T0145380

Studies in Computational Intelligence

Volume 753

Series editor

Janusz Kacprzyk, Polish Academy of Sciences, Warsaw, Poland
e-mail: kacprzyk@ibspan.waw.pl

About this Series

The series "Studies in Computational Intelligence" (SCI) publishes new developments and advances in the various areas of computational intelligence—quickly and with a high quality. The intent is to cover the theory, applications, and design methods of computational intelligence, as embedded in the fields of engineering, computer science, physics and life sciences, as well as the methodologies behind them. The series contains monographs, lecture notes and edited volumes in computational intelligence spanning the areas of neural networks, connectionist systems, genetic algorithms, evolutionary computation, artificial intelligence, cellular automata, self-organizing systems, soft computing, fuzzy systems, and hybrid intelligent systems. Of particular value to both the contributors and the readership are the short publication timeframe and the world-wide distribution, which enable both wide and rapid dissemination of research output.

More information about this series at http://www.springer.com/series/7092

Vladik Kreinovich · Songsak Sriboonchitta
Nopasit Chakpitak
Editors

Predictive Econometrics and Big Data

 Springer

Editors
Vladik Kreinovich
Computer Science Department
University of Texas at El Paso
El Paso, TX
USA

Nopasit Chakpitak
International College
Chiang Mai University
Chiang Mai
Thailand

Songsak Sriboonchitta
International College
Chiang Mai University
Chiang Mai
Thailand

ISSN 1860-949X ISSN 1860-9503 (electronic)
Studies in Computational Intelligence
ISBN 978-3-319-89018-0 ISBN 978-3-319-70942-0 (eBook)
https://doi.org/10.1007/978-3-319-70942-0

Printed on acid-free paper

This Springer imprint is published by Springer Nature
The registered company is Springer International Publishing AG
The registered company address is: Gewerbestrasse 11, 6330 Cham, Switzerland

Preface

Econometrics is a branch of economics that uses mathematical (especially statistical) methods to analyze economic systems, to forecast economic and financial dynamics, and to develop strategies for achieving desirable economic performance.

Traditional econometric techniques have been focused on the quantitative description of economic phenomena. However, the ultimate goal of econometrics—as well as the ultimate goal of science in general—is to predict future development of economics and to develop strategies that optimize the future state of economics. It is therefore desirable to develop techniques that are specifically aimed at predicting economic phenomena. Such predictive econometric techniques—and their applications to real-life economic and financial situations—are one of the main foci of this volume.

Another focus of this book is related to the fact that in the modern world, in which computers are ubiquitous, the amount of economic-related data generated and processed by these computers has grown exponentially. The amount of available economic data is so huge that many traditional statistical data processing algorithms are no longer capable of processing all these data in real time. To process this data, we need to utilize "big data" techniques specifically developed for processing such huge amounts of data and we need to develop big data versions of the state-of-the-art econometric techniques and algorithms. This is a new and promising direction in econometrics. Big data is the main subject of this volume's keynote paper by Dr. Chaitanya Baru from the US National Science Foundation.

In addition to papers on predictive econometric techniques and on big data applications, this book also contains applications of more traditional statistical techniques to econometric problems.

We hope that this volume will help practitioners to learn how to apply new predictive and big data econometric techniques and help researchers to further improve the existing predictive and big data techniques and to come up with new ideas on how econometric techniques can utilize large amounts of data to make more accurate predictions.

We want to thank all the authors for their contributions and all anonymous referees for their thorough analysis and helpful comments.

The publication of this volume is partly supported by the Chiang Mai School of Economics (CMSE), Thailand. Our thanks go to Dean Pirut Kanjanakaroon and CMSE for providing crucial support. Our special thanks go to Prof. Hung T. Nguyen for his valuable advice and constant support.

We would also like to thank Prof. Janusz Kacprzyk (Series Editor) and Dr. Thomas Ditzinger (Senior Editor, Engineering/Applied Sciences) for their support and cooperation in this publication.

September 2017   Vladik Kreinovich
                                                                  Songsak Sriboonchitta
                                                                    Nopasit Chakpitak

Contents

Keynote Address

Data in the 21st Century

Chaitanya Baru[1,2(✉)]

[1] University of California San Diego, San Diego, USA
cbaru@ucsd.edu
[2] US National Science Foundation, Alexandria, USA

Abstract. The past couple of decades have witnessed exponential growth in data, due to the penetration of information technology across all aspects of science and society; the increasing ease with which we are able to collect more data; and the growth of Internet-scale, planet-wide Web-based and mobile services—leading to the notion of "big data". While the emphasis so far has been on developing technologies to manage the volume, velocity, and variety of the data, and to exploit available data assets via machine learning techniques, going forward the emphasis must also be on *translational* data science and the responsible use of all of these data in real-world applications. Data science in the 21st century must provide trust in the data and provide responsible and trustworthy techniques and systems by supporting the notions of transparency, interpretability, and reproducibility. The future offers exciting opportunities for transdisciplinary research and *convergence* among disciplines—computer science, statistics, mathematics, and the full range of disciplines that impact all aspects of society. Econometrics and economics can find an important role in this convergence of ideas.

Keywords: Data science · Translational data science · Convergence
Responsible data management · Trustworthiness

1 Background and Context

The past few decades we have witnessed continuous growth in the amount of data in all spheres across science and society, leading to the idea of the "4th paradigm" of science [9], and giving rise to the phrase, "big data". In the early days of computing, data was managed simply, as "flat files". That was soon followed by "structured files"—files with an internal or external index structure to allow for faster access to the contents of the file [1]. Early applications of such data were in the areas of accounting and finance, in addition to basic scientific and cryptographic calculations. A decade or two later, in the 1980s, with growing use of computers in business, finance, and commerce, and the need to manage increasing amounts of data in multi-user environments, came the new technology of relational databases. Corporate data was converted to structured databases, following principles established by relational database theory [6].

© Springer International Publishing AG 2018
V. Kreinovich et al. (eds.), *Predictive Econometrics and Big Data*, Studies in Computational Intelligence 753, https://doi.org/10.1007/978-3-319-70942-0_1

During that same period, the amount of scientific data was also beginning to grow due to the availability of higher resolution sensors and detectors; launch of science initiatives like the Human Genome Project and supercolliders; and, the ability to build larger supercomputers generating larger amounts of data—all made possible by increasingly cheaper and more efficient processing and storage hardware. In science, the use of computing began to spread across many different research communities. Business applications went beyond transaction processing and simple query processing to decision support and use in strategic planning.

The next knee in the data growth curve occurred with the growth of the Internet and the advent of the Web—beginning in the 1990s and into the 2000s—leading to the creation of "internet-scale" services, businesses, and operations. While data has always been with us, this last inflexion in growth of data was different. Until then, data in the scenarios described above were collected "by design"—through deliberate design of corporate databases and designed data collection efforts in science. The explosion in Web data—and afterwards data from mobile computing and smartphones—gave rise to the notion of *organic data* [8]: observational data that is not collected deliberately, but is available simply as a "by-product" of other actions such as commercial transactions, social media interactions, interactions with online services, location-based information from mobile platforms, and the like. A major component of this huge explosion in data was "unstructured" and "semistructured" data (really, data with varying structure)—text, images, video, weblogs. In addition to the "organic" nature of the data (versus "designed" data), another key characteristic was that these data were being used for multiple purposes—not just for a single, designed application. For example, location-based information from mobile platforms can provide information about instantaneous population density in a given spatial extent. That information could be used for multiple purposes: for traffic control; for drawing up and modifying dynamic disaster response/evacuation plans; as a spot census; for load balancing cell phone traffic across multiple cell towers; or, for planning police and security operations.

The visible successes in data exploitation by large, Internet-scale (indeed, planet-scale) companies, like Google, Facebook, Amazon, Uber, as well as by other companies in other sectors such as, say, finance, intelligence, and others, have led to the current enduring interest in "big data". The Data Deluge has been mentioned as a phenomenon [7]; there is the notion of the "unreasonable effectiveness of data"; and, data is said to be the "new oil". There are also many cautionary messages, for example, in "Weapons of Math Destruction" [16], and concerns regarding bias, fairness, and transparency related to data[1]. As data collection increases, and the power of data analytics begins to dominate every aspect of our lives, one must pay serious attention to consequences from data-driven decision making. This will be a key challenge of data in the 21st century.

---

[1] Data and Society, https://datasociety.net/.

While the 20th century witnessed the invention of digital computing—ending with the introduction of the Web, and the early 21st century witnessed explosion in data—leading to the notion of "big data", the defining character of the data in the 21st century will be in how we manage the data that we are collecting (or, one might even say, "receiving")—and how we use that data for the benefit of science and society. The transition must be made from just thinking about algorithms and technology to developing unified, holistic approaches that bring together a number of seemingly disparate concerns such as, say, data *uncertainty, quality, bias, fairness,* in conjunction with systems that are able to provide *transparency, interpretability,* and *reproducibility* of results. A *convergence* of ideas is needed[2]. From a research standpoint, teams of researchers will need to come together to tackle problems in a holistic fashion. While single investigator-driven research will continue to thrive, it also misses a larger context and the holistic view, and the important tradeoffs involved in that larger context. The issue of ethics and the impacts of data-driven decision-making will be an essential concern.

In short, data in the 21st century will require a "reboot" of a sort. From the deep silos of expertise that we have created over the past 4–5 decades, we now need to move to a convergent approach, where experts from widely different areas and fields come together to solve complex, societal-scale problems. Indeed, the fact that so much data is available at a time when societies and economies appear to be undergoing a huge change is a boon: data-driven approaches will be essential, but they must be holistic in their approach. As noted in a recent Dagstuhl workshop titled, "Data—Responsibly" [2], societies are becoming "data-driven," and large scale data analysis ("big data") now reaches all of us in our private lives, and is a dominant force in commercial domains as varied as manufacturing, e-commerce and personalized medicine. Big data assists in and, in some cases, fully automates decision making in both the public and private sectors. Data-driven algorithms are used in criminal sentencing—determining who goes free and who remains behind bars; in college admissions—granting or denying access to education; and, in employment and credit decisions—offering or withholding economic opportunities. The promise of using these approaches is that they can improve people's lives, accelerate scientific discovery and innovation, and enable broader participation. Yet, if not used responsibly, the big data phenomenon can increase economic inequality and affirm systemic bias, polarize rather than democratize, and deny opportunities rather than improve access. Worse yet, all this can be done in a way that is non-transparent and defies public scrutiny.

While the massive explosion in data, and the "unreasonable effectiveness" of analytics on that data, have brought attention to the notion of "big data", it is important to exercise care and pay equal attention to *all* data—big or small. All data have a lifecycle—from collection to curation, to use in analytics, to preservation. Indeed, there is increasingly a discussion of "small data"—i.e., the data that pertains to individuals and, therefore in some sense, "belongs to" an individual. Thus, the key notions going forward are not just about "big data", but about *data science,* more broadly.

---

[2] Convergence Research at NSF: https://www.nsf.gov/od/oia/convergence/index.jsp.

In the next sections, we will discuss 21st century issues related to foundations of data science; issues in developing software systems for data science to manage and analyze data; and *translational data science*—i.e., the translation of data science concepts, theories, and systems into real-world applications. We will end with some observations about big data and econometrics.

2   Data Science Foundations

In the United States, the BIGDATA research program was launched in 2012, as a cross-agency effort among NSF, NIH, and DARPA [5]. Shortly thereafter, a BIGDATA program was established at NSF. The program funded research in foundations of big data as well as innovative applications of big data [22]. A federal R&D strategic plan for big data was released in the United States in May 2016. Not surprisingly, much of the research in big data generally focused on scaling issues in data management and machine learning, statistical modeling issues with organic data, the so-called "volume, velocity, and variety" problems, modeling of new types of data, especially, graphs, time series, and geospatial and spatiotemporal data, and "needle in the haystack" problems. The explosion in data motivated research on whether and how techniques scaled up with data size (volume), and their performance at scale. For example, a project on *"Analytical Approaches to Massive Data Computation with Applications to Genomics,"* funded by the NSF in 2013 [23], has the goal of designing and testing mathematically well-founded algorithmic and statistical techniques for analyzing large scale, heterogeneous and noisy data. The project was motivated by the challenges in analyzing molecular biology data that are typically large and noisy, given that DNA/RNA sequence data repositories have been growing at a "super-exponential" rate. However, the methods and techniques developed would be broadly applicable to other scientific communities that process massive multi-variant data sets.

Another project on *"Big Tensor Data Mining"* [17], addressed the issue of insufficient theory and methods for big sparse tensor representations of data. It investigated the theory, scalability of algorithms and applications of tensor-based representations of big data, addressing terabyte and petabyte scaling issues; distributed fault-tolerant computations; and datasets with large proportions of missing data. Early projects also addressed the data management aspects of big data. For example, the project on *"Formal Foundation for Big Data Management"* developed a multi-platform software middleware for expressing and optimizing ad hoc data analytics techniques. The goal of is to augment and integrate existing analytics solutions to facilitate and improve methods typically use by a community of users, to make it easier for end users to conduct complex data analyses on big data and on large computer clusters. The project has developed middleware that is accessible as a Web-based service[3].

Another project on the "Theory and Algorithms for Parallel Probabilistic Inference with Big Data, via Big Model, in Realistic Distributed Computing

---

[3] http://myriadb.cs.washington.edu.

Environments,"[4] focuses on scale-up and parallelization of Bayesian machine learning—providing a powerful and theoretically justified framework for modeling a wide variety of datasets. A suite of complementary distributed inference algorithms were developed for hierarchical Bayesian models, covering the most commonly used Bayesian machine learning methods. The project focused on combining speed and scalability with theoretical guarantees that allow one to assess the accuracy of the resulting methods, thereby allowing practitioners to make trade-offs between speed and accuracy. The project developed techniques applicable to a broad spectrum of hierarchical Bayesian models that can be combined as needed for arbitrary probabilistic models, whether they are parametric or nonparametric, discriminative or generative [27].

Foundational data science offers many opportunities for interdisciplinary collaborations among computer scientists, statisticians, and mathematicians, to fully consider the computation-statistics tradeoff. Traditionally, the statistics community has focused mainly on the inferential aspect, while the theoretical computer science and mathematics communities have focused more on the computational aspects. The NSF sponsored a workshop on *Theoretical Foundations of Data Science,* in April 2016, to explore future directions in this area[5]. The meeting brought together computer scientists (experts in algorithmics and theoretical aspects), statisticians, and mathematician, with the notion that future progress in this area requires a convergence (at least) among these disciplines, but perhaps more. A range of topics were discussed at the meeting, and incorporated into a workshop report[6]. There was a recognition that foundational issues must be considered across the full lifecycle of data, rather than just the analysis phase. It is well-known among data scientists that a large fraction of the total data analysis time is spent in data preparation and preprocessing, which is often ignored in theoretical studies. Indeed, data preparation and preprocessing pose many intellectually challenging problems that are related to deep mathematical issues that cannot be easily be formalized. It is not merely "engineering", but rather a critical part of deploying models in production. It constitutes the *before* and *after* of many machine learning problems that have a greater cachet. Also, the report states, "data science involves iterative procedures with a dynamic feedback loop. Typically, the formulations of so-called online algorithms in machine learning are not particularly well-suited to understanding the iterative aspect of data science more generally. Targets can change as more data are acquired; instead of restricting our attention to idealized systems under restrictive assumptions, dynamic data collection is general, heterogeneous, and messy. Latency is an important issue, and there are also humans in the loop. Data science foundations should investigate how data analysis techniques, user behaviors, and the data collection process fit together."

---

[4] https://nsf.gov/awardsearch/showAward?AWD_ID=1447676&HistoricalAwards=false.

[5] TFoDS Workshop, http://www.cs.rpi.edu/TFoDS/.

[6] TFoDS Workshop Report, http://www.cs.rpi.edu/TFoDS/TFoDS_v5.pdf.

A workshop identified that a potentially high impact area of foundational research is the integration of statistical and computational approaches into a united theoretical framework. Topics for interdisciplinary collaboration could include studies of randomized numerical linear algebra [11]; signal processing/harmonic analysis on graphs; nonconvex statistical optimization; combining physical and statistical models; mixed type and multi-modality data; applied representation theory and non-commutative harmonic analysis; topological data analysis and homological algebra; security, privacy, and algorithmic fairness; provenance and reproducibility; and, the development of a common terminology/language to enable interdisciplinary collaborations.

Indeed, NSF's recent program in *Transdisciplinary Research in Principles of Data Science (TRIPODS)* picks up on these recommendations and promotes interdisciplinary research across computer science, statistics, and mathematics. Phase I of this program has recently been funded, and supports twelve collaborative projects that explore research in this direction. For example, one of the funded efforts, "Towards a Unified Theory of Structure, Incompleteness & Uncertainty in Heterogeneous Graphs," brings together researchers from mathematics, statistics, and computer science to develop a unified theory of data science applied to uncertain and heterogeneous graph and network data, with the goal of developing a new foundation for data science to deal with incomplete, noisy, heterogeneous data with multiple modalities and multiple scales in the context of graph and network data. A unified theory will be developed to understand how to quantify the uncertainty in the system that arises from the uncertainty in the relationships among its actors via transdisciplinary collaboration among statisticians, mathematicians, and computer scientists. This project plans to investigate models of algorithms on uncertain network data, combining techniques from sub-linear algorithms with Bayesian methods. Another theme is on how algorithms can benefit from data uncertainty in the context of privacy, disclosure, and robustness to noise. Both of these themes require marrying computational approaches to uncertainty with statistical and mathematical approaches for uncertainty.

Another TRIPODS project, *"Algorithms for Data Science: Complexity, Scalability, and Robustness,"* examines the complexity, scalability and robustness of data science algorithms [10]. Since the challenges that a range of fields now face are no longer easily handled by ideas from a single discipline, a central goal of the project is to provide a common language and unifying methods for addressing contemporary data science challenges. Each of the three disciplines of computer science, mathematics, and statistics has rich theories of complexity and robustness. These theories have influenced the design of the available tools that are used to address real world computational problems. Going forward, there is a need for new algorithms and design principles that unify ideas and provide a common language for addressing contemporary data science challenges. The complexity and algorithmic questions this work seeks to address include: (i) how to unify various notions of complexity (which range from information theoretic to computational to black box oracle models), (ii) how to unify notions of robustness and adaptivity (e.g., how solutions and methods change as oracle models are corrupted by random or adversarial noise), (iii) how to address optimization challenges due to nonconvexity, and (iv) how to use these unified approaches to design more effective scalable tools, in theory and practice.

Foundations of data in the 21st century must embrace not only the technical issues, but also the larger issues of fairness and ethics. A third TRIPODS project, *"Data Science for Improved Decision-Making: Learning in the Context of Uncertainty, Causality, Privacy, and Network Structures,"* will examine issues related to privacy and fairness, learning on social graphs, learning to intervene, uncertainty quantification, and deep learning. As data science becomes pervasive across many areas of society, and as it is increasingly used to aid decision-making in sensitive domains, it becomes crucial to protect individuals by guaranteeing privacy and fairness. The project proposes to research the theoretical foundations to providing such guarantees and to surface inherent limitations. Furthermore, data-driven approaches to learning good interventions (including policies, recommendations, and treatments) inspire challenging questions about the foundations of sequential experimental design, counterfactual reasoning, and causal inference.

In sum, foundational approaches are needed to establish data science on a firm multidisciplinary footing with principles drawn from computer science, statistics, and mathematics, as well as from areas related to privacy, fairness, and ethics.

3   Systems

Data science hardware and software systems must support the needs of data in the 21st century. As mentioned earlier in Sect. 1, the primary focus of systems thus far has been on managing the scale (volume) of data and ensuring that systems continue to perform well as data volumes and velocities increase. Significant work has also been done in integration of data from heterogeneous sources, and in supporting the heterogeneity of data types, from structured data, to images, video, and more recently graph data—a rapidly increasing modality of data. A number of current efforts focus on support for streaming data and real time analytics on data streams.

Hardware systems. In terms of computing hardware, we are in a period of "hardware renaissance", with a number of new technologies becoming available for data storage and for computing. Storage/memory hierarchies have become deeper with the availability of solid state disk (SSD), and a variety of non-volatile memories (NVM). Computing has entered the so-called "post CPU" era—graphical processing units (GPUs) and other specialized hardware, FPGAs, ASICs, and Google's Tensorflow Processing Units [18], are being utilized to speed up machine learning applications. Major software suites are available for machine learning such as MXNet from Amazon Web Services (AWS)[7], Google's Tensor-

---

[7] Apache MXnet, https://aws.amazon.com/mxnet/.

flow software[8], Microsoft's Cognitive Toolkit[9], and IBM's Cognitive Computing suite[10]. Innovation in hardware and software can be expected to continue, including in end-to-end software environments for data science to help increase end-user productivity.

In terms of new directions and vistas for data in the 21st century, there is a pressing need as well as community interest in developing systems to support trustworthiness as an intrinsic feature of data science [14][11]. Systems that support transparency, interpretability and reproducibility of results can promote trustworthiness. The problem becomes quite challenging when data quality can vary—over time and across different sources, and when the data may be dynamic in nature, as in the case of streaming, real time data flows. Appropriate methods and quantification approaches are needed to capture uncertainty in data as well as to ensure reproducibility and replicability of results. These issues become especially important when data are repurposed for a use different than the one for which they were was originally collected, and also when data are integrated from multiple, heterogeneous sources of different quality. Data collected by an original researcher/community for a purpose may be well understood by that community and properly characterized for the original purpose. However, as those data are reused—as is often the case in big data applications—it is important to retain provenance of the information, as well as develop metadata methods by which data can be characterized for uses other than originally planned. Novel techniques are needed for developing and maintain such provenance and metadata information, especially since reuse and repurposing of data may not only be useful and convenient but, in many cases, it may be the only option, since the desired data may be difficult, or too expensive, to collect.

Software systems. As mentioned earlier, relational databases emerged in the 1980s and have since become a dominant technology—underlying a vast number of data-based applications. Relational databases incentivized thoughtful design of enterprise-level data and databases and incorporated a number of key concepts and useful features, including data normalization, ad hoc query processing, and transaction support. In the process, they also developed sophisticated underlying technologies for database query optimization and complex transaction processing in complex, distributed environments. The systems were designed to scale to terabyte and 100s of terabyte scale databases, at the high end, before the cost became prohibitive. However, the Internet-scale, big data explosion of the 1990s and later required a similar set of features to relational database systems, in some ways, but also quite different in other ways. Big data needed to scale to petabyte-size databases at reasonable cost; accommodate unstructured and

---

[8] Google Tensorflow, https://www.tensorflow.org/.

[9] Microsoft Cognitive Toolkit, https://www.microsoft.com/en-us/cognitive-toolkit/.

[10] IBM Cognitive Computing, https://www.ibm.com/it-infrastructure/us-en/cognitive-computing/.

[11] Administration Issues Strategic Plan for Big Data Research and Development, https://obamawhitehouse.archives.gov/blog/2016/05/23/administration-issues-strategic-plan-big-data-research-and-development.

semi-structured data, such as free text, dynamically formatted web log records, images, and video; support basic ad hoc query processing capability, but not necessarily the "full blown" SQL[12] supported by relational databases systems. Thus, query interfaces to these petabyte-scale databases could be much simpler; and, since the system configurations were very large, i.e., 1000s of computing and storage nodes in a typical compute cluster, the software had to function in a distributed environment and operate in the presence of systems faults, without using expensive hardware/software solution. Also, given the scale of the datasets, and the fact that much of the computing involved statistical analysis and machine learning, small variations in the data content from run to run were permissible. In other words, the database was not required to follow strict consistency rules as enforced by standard relational databases. Loose consistency was acceptable [25].

Technologies that have been invented to address these big data issues include noSQL databases[13], Hadoop/MapReduce and all of its variants[14], Spark[15], and systems like Storm for managing and processing data streams[16].

While performance will always remain an important consideration, technologies, systems, and tools for data in the 21st century must also support "responsible" data management and analysis. This means building systems that intrinsically support notions of transparency, interpretability and reproducibility. Many data analytics and predictive analytics algorithms and systems are not transparent to end-users. How the underlying models work, and when and why such models may fail, is not always clear. Dealing at multiple levels with bias in the data—from algorithmic to decision support—is a key issue. Machine learning systems may learn and reinforce pre-existing biases, which can be especially problematic with human-related data, where such systems may lead to unfair treatment of minority sections of a population. Enabling widespread adoption of data science approaches requires assurances that not only will the system operate securely and in a controlled fashion, but also that it will be "fair", by producing transparent, interpretable, dependable, reproducible—and, therefore, trustworthy—solutions. Interdisciplinary, convergent research is needed in the use of machine learning in data-driven decision-making and discovery systems to examine how data can be used to best support and enhance human judgment. From a computing systems point of view, this requires support for the full data lifecycle: from data acquisition, to curation, analysis and use, decision support, and preservation.

Intrinsic to the notion of trustworthiness is reproducibility—it should be possible to reproduce previously produced results. While simple in concept, this can be a complex issue and has received significant attention. There have been a number of discussions in the community at large, as well as in many individual

---

[12] Structured Query Language, https://www.w3schools.com/sql/sql_intro.asp.
[13] noSQL Databases, http://nosql-database.org/.
[14] Apache Hadoop, http://hadoop.apache.org/.
[15] Apache Spark, https://spark.apache.org/.
[16] Apache Storm, http://storm.apache.org/.

science domains/communities, about reproducibility [3, 12]. While there is significant discussion about distinctions among terms like reproducibility, repeatability, and replicability—the key notion here is that there should be some support for the ability to reproduce results previously produced, specifically in the context of computational and data-driven experiments, by the team that produced the original result as well as by others. Even this seemingly simple notion can be quite complex to define and implement—how should one deal with situations where the results are statistical in nature, and there may not be a single definitive numerical answer?; what about big data, where the datasets are indeed very large—does reproducibility require maintaining a copy of the entire original dataset?; what about streaming data, where the data itself may be change quickly over time?; how does one account for variations in hardware and software when computations are run in one environment versus another?; and, how can we determine standards for acceptability of results, accounting for run to run variations?

In sum, with increasing use of data in every sphere of our personal lives and in society at large, systems in the 21st century will be required to support the broad notion of trustworthiness, which has many facets as introduced above.

Data as Infrastructure. To develop and test foundational approaches and systems for data science, there is a need for cyberinfrastructure. In addition to hardware platforms and software tools and environments, infrastructure for data science also includes the data itself, as well as the knowledge structures that may be derived from that data, and enable use of the data. Science involves research _with_ data. Data science involves research _on_ data. The data, and knowledge, structures—and the algorithms and processes that act on them—are themselves the target of study in data science. Arguably, data is to data science as computers are to computer science. Thus, easy access to data is essential for data science.

Interestingly, given all the interest and fascination in big data, easy access to truly big data datasets is not available to most researchers and students in academia. While data scientists in industry and government have access to large amounts of data (though, even in those cases, oftentimes the access is not easily facilitated. Organizations that provide easy access to such data have a competitive advantage over those that don't), academia does not have such ready access to open, big data—other than whatever data that industry or government might make available in the open. It was remarked recently that it is not possible today in academia to do the *"hello, world"* problem for big data[17]. The *"hello, world"* problem refers to a very simple, first programming assignment usually given to students in introductory programming courses, to ensure that they know how to write, compile, and run a very simple program of one, or few, lines. The equivalent for big data would be an assignment that might ask students to, say, "Access a 1 petabyte (or, even, 100 TB) dataset; perform a sub-selection or sampling operation to obtain 1 TB out of the 1 PB dataset; compute a simple

---

[17] Personal communication with R.V. Guha, July 2016.

descriptive statistic on the 1 TB dataset; print the result. And, complete the assignment by next class." A typical student would not know where to go to find a 1PB or 100 TB (or, in many cases, a 10 TB) data sets, and where to get resources to run even a simple sub-selection operation on a petabyte or 100 TB of data. Thus, a key immediate challenge for the community is to make available a larger number of large datasets for education and training.

Knowledge Networks/Knowledge Graphs. Another area where industry has been making rapid progress and academia needs to catch up and keep us, is in creating large *knowledge networks* or *knowledge graphs.* Such graphs typically encode information about real world entities and the relationship among these entities, e.g., sports teams and the cities they are located in, their roster of players, and the arenas they play in. Very large knowledge graphs of this type are being constructed by companies like Amazon, Google, Apple, Microsoft, etc., to support "intelligent" applications like Alexa, Google Assistant, Siri, Cortana, respectively. Extensive machine learning techniques are utilized to create such knowledge graphs from unstructured and structured data on the Web and in corporate databases, and they form the essential semantic information infrastructure for next generation "smart" and "intelligent" applications. Yet, there is no corresponding effort in the open community that is open and accessible to academia. Two meetings have been conducted so far to discuss the notion of an *Open Knowledge Network,* organized by the federal Big Data Interagency Working Group[18], with a third workshop planned for October 4–5, 2017 at the National Library of Medicine, Bethesda, MD [15]. Access to large-scale open datasets and open knowledge networks, as infrastructure, will be essential for innovation in data science, and for research in general.

4   Translational Data Science

The development of foundations of data science and trustworthy systems are, in the end, in service of data-driven applications, across science and society. *Translational data science* is a new term that is being used for an emerging field that applies data science principles, techniques, and technologies to challenging scientific and societal problems that hold the promise of having an important impact on human or societal welfare. The term is also used when data science principles, techniques and technologies are applied to problems in different domains in general, including—but not restricted to—science and engineering research. A workshop on Translational Data Science was held in June 2017 at the University of Chicago[19].

The term "translational research" emerged in a medical research context, referring to the application of findings from basic science to enhance human

---

[18] NITRD Big Data Interagency Working Group (BDIWG), https://www.nitrd.gov/nitrdgroups/index.php?title=Big_Data.

[19] Translational Data Science workshop, https://cdis.uchicago.edu/tds-17/.

health and well-being[20]. The objective is to "translate" findings in fundamental research into medical practice and meaningful health outcomes. As employed in biomedical and medical research context, the term "translational" describes "the process of turning observations in the laboratory, clinic and community into interventions that improve the health of individuals and populations from diagnostics and therapeutics to medical procedures and behavioral interventions"[21]. Correspondingly, "translational science" is "the field of investigation focused on understanding the scientific and operational principles underlying each step of the translational process". The notion of translating findings in a laboratory setting into potential treatments for disease has received significant attention in biomedicine/medicine because (a) it was felt that research findings were not making their way into clinical practice quickly enough, (b) there is always a pressing need to apply the latest findings from fundamental research in basic science towards enhancing human health and well-being, by producing promising new treatments that can be applied in practice, and (c) to ensure not only that research knowledge reaches practice, but that it is implemented correctly in practice.

In data science, the experience thus far has been that tools, techniques, and technologies are being developed rapidly and deployed immediately into practice. Thus, in a sense, research knowledge is already reaching practice! However, that is primarily a technology-driven view of data science. Translational data science requires a partnership between the technology side and the applications side, so that methods are developed and implemented in a fashion that serves the end users and the end applications. Formalizing and systematizing the data science translation activity allows us to address a number of issues that lie at the interface of theories and technologies, and applications. By identifying the translation step as a distinct activity, one is able to focus and make that activity itself the object of study and research. Data science methods and processes are iterative in nature. They should also be closed-loop—lessons learned in practice should inform the theory and modify techniques for future use. Important issues, such as those related to ethics and policy, arise only at the translation phase. Careful thought must be given to the structures and processes of translation and the implications thereof. Thus, there is a need to study and better understand these processes.

Translational medicine identifies several distinct steps or phases or translation encompassing: *basic research*—research on fundamental aspects in science; *pre-clinical research*—research to connect basic science results to human medicine, for example, via use of model organisms and animal models; *clinical research*—moving the research to human subjects, involving clinical trials, for example; *clinical implementation*—implementation in a clinical setting; and, *public health*—studying the impact of the new techniques in terms of population health. The multiple steps indicate the complexity of the process; the need

---

[20] Translational Research, wikipedia, https://en.wikipedia.org/wiki/Translational_research.

[21] National Center for Advancing Translational Science, https://ncats.nih.gov/.

for a careful, considered approach; and the need to consider the diversity of response among the population to the treatment. It has also been recognized that the move towards precision medicine and personalized medicine will make this process even more complex due to the need for "clinical trial that focuses on individual, not average responses to therapy" [19]. Data science has a similar need for care in applying methods and processes, which can be a complex issue. If data driven decisions are being made about every aspect of our lives, then it is important to consider the impact of these on individuals. Furthermore, one of the key promises of data science and big data is the ability to *personalize* responses. That immediately raises a number of issues, related to privacy of information; the micro-level impact of macro-level methods like machine learning on big data; and, similar to the clinical trials issues, how personalization methods could be tested prior to deployment—or, would the deployment itself be the test?

In sum, calling out translational data science as a distinct activity puts focus on the collaborative nature of applying data science foundations and systems to real applications, in practice. Many key issues in privacy and ethics, surface only at the translational step. Translation requires "co-design", where application domain experts work in collaboration with data scientists to implement systems that meet the needs of the application. In the process, lesson learned in on situation, or by one application domain, could be useful and helpful to another application domain, thereby making the process more efficient by reducing duplication and "not making the same mistakes".

5   Data Science and Econometrics

Use of data science methods in any scientific application will require the ability to "explain" the results—black box approaches will typically not suffice—though, they may suffice in some well-known or well-defined parts of a larger data analysis workflow. Workshops on causal inference have pointed out the importance of such inferencing methods [13, 20]. At the same time, black box methods could be useful, for example, machine learning methods are used extensively in, say, astronomy for image processing, in order to identify candidate images that require further analysis [4].

In his article on *"New Tricks for Econometrics,"* Hal Varian provides an overview of a number of big data technologies which he deemed are useful for the community [24]. To his point, some of the technologies mentioned in the paper written in 2014 have already been overtaken by other newer technologies. In the paper, he states "my standard advice to graduate students these days is go to the computer science department and take a class in machine learning." That, indeed, is a reaction to the lack of a Translational Data Science area, and corresponding curriculum. There is a difference between how a topic like machine learning, or any other technical topic, would be taught to computer scientists versus to, say, econometricians. While there may be a common core, there will be significant differences in the nature and style of coverage—including even the examples selected—for computer science versus econometrics. Translational data science can provide the home for such curriculum. While some econometricians (or, researchers/scientists from any field) may be interested in actually taking a computer science course on the topic, for many, what is needed is the translational aspect of the subject matter. Or, even if a foundations course is taken in computer science, the translational course will still be essential.

As described in Sect. 2, there are on-going collaborations among the computer science, statistics, and mathematics community to define foundations of data science. There is now an opportunity for econometrists to join in such collaborations to help develop foundations, but also to develop the translational aspects of data science as it applies to economics and econometrics.

References

1. Abel, P.: Cobol Programming: A Structured Approach. Prentice Hall, Upper Saddle River (1988)
2. Abiteboul, S., Miklau, G., Stoyanovich, J., Weikum, G.: Data, Responsibly. Seminar 16291, Dagstuhl, 17–22 July 2016. http://www.dagstuhl.de/16291
3. ACM: Artifact Review and Badging, June 2016. https://www.acm.org/publications/policies/artifact-review-badging
4. Ball, N.M., Brunner, R.J.: Data Mining and Machine Learning in Astronomy, arxiv.org, August 2010. https://arxiv.org/abs/0906.2173
5. CCC Blog, Obama Administration Unveils $200M Big Data R&D Initiative, 29 March 2012. http://www.cccblog.org/2012/03/29/obama-administration-unveils-200m-big-data-rd-initiative/
6. Codd, E.F.: The Relational Model for Database Management (Version 2 ed.). Addison Wesley Publishing Company (1990). ISBN 0-201-14192-2
7. Economist: The Data Deluge, February 2010. http://www.economist.com/node/15579717
8. Groves, R.: "Designed Data" and "Organic Data", May 2011. https://www.census.gov/newsroom/blogs/director/2011/05/designed-data-and-organic-data.html
9. Hey, T., Tansley, S., Tolle, K.: The Fourth Paradigm: Data-Intensive Scientific Discovery. Microsoft Research (2009). ISBN 978-0-9825442-0-4
10. Kakade, S., Harchaoui, Z., Drusvyatskiy, D., Lee, Y.T., Fazel, M.: Algorithms for data science: complexity, scalability, and robustness (2017). https://nsf.gov/awardsearch/showAwardAWD_ID=1740551&HistoricalAwards=false
11. Mahoney, M.W.: Lecture Notes on Randomized Linear Algebra, arXiv:1608.04481, August 2016
12. National Academy of Sciences, Arthur M. Sackler Colloquia: Reproducibility of research: issues and proposed remedies. http://www.nasonline.org/programs/sackler-colloquia/completed_colloquia/Reproducibility_of_Research.html
13. National Academy of Sciences: Refining the Concept of Scientific Inference When Working With Big Data: A Workshop, June 2016. http://sites.nationalacademies.org/DEPS/BMSA/DEPS_171738
14. NITRD Big Data Interagency Working Group: The Federal Big Data R&D Strategic Plan, May 2016. https://obamawhitehouse.archives.gov/sites/default/files/microsites/ostp/NSTC/bigdatardstrategicplan-nitrd_final-051916.pdf
15. NITRD Big Data Interagency Working Group: 3rd Workshop on an Open Knowledge Network (2017). https://www.nitrd.gov/nitrdgroups/index.php?title=Open_Knowledge_Network

16. O'Neil, C.: Weapons of Math Destruction. Crown Publishing, New York (2016)
17. Papalexakis, E.E., Kang, U., Faloutsos, C., Sidiropoulos, N.D., Harpale, A.: Large scale tensor decompositions: algorithmic developments and applications. IEEE Data Eng. Bull. - Special Issue on Social Media **36**, 59 (2013)
18. Sato, K., Young, C., Patterson, D.: An in-depth look at Google's first Tensor Processing Unit (TPU), May 2017. https://cloud.google.com/blog/big-data/2017/05/an-in-depth-look-at-googles-first-tensor-processing-unit-tpu
19. Schork, N.: Personalized medicine: time for one-person trials. Nature **520**(7549), 609–611 (2015). https://doi.org/10.1038/520609a. https://www.nature.com/news/personalized-medicine-time-for-one-person-trials-1.17411
20. Shiffrin, R.M.: Drawing causal inference from Big Data, vol. 113, no. 27, pp. 7308–7309 (2016). https://doi.org/10.1073/pnas.1608845113
21. Suciu, D., Balazinska, M., Howe, B.: A formal foundation for big data management. https://nsf.gov/awardsearch/showAward?AWD_ID=1247469&HistoricalAwards=false
22. NSF: Core Techniques and Technologies for Advancing Big Data Science & Engineering (BIGDATA) (2012). https://www.nsf.gov/pubs/2012/nsf12499/nsf12499.htm
23. Upfal, E.: Analytical approaches to massive data computation with applications to genomics (2012). https://nsf.gov/awardsearch/showAward?AWD_ID=1247581&HistoricalAwards=false
24. Varian, H.R.: Big data: new tricks for econometrics. J. Econ. Perspect. **28**(2), 3–28 (2014). https://doi.org/10.1257/jep.28.2.3. http://www.aeaweb.org/articles?id=10.1257/jep.28.2.3
25. Viotti, P., Vukolic, M.: Consistency in non-transactional distributed storage systems. ACM Comput. Surv. **49**(1), 19:1–19:34 (2016). https://doi.org/10.1145/2926965
26. Weinberger, K., Strogatz, S., Hooker, G., Kleinberg, J., Shmoys, D.: Data science for improved decision-making: learning in the context of uncertainty, causality, privacy, and network structures (2017). https://nsf.gov/awardsearch/showAward?AWD_ID=1740822&HistoricalAwards=false
27. Xing, E.P., Ho, Q., Dai, W., Kim, J.K., Wei, J., Lee, S., Zheng, X., Xie, P., Kumar, A., Yu, Y.: Petuum: a new platform for distributed machine learning on big data. IEEE Trans. Big Data **1**, 49 (2015). https://doi.org/10.1109/TBDATA.2015.2472014

Fundamental Theory

Model-Assisted Survey Estimation
with Imperfectly Matched Auxiliary Data

F. Jay Breidt[✉], Jean D. Opsomer, and Chien-Min Huang

Colorado State University, Fort Collins, USA
{FJay.Breidt,Jean.Opsomer,Chien-Min.Huang}@colostate.edu

Abstract. Model-assisted survey regression estimators combine auxiliary information available at a population level with complex survey data to estimate finite population parameters. Many prediction methods, including linear and mixed models, nonparametric regression, and machine learning techniques, can be incorporated into such model-assisted estimators. These methods assume that observations obtained for the sample can be matched without error to the auxiliary data. We investigate properties of estimators that rely on matching algorithms that do not in general yield perfect matches. We focus on difference estimators, which are exactly unbiased under perfect matching but not under imperfect matching. The methods are investigated analytically and via simulation, using a study of recreational angling in South Carolina to build a simulation population. In this study, the survey data come from a stratified, two-stage sample and the auxiliary data from logbooks filed by boat captains. Extensions to regression estimators under imperfect matching are discussed.

Keywords: Complex survey · Difference estimator
Probability sampling · Survey regression estimation

1 Introduction

1.1 Probability Sampling and Weighted Estimation

Let $U = \{1, 2, \ldots, N\}$ denote a finite population and let y_k denote the (non-random) value of some variable of interest for element $k \in U$. We are interested in the finite population total $T_y = \sum_{k \in U} y_k$. As a motivating example, which we return to in Sect. 4, suppose U is the set of all recreational angling boat trips on the coast of the state of South Carolina in 2016. Further, suppose y_k is the number of anglers on the kth boat trip, so that T_y is the total number of recreational angler trips on boats in South Carolina waters in 2016; or suppose that y_k is the number of black sea bass caught on the kth boat trip, so that T_y is the total number of black sea bass caught in 2016.

Because it is often impractical to measure y_k for all $k \in U$, we instead estimate T_y based on information obtained for a sample $s \subset U$, which is selected

© Springer International Publishing AG 2018
V. Kreinovich et al. (eds.), *Predictive Econometrics and Big Data*, Studies in Computational Intelligence 753, https://doi.org/10.1007/978-3-319-70942-0_2

via a random mechanism. Following [8], let s denote one of the 2^N possible subsets of U, selected with probability given by $p(s)$, the *sampling design*. The *sample membership indicators* are random variables defined by $I_k = 1$ if $k \in s$, $I_k = 0$ otherwise, so that $s = \{k \in U : I_k = 1\}$. Let $\pi_k = \mathrm{E}\,[I_k] = \mathrm{P}\,[I_k = 1]$ and suppose that these *first-order inclusion probabilities* satisfy $\pi_k > 0$ for all $k \in U$. The sampling design is then said to be a *probability sampling design*, for which it is well known that the *Horvitz-Thompson estimator* [4]

$$\widehat{T}_y = \sum_{k \in U} y_k \frac{I_k}{\pi_k} = \sum_{k \in s} \frac{y_k}{\pi_k} \tag{1}$$

is unbiased for T_y, where the expectation is with respect to the probability distribution $p(\cdot)$. Its variance is given by

$$\mathrm{Var}\left(\widehat{T}_y\right) = \sum_{j,k \in U} \mathrm{Cov}\,(I_j, I_k) \frac{y_j}{\pi_j} \frac{y_k}{\pi_k} = \sum_{j,k \in U} \Delta_{jk} \frac{y_j}{\pi_j} \frac{y_k}{\pi_k}, \tag{2}$$

where $\Delta_{jk} = \pi_{jk} - \pi_j \pi_k$ and $\pi_{jk} = \mathrm{E}[I_j I_k] = \mathrm{P}\,[I_j = 1,\, I_k = 1]$ denotes a *second-order inclusion probability*.

If the second-order inclusion probabilities satisfy $\pi_{jk} > 0$ for all $j, k \in U$, then the design is a *measurable sampling design* and an unbiased variance estimator is given by

$$\widehat{V}(\widehat{T}_y) = \sum_{j,k \in U} \Delta_{jk} \frac{y_j}{\pi_j} \frac{y_k}{\pi_k} \frac{I_j I_k}{\pi_{jk}}; \tag{3}$$

this estimator or closely-related approximations are computed in standard survey software such as `proc surveymeans` in SAS [9] or the `survey` package of R [6,7].

1.2 Auxiliary Information

In addition to observations obtained on the sample, auxiliary information may be available from external records. Let $\mathscr{A} = \{1, 2, \ldots, A\}$ denote the indices for this external database and let $\mathbf{a}_\ell$ denote the vector of auxiliary information available for record $\ell \in \mathscr{A}$. We write $A_x = \sum_{\ell \in \mathscr{A}} x_\ell$ for sums over the database, in particular noting that the size of the database is $A_1 = \sum_{\ell \in \mathscr{A}} 1$.

The auxiliary vector $\mathbf{a}_\ell$ could be used to construct a predictor, $\mu(\mathbf{a}_\ell)$ of y_k provided record $\ell \in \mathscr{A}$ in the database matches element $k \in U$ in the population. We assume for the present that the construction of the prediction method $\mu(\cdot)$ does not involve the sample, s. We write $A_\mu = \sum_{\ell \in \mathscr{A}} \mu(\mathbf{a}_\ell)$.

2 Estimation Under Perfect Matching

2.1 Notation for Perfect Matching

We first consider the case of perfect matching: suppose that every record in the database can be matched to one and only one element in the population, and vice versa. We write

$$M_{k\ell} = \begin{cases} 1, & \text{if database record } \ell \in \mathscr{A} \text{ matches element } k \in U, \\ 0, & \text{otherwise.} \end{cases} \tag{4}$$

The appropriate predictor of y_k would then be denoted $\sum_{\ell \in \mathscr{A}} M_{k\ell}\mu(\mathbf{a}_\ell)$, to reflect the matching step.

2.2   Difference Estimator Under Perfect Matching

Under this perfect matching scenario, $\sum_{k \in U} M_{k\ell} = 1$. It follows that an unbiased estimator of T_y is given by the *difference estimator*,

$$\widehat{T}_{y,\text{diff}} = \sum_{k \in U} \sum_{\ell \in \mathscr{A}} M_{k\ell}\mu(\mathbf{a}_\ell) + \sum_{k \in s} \frac{y_k - \sum_{\ell \in \mathscr{A}} M_{k\ell}\mu(\mathbf{a}_\ell)}{\pi_k}$$

$$= \sum_{\ell \subset \mathscr{A}} \mu(\mathbf{a}_\ell) + \sum_{k \in s} \frac{y_k - \sum_{\ell \in \mathscr{A}} M_{k\ell}\mu(\mathbf{a}_\ell)}{\pi_k}; \tag{5}$$

this is simply a more elaborate notation for a standard estimator (e.g., Eq. (4) of [2]), to account for the matching step. The variance of the perfect-matching difference estimator is

$$\text{Var}\left(\widehat{T}_{y,\text{diff}}\right) = \sum_{j,k \in U} \Delta_{jk} \frac{y_j - \sum_{\ell \in \mathscr{A}} M_{j\ell}\mu(\mathbf{a}_\ell)}{\pi_j} \frac{y_k - \sum_{\ell \in \mathscr{A}} M_{k\ell}\mu(\mathbf{a}_\ell)}{\pi_k}. \tag{6}$$

The unbiased difference estimator will have smaller variance and mean square error than the unbiased Horvitz-Thompson estimator (1) provided the residuals $\{y_k - \sum_{\ell \in \mathscr{A}} M_{k\ell}\mu(\mathbf{a}_\ell)\}_{k \in U}$ in (6) have less variation than the raw values $\{y_k\}_{k \in U}$ in (2).

3   Estimation Under Imperfect Matching

3.1   Notation for Imperfect Matching

In practice, perfect matching may not be possible. The sampled element k might have no corresponding record in the database. It might have a corresponding record ℓ, but fail to match it perfectly due to missing values or inaccuracies in the survey observation, the database record, or both. Similarly, the sampled element might appear to match multiple database records due to agreement on a number of data values.

Hence, we replace the $M_{k\ell} = 0$ or 1 by a possibly-fractional value $m_{k\ell} \in [0, 1]$, computed via a deterministic algorithm that does not depend on the sample. We refer to these values as *match metrics*. Assume that for any sampled element

$k \in U$, the match metrics $\{m_{k\ell}\}_{\ell \in \mathscr{A}}$ for every database record can be computed. For example, sampled element k might match record ℓ_1 perfectly, in which case

$$m_{k\ell} = \begin{cases} 1, & \text{if } \ell = \ell_1, \\ 0, & \text{otherwise.} \end{cases}$$

It might not match any records, in which case

$$m_{k\ell} = 0, \quad \text{for all } \ell \in \mathscr{A};$$

or it might match three records ℓ_1, ℓ_2, ℓ_3 equally well, in which case

$$m_{k\ell} = \begin{cases} 1/3, & \text{if } \ell = \ell_1, \ell = \ell_2 \text{ or } \ell = \ell_3, \\ 0, & \text{otherwise.} \end{cases}$$

If $\sum_{\ell \in \mathscr{A}} m_{k\ell} < 1$, then the matching algorithm has determined that there is a non-trivial possibility that the sampled element does not match any database record. This can occur when there is potential non-overlap between the target population U and the database $\mathscr{A}$. This is of interest in situations such as the application we will describe in Sect. 4, where $\mathscr{A}$ is a possibly-incomplete set of recreational angling trips self-reported by boat captains, while U is the actual population of trips.

3.2   Difference Estimation Under Imperfect Matching

3.2.1   First Difference Estimator

Under imperfect matching, an estimator analogous to (5) is

$$\widetilde{T}_{y,\text{diff1}} = \sum_{\ell \in \mathscr{A}} \mu(\mathbf{a}_\ell) + \sum_{k \in s} \frac{y_k - \sum_{\ell \in \mathscr{A}} m_{k\ell}\mu(\mathbf{a}_\ell)}{\pi_k}. \tag{7}$$

This estimator is no longer unbiased. Instead, its expectation is

$$\mathrm{E}\left[\widetilde{T}_{y,\text{diff1}}\right] = \sum_{\ell \in \mathscr{A}} \mu(\mathbf{a}_\ell) + \sum_{k \in U} \left(y_k - \sum_{\ell \in \mathscr{A}} m_{k\ell}\mu(\mathbf{a}_\ell) \right)$$

$$= T_y + \sum_{\ell \in \mathscr{A}} \left(1 - \sum_{k \in U} m_{k\ell} \right) \mu(\mathbf{a}_\ell). \tag{8}$$

Its variance is

$$\mathrm{Var}\left(\widetilde{T}_{y,\text{diff1}}\right) = \sum_{j,k \in U} \Delta_{jk} \frac{y_j - \sum_{\ell \in \mathscr{A}} m_{j\ell}\mu(\mathbf{a}_\ell)}{\pi_j} \frac{y_k - \sum_{\ell \in \mathscr{A}} m_{k\ell}\mu(\mathbf{a}_\ell)}{\pi_k}. \tag{9}$$

The variance is small if the match-weighted quantities $\{\sum_{\ell \in \mathscr{A}} m_{k\ell}\mu(\mathbf{a}_\ell)\}_{k \in U}$ are good predictors of the response values $\{y_k\}_{k \in U}$. Under a measurable sampling design, an unbiased variance estimator is given by

$$\widehat{V}(\widetilde{T}_{y,\text{diff1}}) = \sum_{j,k \in U} \Delta_{jk} \frac{y_j - \sum_{\ell \in \mathscr{A}} m_{j\ell}\mu(\mathbf{a}_\ell)}{\pi_j} \frac{y_k - \sum_{\ell \in \mathscr{A}} m_{k\ell}\mu(\mathbf{a}_\ell)}{\pi_k} \frac{I_j I_k}{\pi_{jk}} \tag{10}$$

which, like (3), can be computed or closely approximated using standard survey software.

The behavior of the estimator under three extreme cases is of interest. First, if there is no matching at all, so that $m_{k\ell} \equiv 0$ for all $k \in U$, $\ell \in \mathscr{A}$, then $\widetilde{T}_{y,\text{diff1}}$ becomes

$$\sum_{\ell \in \mathscr{A}} \mu(\mathbf{a}_\ell) + \sum_{k \in s} \frac{y_k}{\pi_k} = A_\mu + \widehat{T}_y, \tag{11}$$

with expectation $A_\mu + T_y$ and variance equal to that of the Horvitz-Thompson estimator, (2). Effectively, the estimator regards the sampling design as having failed to cover the complete population, which is actually the disjoint union $\mathscr{A} \cup U$ and not U. It thus separately estimates the totals for the database and the universe and adds them together.

The second extreme case is that of full matching in the sense that $\sum_{k \in U} m_{k\ell} = 1$ for all $\ell \in \mathscr{A}$ (this is not the same as perfect matching). In this case, $\widetilde{T}_{y,\text{diff1}}$ is exactly unbiased for T_y by (8).

The third and final extreme case can occur if a rare characteristic appears in the population but is never encountered in the sample, so that $y_k \equiv 0$ for all $k \in s$. In this case, the estimator (7) becomes

$$\sum_{\ell \in \mathscr{A}} \mu(\mathbf{a}_\ell) - \sum_{k \in s} \sum_{\ell \in \mathscr{A}} \frac{m_{k\ell} \mu(\mathbf{a}_\ell)}{\pi_k}. \tag{12}$$

This behavior may be undesirable, as for a non-negative characteristic with non-negative predictions, the estimator predicts less than what is known to be present in the database. This behavior is better than that of the Horvitz-Thompson estimator, however, which would estimate zero for the population with such a degenerate sample. Nonetheless, other difference-type estimators are worth considering, including the one proposed below.

3.2.2   Second Difference Estimator
An alternative to $\widetilde{T}_{y,\text{diff1}}$ in (7) is obtained by an additional differencing adjustment,

$$\widetilde{T}_{y,\text{diff2}} = \widetilde{T}_{y,\text{diff1}} + \sum_{k \in s} \sum_{\ell \in \mathscr{A}} \frac{m_{k\ell} \{\mu(\mathbf{a}_\ell) - y_k\}}{\pi_k}$$

$$= \sum_{\ell \in \mathscr{A}} \mu(\mathbf{a}_\ell) + \sum_{k \in s} \frac{y_k(1 - \sum_{\ell \in \mathscr{A}} m_{k\ell})}{\pi_k}. \tag{13}$$

The expectation of the estimator is

$$\mathrm{E}\left[\widetilde{T}_{y,\text{diff2}}\right] = \mathrm{E}\left[\widetilde{T}_{y,\text{diff1}}\right] + \sum_{k \in U} \sum_{\ell \in \mathscr{A}} m_{k\ell} \{\mu(\mathbf{a}_\ell) - y_k\}$$

$$= T_y + \sum_{\ell \in \mathscr{A}} \left(1 - \sum_{k \in U} m_{k\ell}\right) \mu(\mathbf{a}_\ell) + \sum_{k \in U} \sum_{\ell \in \mathscr{A}} m_{k\ell} \{\mu(\mathbf{a}_\ell) - y_k\}. \tag{14}$$

Its variance is

$$\mathrm{Var}\left(\widetilde{T}_{y,\mathrm{diff2}}\right) = \sum_{j,k\in U} \Delta_{jk} \frac{y_j(1 - \sum_{\ell\in\mathscr{A}} m_{j\ell})}{\pi_j} \frac{y_k(1 - \sum_{\ell\in\mathscr{A}} m_{k\ell})}{\pi_k}. \tag{15}$$

The variance is small if the matching is good in the sense that $\sum_{\ell\in\mathscr{A}} m_{k\ell} \simeq 1$ for all $k \in U$. Under a measurable sampling design, an unbiased variance estimator is given by

$$\widehat{V}(\widetilde{T}_{y,\mathrm{diff2}}) = \sum_{j,k\in U} \Delta_{jk} \frac{y_j(1 - \sum_{\ell\in\mathscr{A}} m_{j\ell})}{\pi_j} \frac{y_k(1 - \sum_{\ell\in\mathscr{A}} m_{k\ell})}{\pi_k} \frac{I_j I_k}{\pi_{jk}}. \tag{16}$$

Again, like (3) and (10), this estimator can be computed or closely approximated using standard survey software.

We next consider the behavior of $\widetilde{T}_{y,\mathrm{diff2}}$ under the three extreme scenarios described above. First, if there is no matching at all, so that $m_{k\ell} \equiv 0$ for all $k \in U$, $\ell \in \mathscr{A}$, then $\widetilde{T}_{y,\mathrm{diff2}}$ reduces to $\widetilde{T}_{y,\mathrm{diff1}}$ by (13) and has exactly the same behavior.

Second, under full database matching in the sense that $\sum_{k\in U} m_{k\ell} = 1$ for all $\ell \in \mathscr{A}$, the expectation of $\widetilde{T}_{y,\mathrm{diff2}}$ in (14) becomes

$$T_y + \sum_{\ell\in\mathscr{A}} \mu(\mathbf{a}_\ell) - \sum_{k\in U}\sum_{\ell\in\mathscr{A}} m_{k\ell}y_k \tag{17}$$

so that, unlike $\widetilde{T}_{y,\mathrm{diff1}}$ under this scenario, $\widetilde{T}_{y,\mathrm{diff2}}$ is biased. The bias is small if $\sum_{k\in U} m_{k\ell}y_k$ is close to $\mu(\mathbf{a}_\ell)$ for all ℓ.

Third, with $y_k \equiv 0$ for all $k \in s$, the estimate computed from (13) becomes the full database total

$$\sum_{\ell\in\mathscr{A}} \mu(\mathbf{a}_\ell), \tag{18}$$

which may be preferable to either the zero estimate from Horvitz-Thompson or the reduced database total of $\widetilde{T}_{y,\mathrm{diff1}}$ from (12).

4   Simulation Experiments

4.1   Constructing the Population and Database

In the US state of South Carolina, there are about 500 operators of charter boats who take recreational angling trips with paying customers. Each boat can take multiple anglers, and over the course of 2016 there were about 50,000 angler trips on approximately 15,000 boat trips. These boat trips, along with the boat's logbook data on number of anglers and number of fish of each species caught by those anglers, are required to be reported to the South Carolina Department of Natural Resources, though reporting is incomplete. After removing logbook reports with missing values, we took the remaining $N = 10,647$ as the universe

U of actual boat trips to be studied. We then used a stochastic algorithm to simulate a corresponding database $\mathscr{A}$ of logbook records and a set of match metrics, $[m_{k\ell}]_{k\in U, \ell\in\mathscr{A}}$. In keeping with the real match metrics used in South Carolina, at most five of the $\{m_{k\ell}\}_{\ell\in\mathscr{A}}$ are non-zero for a given population element k.

We simulated the database by first sorting the universe in space and time, so that nearby elements in the population tend to be from the same coastal location and from nearby dates. We then used a Markov chain to determine the true (but unobservable) matching state of the population elements: no match, perfect match, high-quality match, or low-quality match. The transition probability matrix of the chain is as follows:

State	0	1	2	3	4	5	6	7	8	9	10	11	12	13	14	15	16
0	ρ_0	ρ_1	ρ_2	0	0	0	0	ρ_3	0	0	0	0	0	0	0	0	0
1	ρ_0	ρ_1	ρ_2	0	0	0	0	ρ_3	0	0	0	0	0	0	0	0	0
2	0	0	0	1	0	0	0	0	0	0	0	0	0	0	0	0	0
3	0	0	0	0	1	0	0	0	0	0	0	0	0	0	0	0	0
4	0	0	0	0	0	1	0	0	0	0	0	0	0	0	0	0	0
5	0	0	0	0	0	0	1	0	0	0	0	0	0	0	0	0	0
6	ρ_0	ρ_1	ρ_2	0	0	0	0	ρ_3	0	0	0	0	0	0	0	0	0
7	0	0	0	0	0	0	0	0	1	0	0	0	0	0	0	0	0
8	0	0	0	0	0	0	0	0	0	1	0	0	0	0	0	0	0
9	0	0	0	0	0	0	0	0	0	0	1	0	0	0	0	0	0
10	0	0	0	0	0	0	0	0	0	0	0	1	0	0	0	0	0
11	0	0	0	0	0	0	0	0	0	0	0	0	1	0	0	0	0
12	0	0	0	0	0	0	0	0	0	0	0	0	0	1	0	0	0
13	0	0	0	0	0	0	0	0	0	0	0	0	0	0	1	0	0
14	0	0	0	0	0	0	0	0	0	0	0	0	0	0	0	1	0
15	0	0	0	0	0	0	0	0	0	0	0	0	0	0	0	0	1
16	ρ_0	ρ_1	ρ_2	0	0	0	0	ρ_3	0	0	0	0	0	0	0	0	0

where $\sum_{i=0}^{3} \rho_i = 1$. This chain determines that an element k has no match (state 0); or determines that element k has a perfect match (state 1); or determines that five successive elements $k, k+1, \ldots, k+4$ are high-quality (HQ) matches (states 2–6); or determines that ten successive elements $k, k+1, \ldots, k+9$ are low-quality (LQ) matches (states 7–16).

In the event of no match, no database record is created, and $m_{k\ell} = 0$ for all $\ell \in \mathscr{A}$.

In the event of a perfect match, a database record that matches element k is created, and $m_{k\ell} = 1$ for $k = \ell$ and zero otherwise.

In the event of five HQ matches, five database records are created: the first record matches element k, the next record matches element $k+1$, and so on until the fifth record matches element $k+4$. Further, we generate five match metric values that sum to one by independently generating $U_k, U_{k+1}, \ldots, U_{k+4}$

as Uniform (0,1) and setting

$$(m_{k+i,k}, m_{k+i,k+1}, \ldots, m_{k+i,k+4}) = \frac{1}{\sum_{i=0}^{4} U_{k+i}} (U_k, U_{k+1}, \ldots, U_{k+4})$$

and $m_{k+i,\ell} = 0$ otherwise for all five elements $i = 0, 1, \ldots, 4$. That is, all five elements have the same match metric values with the same five database records. If we sample one of these five elements, we know that it (in truth) matches one of the five database records with non-zero match metric values, but we do not know which one.

In the event of ten LQ matches, five database records are created: the first record matches element k, the next record matches element $k + 1$, and so on until the fifth record matches element $k + 4$. The remaining five elements have no matching database records. All ten population elements share the same match metric values, constructed similarly to those for the HQ matches, but with match metric values summing to $1/2$ instead of 1: independently generate $U_k, U_{k+1}, \ldots, U_{k+4}$ as Uniform (0,1) and set

$$(m_{k+i,k}, m_{k+i,k+1}, \ldots, m_{k+i,k+4}) = \frac{1}{2\sum_{i=0}^{4} U_{k+i}} (U_k, U_{k+1}, \ldots, U_{k+4})$$

and $m_{k+i,\ell} = 0$ otherwise for all ten elements $i = 0, 1, \ldots, 9$. Thus, if we sample one of these ten elements, we think there might be no match at all (true for half of the ten elements) or there might be a match among the five database records (true for half of the elements), but we do not know which one.

We consider two population/database combinations, determined by the choice of $\rho_0, \rho_1, \rho_2, \rho_3$. The "Poor Match" combination results in simulated proportions of match metric values that closely mirror those in the actual South Carolina data, while the "Better Match" combination has greatly improved matching:

	Records	ρ_0	ρ_1	ρ_2	ρ_3	No match	LQ	HQ	Perfect
South Carolina						11.0%	52.5%	36.5%	0.0%
Poor Match	6836	0.35	0.20	0.25	0.20	8.6%	54.4%	31.7%	5.3%
Better Match	9031	0.10	0.20	0.60	0.10	2.3%	23.3%	69.8%	4.7%

Under the Poor Match combination, there are $6,836$ logbook records, so that many of the $N = 10,647$ population elements have no matching logbook records. Under the Better Match combination, there are $9,031$ logbook records. For each combination, we simulated the database once, and each population/database combination was then fixed for the remainder of the sampling experiment.

4.2 Estimation Properties Under Repeated Sampling

The sampling design used in our simulation study follows closely the design actually used by the Marine Recreational Information Program (MRIP) in

South Carolina. We stratified the population into fifteen strata by crossing three regions (each consisting of contiguous South Carolina counties) and five waves (March–April, May–June, July–August, September–October, November–December). Similar to MRIP, our sampling design selects particular sites on particular days ("site-days") and intercepts all boat trips on those selected site-days. In MRIP, the site-days are selected with probability proportional to a measure of size that is an estimate of fishing activity ("pressure") for the site-day. In our design, we approximate this unequal probability design by allocating an overall sample size of $n = 500$ site-days to the 15 strata using a database

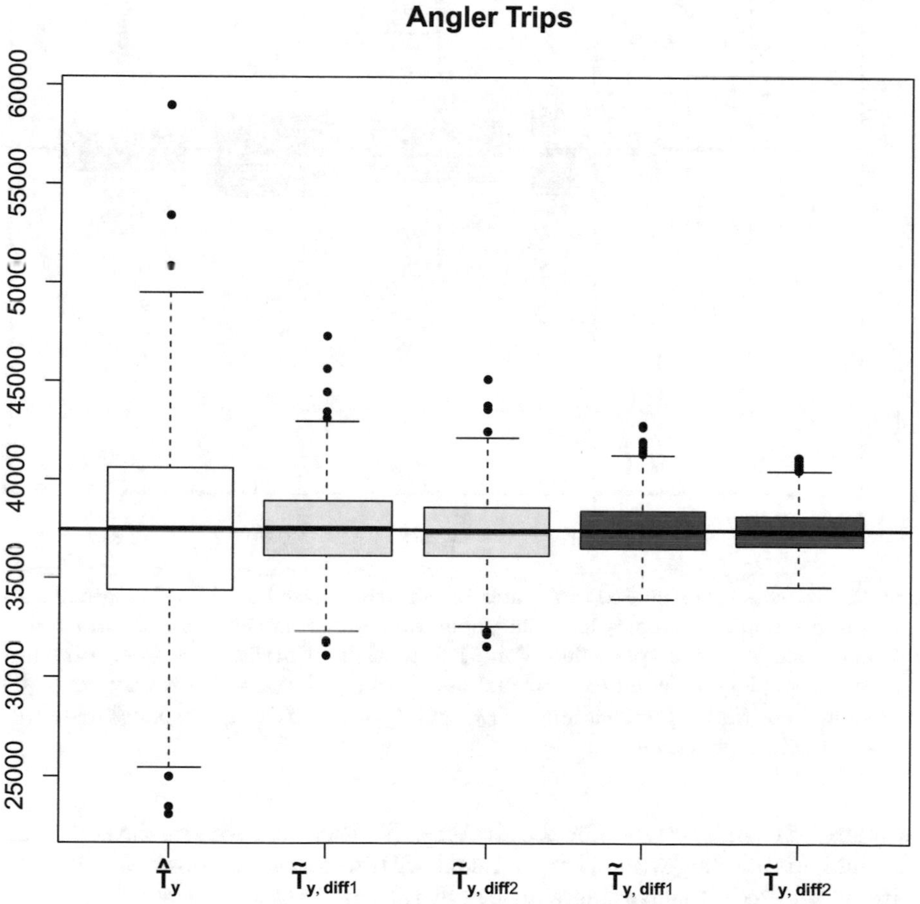

Angler Trips

Fig. 1. Boxplots for estimated total angler trips, based on 1000 simulated stratified simple random samples for each population/database combination. Horizontal reference line is at the true value. From left to right: Horvitz-Thompson estimator $\widehat{T}_y$ (white boxplot) under either combination; $\widetilde{T}_{y,\text{diff1}}$ and $\widetilde{T}_{y,\text{diff2}}$ (light gray boxplots) under the Poor Match combination; $\widetilde{T}_{y,\text{diff1}}$ and $\widetilde{T}_{y,\text{diff2}}$ (dark gray boxplots) under the Better Match combination.

Red Drum Catch

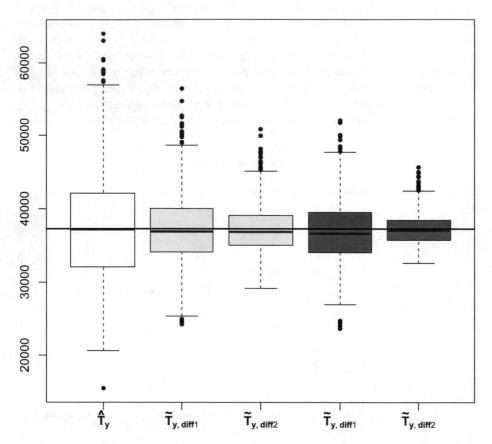

Fig. 2. Boxplots for estimated total catch of red drum, based on 1000 simulated stratified simple random samples for each population/database combination. Horizontal
reference line is at the true value. From left to right: Horvitz-Thompson estimator
$\widehat{T}_y$ (white boxplot) under either combination; $\widetilde{T}_{y,\text{diff1}}$ and $\widetilde{T}_{y,\text{diff2}}$ (light gray boxplots)
under the Poor Match combination; $\widetilde{T}_{y,\text{diff1}}$ and $\widetilde{T}_{y,\text{diff2}}$ (dark gray boxplots) under the
Better Match combination.

estimate of fishing pressure for the stratum. We then selected site-days via simple random sampling without replacement within strata, and observed all boat-
trips on selected site-days (there may, in fact, be no trips for a selected site-day).
We chose $n = 500$ so that the number of selected site-days with non-zero fishing activity closely matches the 109 non-zero site-days for South Carolina in
2016. Site-days are thus the primary sampling units (PSUs), selected via stratified simple random sampling, and boat-trips are the secondary sampling units,
selected with certainty within PSUs. Variance estimation needs to account for
this stratified two-stage structure.

Black Sea Bass Catch

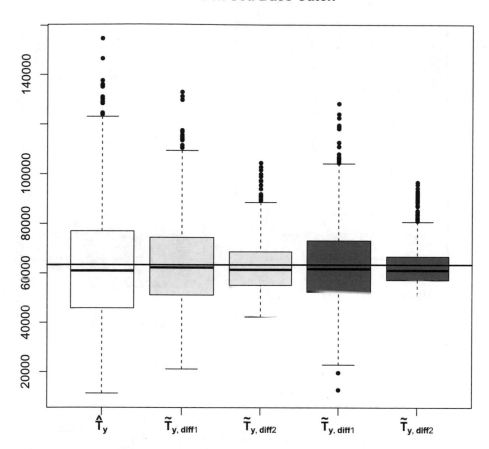

Fig. 3. Boxplots for estimated total catch of black sea bass, based on 1000 simulated stratified simple random samples for each population/database combination. Horizontal reference line is at the true value. From left to right: Horvitz-Thompson estimator $\widehat{T}_y$ (white boxplot) under either combination; $\widetilde{T}_{y,\text{diff1}}$ and $\widetilde{T}_{y,\text{diff2}}$ (light gray boxplots) under the Poor Match combination; $\widetilde{T}_{y,\text{diff1}}$ and $\widetilde{T}_{y,\text{diff2}}$ (dark gray boxplots) under the Better Match combination.

For each sampled boat-trip k in stratum h, the inclusion probability is $\pi_k = n_h/N_h$ where n_h is the number of site-days allocated to stratum h and N_h is the total number of site-days in stratum h, for $h = 1, 2, \ldots, 15$.

For this setting, our vector $\mathbf{a}_\ell$ of auxiliary information available for each element in the database includes time, location, number of anglers, and catch by species for multiple species of fish. Number of anglers and catch by species are of particular interest for estimation, and are observed for the sample of intercepted trips. The predictor $\mu(\mathbf{a}_\ell)$ for a characteristic of interest then simply returns the logbook value of the survey response: $\mu(\mathbf{a}_\ell) = $ logbook number of anglers when

Gag Grouper Catch

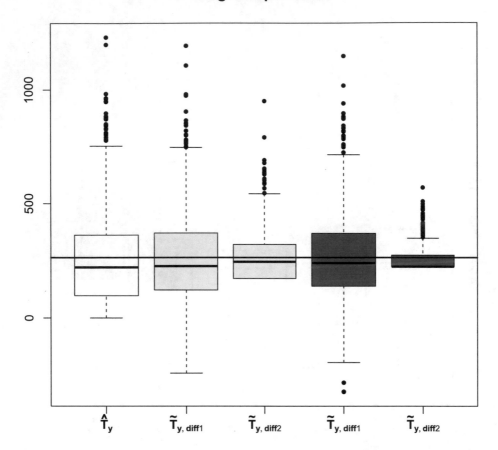

Fig. 4. Boxplots for estimated total catch of gag grouper, based on 1000 simulated stratified simple random samples for each population/database combination. Horizontal reference line is at the true value. From left to right: Horvitz-Thompson estimator $\widehat{T}_y$ (white boxplot) under either combination; $\widetilde{T}_{y,\text{diff1}}$ and $\widetilde{T}_{y,\text{diff2}}$ (light gray boxplots) under the Poor Match combination; $\widetilde{T}_{y,\text{diff1}}$ and $\widetilde{T}_{y,\text{diff2}}$ (dark gray boxplots) under the Better Match combination.

y_k = intercepted number of anglers, $\mu(\mathbf{a}_\ell)$ = logbook number of black sea bass when y_k = intercepted number of black sea bass, etc.

For each population/database combination, we drew 1000 independent stratified simple random samples from the fixed population and constructed the estimators $\widehat{T}_y$, $\widetilde{T}_{y,\text{diff1}}$, and $\widetilde{T}_{y,\text{diff2}}$ for several characteristics, including number of angler trips and total catch for red drum, black sea bass, gag grouper, Atlantic croaker, toadfish, and wahoo. These species were chosen to reflect a variety of reporting behaviors: in particular, they include species that are reported frequently in the database and are common enough to appear frequently in the on-site interviews, and species that are reported regularly but are rare enough to appear infrequently in the interviews. We also computed variance estimates as in (3), (10), and (16), but using the standard approximation of ignoring finite population corrections within strata. We present selected results here, noting that the Horvitz-Thompson estimator $\widehat{T}_y$ does not use the auxiliary information and has the same behavior under either combination.

Side-by-side boxplots for estimated total angler trips are shown in Fig. 1, for estimated red drum catch in Fig. 2, for estimated black sea bass catch in Fig. 3, and for estimated gag grouper catch in Fig. 4. We further summarized the results of the 1000 simulated samples for each estimator with the percent relative bias, root mean square error (RMSE), RMSE ratio (with Horvitz-Thompson estimator in the numerator), average estimated standard error (SE), and coverage of nominal 95% confidence intervals computed assuming approximate normality. Results are presented in Table 1.

Table 1. Summary results for estimated angler trips, red drum catch, black sea bass catch, and gag grouper catch, based on 1000 simulated stratified simple random samples for each population/database combination. Relative RMSE (Root Mean Square Error) is RMSE of the estimator in the denominator and RMSE of $\widehat{T}_y$ in the numerator. Estimated SE (standard error) is for stratified simple random sampling, but ignoring within-stratum finite population corrections. Confidence interval coverage is for nominal 95% coverage under normality, using (estimator) $\pm 1.96 \times$ (estimated SE).

		$\widehat{T}_y$	Poor Match		Better Match	
			$\widetilde{T}_{y,\text{diff1}}$	$\widetilde{T}_{y,\text{diff2}}$	$\widetilde{T}_{y,\text{diff1}}$	$\widetilde{T}_{y,\text{diff2}}$
Angler trips	Mean	37567.7	37600.0	37424.6	37534.3	37471.0
	Percent Relative Bias	0.3	0.4	−0.1	0.2	0.0
	Relative RMSE	1.0	2.1	2.4	3.1	3.9
	RMSE	4427.2	2071.0	1828.9	1443.3	1138.3
	Average Estimated SE	4555.0	2155.3	1915.4	1476.5	1138.3
	Confidence Interval Coverage	95.2	94.9	94.3	94.9	92.8
Red drum catch	Mean	37508.6	37266.3	37197.5	36887.2	37236.7
	Percent Relative Bias	0.6	0.0	−0.2	−1.1	−0.1
	Relative RMSE	1.0	1.5	2.3	1.7	3.5
	RMSE	7417.2	4857.0	3164.3	4301.2	2086.9
	Average Estimated SE	7270.7	4693.0	3042.5	4235.8	1960.8
	Confidence Interval Coverage	93.1	92.6	90.5	92.8	87.6
Black sea bass catch	Mean	63094.5	63915.1	62853.5	63509.8	62806.4
	Percent Relative Bias	−0.5	0.8	−0.9	0.2	−0.9
	Relative RMSE	1.0	1.3	2.2	1.4	3.0
	RMSE	23526.6	18008.0	10533.4	16287.8	7793.4
	Average Estimated SE	22725.9	17063.2	9653.7	15397.5	6473.7
	Confidence Interval Coverage	87.5	91.2	83.1	92.4	76.9
Gag grouper catch	Mean	256.0	260.9	272.1	263.7	258.5
	Percent Relative Bias	−3.4	−1.5	2.7	−0.5	−2.5
	Relative RMSE	1.0	1.1	2.0	1.1	3.8
	RMSE	209.2	196.2	105.0	188.8	54.4
	Average Estimated SE	170.7	170.5	75.5	163.3	29.2
	Confidence Interval Coverage	72.4	79.9	62.2	86.0	45.5

The Horvitz-Thompson estimator is theoretically unbiased, and both difference estimators are also nearly unbiased under each population/database combination and for each quantity of interest. Due to the low bias in all cases, the average estimated standard errors tend to be close to the RMSE's over the 1000 simulations, with the exceptions occurring for rarely-caught species like gag grouper. The sampling distributions of $\widehat{T}_y$ and $\widetilde{T}_{y,\mathrm{diff2}}$, which are nonnegative by construction for nonnegative responses, are then highly skewed, with corresponding poor confidence interval coverage. The sampling distribution of $\widetilde{T}_{y,\mathrm{diff1}}$, which is not constrained to be nonnegative, tends to be more symmetric and hence have better coverage with skewed distributions. This improved coverage comes at the expense of worse RMSE.

Under each population/database combination and for each quantity of interest, both difference estimators are better than the Horvitz-Thompson estimator, in terms of lower RMSE. The first difference estimator $\widetilde{T}_{y,\mathrm{diff1}}$ is sometimes not much better than $\widehat{T}_y$, but the second difference estimator $\widetilde{T}_{y,\mathrm{diff2}}$ is often much better than $\widehat{T}_y$, and is always better than $\widetilde{T}_{y,\mathrm{diff1}}$.

5   Discussion

The difference estimators described here are feasible in practice, given an auxiliary database and a suitable matching algorithm. The methodology offers substantial efficiency gains in a simulation study motivated by a real application in fisheries management. The simulation described here does not reflect any differential reporting, allowing probabilities of the match states to depend on the population characteristics. For example, boat captains catching only Atlantic croaker might be less likely to file a report than captains catching other species. The simulation also does not reflect differential measurement errors between the survey interviews and the logbook reports. In current practice, the boat captain is not required to file a logbook report immediately, and the catch recalled by the captain at the time of reporting may differ from the catch observed by an interviewer at a dockside intercept. These are directions for further study, both analytically and via simulation.

In results not reported here, we have also considered multiplicative adjustments of the Horvitz-Thompson estimator, as opposed to the additive adjustments of the difference-type estimators. These multiplicative adjustments lead to ratio-type estimators that can be considered generalizations of capture-recapture sampling, extending the work of [5]. These multiplicative adjustments, however, seem particularly sensitive to poor matching and can have large biases and variances. Further study on such estimators is necessary.

Two other directions for generalization of the results reported here are (1) allowing the predictor $\mu(\mathbf{a}_\ell)$ to be estimated from the sample and (2) allowing the match metric values $m_{k\ell}$ to depend on the sample. The first of these generalizations is standard in the survey literature (see [2] for an extensive review), but will be novel in this context due to the uncertain matching. The second generalization is also novel; some of the techniques of [1,3] may be relevant in determining suitable variance estimation strategies.

Acknowledgements. We thank Eric Hiltz of the South Carolina Department of Natural Resources (SC DNR) for development of the matching algorithm that motivated this work, and for assistance with the South Carolina logbook records and other data sources; John Foster of the National Oceanic and Atmospheric Administration Fisheries Service for assistance with the MRIP data for South Carolina; and Amy Dukes (SC DNR), Brad Floyd (SC DNR) and Geoffrey White (Atlantic Coastal Cooperative Statistics Program) for useful discussions.

References

1. Breidt, F.J., Opsomer, J.D.: Endogenous post-stratification in surveys: classifying with a sample-fitted model. Ann. Stat. **36**, 403–427 (2008)
2. Breidt, F.J., Opsomer, J.D.: Model-assisted survey estimation with modern prediction techniques. Stat. Sci. **32**(2), 190–205 (2017)
3. Dahlke, M., Breidt, F.J., Opsomer, J.D., Van Keilegom, I.: Nonparametric endogenous post-stratification estimation. Stat Sin, **23**, 189–211 (2013)
4. Horvitz, D.G., Thompson, D.J.: A generalization of sampling without replacement from a finite universe. J. Am. Stat. Assoc. **47**, 663 685 (1952)
5. Liu, B., Stokes, L., Topping, T., Stunz, G.: Estimation of a total from a population of unknown size and application to estimating recreational red snapper catch in Texas. J. Surv. Stat. Methodol. **5**, 350–371 (2017)
6. Lumley, T.: Analysis of complex survey samples. J. Stat. Softw. **9**(1), 1–19 (2004)
7. Lumley, T.: R package: survey: analysis of complex survey samples, version 3.30 (2014). https://www.r-project.org/(05.07.16)
8. Särndal, C.E., Swensson, B., Wretman, J.: Model Assisted Survey Sampling. Springer, New York (1992)
9. SAS Institute: SAS/STAT 14.1 User's Guide. Cary, NC (2015)

COBra: Copula-Based Portfolio Optimization

Marc S. Paolella[1,2(✉)] and Paweł Polak[3]

[1] Department of Banking and Finance, University of Zurich, Zurich, Switzerland
[2] Swiss Finance Institute, Geneva, Lausanne, Lugano, Zurich, Switzerland
marc.paolella@bf.uzh.ch
[3] Department of Statistics, Columbia University, New York, USA

Abstract. The meta-elliptical t copula with noncentral t GARCH univariate margins is studied as a model for asset allocation. A method of parameter estimation is deployed that is nearly instantaneous for large dimensions. The expected shortfall of the portfolio distribution is obtained by combining simulation with a parametric approximation for speed enhancement. A simulation-based method for mean-expected shortfall portfolio optimization is developed. An extensive out-of-sample backtest exercise is conducted and comparisons made with common asset allocation techniques.

Keywords: CCC · Expected shortfall · GARCH · Non-ellipticity
Student's t-copula

JEL classification: C13 · C32 · G11

1 Introduction

The use of copulae for modeling the joint distribution of financial asset returns is now a well-established field, with the seminal and still highly relevant article being Embrechts et al. (2002); see also Jondeau and Rockinger (2006). However, it appears that the literature is moving towards more sophisticated and complex copula constructions that are more able to adequately capture the co-movements and dependency structures in asset returns, particularly for large dimensions; see, e.g., Aas et al. (2009), Ausin and Lopes (2010), Scherer (2011), Christoffersen et al. (2012), Patton (2012), Aas (2016), Fink et al. (2017), and the references therein. As emphasized in McNeil et al. (2015), in the context of modeling asset returns within a copula framework, the choice of the copula is critical for determining the characteristics of the multivariate distribution. On the other hand, as pointed out in Embrechts (2009), there is usually no clear guidelines regarding which copula to use.

Before proceeding, we provide a list of the numerous abbreviations used throughout the paper (and thank an anonymous referee for this useful suggestion).

M.S. Paolella—Financial support by the Swiss National Science Foundation (SNSF) through project #150277 is gratefully acknowledged.

© Springer International Publishing AG 2018
V. Kreinovich et al. (eds.), *Predictive Econometrics and Big Data*, Studies in Computational Intelligence 753, https://doi.org/10.1007/978-3-319-70942-0_3

- APARCH: Refers to an Asymmetric Power ARCH process.
- ALRIGHT: The t-copula model proposed in Paolella and Polak (2015a), with the acronym coming from the title, Asymmetric LaRge-scale (I)GARCH with Hetero-Tails.
- COBra: The acronym for the title of this paper, Copula-Based Portfolio Optimization.
- DCC: Dynamic Conditional Correlation.
- ES: Expected Shortfall.
- GARCH: Generalized Autoregressive Conditional Heteroskedasticity.
- KP: Refers to the method of estimation proposed in the paper of Krause and Paolella (2014).
- MLE: Maximum Likelihood Estimator.
- MSE: Mean Squared Error.
- NCT: The noncentral (Student's) t distribution.
- NCT*: A location-shifted NCT random variable such that the mean is zero, as given in (3) below.
- pdf: Probability density function.
- SIMBA: The Simulation-Based Approximate mean-ES portfolio optimization method proposed in Sect. 3.4.
- VaR: Value at Risk.

   In this paper, we choose to work with the relatively simple Student's t copula, with noncentral-t (NCT) GARCH margins, as studied in Paolella and Polak (2015a), hereafter denoted ALRIGHT. A detailed study of the t copula can be found in Demarta and McNeil (2005) and McNeil et al. (2015). With respect to the margin distribution, note that it is rather flexible: Each is endowed with its own tail thickness parameter (the degrees of freedom of the NCT) and also an asymmetry parameter (via the noncentrality). While allowance for non-Gaussianity via leptokurtic or heavy-tailed distributions for financial asset returns data is virtually understood now, the asymmetry of asset returns is also important, and has been shown to be highly relevant for risk and density forecasting by numerous authors; see, e.g., Jondeau and Rockinger (2003, 2012), Haas et al. (2004), Kuester et al. (2006), Adcock (2010, 2014), Haas et al. (2013), Adcock et al. (2015), and the references therein. The margin distribution is also endowed with a GARCH-type structure for the evolution of the scale term that allows for asymmetries for the effect of the innovation term on the volatility, sometimes referred to as the leverage effect.
   With respect to our choice of the t copula, besides the dispersion matrix, it has only one shape parameter, ν_0. For finite ν_0, the copula exhibits tail dependence, though not in an asymmetric way (see, e.g., Jondeau (2016)). For a given window of data, ν_0 is estimated, as discussed below, and does not vary over time, except in the primitive way of allowing it to vary as the data window moves through time. While this choice appears to be rather limited, particularly in light of the aforementioned trend towards ever more sophisticated copula constructions, the literature is rather silent on, first, how to use copula-based models for financial *asset allocation* and, secondly, its performance in this regard

as compared to more traditional allocation methods. Addressing these issues is the point of this paper.

Compared to other recent methods of modeling multivariate asset returns that are strongly superior to classic Markowitz allocation or use of the DCC-GARCH model (e.g., Paolella and Polak (2015c); Gambacciani and Paolella (2017)), use of a copula-based model has several benefits. First, it has the ability to use a two-step estimation procedure, first estimating the univariate margin distributions, and then the copula parameters, this being much faster than joint maximum likelihood (ML). Second, the copula framework allows the distribution of asset returns to be non-elliptic, this being another accepted stylized fact of asset returns; see, e.g., McNeil et al. (2015), Chicheportiche and Bouchaud (2012), and the references therein. In particular, and differing from other sophisticated non-elliptic models such as Paolella and Polak (2015b), the margins can be specified separately from the copula structure, and thus can accommodate heterogeneous tail behaviors of the assets, which is a definitive stylized fact of asset returns; see, e.g., Paolella and Polak (2015a) for evidence. Third, the model accommodates (possibly time-varying) tail dependence—a potentially crucial aspect of risk management, especially during times of severe market downturns, when correlations among assets tend to substantially increase, and contagion effects arise. Note that Gaussian based models, such as the popular DCC-GARCH model (Engle 2009), cannot exhibit tail dependence.

In ALRIGHT, the marginal NCT-GARCH processes are estimated using ML, but based on a closed-form and "vectorizable" saddlepoint approximation to the NCT density, enabling vastly faster estimation than if the traditional expression for the probability density function (pdf) of the NCT distribution were used. In this paper, we use instead of GARCH the APARCH model of Ding et al. (1993), as it is able to capture asymmetry in the impact on volatility. Also, estimation of the NCT-APARCH univariate margin distributions is conducted using the method presented in Krause and Paolella (2014), which we hereafter denote as KP, and now discuss.

The KP method is unconventional in three respects. First, the APARCH parameters are fixed and not estimated, reminiscent of the suggestion of the 1994 RiskMetrics technical document. It is argued and demonstrated in KP that this is superior to their estimation in terms of value-at-risk (VaR) predictive ability, not to mention speed. Second, the location parameters ($a_{i,0}$; see below) are each estimated via an iterative convergent trimmed-mean procedure where the optimal trimming amount is based on a pre-determined mapping to the estimated degrees of freedom parameter. This has been shown in simulations in KP to be nearly as good as use of the MLE, and significantly outperforming the mean (this being a very poor estimator when the tail thickness is relatively large) and also the median in terms of mean squared error (MSE) accuracy, but obviating the need for (the otherwise slow) joint maximum likelihood estimation of the model parameters. Third, the estimation of the remaining two shape parameters of the NCT distribution (degrees of freedom and noncentrality parameter) is conducted based on a table-lookup procedure using sample quantiles, instead of ML estimation. This is not only extraordinarily fast, but, depending on the granularity of the table employed, (slightly) outperforms the MLE in terms of MSE when using 250 observations (though as the sample size grows, the MLE does, as the theory suggests, become slightly preferable).

The KP method also delivers the VaR and expected shortfall (ES) of the conditional NCT distribution instantaneously, again having used a table-lookup method. This obviates the otherwise slow integration of a heavy-tailed density with a complicated pdf to obtain the ES, and the root-solving required to obtain the quantile function—which involves the slow, repeated evaluation of the NCT cumulative distribution function (cdf).

In ALRIGHT, the emphasis was on assessing and improving the out-of-sample density prediction. This is arguably far more relevant than, say, reporting the statistical significance of certain coefficients based on asymptotically valid p-values (see, e.g., Nguyen, (2016), on the futility of using p-values for model selection), or inspection of in-sample fit measures, when applications are concerned with using the predictive density for volatility and correlation forecasting, and/or portfolio construction. In this paper, we emphasize portfolio performance, this being a strong test of a model that purportedly outperforms other constructs for modeling and predicting the multivariate distribution of asset returns. To this end, we develop a method to significantly expedite the calculation of the mean, variance, and ES of the portfolio distribution. In particular, as the weighted sum of the margins of a t-copula is analytically intractable, extensive simulation would be required for accurate assessment of the variance and, in particular, the ES, of the portfolio distribution, rendering such a naive approach to be too slow for extensive application. We propose use of various parametric approximations, all of which are both highly accurate and fast to estimate, and, crucially, whose mean and ES can be very quickly evaluated.

While our framework supports the use of the variance as a risk measure in portfolio optimization, we concentrate on use of the ES because, in a non-elliptic framework, optimal asset allocation requires the use of left-tail risk measures, as opposed to the variance. The ES is considered in this regard to be ideal; see, e.g., Embrechts et al. (2002), and the consultative document of the Basel Committee on Banking Supervision (2013). Empirical evidence favoring the use of mean-ES instead of mean-variance portfolio optimization in the presence of non-elliptic data is demonstrated, for example, in Paolella and Polak (2015c). See also Righi and Ceretta (2013, 2015) and the references therein.

The remainder of this paper is as follows. Section 2 presents the copula model, discusses the employed estimation methodology, and how it is forecast. Section 3 reviews how simulation is used for portfolio optimization and the concept of ES span. Section 4 walks through some empirical demonstrations using both simulated and real data, in order to assess the viability of the method and why it could fail. Section 5 details a new heuristic algorithm for improved asset allocation, and demonstrates its superior performance. Section 6 provides concluding remarks on the proposed methodology, discusses some of its benefits compared to more common asset allocation techniques, and provides some ideas for future research. Appendix A details the new method for calculating the ES of the portfolio distribution. Appendix B discusses estimation of the Gaussian DCC model, as used in our empirical work.

2   Model and Estimation

2.1   Notation

Let $R_{i,t}$ denote the (percentage log) return on asset i at time t, which we assume to be equally spaced in time (e.g., daily, ignoring weekend effects). To designate an individual time series, we use the notation $\mathbf{R}_{i,\bullet} = (R_{i,1}, \ldots, R_{i,T})'$, $i = 1, \ldots, d$. For a particular point in time t, the set of all returns is $\mathbf{R}_{\bullet,t} = (R_{1,t}, \ldots, R_{d,t})'$, $t = 1, \ldots, T$, while $\mathbf{D} = \mathbf{R}_{\bullet,\bullet} = [\mathbf{R}_{1,\bullet} \mid \mathbf{R}_{2,\bullet} \mid \cdots \mid \mathbf{R}_{d,\bullet}]$ denotes the whole data set.

We will require the (singly) noncentral-t, hereafter NCT, distribution, and a mean-zero version of it. Let $Z \sim \mathrm{N}(\gamma, \sigma^2)$, and $Y \sim \chi^2(\nu)$, independently of Z. Then

$$X = \mu + \frac{Z}{\sqrt{Y/\nu}} \sim \mathrm{NCT}(\mu, \sigma; \nu, \gamma), \tag{1}$$

where μ and γ are location and noncentrality coefficients, respectively, ν denotes the degrees of freedom parameter, which we assume is bounded below by 1, and σ is the scale parameter. With $\zeta = \gamma(\nu/2)^{1/2}\Gamma((\nu-1)/2)/\Gamma(\nu/2)$,

$$\mathbb{E}[X] = \mu + \zeta \quad \text{and} \quad \mathbb{V}(X) = \frac{\nu}{\nu-2}\left(\sigma^2 + \gamma^2\right) - \zeta^2, \tag{2}$$

the variance existing if $\nu > 2$ (see, e.g., Paolella 2007). A location-shifted NCT such that the mean is zero is required; this is denoted as

$$\mathrm{NCT}^*(\nu, \gamma) = \mathrm{NCT}(-\zeta, 1; \nu, \gamma), \quad Z \sim \mathrm{NCT}^*(\nu, \gamma), \quad \mathbb{E}[Z] = 0. \tag{3}$$

2.2   Model: Marginal Specifications

The NCT-APARCH model embodies the two most important stylized facts of conditional heteroskedasticity and leptokurtic tails of the innovations process (in common with the ubiquitous t-GARCH model of Bollerslev 1987), but also asymmetry in the innovations process and the effect of shocks on volatility. In particular, we assume $\mathbf{R}_{i,\bullet}$ follows the location-scale process

$$R_{i,t} = a_{i,0} + \sigma_{i,t} X_{i,t}, \quad X_{i,t} \overset{\mathrm{iid}}{\sim} \mathrm{NCT}^*(\nu_i, \gamma_i), \quad i = 1, \ldots, d, \tag{4}$$

where the evolution of $\sigma_{i,t}$ is governed by an APARCH$(1,1)$ process from Ding et al. (1993), and studied extensively in He and Teräsvirta (1999a, b), Ling and McAleer (2002), Karanasos and Kim (2006), and Francq and Zakoïan (2010, Chap. 10.) It is given by

$$|\sigma_{i,t}|^p = c_0 + c_1\left(|\epsilon_{i,t-1}| - g_1\epsilon_{i,t-1}\right)^p + d_1|\sigma_{i,t-1}|^p, \quad \epsilon_{i,t} = \sigma_{i,t} X_{i,t}, \tag{5}$$

with $c_0, c_1 > 0$, $d_1 \geq 0$, $|g_1| < 1$, and $p > 0$. From (2) and (3), the first two moments of $R_{i,t}$ are given by

$$\mathbb{E}[R_{i,t}] = a_{i,0}, \quad \mathbb{V}(R_{i,t}) = \sigma_{i,t}^2\left(\frac{\nu_i}{\nu_i - 2}(1 + \gamma_i^2) - \zeta_i^2\right), \quad \zeta_i = \gamma_i\sqrt{\frac{\nu_i}{2}}\Gamma\left(\frac{\nu_i - 1}{2}\right)\Gamma\left(\frac{\nu_i}{2}\right)^{-1}, \quad (6)$$

the variance existing if $\nu_i > 2$, $i = 1, \ldots, d$. As mentioned above, we assume $\nu_i > 1$, to ensure existence of the mean, but do not impose that the variance exists. However, in our applications, it was always the case that all $\widehat{\nu}_i > 2$, so we fix the power parameter p in (5) to be two. From an empirical point of view, estimates of p are often between one and two, with large standard errors, and fixing it to any value in $[1, 2]$ has very little effect on the forecasts. For data sets that are suspected of not possessing a conditional variance (see, e.g., Paolella, 2016, and the references therein), p can be taken to be unity in (5). As discussed in Krause and Paolella (2014), the APARCH parameters are fixed to judiciously chosen values given by

$$c_0 = 0.04, \ c_1 = 0.05, \ d_1 = 0.90, \text{ and } g_1 = 0.4. \quad (7)$$

2.3   Model: Student's t Copula

With $\mathbf{y} = (y_1, \ldots, y_d)' \in \mathbb{R}^d$, $\mathbf{a}_0 = (a_{1,0}, \ldots, a_{d,0})' \in \mathbb{R}^d$, and similarly defined for parameter vectors $\boldsymbol{\nu}$, $\boldsymbol{\gamma}$ and $\boldsymbol{\sigma}_t$, the joint density of the asset returns at time t is

$$f_{\mathbf{R}_\bullet,t}(\mathbf{y}; \mathbf{a}_0, \boldsymbol{\nu}, \boldsymbol{\gamma}, \boldsymbol{\sigma}_t, \boldsymbol{\Upsilon}, \nu_0) = \frac{f_{\mathbf{X}}(\mathbf{x}; \boldsymbol{\nu}, \boldsymbol{\gamma}, \boldsymbol{\Upsilon}, \nu_0)}{\sigma_{1,t}\sigma_{2,t}\cdots\sigma_{d,t}}, \quad \mathbf{x} = \left(\frac{y_1 - a_{1,0}}{\sigma_{1,t}}, \ldots, \frac{y_d - a_{d,0}}{\sigma_{d,t}}\right)', \quad (8)$$

where $f_{\mathbf{X}}(\mathbf{x}; \boldsymbol{\nu}, \boldsymbol{\gamma}, \boldsymbol{\Upsilon}, \nu_0)$ is given by

$$f_{\mathbf{X}}(\mathbf{x}; \cdot) = C\left(\Phi_{\nu_0}^{-1}\big(\Psi(x_1; \nu_1, \gamma_1)\big), \ldots, \Phi_{\nu_0}^{-1}\big(\Psi(x_d; \nu_d, \gamma_d)\big); \boldsymbol{\Upsilon}, \nu_0\right)\prod_{i=1}^{d}\psi(x_i; \nu_i, \gamma_i); \quad (9)$$

Ψ and ψ denote, respectively, the NCT* cdf and pdf; and $\Phi_{\nu_0}^{-1}(x)$ denotes the Student's t inverse cdf with $\nu_0 \in \mathbb{R}_{>0}$ degrees of freedom evaluated at $x \in \mathbb{R}$. Function $C(\cdot; \cdot)$ is the d-dimensional t-copula, referred to by Fang et al. (2002) as the *density weighting function*, and given by

$$\begin{aligned}C(\cdot; \cdot) &= C(z_1, z_2, \ldots, z_d; \boldsymbol{\Upsilon}, \nu_0)\\ &= \frac{\Gamma\{(\nu_0 + d)/2\}\{\Gamma(\nu_0/2)\}^{d-1}}{[\Gamma\{(\nu_0 + 1)/2\}]^d|\boldsymbol{\Upsilon}|^{1/2}}\left(1 + \frac{\mathbf{z}'\boldsymbol{\Upsilon}^{-1}\mathbf{z}}{\nu_0}\right)^{-(\nu_0+d)/2}\prod_{i=1}^{d}\left(1 + \frac{z_i^2}{\nu_0}\right)^{(\nu_0+1)/2}, \quad (10)\end{aligned}$$

where $\mathbf{z} = (z_1, z_2, \ldots, z_d)' \in \mathbb{R}^d$ and $\boldsymbol{\Upsilon}$ is a $d \times d$ correlation matrix, i.e.,

$$\boldsymbol{\Upsilon} = \big\{\rho_{ij} : \rho_{ii} = 1, -1 < \rho_{ij} < 1, i \neq j, \rho_{ji} = \rho_{ij}; i, j = 1, \ldots, d\big\}. \quad (11)$$

Thus, the model exhibits constant conditional correlation (CCC), and generalizes the Gaussian CCC model first presented in Bollerslev (1990). When the margins are restricted to being symmetric (central Student's t), we will also refer to the model as FaK, as an abbreviation of the authors in Fang et al. (2002), who considered the copula structure (albeit without GARCH effects in the margins) in detail.

2.4   Predictive Distribution: Simulation and Moments

Based on the obtained parameter estimates (denoted with hats), as discussed below in Sect. 2.5, and the predicted scale terms $\widehat{\sigma}_{i,t+1}$, $i = 1, \ldots, d$, computed as the usual deterministic update of the APARCH recursion (5), the predictive distribution at time $t + 1$ based on information at time t is given by

$$f_{\mathbf{R}_{\bullet,t+1|t}}(\mathbf{y}; \widehat{\mathbf{a}}_0, \widehat{\boldsymbol{\nu}}, \widehat{\boldsymbol{\gamma}}, \widehat{\boldsymbol{\sigma}}_{t+1}, \widehat{\mathbf{\Upsilon}}, \widehat{\nu}_0) = \frac{f_{\mathbf{X}}(\mathbf{x}; \widehat{\boldsymbol{\nu}}, \widehat{\boldsymbol{\gamma}}, \widehat{\mathbf{\Upsilon}}, \widehat{\nu}_0)}{\widehat{\sigma}_{1,t+1|t}\widehat{\sigma}_{2,t+1|t}\cdots\widehat{\sigma}_{d,t+1|t}}, \tag{12}$$

where

$$\mathbf{x} = \left(\frac{y_1 - \widehat{a}_{1,0}}{\widehat{\sigma}_{1,t+1|t}}, \ldots, \frac{y_d - \widehat{a}_{d,0}}{\widehat{\sigma}_{d,t+1|t}}\right)'.$$

As analytic expressions for the convolution of the margins (as required for the portfolio distribution) are not available, simulation will be used, as discussed further below. Generating a realization of $\mathbf{R}_{\bullet,t+1|t}$, denoted with a tilde, is conducted as follows Fang et al., (2002, p. 15): Draw $\mathbf{Y} = (Y_1, \ldots, Y_d)'$ from a d-dimensional multivariate Student's t distribution with location vector zero, correlation matrix $\widehat{\mathbf{\Upsilon}}$ and degrees of freedom parameter $\widehat{\nu}_0$, and set

$$\widetilde{R}_{i,t+1|t} = \widehat{a}_{i,0} + \widehat{\sigma}_{i,t+1}\Psi^{-1}\left(\Phi_{\widehat{\nu}_0}(Y_i); \widehat{\nu}_i, \widehat{\gamma}_i\right), \quad i = 1, \ldots, d, \tag{13}$$

where, as in (9), Ψ and Φ correspond to the NCT* and Student's t cdf, respectively, with the NCT* distribution given in (3).

Below, for portfolio optimization, it is desirable to have the mean and variance of $\mathbf{R}_{\bullet,t+1|t}$. As each margin Y_i is Student's t with location zero, scale one, and degrees of freedom $\widehat{\nu}_0$, it follows that $\Phi_{\widehat{\nu}_0}(Y_i) \sim \text{Unif}(0,1)$. By construction, the NCT* has mean zero. Thus, from the probability integral transform, $\mathbb{E}[\Psi^{-1}(\Phi_{\widehat{\nu}_0}(Y_i); \widehat{\nu}_i, \widehat{\gamma}_i)] = 0$, and

$$\widehat{\mathbb{E}}[\mathbf{R}_{\bullet,t+1|t}] = \widehat{\mathbf{a}}_0. \tag{14}$$

Similarly, from (6), if $\widehat{\nu}_i > 2$ for all $i = 1, \ldots, d$,

$$d_{ii} \equiv \mathbb{V}(R_{i,t+1|t}) = \widehat{\sigma}_{i,t+1}^2\left(\frac{\widehat{\nu}_i}{\widehat{\nu}_i - 2}(1 + \widehat{\gamma}_i^2) - \widehat{\zeta}_i^2\right), \quad i = 1, \ldots, d. \tag{15}$$

Determining the covariance matrix $\mathbb{V}(\mathbf{R}_{\bullet,t+1|t})$ associated with density (8) is not straightforward. Indeed, Fang et al. (2002) and Abdous et al. (2005) are silent on its off-diagonal elements. The simple expression

$$\widehat{\mathbb{V}}(\mathbf{R}_{\bullet,t+1|t}) = \mathbf{D}\widehat{\mathbf{\Upsilon}}\mathbf{D}, \quad \text{if } \widehat{\nu}_i > 2 \,\forall i, \tag{16}$$

suggests itself as a first-order approximation, where $\mathbf{D}$ is the $d \times d$ diagonal matrix with ith diagonal element $d_{ii}^{1/2}$ given in (15). Simulation confirms its viability when used with parameters typical for financial asset returns, with the obtained discrepancies of the variances and covariances having the same order of magnitude.

2.5 Parameter Estimation

Model estimation is conducted via a three-step procedure. First, the parameters of the marginal NCT-APARCH distributions are obtained via the KP method, as discussed in the introduction. Second, using the NCT-APARCH filtered residuals from the first step, $(R_{i,t} - \widehat{a}_{i,0})/\widehat{\sigma}_{i,t}$, the correlation matrix Υ is estimated using the traditional plug-in estimator for correlation, possibly augmented with shrinkage (see below), which differs from conventional wisdom of using the Kendall's tau transform estimator (see, e.g., McNeil et al. 2005, p. 98, 230). The latter is far slower than the simple plug-in correlation estimator, and does not result necessarily in a positive definite matrix. Moreover, as demonstrated in Paolella and Polak (2015a), despite evidence to the contrary, the use of the sample correlation estimator is nearly as efficient as use of the MLE in this context. Third, the single remaining copula parameter is estimated, or actually just assigned, based on the analysis in Paolella and Polak (2015a) showing that its estimation is superfluous with respect to density forecasts. In particular, they show that it is adequate to restrict $\nu_0 = \max_i \nu_i$, $i = 1, \ldots, d$.

Given the near-instantaneous nature of all three steps, the entire model estimation procedure is trivial: For $d = 30$ assets and $T = 250$ observations, we require 0.17 seconds, while with $d = 376$ and $T = 1,000$, we require 2.1 seconds (based on a desktop PC with an Intel Core i7-4790k processor, and using Matlab R2014B).

Because of the proliferation of the parameters as d increases, the estimated correlation matrix $\widehat{\Upsilon}$ will be subject to large estimation error, and shrinkage can be highly beneficial; see, e.g., Wolf (2004) and the references therein. Denote by $\widehat{\Upsilon}_{\mathrm{samp}}$ the sample correlation estimator. Shrinkage towards zero can be applied to its off-diagonal elements by taking the estimator to be $\widehat{\Upsilon} = (1 - s_\Upsilon)\widehat{\Upsilon}_{\mathrm{samp}} + s_\Upsilon \mathbf{I}$, for some $0 \leq s_\Upsilon \leq 1$. Alternatively, shrinkage towards the average of the correlation coefficients can be used, which can be algebraically expressed as, with $a = \mathbf{1}'(\widehat{\Upsilon}_{\mathrm{samp}} - \mathbf{I})\mathbf{1}/(d(d-1))$ and $\mathbf{1}$ a d-vector of ones,

$$\widehat{\Upsilon} = (1 - s_\Upsilon)\widehat{\Upsilon}_{\mathrm{samp}} + s_\Upsilon\left((1-a)\mathbf{I} + a\mathbf{1}\mathbf{1}'\right). \tag{17}$$

Use of (17), with $s_\Upsilon = 0.2$ for $d = 30$ was demonstrated in Paolella and Polak (2015a) to be most effective, in terms of out-of-sample density forecasting.

2.6 Calculating the Portfolio Expected Shortfall

Observe that the distribution of (weighted) sums of the margins of (12) is not analytically tractable, so that simulation is required to conduct the portfolio optimization. In particular, for a valid portfolio weight vector $\mathbf{w} = (w_1, \ldots, w_d)'$, denote by random variable $P_{t+1|t,\mathbf{w}}$ the return at time $t+1$ given information up to time t, and portfolio $\mathbf{w}$. The predictive portfolio returns distribution can be empirically generated by drawing s_1 replications from (12), stored in $d \times s_1$ matrix, say $\mathbf{M}$, and then computing

$$\widetilde{\mathbf{P}}_{t+1|t,\mathbf{w}}^{\mathrm{Emp}} = \mathbf{w}'\mathbf{M}. \tag{18}$$

The ES can be empirically approximated based on (18), though, being a tail measure, a large number of replications s_1 will be necessary. This approach is too time consuming to get the desired accuracy, so we propose to use much smaller samples and approximate its distribution with a flexible parametric form, from which the ES can be analytically calculated. The details are contained in Appendix A, where four such distributions are considered, and a heuristic algorithm for the choice of s_1 is developed that leads to accurate approximation of the ES.

3   Simulation–Based Portfolio Optimization

Even if one uses the same value of $\mathbf{M}$ from (18) during the optimization to select $\mathbf{w}$ (which is what we suggest and do), observe that the ES (irrespective of its calculation) will not be perfectly continuous (let alone differentiable) with respect to the portfolio weights. As such, standard gradient and/or Hessian-based optimization routines will not be effective for obtaining the optimal portfolio vector. This is not a serious drawback however: One can repeat the simulation exercise with, say, s_2 randomly chosen portfolio vectors, and then choose the portfolio vector with the desired characteristics, such as the largest expected return for a given ES, or the smallest ES for a given desired expected return. This simulation-based method is illustrated and discussed in detail in Paolella (2014), and briefly reviewed in Sect. 3.1. Section 3.2 discusses and illustrates the useful concept of ES span, while Sect. 3.3 introduces mean-variance portfolio optimization and the associated ES with respect to the ES span.

3.1   Drawing Random Portfolio Weight Vectors

The natural starting point for sampling a random non-negative portfolio vector $\mathbf{w}$ is to draw values that are uniform on the simplex $\mathbf{w} \in [0, 1]^d$, $\mathbf{w}'\mathbf{1} = 1$. It is important to note that taking $w_i = U_i/S$, for $S = \sum_{i=1}^{d} U_i$, with the U_i being i.i.d. standard uniform, does *not* result in a uniform distribution in the simplex. As detailed in, e.g., Devroye (1986, Chap. 5), to simulate uniformly, it is required to take logs, a point also correctly noted by Shaw (2010). That is, the portfolio vector $\mathbf{w}$ is uniform in the simplex by taking

$$\mathbf{w} = \mathbf{U}^{(\log)}/\mathbf{1}'\mathbf{U}^{(\log)}, \quad \mathbf{U}^{(\log)} = (\log U_1, \ldots, \log U_d)', \quad U_i \overset{\mathrm{iid}}{\sim} \mathrm{Unif}(0, 1). \tag{19}$$

Now consider taking instead

$$\mathbf{w} = \mathbf{U}^{(q)}/\mathbf{1}'\mathbf{U}^{(q)}, \quad \mathbf{U}^{(q)} = (U_1^q, \ldots, U_d^q)', \quad U_i \overset{\text{iid}}{\sim} \text{Unif}(0,1). \quad (20)$$

The non-uniformity corresponding to $q = 1$ is such that there are a disproportionate number of values close to the "$1/N$" (in our case, $1/d$), i.e., equally weighted, portfolio, and too few near the corner solutions. Observe that, as $q \to 0$, (20) collapses to the $1/N$ portfolio. The ability of the $1/N$ portfolio to outperform even sophisticated allocation methods goes back at least to Bloomfield et al. (1977), and is further detailed in DeMiguel et al. (2009), Brown et al. (2013), and the references therein. As such, sampling with $q = 1$ in (20) is useful because it covers the region around the equally weighted portfolio. As $q \to \infty$ in (20), $\mathbf{w}$ will approach a vector of all zeroes, except for a one at the position corresponding to the largest U_i. Thus, a large value of q is very useful for exploring corner solutions.

We sample by mixing over several types. In particular, out of the s_2 values, we take 40% from the uniform, 10% using $q = 1$, 40% using $q = 64$, and 10% using $q = 1024$. We denote this as "mixed $\mathbf{w}$ sampling". A data-driven heuristic method for improved sampling in conjunction with a related method for asset allocation is developed in Paolella (2017).

3.2   The ES Span

The differences in sampling techniques can be effectively illustrated using the so called ES-span, as introduced in Paolella (2014). With $\mathbf{D} = [\mathbf{R}_{1,\bullet} \mid \mathbf{R}_{2,\bullet} \mid \cdots \mid \mathbf{R}_{d,\bullet}]$ denoting a particular data set consisting of d assets, and a specified probability ξ for the ES level, we define the distribution of possible values that the ES can take on, over the set of all $\mathbf{w}$, under uniform $\mathbf{w}$ sampling, and conditional on a chosen model $\mathcal{M}$, to be $\text{span}_{\text{ES}}(\mathbf{D}, \mathcal{M}, \xi)$. When not otherwise specified, we use $\xi = 0.01$, i.e., 1% ES values. For sampling via (20) with a particular value of q, we write $\text{span}_{\text{ES}}(\mathbf{D}, \mathcal{M}, \xi, q)$, and for the mixed $\mathbf{w}$ sampling method, we write $\text{span}_{\text{ES}}(\mathbf{D}, \mathcal{M}, \xi, \text{Mix})$.

The values obtained from simulation can be plotted as a histogram or kernel density estimate, and convey knowledge of the distribution of the ES corresponding to $\mathbf{D}$ (and $\mathcal{M}$ and ξ). Observe that use of optimization algorithms gives no such information—they just return a single value (also dependent on $\mathbf{D}$, $\mathcal{M}$ and ξ) that one hopes is the global optimum. The spread of the ES values, measured as, say, the (sample) variance or interquartile range of $\text{span}_{\text{ES}}(\mathbf{D}, \mathcal{M}, \xi)$, or other measures, such as the distance from the minimal ES value to, say, the ES corresponding to the equally weighted portfolio, contain information about the time-varying nature of the data and also indicates under what conditions the $1/N$ portfolio is expected to do well. In general, the $1/N$ portfolio would be expected to be close to the minimum ES (or minimum variance) portfolio during relatively calm market periods, such that the returns are not very heavy-tailed, in which case the central limit theorem is applicable and results in the $1/N$

portfolio being close to Gaussian, i.e., a distribution with exponential (and not heavy) tails, and thus whose ES is relatively small, compared to other portfolios.

This aforementioned effect can be seen by comparing the ES-spans associated with the predictive density corresponding to the next trading day, based on the year of daily data for years 2005 and 2008, and taking $\mathcal{M}$ to be the the the copula model discussed above in Sect. 2. The year 2005 was rather calm, while 2008 corresponds to the global financial crisis. This is shown in Fig. 1. Indeed, the ES corresponding to $1/N$ is smaller (larger) than the mode of $\text{span}_{ES}(\mathbf{D}, \mathcal{M})$ for 2005 (2008).

3.3   ES Span for Mean–Variance Optimization

In the classic portfolio optimization framework going back to the seminal work of Markowitz (1952), the returns are assumed to be an i.i.d. multivariate normal process (or, more generally, from an elliptic distribution with existing second moments). One wishes to determine the portfolio weight vector, say $\mathbf{w}^*$, that yields the lowest variance of the predictive portfolio distribution. We choose to impose a no-short-selling constraint because it is common in practice, required by some investors (such as pension funds), and also because it can be interpreted as a type of shrinkage estimator that reduces the risk in estimated optimal portfolios; see Jagannathan and Ma (2003). Let $\widehat{\mathbf{w}}_M(\mathbf{D})$ denote the minimum variance portfolio without short selling, and based on the data $\mathbf{D}$ treated as an i.i.d. set of realizations, computed based on the sample plug-in mean and variance-covariance estimators. Further, let $\widehat{\mathbf{w}}_M(\mathbf{D}, \tau)$ be the same, but subject to an expected percentage annual mean return of $\tau = \tau_{\text{annual}}$, i.e.,

$$\widehat{\mathbf{w}}_M(\mathbf{D}, \tau) = \arg\min_{\mathbf{w} \in \mathcal{W}} \mathbf{w}'\widehat{\mathbb{V}}(\mathbf{D})\mathbf{w} \quad \text{such that} \quad \mathbf{w}'\widehat{\mathbb{E}}[\mathbf{D}] \geq \tau_{\text{daily}}, \quad (21)$$

where $\widehat{\mathbb{E}}$ and $\widehat{\mathbb{V}}$ denote plug-in estimators of their arguments,

$$\mathcal{W} = \{\mathbf{w} \in [0, 1]^d : \mathbf{1}'_d \mathbf{w} = 1\}, \quad (22)$$

and, with discrete compounding,

$$\tau_{\text{daily}} = 100\left(\left(1 + \frac{\tau}{100}\right)^{1/250} - 1\right), \quad \tau = 100\left(\left(1 + \frac{\tau_{\text{daily}}}{100}\right)^{250} - 1\right), \quad (23)$$

here calculated assuming 250 business days per year.

Observe that calculation of $\widehat{\mathbf{w}}_M(\mathbf{D})$ and $\widehat{\mathbf{w}}_M(\mathbf{D}, \tau)$ requires numeric optimization, but these are convex programming problems whose solutions are fast and numerically reliable. The ES values for $\widehat{\mathbf{w}}_M(\mathbf{D})$ and $\widehat{\mathbf{w}}_M(\mathbf{D}, 0.10)$ are also displayed in Fig. 1. As expected, the ES of $\widehat{\mathbf{w}}_M(\mathbf{D}, 0.10)$ exceeds that of $\widehat{\mathbf{w}}_M(\mathbf{D})$. It appears that $\widehat{\mathbf{w}}_M(\mathbf{D}, \tau)$ can be used as a good reference point from which random $\mathbf{w}$ should be generated. In particular, we suggest generating s_5 values of $\mathbf{w}$ according to

$$\mathbf{w} = s_6\widehat{\mathbf{w}}_M + (1 - s_6)\widetilde{\mathbf{w}}, \quad (24)$$

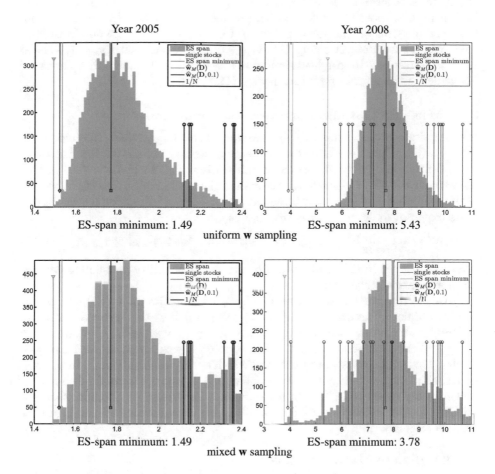

Fig. 1. Histogram of 10,000 1% ES values of $\mathrm{span}_{\mathrm{ES}}(\mathbf{D}, \mathcal{M}, 0.01)$ (top) and $\mathrm{span}_{\mathrm{ES}}(\mathbf{D}, \mathcal{M}, 0.01, \mathrm{Mix})$ (bottom), for the predictive density of January 3rd, 2006, based on $\mathbf{D}$ being the $d = 30$ constituents (as of April 2013) of the Dow Jones Industrial Average index (obtained from Wharton/CRSP) for the 252 trading days in year 2005 (left panels) and for January 2nd, 2009, based on the 253 trading days in 2008 (right panels); and such that model $\mathcal{M}$ is the copula model discussed herein, and using the NCT method for the parametric approximation of the simulated portfolio returns $\widetilde{\mathbf{P}}_{t+1|t,\mathbf{w}}$ to compute the ES, as discussed in Appendix A.

The ES of the 30 single stocks are shown as black vertical lines, though the graphs were truncated to improve readability, so that not all values appear. $1/N$ refers to the equally weighted portfolio, while $\widehat{\mathbf{w}}_M(\mathbf{D})$ and $\widehat{\mathbf{w}}_M(\mathbf{D}, 0.10)$ refer to "Markowitz"-type portfolio vectors, obtained by treating the data as i.i.d. and using the sample mean and variance-covariance matrix to determine the minimum variance, and 10%-mean return constrained minimum variance portfolios, respectively, and such that short selling is not allowed.

where $\widetilde{\mathbf{w}}$ is generated via uniform $\mathbf{w}$ sampling and s_6 is a tuning parameter indicating how close the new random portfolios are to $\widehat{\mathbf{w}}_M$ (as measured, for example, in terms of Euclidean distance). Use of this method to obtain the ES-span, which we denote by $\mathrm{span}_{\mathrm{ES}}(\mathbf{D}, \mathcal{M}, \xi, \mathrm{Mix}+M)$, and values $s_5 = 1,000$ and $s_6 = 0.90$, resulted in graphics similar to those in Fig. 1, but with a cluster of ES values in the neighborhood of the $\widehat{\mathbf{w}}_M(\mathbf{D}, \tau)$ solution.

3.4 Mean–ES Optimization Based on Simulation

We propose a *simulation-based approximate mean-ES portfolio* method, hereafter SIMBA. There are numerous benefits of using pure simulation instead of an optimization algorithm; see the detailed discussion in Paolella (2014). The objective is to deliver a portfolio vector that yields the lowest expected shortfall of the predictive portfolio distribution at time $t+1$, conditional on a lower bound of its expected daily percentage return τ_{daily},

$$\mathbf{w}^\star = \arg \min_{\mathbf{w} \in \mathcal{W}} \mathrm{ES}\big(P_{t+1|t,\mathbf{w}}, \xi\big) \quad \text{such that} \quad \mathbb{E}\big[P_{t+1|t,\mathbf{w}}\big] \geq \tau_{\mathrm{daily}}, \qquad (25)$$

where $\mathcal{W}$ is given in (22) and ξ is a pre-specified probability associated with the ES (for which we take 0.01).

Observe that, by the nature of simulated-based estimation, (25) will not be exactly obtained, but only approximated. We argue that this is not a drawback: All models, including ours, are wrong w.p.1; are anyway subject to estimation error; and the portfolio delivered will depend on the chosen data set $\mathbf{D}$, in particular, how much past data to use, and which assets to include; and, in the case of non-ellipticity, also depends on the choice of ξ. As such, the method should be judged not on how well (25) can be evaluated, but rather on the out-of-sample portfolio performance, conditional on all tuning parameters and the heuristics used to calculate (25).

4   Empirical Demonstration

Before applying our methods to real data, we investigate its performance using simulated data and based on the true model parameters. This obviously unrealistic setting serves as a check on the methodology and also (assuming the method is programmed correctly), will illustrate the large variation in the performance of the methods due strictly to the nature of statistical sampling. We begin with the simplified case of the usual multivariate Student's t distribution which, being elliptic, implies that standard Markowitz allocation is optimal, provided that second moments exist.

4.1 Use of the Multivariate t Distribution

We simulate first from the multivariate t distribution (hereafter MVT), this being a special case of our general model with all margins having a (central) Student's t distribution with the same degrees of freedom parameter, and no GARCH effects. We begin with a summary of important facts on the MVT distribution.

4.1.1 Distribution Theory

Recall that the d-dimensional, zero-location vector, identity-scale matrix, multivariate Student's t distribution with $v > 0$ degrees of freedom has density

$$f_{\mathbf{X}}(\mathbf{x}; v) = \frac{\Gamma\left(\frac{v+d}{2}\right)}{\Gamma\left(\frac{v}{2}\right)(v\pi)^{d/2}} \left(1 + \frac{\mathbf{x}'\mathbf{x}}{v}\right)^{-(v+d)/2}, \tag{26}$$

for $\mathbf{x} = (x_1, \dots, x_d)'$. The characteristic function (c.f.) corresponding to (26) was first (correctly) given by Sutradhar (1986) (without use of the Bessel function, but with different expressions for when v is odd, even, and fractional), while Song et al. (2014) derive it (and that of a type of generalized multivariate t) by extending the method for the univariate case from Hurst (1995), resulting in a much more compact expression in terms of the Bessel function. With $\mathbf{t} = (t_1, \dots, t_d)' \in \mathbb{R}^d$, it is given by

$$\varphi_{\mathbf{X}}(\mathbf{t}; v) = \frac{K_{v/2}\left(\|\sqrt{v}\mathbf{t}\|\right)\left(\|\sqrt{v}\mathbf{t}\|\right)^{v/2}}{\Gamma(v/2)\, 2^{v/2-1}}, \quad \|\mathbf{t}\| = \sqrt{\mathbf{t}'\mathbf{t}}. \tag{27}$$

For vector $\boldsymbol{\mu} = (\mu_1, \dots, \mu_d)' \in \mathbb{R}^d$ and $d \times d$ dispersion matrix $\boldsymbol{\Sigma} > 0$ with typical entry denoted σ_{ij} and diagonal elements denoted σ_j^2, $j = 1, \dots, d$, the location-scale version of (26) is given by

$$f_{\mathbf{X}}(\mathbf{x}; \boldsymbol{\mu}, \boldsymbol{\Sigma}, v) = \frac{\Gamma\left(\frac{v+d}{2}\right)}{\Gamma\left(\frac{v}{2}\right)(v\pi)^{d/2} |\boldsymbol{\Sigma}|^{1/2}} \left(1 + \frac{(\mathbf{x}-\boldsymbol{\mu})' \boldsymbol{\Sigma}^{-1} (\mathbf{x}-\boldsymbol{\mu})}{v}\right)^{-(v+d)/2}, \tag{28}$$

denoted $\mathbf{X} \sim t_v(\boldsymbol{\mu}, \boldsymbol{\Sigma})$. This distribution arises as follows: Let $X \sim \mathrm{IGam}(\alpha, \beta)$ with density

$$f_X(x; \alpha, \beta) = [\beta^\alpha / \Gamma(\alpha)]\, x^{-(\alpha+1)} \exp\{-\beta/x\}\, \mathbb{I}_{(0,\infty)}(x), \quad \alpha > 0, \beta > 0, \tag{29}$$

and trivial calculations confirming that

$$\mathbb{E}[X^r] = \frac{\Gamma(\alpha - r)}{\Gamma(\alpha)} \beta^r, \quad \alpha > r, \tag{30}$$

so that

$$\mathbb{E}[X] = \frac{\beta}{\alpha - 1}, \quad \mathbb{V}(X) = \frac{\beta^2}{(\alpha-1)^2(\alpha-2)}, \tag{31}$$

if $\alpha > 1$ and $\alpha > 2$, respectively. Let $G \sim \mathrm{IGam}(v/2, v/2)$, $v \in \mathbb{R}_{>0}$ and let $\mathbf{Z} = (Z_1, Z_2, \dots, Z_d)' \sim \mathrm{N}_d(\mathbf{0}, \boldsymbol{\Sigma})$. Then

$$\mathbf{X} = (X_1, X_2, \dots, X_d)' = \boldsymbol{\mu} + \sqrt{G}\mathbf{Z} \tag{32}$$

follows a d-variate $t_v(\boldsymbol{\mu}, \boldsymbol{\Sigma})$ distribution. From (30),

$$\mathbb{E}\left[G^{1/2}\right] = \sqrt{v/2}\frac{\Gamma\left(\frac{v-1}{2}\right)}{\Gamma\left(\frac{v}{2}\right)}, \quad v > 1, \tag{33}$$

so that, for $v > 1$, $\mathbb{E}[\mathbf{X}]$ exists, and, from (32), $\mathbb{E}[\mathbf{X}] = \boldsymbol{\mu} + \mathbb{E}\left[G^{1/2}\right]\mathbb{E}[\mathbf{Z}] = \boldsymbol{\mu}$. From (31),

$$\mathbb{E}[G] = v/(v-2), \text{ if } v > 2, \tag{34}$$

implying the well-known results that

$$\mathbb{E}[\mathbf{X}] = \boldsymbol{\mu}, \text{ if } v > 1, \qquad \mathbb{V}(\mathbf{X}) = \frac{v}{v-2}\boldsymbol{\Sigma}, \text{ if } v > 2. \tag{35}$$

Expression (32) is equivalent to saying that $(\mathbf{X} \mid G = g) \sim \mathrm{N}\left(\boldsymbol{\mu}, g\boldsymbol{\Sigma}\right)$, so that

$$f_{\mathbf{X}}\left(\mathbf{x}; \boldsymbol{\mu}, \boldsymbol{\Sigma}, v\right) = \int_0^\infty f_{\mathbf{X}\mid G}\left(\mathbf{x}; g\right) f_G\left(g; v/2, v/2\right) \mathrm{d}g. \tag{36}$$

The c.f. corresponding to (28) is

$$\varphi_{\mathbf{X}}(\mathbf{t}; \boldsymbol{\mu}, \boldsymbol{\Sigma}, v) = \mathbb{E}\left[e^{i\mathbf{t}'\mathbf{X}}\right] = e^{i\mathbf{t}'\boldsymbol{\mu}}\frac{K_{v/2}\left(\left\|\sqrt{v}\boldsymbol{\Sigma}^{1/2}\mathbf{t}\right\|\right)\left(\left\|\sqrt{v}\boldsymbol{\Sigma}^{1/2}\mathbf{t}\right\|\right)^{v/2}}{\Gamma\left(v/2\right)2^{v/2-1}}. \tag{37}$$

Let $j \in \{1, 2, \ldots, n\}$ and define $\mathbf{t} = (0, \ldots, 0, t, 0, \ldots, 0)'$, where t appears in the jth position. The marginal c.f. corresponding to X_j is Student's t, seen as follows. Observe that, with $\boldsymbol{\Sigma}^{1/2}$ symmetric (as can be obtained via the spectral decomposition method of calculating it), $\left\|\sqrt{v}\boldsymbol{\Sigma}^{1/2}\mathbf{t}\right\| = \sqrt{v}\sqrt{\mathbf{t}'\boldsymbol{\Sigma}\mathbf{t}} = \sqrt{v}|t|\sigma_j$. Thus,

$$\varphi_{X_j}(t) = e^{it_j\mu_j}\frac{K_{v/2}\left(v^{1/2}|t|\sigma_j\right)\left(v^{1/2}|t|\sigma_j\right)^{v/2}}{\Gamma(v/2)2^{v/2-1}}, \tag{38}$$

so that, from the uniqueness theorem of characteristic functions, $X_j \sim t_v(\mu_j, \sigma_j)$. See Ding (2016) for a simple derivation of (and corrections to mistakes in previous literature) of the conditional distribution of subsets of $\mathbf{X}$ given a different subset, paralleling the well-known result for the multivariate normal.

The following result is necessary for calculating the portfolio distribution when the assets follow a MVT distribution. Let $\mathbf{X} = (X_1, \ldots, X_d)' \sim t_v(\boldsymbol{\mu}, \boldsymbol{\Sigma})$ with p.d.f. (28), and define $S = \sum_{j=1}^d a_i X_i = \mathbf{a}'\mathbf{X}$ for $\mathbf{a} = (a_1, \ldots, a_d)' \neq \mathbf{0}$, i.e., S is a non-zero weighted sum of the univariate margins. Then

$$\varphi_S(t) = \mathbb{E}_S\left[e^{itS}\right] = \mathbb{E}_{\mathbf{X}}\left[e^{it\mathbf{a}'\mathbf{X}}\right] = \mathbb{E}_{\mathbf{X}}\left[e^{i(t\mathbf{a})'\mathbf{X}}\right] = \varphi_{\mathbf{X}}(t\mathbf{a}; \boldsymbol{\mu}, \boldsymbol{\Sigma}, v)$$

$$= e^{it\mathbf{a}'\boldsymbol{\mu}}\frac{K_{v/2}\left(\left\|\sqrt{v}t\boldsymbol{\Sigma}^{1/2}\mathbf{a}\right\|\right)\left(\left\|\sqrt{v}t\boldsymbol{\Sigma}^{1/2}\mathbf{a}\right\|\right)^{v/2}}{\Gamma\left(v/2\right)2^{v/2-1}}$$

$$= e^{it\mu_S}\frac{K_{v/2}\left(v^{1/2}|t|\kappa\right)\left(v^{1/2}|t|\kappa\right)^{v/2}}{\Gamma\left(v/2\right)2^{v/2-1}} = e^{it\mu_S}\varphi_T(\kappa t; v), \tag{39}$$

where $T \sim t_v$, $\mu_S = \mathbf{a}'\boldsymbol{\mu}$ and $\kappa = \left\|\boldsymbol{\Sigma}^{1/2}\mathbf{a}\right\| = \sqrt{\mathbf{a}'\boldsymbol{\Sigma}\mathbf{a}} > 0$. Thus, $S \overset{d}{=} \mu_S + \kappa T$, a location-scale Student's t with v degrees of freedom, where $\overset{d}{=}$ means equality in distribution (and is not to be confused with the dimension d of $\mathbf{X}$).

This method of proof can be extended to show a more general result encompassing (38) and (39). Let $\mathbf{X} \sim t_v(\boldsymbol{\mu}, \boldsymbol{\Sigma})$. For $1 \leq k \leq d$, $\mathbf{c} \in \mathbb{R}^k$, and $\mathbf{B}$ a $k \times d$ real matrix,

$$\mathbf{c} + \mathbf{B}\mathbf{X} \sim t_v\left(\mathbf{c} + \mathbf{B}\boldsymbol{\mu}, \mathbf{B}\boldsymbol{\Sigma}\mathbf{B}'\right). \tag{40}$$

This also follows from the more general statement for elliptic random variables; see Kelker (1970), Cambanis et al. (1981), Fang et al. (1989), and McNeil et al. (2015) for further details on elliptic distribution theory.

4.1.2   Empirical Performance Based on the MVT

We use $d = 10$, $v = 4$ degrees of freedom, and take the mean vector to be i.i.d. $N(0, 0.1^2)$, scale terms to be i.i.d. $Exp(1, 1)$ (i.e., scale one, and location one), and the off-diagonal elements of the correlation matrix of the MVT to be i.i.d. $Beta(4, 9)$, with mean $4/13$ and such that the resulting matrix is positive definite.

The next step is to compute the optimal portfolio vectors and the associated realized returns, over moving windows of (arbitrary) length $1,000$ and $\tau = 10\%$, using the long-only Markowitz method, and the allocation method using the simulation method (19) and knowledge of the true MVT parameters. This simulation exercise runs quickly, and it was conducted several times, with six representative results shown in Fig. 2. Observe that the portfolio ES is trivially calculated in this situation, as the MVT is in the elliptic class, as discussed above, so that (weighted) sums of its margins remains in the class, i.e., is univariate Student's t. We plot the cumulative realized returns from the Markowitz and MVT method via simulation from (19) based on $s_2 = 10,000$ replications, as well as the performance of the equally weighted ("$1/N$") portfolio. Overlaid are 100 cumulative returns based on randomly selecting the portfolio weights at each point in time to give a sense of if the computational methods are genuinely outperforming pure luck.

In each of the six cases, the true MVT parameters are different, but come from the same underlying distribution, as discussed above. The fact that the MVT case is using the true parameter values gives it an edge in terms of total returns, as seen in the middle and lower left panels, though in other cases, it does not perform better in finite-time experiments, such as in the middle right panel. The take-away message is that, even over a period using $2,000$ days of trading, allocation based on the true model and true parameters may not outperform the somewhat naive Markowitz approach (at least in terms of total return), and that the latter can even be beaten by the very naive $1/N$ strategy.

4.2   Use of the Student's t Copula Model with Known Parameters

We next take one step (of several) towards reality and leave the elliptic world, using instead the t copula model with heterogeneous degrees of freedom, such that, instead of the fixed degrees of freedom $v = 4$ used above, we take the ν_i in (4) and (8) to be realizations from a simulated $Unif(2, 7)$ distribution. There are

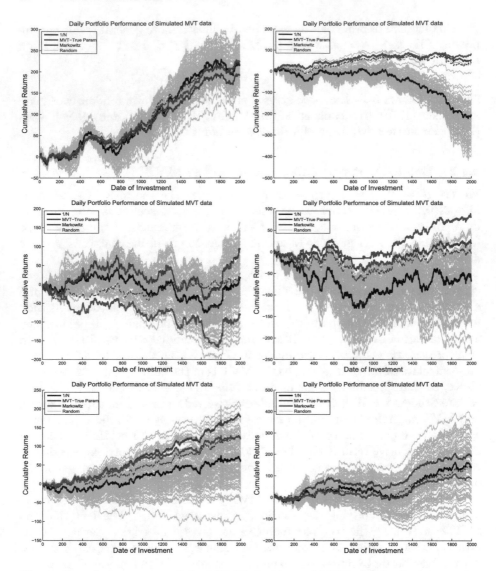

Fig. 2. Cumulative returns of the $1/N$, Markowitz and MVT models, the latter using the true parameter values. The thinner, dashed (red) line uses $s_2 = 1,000$ instead of $s_2 = 10,000$. In all but the top left case, use of $s_2 = 10,000$ is at least as good as $s_2 = 1,000$ and in some cases, such as the last four panels, leads to substantially better results.

still no GARCH effects, and we still assume the model, and the true parameters, are known. We also keep the restriction that the margins are symmetric, so that this coincides with the FaK model from Sect. 2.3. As the distribution of the (weighted) sum of margins in this case is not analytically tractable, computation of the ES is done via the method in Sect. 2.6, namely using the empirical VaR and ES, obtained from $s_1 = 10,000$ draws.

Results for four runs are shown in Fig. 3, with other runs (not shown) being similar. We obtain our hoped-for result that the t-copula model outperforms Markowitz (which is designed for elliptic data), and does so particularly when the set of ν_i tended to have smaller (heavier-tail) values. The $1/N$ portfolio is also seen to be inferior in this setting, particularly in the last of the four shown runs. The graphs corresponding to the t copula model are also such that they systematically lie near or above the top of the cloud of cumulative returns obtained from random portfolio allocations, indicating that accounting for the heavy-tailed and heterogeneous-tailed nature of the data indeed leads to superior asset allocation. This exercise also adds confirmation to the fact that allocations differ in the non-elliptic case, particularly amid heavy tails, and also that the algorithm for obtaining the optimal portfolio, and the method of calculating the ES for a given portfolio vector, are working.

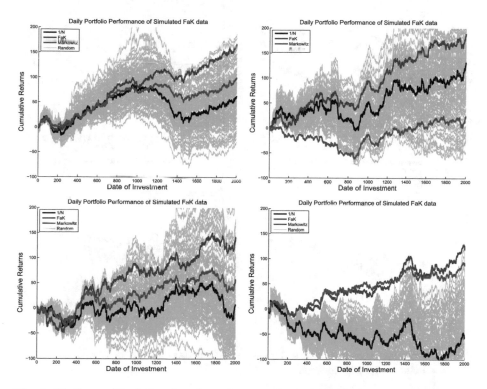

Fig. 3. Similar to Fig. 2, but based on the t copula model with symmetric Student's t margins without GARCH and with different degrees of freedom, using the true parameter values (labeled FaK). All plots were truncated in order to have the same y-axis.

4.3   Use of the Student's t Copula Model with Unknown Parameters

The crucial next step is to still use knowledge that the data generating process is the t copula (still without GARCH), but use parameter estimates instead of the true values, based on the estimation method discussed in Sect. 2, along with applying shrinkage to the estimated correlation matrix from (17), with the two cases $s_\Upsilon = 0$ and $s_\Upsilon = 0.30$.

Figure 4 is similar to Fig. 3, and uses the same generated data, so that the two figures can be directly compared. The degradation in performance of the copula model is apparent: The realistic necessity of parameter estimation when using parametric models takes a strong toll for all of the four runs shown, and also shrinkage of the estimated correlation matrix does not help, but rather, at least for the cases shown and the choice of $s_\Upsilon = 0.30$, predominantly hurts.

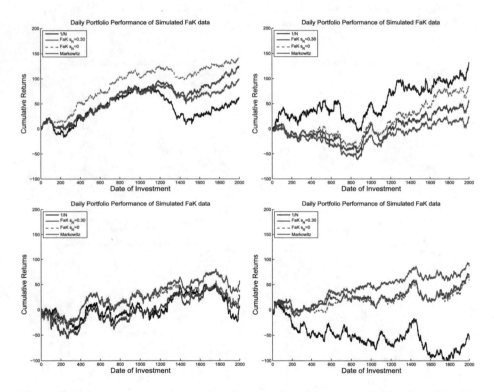

Fig. 4. Performance comparison using the same four data sets as in Fig. 3, and having estimated the model parameters.

Note that, if the true multivariate predictive density were somehow available, then the optimal portfolio can be elicited from it. However, this is probabilistic, and thus only with repeated investment over very many time periods would it be the case that, on average, the desired return is achieved with respect to the specified risk. As (i) the true predictive density is clearly not attainable (because the specified model is wrong w.p.1, along with the associated parameter estimation error); and (ii) backtest exercises necessarily involve a finite amount of data (so that the real long-term performance cannot be assessed with great accuracy), assessment of genuine asset allocation performance is difficult.

4.4 Alternative Investment Strategy

In light of the previous humbling results based on simulated data and use of the FaK model, we now consider an alterative investment strategy that capitalizes on the nature of how the optimal portfolio is determined. In particular, as we use random sampling instead of a black-box optimization algorithm to determine (25), we have access to $s_2 = 10,000$ portfolios. We apply the following algorithm for a given expected return τ, for which we use 10%:

1. For a given data set of dimension d, window length T, and expected return τ, estimate the copula model. In the below exercise, we still use simulated data, and so, for simplicity, omit GARCH effects.
2. Attempt s_2 random portfolios (we use $s_2 = 10,000$ for $d = 10$), and if after $s_2/10$ generations, no portfolio reaches the desired expected annual return (the τ-constraint), give up (and trading does not occur).
3. Assuming the exit in step 1 is not engaged, from the s_2 portfolios, store those that meet the τ-constraint, amassing a total of v valid portfolios.
4. If $v < s_2/100$, then do not trade. The idea is that, if so few portfolios meet the τ-constraint, then, taking the portfolio parameter uncertainty into account, it is perhaps unlikely that the expected return will actually be met.
5. Assuming $v \geq s_2/100$, keep the subset consisting of the $s_2/10$ with the lowest ES. (This requires that the stored ES, and the associated stored expected returns and portfolio vectors, are sorted.)
6. From this remaining subset, choose that portfolio with the highest expected return.

The core idea is to collect the 10% of portfolios yielding the lowest ES, and then choose among them the one with the highest expected return. Observe how this algorithm could also be applied to the Markowitz setting, using variance as a risk measure, but then the sampling algorithm would need to be used, as opposed to a direct optimization algorithm, as is applicable with (25). This alternative method contains several tuning parameters, such as the choice of τ, the window size, s_2, the shrinkage amount for the correlation matrix, and the (arbitrary) values of $s_2/100$ in step 4, and $s_2/10$ in step 5. In light of the risks and pitfalls of backtest overfitting (see, e.g., Bailey et al. 2014; 2016), one is behooved to investigate its performance for a range of such values (and data sets), and confirm that the results are reasonably robust with respect to their choices around an optimal range.

Figure 5 shows the resulting graphs based again on the same four simulated data sets as used above. There now appears to be some space for optimism, and tweaking the—somewhat arbitrarily chosen—tuning parameters surely will lead to enhanced performance. Realistic application would entail (i) applying the method also to real data; (ii) computing additional performance measures such as the Sharpe and related ratios; and (iii) taking into account transaction costs (these can be approximated by the simple but reasonably accurate method in DeMiguel et al. 2013). Finally, for real daily returns data, it would be wise to use the APARCH filter as discussed in Sect. 2.2.

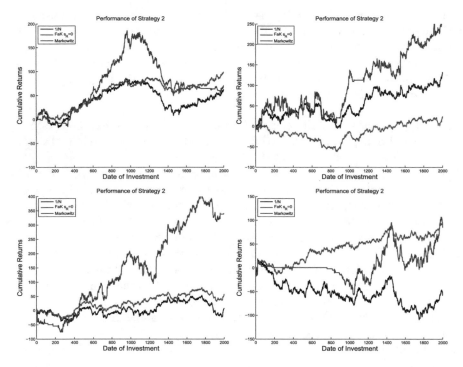

Fig. 5. Similar to Fig. 4, with estimated parameters and not using shrinkage on the correlation matrix, but having used the alternative investment strategy based on choosing among the 10% of generated portfolios with the lowest ES the one with the highest expected return.

4.5   Application to Real Data

We finally turn to the use of real data, using the daily (percentage log) returns based on the closing prices of the 30 stocks associated with the DJIA index, from Jan 2, 2001 to July 21, 2016 (conveniently including the market turmoil associated with the Brexit event). We use the alternative investment strategy from Sect. 4.4 and continue to use the FaK model instead of the more general case with noncentral t margins because the calculation of the portfolio ES is much faster in this case. Figure 6 shows the results in terms of the cumulative returns. The results are not encouraging, to put it mildly. Based on the previous progression of results, we can surmise two reasons for this: First, the fact that the parameters need to be estimated contributes significantly to the bad performance. Second, having moved from simulated to real data, we can conclude that the true data generating process is not being adequately captured by the FaK model.

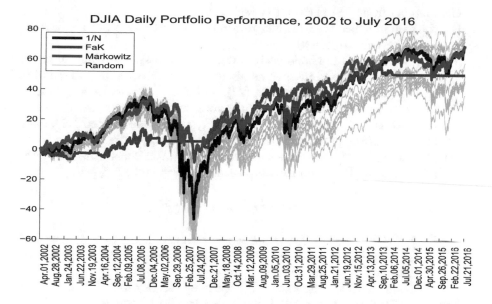

Fig. 6. Cumulative returns for the DJIA data using the FaK model, the $1/N$ allocation, and 400 random portfolios (showing only the most extreme ones).

To help address these issues, we turn to a more general alternative investment strategy that is applicable to all copula-based models under the assumption that they can be simulated from, and also use the t copula model with NCT-APARCH innovations, this being computationally more expensive because of the noncentrality parameters and simulation from the t copula.

5   SIMBA: Simulation-Based Approximate Mean-ES Optimized Portfolio Selection

Based on the disappointing performance when applied to real data of the FaK model in conjunction with simple mean-ES optimization as well as the alternative allocation method, we attempt to generalize the latter, capitalizing on the nature of the simulation-based method of portfolio optimization. The proposed heuristic algorithm, which we term SIMBA, addresses two situations: The first, as applies to our model, is such that the expected return of a candidate portfolio is known; in our case, it is given as $\mathbf{w}'\widehat{\mathbf{a}}_0$ using (14). The second applies to situations in which more sophisticated copula structures are used, and the expectation is not analytically known. Both cases assume that the ES is not analytically available.

5.1 The SIMBA Algorithm

The SIMBA algorithm is given as follows, applicable to an h-step ahead optimal asset allocation, for which we always use $h = 1$ in this paper. Let $E_i \equiv \mathbb{E}[P_{t+1|t,\mathbf{w}_i}]$.

1. Compute $\mathbf{M}$ from (18) based on s_1 replications.
2. For the case in which the expected return of the portfolio is not analytically known: (Alternate to step 3)
   (a) Draw s_2 random portfolio vectors $\mathbf{w}_i$, $i = 1, \ldots, s_2$, according to either the uniform or mixed method as discussed in Sect. 3.1 above, and add to this set the s_5 values drawn via (24), and the s_4 candidate vectors from the remarks below. Denote by $\check{s}_2$ the total number of vectors.
   (b) Approximate the $\check{s}_2$ expected returns E_i, and the associated ES values, using one of the parametric forms discussed in Appendix A.
   (c) Discard all portfolio vectors such that $E_i < \tau_{\text{daily}}$. Denote the remaining number as $\check{s}_2^*$. For given tuning parameters $\check{s}_2^{\min}$ and $\check{s}_2^{\max}$, set

$$\check{s}_2^\tau = \max\left(\check{s}_2^{\min}, \min\left(\check{s}_2^*, \check{s}_2^{\max}\right)\right).$$

   If $\check{s}_2^{\min} = 0$ and $\check{s}_2^\tau = 0$, trading is not conducted. If $\check{s}_2^\tau < \check{s}_2^*$, then sort the $\check{s}_2^*$ according to expected return, highest first, and keep only the first $\check{s}_2^\tau$ associated portfolio vectors.
3. For the case in which the expected return of the portfolio is analytically known: (Alternative to step 2) Sample s_2 portfolio vectors such that $E_i \geq \tau_{\text{daily}}$, noting that, for each, the ES does not need to be calculated. To this set, add the s_4 and s_5 candidate vectors, as discussed in step 2a, that satisfy the mean constraint. Let $\check{s}_2^\tau$ denote the total number of resulting vectors.
4. Calculate $\widehat{\mathbf{w}}_0^\star$, the (approximate) min-ES or mean-ES portfolio:
   (a) If step 2 is conducted and $\tau_{\text{daily}} = -\infty$, then from the $\check{s}_2^\tau$ ES values, deliver the portfolio vector that leads to the portfolio with the lowest ES.
   (b) If step 2 is conducted and $\tau_{\text{daily}} > -\infty$, or if step 3 is conducted, then from the $\check{s}_2^\tau$ ES values, deliver the portfolio vector that leads to the portfolio with the lowest ES if $E_i \geq \tau_{\text{daily}}$ holds for some $i = 1, \ldots, \check{s}_2$; otherwise, return $\arg\max_i E_i$, the portfolio corresponding to the highest return.
5. Calculate $\widehat{\mathbf{w}}_1^\star$: Based on the set of $\check{s}_s^\tau$ portfolio vectors resulting from step 2 or 3, deliver

$$\widehat{\mathbf{w}}_1^\star = \arg\max_{\mathbf{w}_i} \mathbb{E}\left[\widetilde{\mathbf{P}}_{t+1|t,\mathbf{w}_i}\right] \Big/ \text{ES}\left(\widetilde{\mathbf{P}}_{t+1|t,\mathbf{w}_i}\right), \tag{41}$$

this fraction being an obvious analog to many ratios, such as Sharpe, with expected return divided by a risk measure, but computed not for a sample of realized returns, but rather across different portfolio vectors, for a particular single point in time. We deem this object the *performance ratio of individual time forecasts*, or PROFITS measure.

6. Calculate $\widehat{\mathbf{w}}_2^\star$: Sort the $\breve{s}_2^T$ portfolio vectors obtained in step 2 or 3 according to expected return, highest first, and select the first $s_3\%$ of them. From these vectors, return $\widehat{\mathbf{w}}_2^\star(s_3)$ as the vector corresponding to the smallest portfolio ES value. For example, $\widehat{\mathbf{w}}_2^\star(1)$ corresponds to use of 1% of the $\breve{s}_2^T$ values. For $s_3 = 100$, $\widehat{\mathbf{w}}_2^\star$ and $\widehat{\mathbf{w}}_0^\star$ coincide.

Remarks:

1. Observe that, if the expectation of the predictive portfolio distribution is known (as a function of the estimated location terms $\mathbf{a}_0$ and any other relevant parameters), then use of step 3 instead of step 2 will be faster.
2. By construction of $\widehat{\mathbf{w}}_2^\star$, larger (smaller) values of ES are obtained for smaller (larger) values of s_3. This relates directly to an approximation of the efficient portfolio frontier. By taking advantage of sorting values in $\widehat{\mathbf{w}}_2^\star$ (step 6), different values of s_3 can be processed at once without noticeable additional computational effort, so that the efficient frontier based on $\widehat{\mathbf{w}}_2^\star$ could be plotted.
3. For use with rolling windows exercises, we advocate augmenting the simulated set of candidate portfolio vectors for estimators $\widehat{\mathbf{w}}_i^\star$, $i = 0, 1, 2$, by carrying over, say, s_4 of the best performing vectors, e.g., in terms of expected return, from the previous window. If not otherwise specified, we use $s_4 = \breve{s}_2^T/100$.
4. Similar to (24), we suggest augmenting the set of vectors to be carried over to the subsequent rolling window as follows. Let $\widehat{\mathbf{w}}^*$ denote some particular portfolio vector from the current window. For the subsequent window, select, say, s_7 vectors among the set of current portfolio vectors, $\mathbf{w}_j$, $j = 1, \ldots, s_7$, which are at most in a fixed L^2 distance from $\widehat{\mathbf{w}}^*$, i.e., such that $\|\mathbf{w}_j - \widehat{\mathbf{w}}^*\|_2^2 < s_8$ for all j, where s_8 is another tuning parameter related to a portfolio *turnover constraint*. This can serve to lower transaction costs and provide a speed increase to SIMBA by reducing the number of sampled portfolios s_2.

5.2   Second Empirical Portfolio Performance Comparison

We use for this empirical exercise the daily data for the period Jan 4, 1993, to Dec 31, 2012, of the $d = 30$ constituents (as of April 2013) of the Dow Jones Industrial Average index, as obtained from Wharton/CRSP, and moving windows of length $T = 250$. In particular, the first window spans Jan 4, 1993, to Dec 28, 1993, and delivers estimated portfolio vectors for the subsequent trading day, Dec 29, 1993. The window is then moved one period ahead (and thus changes by two data points). As such, forecast dates comprise all $4,787$ business days between Dec 29, 1993, and Dec 31, 2012. The SIMBA tuning parameters used are $s_1 = 1e3$, $s_2 = 1e4$, $s_3 = 1$, $s_4 = 0.01\breve{s}_2^T$, and the default values for the other s_i as stated above, in particular, $s_5 = 1,000$, $s_6 = 0.90$, $s_7 = 1,000$, $s_8 = 0.1$, $\tau_{\text{annual}} = 0.1$ and $\xi = 0.01$. For drawing the s_2 portfolio vectors, "mixed $\mathbf{w}$ sampling" is used, as described in Sect. 3.1.

The left panel of Fig. 7 shows the cumulative returns obtained from $\widehat{\mathbf{w}}_i^\star$, $i = 0, 1, 2$, when the NCT-APARCH marginals are estimated using the fast KP method, and based on the stable Paretian methods for the ES; see Appendix A.4. These are compared to the following benchmark models: (i) the equally weighted $(1/N)$ portfolio; (ii) the portfolio according to (21); and (iii) the portfolio similar to (21), but such that the sample mean and variance are replaced by the values obtained by using the standard Gaussian DCC-GARCH model of Engle (2002, 2009), which we denote by DCC, and whose calculations are described in detail in Appendix B.

We see that the use of $\widehat{\mathbf{w}}_2^\star(1)$ outperforms the usual competition by a wide margin in terms of total wealth. The exercise was repeated, but having used maximum likelihood estimation for the NCT-APARCH marginals. In addition to being far slower to estimate, the performance was blatantly lower, with a terminal wealth of 250 instead of 300, as obtained using the fast KP-method of estimation. This adds further evidence to the benefit of fixing the APARCH parameters instead of estimating them.

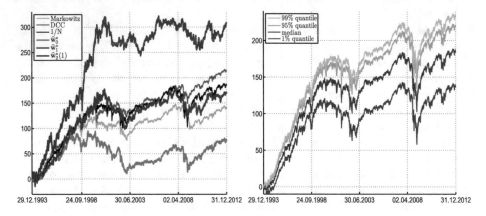

Fig. 7. Left: Cumulative realized portfolio returns for the SIMBA mean-ES portfolios $\widehat{\mathbf{w}}_0^\star$, $\widehat{\mathbf{w}}_1^\star$, and $\widehat{\mathbf{w}}_2^\star(1)$, estimated by the fast KP method, and using the stable Paretian distribution parametric approximation for determining the ES, overlaid with the performance of the three benchmark models.
Right: Quantile values over time based on $100,000$ paths of cumulative returns of random portfolios. Each path consists of 505 portfolio returns, obtained from a random portfolio characterized by a portfolio vector drawn uniformly according to (19). Note the change of the y-axis, for improved reading.

It is of interest to assess which models, if any, beat randomly chosen portfolios. It is well known that many active strategies do not perform better than random allocations; see, for example Edwards and Lazzara (2014), Hough (2014), and Clare et al. (2015). In particular, we wish to take the null hypothesis to be that a proposed method does not outperform purely random asset allocations. To obtain the distribution of the null, the portfolio vector obtained by drawing uniformly according to (19) is used at each point in time, and this is conducted $B = 100,000$ times. Instead of plotting the resulting B cumulative returns graphs, we plot a set of quantiles, computed at each point in time. This "monkey plot" is shown in the right panel of Fig. 7. The terminal wealth of the copula-based strategy is well above the 99% quantile (with a terminal wealth of about 220), lending evidence to the claim that the copula strategy is not performing well by chance. Interestingly, the next best performer, as shown in the left panel of Fig. 7, is the DCC model, which results in a terminal wealth just below that of the 95% quantile of the random portfolios, suggesting that it is not significantly outperforming "lucky random allocations". All other models lie well below the 95% quantile.

6   Conclusions

The goal of this paper is to operationalize the use of a basic t-copula structure with flexible and quickly-estimated NCT-APARCH margins, for portfolio optimization. It embodies the proposed method of simulation for eliciting optimal mean-ES portfolios based on a parametric approximation to the simulated portfolio density for speed enhancement. Future work includes further refinement of the algorithm, application to a variety of data sets, and accounting for transaction costs.

The better performance of the $\widehat{\mathbf{w}}_2^\star(1)$ estimator via the ALRIGHT-SIMBA model (hereafter, just A-S) compared to the Gaussian DCC-GARCH model (hereafter, just DCC) can be attributed to several factors.

1. A-S uses a copula framework, so that a basic form of tail dependence is captured. Tail dependence is not possible in the DCC model. Via use of short windows of estimation (we use 250) and re-estimation at every window, the copula parameter ν_0 is allowed to change through time, albeit in a primitive manner. More advanced copula structures could be used that exhibit richer tail dependency structures, possibly in conjunction with methods that are explicitly designed to detect changes in tail dependence, such as Bücher et al. (2015).
2. The A-S model is non-elliptic and heavy tailed; the DCC is not only elliptic, but also Gaussian.
3. The marginals of A-S use an NCT-APARCH model, allowing for fat-tailed, asymmetric innovations and volatility asymmetry. Moreover, estimation of the predictive means of the constituent returns are more accurate, owing to the use of the iterative trimmed mean procedure in KP, as opposed to the sample mean, which is non-optimal for heavy-tailed data, or just setting the mean to zero, in which case, only the minimum ES portfolio can be obtained, this being of far less interest to investors.
4. The APARCH model in A-S uses fixed parameters, instead of (ML) estimation, and as argued and demonstrated in KP, is superior in terms of VaR prediction and density forecasting.
5. While not related to performance, observe that model estimation via the KP-method is much faster, despite using a NCT-APARCH model for the margins, compared to DCC, which estimates the GARCH model for each asset.

Portfolio optimization, however, is not faster, because of the intractability of the portfolio distribution, though the SIMBA method, with the parametric approximation to the portfolio distribution and its fast estimation and calculation of mean and ES, does allow for reasonable computation time.

6. A-S uses only a CCC structure within the copula framework, but makes use of shrinkage applied to the dispersion matrix. Weighted likelihood could also be used, allowing the window size T to be larger while still mitigating the effect of a changing data generating process through time. We note that both techniques of shrinkage and weighted likelihood could also be used in the DCC framework. A possible avenue for future research is to incorporate a Markov-switching structure into the ALRIGHT model, similar to Pelletier (2006), thus allowing for a time-varying dispersion matrix Υ. In such a case, larger sample sizes can and should be used, instead of $T = 250$.

7. Note that asymmetric extensions of DCC exist; namely that of Cappiello et al. (2006) and Asai (2013); and presented in a general multivariate matrix structure, allowing each asset to have its own asymmetric DCC parameterization; though estimation of the matrix version of that model is subject to the curse of dimensionality, and so not applicable for direct use, even with a modest number of assets. Furthermore, it has yet to be demonstrated that such extensions lead to improved portfolio performance and, if still conducted under a Gaussian assumption, will most likely not be able to outperform the A-S or other models that use non-Gaussian structures, such as Paolella (2014) and Paolella and Polak (2015c).

A   Parametric Forms for Approximating the Distribution of $\widetilde{\mathbf{R}}_P$

We detail here the four candidate parametric structures mentioned in Sect. 2.6.

A.1   The Noncentral Student's t

The first is the location-scale NCT* distribution (3). As location μ and scale σ parameters need to be estimated along with the NCT* shape parameters, we compute

$$\arg\max_{\mu,\sigma} f_{\mathrm{NCT}}\left(P_{t+1|t,\mathbf{w}}; \widetilde{\nu}, \widetilde{\gamma}, \mu, \sigma\right), \quad \widetilde{\nu}, \widetilde{\gamma} = \mathrm{KP}\left(Z_{t+1|t,\mathbf{w}}\right), \quad Z_{t+1|t,\mathbf{w}} = \frac{P_{t+1|t,\mathbf{w}} - \mu}{\sigma}. \quad (42)$$

Starting values are taken to be the 50% trimmed mean for μ (i.e., the lower and upper 25% of the sorted sample are ignored) and, using (6) with $\nu = 4$ and $\gamma = 0$, gives $(s^2/2)^{1/2}$ for σ, where s^2 denotes the sample variance. Two box constraints $q_{0.25} < \widehat{\mu} < q_{0.75}$ and $(s^2/10)^{1/2} < \widehat{\sigma} < s$ are imposed during estimation, where q_ξ denotes the ξth sample quantile. The mean and variance are then determined from (6), while the ES is, via a table-lookup procedure, given essentially instantaneously from the KP method, noting that, for any probability $0 < \xi < 1$, $\mathrm{ES}(P_{t+1|t,\mathbf{w}}; \xi) = \mu + \sigma\mathrm{ES}(Z_{t+1|t,\mathbf{w}}; \xi)$.

A.2   The Generalized Asymmetric t

The second candidate is the five-parameter generalized asymmetric t, or GAt distribution. The pdf is

$$
f_{\mathrm{GA}t}(z; d, \nu, \theta) = K \times
\begin{cases}
\left(1 + \dfrac{(-z\theta)^d}{\nu}\right)^{-(\nu + 1/d)} & , \text{if } z < 0, \\[4mm]
\left(1 + \dfrac{(z/\theta)^d}{\nu}\right)^{-(\nu + 1/d)} & , \text{if } z \geq 0,
\end{cases}
\tag{43}
$$

where $d, \nu, \theta \in \mathbb{R}_{>0}$, and $K^{-1} = (\theta^{-1} + \theta)d^{-1}\nu^{1/d}B(1/d, \nu)$. It is noteworthy because limiting cases include the generalized exponential (GED), and hence the Laplace and normal, while the Student's t (and, thus, the Cauchy) distributions are special cases. For $\theta > 1$ ($\theta < 1$) the distribution is skewed to the right (left), while for $\theta = 1$, it is symmetric. See Paolella (2007, p. 273) for further details. The rth moment for integer r such that $0 \leq r < \nu d$ is

$$
\mathbb{E}[Z^r] = \frac{I_1 + I_2}{K^{-1}} - \frac{(-1)^r \theta^{-(r+1)} + \theta^{r+1}}{\theta^{-1} + \theta} \frac{B\big((r+1)/d, \nu - r/d\big)}{B(1/d, \nu)} \nu^{r/d},
$$

i.e., the mean is

$$
\mathbb{E}[Z] = \frac{\theta^2 - \theta^{-2}}{\theta^{-1} + \theta} \frac{B\big(2/d, \nu - 1/d\big)}{B(1/d, \nu)} \nu^{1/d}
\tag{44}
$$

when $\nu d > 1$, and the variance is computed in the obvious way. The cumulative distribution function (cdf) of $Z \sim \mathrm{GA}t(d, \nu, \theta)$ is

$$
F_Z(z; d, \nu, \theta) =
\begin{cases}
\dfrac{\bar{B}_L\big(\nu, 1/d\big)}{1 + \theta^2}, & \text{if } z \leq 0, \\[4mm]
\dfrac{\bar{B}_U\big(1/d, \nu\big)}{1 + \theta^{-2}} + \big(1 + \theta^2\big)^{-1}, & \text{if } z > 0,
\end{cases}
\tag{45}
$$

where $\bar{B}$ is the incomplete beta ratio,

$$
L = \frac{\nu}{\nu + \big(-z\theta\big)^d}, \quad \text{and} \quad U = \frac{\big(z/\theta\big)^d}{\nu + \big(z/\theta\big)^d}.
$$

For computing the ES, we require $\mathbb{E}[Z^r \mid Z < c]$ for $r = 1$. For $c < 0$, this is given by

$$
S_r(c) = (-1)^r \nu^{r/d} \frac{(1 + \theta^2)}{(\theta^r + \theta^{r+2})} \frac{B_L\big(\nu - r/d, (r+1)/d\big)}{B_L\big(\nu, 1/d\big),} \quad L = \frac{\nu}{\nu(-c\theta)^d}.
\tag{46}
$$

The existence of the mean and the ES requires $\nu d > 1$.

A.3   The Two-Component Mixture GA*t*

With five parameters (including location and scale), the GA*t* is a rather flexible distribution. However, as our third choice, greater accuracy can be obtained by using a two-component mixture of GA*t*, with mixing parameters $0 < \lambda_1 < 1$ and $\lambda_2 = 1 - \lambda_1$. This 11 parameter construction is extraordinarily flexible, and should be quite adequate for modeling the portfolio distribution. We also assume that the true distribution is not (single component) GA*t*, and that the distributional class of two-component mixtures of GA*t* is identified. Its pdf and cdf are just weighted sums of GA*t* pdfs and cdfs respectively, so that evaluation of the cdf is no more involved than that of the GA*t*. Let P denote a K-component mixGA*t* distribution, where each component has the three aforementioned shape parameters, as well as location u_i and scale c_i, $i = 1, \ldots, K$. First observe that the cdf of the mixture is given by

$$F_P(z) = \sum_{j=1}^{K} \lambda_j F_{Z_j}\left(\frac{z - u_j}{c_j}; d_j, \nu_j, \theta_j\right), \quad 0 < \lambda_j < 1, \quad \sum_{j=1}^{K} \lambda_j = 1, \quad (47)$$

where the ith cdf mixture component is given as the closed-form expression in (45), so that a quantile can be found by simple one-dimensional root searching. Similar to calculations for the ES of mixture distributions in Broda and Paolella (2011), the ES of the mixture is given by

$$
\begin{aligned}
\text{ES}_\xi(P) &= \frac{1}{\xi} \int_{-\infty}^{q_{P,\xi}} x f_P(x) \mathrm{d}x = \frac{1}{\xi} \sum_{j=1}^{K} \lambda_j \int_{-\infty}^{q_{P,\xi}} x c_j^{-1} f_{Z_j}\left(\frac{x - u_j}{c_j}\right) \mathrm{d}x \\
&= \frac{1}{\xi} \sum_{j=1}^{K} \lambda_j \int_{-\infty}^{\frac{q_{P,\xi} - u_j}{c_j}} (c_j z + u_j) c_j^{-1} f_{Z_j}(z) c_j \mathrm{d}z \\
&= \frac{1}{\xi} \sum_{j=1}^{K} \lambda_j \left[c_j \int_{-\infty}^{\frac{q_{P,\xi} - u_j}{c_j}} z f_{Z_j}(z) \mathrm{d}z + u_j \int_{-\infty}^{\frac{q_{P,\xi} - u_j}{c_j}} f_{Z_j}(z) \mathrm{d}z \right] \\
&= \frac{1}{\xi} \sum_{j=1}^{K} \lambda_j \left[c_j S_{1,Z_j}\left(\frac{q_{P,\xi} - u_j}{c_j}\right) + u_j F_{Z_j}\left(\frac{q_{P,\xi} - u_j}{c_j}\right) \right], \quad (48)
\end{aligned}
$$

where $q_{P,\xi}$ is the ξ-quantile of P, S_{1,Z_j} is given in (46), and F_{Z_j} is the cdf of the GA*t* random variable given in (45), both functions evaluated with the parameters d_j, ν_j, and θ_j from the mixture components, Z_j, for $j = 1, \ldots, K$.

While estimation of the two-component mixture GA*t* is straightforward using standard ML estimation, it was found that this occasionally resulted in an inferior, possibly bi-modal fit that optically did not agree well with a kernel-density estimate. This artefact arises from the nature of mixture distributions and the problems associated with the likelihood. We present a method that leads, with far higher probability, to a successful model fit, based on a so-called augmented likelihood procedure. The technique was first presented in Broda et al. (2013) and is adapted for the mixture GA*t* as follows.

Let $f(x; \boldsymbol{\theta}) = \sum_{i=1}^{K} \lambda_i f_i(x; \boldsymbol{\theta}_i)$ be the univariate pdf of a K-component (finite) mixture distribution with component weights $\lambda_1, \ldots, \lambda_K$ positive and summing to one. The likelihood function is

$$\ell^\star(\boldsymbol{\theta}; \mathbf{x}) = \sum_{t=1}^{T} \log \sum_{i=1}^{K} \lambda_i f_i(x_t; \boldsymbol{\theta}_i), \tag{49}$$

where $\mathbf{x} = (x_1, \ldots, x_T)'$ is the sequence of evaluation points, and $\boldsymbol{\theta} = (\boldsymbol{\lambda}, \boldsymbol{\theta}_1, \ldots, \boldsymbol{\theta}_K)'$ is the vector of all model parameters. Assuming that the $\boldsymbol{\theta}_i$ include location and scale parameters, $\ell^\star$ is plagued with "spikes"—it is an unbounded function with multiple maxima, see, e.g., Kiefer and Wolfowitz (1956). Hence, numerical maximization of (49) is prone (depending on factors like starting values and the employed numerical optimization method) to result in inaccurate, if not arbitrary, estimates. To avoid this problem, an augmented likelihood function is proposed in Broda et al. (2013). The idea is to remove unbounded states from the likelihood function by introducing a smoothing (shrinkage) term that, at maximum, drives all components to act as one (irrespective of their assigned mixing weight) such that the mixture loses its otherwise inherently large flexibility. The suggested augmented likelihood function is given by

$$\widetilde{\ell}(\boldsymbol{\theta}; \mathbf{x}) = \ell^\star(\boldsymbol{\theta}; \mathbf{x}) + \kappa \sum_{i=1}^{K} \frac{1}{T} \sum_{t=1}^{T} \log f_i(x_t; \boldsymbol{\theta}_i), \tag{50}$$

where κ, $\kappa \geq 0$, controls the shrinkage strength. If all component densities f_i are of the same type, larger values of κ lead to more similar parameter estimates across components, with identical estimates in the limit, as $\kappa \to \infty$. At $\kappa = 0$, (50) reduces to (49). The devised estimator,

$$\widehat{\boldsymbol{\theta}}_{\text{ALE}} = \arg\max_{\boldsymbol{\theta}} \widetilde{\ell}(\boldsymbol{\theta}; \mathbf{x}),$$

is termed the augmented likelihood estimator (ALE) and is asymptotically consistent, as $T \to \infty$. By changing κ, smooth density estimates can be enforced, even for small sample sizes. For mixGAt with $K = 2$ and 250 observations, we obtain $\kappa = 10$ as an adequate choice, which, in our empirical testing, guaranteed unimodal estimates in all cases, while still offering enough flexibility for accurate density fits, significantly better than those obtained with the single component GAt.

A.4   The Asymmetric Stable Paretian

The fourth candidate we consider is the use of the asymmetric non-Gaussian stable Paretian distribution, hereafter stable, with location μ, scale c, tail index α, and asymmetry parameter β. We use the parametrization such that the mean, assuming $\alpha > 1$, is given by μ. (In Nolan 2015, and the use of his software, this corresponds to his first parametrization; see also Zolotarev 1986; and Samorodnitsky and Taqqu 1994.)

This might at first seem like an odd candidate, given the historical difficulties in its estimation and the potentially problematic calculation of the ES, given the extraordinary heavy-tailed nature of the distribution and the problems associated with the calculation of the density far into the tails; see, e.g., Paolella (2016) and the references therein. We circumvent both of these issues as follows. We make use of the estimator based on the sample characteristic function of Kogon and Williams (1998), which is fast to calculate, and results in estimates that are very close in performance to the MLE. We use the function provided in John Nolan's STABLE toolbox, saving us the implementation, and easily confirming via simulation that his procedure is correct (and very fast). For the ES calculation, we first need the appropriate quantile, which is also implemented in Nolan's toolbox. The ES integral can then be computed using the integral expression given in Stoyanov et al. (2006), which cleverly avoids integration into the tail.

This procedure, while feasible, is still too time consuming for our purposes. Instead, we use the same procedure employed in Krause and Paolella (2014) to generate a (massive) table in two dimensions (α and β) to deliver the VaR (the required quantile) and the ES, essentially instantaneously and with very high accuracy. There is one caveat with its use that requires remedying. It is well-known, and as simulations quickly verify, that estimation of the asymmetry parameter β is subject to the most variation, for any particular sample size. The nature of the stable distribution, with its extremely heavy tails, relative to asymmetric Student's t distributions, will induce observations in small samples that have a relatively large impact on the estimation of β. This is particularly acute when using a relatively small sample size of $T = 250$. As such, we recommend use of a simple shrinkage estimator, with target zero and weight s_β, namely delivering $\widehat{\beta} = s_\beta \widehat{\beta}_{\text{MLE}}$. Some trial and error suggests $s_\beta = 0.3$ to be a reasonable choice for $T = 250$.

The motivation for using the stable is the conservative nature of the delivered ES. In particular, the first three methods we discussed are all based on asymmetric variations of the Student's t distribution which, while clearly heavy-tailed (it does not possess a moment generating function on an open neighborhood around zero), still potentially possesses a variance; as opposed to the stable, except in the measure-zero case of $\alpha = 2$. As such, and because estimation is based on a finite amount of data, the ES delivered from the stable will be expected to be larger than those from the t-based models. This might be desirable when more conservative estimates of risk should be used, and will also be expected to affect the optimized portfolio vectors and the performance of the method.

A.5   Discussion of Portfolio Tail Behavior and ES

It is worth mentioning that the actual tail behavior of financial assets is *not* necessarily heavy-tailed; the discussion in Heyde and Kou (2004) should settle this point. This explains why, on the one hand, exponential-tailed distributions, such as the mixed normal, can deliver excellent VaR predictions; see, e.g., Haas et al. (2004), Haas et al. (2013), and Paolella (2013); while, on the other hand, stable-Paretian GARCH models also work admirably well; see e.g., Mittnik et al. (2002) and Mittnik and Paolella (2003).

Further, observe that the tail behavior associated with the $P_{t+1|t,\mathbf{w}}$, given the model and the parameters, is not subject to debate: by the nature of the model we employ, it involves convolutions of (dependent) random variables with power tails, and, as such, will also have power tails, and will (presumably) be in the domain of attraction of a stable law. It is, however, analytically intractable. Observe that it is fallacious to argue that, as our model involves use of the (noncentral) Student's t, with estimated degrees of freedom parameters (after application of the APARCH filter) above two, the convolution will have a finite variance, and so the stable distribution cannot be considered. It is crucial to realize first that *the model we employ is wrong w.p.1* (and also subject to estimation error) and, second, recalling that, if an i.i.d. set of stable data with, say, $\alpha = 1.7$ is estimated as a location-scale Student's t model, the resulting estimated degrees of freedom will not be below two, but rather closer to four.

As such, we believe it makes sense to consider several methods of determining the ES, and compare them in terms of portfolio performance.

A.6   Comparison of Methods

The computation times for estimating the model and evaluating mean and ES for each of the four methods discussed above were compared. Based on a sample size of $s_1 = 1e3$, the NCT method requires, on average, 0.20 seconds. The GAt and mixGAt require 0.23 and 1.96 seconds, respectively, while the stable requires 0.00064 seconds. Generation of $s_1 = 1e6$ ($1e3$) samples requires approximately 2769.34 (2.91) seconds, and the empirical calculation of the mean and ES based on $s_1 = 1e6$ requires approximately 0.35 seconds. The bottleneck in the generation of samples is the evaluation of the NCT quantile function in (13). In summary, it is fastest to use $s_1 = 1e3$ samples and one of the parametric methods to obtain the mean and ES.

We now wish to compare the ES values delivered by each of the methods. For this, we fix the portfolio vector $\mathbf{w}$ to be equally weighted, and use 100 moving windows of data, each of length 250, and compute, for each method, the ES corresponding to the one-day-ahead predictive distribution and the fixed equally weighted portfolio. All the ES values (the empirically determined ones as well as the parametric ones) are based on (the same) 1e5 replications. The 100 windows have starting dates 8 August 2012 to 31 December 2012 and use the $d = 30$ constituents (as of April 2013) of the Dow Jones Industrial Average index from Wharton/CRSP. The values are shown in Fig. 8, and have been smoothed to enhance visibility. As expected, the stable ES values are larger than those delivered from the t-based models and also the empirically determined ES values. The mixGAt is the most flexible distribution and approximates the empirical ES nearly exactly, though takes the longest time to compute of the four parametric methods.

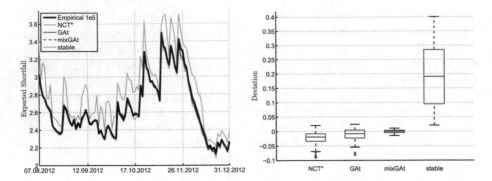

Fig. 8. Comparison of five methods of estimating ES, as discussed above, for a sequence of 100 rolling windows and based on the equally weighted portfolio. Each point was calculated based on (across the methods, the same) 1e5 replications. **Left**: The 100 values, for each method, plotted as a function of time. **Right**: The deviations of the four parametric methods from the empirical one.

A.7   Calibrating the Number of Samples s_1

As stated in Sect. 2.6, we wish to determine a heuristic for selecting the number of samples, s_1, from the predictive copula distribution, in order to obtain the ES. This is conducted as follows. The copula model is estimated for all non-overlapping windows of length $T = 250$ based on the 30 components of the DJIA returns available from 4 Jan. 1993 to 31 Dec. 2012 and the ES of the predictive returns distribution for the equally weighted portfolio is computed. The goal is to determine an approximation to the smallest value of s_1, say s_1^*, such that the sampling variance of the ES determined from the parametric methods is less than some threshold. This value s_1^* is then linked to the tail thickness of the various predictive returns distributions over the non-overlapping windows.

To compute s_1^* for a particular data set, the ES is calculated $n = 50$ times for a fixed s_1, based on simulation of the predictive returns distribution, and having used the NCT and stable parametric forms for its approximation. This is conducted for a range of s_1 values, and s_1^* is taken to be the smallest number such that the sample variance is less than a threshold value, For the NCT and stable estimators, Fig. 9 shows the results for selected values of s_1 for the NCT case. As expected, ES variances across rolling windows decrease with s_1 increasing. As can be seen from the middle right panels, a roughly linear relationship is obtained for the logarithm of ES variance. The analysis was also conducted for the stable Paretian distribution, resulting in a similar plot (not shown).

A simple regression approach then yields the following. For a threshold of $\exp(-2)$,

$$s_1(\widehat{\nu}_P) = \lceil 100 + (49.5 - 3.8\,\widehat{\nu}_P + 100.5\,\widehat{\nu}_P^{-1})^2 \rceil \mathbb{I}_{\{\widehat{\nu}_P \leq 15\}} + 100\mathbb{I}_{\{\widehat{\nu}_P > 15\}}$$

$$s_1(\widehat{\alpha}_P) = \lceil 100 + (-54800 - 13612\,\widehat{\alpha}_P + 55151\,\widehat{\alpha}_P^{-1})^2 \rceil \mathbb{I}_{\{\widehat{\alpha}_P \leq 1.97\}} + 100\mathbb{I}_{\{\widehat{\alpha}_P > 1.97\}} \cdot$$

$$(51)$$

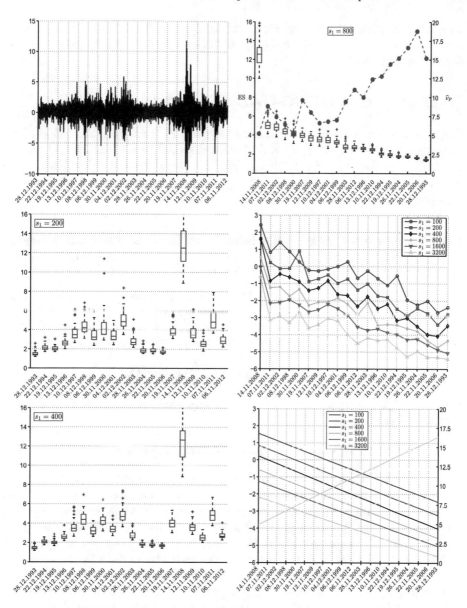

Fig. 9. Upper left: Percentage log returns of the equally weighted portfolio as used in the other panels. Mid and lower left: Boxplots of 1% ES values obtained from 50 simulations based on s_1 draws from the fitted copula for different non-overlapping rolling windows of size 250, spanning Jan 4, 1993, to Dec 31, 2012. Timestamps denote the most recent date included in a data window. All values are obtained via the NCT estimator. Upper right: Boxplots of 1% ES values sorted in descending order by the average ES value, overlayed by the average of the estimated degrees of freedom parameters. Mid right: ES variances in log scale across rolling windows for different samples sizes s_1, sorted by the average ES value per window. Lower right: Linear approximation of the above panel using ordinary least squares regression, overlayed by another linear approximation for the estimated degrees of freedom for the largest sample size $s_1 = 3,200$ under study.

The resulting procedure is then: From an initial set of 300 copula samples, the ES is evaluated, s_1 is computed from (51), and if $s_1 > 300$, an additional $s_1 - -300$ samples are drawn.

B   The Gaussian DCC-GARCH Model

Consider a d-dimensional vector of asset returns, $\mathbf{Y}_t = (Y_{t,1}, Y_{t,2}, \ldots, Y_{t,d})'$. The ith univariate series, $i = 1, \ldots, d$, is assumed to follow a GARCH(1,1) model, which is a special case of (5). We assume an unknown mean μ_i, so that $Y_{t,i} - \mu_i = \epsilon_{t,i} = Z_{t,i}\sigma_{t,i}^2$, $\sigma_{t,i}^2 = c_{0,i} + c_{1,i}(Y_{t-1,i} - \mu_i)^2 + d_{1,i}\sigma_{t-1,i}^2$, and $Z_{t,i}$ are i.i.d. standard normal.

B.1   Estimation Using Profile Likelihood for Each GARCH Margin

The DCC multivariate structure can be expressed as

$$\mathbf{Y}_{t|t-1} \sim \mathrm{N}_d(\boldsymbol{\mu}, \mathbf{H}_t), \quad \mathbf{H}_t = \mathbf{D}_t \mathbf{R}_t \mathbf{D}_t, \tag{52}$$

with $\boldsymbol{\mu} = (\mu_1, \ldots, \mu_d)'$, $\mathbf{D}_t^2 = \mathrm{diag}([\sigma_{t,1}^2, \ldots, \sigma_{t,d}^2])$, and $\{\mathbf{R}_t\}$ the set of $d \times d$ matrices of time varying conditional correlations with dynamics specified by

$$\mathbf{R}_t = \mathbb{E}_{t-1}[\epsilon_t \epsilon_t'] = \mathrm{diag}(\mathbf{Q}_t)^{-1/2} \mathbf{Q}_t \mathrm{diag}(\mathbf{Q}_t)^{-1/2}, \tag{53}$$

$t = 1, \ldots, T$, where $\epsilon_t = \mathbf{D}_t^{-1}(\mathbf{Y}_t - \boldsymbol{\mu})$. The $\{\mathbf{Q}_t\}$ form a sequence of conditional matrices parameterized by

$$\mathbf{Q}_t = \mathbf{S}(1 - a - b) + a(\epsilon_{t-1}\epsilon_{t-1}') + b\mathbf{Q}_{t-1}, \tag{54}$$

with $\mathbf{S}$ the $d \times d$ unconditional correlation matrix (Engle 2002, p. 341) of the ϵ_t, and parameters a and b are estimated via maximum likelihood conditional on estimates of all other parameters, as discussed next. Matrices $\mathbf{S}$ and $\mathbf{Q}_0$ can be estimated with the usual plug-in sample correlation based on the filtered ϵ_t; see also Bali and Engle (2010) and Engle and Kelly (2012) on estimation of the DCC model. Observe that the resulting $\mathbf{Q}_t$ from the update in (54) will not necessarily be precisely a correlation matrix; this is the reason for the standardization in (53). See Caporin and McAleer (2013) for several critiques of this DCC construction; and Aielli (2013) for a modified DCC model, termed cDCC, with potentially better small-sample properties. The CCC model is a special case of (52), with $a = b = 0$ in (54).

The mean vector, $\boldsymbol{\mu}$, can be set to zero, or estimated using the sample mean of the returns, as in Engle and Sheppard (2001) and McAleer et al. (2008), though in a more general non-Gaussian context, is best estimated jointly with the other parameters associated with each univariate return series; see Paolella and Polak (2017). Let $\mathbf{Y} = [\mathbf{Y}_1, \ldots, \mathbf{Y}_T]'$, and denote the set of parameters as $\boldsymbol{\theta}$. The log-likelihood of the remaining parameters, conditional on $\boldsymbol{\mu}$, is given by

$$\ell(\boldsymbol{\theta}; \mathbf{Y}, \boldsymbol{\mu}) = -\frac{1}{2} \sum_t \left(d\ln(2\pi) + \ln(|\mathbf{H}_t|) + (\mathbf{Y}_t - \boldsymbol{\mu})' \mathbf{H}_t^{-1} (\mathbf{Y}_t - \boldsymbol{\mu}) \right)$$

$$= -\frac{1}{2} \sum_t \left(d\ln(2\pi) + 2\ln(|\mathbf{D}_t|) + \ln(|\mathbf{R}_t|) + \boldsymbol{\epsilon}_t' \mathbf{R}_t^{-1} \boldsymbol{\epsilon}_t \right).$$

Then, as in Engle (2002), adding and subtracting $\boldsymbol{\epsilon}_t' \boldsymbol{\epsilon}_t$, ℓ can be decomposed as the sum of volatility and correlation terms, $\ell = \ell_V + \ell_C$, where

$$\ell_V = -\frac{1}{2} \sum_t \left(d\ln(2\pi) + 2\ln(|\mathbf{D}_t|) + \boldsymbol{\epsilon}_t' \boldsymbol{\epsilon}_t \right), \quad \ell_C = -\frac{1}{2} \sum_t \left(\ln(|\mathbf{R}_t|) + \boldsymbol{\epsilon}_t' \mathbf{R}_t^{-1} \boldsymbol{\epsilon}_t - \boldsymbol{\epsilon}_t' \boldsymbol{\epsilon}_t \right),$$

so that a two-step maximum likelihood estimation procedure can be applied: First, estimate the GARCH model parameters for each univariate returns series and construct the standardized residuals; second, maximize the conditional likelihood with respect to parameters a and b in (54) based on the filtered residuals from the previous step. We now discuss this first step in more detail.

While Francq and Zakoïan (2004) prove the consistency and asymptotic normality of the GARCH model parameters, interest centers on their numeric estimation. Dropping the subscript i, the choice of starting values for $\hat{c}_0$, $\hat{c}_1$, and $\hat{d}_1$ are important, as the log-likelihood can exhibit more than one local maxima. This issue of multiple maxima has been noted by Ma et al. (2006), Winker and Maringer (2009), and Paolella and Polak (2015b), though seems to be often ignored, and can lead to inferior forecasts and jeopardize results in applied work. This unfortunate observation might help explain the results of Brooks et al. (2001, p. 54) in their extensive comparison of econometric software. In particular, they find that, with respect to estimating just the simple normal GARCH model, "the results produced using a default application of several of the most popular econometrics packages differ considerably from one another". Another reason for discrepant results is the choice of ϵ_0 and σ_0 to start the GARCH(1,1) recursion, for which several suggestions exist in the literature. We take $\hat{\sigma}_0^2$ to be the sample unconditional variance of the R_t, and $\hat{\epsilon}_0^2 = \kappa \hat{\sigma}_0^2$, where

$$\kappa := \mathbb{E}\left[(|Z| - gZ)^\delta \right] \tag{55}$$

depends on the density specification $f_Z(\cdot)$ and is stated for the more general APARCH model (5). For $Z \sim N(0, 1)$, a trivial calculation yields

$$\mathbb{E}\left[(|Z| - gZ)^\delta \right] = \frac{1}{\sqrt{2\pi}} \left[(1 + g)^\delta + (1 - g)^\delta \right] 2^{(\delta - 1)/2} \Gamma\left(\frac{\delta + 1}{2} \right).$$

In our case, with $\delta = 2$ and $g = 0$, this reduces to $\kappa = \mathbb{E}\left[|Z|^2 \right] = 1$.

Paolella and Polak (2015b) demonstrate the phenomenon of multiple maxima with a real (and typical) data set, and propose a solution that is simple to implement, making use of the profile log-likelihood (p.l.) obtained by fixing the value of c_0, and using a grid of points of c_0 between zero and 1.1 times the sample variance of the series. That is, for a fixed value of c_0, we compute

$$\widehat{\boldsymbol{\theta}}_{\text{p.l.}}(c_0) = \arg\max_{\boldsymbol{\theta}_{\text{p.l.}}} \ell(\boldsymbol{\theta}_{\text{p.l.}}; \mathbf{R}), \qquad \boldsymbol{\theta}_{\text{p.l.}} = (c_1, d_1)'. \tag{56}$$

To obtain (with high probability) the global maximum, the following procedure suggests itself: (i) Based on a set of c_0 values, compute (56); (ii) take the value of c_0 from the set, say c_0^*, and its corresponding $\widehat{\boldsymbol{\theta}}_{\text{p.l.}}(c_0^*)$ that results in the largest log-likelihood as starting values, to (iii) estimate the full model. The finer the grid, the higher the probability of reaching the global maximum; some trials suggest that a grid of length 10 is adequate. The use of more parameters, as arise with more elaborate GARCH structures such as the APARCH formulation, or additional shape parameter(s) of a non-Gaussian distribution such as the NCT or stable Paretian, can further exacerbate the problem of multiple local maxima of the likelihood.

B.2   Remarks on DCC

One might argue that only two parameters for modeling the evolution of an entire correlation matrix will not be adequate. While this is certainly true, the models of Engle (2002) and Tse and Tsui (2002) have two strong points: First, their use is perhaps better than no parameters (as in the CCC model), and second, it allows for easy implementation and estimation. Generalizations of the simple DCC structure that allow the number of parameters to be a function of d, and also introducing asymmetric extensions of the DCC idea, are considered in Engle (2002) and Cappiello et al. (2006), though with a potentially very large number of parameters, the usual estimation and inferential problems arise.

Bauwens and Rombouts (2007) consider an approach in which similar series are pooled into one of a small number of clusters, such that their GARCH parameters are the same within a cluster. A related idea is to group series with respect to their correlations, generalizing the DCC model; see, e.g., Vargas (2006), Billio et al. (2006), Zhou and Chan (2008), Billio and Caporin (2009), Engle and Kelly (2012), So and Yip (2012), Aielli and Caporin (2013), and the references therein.

An alternative approach is to assume a Markov switching structure between two (or more) regimes, each of which has a CCC structure, as first proposed in Pelletier (2006), and augmented to the non-Gaussian case in Paolella et al. (2017). Such a construction implies many additional parameters, but their estimation makes use of the usual sample correlation estimator, thus avoiding the curse of dimensionality, and shrinkage estimation can be straightforwardly invoked to improve performance. The idea that, for a given time segment, the correlations are constant, and take on one set (of usually two, or at most three sets) of values. This appears to be better than attempting to construct a model that allows for their variation at every point in time. The latter might be "asking too much of the data" and inundated with too many parameters. Paolella et al. (2017) demonstrate strong out-of-sample performance of their non-Gaussian Markov switching CCC model with two regimes, compared to the Gaussian CCC case, the Gaussian CCC switching case, the Gaussian DCC model, and the non-Gaussian single component CCC of Paolella and Polak (2015b).

References

Aas, K.: Pair-copula constructions for financial applications: a review. Econometrics 4(4), 1–15 (2016). Article 43

Aas, K., Czado, C., Frigessi, A., Bakken, H.: Pair-Copula Constructions of Multiple Dependence. Insur. Math. Econ. 44, 182–198 (2009)

Abdous, B., Genest, C., Rémillard, B.: Dependence Properties of Meta-Elliptical Distributions. In: Duchesne, P., Rémillard, B. (eds.) Statistical Modeling and Analysis for Complex Data Problems. Springer Verlag, New York (2005). Chapter 1

Adcock, C.J.: Asset pricing and portfolio selection based on the multivariate extended skew-student-t distribution. Ann. Oper. Res. 176(1), 221–234 (2010)

Adcock, C.J.: Mean-variance-skewness efficient surfaces, Stein's lemma and the multivariate extended skew-student distribution. Eur. J. Oper. Res. 234(2), 392–401 (2014)

Adcock, C.J., Eling, M., Loperfido, N.: Skewed distributions in finance and actuarial science: a preview. Eur. J. Financ. 21(13–14), 1253–1281 (2015)

Aielli, G.P.: Dynamic conditional correlation: on properties and estimation. J. Bus. Econ. Stat. 31(3), 282–299 (2013)

Aielli, G.P., Caporin, M.: Fast clustering of GARCH processes via gaussian mixture models. Math. Comput. Simul. 94, 205–222 (2013)

Asai, M.: Heterogeneous asymmetric dynamic conditional correlation model with stock return and range. J. Forecast. 32(5), 469–480 (2013)

Ausin, M.C., Lopes, H.F.: Time-varying joint distribution through copulas. Comput. Stat. Data Anal. 54, 2383–2399 (2010)

Bailey, D.H., Borwein, J.M., López de Prado, M., Zhu, Q.J.: Pseudo-mathematics and financial charlatanism: the effects of backtest overfitting on out-of-sample performance. Not. Am. Math. Soc. 61(5), 458–471 (2014)

Bailey, D.H., Borwein, J.M., López de Prado, M., Zhu, Q.J.: The probability of backtest overfitting. J. Comput. Finan. (2016). https://papers.ssrn.com/sol3/papers.cfm?abstract_id=2840838

Bali, T.G., Engle, R.F.: The intertemporal capital asset pricing model with dynamic conditional correlations. J. Monetary Econ. 57(4), 377–390 (2010)

Fundamental Review of the Trading Book: A Revised Market Risk Framework. Consultative document, Bank for International Settlements, Basel (2013)

Bauwens, L., Rombouts, J.V.K.: Bayesian clustering of many GARCH models. Econometric Rev. 26(2), 365–386 (2007)

Billio, M., Caporin, M.: A generalized dynamic conditional correlation model for portfolio risk evaluation. Math. Comput. Simul. 79(8), 2566–2578 (2009)

Billio, M., Caporin, M., Gobbo, M.: Flexible dynamic conditional correlation multivariate GARCH models for asset allocation. Appl. Financ. Econ. Lett. 2(2), 123–130 (2006)

Bloomfield, T., Leftwich, R., Long, J.: Portfolio strategies and performance. J. Financ. Econ. 5, 201–218 (1977)

Bollerslev, T.: A conditional heteroskedastic time series model for speculative prices and rates of return. Rev. Econ. Stat. 69, 542–547 (1987)

Bollerslev, T.: Modeling the coherence in short-run nominal exchange rates: a multivariate Generalized ARCH approach. Rev. Econ. Stat. 72, 498–505 (1990)

Broda, S.A., Haas, M., Krause, J., Paolella, M.S., Steude, S.C.: Stable mixture GARCH models. J. Econometrics 172(2), 292–306 (2013)

Broda, S. A., Paolella, M. S:. Expected Shortfall for Distributions in Finance. In: Čížek, P., Härdle, W., and Rafał W. (eds.) Statistical Tools for Finance and Insurance (2011)

Brooks, C., Burke, S.P., Persand, G.: Benchmarks and the accuracy of GARCH model estimation. Int. J. Forecast. **17**(1), 45–56 (2001)

Brown, S. J., Hwang, I., In, F.: Why Optimal Diversification Cannot Outperform Naive Diversification: Evidence from Tail Risk Exposure (2013)

Bücher, A., Jäschke, S., Wied, D.: Nonparametric tests for constant tail dependence with an application to energy and finance. J. Econometrics **1**(187), 154–168 (2015)

Cambanis, S., Huang, S., Simons, G.: On the theory of elliptically contoured distributions. J. Multivar. Anal. **11**(3), 368–385 (1981)

Caporin, M., McAleer, M.: Ten things you should know about the dynamic conditional correlation representation. Econometrics **1**(1), 115–126 (2013)

Cappiello, L., Engle, R.F., Sheppard, K.: Asymmetric dynamics in the correlations of global equity and bond returns. J. Financ. Econometrics **4**(4), 537–572 (2006)

Chicheportiche, R., Bouchaud, J.-P.: The joint distribution of stock returns is not elliptical. Int. J. Theor. Appl. Financ. **15**(3), 1250019 (2012)

Christoffersen, P., Errunza, V., Jacobs, K., Langlois, H.: Is the potential for international diversification disappearing? a dynamic copula approach. Rev. Financ. Stud. **25**, 3711–3751 (2012)

Clare, A., O'Sullivan, N., and Sherman, M.: Benchmarking UK mutual fund performance: the random portfolio experiment. Int. J. Financ. (2015). https://www.ucc.ie/en/media/research/centreforinvestmentresearch/RandomPortfolios.pdf

Demarta, S., McNeil, A.J.: The t copula and related copulas. Int. Stat. Rev. **73**(1), 111–129 (2005)

DeMiguel, V., Garlappi, L., Uppal, R.: Optimal versus naive diversification: how inefficient is the $1/N$ portfolio strategy? Rev. Financ. Stud. **22**(5), 1915–1953 (2009)

DeMiguel, V., Martin-Utrera, A., Nogales, F.J.: Size matters: optimal calibration of shrinkage estimators for portfolio selection. J. Bank. Financ. **37**(8), 3018–3034 (2013)

Devroye, L.: Non-Uniform Random Variate Generation. Springer Verlag, New York (1986)

Ding, P.: On the conditional distribution of the multivariate t distribution. Am. Stat. **70**(3), 293–295 (2016)

Ding, Z., Granger, C.W.J., Engle, R.F.: A long memory property of stock market returns and a new model. J. Empir. Financ. **1**(1), 83–106 (1993)

Edwards, T., Lazzara, C.J.: Equal-Weight Benchmarking: Raising the Monkey Bars. Technical report, McGraw Hill Financial (2014)

Embrechts, P.: Copulas: a personal view. J. Risk Insur. **76**, 639–650 (2009)

Embrechts, P., McNeil, A., Straumann, D.: Correlation and dependency in risk management: properties and pitfalls. In: Dempster, M.A.H. (ed.) Risk Management: Value at Risk and Beyond, pp. 176–223. Cambridge University Press, Cambridge (2002)

Engle, R.: Anticipating Correlations: A New Paradigm for Risk Management. Princeton University Press, Princeton (2009)

Engle, R., Kelly, B.: Dynamic equicorrelation. J. Bus. Econ. Stat. **30**(2), 212–228 (2012)

Engle, R.F.: Dynamic conditional correlation: a simple class of multivariate generalized autoregressive conditional heteroskedasticity models. J. Bus. Econ. Stat. **20**, 339–350 (2002)

Engle, R.F., Sheppard, K.: Theoretical and Empirical Properties of Dynamic Conditional Correlation Multivariate GARCH. NBER Working Papers 8554, National Bureau of Economic Research Inc (2001)

Fang, H.B., Fang, K.T., Kotz, S.: The meta-elliptical distribution with given marginals. J. Multivar. Anal. **82**, 1–16 (2002)

Fang, K.-T., Kotz, S., Ng, K.-W.: Symmetric Multivariate and Related Distributions. Chapman & Hall, London (1989)

Fink, H., Klimova, Y., Czado, C., Stöber, J.: Regime switching vine copula models for global equity and volatility indices. Econometrics **5**(1), 1–38 (2017). Article 3

Francq, C., Zakoïan, J.-M.: Maximum likelihood estimation of pure GARCH and ARMA-GARCH processes. Bernoulli **10**(4), 605–637 (2004)

Francq, C., Zakoïan, J.-M.: GARCH Models: Structure Statistical Inference and Financial Applications. John Wiley & Sons Ltd., Chichester (2010)

Gambacciani, M., Paolella, M.S.: Robust normal mixtures for financial portfolio allocation. Forthcoming. In: Econometrics and Statistics (2017)

Haas, M., Krause, J., Paolella, M.S., Steude, S.C.: Time-varying mixture GARCH models and asymmetric volatility. North Am. J. Econ. Financ. **26**, 602–623 (2013)

Haas, M., Mittnik, S., Paolella, M.S.: Mixed normal conditional heteroskedasticity. J. Financ. Econometrics **2**(2), 211–250 (2004)

He, C., Teräsvirta, T.: Properties of moments of a family of GARCH processes. J. Econometrics **92**(1), 173–192 (1999a)

He, C., Teräsvirta, T.: Statistical properties of the asymmetric power ARCH model. In: Engle, R.F., White, H. (eds) Cointegration, Causality, and Forecasting. Festschrift in Honour of Clive W. J. Granger, pp. 462–474. Oxford University Press (1999b)

Heyde, C.C., Kou, S.G.: On the controversy over tailweight of distributions. Oper. Res. Lett. **32**, 399–408 (2004)

Hough, J.: Monkeys are better stockpickers than you'd think. Barron's magazine (2014)

Hurst, S.: The characteristic function of the student t distribution. Financial Mathematics Research Report FMRR006-95, Australian National University, Canberra (1995). http://wwwmaths.anu.edu.au/research.reports/srr/95/044/

Jagannathan, R., Ma, T.: Risk reduction in large portfolios: why imposing the wrong constraints helps. J. Financ. **58**(4), 1651–1683 (2003)

Jondeau, E.: Asymmetry in tail dependence of equity portfolios. Computat. Stat. Data Anal. **100**, 351–368 (2016)

Jondeau, E., Rockinger, M.: Conditional volatility, skewness, and kurtosis: existence, persistence, and comovements. J. Econ. Dyn. Control **27**, 1699–1737 (2003)

Jondeau, E., Rockinger, M.: The Copula-GARCH model of conditional dependencies: an international stock market application. J. Int. Money Financ. **25**, 827–853 (2006)

Jondeau, E., Rockinger, M.: On the importance of time variability in higher moments for asset allocation. J. Financ. Econometrics **10**(1), 84–123 (2012)

Karanasos, M., Kim, J.: A re-examination of the asymmetric power ARCH model. J. Empir. Financ. **13**, 113–128 (2006)

Kelker, D.: Distribution theory of spherical distributions and a location-scale parameter generalization. Sankhyā, Series A **32**(4), 419–430 (1970)

Kiefer, J., Wolfowitz, J.: Consistency of the maximum likelihood estimator in the presence of infinitely many incidental parameters. Ann. Math. Stat. **27**(4), 887–906 (1956)

Kogon, S.M., Williams, D.B.: Characteristic function based estimation of stable parameters. In: Adler, R.J., Feldman, R.E., Taqqu, M.S. (eds) A Practical Guide to Heavy Tails, pp. 311–335. Birkhauser Boston Inc. (1998)

Krause, J., Paolella, M.S.: A fast, accurate method for value at risk and expected shortfall. Econometrics **2**, 98–122 (2014)

Kuester, K., Mittnik, S., Paolella, M.S.: Value-at-risk prediction: a comparison of alternative strategies. J. Financ. Econometrics **4**, 53–89 (2006)

Ling, S., McAleer, M.: Necessary and sufficient moment conditions for the garch(r, s) and asymmetric power garch(r, s) models. Econometric Theor. **18**(3), 722–729 (2002)

Ma, J., Nelson, C.R., Startz, R.: Spurious inference in the GARCH(1,1) model when it is weakly identified. Stud. Nonlinear Dyn. Econometrics **11**(1), 1–27 (2006). Article 1

Markowitz, H.: Portfolio Selection. J. Financ. **7**(1), 77–91 (1952)

McAleer, M., Chan, F., Hoti, S., Lieberman, O.: Generalized autoregressive conditional correlation. Econometric Theor. **24**(6), 1554–1583 (2008)

McNeil, A.J., Frey, R., Embrechts, P.: Quantitative Risk Management: Concepts, Techniques, and Tools. Princeton University Press, Princeton (2005)

McNeil, A.J., Frey, R., Embrechts, P.: Quantitative Risk Management: Concepts, Techniques, and Tools. Princeton University Press, Princeton (2015). Revised edition

Mittnik, S., Paolella, M.S.: Prediction of financial downside risk with heavy tailed conditional distributions. In: Rachev, S.T. (ed.) Handbook of Heavy Tailed Distributions in Finance. Elsevier Science, Amsterdam (2003)

Mittnik, S., Paolella, M.S., Rachev, S.T.: Stationarity of stable power-GARCH processes. J. Econometrics **106**, 97–107 (2002)

Nguyen, H.T.: On evidential measures of support for reasoning with integrate uncertainty: a lesson from the ban of P-values in statistical inference. In: Huynh, V.-N., Inuiguchi, M., Le, B., Le, B.N., Denoeux, T. (eds.) 5th International Symposium on Integrated Uncertainty in Knowledge Modeling and Decision Making IUKM 2016, pp. 3–15. Springer, Cham (2016)

Nolan, J. P.: Stable Distributions - Models for Heavy Tailed Data. Birkhäuser, Boston (2015, forthcoming). Chapter 1 online

Paolella, M.S.: Intermediate Probability: A Computational Approach. John Wiley & Sons, Chichester, West Sussex, England (2007)

Paolella, M.S.: Multivariate asset return prediction with mixture models. Eur. J. Financ. **21**, 1–39 (2013)

Paolella, M.S.: Fast methods for large-scale non-elliptical portfolio optimization. Ann. Financ. Econ. **09**(02), 1440001 (2014)

Paolella, M.S.: Stable-GARCH models for financial returns: fast estimation and tests for stability. Econometrics **4**(2), 25 (2016). Article 25

Paolella, M.S.: The univariate collapsing method for portfolio optimization. Econometrics **5**(2), 1–33 (2017). Article 18

Paolella, M.S., Polak, P.: ALRIGHT: Asymmetric LaRge-Scale (I)GARCH with hetero-tails. Int. Rev. Econ. Financ. **40**, 282–297 (2015a)

Paolella, M.S., Polak, P.: COMFORT: A common market factor non-gaussian returns model. J. Econometrics **187**(2), 593–605 (2015b)

Paolella, M.S., Polak, P.: Portfolio Selection with Active Risk Monitoring. Research paper, Swiss Finance Institute (2015c)

Paolella, M.S., Polak, P.: Density and Risk Prediction with Non-Gaussian COMFORT Models (2017). Submitted

Paolella, M.S., Polak, P., Walker, P.: A Flexible Regime-Switching Model for Asset Returns (2017). Submitted

Patton, A.J.: A review of copula models for economic time series. J. Multivar. Anal. **110**, 4–18 (2012)

Pelletier, D.: Regime switching for dynamic correlations. J. Econometrics **131**, 445–473 (2006)

Righi, M.B., Ceretta, P.S.: Individual and flexible expected shortfall backtesting. J. Risk Model Valid. **7**(3), 3–20 (2013)

Righi, M.B., Ceretta, P.S.: A comparison of expected shortfall estimation models. J. Econ. Bus. **78**, 14–47 (2015)

Samorodnitsky, G., Taqqu, M.S.: Stable Non-Gaussian Random Processes: Stochastic Models with Infinite Variance. Chapman & Hall, London (1994)

Scherer, M.: CDO pricing with nested archimedean copulas. Quant. Financ. **11**, 775–787 (2011)

Shaw, W.T.: Monte Carlo Portfolio Optimization for General Investor Risk-Return Objectives and Arbitrary Return Distributions: a Solution for Long-only Portfolios (2010)

So, M.K.P., Yip, I.W.H.: Multivariate GARCH models with correlation clustering. J. Forecast. **31**(5), 443–468 (2012)

Song, D.-K., Park, H.-J., Kim, H.-M.: A note on the characteristic function of multivariate t distribution. Commun. Stat. Appl. Methods **21**(1), 81–91 (2014)

Stoyanov, S., Samorodnitsky, G., Rachev, S., Ortobelli, S.: Computing the portfolio conditional value-at-risk in the alpha-stable case. Probab. Math. Statistics **26**, 1–22 (2006)

Sutradhar, B.C.: On the characteristic function of multivariate student t-distribution. Can. J. Stat. **14**(4), 329–337 (1986)

Tse, Y.K., Tsui, A.K.C.: A multivariate generalized autoregressive conditional heteroscedasticity model with time-varying correlations. J. Bus. Econ. Stat. **20**(3), 351–362 (2002)

Vargas, G.A.: An asymmetric block dynamic conditional correlation multivariate GARCH model. Philippine Stat. **55**(1–2), 83–102 (2006)

Winker, P., Maringer, D.: The convergence of estimators based on heuristics: theory and application to a GARCH model. Comput. Stat. **24**(3), 533–550 (2009)

Wolf, O.L.M.: Honey, I shrunk the sample covariance matrix: problems in mean-variance optimization. J. Portfolio Management **30**(4), 110–119 (2004)

Zhou, T., Chan, L.: Clustered dynamic conditional correlation multivariate garch model. In: Song, I.-Y., Eder, J., Nguyen, T. M. (eds) Proceedings of the 10th International Conference Data Warehousing and Knowledge Discovery, DaWaK 2008, Turin, Italy, 2–5 September 2008, pp. 206–216 (2008)

Zolotarev, V.M.: One Dimensional Stable Distributions (Translations of Mathematical Monograph, Vol. 65). American Mathematical Society, Providence, RI (1986). Translated from the original Russian verion (1983)

Multiple Testing of One-Sided Hypotheses: Combining Bonferroni and the Bootstrap

Joseph P. Romano[1] and Michael Wolf[2]($\boxtimes$)

[1] Departments of Economics and Statistics, Stanford University, Stanford, USA
romano@stanford.edu
[2] Department of Economics, University of Zurich, Zurich, Switzerland
michael.wolf@econ.uzh.ch

Abstract. In many multiple testing problems, the individual null hypotheses (i) concern univariate parameters and (ii) are one-sided. In such problems, power gains can be obtained for bootstrap multiple testing procedures in scenarios where some of the parameters are 'deep in the null' by making certain adjustments to the null distribution under which to resample. In this paper, we compare a Bonferroni adjustment that is based on finite-sample considerations with certain 'asymptotic' adjustments previously suggested in the literature.

1 Introduction

Multiple testing refers to any situation that involves the simultaneous testing of several hypotheses. This scenario is quite common in empirical research in just about any field, including economics and finance. Some examples are: one fits a multiple regression model and wishes to decide which coefficients are different from zero; one compares several forecasting strategies to a benchmark and wishes to decide which strategies are outperforming the benchmark; and one evaluates a policy with respect to multiple outcomes and wishes to decide for which outcomes the policy yields significant effects.

If one does not take the multiplicity of tests into account, then the probability that some of the true null hypotheses will be rejected by chance alone is generally unduly large. Take the case of $S = 100$ hypotheses being tested at the same time, all of them being true, with the size and level of each test exactly equal to α. For $\alpha = 0.05$, one then expects five true hypotheses to be rejected. Furthermore, if all test statistics are mutually independent, then the probability that at least one true null hypothesis will be rejected is given by $1 - 0.95^{100} \approx 0.994$.

The most common solution to multiple testing problems is to control the *familywise error rate* (FWE), which is defined as the probability of rejecting at least one of the true null hypotheses. In other words, one uses a *global* error rate that combines all tests under consideration instead of an *individual* error rate that only considers one test at a time.

Controlling the FWE at a pre-specified level α corresponds to controlling the probability of a Type I error when carrying out a single test. But this is only one side of the testing problem — and it can be achieved trivially by rejecting

© Springer International Publishing AG 2018
V. Kreinovich et al. (eds.), *Predictive Econometrics and Big Data*, Studies in Computational Intelligence 753, https://doi.org/10.1007/978-3-319-70942-0_4

a particular hypothesis under test with probability α without even looking at data. The other side of the testing problem is 'power', that is, the ability to reject a false null hypothesis.

In this paper, we shall study certain adjustments to 'null sampling distributions' with the hope of power gains in the setting where the individual null hypotheses (i) concern univariate parameters and (ii) are one-sided.

2   Testing Problem

Suppose data X are generated from some unknown probability mechanism $\mathbb{P}$. A model assumes that $\mathbb{P}$ belongs to a certain family of probability distributions, though we make no rigid requirements for this family; it may be a parametric, semiparametric, or nonparametric model.

We consider the following generic multiple testing problem:

$$H_s : \theta_s \leq 0 \quad \text{vs.} \quad H'_s : \theta_s > 0 \qquad \text{for} \quad s = 1, \ldots, S, \tag{1}$$

where the $\theta_s := \theta_s(\mathbb{P})$ are real-valued, univariate parameters and the values under the null hypotheses are always zero without loss of generality. We also denote $\theta := (\theta_1, \ldots, \theta_S)'$.

Remark 1 (Arbitrary Null Parameters). Of course, in practice the values of the parameters under the null hypotheses ("null parameters") may not always be zero. But this situation can easily be handled by our framework as well. To see how, denote the 'original' parameters of interest by γ_s and consider the multiple testing problem

$$H_s : \gamma_s \leq \gamma_{0,s} \quad \text{vs.} \quad H'_s : \gamma_s > \gamma_{0,s} \qquad \text{for} \quad s = 1, \ldots, S, \tag{2}$$

where the null parameters $\gamma_{0,s}$ can take on any value. In such a case, simply define $\theta_s := \gamma_s - \gamma_{0,s}$, for $s = 1, \ldots, S$. □

The familywise error rate (FWE) is defined as

$$\text{FWE}_\mathbb{P} := \mathbb{P}\{\text{Reject at least one hypothesis } H_s : \theta_s \leq 0\}.$$

The goal is to control the FWE rate at a pre-specified level α while at the same time to achieve large 'power', which is loosely defined as the ability to reject false null hypotheses, that is, the ability to reject null hypotheses H_s for which $\theta_s > 0$. For example, particular notions of 'power' can be the following:

- The probability of rejecting at least one of the false null hypotheses
- The probability of rejecting a particular false null hypothesis
- The expected number of the false null hypotheses that will be rejected
- The probability of rejecting all false null hypotheses

Control of the FWE means that, for a given significance level α,

$$\text{FWE}_{\mathbb{P}} \leq \alpha \quad \text{for any } \mathbb{P}. \tag{3}$$

Control of the FWE allows one to be $1 - \alpha$ confident that there are no false discoveries among the rejected hypotheses.

Control of the FWE is generally equated with 'finite-sample' control: (3) is required to hold for any given sample size n. However, such a requirement can often only be achieved under strict parametric assumptions or for special permutation set-ups. Instead, we settle for *asymptotic* control of the FWE:

$$\limsup_{n \to \infty} \text{FWE}_{\mathbb{P}} \leq \alpha \quad \text{for any } \mathbb{P}. \tag{4}$$

Note here that the statement "for any $\mathbb{P}$" is meant to mean any $\mathbb{P}$ in the underlying assumed model for the family of distributions generating the data; for example, often one would assume the existence of some moments.

3   Multiple Testing Procedures

We assume that individual test statistics are available of the form

$$T_{n,s} := \frac{\hat{\theta}_{n,s}}{\hat{\sigma}_{n,s}},$$

where $\hat{\theta}_{n,s}$ is an estimator of θ_s based on a sample of size n and $\hat{\sigma}_{n,s}$ is a corresponding standard error.[1] We also denote $\hat{\theta}_n := (\hat{\theta}_{n,1}, \dots, \hat{\theta}_{n,s})'$. We further assume that these test statistics are 'proper' t-statistics in the sense that $T_{n,s}$ converges in distribution to the standard normal distribution under $\theta_s = 0$, for $s = 1, \dots, S$.

There exists by now a sizeable number of multiple testing procedures (MTPs) designed to control the FWE, at least asymptotically. The oldest and best-known such procedure is the Bonferroni procedure that rejects hypothesis H_s if $\hat{p}_{n,s} \leq \alpha/S$, where $\hat{p}_{n,s}$ is a p-value for H_s. Such a p-value can be obtained via asymptotic approximations or alternatively via resampling methods; for example, an 'asymptotic' p-value is obtained as $\hat{p}_{n,s} := 1 - \Phi(T_{n,s})$, where $\Phi(\cdot)$ denotes the c.d.f. of the standard normal distribution. Although the Bonferroni procedure controls the FWE asymptotically under weak regularity conditions, it is generally suboptimal in terms of 'power'.

There are two main avenues of increasing 'power' while maintaining (asymptotic) control of the FWE. The first avenue, dating back to [Hol79], is to use stepwise procedures where the threshold for rejecting hypotheses becomes less lenient in subsequent steps in case some hypotheses have been rejected in a first step. The second avenue, dating back to [Whi00], at least in nonparametric settings, is to take the dependence structure of the individual test statistics $T_{n,s}$

---

[1] This means that $\hat{\sigma}_{n,s}$ is an estimator of the standard deviation of $\hat{\theta}_{n,s}$.

into account rather than assuming a 'worst-case' dependence structure as the Bonferroni procedure does; taking this true dependence structure into account — in the absence of strict assumptions — requires the use of resampling methods, such as the bootstrap, subsampling, and permutation methods. [RW05] suggest to combine both avenues, resulting in resampling-based stepwise MTPs.

We start by discussing a bootstrap-based single-step method. An idealized method would reject all H_s for which $T_{n,s} \geq d_1$ where d_1 is the $1 - \alpha$ quantile under the true probability mechanism $\mathbb{P}$ of the random variable $\max_s(\hat{\theta}_{n,s} - \theta_s)/\hat{\sigma}_{n,s}$. Naturally, the quantile d_1 not only depends on the marginal distributions of the centered statistics $(\hat{\theta}_{n,s} - \theta_s)/\hat{\sigma}_{n,s}$ but, crucially, also on their dependence structure.

Since the true probability mechanism $\mathbb{P}$ is unknown, the idealized critical value d_1 is not available. But it can be estimated consistently under weak regularity conditions as follows. Take $\hat{d}_1$ as the $1 - \alpha$ quantile under $\hat{\mathbb{P}}_n$ of $\max_s(\hat{\theta}_{n,s}^* - \hat{\theta}_{n,s})/\hat{\sigma}_{n,s}^*$. Here, $\hat{\mathbb{P}}_n$ is an *unrestricted* estimate of $\mathbb{P}$. For example, if $X = (X_1, \ldots, X_n)$ with $X_i \overset{\text{iid}}{\sim} \mathbb{P}$, then $\hat{\mathbb{P}}_n$ is typically the empirical distribution of the X_i. Furthermore, $\hat{\theta}_{n,s}^*$ is the estimator of $\hat{\theta}_s$ and $\hat{\sigma}_{n,s}^*$ is the corresponding standard error, both computed from X^* where $X^* \sim \hat{\mathbb{P}}_n$. In other words, we use the bootstrap to estimate d_1. The particular choice of $\hat{\mathbb{P}}_n$ depends on the situation. In particular, if the data are collected over time a suitable time series bootstrap needs to be employed; for example, see [DH97, Lah03].

We have thus described a single-step MTP. However, a stepwise improvement is possible.[2] In any given step j, one simply discards the hypotheses that have been rejected so far and applies the single-step MTP to the remaining universe of non-rejected hypotheses. The resulting critical value $\hat{d}_j$ necessarily satisfies $\hat{d}_j \leq \hat{d}_{j-1}$, and typically satisfies $\hat{d}_j < \hat{d}_{j-1}$, so that new rejections may result; otherwise the method stops with no further rejections.

This bootstrap stepwise MTP provides asymptotic control of the FWE under remarkably weak regularity conditions. Mainly, it is sufficient that (i) $\sqrt{n}(\hat{\theta}_n - \theta)$ converges in distribution to a (multivariate) continuous limit distribution and that the bootstrap consistently estimates this limit distribution; and that (ii) the 'scaled' standard errors $\sqrt{n}\hat{\sigma}_{n,s}$ and $\sqrt{n}\hat{\sigma}_{n,s}^*$ converge to the same, non-zero limiting values in probability, both in the 'real world' and in the 'bootstrap world'. Under even weaker regularity conditions, a subsampling approach could be used instead; see [RW05]. Furthermore, when a randomization setup applies, randomization methods can be used as an alternative; see [RW05] again.

4   Adjustments for Power Gains

As stated before, the bootstrap stepwise MTP of the previous section provides asymptotic control of the FWE under weak regularity conditions. But in the one-sided setting (1) considered in this paper, it might be possible to obtain

---

[2] More precisely, the improvement is a stepdown method.

further power gains by making adjustments for null hypotheses that are 'deep in the null', an idea going back to [Han05].

To motivate such an idea, it is helpful to first point out that for many parameters of interest, θ, there is a one-to-one relation between the bootstrap stepwise MTP of the previous section, which is based on an *unrestricted* estimate $\hat{\mathbb{P}}_n$ of $\mathbb{P}$, and a bootstrap stepwise MTP that is based on a *restricted* estimate $\hat{\mathbb{P}}_{0,n}$ of $\mathbb{P}$, satisfying the constraints of the S null hypotheses. In the latter approach the critical value $\hat{d}_1$ in the first step is obtained as the $1 - \alpha$ quantile under $\hat{\mathbb{P}}_{0,n}$ of $\max_s \hat{\theta}^*_{n,s}/\hat{\sigma}^*_{n,s}$. Here, $\hat{\theta}^*_{n,s}$ is the estimator of θ_s and $\hat{\sigma}^*_{n,s}$ is the corresponding standard error, both computed from X^* where $X^* \sim \hat{\mathbb{P}}_{0,n}$. Note that in this latter approach, there is no (explicit) centering in the numerator of the bootstrap test statistics, since the centering already takes place implicitly in the restricted estimator $\hat{\mathbb{P}}_{0,n}$ by incorporating the constraints of the null hypotheses.

For many parameters of interest, the unrestricted bootstrap stepwise MTP of the previous section is equivalent to the restricted bootstrap stepwise MTP of the previous paragraph based on an estimator $\hat{\mathbb{P}}_{0,n}$ that satisfies $\theta_s(\hat{\mathbb{P}}_{0,n}) = 0$ for $s = 1, \ldots, S$. In statistical lingo, such a null parameter $\theta(\hat{\mathbb{P}}_{0,n})$ corresponds to a *least favorable configuration* (LFC), since all the components $\theta_s(\hat{\mathbb{P}}_{0,n})$ lie on the boundary of the respective null hypotheses H_s.

Remark 2 (Example: Testing Means). To provide a specific example of a null-restricted estimator $\hat{\mathbb{P}}_{0,n}$, consider the setting where $X = (X_1, \ldots, X_n)$ with $X_i \overset{iid}{\sim} \mathbb{P}$, $X_i \in \mathbb{R}^S$, and $(\theta_1, \ldots, \theta_S)' = \theta := \mathbb{E}(X_i)$. Then an unrestricted estimator $\hat{\mathbb{P}}_n$ is given by the empirical distribution of the X_i whereas a null-restricted estimator $\hat{\mathbb{P}}_{0,n}$ is given by the empirical distribution of the $X_i - \hat{\theta}_n$, where $\hat{\theta}_n$ is the sample average of the X_i. In other words, $\hat{\mathbb{P}}_{0,n}$ is obtained by suitably shifting $\hat{\mathbb{P}}_n$ to achieve mean zero for all components.   □

[Han05] argues that such an approach is overly conservative when some of the θ_s lie 'deep in the null', that is, for $\theta_s \ll 0$. Indeed, it can easily be shown that asymptotic control of the FWE based on the restricted bootstrap stepwise MTP could be achieved based on an infeasible 'estimator' $\hat{\mathbb{P}}_{0,n}$ that satisfies

$$\theta_s(\hat{\mathbb{P}}_{0,n}) = \min\{\theta_s, 0\}.$$

(We use the term 'estimator' here, since such an $\hat{\mathbb{P}}_{0,n}$ is infeasible in practice because one does not know the true values θ_s.) Clearly, when some of the θ_s are smaller than zero, one would obtain smaller critical values $\hat{d}_j$ in this way compared to using the LFC.

The idea then is to adjust $\hat{\mathbb{P}}_{0,n}$ in a *feasible*, data-dependent fashion such that $\theta_s(\hat{\mathbb{P}}_{0,n}) < 0$ for all θ_s 'deep in the null'.

4.1 Asymptotic Adjustments

Based on the law of iterated logarithm, [Han05] proposes an adjustment $\hat{\mathbb{P}}^{A}_{0,n}$ that satisfies

$$\theta_s(\hat{\mathbb{P}}^{A}_{0,n}) := \hat{\theta}_{n,s}\mathbb{1}_{\{T_{n,s}<-\sqrt{2\log\log n}\}}, \tag{5}$$

where $\mathbb{1}_{\{\cdot\}}$ denotes the indicator function of a set. Therefore, if the t-statistic $T_{n,s}$ is sufficiently small, the parameter of the restricted bootstrap distribution is adjusted to the sample-based estimator $\hat{\theta}_s$, and otherwise it is left unchanged at zero. How one can construct such an estimator $\hat{\mathbb{P}}^{A}_{0,n}$ depends on the particular application. In the example of Remark 2, say, $\hat{\mathbb{P}}^{A}_{0,n}$ can be constructed by suitably shifting the empirical distribution $\hat{\mathbb{P}}_n$.

[Han05] only considers a bootstrap single-step MTP. [HHK10] propose the same adjustment (5) in the context of a bootstrap stepwise MTP in the spirit of [RW05].

The adjustment (5) is of asymptotic nature, since one does not have to pay any 'penalty' in the proposals of [Han05,HHK10]. In other words, the MTP procedure proceeds as if $\theta_s(\hat{\mathbb{P}}^{A}_{0,n}) = \theta_s$ in case $\theta_s(\hat{\mathbb{P}}^{A}_{0,n})$ has been adjusted to $\hat{\theta}_{n,s} < 0$. The point here is that in finite samples, it may happen that $T_{n,s} < -\sqrt{2\log\log n}$ even though $\theta_s \geq 0$ in reality; in such cases, the null distribution $\hat{\mathbb{P}}_{0,n}$ is generally too 'optimistic' and results in critical values $\hat{d}_j$ that are too small. As a consequence, control of the FWE in finite samples will be negatively affected.

Also note that the cutoff $-\sqrt{2\log\log n}$ is actually quite arbitrary and could be replaced by any multiple of it, however big or small, without affecting the asymptotic validity of the method.

Remark 3 (Related Problem: Testing Moment Inequalities). The literature on moment inequalities is concerned with the related testing problem

$$H : \theta_s \leq 0 \quad \text{for all } s \quad \text{vs.} \quad H' : \theta_s > 0 \quad \text{for at least one } s. \tag{6}$$

This is not a multiple testing problem but the multivariate hypothesis H, which is a single hypothesis, also involves an S-dimensional parameter θ and is one-sided in nature. For this testing problem, [AS10] suggest an adjustment to $\hat{\mathbb{P}}_{0,n}$ that is of asymptotic nature and corresponds to the adjustment of [Han05] for testing problem (1). But then, in a follow-up paper, [AB12] propose an alternative method based on finite-sample considerations that incorporates an explicit 'penalty' for making adjustments to the LFC. The proposal of [AB12] is computationally quite complex and also lacks a rigorous proof of validity. [RSW14] suggest a Bonferroni adjustment as an alternative, which is simpler to implement and also comes with a rigorous proof of validity.   □

4.2 Bonferroni Adjustments

We now 'translate' the Bonferroni adjustment of [RSW14] for testing problem (6) to the multiple testing problem (1).

In the first step, we adjust $\hat{\mathbb{P}}_{0,n}$ based on a nominal $1 - \beta$ upper rectangular joint confidence region for θ of the form

$$(-\infty, \hat{\theta}_{n,1} + \hat{c}\,\hat{\sigma}_{n,1}] \times \cdots \times (-\infty, \theta_{n,S} + \hat{c}\,\hat{\sigma}_{n,S}]. \tag{7}$$

Here, $0 < \beta < \alpha$ and $\hat{c}$ is a bootstrap-based estimator of the $1 - \beta$ quantile of the sampling distribution of the statistic

$$\max_s \frac{\theta_s - \hat{\theta}_{n,s}}{\hat{\sigma}_{n,s}}.$$

For notational compactness, denote the upper end of a generic joint confidence interval in (7) by

$$\hat{u}_{n,s} := \hat{\theta}_{n,s} + \hat{c}\,\hat{\sigma}_{n,s}. \tag{8}$$

Then we propose an adjustment $\hat{\mathbb{P}}^{B}_{0,n}$ that satisfies

$$\theta_s(\hat{\mathbb{P}}^{B}_{0,n}) := \min\{\hat{u}_{n,s}, 0\}. \tag{9}$$

How one can construct such an estimator $\hat{\mathbb{P}}^{B}_{0,n}$ depends on the particular application. In the example of Remark 2, say, $\hat{\mathbb{P}}^{B}_{0,n}$ can be constructed by suitably shifting the empirical distribution $\hat{\mathbb{P}}_n$.

In the second step, the restricted bootstrap stepwise MTP (i) uses $\theta(\hat{\mathbb{P}}^{B}_{0,n})$ defined by (9) and (ii) is carried out at nominal level $\alpha - \beta$ as opposed to at nominal level α. Feature (ii) is a finite-sample 'penalty' that accounts for the fact that with probability β, the true θ will not be contained in the joint confidence region (7) in the first step and, consequently, the adjustment in (i) will be overly optimistic.

As reasonable 'generic' choice for β is $\beta := \alpha/10$, as per the suggestion of [RSW14].

It is clear that the Bonferroni adjustment is necessarily less powerful compared to the asymptotic adjustment for two reasons. First, typically $\theta_s(\hat{\mathbb{P}}^{A}_{0,n}) \leq \theta_s(\hat{\mathbb{P}}^{B}_{0,n})$ for all $s = 1, \ldots, S$. Second, the asymptotic adjustment uses the full nominal level α in the stepwise MTP whereas the Bonferroni adjustment only uses the reduced level $\alpha - \beta$. On the other hand, it can be expected that the asymptotic adjustments will be liberal in terms of the finite-sample control of the FWE in some scenarios.

4.3 Adjustments for Unrestricted Bootstrap MTPs

We have detailed the asymptotic and Bonferroni adjustments in the context of the restricted bootstrap stepwise MTPs, since they are conceptually somewhat easier to understand.

But needless to say, these adjustments carry over one-to-one to the unrestricted bootstrap stepwise MTPs of [RW05].

Focusing on the first step to be specific, the asymptotic adjustment takes $\hat{d}_1$ as the $1 - \alpha$ quantile under $\hat{\mathbb{P}}_n$ of $\max_s(\hat{\theta}_{n,s}^* - \hat{\theta}_{n,s}^{\mathrm{A}})/\hat{\sigma}_{n,s}^*$. Here, $\hat{\mathbb{P}}_n$ is an unrestricted estimator of $\mathbb{P}$ and

$$\hat{\theta}_{n,s}^{\mathrm{A}} := \begin{cases} \hat{\theta}_{n,s} & \text{if } T_{n,s} < -\sqrt{2 \log \log n} \\ 0 & \text{otherwise} \end{cases}$$

Furthermore, $\hat{\theta}_{n,s}^*$ and $\hat{\sigma}_{n,s}^*$ are the estimator of θ_s and the corresponding standard error, respectively, computed from X^*, where $X^* \sim \hat{\mathbb{P}}_n$.

On the other hand, the Bonferroni adjustment takes $\hat{d}_1$ as the $1 - \alpha + \beta$ quantile under $\hat{\mathbb{P}}_n$ of $\max_s(\hat{\theta}_{n,s}^* - \hat{\theta}_{n,s}^{\mathrm{B}})/\hat{\sigma}_{n,s}^*$. Here, $\hat{\mathbb{P}}_n$ is an unrestricted estimator of $\mathbb{P}$ and

$$\hat{\theta}_{n,s}^{\mathrm{B}} := \hat{\theta}_{n,s} - \min\{\hat{u}_{n,s}, 0\}, \tag{10}$$

with $\hat{u}_{n,s}$ defined as in (8). Furthermore, $\hat{\theta}_{n,s}^*$ and $\hat{\sigma}_{n,s}^*$ are the estimator of θ_s and the corresponding standard error, respectively, computed from X^* where $X^* \sim \hat{\mathbb{P}}_n$.

The computation of the critical constants $\hat{d}_j$ in subsequent steps $j > 1$ is analogous for both adjustments.

Remark 4 (Single Adjustment versus Multiple Adjustments). In principle, the Bonferroni adjustments (10) could be updated in each step of the bootstrap step-wise MTP by updating the joint confidence region for the remaining part of θ in each step, that is, for the elements θ_s of θ for which the corresponding null hypotheses H_s have not been rejected in previous steps. This approach can be expected to lead to small further power gains, though at additional computational (and software coding) costs.   □

5   The Gaussian Problem

5.1   Single-Step Method

In this section, we derive an exact finite-sample result for the multivariate Gaussian model, which motivates the method proposed in the paper. Assume that $W := (W_1, \ldots, W_S)' \sim \mathbb{P} \in \mathbf{P} := \{N(\theta, \Sigma) : \mu \in \mathbb{R}^S\}$ for a known covariance matrix Σ. The multiple testing problem consists of S one-sided hypotheses

$$H_s : \theta_s \leq 0 \quad \text{vs.} \quad H_s' : \theta_s > 0 \qquad \text{for} \quad s = 1, \ldots, S. \tag{11}$$

The goal is to control the FWE exactly at nominal level α in this model, for any possible choice of the θ_s, for some pre-specified value of $\alpha \in (0, 1)$. Note further that, because Σ is assumed known, we may assume without loss of generality that its diagonal consists of ones; otherwise, we can simply replace W_s by W_s divided by its standard deviation. This limiting model applies to the nonparametric problem in the large-sample case, since standardized sample means are asymptotically multivariate Gaussian with a covariance matrix that can be estimated consistently.

First, if instead of the multiple testing problem, we were interested in the single multivariate joint hypothesis that all θ_s satisfy $\theta_s \leq 0$, then we are in the moment inequalities problem; see Remark 3. For such a problem, there are, of course, many ways in which to construct a test that controls size at level α. For instance, given any test statistic $T := T(W_1, \ldots, W_S)$ that is nondecreasing in each of its arguments, we may consider a test that rejects H for large values of T. Note that, for any given fixed critical value c, $\mathbb{P}_\theta\{T(W_1, \ldots, W_S) > c\}$ is a nondecreasing function of each component θ_s in θ. Therefore, if $c := c_{1-\alpha}$ is chosen to satisfy

$$\mathbb{P}_0\{T(W_1, \ldots, W_S) > c_{1-\alpha}\} \leq \alpha,$$

then the test that rejects H_0 when $T > c_{1-\alpha}$ is a level α test. A reasonable choice of test statistic T is the likelihood ratio statistic or the maximum statistic $\max(W_1, \ldots, W_S)$. For this latter choice of test statistic, $c_{1-\alpha}$ may be determined as the $1 - \alpha$ quantile of the distribution of $\max(W_1, \ldots, W_S)$ when $(W_1, \ldots, W_S)'$ is multivariate normal with mean 0 and covariance matrix Σ. Unfortunately, as S increases, so does the critical value, which can make it difficult to have any reasonable power against alternatives. The same issue occurs in multiple testing, as described below. The main idea of our procedure is to essentially remove from consideration those θ_s that are 'negative'.[3] If we can eliminate such θ_s from consideration, then we may use a smaller critical value with the hope of increased power against alternatives.

In the multiple testing problem using the max statistic, one could simply reject any θ_s for which $X_s > c_{1-\alpha}$. But as in the single testing problem above, $c_{1-\alpha}$ increases with S and therefore it may be helpful to make certain adjustments if one is fairly confident that a hypothesis H_s satisfies $\theta_s < 0$. Using this reasoning as a motivation, we may use a confidence region to help determine which θ_s are 'negative'. To this end, let $M(1 - \beta)$ denote an upper rectangular joint confidence region for θ at level $1 - \beta$. Specifically, let

$$M(1 - \beta) := \left\{ \theta \in \mathbb{R}^S : \max_{1 \leq s \leq S} (\theta_s - W_s) \leq K^{-1}(1 - \beta) \right\} \tag{12}$$

$$= \left\{ \theta \in \mathbb{R}^S : \theta_s \leq W_s + K^{-1}(1 - \beta) \text{ for all } 1 \leq s \leq S \right\},$$

where $K^{-1}(1 - \beta)$ is the $1 - \beta$ quantile of the distribution (function)

$$K(x) := \mathbb{P}_\theta \left\{ \max_{1 \leq s \leq S} (\theta_s - W_s) \leq x \right\}.$$

Note that $K(\cdot)$ depends only on the dimension S and the underlying covariance matrix Σ. In particular, it does not depend on the θ_s, so it can be computed under the assumption that all $\theta_s = 0$. By construction, we have for any $\theta \in \mathbb{R}^S$, that

$$\mathbb{P}_\theta\{\theta \in M(1 - \beta)\} = 1 - \beta.$$

---

[3] Such a program is carried out in the moment inequality problem by [RSW14].

The idea now is that with probability at least $1 - \beta$, we may assume that θ will lie in $\Omega_0 \cap M(1 - \beta)$ rather than just in Ω_0, where Ω_0 is the 'negative quadrant' given by $\{\theta : \theta_s \leq 0, \ s = 1, \ldots, S\}$. Instead of computing the critical value under $\theta = 0$, the 'largest' value of θ in Ω_0 (or the value under the LFC), we may therefore compute the critical value under $\tilde{\theta}$, the 'largest' value of θ in the (data-dependent) set $\Omega_0 \cap M(1 - \beta)$. It is straightforward to determine $\tilde{\theta}$ explicitly because of the simple shape of the joint confidence region for θ. In particular, $\tilde{\theta}$ has sth component equal to

$$\tilde{\theta}_s := \min\{W_s + K^{-1}(1 - \beta), 0\}. \tag{13}$$

But, to account for the fact that θ may not lie in $M(1-\beta)$ with probability β, we reject any H_s for which W_s exceeds the $1 - \alpha + \beta$ quantile of the distribution of $T := \max(W_1, \ldots, W_S)$ under $\tilde{\theta}$ rather than the $1 - \alpha$ quantile of the distribution of T under $\tilde{\theta}$. The following result establishes that this procedure controls the FWE at level α.

Theorem 1. *Let $T := \max(W_1, \ldots, W_S)$. For $\theta \in \mathbb{R}^S$ and $\gamma \in (0, 1)$, define*

$$b(\gamma, \theta) := \inf\{x \in \mathbb{R} : \mathbb{P}_\theta\{T(W_1, \ldots, W_k) \leq x\} \geq \gamma\},$$

that is, as the γ quantile of the distribution of T under θ. Fix $0 < \beta < \alpha$. The multiple testing procedure that rejects any H_s for which $W_s > b(1 - \alpha + \beta, \tilde{\theta})$ controls the FWE at level α.

Remark 5. As emphasized above, an attractive feature of the procedure is that the 'largest' value of θ in $\Omega_0 \cap M(1 - \beta)$ may be determined explicitly. This follows from our particular choice of the initial joint confidence region for θ. If, for example, we had instead chosen $M(1 - \beta)$ to be the usual Scheffé confidence ellipsoid, then there may not even be a 'largest' value of θ in $\Omega_0 \cap M(1 - \beta)$. $\square$

Proof of Theorem 1. First note that $b(\gamma, \theta)$ is nondecreasing in θ, since T is nondecreasing in its arguments. Fix any θ. Let $I_0 := I_0(\theta)$ denote the indices of true null hypotheses, that is,

$$I_0 := \{s : \theta_s \leq 0\}.$$

Let $\theta_s^* := \min(\theta_s, 0)$ and let E be the event that $\theta \in M(1 - \beta)$. Then, the familywise error rate (FWE) satisfies

$$\mathbb{P}_\theta\{\text{reject any true } H_s\} \leq \mathbb{P}_\theta\{E^c\} + \mathbb{P}_\theta\{E \cap \{\text{reject any } H_s \text{ with } s \in I_0\}\}$$
$$= \beta + \mathbb{P}_\theta\{E \cap \{\text{reject any } H_s \text{ with } s \in I_0\}\}.$$

But when the event E occurs and some true H_s is rejected — so that $\max_{s \in I_0} W_s > b(1 - \alpha + \beta, \tilde{\theta})$ — then the event $\max_{s \in I_0} W_s > b(1 - \alpha + \beta, \theta^*)$

must occur, since $b(1 - \alpha + \beta, \theta)$ is nondecreasing in θ and $\theta \leq \tilde{\theta}$ when E occurs. Hence, the FWE is bounded above by

$$\beta + \mathbb{P}_\theta \Big\{ \max_{s \in I_0} W_s > b(1 - \alpha + \beta, \theta^*) \Big\} \leq \beta + \mathbb{P}_{\theta^*} \Big\{ \max_{s \in I_0} W_s > b(1 - \alpha + \beta, \theta^*) \Big\}$$

because the distribution of $\max_{s \in I_0} W_s$ only depends on those θ_s in I_0. Therefore, the last expression is bounded above by

$$\beta + \mathbb{P}_{\theta^*} \Big\{ \max_{\text{all } s} W_s > b(1 - \alpha + \beta, \theta^*) \Big\} = \beta + 1 - (1 - \alpha + \beta) = \beta + (\alpha - \beta) = \alpha.$$

$\square$

5.2   Stepwise Method

One can improve upon the single-step method in Theorem 1 by a stepwise method.[4] More specifically, consider the following method. Begin with the method described above, which rejects any H_s for which $W_s > b(1 - \alpha + \beta, \tilde{\theta})$. Basically, one applies the closure method to the above and show that it may be computed in a stepwise fashion. To do this, we first need to describe the situation when testing only a subset of the hypotheses. So, let I denote any subset of $\{1, \ldots, S\}$ and let $b_I(\gamma, \theta)$ denote the γ quantile of the distribution of $\max(T_s : s \in I)$ under θ. Also, let $\tilde{\theta}(I) := \{\tilde{\theta}_s(I) : s \in I\}$ with $\tilde{\theta}_s(I)$ be defined as in (13) except that $K^{-1}(1 - \beta)$ is replaced by $K_I^{-1}(1 - \beta)$, defined to be the $1 - \beta$ quantile of the distribution (function)

$$K_I(x) := \mathbb{P}_\theta \Big\{ \max_{s \in I}(\theta_s - W_s) \leq x \Big\}.$$

The stepwise method can now be described. Begin by testing all H_s with $s \in \{1, \ldots, S\}$ as described in the single-step method. If there are any rejections, remove the rejected hypotheses from consideration and apply the single-step method to the remaining hypotheses. That is, if I is the set of indices of the remaining hypotheses not previously rejected, then reject any such H_s if $W_s > b_I(1 - \alpha + \gamma, \tilde{\theta}(I))$. And so on. (Note that at each step of the procedure, a new joint confidence region is computed to determine $\tilde{\theta}(I)$, but β remains the same in each step.)

Theorem 2. *Under the Gaussian setup of Theorem 1, the above stepwise method controls the FWE at level α.*

Proof of Theorem 2. We just need to show that the closure method applied to the above tests results in the stepwise method as described. To do this, it suffices to show that if H_s is rejected by the stepwise method, $s \in I$, and $I \subset J$, then when J is tested (meaning the H_s with $s \in J$ are jointly tested) and the method rejects the joint (intersection) hypothesis, then it also rejects the particular joint (intersection) hypothesis when just I is tested.

---

[4] More precisely, the improvement is a stepdown method.

First, the distribution of $\max_{\theta_s \in I} W_s$ is stochastically dominated by that of $\max_{\theta_s \in J} W_s$ (since we are just taking the max over a larger set), under any θ and in particular under $\tilde{\theta}(I)$. But the distribution of the maximum statistic $\max_{\theta_s \in J} W_s$ is monotone increasing with respect to θ_s because of the important fact that, component wise,

$$\tilde{\theta}(I) \leq \tilde{\theta}(J).$$

Hence, the distribution of $\max_{\theta_s \in J}$ under $\tilde{\theta}(I)$ is further dominated by the distribution of $\max_{\theta_s \in J} W_s$ under $\tilde{\theta}(J)$. Therefore, the critical values satisfy

$$b_I(1 - \alpha + \beta, \tilde{\theta}(I)) \leq b_J(1 - \alpha + \beta, \tilde{\theta}(J)),$$

which is all we need to show, since then any H_s for which W_s exceeds $b_J(1 - \alpha + \beta, \tilde{\theta}(J))$ will satisfy that W_s also exceeds $b_I(1 - \alpha + \beta, \tilde{\theta}(I))$.   □

6 Monte Carlo Simulations

The data are of the form $X := (X_1, X_2, \ldots, X_n)$ with $X_i \overset{\text{iid}}{\sim} N(\theta, \Sigma)$, $\theta \in \mathbb{R}^S$, and $\Sigma \in \mathbb{R}^{S \times S}$. We consider $n = 50, 100$.

For $n = 50$, we consider $S = 25, 50, 100$ and the following mean vectors $\theta = (\theta_1, \ldots, \theta_S)'$:

- All $\theta_s = 0$
- Five of the $\theta_s = 0.4$
- Five of the $\theta_s = 0.4$ and $S/2$ of the $\theta_s = -0.4$
- Five of the $\theta_s = 0.4$ and $S/2$ of the $\theta_s = -0.8$

For $n = 100$, we consider $S = 50, 100, 200$ and the following mean vectors $\theta = (\theta_1, \ldots, \theta_S)'$:

- All $\theta_s = 0$
- Ten of the $\theta_s = 0.3$
- Ten of the $\theta_s = 0.3$ and $S/2$ of the $\theta_s = -0.3$
- Ten of the $\theta_s = 0.3$ and $S/2$ of the $\theta_s = -0.6$

For $S = 50, 100$, the covariance matrix Σ is always a constant-correlation matrix with constant variance one on the diagonal and constant covariance $\rho = 0$, 0.5 on the off-diagonal.

The test statistics $T_{n,s}$ are the usual t-statistics based on the individual sample means and sample standard deviations.

The multiple testing procedure is always the bootstrap stepwise MTP of [RW05] and we consider three variants:

- **LFC:** No adjustment at all
- **Asy:** Asymptotic Adjustment
- **Bon:** Bonferroni Adjustment

Note that for computational simplicity, Bon is based on a single adjustment throughout the stepwise MTP; see Remark 4.

The nominal level for FWE control is $\alpha = 10\%$ and the value of β for the Bonferroni adjustment is chosen as $\beta = 1\%$ following the 'generic' suggestion $\beta := \alpha/10$ of [RSW14].

We consider two performance measures:

- **FWE:** Empirical FWE
- **Power:** Average number of rejected false hypotheses

The number of Monte Carlo repetitions is $B = 50,000$ in each scenario and the bootstrap à la [Efr79] is based on 1,000 resamples always.

The results for $n = 50$ are presented in Sect. A.1 and the results for $n = 100$ are presented in Sect. A.2. They can be summarized as follows.

- As pointed out before, Asy is always more powerful than Bon necessarily.
- There are some scenarios where Asy fails to control the FWE, though the failures are never grave: In the worst case, the empirical FWE is 10.6%.
- Bon can actually be less powerful than LFC (though never by much). This is not surprising: When null parameters are on the boundary or close to the boundary, then the 'minor' adjustment in the first stage of Bon does not offset the reduction in the nominal level (from α to $\alpha - \beta$) in the second stage.
- When null parameters are 'deep in the null', also the power gains of Bon over LFC are noticeable (though never quite as large as the power gains of Asy over LFC). Of course, such power gains would even be greater by increasing the proportion of null parameters 'deep in the null' and/or the distance away from zero of such null parameters.

7   Conclusion

In many multiple testing problems, the individual null hypotheses (i) concern univariate parameters and (ii) are one-sided. In such problems, power gains can be obtained for bootstrap multiple testing procedures in scenarios where some of the parameters are 'deep in the null' by making certain adjustment to the null distribution under which to resample. In this paper we have compared a Bonferroni adjustment that is based on finite-sample considerations to certain 'asymptotic' adjustments previously suggested in the literature. The advantage of the Bonferroni adjustment is that it guarantees better finite-sample control of the familywise error rate. The disadvantage is that it is always somewhat less powerful than the asymptotic adjustments.

A   Detailed Monte Carlo Results

A.1   Results for $n = 50$

See Tables 1, 2, 3 and 4.

Table 1. All $\theta_s = 0$: FWE.

S	LFC	Asy	Bon
$\rho = 0$			
25	9.8	10.5	8.8
50	9.7	10.2	8.7
100	9.5	10.1	8.5
$\rho = 0.5$			
25	10.0	10.0	9.0
50	9.8	9.8	8.8
100	9.9	9.9	8.9

Table 2. Five of the $\theta_s = 0.4$: FWE | Power.

S	LFC	Asy	Bon	LFC	Asy	Bon
$\rho = 0$						
25	8.9	9.4	7.9	2.7	2.8	2.6
50	9.1	9.6	8.2	2.2	2.2	2.1
100	9.4	10.0	8.4	1.7	1.8	1.6
$\rho = 0.5$						
25	9.9	9.9	8.9	3.2	3.2	3.1
50	9.8	9.8	8.8	2.8	2.8	2.7
100	10.0	10.1	9.2	2.4	2.4	2.3

Table 3. Five of the $\theta_s = 0.4$ and $S/2$ of the $\theta_s = -0.4$: FWE | Power.

S	LFC	Asy	Bon	LFC	Asy	Bon
$\rho = 0$						
25	3.7	7.7	3.6	2.7	3.3	2.7
50	4.2	8.2	4.0	2.2	2.7	2.2
100	4.7	8.8	4.4	1.7	2.1	1.7
$\rho = 0.5$						
25	5.3	7.6	4.7	3.1	3.6	3.2
50	5.9	8.0	5.3	2.8	3.2	2.7
100	6.6	8.4	5.9	2.4	2.8	2.3

Table 4. Five of the $\theta_s = 0.4$ and $S/2$ of the $\theta_s = -0.8$: FWE | Power.

S	LFC	Asy	Bon	LFC	Asy	Bon
			$\rho = 0$			
25	3.7	8.8	6.9	2.7	3.4	3.2
50	4.2	9.3	7.1	2.2	2.8	2.6
100	4.7	9.8	7.4	1.7	2.2	2.0
			$\rho = 0.5$			
25	5.3	9.8	7.8	3.1	3.7	3.5
50	5.9	9.8	7.7	2.8	3.2	3.1
100	6.6	10.0	7.8	2.4	2.8	2.6

A.2   Results for $n = 100$

See Tables 5, 6, 7 and 8.

Table 5. All $\theta_s = 0$: FWE.

S	LFC	Asy	Bon
		$\rho = 0$	
50	9.8	10.2	8.8
100	10.0	10.4	8.9
200	10.0	10.6	8.9
		$\rho = 0.5$	
50	10.0	10.0	9.0
100	10.1	10.1	9.0
200	10.1	10.1	9.1

Table 6. Ten of the $\theta_s = 0.3$: FWE | Power.

S	LFC	Asy	Bon	LFC	Asy	Bon
			$\rho = 0$			
50	8.8	9.1	7.9	5.4	5.5	5.3
100	9.5	9.8	8.5	4.5	4.5	4.3
200	9.7	10.2	8.7	3.6	3.7	3.5
			$\rho = 0.5$			
50	9.9	9.9	8.9	6.5	6.5	6.3
100	10.0	10.0	9.0	5.8	5.8	5.6
200	10.1	10.1	9.1	5.1	5.1	4.9

Table 7. Ten of the $\theta_s = 0.3$ and $S/2$ of the $\theta_s = -0.3$: FWE | Power.

S	LFC	Asy	Bon	LFC	Asy	Bon
			$\rho = 0$			
50	3.3	7.3	3.4	5.4	6.4	5.4
100	4.3	8.4	4.1	4.5	5.3	4.4
200	4.7	8.9	4.4	3.6	4.4	3.6
			$\rho = 0.5$			
50	5.3	7.7	4.7	6.5	7.3	6.5
100	6.2	8.4	5.6	5.8	6.5	5.7
200	6.9	8.8	6.1	5.1	5.7	5.0

Table 8. Ten of the $\theta_s = 0.3$ and $S/2$ of the $\theta_s = -0.6$: FWE | Power.

S	LFC	Asy	Bon	LFC	Asy	Bon
			$\rho = 0$			
50	3.4	8.4	6.9	5.4	6.6	6.4
100	4.3	9.4	7.7	4.5	5.5	5.2
200	4.7	9.9	7.9	3.6	4.5	4.3
			$\rho = 0.5$			
50	5.3	9.9	8.3	6.5	7.4	7.1
100	6.2	10.0	8.4	5.8	6.5	6.3
200	6.9	10.2	8.6	5.1	5.9	5.6

References

[AB12] Andrews, D.W.K., Barwick, P.J.: Inference for parameters defined by moment inequalities: a recommended moment selection procedure. Econometrica **80**(6), 2805–2826 (2012)

[AS10] Andrews, D.W.K., Soares, G.: Inference for parameters defined by moment inequalities using generalized moment selection. Econometrica **78**(1), 119–157 (2010)

[DH97] Davison, A.C., Hinkley, D.V.: Bootstrap Methods and their Application. Cambridge University Press, Cambridge (1997)

[Efr79] Efron, B.: Bootstrap methods: another look at the jackknife. Ann. Stat. **7**, 1–26 (1979)

[Han05] Hansen, P.R.: A test for superior predictive ability. J. Bus. Econ. Stat. **23**, 365–380 (2005)

[HHK10] Hsu, P.-H., Hsu, Y.-C., Kuan, C.-M.: Testing the predictive ability of technical analysis using a new stepwise test with data snooping bias. J. Empir. Finance **17**, 471–484 (2010)

[Hol79] Holm, S.: A simple sequentially rejective multiple test procedure. Scand. J. Stat. **6**, 65–70 (1979)

[Lah03] Lahiri, S.N.: Resampling Methods for Dependent Data. Springer, New York (2003)

[RSW14] Romano, J.P., Shaikh, A.M., Wolf, M.: A practical two-step method for testing moment inequalities. Econometrica **82**(5), 1979–2002 (2014)

[RW05] Romano, J.P., Wolf, M.: Exact and approximate stepdown methods for multiple hypothesis testing. J. Am. Stat. Assoc. **100**(469), 94–108 (2005)

[Whi00] White, H.L.: A reality check for data snooping. Econometrica **68**(5), 1097–1126 (2000)

Exploring Message Correlation in Crowd-Based Data Using Hyper Coordinates Visualization Technique

Tien-Dung Cao[✉], Dinh-Quyen Nguyen, and Hien Duy Tran

School of Engineering, Tan Tao University, Duc Hoa, Long An, Vietnam
dung.cao@ttu.edu.vn

Abstract. Analytical exploration for necessary information and insights from heterogeneous and multivariate dataset is challenging in visual analytics research due to the complexity of data and tasks. One of the data analytics target is to examine the relationship in the dataset, such as considering how the data elements and subsets are connected together. This work takes into account the *direct* and *indirect* connection relations: elements and subsets of elements might not only be directly linked together, but also possibly be indirectly associated via the relationships from other elements/subsets as well. Stream of messages instantly put on the cyberspace from the crowd is an example for such kind of dataset. In this paper, we present an approach to estimate the correlation between streaming messages collection in terms of large scale data processing, whilst the *Hyper Coordinates* visualization technique is designed to support those correlations exploration. The prototype tool is built to demonstrate the concepts for crowd-based data in the financial market domain.

Keywords: Hyper Coordinates · Multivariate data visualization
Message correlation · Direct/indirect relationship

1 Introduction

In this age of big data and internet of things, more than ever before, data are everywhere thanks to the advances in ubiquitous devices, networking and data management. This flushes a wealth of valuable information that can be exploited in many application scenarios from everyday personal lives, to businesses, and to governments. But because big data are normally heterogeneous, from various kinds of sources, and typically contains different types of information, the task of identifying and extracting valuable information necessary for a specific usage scenario is complicated and challenging. With complex and multivariate datasets, source-origin big data are usually chaotic and unwell-modeled for a specific usage. Therefore, they must be normally cleansed (removed the duplicate and similarity or unnecessary information), aggregated or augmented with expertise knowledge, and transformed into well-modeled structures for further

© Springer International Publishing AG 2018
V. Kreinovich et al. (eds.), *Predictive Econometrics and Big Data*, Studies in Computational Intelligence 753, https://doi.org/10.1007/978-3-319-70942-0_5

analyses. The advances of today machine learning and artificial intelligence make those steps more realistic but not always fulfilled. In many cases, analysts can be overloaded in the exploration and examination process for data selection and transformation. To deal with it, visualization supports in terms of visual analytics have been being taken into account [27].

One of the subjects for visual analytics research on multivariate data visualization is to investigate the relationships of data in the datasets. There are various types of relationships that can be taken into consideration, including *set* relations – data elements of categories, *connection* relations – linking pairs of data elements or sets, *ordered* relations – such as the orders of connections that form the relations, *quantitative* relations – quantitative information of elements in a set or connection, and *spatially explicit* relations – such as positions on geographic maps [7]. Except the spatially explicit relations which are geometrically constrained, all the remaining just-listed relationships are representable in the *Hyper Word Clouds* visualization technique [20]. It is proved as those types of relationships are relatively constituted to each others. Based on that technique, this paper additionally takes the *direct* and *indirect* aspects of connection relations into account.

We investigate on that visual analytics problem with regard to the scenario of financial market domain. In the domain of financial markets, understanding the market influences from big data so that to predict the market trends is an expectation from the business investors. The software company Sentifi,[1] in our showcase, investigates big data from the crowd to develop products for global financial markets, i.e., identifying and ranking the market influences and predicting market trends. Sentifi has developed a tool for cleansing and augmenting the crowd data around messages (pieces of text posted or discussed on the cyberspace, mainly from Twitter). As mentioned above, the processed data may still contain duplicated, similarity, or associated data. And thus, analysts at Sentifi have to analyze the relationships in the data to cleanse and structurize the dataset. To that end, we develop to support Sentifi in spotting out how messages streamed from the crowd are correlated together, considered as the problem in aggregating and selecting important messages in clusters (see more in Sect. 2).

Although this also concerns a classic problem in data mining, clustering multivariate and real-time updated big data is still the challenge due to the complexity of data as well as the limitation of automatic analytical solutions. Several approaches have been proposed to examine the correlation of messages to either reduce the duplicate/similarity information [21,31] or connect messages in a log file into a business process that allows users to effectively navigate through the process space and identify the ones of interest [19,22,25]. Ye *et al.* [34] also introduced a data correlation for a stream based similarity clustering. However, it is challenging because we here care about the correlation of message that does not come only from direct relationship among messages – i.e., two messages share a common content –, but also from indirect relationship as well. It means that

---

[1] https://sentifi.com.

hidden/related information and/or domain expert's knowledge may also be useful information to make two messages cohere together. Unfortunately the information like this usually is stored in different datasets and the mentioned approaches thus do not take into account this information. To this end, this paper reaches the following contributions:

- An approach in which we do not only consider the direct connection relations of messages in a set for correlation estimation, but also their indirect connection relations, i.e., the additional information that comes from other associated elements/sets. At this point, we leverage the capability of today parallel and cluster computing in computing correlation weights among messages because it is a large scale data processing problem.
- The *Hyper Coordinates* visualization technique, which is the extension of *Hyper Word Clouds*, that supports the users in understanding, interacting, and verifying relationships in multivariate dataset. This reduces the limitation of solely automatic data computation.
- A prototype tool as a proof of concepts discussed in our novel approach.

The remaining sections of the paper is organized as follows. In Sect. 2, the working scenario at Sentifi is outlined for general problem identification. Following that, we discuss our main development in terms of message correlation computation and visualization support in Sects. 3 and 4, respectively. Section 5 will then showcase the developed technique for the exploration of messages regarding Sentifi scenario. Section 6 compares our development with existing research. Finally, we conclude our work and shortlist future work in Sect. 7.

2   Working Scenario

2.1   Crowd-Based Data Analytics at Sentifi

The software company Sentifi aims at developing products for global financial markets. By collecting and analyzing crowd-based data available on the cyberspace, Sentifi identifies and ranks market influences, and then provides predictions for market trends. This section describes a scenario for that development, as outlined in Fig. 1:

Firstly, data are selectively collected and preprocessed with three major parallel activities:

1. Everyday, workers at Sentifi – with the help of a software engine – collect and verify social network profiles and store in their database a list of *publishers* who talk about something on the Internet;
2. They also process to collect other information which are categories about *topic*, name of *entity*, *lexicon*, *source*, *impact* and *event*. Here, Sentifi's domain expertise is needed in manually defining and refining relevant information extracted from the software engine;

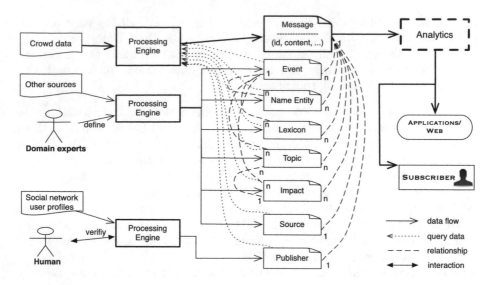

Fig. 1. A general working scenario at Sentifi

3. And, from the crowd data such as news articles, blogs, and Twitter tweets, etc. – many of which are chaotic and contain tremendous stuffs which are meaningless for financial domain – Sentifi selects *messages* in terms of financial contexts by restructurizing and augmenting them with information from the above-mentioned categories with the aim of a semantic engine. For instance, a message is analyzed to be about someone/organizations (i.e., name of entity) on one or some topics (e.g., "stock market", "downing of oil price", etc.), from or related to one or some events (e.g., "Obama recommends Apple to produce their product in US instead of China or other countries"), and indicates how its information impacts to a list of topics, and so on.

To this point, *messages* are central objects in analyzing data streaming on the Internet. However, the problem of redundancy and similarity exists amongst the messages. In that regard, analytical support is expected as a next step in structurizing verification, message similarity detection, important messages identification, data augmenting correction before the messages are sent to subscribers or for reporting. Currently, Senfiti employs a lot of staffs/experts in this phase, therefore they need a tool to simplify the work. In the next section, the challenges for this analytical support is discussed.

2.2 Problem Analysis

Message similarity detection is the most important requirement at the Analytics phase of Sentifi's scenario. This is because they want to avoid the case of sending a group of messages to their subscribers in which those messages have similar content. This is not a new problem since it relates to message correlation problem which estimates a distance or weight between two messages. Several approaches have been proposed for similar issues such as [10,19,21,31,34]. Unfortunately all those approaches were proposed for a concrete context and purpose with limitation of data available or unwell-structured of data. Therefore, the hidden/related information – which is usually stored in different datasets and linked to processing data via direct and/or indirect relationships – does not take into account. In financial domain, considering all direct as well as indirect information for a decision making is very important. For instance, since two messages talk about two different events, they usually have no correlation if considering only direct information about events. However, if some additional information of events such as: (i) two events are organized by the same community, (ii) they talk about the same topics, or (iii) in the same context, and so on – then the messages still somehow have a correlation. We call this type of correlation as indirect connection relationship. In other words, as examining correlation between messages, all indirect relationships through many augmented data must take into account beside the direct ones.

As mentioned earlier, the analytics phase do not only cover the message similarity problem, it must also take other problems into consideration as well. We highlight all those related problems in the following question list:

1. *How two messages correlate together?*
2. *Which ones are important in a group of correlated messages?*
3. *Could an important message represent to all others in a group?*
4. *If yes, which missing information of the group that a message must cover?*
5. *Does the processing engine map correctly financial terms?*

All those questions raise us two following problems: (i) message clustering as well as recognizing the representatives, (ii) visualizing message correlation that allows domain experts to explore the message clusters, confirming the representatives as well as interacting with data by correcting message value or merging some of them together. Two those major problems are really necessary in coping with streaming messages from the crowd. In the two next sections, we will introduce in details our solutions for those two problems.

3   Message Correlations

We present in this section an approach to measure the correlations between messages that works as a back-end system in supporting the visualization technique described in Sect. 4. We first define our supported data model in Sect. 3.1 in which the concepts of *direct* and *indirect* connection relations are described. Section 3.2 presents a method to calculate *weight* – the parameter in measuring correlation between two messages. Section 3.3 introduces the algorithms used for: (i) calculating correlation weights of all pairs of messages; (ii) clustering messages into clusters and extracting their cluster representatives as well as their properties; and (iii) calculating correlation weight for a new incoming message based on the already calculated messages.

3.1   Data Model and Relationships

Before going further to discuss on message correlation, we describe in this section the supported data model as well as the analysis on data relationships. We organize data into tables centered around a *focus table* (denoted, ft), in which the correlation among rows of ft is measured based on its relationship with other tables (denoted, $rt = relation\ table$). For instance, take a look into the Sentifi's scenario, ft is the *message* table because we want to measure the correlation among the rows of messages in this table. Whereas the other tables such as *event, topic, lexicon, publisher*, etc. are the relation tables. Based on these concepts (i.e., focus table and relation table), we define that a *direct* connection relation is a relationship established by ft table and a relation table rt. An *indirect* connection relation is a relationship established by two relation tables. Definition 1 defines a formal definition of *direct* and *indirect* connection relations.

Definition 1. Let ft be a focus table in which $A_{ft} = (a^1, a^2, ..., a^n)$ is a set of attributes of ft and $RT = \{rt_1, rt_2, ..., rt_m\}$ be a set of relation tables where $A_{rt_i} = (a_i^1, a_i^2, ..., a_i^{k_i})$ is a set of attributes of rt_i. A set of relationships[2] (based on ft and RT) $R = R_d \cup R_u$ is defined as follows:

- Let $RT_d \subseteq RT$, $R_d = \{r_1^d, ..., r_{|RT_d|}^d\}$ is a relationship set between ft and RT_d (called, direct connection relation), in which $r_i^d = [a^k, a_i^l] \mid a^k \in A_{ft}, a_i^l \in A_{rt_i}$.
- $R_u = \{r_1^u, ..., r_h^u\}$ is a relationship set in RT (called, indirect connection relation), in which $r_i^u = [a_p^k, a_q^l] \mid a_p^k \in A_{rt_p}, a_q^l \in A_{rt_q}$ (A_{rt_p} and A_{rt_q} are the set of attributes of relation table rt_p and $rt_q \in RT$).

Constraints on Relationship: Basically, there are 4 types of constraints on a binary *connection* relationship between elements on two tables, which are *one-to-one, one-to-many, many-to-one* and *many-to-many*. However, we target them here through *ordered* relations which indicates the sequence of connections in the order from a source to a destination. Besides, *set* relation is used to explain a data element belongs to a set (typically a category), and *quantitative* relation is used to represent the quantitative information of data elements in a set (see more in Sect. 4.1). For instance, take a look back to Sentifi's scenario, there can be ordered relations which are directly –by meaning of direct connection relation– from *message* to *event*, *message* to *topic*, *message* to *name entity*, *message* to *lexicon*, *message* to *impact*, *message* to *publisher* and *message* to *source*. While there are also ordered relations indirectly –by meaning of indirect connection relation– via *event* such as *event* to *topic*, *event* to *name entity*, *event* to *lexicon* and *event* to *impact* or via *impact* to *topic*.

---

[2] We define here a binary relationship between two tables, i.e., we pick one attribute up from each table to make a relationship. In real database, a relationship may be created from several attributes. However, we can transform them into a single attribute by specifying representative identifier.

3.2   Correlation Weight

Correlation weight is a parameter to measure the relationship between two messages. Two messages have a strong relationship if their correlation weight is high and a weak relationship if the weight is low. If the weight is zero, there is no relation between two messages. In our approach, the weight is calculated based on two types of connection relation defined in Definition 1, which are R_d and R_u. Suppose that all ordered relations from the focus table to relation tables or between relation tables are set relations, a constraint *one-to-one* is considered as a set with only one element. Starting from the focus table, each direct connection relation (denoted r^d) will contribute a direct weight (denoted w^d) to the total weight. This direct weight is calculated by getting the size of intersection set (denoted IS) of value sets of two rows m_i and m_j projected on attribute a_k of direct connection relation r_k^d, multiplied with a factor w_k^d defined by user. In addition, if both value sets of two rows m_i and m_j projected on attribute a_k of direct connection relation r_k^d (denoted $VS_{m_i}^k$ and $VS_{m_j}^k$) subtract to intersection set (i.e., IS) are not empty, and the attribute a_k is an object defined by an indirect connection relations, an indirect weight (denoted w^u) of two value set $VS_{m_i}^k$ and $VS_{m_j}^k$ is calculated by sum of correlation weights of all element pairs $(c_i, c_j) \in (VS_{m_i}^k \setminus IS \times VS_{m_j}^k \setminus IS)$. While calculating the correlation weight of element pair (e_i, e_j), if the indirect connection relation is defined from current relation table to others, we recursively calculate the correlation weight of those indirect connection relation until the end of ordered relation path. For instance, in the Sentifi's scenario, we first calculate the direct weight of two messages using direct connection relations such as *message-to-event, message-to-topic, message-to-lexicon*, etc. While considering connection relation *message-to-event*, suppose that two messages m_i, m_j do not share the same event (i.e., $w_e^d = 0$), we calculate an indirect weight of two events e_{m_i} and c_{m_j} (i.e., w_e^u). To calculate the indirect weight w_e^u, we use the indirect connection relations via *event* such as *event-to-topic, event-to-impact, event-to-lexicon*, etc. If e_{m_i} and e_{m_j} do not share any common value set of topic or lexicon, it means topic and lexicon do not contribute a factor to w_e^u. However, suppose that e_{m_i} and e_{m_j} do not share a common value set of impact, we continue to calculate an indirect factor of two impacts, which e_{m_i} and e_{m_j} hold, because there is an indirect connection relation defined from impact to topic. For each recursive step, a function is defined to scale an indirect factor to be smaller than a direct one. For instance, we define the following function is defined to scale the indirect factor of events.

$$f(x) = x/(max + min)$$

where $max = \mathrm{MAX}(\text{weight}(e_{m_i}, e_{m_j}))$ and $min = \mathrm{MIN}(\text{weight}(e_{m_i}, e_{m_j}))$.

Example 1: For the data model provided in Fig. 1, without considering any unrelated fields, we suppose to have the two following messages:

- $m_1 = (id_1, \{event_1, event_2\}, \{topic_1\}, \{lexicon_1, lexicon_2\}, \{name\_entity_1\}, \{impact_1\}, publisher_1, source_1)$

- $m_2 = (id_2, \{event_2, event_3\}, \{topic_2\}, \{lexicon_1, lexicon_3\}, \{name\_entity_1\},$
  $\{impact_2\}, publisher_1, source_2)$

To compute the correlation weight w between m_1 and m_2, we also suppose to have *factors* already specified for each property as follows (*event = 2, topic = 1, name_entity = 0.5, lexicon = 0.7, impact = 0.5, publisher = 0.3, source = 0.05*), and thus w will be computed as $w = 2*w_e + 1*w_t + 0.7*w_l + 0.5*w_n + 0.5*w_i + 0.3*w_p + 0.05*w_s$, where:

- $w_e = w_e^d + f(w_e^i)$ is the weight of event in which $w_e^d = |\ \{event_1, event_2\} \cap \{event_2, event_3\}\ | = 1$ is the direct weight, and w_e^i is the correlation weight between the non-intersected set $\{event_1, event_3\}$ which is calculated based on their properties such as name entity, topic, lexicon and impact (i.e., indirect connection relations of message via event). w_e^i is calculated with the same method applied for messages, so we do not present them in details but just assume that w_e^i is calculated $= 1.5$ and $f(w_e^i) = 0.65$. As a result, we have $w_e = 1 + 0.65 = 1.65$.
- $w_t = |\ \{topic_1\} \cap \{topic_2\}\ | = 0$ is the weight of topic.
- $w_l = |\ \{lexicon_1, lexicon_2\} \cap \{lexicon_1, lexicon_3\}\ | = 1$ is the weight of lexicon.
- $w_n = |\ \{name\_entity_1\} \cap \{name\_entity_1\}\ | = 1$ is the weight of name entity.
- $w_i = w_i^d = |\ \{impact_1\} \cap \{impact_2\}\ | = 0$ is the direct weight of impact, suppose that the indirect weight of $impact_1$ and $impact_2$ is zero.
- $w_p = |\ \{publisher_1\} \cap \{publisher_1\}\ | = 1$ is the weight of publisher.
- $w_s = |\ \{source_1\} \cap \{source_2\}\ | = 0$ is the weight of source.

Finally, we have $w = 2*1.65 + 0 + 0.7 + 0.5 + 0 + 0.3 + 0 = 4.8$.

3.3   Computing Correlation Weight

As discussed in Sect. 2.2, to help domain experts to explore the crowd-data such as identifying message similarity, recognizing important messages, correcting relations, visualization supports are needed. However, to visualize a data set, we need a back-end system to estimate the correlation among messages as well as other related properties. This section presents three algorithms for the following purposes: (i) calculating correlation weights of all pairs of the available messages; (ii) based on these weights, classifying messages into clusters that allow us to overview the relationship of the whole dataset; (iii) calculating the correlation weight of a new incoming message against the existing ones. Because of big data, all the algorithms are designed to run on a parallel processing platform such as Spark [1].

Algorithm 1 presents the algorithm to calculate correlation weight of all pairs of messages. This algorithm requires 4 additional parameters beside the set of messages. The first and the second ones are two set of direct/indirect connection relations. The third and the fourth ones are the impact factors corresponding to the direct/indirect connection relations. In this algorithm, we first calculate a Cartesian product of all rows of the *message* table (line 2). This operation gives

---

Algorithm 1. Correlation weights of all pairs of messages

Data: M is a message table;
R^d is a set of direct connection relations on M;
R^u is a set of indirect connection relations on attributes of M;
W^d is a set of impact weights of R^d;
F is a set of scale functions.
Result: Correlation weight of all pairs in M.

```
 1  begin
 2  │   CP ⟵ M × M;
 3  │   FCP ⟵ Remove all pairs (m_i, m_j) ∈ CP such that m_i.id after m_j.id;
 4  │   Load all weights of indirect connection relation into buffer;
 5  │   for (m_i, m_j) ∈ FCP do
 6  │   │   w ⟵ 0;
    │   │   // each relation corresponds to an attribute of M
 7  │   │   for r ∈ R^d do
 8  │   │   │   A_i ⟵ set of values of m_i projected on attribute r;
 9  │   │   │   A_j ⟵ set of values of m_j projected on attribute r;
10  │   │   │   I ⟵ A_i ∩ A_j;
11  │   │   │   w_r^d ⟵ | I |;
12  │   │   │   w_r^u ⟵ 0;
13  │   │   │   if A_i \ I ≠ ∅ and A_j \ I ≠ ∅ and ∃r_u ∈ R^u defined on attribute r then
14  │   │   │   │   A_i^* ⟵ A_i \ I;
15  │   │   │   │   A_j^* ⟵ A_j \ I;
16  │   │   │   │   for (a_p, a_q) ∈ (A_i^* × A_j^*) do
17  │   │   │   │   │   w_r^u ⟵ w_r^u + f_r(weight(a_p, a_q));
    │   │   │   │   │   // f_r ∈ F and weight(a_p, a_q) is queried in buffer
18  │   │   └   w ⟵ w + W^d[r] * (w_r^d + w_r^u);
19  │   └   Store tuple (m_i, m_j, w) into result;
```

---

us two relation pairs between row r_i and row r_j, i.e., $<r_i, r_j>$ and $<r_j, r_i>$. However, these relations are equivalent, so we remove one of them by defining a rule "r_j after r_i" then removing $<r_j, r_i>$ (line 3). From line 5 to line 18, we calculate the weights for the remainder pairs. For each pair, we examine all attributes (line 7), which are defined in direct connection relations, of message table to compute the factors they contribute to the total weight of the pair (i.e., variable w in line 6). For each attribute (denoted at), the factor of direct connection relation is computed by counting the size of intersection set between two value sets of at (line 8–11). If both value sets have remainder values (after removing all common elements) and the indirect connection relation are defined on table referring to at (line 13), we recursively compute the correlation weights for all pairs in these two remainder value sets. However, to speed up the performance, based on the indirect connection relation set and the ordered relation property, we apply the same method to compute the weight of all pair values of at before this processing, and we just query its value from buffer or database (line 17). We next sum up all weights of each pair to have a weight of indirect connection relation. Finally, the correlation weight of two messages is the sum of all attribute factors (line 18).

Using the output of Algorithm 1, we propose Algorithm 2 to group messages into clusters. The output of Algorithm 1 can be considered as a weighted graph, in which each message is a vertex and the edges are the correlation message pairs. We define clusters to be the connected components of this graph. Therefore, by setting up a threshold and removing all edges with weak correlation, the remainder graph is then a set of clusters (line 2–4) – where every cluster contains messages that correlate strongly together. However, on the one side, from the view of visualization, user usually want to see an overview of data before going further to detail. On the other side, the domain experts also want to recognize which ones are the important messages in every cluster. This requires us to select the representatives for each cluster. We define that an important message is the one which is correlated to many others. In the case that two messages have the same number of correlations with the others, the one with stronger correlation weight is more important. Using this rule, we rank the messages as follows:

$$rank(m) = \sum_{m_i \in C_m} weight(m, m_i) \tag{1}$$

where C_m is a set of messages that m has a correlation.

Although two messages belong to a cluster, they might still not share a common set of values because they do not have the direct correlation. For example, A has correlation with B (denoted $A \longleftrightarrow B$), then $B \longleftrightarrow C$ and $C \longleftrightarrow D$. However, A may not have a correlation with D because they may have different value sets. Since both A and D have the highest rank in a cluster, we select both as the representatives. After sorting messages by their rank, the user will decide how many messages he/she wants to get as the representatives for each cluster by indicating a percentage (line 7). Finally, the values of representative set are extracted and stored for further calculation (line 8–9).

At this point, all calculations necessary for the visualization are processed. However, there is another problem in the case that data are still being collected from the crowd. It means that we need to continue to measure the correlation of any new incoming message (against the existing ones). This problem is usually carried out on-the-fly where the calculated results are provided instantly to the visualization component. In this case, using Algorithm 1 is not appropriate since the available messages have been grouped into clusters and their representatives and properties have been extracted, where the incoming message usually has no correlation to all clusters but just a few. Therefore, Algorithm 3 is proposed for this purpose.

We first find the cluster(s) where the new message m is close to (line 6), then we only compute the correlation weight of this message against the messages in the clusters that the new message close to (line 10). This allows us to reduce the computation. If m belongs to some clusters, we remove all weak correlations to identify which cluster that m really belongs to (line 13). If m still belongs to more than one cluster, we merge those clusters into one and identify its representatives

---

Algorithm 2. Clustering message

Data: CW: a set of tuple (m_i, m_j, w);
t is a threshold of correlation weight;
p top percent of message in cluster (i.e., representatives of cluster).
Result: A set of cluster and their representatives as well as values.

```
1 begin
        // remove all weak correlations then we have clusters
2       C ⟵ Remove all tuple (mᵢ, mⱼ, w) in CW where w < t;
3       G ⟵ Build a graph from C;
4       CS ⟵ Get a set of connected components of G ;
        // Note: each component is a cluster, then we identify
            their representatives
5       for c ∈ CS do
6           Rank messages in c using formula (1);
7           T_c ⟵ Get top p percent of messages in c;
8           At_c ⟵ Get all attribute values of messages in T_c;
9           Store tuple (c, T_c, At_c) into result;
```

Algorithm 3. Computing correlation weight for new incoming message

---

Data: CS: a set of clusters and their representative values;
m: incoming message;
R^d is a set of direct connection relations.
Result: Set of messages that m has correlation as well as m's cluster.

```
1  begin
2      flag₁ ⟵ false;
3      for c ∈ CS do
            // c consists of id, value sets of attribute i ∈ 1,...,n
                where each attribute i represents a direct
                relationship.
4          flag₂ ⟵ false;
5          for r ∈ Rᵈ do
6              if value of m projected on attribute r belong to value set of attribute r of c then
7                  flag₂ ⟵ true;
8                  break;
9          if flag₂ = true then
               // m belongs to c
10             Compute the correlation weight of m and messages in c (Similarly with lines 5 - 19 in
                   Algorithm 1);
11             flag₁ ⟵ true;
12     if flag₁ = true then
13         Remove all weak correlations then identify which cluster m really belongs to;
14         if m belong to more than one clusters then
15             Merge those clusters into one;
16             Re-identify the representatives of cluster;
17     else
           // m does not belong to any cluster
18         A new cluster is added into CS where m is the representative;
```

(line 15–16). Finally, if m does not belong to any cluster, a new cluster is created and this message is the representative of that new cluster (line 18).

4   Visualization Design

In this section, we continue with the *Hyper Coordinates* visualization technique which is designed to support the analysts in visually exploring to understand how the messages calculated in Sect. 3 are correlated and clustered together, verifying their representatives, or updating the irrelevant data. Section 4.1 begins with the general analysis on data and data relationships as well as the expected interaction tasks. Following that, the details on visualization design in terms of data relationships and interaction tasks are then provided in Sects. 4.2 and 4.3, respectively.

4.1   Data Relationships and Tasks Analysis

Section 3.1 already presented the data model where the concepts of *direct* and *indirect connection* relations and other types of data relationships in a dataset are mentioned. In this section, in terms of visualization design, we further discuss on those relationships – which are *connection* relations, *set* relations, *quantitative* relations, and *order* relations. We consider all of those relationship types here because they are relatively constituted to each others in communicating data characteristics and tasks at hand.

When examining a data element in a relation table associated with a message in the focus table, definitely we need to show there is a connection between that associated element and the message, and thus the representation of such connection in the visualization is requested. Additionally, it is expected to visually differentiate the *direct* and *indirect connection* relations: visual *direct* connection is provided if the element is directly associated with the message, while *indirect* connection is provided when the element is indirectly associated via the relationships from other elements or other sets on other relation tables.

With a connection relation between a message and an associated element, it obviously is also the connection between the focus table and a relation table (see Definition 1). Therefore, one might easily understand that we also need ways to represent the connection between the tables. This is a kind of *set* relations visualization. Further on, by visually representing set relations, we can also communicate the situation that messages belonging to clusters (referring Algorithm 2).

Next, we seek to communicate the weights that constitute the correlation between messages or between any data elements or sets on tables. By that request, it is meant that *quantitative* relations visualization are desired. Finally, as introduced in Algorithm 1, correlation weights between messages are aggregated from different direct and indirect connections, such as from *topic* to *impact*, then to *event*, and then to *message* tables. In that regard, the *order* of such connection relations are expected as well.

Regarding tasks analysis, we concentrate on low-level visualization tasks – which are *visual summarization, visual identification,* and *visual comparison* (in terms of the typology defined [4]). Firstly, we need interactions to support the users in detecting important messages from correlated messages clusters. But because the dataset is large, it is impracticable to visually display all the data elements and relationships in details. Therefore, a *visual summarization* about how messages are distributed over the clusters should be provided first so that the users will have a general overview. Secondly, from the overview of message clusters, users normally want to dig into the view to find out further information, such as the messages and correlations that they are interested in. In other words, we need supports for *visual identification* of data of interest. And finally, since messages and their correlations need to be verified by the analysts, we also need supports for *visual comparison* on different combinations of relationships for correlation computation.

4.2 Design in Terms of Data Relationships

With the above discussion on data relationships analysis, visualization development works around how *direct* and *indirect connection* relations, *set* relations, *quantitative* relations, and *order* relations are visually represented. In our design, it is accomplishable through the ways visual objects showing in parallel coordinates for *set*, anchor path for *connection* and *order*, node size for *quantitative*, and link for *connection* relations, amongst some other encodings using common visual primitives.

4.2.1 Parallel Coordinates for Set Relations

The crucial idea behind the *Hyper Coordinates* is to represent the dimensions of a multivariate dataset as parallel coordinates so that their variables and relationships can be mapped to. Every coordinate, which is a parallel vertical column on the screen (Fig. 2a) is designed to communicate a data table. Accordingly, when data elements are shown on this column (referring Sect. 4.2.3), they communicate the *set* relations about the table in which they belong to. The name of the data set, such as table name, is placed on top of the column for table identification. To support *connection* relations between sets or their member elements, we also use colors to differentiate the tables.

4.2.2 Anchor Paths for Connection and Order Relations

Anchor path is the concept developed to indicate the connection between the tables, or set of elements in the column that the table represents for. It is named as *anchor* because through it, the elements or links from the elements on the columns are identifiable. Instead of looking at the detailed connection between elements on the columns, looking at the anchor path is more vivid in examining those *connection* and *order* relations (Fig. 2b). We simply draw curve path to link pairs of tables for connection relations. The color of the path indicates the direction from a table to another, and thus it is colored by the hue value of the

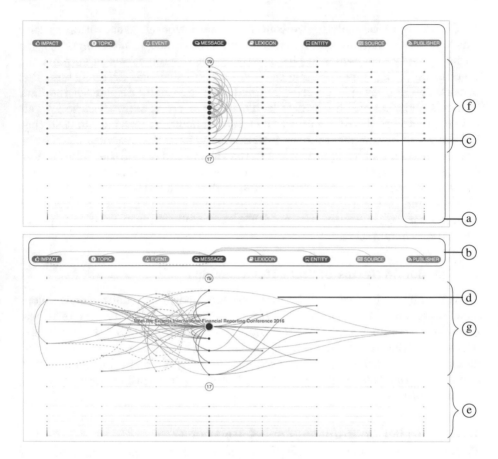

Fig. 2. *Hyper Coordinates* visualization design: (a) each column communicates *set relation* of elements in a table, where (b) table names on top are connected through the *anchor paths* for *connection* and *order relations*; (c) size of node indicates its *quantitative relation* in a set, while (d) types of *connection relation* between nodes are encoded by color and solidity. During interaction, nodes and links are shown based on the level of overview-detail: (e) overview of clusters, (f) overview of messages in a cluster of interest, and (g) detail of nodes and connections between elements associated with a message of interest.

source table. To differentiate the *direct* and indirect *connection* relations, the solidity of line is used: solid line is for the direct connection, and dashed line is for the indirect one. To avoid the matter of cluttering if the direct and indirect paths are too close together, especially in the case they are coincident, they have to be positioned further from each others, such as in two opposite sides above and below the baseline of the table name.

4.2.3   Nodes and Links for Quantitative and Connection Relations

Node-and-link is a basic way to represent connected elements in a set, especially in graph. In Fig. 2c, sets of nodes are placed along the columns and they communicate well the sets that they belong to. In that way, the simplest encoding is to use a dot for a node. Since node's position is associated with set relations (nodes have to be on the columns), size of the node is designed for *quantitative* relations. To encode *connection* relations, path line linked every pair of nodes is provided. The color and solidity of line indicate the source of the associated table, as well as whether the connection is direct or indirect between nodes – similarly to the design of anchor paths (Sect. 4.2.2). An indirect connection is communicated through the two dashed lines in between two relation tables with a solid line between the nodes on the same *via* table. Since the amount of nodes and links are typically tremendous, linking lines have to be smoothly blended as curves to minimize negative affects as well as visual clutters (Fig. 2d). In Sect. 4.3, we will present different visual encodes designed for nodes and links for different levels of overview + detail exploration.

Finally, it is also possible to create node as a glyph with further visual cues. Text can be added to nodes to form the parallel word clouds, where visual cues are browsed and explored through interaction. However, this has to be carefully examined since the dataset is typically so big and complex.

4.3   Design in Terms of Interaction Tasks

The three low-level tasks mentioned in Sect. 4.1 in fact follow the sequence of tasks carried out in the typical *visual information seeking mantra*, which are "overview first, zoom and filter, then details on demand – together with relate, history, and extract" [26]. This section describes how those interaction tasks affect our visualization development. Since the display space is restricted but data are numerous, different data have to be selected to be represented on the *Hyper Coordinates* interface following user interactions. We design here three levels of overview + detail interactions: overview of clusters (first level), overview of messages on a cluster (second level), and details of a message of interest (third level). These three levels of overview+detail exploration are discussed with regard to the management of visual primitives of nodes, links, anchor paths, and parallel columns.

4.3.1   Overview of Message Clusters

With the large stream of messages from the crowd, they are clustered into groups as presented in Sect. 3.3. Therefore, to examine the messages, the straightforward strategy is to show them in clusters. Typically, users want to get general information such as: for each cluster, how many messages are there, what is the highest correlation weight of a message in that cluster, what relation tables contribute to the correlation weights of the messages in that cluster, etc. This is because the data are big and complex and thus it is impractical to show all their messages as well as how the messages are correlated at the same time.

To that end, at the first level of overview, we design to show only simple information indicating the tables constitute to each cluster which give hints for further interaction. Because messages are clustered together, the *message* column in the parallel coordinates is used to communicate the clusters, while the other columns are still for the other tables (users can interact to drag each column to any horizontal position on the screen). For every cluster, straight lines are used to represent *connection* relations between the relation tables and the focus table, while small dots are shown on each relevant table columns to emphasize the connections (in fact, these forms a group of overlapping lines from relation table columns to the cluster column). To provide a coherent view on the interaction, the colors of the nodes are unique with the colors of the tables. However, we want to discern the connection lines with the similar ones in the second level for messages overview – as presented next, thus we use monotonic color (grey) for line connections (Fig. 2e).

Now, with numerous clusters shown as parallel lines on this overview level, we employ fisheye lens [24] to support the users in focusing on a cluster of interest for further interactions.

4.3.2   Overview of Messages in a Cluster of Interest

Following the above overview level for clusters, when a user focuses on a cluster, s/he is supported to explore its messages of interest. Therefore, we allow the user to select a cluster of interest that leads to this second level of overview + detail interaction.

Most of the display space is now reserved for the messages, while the general context of the fisheye clusters reduces. At this point, we only show the overview of messages so that the user can skim and decide to continue to explore or move to another cluster (we give hints about the clusters above and below the selected cluster with information such as the number of messages in those clusters). The message nodes and their connection tables are encoded completely like in the first level of overview, except two main things: (i) the nodes on the message column is for messages, and thus we only encode information of message nodes, which is its weight in the cluster, and (ii) connection lines are colorized by connection columns' colors to distinguish with the line of clusters (Fig. 2f).

Furthermore, at this level, to support the user to pay more attention to a message, as the user hovers on it, we highlight the message of interest (without changing visual encoding) while dimming the other messages (including their connection links and colors). The anchor path is also updated with detailed relation tables (direct vs indirect).

With such overview, users are given with hints about messages and correlations, but they do not know how detailed the weights are aggregated. Therefore, the next step is to support the user to zoom more and filter into details for a selected message.

4.3.3   Examine Messages of Interest in Details

With a selected message is now all related messages (they are all from the same cluster) that contribute to the aggregated correlation weight of the selected message to be explored. To support the users to perform this examination, all messages and associated elements on other tables are to be detailed. The display space is the space used for the messages overview in the second level (but all old messages and their associated nodes on other table columns as well as the existing connection links are removed).

Then, we show detailed nodes for elements on their table columns and detailed links for connection relations following the design provided in Sect. 4.2.3. The colors of nodes and links indicate the associated tables. The size of a node indicates how quantitative the correlation weight is. For connection relations, there are three types of lines between the nodes: the solid paths for direct connection relations between elements on two different tables, and the solid paths between nodes on the same table –which represents a correlation of messages– and dashed paths between different tables for indirect connection relations (Fig. 2g).

With that encoding, the user can interact by hovering on and selecting the node or link that s/he wants to examine, where more visual cues are added to the associated nodes and connection links. This supports the user in understanding the detailed correlations between messages. The node which is hovered is highlighted bigger and with a tooltip. To filter the nodes and connections of interest, e.g. in checking their relationships, the user can click on a node to include/exclude them from the examination – the excluded node is dimmed with low opacity, while the included one will show with full opacity. To include/exclude a node and all of its connected nodes, the user can press a predefined key while clicking on the node, where the connection links are highlighted as well. S/he can also drag any node to any place on the associated column for a clear understanding of relations between nodes (in addition to dragging columns). By default, we show only the nodes and connections those contribute to the correlation weights which form the cluster. However, the user might want to check the unrelated data elements (properties) of messages as well (e.g., so that s/he can decide whether or not a message should be a representative for another one). As a result, we also design to include the unrelated data elements when two messages are in comparison (but to avoid using so much ink on the screen, they are only included as the user clicks on the connection link of interest between two messages). To this point, tasks such as identifying, relating information, and comparing nodes and links are easily carried out. In the next section, we will present examples that analysts can do in exploring and verifying messages through their correlations.

5   Showcase

In this section, we first outline the overview of the system prototype developed to actualize our proposed approach, in Sect. 5.1. Then, Sect. 5.2 describes the

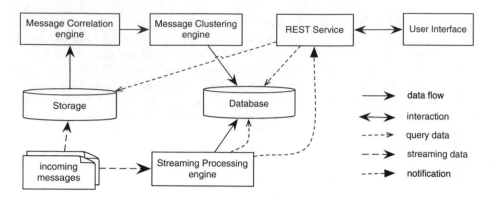

Fig. 3. System prototype overview

visualization interface, where sample cases are demonstrated in Sect. 5.3. The results of user interactions are discussed in Sect. 5.4.

5.1   System Overview

We implement a system prototype based on the message correlation approach and the visualization technique described above. Its aim is to provide an interface for message correlations examination and verification, as a part of the Analytics component mentioned in Fig. 1 of Sect. 2. Figure 3 shows our prototype overview. Ignoring data collection and pre-processing steps, starting from storage, the prototype system measures the correlation of all message pairs by implementing Algorithm 1. We use Spark framework [1] to implement this component. The result is then passed to the next component, named Message Clustering, which implements Algorithm 2 using Spark GraphX[3] library for clustering. Because the output of Message Clustering component is unwell-formatted and lacks of related information for visualization, we store it on a database. We next implement a REST Service to transform the previous output into a well-formatted structure which augments them with additional information before sending to the client-side User Interface. In parallel with those components, we also implement Algorithm 3 for Streaming Processing component, which receives incoming messages from a REST service of pre-processing engine and calculates the correlation of the incoming message with the existing ones in the database. It then notifies to REST service for visualization updating. The User Interface is implemented in Javascript, using D3js[4], jQuery[5], Bootstrap[6], and Font Awesome[7] for visualization and interaction handling.

---

[3] http://spark.apache.org/graphx/.

[4] https://d3js.org/.

[5] http://jquery.com/.

[6] http://getbootstrap.com/.

[7] http://fontawesome.io/.

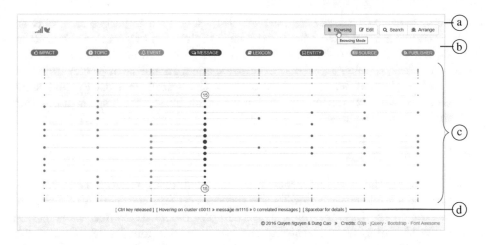

Fig. 4. *Hyper Coordinates* user interface: (a) toolbar, (b) anchor bar, (c) visualization area, and (d) notification bar.

5.2   The User Interface

The User Interface, as screenshotted in Fig. 4, includes a *toolbar* on top; then the *anchor bar*; then the *visualization area* for clusters, messages, and their related elements and tables visualization; and at the bottom of the visualization is a *notification bar*.

The *toolbar* is designed to contain a collection of controls to support the users in manipulating the visualization elements in focus. There are four option buttons to support the modes of *browsing, editing, searching*, and *arranging* data of interest. The default mode is *browsing* which supports the users in exploring how messages are correlated in clusters. If switching to *editing* mode, the user can adjust the connections between nodes or update their factors to examine the updated correlations and clusters. For *searching* or *arranging* mode, more controls will be added: with *searching*, there will include a search box and a multiple-select box that lists the tables for data searching or filtering; while with *arranging*, there includes a dropdown for selecting a table and criteria for arranging on that table (such as arranging clusters by the number of messages, or by highest correlation weight of the centroid message, or by the number of connected tables, etc.).

The *anchor bar* is immediately above the *visualization area* for table names and anchor paths manipulation. It is separately mentioned (along with the *visualization area*) because we want to differentiate interaction tasks on it and tasks on *visualization area*. The anchor bar is implemented following the design in Sect. 4.2.2.

The *visualization area* is the main space where users interact with the data elements (such as clusters, messages, ...) and connection links following the design with three levels of overview + detail interactions discussed in Sect. 4.3. In the default mode of *browsing*, the user can keep Ctrl key for overview browsing with clusters, releasing Ctrl key for overview browsing with messages on a cluster, and pressing the Spacebar to enter the level of detailed exploration. For *editing* mode, the visualization does not change the level of zooming, while the user can click on a link, a node, a parallel coordinate, or an anchor path to edit its value.

The *notification bar* is provided so that the system can briefly text some messages to the users during interaction while not interfering his/her interaction on the *visualization area*.

5.3 Sample Cases

Now, ignoring the back-end processing (i.e., computing message correlations), we provide sample cases regarding the exploration and verification of messages of interest on the User Interface. These samples demonstrate how visualization supports the analysts in answering the question mentioned in Sect. 2.2. Let's start with the situation that Jane, an analyst at Sentifi, wants to explore just the messages about "oil price". To accomplish that action, she uses the search/filter function to limit the number of clusters as well as messages to be shown on the screen. With the list of filtered clusters, she presses Ctrl key and hovers on each cluster to see the overview of the messages on that cluster of interest, as already provided in Fig. 2f. At this point, she can easily notice the important messages in the cluster under exploration by looking at the size of the node as well as the number of connections with other nodes on that cluster, for instance, as arranging the ranked messages from the center of the cluster. To this point, question Q2 is answered.

After that, Jane hovers on an overview message and press Spacebar key to explore in details how a message of interest correlates with the other ones in that cluster, as shown in Fig. 2g. This answers question Q1. To check whether a message of interest can be the representative for other messages in the cluster or not (question Q3), Jane continues to examine the correlations of that message with every other message, accomplishable when she clicks to select every pair of messages to look at their detail direct/indirect connections, as presented on Fig. 5a. In addition to finding the answer for question Q3, Jane also gets the answer for question Q4 by checking the differences between any two messages. This is done when Jane hovers on a correlation link of the two messages, where separated nodes connected on the other relation tables are highlighted (Fig. 5b). With her domain expertise, Jane will decide whether to merge those messages together or not, or whether a missing information should be added into the representative message.

Finally, when with a message of interest, Jane may hover on any node to examine the detailed data – such as through tooltip as in Fig. 5c). There, she can check whether the semantic processing engine correctly mapped financial terms or not. To that end, question Q5 is also answered. From the interface in Fig. 5b and c, choosing the edit mode, this analyst can manually correct the mapping work of semantic processing engine by editing the financial terms, making a new connection relation or removing them from the list.

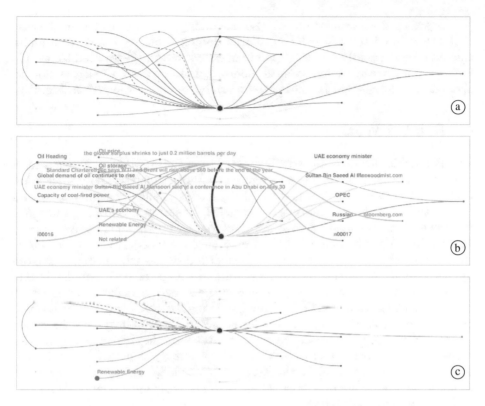

Fig. 5. Exploring detail correlation of a message of interest: (a) direct/indirect connection relation, (b) two messages comparison, (c) semantic engine verification.

Besides, when having a new message streamed to the system, the Streaming Processing component calculates its correlation against the existing messages in the database. As having a major change, it is prompted on a cluster (if it is not on the current cluster of interest) or as a (updated) node on the current cluster of interest. It is then provided to Jane as blinking on the interface. Got noticed, Jane might browse to that message to examine in details when she wants.

5.4 Discussion

We have shown on the above showcase the situation that a domain expert can easy explore streaming messages in terms of their correlations and relationships, answering the questions raised in Sect. 2.2. However, our current prototype system does not allow updating data in editing mode back to the data storage system yet. This is a challenge because we need mechanism to identify the effected messages as well as the solution for re-computing the correlations if the analysts make change on the data. This is also a limitation when a message needs re-computation by the semantic processing engine to improve the correctness of the mapping terms.

Regarding the visualization, we also have not provided solutions in the case that (1) we need to show a lot of information on a message or a node under examination (currently we just support showing tooltip), (2) we need to provide a clearer view when having so much nodes or links on screen (such as how to arrange the tooltips well instead of the overlapping ones presented on Fig. 5b).

6   Related Work

Visual analytics for real-time streaming text from the crowd is an active area of research. Marcus et al. develop a system for sentiment analysis and visualization from Twitter tweets [18]. The goal of that system is to identify significant events and sentiment in the streaming text. Frequency of searched keywords and sentiment classifiers over time windows are employed for peaks of high tweet activity aggregation. And events are displayed on a timeline-based visualization, where the users can interactively focus + context drill down to subevents and explore further relevant data, such as from geographical maps or with popular URLs. Ren et al. meanwhile leverage supports from the masses in analyzing events from Weibo microblog streams [23]. The crowd sourcing feedbacks and comments from users are collected using predefined criteria, which are combined with the statistics on how tweets and retweets are propagated to form the clusters of events. The clusters are then tree-based displayed, such as through connected graphs, and combined with other views for the experts to examine. They are some amongst common work focusing on statistical analysis and visually support information exploration using multiple views.

Gansner et al. [10], however, follow a different approach. They compute the similarity of Twitter messages based on semantic analysis – using Latent Dirichlet Allocation model, then cluster and display the message relations as clusters of nodes on a dynamic graph layout and graph-based stable packing algorithm. Another work is the TextWheel system developed by Cui et al. [8]. Sentiment and semantic relations between selected keywords and news articles are analyzed; then, patterns about those relations in terms of keywords are visualized as a wheel, connected to a so-called document transportation belt for articles, together with a trend chart for sentiment changes. The approach of these two techniques are quite similar to ours: focusing on semantics of messages or keywords relations to find out the important information, and visualizations are developed to communicate those relations. The difference is that they focus only on few direct relations (i.e., between semantic relations of messages to messages, or between keywords and articles), while ours examine many types and impacts of relationships between messages and their associated user-defined tables, especially the aggregated direct + indirect connection relations.

In the next two subsections, we present related work regarding our algorithms for message correlation computation and visualization development.

6.1   Message Correlation

Message correlation is the process of finding relationships between messages that belong to the same group with several purposes such as: finding messages belong to the same process execution instance, classifying faults in a systems, reducing storage data volume, or cohering many log files of a distributed system, etc. In the last years, several approaches of message/event correlation have been proposed. Some works such as [19, 22, 25] focus on finding correlation of events/messages for a process of business or protocol interaction that allows users to effectively navigate through the process space and identify the ones of interest. While the other works [21, 31] analyze correlation of messages of log files of distributed systems or clouds to minimize the log number without missing necessary information for later analysis. Jakobson and Weissman [16] also focus on the same problem by proposing a framework to find the correlation among alarms of telecommunication network for class hierarchy and reducing the number of alarms displayed on operators' terminals. Focusing on data replication among datacenters in geo-cloud environment and difficulty of determination of the replicas location, Ye et al. [34] also propose a data correlation for a stream based similarity clustering, which uses a small number of micro clusters to represent huge number of users and thus significantly reducing the cost of replica placement algorithm. In these works, although their approaches are either text-based search or semantic analysis of semi-structure format, the correlation condition is computed using direct relationship among messages. It means if two messages have a correlation, they must share at least a common content (e.g., a word). However, in many case, if the structure of message is well defined with additional information, they still have a correlation (i.e., indirect relation) even they do not share any common content. This problem is not considered in these works.

6.2   Visualization

In our visualization development, we design to communicate (i) the overview of correlations of messages in a collection, and (ii) the details of relationships between messages and elements on relation tables that form the correlations.

There are various techniques dealing with the first issue about how interrelated elements in a collection are visually communicated. Since the elements are connected together as a network, node-link graph is the classical mean to represent the elements and the connections. The visualizations developed in the above mentioned systems [18, 23] use this common representation. Basically, size or color of the nodes are used to represent node's importance in the network, while the positions are calculated based on the relative distances between the nodes. Besides, directions of connections are also representable as link arrows [30]. The elements can be grouped into or expanded from subsets for different zooming levels as well, which typically used when the groups of clusters of elements are separable [17], or augmented with associated data through node extensions as in [29]. However, this is not applicable for elements belonging to different clusters due to different types of relationships as in our case.

The second classical technique is to encode the connection between every pair of network elements as a cell in a matrix (elements are indicated twice as rows and cells of the matrix). Visual matrices are introduced in the "Semiology of graphics" of Bertin [3]. Ghoniem et al. [11] show that matrices outperform node-link diagrams for large graphs or dense graphs in various low-level reading tasks, except path finding. This technique therefore communicate well the overview of message correlations, but multiple views are needed when additional data are explored – using brushing and linking technique. A modification is to show half matrix to reduce visual ink and display space. Elements can also be placed along an axis [32] or a circle [14], and connected by arc paths. With the axis-based arc diagram, it completely satisfies as a coordinate in the parallel coordinates that we develop.

The second issue in our development deals with multivariate data representation. One of the most popular techniques is the scatterplots matrix, which is an array of scatterplots displaying all possible pairwise combinations of variables or coordinates. This technique has been in use long before its publication [2]. It is popular for multidimensional visualization due to the relative simplicity in comparison to other multidimensional visualization techniques, familiarity among users, and for high visual clarity, as discussed in [9]. Further on multivariate visualization techniques can be found in [12,33], where parallel coordinates [15] is another popular technique that suits well our development.

Parallel coordinates maps points in Cartesian coordinates into lines connected the vertical parallel line coordinates. This technique is used vastly for correlation representation among variables in multivariate data analysis, since it leverages the geometrical aspects of the presentation: the convergence or flow of lines communicate well relationships between many pairs of variables. From that, various extension techniques has been developed for specific contexts. Tominski et al. outline the parallel axes into the radical axes to support additional characteristics of data and tasks [28]. Collins and Carpendale suggest to links visualized 2D planes instead of just 1D coordinates [5]. In addition, local relationships between connected lines can also be grouped for clusters examination [13]. For text visualization, parallel word clouds [6] is the extension of parallel coordinates for facets of text corpora analysis, and hyper word clouds [20] is developed based on the parallel word clouds for text-based collections with four different types of data relationships.

7   Conclusion

This paper targeted a problem of visual analytics in financial markets domain, in which data are collected from the crowd with chaotic and unwell-modeled for its specific usage. Though machine learning and artificial intelligence so far have made the problem of analyzing such crowd-based data to be more realistic, it is not convenient for the analysts to explore them without visualization. In that context, we have developed the visualization technique *Hyper Coordinates* for correlation messages examination and verification. Supporting for this technique, we defined a message correlation computation approach based on the definition of *direct* and *indirect* connection relations between multivariate items in complex data set, as well as mapping those relationship concepts together with the four types of data relationships (i.e., connection relation, set relation, ordered relation and quantitative relation) for the visualization. Our showcase has shown that domain experts can explore their data from overview of clusters to details of messages as well as interact with them to perform other tasks. The showcase also shows us how the correlation is established among message as well as how direct and indirect relations are distinguishable.

In the showcase, we demonstrated most of visualization development provided in Sect. 4. However, our visualization framework still hardcodes for our specific case study and thus some functions are still missing. In the future we thus plan to complete it and release a general framework for users to use with any dataset that follows our definition.

Acknowledgment. The authors would like to thank Quang M. Le and Tuan A. Ta at Sentifi AG (Ho Chi Minh City office, Vietnam) for their supports and discussions on working scenario.

References

1. Spark: Lightning-fast cluster computing. http://spark.apache.org
2. Andrews, D.F.: Plots of high-dimensional data. Biometrics **28**(1), 125–136 (1972)
3. Bertin, J.: Semiology of Graphics: Diagrams, Networks, Maps. University of Wisconsin Press, Madison (1983)
4. Brehmer, M., Munzner, T.: A multi-level typology of abstract visualization tasks. IEEE Trans. Vis. Comput. Graph. **19**(12), 2376–2385 (2013)
5. Collin, C., Carpendale, S.: VisLink: revealing relationships amongst visualizations. IEEE Trans. Vis. Comput. Graph. **13**(6), 1192–1199 (2007)
6. Collins, C., Viegas, F.B., Wattenberg, M.: Parallel tag clouds to explore and analyze faceted text corpora. In: IEEE Symposium on VAST 2009, pp. 91–98, October 2009
7. Collins, C., Penn, G., Carpendale, S.: Bubble sets: revealing set relations with isocontours over existing visualizations. IEEE Trans. Vis. Comput. Graph. **15**(6), 1009–1016 (2009)
8. Cui, W., Qu, H., Zhou, H., Zhang, W., Skiena, S.: Watch the story unfold with textWheel: visualization of large-scale news streams. ACM Trans. Intell. Syst. Technol. **3**(2), 20:1–20:17 (2012)
9. Elmqvist, N., Dragicevic, P., Fekete, J.D.: Rolling the dice: multidimensional visual exploration using scatterplot matrix navigation. IEEE Trans. Vis. Comput. Graph. **14**(6), 1148–1539 (2008)
10. Gansner, E.R., Yifan, H., North, S.C.: Interactive visualization of streaming text data with dynamic maps. J. Graph Algorithms Appl. **17**(4), 515–540 (2013)
11. Ghoniem, M., Fekete, J.-D., Castagliola, P.: On the readability of graphs using node-link and matrix-based representations: a controlled experiment and statistical analysis. Inf. Vis. **4**(2), 114–135 (2005)
12. Grinstein, G., Trutschl, M., Cvek, U.: High-dimensional visualizations. In: Proceedings of the Visual Data Mining Workshop, KDD (2001)

13. Guo, P., Xiao, H., Wang, Z., Yuan, X.: Interactive local clustering operations for high dimensional data in parallel coordinates. In: 2010 IEEE Pacific Visualization Symposium (PacificVis), pp. 97–104, March 2010
14. Holten, D.: Hierarchical edge bundles: visualization of adjacency relations in hierarchical data. IEEE Trans. Vis. Comput. Graph. **12**(5), 741–748 (2006)
15. Inselberg, A.: The plane with parallel coordinates. Vis. Comput. **1**(2), 69–91 (1985)
16. Jakobson, G., Weissman, M.: Alarm correlation. IEEE Netw. **7**(6), 52–59 (1993)
17. Kim, K., Ko, S., Elmqvist, N., Ebert, D.S.: WordBridge: using composite tag clouds in node-link diagrams for visualizing content and relations in text corpora. In: 2011 44th Hawaii International Conference on System Sciences (HICSS), pp. 1–8, January 2011
18. Marcus, A., Bernstein, M.S., Badar, O., Karger, D.R., Madden, S., Miller, R.C.: TwitInfo: aggregating and visualizing microblogs for event exploration. In: Proceedings of the SIGCHI Conference on Human Factors in Computing Systems (CHI 2011), pp. 227–236, New York. ACM (2011)
19. Motahari-Nezhad, H.R., Saint-Paul, R., Casati, F., Benatallah, B.: Event correlation for process discovery from web service interaction logs. Int. J. Very Large Data Bases **20**(3), 417–444 (2011)
20. Nguyen, D.Q., Le, D.D.: Hyper word clouds: a visualization technique for last.fm data and relationships examination. In: Proceedings of the 10th International Conference on Ubiquitous Information Management and Communication (IMCOM 2016), pp. 66:1–66:7, New York. ACM (2016)
21. Pape, C., Reissmann, S., Rieger, S.: Restful correlation and consolidation of distributed logging data in cloud environments. In: The 8th International Conference on Internet and Web Applications and Services, pp. 194–199 (2013)
22. Reguieg, H., Toumani, F., Motahari-Nezhad, H.R., Benatallah, B.: Using Mapreduce to scale events correlation discovery for business processes mining. In: 10th International Conference Business Process Management, pp. 279–284 (2012)
23. Ren, D., Zhang, X., Wang, Z., Li, J., Yuan, X.: WeiboEvents: a crowd sourcing Weibo visual analytic system. In: Proceedings of the 2014 IEEE Pacific Visualization Symposium (PACIFICVIS 2014), Washington, DC, pp. 330–334. IEEE Computer Society (2014)
24. Sarkar, M., Brown, M.H.: Graphical fisheye views of graphs. In: Proceedings of the SIGCHI Conference on Human Factors in Computing Systems (CHI 1992), New York, pp. 83–91. ACM (1992)
25. Serrour, B., Gasparotto, D.P., Kheddouci, H., Benatallah, B.: Message correlation and business protocol discovery in service interaction logs. In: 20th International Conference Advanced Information Systems Engineering, pp. 405–419 (2008)
26. Shneiderman, B.: The eyes have it: a task by data type taxonomy for information visualizations. In: Proceedings of the 1996 IEEE Symposium on Visual Languages, Washington, DC, USA, pp. 336–343. IEEE Computer Society, September 1996
27. Thomas, J.J., Cook, K.A. (eds.): Illuminating the Path: The Research and Development Agenda for Visual Analytics. IEEE CS Press, Los Alamitos (2005)
28. Tominski, C., Abello, J., Schumann, H.: Axes-based visualizations with radial layouts. In: Proceedings of the 2004 ACM Symposium on Applied Computing (SAC 2004), New York, pp. 1242–1247. ACM (2004)
29. van Ham, F., Perer, A.: Search, show context, expand on demand: supporting large graph exploration with degree-of-interest. IEEE Trans. Vis. Comput. Graph. **15**(6), 953–960 (2009)
30. van Ham, F., Wattenberg, M., Viegas, F.B.: Mapping text with phrase nets. IEEE Trans. Vis. Comput. Graph. **15**(6), 1169–1176 (2009)

31. Wang, M., Holub, V., Parsons, T., Murphy, J., OSullivan, P.: Scalable run-time correlation engine for monitoring in a cloud computing environment. In: 17th IEEE International Conference and Workshops on Engineering of Computer Based Systems, pp. 29–38 (2010)
32. Wattenberg, M.: Arc diagrams: visualizing structure in strings. In: Proceedings of the IEEE Symposium on Information Visualization (InfoVis 2002), Washington, DC, pp. 110–116. IEEE Computer Society (2002)
33. Wong, P.C., Bergeron, R.D.: 30 years of multidimensional multivariate visualization. In: Scientific Visualization, Overviews, Methodologies, and Techniques, Washington, DC, pp. 3–33. IEEE Computer Society (1997)
34. Ye, Z., Li, S., Zhou, X.: GCplace: geo-cloud based correlation aware data replica placement. In: The 28th Annual ACM Symposium on Applied Computing, pp. 371–376 (2013)

Bayesian Forecasting for Tail Risk

Cathy W. S. Chen[✉] and Yu-Wen Sun

Department of Statistics, Feng Chia University, Taichung, Taiwan
chenws@mail.fcu.edu.tw, sunyuwun@gmail.com

Abstract. This paper evaluates the performances of Value-at-Risk (VaR) and expected shortfall, as well as volatility forecasts in a class of risk models, specifically focusing on GARCH, integrated GARCH, and asymmetric GARCH models (GJR-GARCH, exponential GARCH, and smooth transition GARCH models). Most of the models incorporate four error probability distributions: Gaussian, Student's t, skew Student's t, and generalized error distribution (GED). We employ Bayesian Markov chain Monte Carlo sampling methods for estimation and forecasting. We further present backtesting measures for both VaR and expected shortfall forecasts and implement two loss functions to evaluate volatility forecasts. The empirical results are based on the S&P500 in the U.S. and Japan's Nikkei 225. A VaR forecasting study reveals that at the 1% level the smooth transition model with a second-order logistic function and skew Student's t error compares most favorably in terms of violation rates for both markets. For the volatility predictive abilities, the EGARCH model with GED error is the best model in both markets.

Keywords: Backtesting · Expected shortfall
Skew Student's t distribution · Smooth transition GARCH model
Second-order logistic function · Markov chain Monte Carlo methods
Value-at-Risk

1 Introduction

Investors regard risk management performance as one of the main criteria for investment. Value-at-Risk (VaR) is a common risk measurement used for subsequent capital allocation for financial institutions worldwide, as chosen by the Basel Committee on Banking Supervision. The benchmark of VaR, first proposed by J. P. Morgan [30], helps investors compare risk under different portfolios (see Duffie and Pan [14]) and also increases their investment confidence.

According to regulations, central banks require that financial institutions deposit a certain amount of reserves in order to ensure that they have enough money to cover any risk in the future and to prevent the bank from going bankrupt. For insurance companies, VaR denotes the threshold of the amounts under claim, whereby the claims less than VaR represent the amount of compensation a company can afford; otherwise, we take it as the amount of reinsurance. For a given portfolio, time horizon, and probability p, VaR denotes as a threshold

© Springer International Publishing AG 2018
V. Kreinovich et al. (eds.), *Predictive Econometrics and Big Data*, Studies in Computational Intelligence 753, https://doi.org/10.1007/978-3-319-70942-0_6

loss value, such that the probability that the loss on the portfolio over the given time horizon exceeds this value is p. Although VaR is a popular measurement for determining regulatory capital requirements, it suffers a number of weaknesses, including its inability to capture "tail risk".

The literature classifies the method for calculating the VaR value into three types. First, some financial institutions employ sample return quantiles, called historical simulation (HS), for nonparametric estimation of VaR (see Hendricks [26]). Second, Engle and Manganelli [17] propose conditional autoregressive value at risk (CAViaR), which is based on dynamic quantile regression. Gerlach, Chen and Chan [22] further investigate the nonlinear CAViaR model to estimate quantiles (VaR) directly. Third, parametric GARCH-type models, such as the autoregressive conditional heteroscedastic (ARCH) model of Engle [16] and the generalized ARCH (GARCH) model of Bollerslev [5] have specified error distributions that are well suited to quantile forecasting. VaR studies also use an IGARCH (integrated generalized autoregressive conditional heteroscedastic) model and RiskMetrics as benchmarks. However, it is widely known that volatility reacts differently to a big price increase and a big price drop with the latter having a greater impact. This phenomenon describes the leverage effect. Volatility asymmetry, as presented by Black [4], and other types of nonlinearity in financial data have received much attention recently. Nelson [31] proposes a model to capture asymmetric volatility, called the exponential GARCH model (EGARCH). Glosten et al. [20] offer another asymmetric GARCH model, popularly known as the GJR-GARCH model, to capture asymmetric volatility via an indicator term in the GARCH equation. Chen et al. [11] focus on nonlinear GARCH model with variant smooth transition functions: first- and second-order logistic functions, and the exponential function.

The first objective of this study is to employ the above-mentioned parametric models and HS approaches for forecasting VaR. We consider the family of GARCH-type model with four error probability normal, Student's t, the skew Student's t, and generalized error distribution (GED). Additionally, we forecast tail risk based on a smooth transition (ST) GARCH model with a second-order logistic function and both Student's t errors. Chen et al. [11] show at the 1% level that the ST model with a second-order logistic function and skew Student's t error is a worthy choice for VaR, when compared to a range of existing alternatives. This model is worth our attention, and thus should include it in our risk model.

There are several limitations from using VaR for determining regulatory capital requirements, including its inability to catch tail risk. For this reason, the Basel Committee on Banking Supervision has considered alternative risk metrics, in particular expected shortfall (ES). ES denotes the expected value of the return less than VaR with a significant α at time t; it is also called conditional tail expectation (see Acerbi and Tasche [1]. Artzner et al. [2] show that VaR lacks a subadditive property, because the risk scatters over a portfolio, and thus we can see that VaR is not coherent. Fllmer and Schied [19] note that ES is coherent, and so ES is more useful than VaR. Many financial institutions and practitioners reply on VaR to evaluate the quantile forecast of financial returns, but it only measures a quantile of the distribution, while ignoring the important facts contained in the tail beyond that quantile.

In order to make a better assessment of potential losses, the second objective of this study is to calculate ES. Yamai and Yoshiba [35] compare VaR with ES and prove that ES is a better risk measure than VaR in terms of tail risk. Harmantzis et al. [25] also emphasize the tail factor with VaR and ES. Gerlach and Chen [23] consider dynamic expectile and ES modeling and forecasting, incorporating information from the daily range.

The problem of computing the integrals of ES when the error term follows a non-Gausian distribution is quite complicated. To solve this problem efficiently we adopt Bayesian Markov chain Monte Carlo (MCMC) methods. We employ Bayesian methods to estimate model parameters via the GARCH-type models, because they describe uncertainty of a statement about an unknown parameter in terms of probability. Bayesian qunatile forecasting provides adequate VaR forecasts (see Chen et al. [8,11,12]; Gerlach et al. [22]). Hoogerheide and van Dijk [27] compare VaR and ES in a Bayesian framework and consider the Bayesian predictive density. Within the Bayesian framework, Chen et al. [9] forecat VaR depending on a range of parametric models. Chen, Weng, and Watanabe [11] evaluate VaR based on various ST GARCH models, but they do not deal with ES in their studies. Therefore, the target of this study is to evaluate both performances of VaR and ES forecasting.

The MCMC methods enable us to solve complicated high dimensional models and more effectively estimate parameters. The advantages of the Bayesian approach using MCMC methods are as follows. (i) The likelihood function is conditional on the unobserved variables to compute the posterior distribution, making the calculation of the parameters faster than the traditional maximum likelihood estimation (see Papp [32]). (ii) This approach solves complex high dimensional models and more effectively estimate parameters. (iii) Bayesian MCMC methods allow simultaneous inferences for all unknown parameters.

Common criteria to compare VaR models are the rate of violation and backtesting. The rate of violation is the proportion of observations for which the actual return is more extreme than the forecasted VaR level over a forecast period. A forecast model's violation rate (VRate) should be close to the nominal level α. We study three hypothesis-testing methods for evaluating and testing the accuracy of VaR models: (1) the unconditional coverage (UC) test of Kupiec [29]: a likelihood ratio test whereby the true violation rate equals α; (2) the conditional coverage (CC) test of Christoffersen [13]: a joint test combining a likelihood ratio test for independence of violations and the UC test; and (3) the Dynamic Quantile (DQ) test of Engle and Manganelli [17]. Both CC and DQ are joint tests of the independence of a model's violations and a proof of the UC test, whereby the true violation equals the significant α. The DQ test is generally more powerful than the CC test (see Berkowitz et al. [3]).

We evaluate ES following Embrechts et al. [15], in order to deal with the weakness of the violations that depend strongly on the VaR estimates without sufficiently reflecting the goodness or badness of these values. We apply the model to daily returns of the S&P500 in the U.S. and Japan's Nikkei 225. We evaluate the volatility forecasts of all parametric models by mean squared error (MSE) and quasi-likelihood (QLIKE). The prediction results show that the asymmetric model with skew and fat tailed distribution improves quantile forecasts.

We organize this paper as follows. Section 2 illustrates a smooth transition heteroskedastic model. Section 3 demonstrates the Bayesian Approach via the MCMC method. Section 4 describes the process of VaR and ES forecasts. Section 5 mentions several methods to evaluate the volatility forecasts and quantile forecasts. Section 6 lists analytic results, using the U.S. and Japanese stock indices. Section 7 provides concluding remarks.

2   The Smooth Transition Heteroskedastic Model

This section describes a double ST GARCH model proposed by Chen et al. [11] since it is the most complicated model in terms of estimation. This model is capable of capturing mean and volatility asymmetry in financial markets. We use a second-order ST function that ensures the mean and volatility parameters are smooth functions of past news or volatility. Chen et al. [11] show that the ST model with a second-order logistic function and skew Student's t error is a worthwhile choice for VaR forecasting.

Suppose that $\{r_t\}$ represents the observation data. To incorporate different speeds of smooth transition functions for the mean and variance, we present the time-varying ST GARCH model as:

$$r_t = \mu_t^{(1)} + F(z_{t-d}; \gamma_1, \boldsymbol{c})\mu_t^{(2)} + a_t \tag{1}$$

$$a_t = \sqrt{h_t}\varepsilon_t, \qquad \varepsilon_t \overset{\text{i.i.d.}}{\sim} D(0,1),$$

$$h_t = h_t^{(1)} + F(z_{t-d}; \gamma_2, \boldsymbol{c})h_t^{(2)}$$

$$\mu_t^{(l)} = \phi_0^{(l)} + \sum_{i=1}^{p} \phi_i^{(l)} r_{t-i}$$

$$h_t^{(l)} = \alpha_0^{(l)} + \sum_{i=1}^{g} \alpha_i^{(l)} a_{t-i}^2 + \sum_{i=1}^{q} \beta_i^{(l)} h_{t-i}, \quad l = 1, 2,$$

where $\mu_t^{(l)}$ and $h_t^{(l)}$ are the respective conditional mean and volatility at regime l; z_t is the threshold variable; d is the delay lag; and $D(0,1)$ is an error distribution with mean 0 and variance 1. We can choose lagged returns, or an exogenous variable, e.g. other asset return, as the threshold z_t. The time-varying ST GARCH model in (1) highlights the model's characteristic, which allows for an ST function with varying speed in the mean and variance. We consider a specification of the second-order logistic function in van Dijk et al. [34].

$$F(z_{t-d}; \gamma_i, \boldsymbol{c}) = \frac{1}{1 + \exp\left\{\frac{-\gamma_i(z_{t-d}-c_1)(z_{t-d}-c_2)}{s_z}\right\}}, \quad i = 1, 2 \quad c_1 < c_2,$$

where $c = (c_1, c_2)'$, as proposed by Jansen and Teräsvirta [28]. Given a point in time t, r_t corresponds to a weighted average of two AR-GARCH models, where the weights assigned to the two models depend on the values taken by the transition functions $F(z_{t-d}; \gamma_i, c)$, $i = 1, 2$. The parameter γ_i determines the smoothness of the change in the value of the $F(z_{t-d}; \gamma_i, c)$ function and the smoothness of the transition from one regime to the other.

Chen et al. [11] discuss some properties of the ST GARCH model. For example, when $\gamma_1 \to 0$, the logistic functions is equal to a constant (equal to 0.5), and the ST GARCH model reduces to a linear AR-GARCH model. When γ_2 in the transition function is moderately large (not necessarily going to infinity) and $c_1 \neq c_2$, then the ST GARCH model becomes the three-regime threshold GARCH model of Chen et al. [10].

Note that the ST GARCH model contains the lagged AR(1) (i.e. $p = 1$) effect in each regime, which allows one to recognize whether the return series exhibits asymmetry mean reversion or market efficiency. For the ST GARCH(1,1) model in (1), we allow an explosive lower regime for the model. The restrictions for positiveness and covariance stationarity are as follows:

$$\alpha_0^{(i)}, \ \alpha_1^{(i)}, \ \beta_1^{(i)} > 0, \ i = 1, 2 \tag{2}$$

$$\left(\alpha_1^{(1)} + \alpha_1^{(2)}\right) > 0, \ \left(\beta_1^{(1)} + \beta_1^{(2)}\right) > 0,$$

$$\left(\alpha_1^{(1)} + 0.5\alpha_1^{(2)}\right) + \left(\beta_1^{(1)} + 0.5\beta_1^{(2)}\right) < 1, \tag{3}$$

$$\alpha_0^{(1)} < b_1, \ \beta_1^{(1)} < b_2, \ \alpha_1^{(1)} + \beta_1^{(1)} < b_3, \tag{4}$$

where b_1, b_2, and b_3 are user-specified. In this study, we let b_2, $b_3 \geq 1$ to allow for explosive behavior.

For the purpose of VaR and ES studies, we also consider standard GARCH model, IGARCH model, and RiskMetrics, as well as asymmetric GARCH models - GJR-GARCH and EGARCH models. Appendix A gives the descriptions of these models.

3   Bayesian Approach

We present a general likelihood functional form for any GARCH model, which is:

$$\mathscr{L}(r \mid \theta) = \prod_t^n \left\{ \frac{1}{\sqrt{h_t}} p_\varepsilon \left(\frac{(r_t - \mu_t)}{\sqrt{h_t}} \right) \right\}, \tag{5}$$

where $r = (r_1, \ldots, r_n)'$, θ is the full parameter vector for any of the combinations of model and error distribution considered, (μ_t, h_t) are the mean and volatility at time t, and $p_\varepsilon(\cdot)$ is the error density function for ε_t.

There are many versions of skew Student's t distribution, such as in Hansen [24] and Fernández and Steel [18]. Both versions introduce skewness into any continuous unimodal and symmetric (with respect to 0) univariate distribution.

We adopt the approach of Hansen [24], denoted by $St(\eta, \nu)$, which has zero mean and unit variance. The probability density function of skew Student's t defined by Hansen [24] is as follows:

$$
p_\varepsilon(\varepsilon_t | \nu, \eta) = \begin{cases} bc \left[1 + \frac{1}{\nu-2} \left(\frac{b\varepsilon_t + a}{1-\eta} \right)^2 \right]^{-(\nu+1)/2} & \text{if } \varepsilon_t < -\frac{a}{b} \\ bc \left[1 + \frac{1}{\nu-2} \left(\frac{b\varepsilon_t + a}{1+\eta} \right)^2 \right]^{-(\nu+1)/2} & \text{if } \varepsilon_t \geq -\frac{a}{b}, \end{cases}
\tag{6}
$$

where degrees of freedom ν and skewness parameter η satisfy $2 < \nu < \infty$, and $-1 < \eta < 1$, respectively. We set the constants a, b, and c as:

$a = 4\eta c \left(\frac{\nu-2}{\nu-1} \right)$, $b^2 = 1 + 3\eta^2 - a^2$, and $c = \Gamma \left(\frac{\nu+1}{2} \right) \left[\Gamma(\nu/2) \sqrt{\pi(\nu-2)} \right]^{-1}$, where Γ denotes the gamma function. The skew Student's t distribution of Hansen [24] includes Gaussian and Student's t distributions as special cases.

GED is known as generalized normal distribution, which is a symmetric family of distributions used in risk models, e.g. Chen et al. [9]. The shape of GED is different from a normal distribution. The GED can include the normal distribution and the Laplace distribution as special cases. The density function for ε_t, a standardized GED with scale parameter σ, is:

$$
p_\varepsilon(\varepsilon_t) = \frac{\lambda}{2\sigma \Gamma(1/\lambda)} \exp \left\{ -\left| \frac{\varepsilon_t}{\sigma} \right|^\lambda \right\},
\tag{7}
$$

where $\sigma = [\Gamma(\frac{1}{\lambda})/\Gamma(\frac{3}{\lambda})]^{0.5}$, and $\lambda \in (0, \infty)$ is the tail-behaviour determining parameter. When $\lambda > 2$, the distribution has thinner tails than a normal distribution. When $\lambda = 2$, it is exactly a normal distribution with mean 0 and standard error σ; while for $\lambda < 2$, the distribution has positive excess kurtosis relative to the normal distribution; see Figure 1 for this information. For real asset return data, we expect $\lambda < 2$.

3.1 The MCMC Methods

We assume all parameters $\boldsymbol{\theta} = (\boldsymbol{\phi}_1, \boldsymbol{\phi}_2, \boldsymbol{\alpha}_1, \boldsymbol{\alpha}_2, c, \gamma, \nu, d)$ of the ST GARCH model as in Eq. (1), which are *a priori* independent. Following the idea of Chen et al. [11], we set up the priors as follows. We define the latent variable $\delta_j^{(i)}$, which determines the prior distribution of $\phi_j^{(i)}$, via a mixture of two normals:

$$
\phi_j^{(i)} | \delta_j^{(i)} \sim (1 - \delta_j^{(i)}) N(0, k^2 \tau_j^{(i)^2}) + \delta_j^{(i)} N(0, \tau_j^{(i)^2}), \quad j = 1, \ldots, p
$$

$$
\delta_j^{(i)} | \gamma_1 = \begin{cases} 1, \text{ if } i = 1 \text{ or } \gamma_1 > \xi \\ 0, \text{ if } i = 2 \text{ and } \gamma_1 \leq \xi, \end{cases}
\tag{8}
$$

where $i = 1, 2$ denotes the regime, and $j = 1, \ldots, p$ denotes the lag order of the AR mean terms in ϕ_j. Here, ξ is a specified threshold, and $\gamma_1 \leq \xi$ indicates that

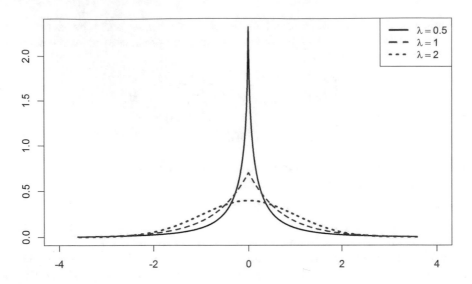

Fig. 1. Shape of GED for $\lambda = (0.5, 1, 2)$.

$F(z_{t-d}; \gamma_1, c) \to 0.5$; that is, an AR-GARCH model. As Gerlach and Chen [21] suggest, we choose k to be a small positive value, so that if $\gamma_1 \le \xi$ and $\delta_j^{(2)} = 0$, then the posterior value for the parameters $\phi_j^{(2)}$ will be weighted by the prior value towards 0.

We take a constrained uniform prior for $p(\boldsymbol{\alpha})$, with the constraint defined by the indicator $I(S)$, where S defines the constraints in Eqs. (2), (3), and (4). We suggest the log-normal prior for γ_i, where $\gamma_i \sim LN(\mu_{\gamma_i}, \sigma_{\gamma_i}^2)$. We choose a discrete uniform prior for d, $\Pr(d) = 1/d_0$, where $d = 1, \dots, d_0$. For ν degrees of freedom, we define $\rho = \nu^{-1}$ and set it to $I(\rho \in [0, 0.25])$ (see Chen et al. [7], for more details). For the skew parameter, we set a flat prior over $\eta \in (-1, 1)$. The general priors for c_1 and c_2 are:

$$c_1 \sim \text{Unif}(lb_1, ub_1);$$
$$c_2 | c_1 \sim \text{Unif}(lb_2, ub_2),$$

where lb_1 and ub_1 are the $\wp_{h_1}$ and $\wp_{1-h_1-h_2}$ percentiles of z_t, respectively. For example, if $h_1 = h_2 = 0.1$, then $c_1 \in (\wp_{0.1}, \wp_{0.8})$. Furthermore, we set $ub_2 = \wp_{(1-h_2)}$ and $lb_2 = c_1 + c^*$, where c^* is a selected number that ensures $c_1 + c^* \le c_2$ and at least $100h_2\%$ of observations are in the range (c_1, c_2).

The conditional posterior distribution for each group is proportional to the product of the likelihood function and the priors:

$$p(\boldsymbol{\theta}_l \mid \boldsymbol{y}^{s+1, n}, \boldsymbol{\theta}_{\neq l}) \propto p(\boldsymbol{y}^{s+1, n} \mid \boldsymbol{\theta}) \cdot p(\boldsymbol{\theta}_l \mid \boldsymbol{\theta}_{\neq l}),$$

where $\boldsymbol{\theta}_l$ is a parameter group, $p(\boldsymbol{\theta}_l)$ is its prior density, and $\boldsymbol{\theta}_{\neq l}$ is the vector of all model parameters, except for $\boldsymbol{\theta}_l$.

Except for the delay parameter d, the rest of the conditional posterior distributions are not standard. We use the adaptive Metropolis-Hastings (MH) algorithm, which combines the random walk MH (RW-MH) algorithm (before burn-in period) and the independent kernel MH (IK-MH) algorithm (after burn-in period) (see Chen and So [6]). Appendix B gives the description of adaptive MH algorithm. Convergence is monitored heavily using trace and ACF plots. We set iteration frequency N samples and burn-in frequency M samples, but take only every second iterate in the sample period for inference.

4   Bayesian Forecasting of Value-at-Risk and Expected Shortfall

This section generates the one-day-ahead volatility and predicted return based on the MCMC algorithm. We use the ST GARCH model with skew Student's t error as an example. The procedure to compute the one-day-ahead volatility h_{n+1} and return r_{n+1} is given as follows.

1. Obtain $\mu_{n+1}^{[j]}|\boldsymbol{\theta}^{[j]}$ and $h_{n+1}^{[j]}|\boldsymbol{\theta}^{[j]}$ using the in-sample data $\boldsymbol{r}$.
2. Generate $\varepsilon_{n+1} \sim St(\eta, \nu)$.
3. Calculate $r_{n+1}^{[j]} = \mu_{n+1}^{[j]} + \sqrt{h_{n+1}^{[j]}}\varepsilon_{n+1}$ and go to Step 1.

VaR is a very popular risk measure, which generally denotes the amount of capital that the banks need to prepare. We assume a long position, such that VaR forecast satisfies:

$$P(r_{n+1} < -\text{VaR}_{n+1}|\mathscr{F}_n) = \alpha, \tag{9}$$

where $\mathscr{F}_n$ is the information available up to n; we typically let α be one percentile or five percentile. A one-step-ahead VaR is the α-level quantile of the conditional distribution $r_{n+1}|\mathscr{F}_n \sim D(\mu_{n+1}, h_{n+1})$, where D is the relevant error distribution. We compute the predictive distribution via the MCMC simulation and give the quantile VaR by:

$$\text{VaR}_{n+1}^{[j]} = -\left[\mu_{n+1}^{[j]} + D_\alpha^{-1}(\boldsymbol{\theta}^{[j]})\sqrt{h_{n+1}^{[j]}}\right], \tag{10}$$

where D_α^{-1} is the inverse CDF of the distribution D. For standardized Student's t errors:

$$D_\alpha^{-1} = \frac{t_\alpha(\nu^{[j]})}{\sqrt{\nu^{[j]}/(\nu^{[j]} - 2)}},$$

where $t_\alpha(\nu^{[j]})$ is a value that follows Student's t distribution with degrees of freedom $\nu^{[j]}$, and $\sqrt{\nu^{[j]}/(\nu^{[j]} - 2)}$ is an adjustment term for a standardized Student's t. Alternatively, one can compute VaR_{n+1} as the α-percentile of the

MCMC sample of r_{n+1} (see Takahashi et al. [33]). The forecasted one-step-ahead VaR is the Monte Carlo posterior mean estimate

$$\text{VaR}_{n+1} = \frac{1}{N - M} \sum_{j=M+1}^{N} \text{VaR}_{n+1}^{[j]}. \tag{11}$$

Because we do not know the severity of the tail of r_{n+1}, we compute ES for evaluating the average of the worst outcomes of a probability distribution. The one-step-ahead forecast of the ES with probability α satisfies

$$\text{ES}_{n+1} = \text{E}\left[r_{n+1} | r_{n+1} < -\text{VaR}_{n+1}\right]. \tag{12}$$

When r_{n+1} follows a normal distribution, we can easily compute ES with a closed form, but the reality is that r_{n+1} is typically skewed and fat tailed. The integral part in (12) would undoubtedly be much more challenging in nonlinear dynamic models with skewed errors. As such, a Bayesian approach can overcome these difficulties. This procedure executes the ES that is approximately obtainable by using a MCMC technique.

To compute the tail of r_{n+1}, we find r_{n+1} that satisfies the condition and then compute its expected value. To compute ES in (12), we provide a numerical approximation of ES by using an adaptive MCMC scheme.

Step 1. Compute either VaR_{n+1} based on (11) or VaR_{n+1} as the α percentile of r_{n+1}.

Step 2. Generate one-step-ahead return $r_{n+1}^{[j]} | \mathscr{F}_n$, $\boldsymbol{\theta}^{[j]}$ from $D(\mu_{n+1}^{[j]}, h_{n+1}^{[j]})$, where D is the relevant error distribution.

Step 3. If $\left(r_{n+1}^{[j]} < -\text{VaR}_{n+1}\right)$ is true, then we save $r_{n+1}^{[j]}$, for $j = M + 1, \dots, N$.

Step 4. Compute the average of $r_{n+1}^{[j]}$, which satisfies the condition in Step 3.

Let n and m be the numbers of samples for the estimation and prediction, respectively. We then estimate the one-day-ahead forecasts of VaR and ES based on the rolling window approach.

5   Evaluation of Volatility and Quantile Forecasts

We evaluate the volatility forecasts of all models using two loss functions: mean squared error (MSE) and quasi-likelihood (QLIKE). Let $\hat{\sigma}_t^2$ and h_t be a volatility proxy and a volatility forecast, respectively, and consider the two loss functions

$$L_t^{\text{MSE}} = \frac{\left(\hat{\sigma}_t^2 - h_t\right)^2}{2}, \quad \text{MSE} = \sum_{i=1}^{m} L_t^{\text{MSE}}, \tag{13}$$

$$L_t^{\text{QLIKE}} = \frac{\hat{\sigma}_t^2}{h_t} - \log\frac{\hat{\sigma}_t^2}{h_t} - 1, \quad \text{QLIKE} = \sum_{i=1}^{m} L_t^{\text{QLIKE}}, \tag{14}$$

where $\hat{\sigma}_t^2$ is a realized kernel that we download from the Oxford-Man Institute "Realised Library", and m is the out-of sample size.

Evaluation of Value-at-Risk

As a simple way to evaluate VaR, we compute the violation rate in the form:

$$\text{VRate} = \frac{1}{m} \sum_{t=n+1}^{n+m} I\left(r_t < -\text{VaR}_t\right), \tag{15}$$

where n is the in-sample size, and m is the out-of-sample size. The result of VRate should be close to the nominal level α. Furthermore, we prefer a model with overestimated risk than one that underestimates risk.

Backtesting methods

We further consider three backtesting methods for evaluating and testing the accuracy of the VaR models. We describe these three hypothesis-testing methods as follows.

The UC test of Kupiec [29]: With null hypothesis $H_0 : \alpha = \alpha_0$, the likelihood ratio test proposed by Kupiec [29] has the form:

$$LR_{uc} = 2 \cdot \log\left[\frac{\hat{\alpha}^X(1-\hat{\alpha})^{m-X}}{\alpha^X(1-\alpha)^{m-X}}\right] \sim \chi_1^2,$$

where X = number of violations, m = forecast period size, and $\hat{\alpha} = X/m$.

The CC test of Christoffersen [13]: The CC test is a joint test that combines a likelihood ratio test for independence of violations and the UC test, where the independence hypothesis stands for VaR violations observed at two different dates being independently distributed.

$$LR_{ind} = 2 \cdot \log\left(\frac{L_1}{L_0}\right) \; ; \; LR_{ind} \sim \chi_1^2.$$

We define T_{ij} as the number of days when condition j occurred under present status, and assuming that condition i occurred on the previous day, we get:

$$i,j = \begin{cases} 1, \text{ if violation occurs} \\ 0, \text{ if no violation occurs,} \end{cases}$$

$$L_1 = \prod_{i=0}^{1}(1-\pi_{i1})^{T_{i0}}\pi_{i1}^{T_{i1}},$$

$$L_0 = (1-\pi)^{\sum_{i=0}^{1}T_{i0}}\pi^{\sum_{i=0}^{1}T_{i1}},$$

$$\pi_{i1} = \frac{T_{i1}}{(T_{i0}+T_{i1})}, \text{ and } \pi = \frac{(T_{01}+T_{11})}{m},$$

with m being the forecast period size. Thus, the joint CC test is a chi-square test, in which $LR_{cc} = LR_{uc} + LR_{ind}$, when $LR_{cc} \sim \chi_2^2$.

The DQ test of Engle and Manganelli [17]: The DQ test is based on a linear regression model of the hits variable on a set of explanatory variables, including a constant, the lagged values of the hit variable, and any function of the past information set suspected of being informative. H_0: $H_t = I(y_t < \text{-VaR}_t) - \alpha$ is independent of $\boldsymbol{W}$. The test statistic is:

$$\text{DQ}(q) = \frac{\boldsymbol{H'W}\left(\boldsymbol{W'W}\right)^{-1}\boldsymbol{W'H}}{\alpha(1-\alpha)},$$

where $\boldsymbol{W}$ is lagged hits, lagged VaR forecasts, or other relevant regressors over time that are discussed in detail by Engle and Manganelli [17]. Under H_0, $\text{DQ}(q) \sim \chi_q^2$. The DQ test is recognized to be more powerful than the CC test.

Evaluation of expected shortfall

To evaluate the ES forecast models, Chen et al. [12] consider the ES rate, which is the analogue of VRate. We implement the measure proposed by Embrechts et al.[15], who evaluates ES based on two measures. $V_1(\alpha)$ gives the standard backtesting measure using the VaR estimates. Takahashi et al. [33] point out that this measure depends strongly on the VaR estimates, without adequately reflecting the correctness of these values. To correct this weakness, we combine a penalty term $V_2(\alpha)$, which evaluates the values that should happen once every $1/\alpha$ days with $V_1(\alpha)$.

Let $\delta_t(\alpha) = r_t - ES_t(\alpha)$ and $q(\alpha)$ be the α-quantile of $\delta_t(\alpha)$. Here, $\kappa(\alpha)$ is a set of time points when a violation happens, and $\tau(\alpha)$ is a set of time points when $\delta_t(\alpha) < q(\alpha)$ occurs, where $q(\alpha)$ is the empirical α-quantile of $\delta_t(\alpha)$. We define the measure as:

$$V(\alpha) = \frac{|V_1(\alpha)| + |V_2(\alpha)|}{2}, \tag{16}$$

$$\text{where} \quad V_1(\alpha) = \frac{1}{T_1}\sum_{t\in\kappa(\alpha)}\delta_t(\alpha), \quad V_2(\alpha) = \frac{1}{T_2}\sum_{t\in\tau(\alpha)}\delta_t(\alpha),$$

and T_1 and T_2 are the numbers of time points in $\kappa(\alpha)$ and $\tau(\alpha)$, respectively. Note that better ES estimates provide lower values of both $|V_1(\alpha)|$ and $|V_2(\alpha)|$, and therefore of $V(\alpha)$.

6   Analytic Results

In this study we consider a very long period for the VaR and ES forecast performance over 19 risk models and 2 HS methods. From the R program "quantmod" package, we download the daily closing prices of (i) the S&P500 (U.S.) and (ii) Nikkei 225 (Japan). We analyze all data with the daily return r_t, $r_t = [\ln(P_t) - \ln(P_{t-1})] \times 100$, where P_t is the closing index price on day t.

The full data period covers January 3, 2006 to March 31, 2017. We consider a learning (in-sample) period from January 4, 2006 to December 31, 2013 and a validation (out-of-sample) period of January 2, 2014 to March 31, 2017. We employ a rolling window approach to produce a one-step-ahead forecasting of h_{n+1}, VaR, and ES over all risk models. There are 817 and 793 prediction samples for S&P500 and Nikkei 225, respectively, in the out-of-sample period.

Table 1 describes summary statistics for the in-sample period of the log returns of the market indices, including sample mean, standard deviation, skewness, excess kurtosis, extreme values, the Jarque-Bera normality (JB) test, and the Ljung-Box values for both returns and squared returns in order to test the null hypothesis of no autocorrelation up to the 5th lag.

For Nikkei 225, the mean of return is not statistically significantly different from zero, and its Ljung-Box statistic does not reject the null hypothesis of no autocorrelation up to the 5th lag, while autocorrelation exists in the S&P500 return. We decide to include the lagged AR(1) effect for all risk models and both markets during the rolling window approach in order to adjust autocorrelation in the mean equation.

The kurtosis reveals that its distribution is leptokurtic, as is observed commonly in financial returns. The normality test indicates a clear rejection for both markets by the JB normality test under a 1% significant level. The skewness is obviously negative for both markets. In summary, the daily returns of both markets have heavy tails and are negatively skewed.

Table 1. Summary statistics of market returns for the in-sample period (January 3, 2006 to December 31, 2013)

Returns	Mean	SD	Skewness	Excess Kurtosis	Min	Max	JB test^a	$Q(5)^b$ p-value	$Q^2(5)^b$ p-value
S&P500	0.0195	1.400	−0.315	9.509	−9.470	10.957	0.000	0.000	0.000
Nikkei 225	0.0006	1.678	−0.554	7.750	−12.111	13.235	0.000	0.235	0.000

^{a} Jarque-Bera normality test.
^{b} $Q(5)$ and $Q^2(5)$ are the p-values of the Ljung-Box test for autocorrelation in the level of returns and the squared returns up to the 5th lag.

The initial values for each parameter in the ST GARCH model are $\phi_i = (0,0)$, $\alpha_i = (0.1, 0.1, 0.1)$, $\gamma_i = 30$, $i = 1, 2$, $\nu = 100$, and $(c_1, c_2) = (0, 0.1)$. We choose the maximum delay, d_0, to be 3. Regarding the GJR-GARCH model, the set-up of initial values is $\phi = (0, 0.0)$, $\alpha = (0.05, 0.05, 0.1, 0.1)$, $\nu = 200$, and $\eta = 0$ for skew Student's t errors, that is, there are no skewness and Gaussian for the initial guess. We perform 20000 MCMC iterations and discard the first 10000 iterates as a burn-in sample for each analyzed data series. To save space, we do not provide the Bayesian estimation for all risk models.

In the out-of-sample period, we use a rolling window estimation scheme with the window size fixed to produce a one-step-ahead forecasting of h_{n+1}, VaR, and ES at the 1% and 5% levels cross 19 risk models. We also apply two HS methods for VaR forecasting. The HS methods encompass the short-term HS with an observation window of 25 days (HS-ST) and the long-term HS with an

Table 2. Volatility forecasts using two loss functions: MSE and QLIKE

	S&P500		Nikkei 225	
	MSE	QLIKE	MSE	QLIKE
HS-ST	NA	NA	NA	NA
HS-LT	NA	NA	NA	NA
RiskMetrics	0.3725	0.3660	3.2264	0.7040
GARCH-n	0.3940	0.4069	4.4367	0.7103
GARCH-t	0.4052	0.4085	3.9871	0.7168
GARCH-st	0.3993	0.3969	3.8921	0.7089
GARCH-ged	0.3943	0.4079	4.2898	0.7170
IGARCH-n	0.4300	0.4093	5.7821	0.7250
IGARCH-t	0.4305	0.4161	5.2880	0.7392
IGARCH-st	0.4263	0.4055	5.2091	0.7317
IGARCH-ged	0.4263	0.4126	5.6719	0.7340
GJR-n	0.3960	0.3862	4.0685	0.7032
GJR-t	0.4392	0.3894	3.8291	0.7077
GJR-st	0.4359	0.3812	3.7680	0.7034
GJR-ged	☐ 0.3670	☐ 0.3445	3.9301	0.6945
EGARCH-n	0.4132	0.3954	2.7702	☐ 0.6552
EGARCH-t	0.4599	0.4056	2.6340	0.6600
EGARCH-st	0.4542	0.3982	☐ 2.5635	0.6569
EGARCH-ged	☐ 0.3673	☐ 0.3232	☐ 2.6075	☐ 0.6425
ST GARCH-t	0.4485	0.4287	3.3069	0.6926
ST GARCH-st	0.4413	0.4276	3.2165	0.6869

The datasets consist of 817 and 793 prediction samples from January 2, 2014 to March 31, 2017, for S&P500 and Nikkei 225, respectively. The values in boxes indicate the best two favored models.

observation window of 100 days (HS-LT), which have been used for nonparametric estimation of VaR (see Gerlch et al. [22]). Eventually, we obtain 817 and 793 prediction samples from January 2, 2014 to March 31, 2017 for S&P500 and Nikkei 225, respectively.

Table 2 presents the results of evaluating the one-day-ahead volatility forecasts by MSE and QLIKE. We obtain the realized kernel, $\hat{\sigma}_t^2$ as in (13) and (14), from the Oxford-Man Institute. The GJR-GARCH model with GED error performs better than the others for the U.S. market based on MSE criteria, while the EGARCH model with skew Student's t error outperforms the others for the Japan market. Based on QLIKE, the measure reveals that for the predictive abilities, the EGARCH model with GED error is the best model in both markets. We can see that the GED assumption plays a great role in volatility forecasts.

Table 3. Evaluating VaR prediction performance based on the U.S. and Japan stock markets.

	Violation	VRate	Violation	VRate	Violation	VRate	Violation	VRate
	No	1%	No	5%	No	1%	No	5%
	S&P500				Nikkei 225			
HS-ST	38	4.65%	67	8.20%	42	5.30%	79	9.96%
HS-LT	14	1.71%	43	5.26%	21	2.65%	48	6.05%
RiskMetrics	19	2.33%	44	5.39%	23	2.90%	47	5.93%
GARCH-n	16	1.96%	40	4.90%	22	2.77%	47	5.93%
GARCH-t	9	1.10%	47	5.75%	14	1.77%	50	6.31%
GARCH-st	9	1.10%	41	5.02%	13	1.64%	46	5.80%
GARCH-ged	10	1.22%	41	5.02%	16	2.02%	48	6.05%
IGARCH-n	15	1.84%	41	5.02%	20	2.52%	49	6.18%
IGARCH-t	9	1.10%	48	5.88%	16	2.02%	51	6.43%
IGARCH-st	8	0.98%	41	5.02%	12	1.51%	44	5.55%
IGARCH-ged	10	1.22%	42	5.14%	16	2.02%	48	6.05%
GJR-n	12	1.47%	38	4.65%	22	2.77%	43	5.42%
GJR-t	8	0.98%	49	6.00%	16	2.02%	45	5.67%
GJR-st	6	0.73%	36	4.41%	13	1.64%	38	4.79%
GJR-ged	12	1.47%	48	5.88%	18	2.27%	46	5.80%
EGARCH-n	10	1.22%	38	4.65%	22	2.77%	47	5.93%
EGARCH-t	8	0.98%	50	6.12%	20	2.52%	47	5.93%
EGARCH-st	5	0.61%	36	4.41%	13	1.64%	45	5.67%
EGARCH-ged	13	1.59%	57	6.98%	19	2.40%	47	5.93%
ST GARCH-t	9	1.10%	37	4.53%	14	1.77%	47	5.93%
ST GARCH-st	8	0.98%	40	4.90%	11	1.39%	44	5.55%

The out-of-sample period: January 2, 2014 to March 31, 2017. A rolling window approach to produce a one-step-ahead VaR and ES over all risk models. There are 817 and 793 prediction samples for S&P500 and Nikkei 225, respectively, in the out-of-sample period.
The values in boxes indicate the best favored models.

To evaluate VaR, Table 3 displays the results of the violation rate at the α significant levels. When VRate is less than α, risk and loss estimates are conservative; while when VRate is greater than α, risk estimates are lower than actuality, and financial institutions may not allocate sufficient capital to cover likely future losses. We prefer that solvency outweighs profitability. In other words, for models where VRate/α are equidistant from 1, lower or conservative rates are preferable.

For the 1% level in the U.S. market, 4 models rank first with a tie: IGARCH-st, GJR-GARCH-t, EGARCH-t, and ST GARCH-st models. For the Japan market, all models and HS methods under-estimate 1% risk levels. The ST GARCH-st model performs best among the 19 risk models. A different scenario applies at the 5% level, where GARCH-n and ST GARCH-st dominate for the U.S. market, while GJR-GARCH-n is favorable for the Japan market. The performance HS methods are very poor and consistently underestimate risk.

Table 4. Evaluating VaR prediction performance for the U.S. stock market

S&P500	UC		CC		DQ	
	1%	5%	1%	5%	1%	5%
HS-ST	0.0000	0.0001	0.0000	0.0002	0.0000	0.0000
HS-LT	0.0628	0.7321	0.0117	0.2289	0.0000	0.0001
RiskMetrics	0.0012	0.6173	0.0010	0.8097	0.0000	0.0300
GARCH-n	0.0149	0.8911	0.0057	0.9902	0.0000	0.1912
GARCH-t	0.7740	0.3345	0.0102	0.6169	0.0000	0.2213
GARCH-st	0.7740	0.9808	0.0102	0.9987	0.0000	0.1819
GARCH-ged	0.5342	0.9808	0.0138	0.9987	0.0000	0.1630
IGARCH-n	0.0315	0.9808	0.0085	0.9987	0.0000	0.1874
IGARCH-t	0.7740	0.2635	0.0102	0.5319	0.0000	0.1850
IGARCH-st	0.9522	0.9808	0.1759	0.9987	0.0436	0.1819
IGARCH-ged	0.5342	0.8542	0.0138	0.9765	0.0000	0.1548
GJR-n	0.2081	0.6436	0.1716	0.8843	0.0878	0.5710
GJR-t	0.9522	0.2040	0.1759	0.4451	0.0623	0.1300
GJR-st	0.4232	0.4272	0.6941	0.6911	0.9528	0.4280
GJR-ged	0.2081	0.2635	0.1716	0.5293	0.0795	0.1359
EGARCH-n	0.5342	0.6436	0.0138	0.8777	0.0000	0.1178
EGARCH-t	0.9522	0.1552	0.1759	0.2946	0.0763	0.0254
EGARCH-st	0.2298	0.4272	0.4715	0.6818	0.8973	0.2042
EGARCH-ged	0.1179	0.0142	0.1281	0.0493	0.0470	0.0134
ST GARCH-t	0.7740	0.5303	0.0102	0.7952	0.0000	0.1135
ST GARCH-st	0.9522	0.8911	0.0063	0.9902	0.0000	0.1030

Tables 4 and 5 provide p-values of three backtests: UC, CC, and DQ tests. We use four lags, as in Engle and Manganelli [17]. At $\alpha = 1\%$, most of the models are rejected in the U.S. market, mainly by the DQ test. The GJR-GARCH models fare best as a group, and EGARCH with skew Student's t error is also adequate in the results of the three tests. Under a similar situation for the Japan market at $\alpha = 1\%$, the DQ test rejects most of the models. The ST GARCH models represent "survival" as a group, whereas GARCH, GJR-GARCH, and EGARCH models with skew Student's t perform well in terms of surviving from the three backtests. In summary, the DQ test is more powerful than the CC test, and the DQ test prefers asymmetric models with skew tails.

Figures 2 and 3 illustrate the VaR forecasts under the 1% significant level compared to log-returns in three models: GJR-GARCH-t, EGARCH-t, and ST GARCH-st. These three models perform best in terms of VRate, but the DQ test rejects the ST GARCH model in the U.S. market. We observe that three consecutive violations (August 20, 21, and 24, 2015) occur in the ST model for the U.S. market. On Sunday night, August 23 2015, a large drop in equities in

Table 5. Evaluating VaR prediction performance for the Japan stock market.

Nikkei 225	UC		CC		DQ	
	1%	5%	1%	5%	1%	5%
HS-ST	0.0000	0.0000	0.0000	0.0000	0.0000	0.0000
HS-LT	0.0001	0.1872	0.0002	0.3305	0.0000	0.0000
RiskMetrics	0.0000	0.2440	0.0000	0.4499	0.0000	0.0050
GARCH-n	0.0000	0.2440	0.0002	0.4499	0.0000	0.0501
GARCH-t	0.0506	0.1045	0.0745	0.2672	0.0684	0.0162
GARCH-st	0.0976	0.3124	0.1130	0.5501	0.0761	0.0258
GARCH-ged	0.0114	0.1872	0.0251	0.3579	0.0097	0.0090
IGARCH-n	0.0003	0.1411	0.0012	0.2771	0.0000	0.0712
IGARCH-t	0.0114	0.0760	0.0251	0.1535	0.0030	0.0123
IGARCH-st	0.1769	0.4858	0.1562	0.7556	0.0362	0.1344
IGARCH-ged	0.0114	0.1872	0.0251	0.3579	0.0025	0.0631
GJR-n	0.0000	0.5900	0.0002	0.5259	0.0000	0.7426
GJR-t	0.0114	0.3930	0.0251	0.3668	0.0435	0.7207
GJR-st	0.0976	0.7867	0.2041	0.7748	0.6223	0.9567
GJR-ged	0.0021	0.3124	0.0063	0.5501	0.0065	0.3709
EGARCH-n	0.0000	0.2440	0.0002	0.4499	0.0000	0.5752
EGARCH-t	0.0003	0.2440	0.0012	0.4499	0.0003	0.5849
EGARCH-st	0.0976	0.3930	0.2041	0.6540	0.5897	0.5317
EGARCH-ged	0.0008	0.2440	0.0028	0.4499	0.0005	0.5458
ST GARCH-t	0.0506	0.2440	0.0745	0.4499	0.1299	0.0565
ST GARCH-st	0.3006	0.4858	0.1948	0.7556	0.1901	0.3360

Asia (Monday time) triggered a drop in index futures in the U.S. and Europe. U.S. stock futures went down (7%) prior to the U.S. open. If we incorporate this external news in our model, then we might avoid this violation. August 24, 2015 was the S&P 500's worst day since 2011 and followed an 8.5% slump in China markets.

We evaluate ES based on Embrechts et al. [15] and choose the smallest value as the best one. The values in boxes of Table 6 indicate the best two favored models. Regarding the backtesting measure for ES at the 1% level, IGARCH-st and EGARCH-t perform as the best two favored models for the U.S. market, while IGARCH-st and ST GARCH-st are the best two favored models for the Japan market. The IGARCH-st performs the best for both markets at the 1% and 5% significant levels, and ST GARCH-st performs the best in the Japan market at both 1% and 5% significant levels. At the 5% level, the parsimonious IGARCH-st model behaves as good as the nonlinear ST GARCH-st model at the 5% level.

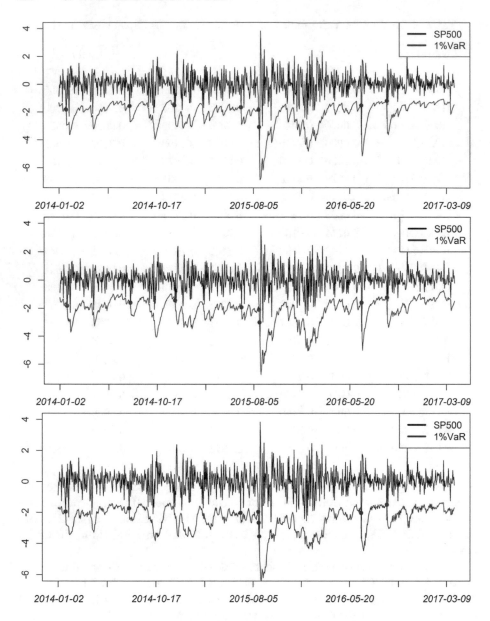

Fig. 2. VaR forecasts (blue line) and daily returns for S&P500 (817 prediction samples from January 2, 2014 to March 31, 2017) at the 1% level. Upper panel: GJR-GARCH-t model, middle panel: EGARCH-t, lower panel: ST GARCH-st.

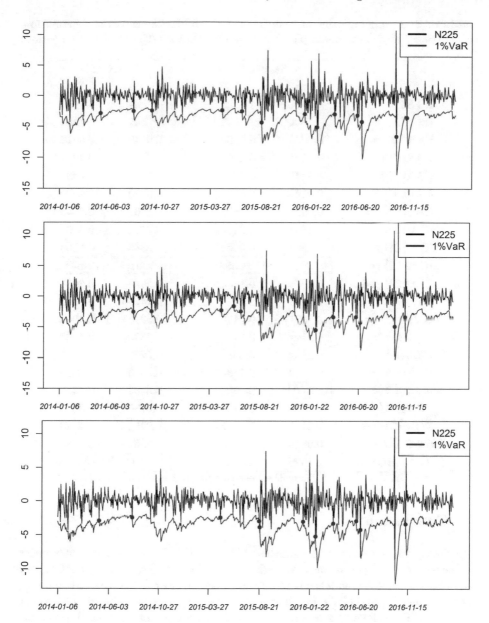

Fig. 3. VaR forecasts (blue line) and daily returns for Nikkei 225 (793 prediction samples from January 2, 2014 to March 31, 2017) at the 1% level. Upper panel: GJR-GARCH-st model, middle panel: EGARCH-st, lower panel: ST GARCH-st.

Table 6. Backtesting measure of Embrechts et al. [15] for the expected shortfall forecasts of S&P500 (817 out-of-sample period) and Nikkei 225 (793 out-of-sample period).

	S&P500		Nikkei 225	
	1%	5%	1%	5%
HS-ST	NA	NA	NA	NA
HS-LT	NA	NA	NA	NA
RiskMetrics	0.6565	0.2990	0.9530	0.5912
GARCH-n	0.4516	0.2400	0.7530	0.4911
GARCH-t	0.0841	0.0963	0.5187	0.3421
GARCH-st	0.0529	0.0211	0.3140	0.2285
GARCH-ged	0.2952	0.1520	0.5702	0.3542
IGARCH-n	0.4667	0.2351	0.7744	0.4305
IGARCH-t	0.0582	0.0603	0.2880	0.2596
IGARCH-st	[0.0315]	[0.0039]	[0.2800]	[0.1684]
IGARCH-ged	0.3075	0.1345	0.4926	0.3031
GJR-n	0.3788	0.1480	0.7423	0.5362
GJR-t	0.1021	0.0450	0.5813	0.4344
GJR-st	0.1192	0.1117	0.3525	0.3665
GJR-ged	0.2489	0.0986	0.6287	0.4187
EGARCH-n	0.3712	0.1230	0.9340	0.5303
EGARCH-t	[0.0140]	0.0523	0.6384	0.4418
EGARCH-st	0.2085	0.1196	0.5736	0.3004
EGARCH-ged	0.2897	0.1066	0.8555	0.4876
ST GARCH-t	0.0819	[0.0196]	0.3331	0.2986
ST GARCH-st	0.1626	0.0502	[0.1769]	[0.1760]

The values in boxes indicate the best two favored models.

7   Conclusion

This study evaluates the performances of VaR and ES forecasts using parametric statistical and HS approaches. We adopt RiskMetrics, GARCH, IGARCH, and asymmetric GARCH models with four distributions in error: normal, Student's t, skew Student's t, and GED. Furthermore, this study considers the smooth transition GARCH model with a second-order logistic function of Chen et al. [11]. This extension makes it possible to estimate quantile forecasts, VaR and ES together. We use Bayesian MCMC methods on these GARCH-type models for estimation, inference, and quantile forecasts and evaluate the one-day-ahead volatility and quantile forecasts.

In volatility forecasts, the GJR-GARCH and EGARCH models with GED perform best for the U.S. market, whereas the EGARCH models with GED or skew Student's t errors are good choices to predict volatility for the Japan market. A VaR forecasting reveals that the ST GARCH model with the second logistic function and skew Student's t compares most favorably in terms of violation rates for both markets. Both asymmetric models, GJR-t and EGARCH-t models, are favorable by independence of violations at the 1% level for the U.S. market. For the Japan market, a range of well-known GARCH models, including RiskMetrics, and historical simulation are not competitive to the ST GARCH-st model at the 1% level. Regarding the backtesting measure of ES, the IGARCH-st performs the best for both markets at the 1% and 5% significant levels, and ST GARCH-st performs the best in Japan market at both 1% and 5% significant levels. The skew and heavy-tailed error distributions are important when evaluating the quantile forecast of financial returns.

The findings of this research contribute to a better understanding in the performance of Bayesian forecasting of VaR and ES based on GARCH-type models. Our Bayesian methodology also makes it possible to forecast the volatility and quantiles by sampling from their predictive distributions.

Acknowledgments. Cathy W.S. Chen's research is funded by the Ministry of Science and Technology, Taiwan (MOST 105-2118-M-035-003-MY2).

A Appendix

All GARCH-type models contain the lagged AR(1) effect, which allows one to recognize whether the return series exhibits asymmetry mean reversion or market efficiency. The AR(1) coefficient may not be significant from time to time. The risk models considered in the empirical study are given in detail below.

AR(1)-GARCH(1,1) model

$$r_t = \phi_0 + \phi_1 r_{t-1} + a_t, \tag{17}$$
$$a_t = \sqrt{h_t}\varepsilon_t, \quad \varepsilon_t \overset{\text{i.i.d.}}{\sim} D(0,1),$$
$$h_t = \alpha_0 + \alpha_1 a_{t-1}^2 + \beta_1 h_{t-1}.$$

To ensure the AR(1)-GARCH(1,1) model is finite and positive, we restrict the parameters by:

$$S_1: \quad |\phi_1| < 1, \tag{18}$$
$$S_2: \quad \alpha_0 > 0, \alpha_1, \beta_1 \geq 0, \alpha_1 + \beta_1 < 1. \tag{19}$$

AR(1)-IGARCH(1,1)
If the AR polynomial of the GARCH representation in Eq. (17) has a unit root, then we have an IGARCH model. Thus, $\alpha_1 + \beta_1 = 1$, and IGARCH models are unit-root GARCH models.

RiskMetrics
Developed by J. P. Morgan [30], still a popular method, specifically used for VaR calculation. It is a special case of the IGARCH model with normal errors,

where $\alpha_0 = 0$, and is thus an exponentially weighted moving average of squared shocks. The model form is:

$$r_t = a_t, \tag{20}$$

$$a_t = \sqrt{h_t}\varepsilon_t, \quad \varepsilon_t \overset{\text{i.i.d.}}{\sim} N(0,1),$$

$$h_t = \delta h_{t-1} + (1-\delta)a_{t-1}^2,$$

where δ is a decay factor belonging to $(0.9,1)$, since J.P. Morgan recommended $\delta = 0.94$ for computing daily volatility.

AR(1)-GJR-GARCH(1,1) model

$$r_t = \phi_0 + \phi_1 r_{t-1} + a_t, \tag{21}$$

$$a_t = \sqrt{h_t}\varepsilon_t, \quad \varepsilon_t \overset{\text{i.i.d.}}{\sim} D(0,1),$$

$$h_t = \alpha_0 + (\alpha_1 + \gamma_1 I_{t-1})a_{t-1}^2 + \beta_1 h_{t-1},$$

$$\text{where } I_{t-1} = \begin{cases} 1, \text{ if } a_{t-1} \le 0 \\ 0, \text{ if } a_{t-1} > 0. \end{cases}$$

We give some constraints for the GJR-GARCH(1,1) model to ensure positiveness and covariance stationarity.

$$S_3: \quad \alpha_0 > 0, \alpha_1, \beta_1 \ge 0, \alpha_1 + \gamma_1 \ge 0, \alpha_1 + \beta_1 + 0.5\gamma_1 < 1. \tag{22}$$

Equations (18) and (22) guarantee stationarity and positivity of the AR(1)-GJR-GARCH(1,1) model.

AR(1)-EGARCH(1,1) model

$$r_t = \phi_0 + \phi_1 r_{t-1} + a_t, \tag{23}$$

$$a_t = \sqrt{h_t}\varepsilon_t, \quad \varepsilon_t \overset{\text{i.i.d.}}{\sim} D(0,1),$$

$$\ln h_t = \alpha_0 + \alpha_1 \left(\frac{|a_{t-1}| + \gamma_1 a_{t-1}}{\sqrt{h_{t-1}}} \right) + \beta_1 \ln h_{t-1}.$$

The stationary condition for the mean part is $|\phi_1| < 1$ as in (18). The model with a positive a_{t-1} contributes $\alpha_1(1+\gamma_1)|\varepsilon_{t-1}|$ to log volatility, whereas a negative a_{t-1} gives $\alpha_1(1-\gamma_1)|\varepsilon_{t-1}|$. The γ_1 parameter thus signifies the leverage effect of a_{t-1}. Therefore, we expect γ_1 to be negative.

B   Appendix

Adaptive MH Algorithm

In our past experience with GARCH-type models, the RW Metropolis algorithm produces very high correlations and slow mixing among the iterates of the parameters $\boldsymbol{\alpha}$. The effect is enhanced when true volatility persistence is high, i.e.

$\alpha_1 + \beta_1$ is close to 1, which is the usual case with financial time series. To speed up mixing and to reduce the dependence, we revise the sampling method for $\boldsymbol{\alpha}$ after the burn-in period and switch to the independent kernel IK MH algorithm for the sampling period (see Chen and So [6]).

RW-MH algorithm

1. Generate the new parameter group $\boldsymbol{\alpha}^*$ from the random walk kernel with the form:
$$\boldsymbol{\alpha}^* = \boldsymbol{\alpha}^{[j-1]} + \varepsilon, \quad \varepsilon \sim N(0, \epsilon \boldsymbol{\Sigma}),$$

where $\boldsymbol{\Sigma}$ is the diagonal variance-covariance matrix.
2. RW chains: Randomly generate p from U(0,1) and compare it with:

$$p^* = \min\left\{1, \frac{\pi(\boldsymbol{\alpha}^*)}{\pi(\boldsymbol{\alpha}^{[j-1]})}\right\}.$$

If $p^* > p$, then accept $\boldsymbol{\alpha}^*$ as the new estimate of the parameter group and then go back to Step 1.

A suitable value of ϵ with good convergence proportion can usually be selected by having an acceptance probability of 25% to 50%.

IK-MH algorithm

1. Generate the new parameter group $\boldsymbol{\alpha}^*$ from the independent walk kernel with the form:
$$\boldsymbol{\alpha}^* = \boldsymbol{\mu}_\alpha + \varepsilon, \quad \varepsilon \sim N(0, \boldsymbol{\Sigma}_\alpha),$$

where mean $\boldsymbol{\mu}_\alpha$ and covariance matrix $\boldsymbol{\Sigma}_\alpha$ are obtained from the burn-in period.
2. IK chains: Randomly generate p from U(0,1) and compare it with:

$$p^* = \min\left\{1, \frac{\pi(\boldsymbol{\alpha}^*)q(\boldsymbol{\alpha})}{\pi(\boldsymbol{\alpha}^{[j-1]})q(\boldsymbol{\alpha}^*)}\right\},$$

where $\boldsymbol{\alpha}^{[j]}$ is the ith iterate of $\boldsymbol{\alpha}$ and $q(.)$ is a Gaussian proposal density. If $p^* > p$, then accept $\boldsymbol{\alpha}^*$ as the new estimate of the parameter group and then go back to Step 1. Repeat until the MCMC iterates converge.

References

1. Acerbi, C., Tasche, D.: On the coherence of expected shortfall. J. Bank. Finance **26**, 1487–1503 (2002)
2. Artzner, P., Delbaen, F., Eber, J.-M., Heath, D.: Coherent measures of risk. Math. Finance **9**, 203–228 (1999)
3. Berkowitz, J., Christoffersen, P.F., Pelletier, D.: Evaluating Value-at-Risk models with desk-level data. Manage. Sci. **57**, 2213–2227 (2011)

4. Black, F.: Studies of stock market volatility changes. In: Proceedings of the American Statistical Association, Business and Economic Statistics Section, pp. 177–181 (1976)
5. Bollerslev, T.: Generalized autoregressive conditional heteroskedasticity. J. Econ. **31**, 307–327 (1986)
6. Chen, C.W.S., So, M.K.P.: On a threshold heteroscedastic model. J. Forecast. **22**, 73–89 (2006)
7. Chen, C.W.S., Chiang, T.C., So, M.K.P.: Asymmetrical reaction to US stock-return news: evidence from major stock markets based on a double-threshold model. J. Econ. Bus. **55**, 487–502 (2003)
8. Chen, C.W.S., Gerlach, R., Hwang, B.B.K., McAleer, M.: Forecasting Value-at-Risk using nonlinear regression quantiles and the intra-day range. Int. J. Forecast. **28**, 557–574 (2012a)
9. Chen, C.W.S., Gerlach, R., Lin, E.M.H., Lee, W.C.W.: Bayesian forecasting for financial risk management, pre and post the global financial crisis. J. Forecast. **31**, 661–687 (2012b)
10. Chen, C.W.S., Gerlach, R., Lin, M.H.: Falling and explosive, dormant, and rising markets via multiple-regime financial time series models. Appl. Stochast. Models Bus. Ind. **26**, 28–49 (2010)
11. Chen, C.W.S., Weng, M.C., Watanabe, T.: Bayesian forecasting of Value-at-Risk based on variant smooth transition heteroskedastic models. Stat. Interface **10**, 451–470 (2017)
12. Chen, Q., Gerlach, R., Lu, Z.: Bayesian Value-at-Risk and expected shortfall forecasting via the asymmetric Laplace distribution. Comput. Stat. Data Anal. **56**, 3498–3516 (2012)
13. Christoffersen, P.: Evaluating interval forecasts. Int. Econ. Rev. **39**, 841–862 (1998)
14. Duffie, D., Pan, J.: An overview of value at risk. J. Deriv. **4**, 7–49 (1997)
15. Embrechts, P., Kaufmann, R., Patie, P.: Strategic Long-term financial risks: single risk factors. Comput. Optim. Appl. **32**, 61–90 (2005)
16. Engle, R.F.: Autoregressive conditional heterosedasticity with estimates of variance of United Kingdom inflation. Econometrica **50**, 987–1008 (1982)
17. Engle, R.F., Manganelli, S.: CAViaR: conditional autoregressive value at risk by regression quantiles. J. Bus. Econ. Stat. **22**, 367–381 (2004)
18. Fernández, C., Steel, M.F.J.: On Bayeasian modeling of fat tail and skewness. J. Am. Stat. Assoc. **93**, 359–371 (1998)
19. Fllmer, H., Schied, A.: Convex measures of risk and trading constraints. Finance Stochast. **6**, 429–447 (2002)
20. Glosten, L.R., Jagannathan, R., Runkle, D.E.: On the relation between the expected value and the volatility of the nominal excess return on stock. J. Finance **48**, 1779–1801 (1993)
21. Gerlach, R., Chen, C.W.S.: Bayesian inference and model comparison for asymmetric smooth transition heteroskedastic models. Stat. Comput. **18**, 391–408 (2008)
22. Gerlach, R., Chen, C.W.S., Chan, N.C.Y.: Bayesian time-varying quantile forecasting for value-at-risk in financial markets. J. Bus. Econ. Stat. **29**, 481–492 (2011)
23. Gerlach, R., Chen, C.W.S.: Bayesian expected shortfall forecasting incorporating the intraday range. J. Finan. Econ. **14**, 128–158 (2015)
24. Hansen, B.E.: Autoregressive conditional density estimation. Int. Econ. Rev. **35**, 705–730 (1994)
25. Harmantzis, F.C., Miao, L., Chien, Y.: Empirical study of valueatrisk and expected shortfall models with heavy tails. J. Risk Finance **7**, 117–135 (2006)

26. Hendricks, D.: Evaluation of Value-at-Risk models using historical data. Econ. Policy Rev. **2**, 39–67 (1996)
27. Hoogerheide, L., van Dijk, H.K.: Bayesian forecasting of Value at Risk and expected shortfall using adaptive importance sampling. Int. J. Forecast. **26**, 231–247 (2010)
28. Jansen, E.S., Teräsvirta, T.: Testing parameter constancy and super exogeneity in econometric equations. Oxford Bull. Econ. Stat. **58**, 735–763 (1996)
29. Kupiec, P.H.: Techniques for verifying the accuracy of risk measurement models. J. Deriv. **3**, 73–84 (1995)
30. Morgan, J.P.: RiskMetrics: J. P. Morgan Technical Document, 4th edn. J. P. Morgan, New York (1996)
31. Nelson, D.B.: Conditional heteroscedasticity in asset returns: a new approach. Econometrica **59**, 347–370 (1991)
32. Papp, R.: What are the advantages of MCMC based inference in latent variable models? Stat. Neerl. **56**, 2–22 (2002)
33. Takahashi, M., Watanabe, T., Omori, Y.: Volatility and quantile forecasts by realized stochastic volatility models with generalized hyperbolic distribution. J. Forecast. **2**, 437–457 (2016)
34. van Dijk, D., Teräsvirta, T., Franses, P.H.: Smooth transition autoregressive models - a survey of recent developments. Econ. Rev. **21**, 1–47 (2002)
35. Yamai, Y., Yoshiba, T.: Value-at-risk versus expected shortfall: a practical perspective. J. Bank. Finance **29**, 997–1015 (2005)

Smoothing Spline as a Guide to Elaborate Explanatory Modeling

Chon Van Le[✉]

School of Business, International University - VNU HCMC, Quarter 6, Linh Trung,
Thu Duc District, Ho Chi Minh City, Vietnam
lvchon@hcmiu.edu.vn

Abstract. Although there are substantial theoretical and empirical differences between explanatory modeling and predictive modeling, they should be considered as two dimensions. And predictive modeling can work as a "fact check" to propose improvements to existing explanatory modeling. In this paper, I use smoothing spline, a nonparametric calibration technique which is originally designed to intensify the predictive power, as a guide to revise explanatory modeling. It works for the housing value model of Harrison and Rubinfeld (1978) because the modified model is more meaningful and fits better to actual data.

Keywords: Predictive econometrics · Calibration · Smoothing spline

1 Introduction

Econometric modeling has two main purposes: causal explanation and empirical prediction. Causal analysis determines whether an independent variable really affects the dependent variable and estimates the magnitude of the effect. And the second goal is to make predictions about the dependent variable, given the observed values of the independent variables.

However, the two objectives have received unequal treatment among academic researchers, especially those who work with cross section and panel data. The bulk of most econometric textbooks is given to issues relevant to causal explanation. Shmueli (2010) attributes this to the assumption that models with high explanatory power are expected to have high predictive power. He argues that there is clear distinction, both theoretically and empirically, between explanatory modeling and predictive modeling.

Although many academic statisticians may consider prediction as unacademic (for example, Kendall and Stuart (1977) and Parzen (2001)), it has played an increasingly important role in banking and financial services and other industries. Recent emergence of Big Data has fueled explosive growth of predictive modeling in the last decade. Large volumes, high velocity inflow, and a wide variety of data, including semi-structured and unstructured data such as text and images, have provided organizations with various opportunities for keeping

© Springer International Publishing AG 2018
V. Kreinovich et al. (eds.), *Predictive Econometrics and Big Data*, Studies in Computational
Intelligence 753, https://doi.org/10.1007/978-3-319-70942-0_7

track of and making prompt adjustments to their performance (Laney 2001). Competent predictive modeling to analyze big data has become a key to gaining insights, forecasting the future, and automating non-routine decision making which enable them to make big improvements such as better risk management, smoothened operations, personalized services, etc. (ICAEW 2014).

On the other hand, heavy emphasis placed on explanatory modeling has led to a widespread acceptance of regression results in many academic papers with highly significant/insignificant coefficients but low R^2. A large proportion of variation in the dependent variable left unexplained may imply that several important independent variables are missing or the functional form that represents the relationship between the regressand and regressor(s) is not appropriate. In addition, since the linear regression model which is used most popularly is an approximation to some unknown, underlying function, such an approximation is likely to be useful only over a small range of variation of the independent variables around their means. It can be asserted that whether or not a factor has an impact on the dependent variable, but the magnitude of its impact, if any, cannot be asserted because it may vary across different observations.

Therefore, calibration techniques under predictive modeling should be used to improve model specification. Although they cannot explain what is wrong in the orginal model, they can work as a guide to elaborate explanatory modeling. In this paper, I focus particularly on smoothing spline and apply it to Harrison and Rubinfeld's (1978) study on housing prices for the Boston area. It is found that their model, though already good, can be fine-tuned to be more meaningful and to fit better to actual data.

The paper is structured as follows. Section 2 reviews predictive modeling in comparison with explanatory modeling. Section 3 presents calibration techniques, including smoothing splines. Section 4 outlines the data, the revised model and its regression results. Conclusions follow in Sect. 5.

2 Predictive Modeling[1]

In economics, statistical methods are used mainly to test economic theory. The theory specifies causal and effect relationships between variables. Explanatory modeling applies statistical models to data for testing these relationships, that is, whether a particular independent variable does influence the dependent variable. In this type of modeling, the theory takes a governing role and the application of statistical methods is done strictly through the lens of the theory (Shmueli 2010).

In contrast, predictive modeling refers to the application of statistical models or data mining algorithms to data for predicting the dependent variable. According to Shmueli (2010), prediction involves point or interval prediction, predictive distribution, or ranking of new or future observations. There is a disagreement among statisticians on the value of predictive modeling. Many see it as unacademic. Berk (2008) indicates that researchers in social sciences only do "causal

---

[1] This part is to a large extent based on Shmueli (2010).

econometric style". As a consequence, many statistics textbooks write very little on predictive modeling. Others such as Geisser (1975), Aitchison and Dunsmore (1975), Friedman (1997) consider prediction as the foremost statistical application. This is true for most organizations and individuals outside of academia that are overwhelmed with exponential growth of data as a result of digital technology, mobile technology, social media, public sector open data, computer chips and sensors implanted in physical assets, etc. Increasingly large and rich datasets often consist of complex patterns and relationships that are beyond the reach of existing theories. Predictive modeling can help reveal new hypotheses and causality or propose improvements to existing explanatory models.

Shmueli (2010) clarifies that explanatory power can testify the strength of a causal hypothesis but cannot assess the distance between theory and empirics. Predictive modeling can work as a "fact check" to evaluate the relevance of theories in the light of actual data. Hence competing theories can be compared by examining their respective predictive power. Ehrenberg and Bound (1993) state that predictive modeling may create benchmarks of predictive accuracy under which scientific development can lead to substantial theoretical and practical gains.

The key element leading to the difference between explanatory and predicting modeling is errors of measurement. Observed data normally do not measure accurately their underlying constructs or variables. There has been considerable discussion in empirical studies on how to obtain reasonable measures of interest rates, profits, services from capital stocks, etc. For this reason, "the operationalization of theories and constructs into econometric models and measurable data creates a disparity between the ability to explain phenomena at the conceptual level and the ability to generate predictions at the measurable level" (Shmueli 2010).

The disparity justifies important differences in several aspects. Firstly, in explanatory modeling, a statistical model "usually begins with a statement of a theoretical proposition" (Greene 2012). It is built based on an economic model that consists of mathematical equations that describe deterministic relationships between independent variables and the dependent variable. The statistical model represents a causal relationship, and independent variables are assumed to cause the dependent variable. In predictive modeling, the statistical model is often built from the data. It shows the association between independent variables and the dependent variable. Interpreting the relationship between independent variables and the dependent variable is not required.

Secondly, Shmueli (2010) claims that explanatory modeling is backward-looking. A statistical model is used to test already existent hypotheses. On the contrary, predictive modeling is forward-looking. The statistical model is built to predict new observations.

Thirdly, explanatory modeling focuses on minimizing bias to secure reliable estimates of the "true" model coefficients. In contrast, since the goal of predictive modeling is to obtain optimal predictions of the dependent variable based on a regression model of whatever variables are available, it seeks to minimize the sum of bias and estimation variance (Shmueli 2010). Therefore, in predictive modeling, bias is not a big problem and can be tolerated as long as empirical precision is improved. However, bias is a crucial issue in explanatory modeling. Several methods have been proposed to deal with the omission of variables that affect the dependent variable and are correlated with independent variables that are present in the statistical model.

Fourthly, explanatory modeling aims to estimate the theory-based statistical model with adequate statistical power for hypothesis testing. Consequently, multicollinearity is often a major concern in causal explanation. When two or more independent variables are highly correlated, the estimator is still unbiased but less precise. In predictive modeling, all these independent variables should be included if each variable contributes significantly to the predictive power of the model.

In addition, sufficient data are required for statistical inference. A variety of data imputation methods such as zero-order method, data augmentation and multiple imputation techniques, inverse probability weighting, etc. have been used to fill gaps in data sets. These methods seem to be constructed for parameter estimation and hypothesis testing. However, according to Shmueli (2010), "beyond a certain amount of data, extra precision is negligible for purposes of inference". Predictive modeling needs a larger sample to reduce bias and variance and to create holdout datasets for prediction testing.

Although explanatory modeling and predictive modeling substantially differ theoretically and empirically, they should be considered as two dimensions. And a statistical model should be evaluated based on its explanatory power and predictive power, whose weights depend on the question of interest. The predictive power can be intensified by calibration techniques which are discussed in the next section.

3 Calibration and Splines

Calibration dates back to the early 1980s in dynamic computable general equilibrium models (Kydland and Prescott, 1982). It is a procedure to select numerical values for the parameters of a model because sometimes no data are available to estimate its parameters. Canova (1994) points out that econometric estimation and calibration are two approaches to exposing general equilibrium models to data. They both begin with formulating a general equilibrium model and selecting funtional forms for production, utility, and exogenous factors. But they are different in choosing the parameters. The estimation approach believes that the model provides an accurate description of the data, or that the model is true, is a data-generating process, and tests what attributes of the model are false, diverge significantly from the data. The calibration approach assumes the opposite view as Box (1976) states that "all models are wrong". As a result, the theoretical model should be modified or calibrated to gain a better approximation of the observed data. Otherwise, the model can produce systematically biased predictions, either too high or too low on average, so it cannot be used for economic decisions.

According to Stine (2011), "a model is calibrated if its predictions are correct on average," or

$$E(y|\hat{y}) = \hat{y}.$$

The adjustment procedure starts with the non-calibrated predicted value from a regression:

$$\hat{y} = \hat{\beta}_0 + \hat{\beta}_1 x_1 + \cdots + \hat{\beta}_k x_k.$$

The predictive ability of the model can be improved if $\hat{y}$ can be transformed to a better predictor, for example, $\hat{\hat{y}} = h(\hat{y})$ where h is a smooth function.

A popular smooth function is a spline function that was first applied by Poirier and Garber (1974). In order to be continuous and continuously differentiable at the 'joins' or 'knots', the spline must take a form of quadratic, cubic or a higher-degree polynomial. Suppose that there are two known knots $\hat{y}_a$ and $\hat{y}_b$ and that we use a cubic spline:

$$\hat{\hat{y}} = \alpha_{i0} + \alpha_{i1}\hat{y} + \alpha_{i2}\hat{y}^2 + \alpha_{i3}\hat{y}^3 + \varepsilon, \tag{1}$$

where the subsets are defined by

$$i = \begin{cases} 1 & \text{if } \hat{y} \leq \hat{y}_a, \\ 2 & \text{if } \hat{y}_a < \hat{y} \leq \hat{y}_b, \\ 3 & \text{if } \hat{y}_b < \hat{y}. \end{cases}$$

Continuity of $\hat{\hat{y}}$ and the first derivatives at the knots requires:

$$\begin{aligned} \alpha_{10} + \alpha_{11}\hat{y}_a + \alpha_{12}\hat{y}_a^2 + \alpha_{13}\hat{y}_a^3 &= \alpha_{20} + \alpha_{21}\hat{y}_a + \alpha_{22}\hat{y}_a^2 + \alpha_{23}\hat{y}_a^3, \\ \alpha_{20} + \alpha_{21}\hat{y}_b + \alpha_{22}\hat{y}_b^2 + \alpha_{23}\hat{y}_b^3 &= \alpha_{30} + \alpha_{31}\hat{y}_b + \alpha_{32}\hat{y}_b^2 + \alpha_{33}\hat{y}_b^3, \\ \alpha_{11} + 2\alpha_{12}\hat{y}_a + 3\alpha_{13}\hat{y}_a^2 &= \alpha_{21} + 2\alpha_{22}\hat{y}_a + 3\alpha_{23}\hat{y}_a^2, \\ \alpha_{21} + 2\alpha_{22}\hat{y}_b + 3\alpha_{23}\hat{y}_b^2 &= \alpha_{31} + 2\alpha_{32}\hat{y}_b + 3\alpha_{33}\hat{y}_b^2. \end{aligned} \tag{2}$$

Additional restrictions should be imposed on the spline (1) before estimation if we want the second derivatives to be continuous:

$$\begin{aligned} 2\alpha_{12} + 6\alpha_{13}\hat{y}_a &= 2\alpha_{22} + 6\alpha_{23}\hat{y}_a, \\ 2\alpha_{22} + 6\alpha_{23}\hat{y}_b &= 2\alpha_{32} + 6\alpha_{33}\hat{y}_b. \end{aligned}$$

The cubic spline though allows discontinuities in the third derivatives at the knots.

So far we have assumed that we have specified the locations of the knots in advance. But in most cases, without further information on abrupt changes over time or size thresholds, it is impossible to determine the knots beforehand. Then we can resort to nonparametric smoothers. The oldest and simplest one is the smoothing spline that connects the medians of equal-width intervals. The number of intervals can be chosen by

$$\text{Number of intervals} = \max\{\min(b_1, b_2), b_3\},$$

where $b_1 = \text{round}\{10 \times \ln 10(N)\}$, $b_2 = \text{round}(\sqrt{N})$, $b_3 = \min(2, N)$, and N is the number of observations (StataCorp 2011). In each interval, the median of y and the median of $\hat{y}$ are calculated. A spline is fit to these medians. If the spline appears to deviate much from the $45°$ line, then the original model can be modified to better capture the "true", complex relationships between the dependent variable and independent variables. Consequently, the model would be more effective as an approximating function, that is, providing more rigorous hypothesis testing of a regressor's impact and an expectedly more accurate estimate of the magnitude of that impact. I suggest using smoothing splines as a guide to elaborate explanatory modeling. An example is presented in the next section.

4 Example of Harrison and Rubinfeld (1978)

In a study of the willingness to pay for air quality improvements in Boston, Harrison and Rubinfeld (1978) use data for 506 census tracts in the Boston Standard Metropolitan Statistical Area (SMSA) in 1970 to estimate a housing value equation

$$
\begin{aligned}
\text{Ln(Medianvalue)} = {} & \alpha_0 + \alpha_1 \text{Room}^2 + \alpha_2 \text{Age} + \alpha_3 \text{Ln(Distance)} + \alpha_4 \text{Ln(Highway)} \\
& + \alpha_5 \text{Tax} + \alpha_6 \text{Pupil/Teacher} + \alpha_7 (\text{Black} - 0.63)^2 \qquad (3) \\
& + \alpha_8 \text{Ln(Lowstatus)} + \alpha_9 \text{Crime} + \alpha_{10} \text{Zoning} + \alpha_{11} \text{Industry} \\
& + \alpha_{12} \text{Charles} + \alpha_{13} \text{Nox}^2 + \epsilon,
\end{aligned}
$$

where variables are defined in Table 1. The results which are reported in the second and third columns of Table 2 seemingly provide strong evidence on the impacts of the independent variables, except Age, Zoning, and Industry. And R^2 is relatively high. However, Fig. 1 indicates that the spline diverges considerably from the $45°$ line, especially at the small predicted values of the dependent variable.

The spline can be approximated by a 3rd degree polynomial of the form:

$$
y = X\boldsymbol{\alpha} + \gamma_2(\hat{y} - \bar{y})^2 + \gamma_3(\hat{y} - \bar{y})^3 + \xi, \qquad (4)
$$

where X and $\hat{y}$ are the set of independent variables and the predict values of the original model.

The second and third columns of Table 3 show that the second term in Eq. (4) that approximates the spline of the Harrison and Rubinfeld's model is significant at 5% level. Moreover, the last two terms in (4) are jointly significant at 10% level (their F-statistics is 2.40), implying that the polynomial (4) fits the data better than the Harrison and Rubinfeld's model. It suggests that the model can be revised.

A closer look at the data set proposes several modifications as follows. Firstly, since the index of accessibility to radial highways takes nine discrete values, namely, 1, 2, 3, 4, 5, 6, 7, 8, and 24, the Highway index should be used instead

Table 1. Variable definitions

Variable	Description
Medianvalue	Median value of owner-occupied homes
Room	Average number of rooms in owner-occupied homes
Age	Proportion of owner-occupied homes built before 1940
Black	Black proportion of population in the community
Lowstatus	Proportion of population that is lower status $= \frac{1}{2}$ (proportion of adults without some high school education and proportion of male workers classified as laborers)
Crime	Crime rate by town
Zoning	Proportion of a town's residential land zoned for lots greater than 25,000 square feet
Industry	Proportion of nonretail business acres per town
Tax	Full value property tax rate ($/$10,000)
Pupil/Teacher	Pupil-teacher ratio by town school district
Charles	Charles River dummy equals 1 if tract bounds the Charles River and 0 otherwise
Distance	Weighted distance to 5 employment centers in the Boston area
Highway	Highway access index
Nox	Annual average nitrogen oxide concentration in pphm

Source: Harrison and Rubinfeld (1978).

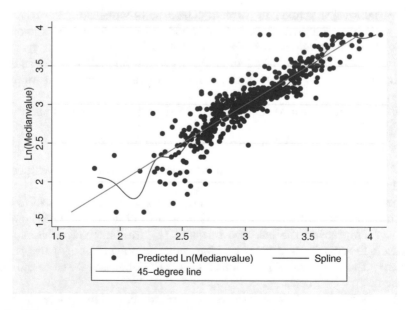

Fig. 1. Smoothing spline based on Harrison and Rubinfeld's (1978) model

Table 2. Housing value models

	Harrison and Rubinfeld's (1978) model[a] (3)		Revised model (5)	
Constant	4.558***	(0.1544)	4.595***	(0.2921)
Room^2	0.0063***	(0.0013)	0.0231***	(0.0025)
Age	0.00009	(0.0005)	0.0052*	(0.0030)
Ln(Distance)	−0.1913***	(0.0334)	−0.2382***	(0.0301)
Ln(Highway)	0.0957***	(0.0191)		
Highway			0.0139*	(0.0076)
Tax	−0.0004***	(0.0001)		
Ln(Tax)			−0.1963***	(0.0427)
Tax600			−0.3011***	(0.0791)
Pupil/Teacher	−0.0311***	(0.0050)	−0.0316***	(0.0045)
(Black − 0.63)^2	0.3637***	(0.1031)	0.1240	(0.0934)
Ln(Lowstatus)	−0.3712***	(0.0250)		
Lowstatus			0.0547***	(0.0095)
Crime	−0.0119***	(0.0012)	−0.0225***	(0.0037)
Crime^2			0.00016***	(0.00005)
Zoning	0.00008	(0.0005)	0.0004	(0.0005)
Industry	0.0002	(0.0024)	−0.0074***	(0.0027)
Charles	0.0914***	(0.0332)	0.0474^b	(0.0292)
Nox^2	−0.6380***	(0.1131)	−0.7763***	(0.1016)
Room×Age			−0.0009**	(0.00047)
Room×Lowstatus			−0.0118***	(0.0015)
Highway×Lowstatus			−0.0012***	(0.00016)
Highway×Industry			0.0019***	(0.0005)
Number of observations	506		506	
R^2	0.806		0.855	

Notes: ***, **, * significant at the 1%, 5%, 10% levels, respectively.
Standard errors in parentheses.
[a] Results are slightly different from those in Harrison and Rubinfeld (1978).
[b] Charles River dummy has a p-value of 10.5.
Source: Author's calculation.

of its log which may be less meaningful. Secondly, the Tax variable is replaced by its log. While most of the observations in the sample have tax rates ranging from \$187 to \$469 per \$10,000, 137 tracts have unusually high tax rates of \$666 and \$711. A dummy variable, Tax600 which equals 1 if the tax rate is over \$600 and 0 otherwise, is included. Thirdly, as the lower status is measured as a percentage of the population in the community, the Lowstatus variable should be in original form, not in logarithmic form. Fourthly, squared crime rate, Crime^2, is added to allow for a changing impact of the crime rate on the housing value. Fifthly, house prices depend on various attributes which are considered not only separately but also together. Therefore, four interaction terms are included. The modified housing value equation is

$$\begin{aligned}
\text{Ln(Medianvalue)} = \beta_0 &+ \beta_1 \text{Room}^2 + \beta_2 \text{Age} + \beta_3 \text{Ln(Distance)} + \beta_4 \text{Highway} \\
&+ \beta_5 \text{Ln(Tax)} + \beta_6 \text{Tax600} + \beta_7 \text{Pupil/Teacher} + \beta_8 (\text{Black} - 0.63)^2 \\
&+ \beta_9 \text{Lowstatus} + \beta_{10} \text{Crime} + \beta_{11} \text{Crime}^2 + \beta_{12} \text{Zoning} \qquad (5) \\
&+ \beta_{13} \text{Industry} + \beta_{14} \text{Charles} + \beta_{15} \text{Nox}^2 + \beta_{16} \text{Room} \times \text{Age} \\
&+ \beta_{17} \text{Room} \times \text{Lowstatus} + \beta_{18} \text{Highway} \times \text{Lowstatus} \\
&+ \beta_{19} \text{Highway} \times \text{Industry} + \varepsilon.
\end{aligned}$$

Table 3. Polynomials approximating splines of the two models

	Harrison and Rubinfeld's (1978)		Revised model	
Constant	4.635***	(0.1581)	4.531***	(0.2951)
Room^2	0.0072***	(0.0014)	0.0271***	(0.0033)
Age	0.0002	(0.0005)	0.0059**	(0.0030)
Ln(Distance)	−0.2048***	(0.0342)	−0.2519***	(0.0310)
Ln(Highway)	0.0997***	(0.0196)		
Highway			0.0151**	(0.0076)
Tax	−0.00046***	(0.0001)		
Ln(Tax)			−0.2048***	(0.0429)
Tax600			−0.2944***	(0.0820)
Pupil/Teacher	−0.0329***	(0.0051)	−0.0322***	(0.0045)
(Black − 0.63)^2	0.3314***	(0.1055)	0.1108	(0.0942)
Ln(Lowstatus)	−0.3880***	(0.0274)		
Lowstatus			0.0651***	(0.0111)
Crime	−0.0103***	(0.0023)	−0.0230***	(0.0040)
Crime^2			0.00016***	(0.00005)
Zoning	0.0003	(0.0005)	0.00055	(0.00046)
Industry	0.0006	(0.0024)	−0.0077***	(0.0028)
Charles	0.0985***	(0.0334)	0.0539*	(0.0295)
Nox^2	−0.6550***	(0.1139)	−0.7926	(0.1025)
Room×Age			−0.0010**	(0.00047)
Room×Lowstatus			−0.0134***	(0.0018)
Highway×Lowstatus			−0.0013***	(0.0002)
Highway×Industry			0.0019***	(0.0005)
$(\widehat{\text{Ln(MV)}} - \overline{\text{Ln(MV)}})^2$	−0.1371**	(0.0672)	−0.0913	(0.0675)
$(\widehat{\text{Ln(MV)}} - \overline{\text{Ln(MV)}})^3$	−0.0396	(0.0858)	−0.1096	(0.0668)
Number of observations	506		506	
R^2	0.808		0.856	

Notes: ***, **, * significant at the 1%, 5%, 10% levels, respectively.
Standard errors in parentheses.
Source: Author's calculation.

Figure 2 shows that the spline based on the revised model does not deviate much from the 45° line, even at the extreme predicted values. This is confirmed by the fact that the last two terms in the spline-approximating polynomial in the fourth and fifth columns of Table 3 are individually and jointly insignificant (their F-statistics is 1.68).

The regression results which are presented in the fourth and fifth columns of Table 2 differ in several aspects from those of Harrison and Rubinfeld. Age and Industry are now significant. Externalities associated with industrial activities such as noise, pollution, heavy traffic, and awful view negatively affect housing values as expected. Black proportion of population no longer has a positive impact on housing values, which makes more sense as black neighbors are often regarded as undesirable. In addition, the explanatory power of the revised model, though already good, still improves since R^2 increases by 5%. The example of Harrison and Rubinfeld demonstrates that smoothing spline helps elaborate modeling.

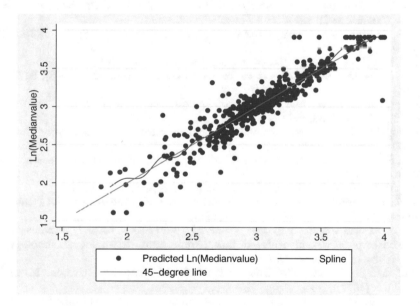

Fig. 2. Smoothing spline based on revised model

5   Conclusions

Greater weight has so far been put on causal explanation than on empirical prediction due to the wrong assumption that models having high explanatory power are supposed to have high predictive power. Explanatory modeling and predictive modeling substantially differ theoretically and empirically, but they should be considered as two dimensions. And predictive modeling can work as a "fact check" to propose improvements to existing explanatory modeling.

In this paper, I use smoothing spline, a nonparametric calibration technique which is originally designed to intensify the predictive power, as a guide to revise explanatory modeling. It works for the housing value model of Harrison and Rubinfeld (1978) as the modified model is more meaningful and fits better to actual data.

The world is producing enormous and complex amounts of data that often contain sophisticated patterns and relationships beyond the reach of current theories. Calibration techniques such as smoothing splines can help incorporate new data, new variables into explanatory models so that new hypotheses and causality can be revealed and tested. Furthermore, organizations and firms can capture and exploit new implications of big data for their own sake.

References

Aitchison, J., Dunsmore, I.R.: Statistical Prediction Analysis. Cambridge University Press, Cambridge (1975)

Berk, R.A.: Statistical Learning from a Regression Perspective. Springer, New York (2008)

Box, G.E.P.: Science and statistics. J. Am. Stat. Assoc. **71**, 791–799 (1976)

Canova, F.: Statistical inference in calibrated models. J. Appl. Econ. **9**(S), S123–S144 (1994)

Ehrenberg, A., Bound, J.: Predictability and prediction. J. Roy. Stat. Soc. Ser. A **156**(2), 167–206 (1993)

Friedman, J.H.: On bias, variance, 0/1-loss, and the curse-of-dimensionality. Data Min. Knowl. Discov. **1**, 55–77 (1997)

Geisser, S.: The predictive sample reuse method with applications. J. Am. Stat. Assoc. **70**, 320–328 (1975)

Greene, W.H.: Econometric Analysis, 7th edn. Pearson, London (2012)

Harrison, D., Rubinfeld, D.L.: Hedonic housing prices and the demand for clean air. J. Environ. Econ. Manage. **5**(1), 81–102 (1978)

ICAEW. Big Data and Analytics—What's New? (2014). https://www.icaew.com/-/media/corporate/archive/files/technical/information-technology/technology/what-is-new-about-big-data-v2.ashx

Kendall, M., Stuart, A.: The Advanced Theory of Statistics, 4th edn. Macmillan, New York (1977)

Kydland, F.E., Prescott, E.C.: Time to build and aggregate fluctuations. Econometrica **50**(6), 1345–1370 (1982)

Laney, D.: 3D data management: controlling data volume, velocity and variety, Gartner (2001). http://blogs.gartner.com/doug-laney/files/2012/01/ad949-3D-Data-Management-Controlling-Data-Volume-Velocity-and-Variety.pdf

Parzen, E.: Comment on statistical modeling: the two cultures. Stat. Sci. **16**(3), 224–226 (2001)

Poirier, D.J., Garber, S.G.: The determinants of aerospace profit rates 1951–1971. South. Econ. J. **41**(2), 228–238 (1974)

Shmueli, G.: To explain or to predict? Stat. Sci. **25**(3), 289–310 (2010)

StataCorp. Stata Release 12: Statistical Software, StataCorp LP (2011)

Stine, R.A.: Lecture Notes on Advanced Quantitative Modeling, The Wharton School at the University of Pennsylvania (2011)

Quantifying Predictive Uncertainty Using Belief Functions: Different Approaches and Practical Construction

Thierry Denœux[(⊠)]

Sorbonne Universités, Université de Technologie de Compiègne, CNRS,
UMR 7253 Heudiasyc, Compiègne, France
`thierry.denoeux@utc.fr`

Abstract. We consider the problem of quantifying prediction uncertainty using the formalism of belief functions. Three requirements for predictive belief functions are reviewed, each one of them inducing a distinct interpretation: compatibility with Bayesian inference, approximation of the true distribution, and frequency calibration. Construction procedures allowing us to build belief functions meeting each of these three requirements are described and illustrated using simple examples.

1 Introduction

Statistical prediction is the task of making statements about a not-yet-observed realization y of a random variable Y, based on past observations x. An important issue in statistical prediction is the quantification of uncertainty. Typically, prediction uncertainty has two components:

1. *Estimation uncertainty*, arising from the partial ignorance of the probability distribution of Y, and
2. *Random uncertainty*, due to the variability of Y.

If the distribution of Y is completely known, there is no estimation uncertainty. If Y is a constant, there is no random uncertainty: this is the case in parameter estimation problems. In all practical problems of interest, both sources of uncertainty coexist, and should be accounted for in the prediction method.

In this paper, we assume the past data X and the future data Y to be independent, and we consider sampling models $X \sim P_X(\cdot; \theta)$ and $Y \sim P_Y(\cdot; \theta)$, where θ is a parameter known only to belong to some set Θ. The sample spaces of X and Y will be denoted by $\mathscr{X}$ and $\mathscr{Y}$, respectively. To keep the exposition simple, we will assume Y to be a real random variable, with $\mathscr{Y} \subseteq \mathbb{R}$.

The statistical prediction problem is treated differently in the Bayesian and frequentist frameworks. Here, we briefly outline the main approaches within each of these two frameworks.

© Springer International Publishing AG 2018
V. Kreinovich et al. (eds.), *Predictive Econometrics and Big Data*, Studies in Computational Intelligence 753, https://doi.org/10.1007/978-3-319-70942-0_8

Bayesian Approach

In the Bayesian framework, X, Y and θ are considered as random variables. A Bayesian posterior predictive distribution $F_B(y|x)$ can then be computed from the conditional distribution $F(y|x; \theta) = F(y|\theta)$ by integrating out θ,

$$F_B(y|x) = \int F(y|\theta)p(\theta|x)d\theta, \tag{1}$$

where $p(\theta|x)$ is the posterior density of θ. The main limitation of this approach is the necessity to specify a prior distribution $p(\theta)$ on θ. In many cases, prior knowledge on θ is either nonexistent, or too vague to be reliably described by a single probability distribution.

Frequentist Approach

In the frequentist framework, the prediction problem can be addressed in several ways. The so-called *plug-in* approach is to replace θ in the model by a point estimate $\widehat{\theta}$ and to estimate the distribution of Y by $P_Y(\cdot; \widehat{\theta})$. This approach amounts to neglecting estimation uncertainty. Consequently, it will typically underestimate the prediction uncertainty, unless the sample size is very large. Another approach is to consider *prediction intervals* $[L_1(X), L_2(X)]$ such that the coverage probability

$$CP(\theta) = P_{X,Y}(L_1(X) \leq Y \leq L_2(X); \theta) \tag{2}$$

has some specified value, perhaps approximately. The coverage probability can take any value only if Y is continuous; consequently, we often make this assumption when using this approach. Confidence intervals do account for estimation and prediction uncertainty, but they do not provide any information about the relative plausibility of values inside or outside that set. To address the issue, we may consider one-sided confidence intervals $(-\infty, L_\alpha(X)]$ indexed by $\alpha \in (0, 1)$, such that

$$CP(\theta) = P_{X,Y}(Y \leq L_\alpha(X); \theta) \tag{3}$$

is equal to α, at least approximately. Then, we may treat α-prediction limits $L_\alpha(x)$ as the α-quantiles of some *predictive distribution function* $\widetilde{F}_{p(y|x)}$ [2, 16]. Such a predictive distribution is not a frequentist probability distribution; rather, it can be seen as a compact way of describing one or two-sided (perhaps, approximate) prediction intervals on Y at any level.

In all the approaches summarized above, uncertainty about Y is represented either as a set (in the case of prediction intervals), or as a probability distribution (such as a frequentist or Bayesian predictive distribution). In this paper, we consider approaches to the prediction problem where uncertainty about Y is represented by a *belief function*. In Dempster-Shafer theory, belief functions are expressions of *degrees of support* for statements about the unknown quantity under consideration, based on evidence. Any subset $A \subseteq \mathcal{Y}$ can be canonically represented by a belief function, and any probability measure is also a particular belief function: consequently, the Dempster-Shafer formalism is more general and flexible than the set-membership or probabilistic representations. The problem addressed in this paper is to exploit this flexibility to represent the prediction uncertainty on Y based on the evidence of observed data x.

The interpretation of a *predictive belief function* will typically depend on the requirements imposed on the construction procedure. There is, however, no general agreement as to which properties should be imposed. The purpose of this paper is to review some desired properties, and to describe practical construction procedures allowing us to build predictive belief functions that verify these properties. As we shall see, three main properties have been proposed in previous work, resulting in three main types of predictive belief functions.

The rest of this paper is organized as follows. Some general definitions and results related to belief functions are first recalled in Sect. 2. The requirements are then presented in Sect. 3, and construction procedures for the three types of predictive belief functions considered in this paper are described in Sect. 4. Section 5 contains conclusions.

2   Background on Belief Functions

In this section, we provide a brief reminder of the main concepts and results from the theory of belief functions that will be used in this paper. The definitions of belief and plausibility functions will first be recalled in Sect. 2.1. The connection with random sets will be explained in Sect. 2.2, and Dempster's rule will be introduced in Sect. 2.4.

2.1   Belief and Plausibility Functions

Let Ω be a set, and $\mathscr{B}$ an algebra of subsets of Ω. A belief function on $(\Omega, \mathscr{B})$ is a mapping $Bel : \mathscr{B} \to [0, 1]$ such that $Bel(\emptyset) = 0$, $Bel(\Omega) = 1$, and for any $k \geq 2$ and any collection $B_1, \ldots, B_k$ of elements of $\mathscr{B}$,

$$Bel \left(\bigcup_{i=1}^{k} B_i \right) \geq \sum_{\emptyset \neq I \subseteq \{1,\ldots,k\}} (-1)^{|I|+1} Bel \left(\bigcap_{i \in I} B_i \right). \tag{4}$$

Given a belief function Bel, the dual *plausibility function* $Pl : \mathscr{B} \to [0, 1]$ is defined by $Pl(B) = 1 - Bel(\overline{B})$, for any $B \in \mathscr{B}$. In the Dempster-Shafer theory of belief functions [22], $Bel(B)$ is interpreted as the degree of support in the proposition $Y \in B$ based on some evidence, while $Pl(B)$ is a degree of consistency between that proposition and the evidence.

If the inequalities in (4) are replaced by equalities, then Bel is a finitely additive probability measure, and $Pl = Bel$. If the evidence tells us that $Y \in A$ for some $A \in \mathscr{B}$, and nothing more, then it can be represented by a function Bel_A that gives full degree of support to any $B \in \mathscr{B}$ such that $B \subseteq A$, and zero degree of support to any other subset. It can easily be verified that Bel_A is a belief function. If $A = \Omega$, the belief function is said to be *vacuous*: it represent complete ignorance on Y.

Given two belief functions Bel_1 and Bel_2, we say that Bel_1 is *less committed* than Bel_2 if $Bel_1 \leq Bel_2$; equivalently, $Pl_1 \geq Pl_2$. The meaning of this notion is that Bel_1 represents a weaker state of knowledge than that represented by Bel_2.

2.2   Connection with Random Sets

A belief function is typically induced by a *source*, defined as a four-tuple $(\mathscr{S}, \mathscr{A}, P, \Gamma)$, where $\mathscr{S}$ is a set, $\mathscr{A}$ an algebra of subsets of $\mathscr{S}$, P a finitely additive probability measure on $(\mathscr{S}, \mathscr{A})$, and Γ a mapping from $\mathscr{S}$ to 2^Ω. The mapping Γ is strongly measurable with respect to $\mathscr{A}$ and $\mathscr{B}$ if, for any $B \in \mathscr{B}$, we have

$$\{s \in \mathscr{S} | \Gamma(s) \neq \emptyset, \Gamma(s) \subseteq A\} \in \mathscr{A}.$$

We can then show [19], that the function Bel defined by

$$Bel(B) = \frac{P(\{s \in \mathscr{S} | \Gamma(s) \neq \emptyset, \Gamma(s) \subseteq B\})}{P(\{s \in \mathscr{S} | \Gamma(s) \neq \emptyset\})}, \tag{5}$$

for all $A \subseteq \mathscr{B}$ is a belief function. The dual plausibility function is

$$Pl(B) = \frac{P(\{s \in \mathscr{S} | \Gamma(s) \cap B \neq \emptyset\})}{P(\{s \in \mathscr{S} | \Gamma(s) \neq \emptyset\})}. \tag{6}$$

The mapping Γ is called a *random set*. We should not, however, get abused by the term "random": most of the time, the probability measure P defined on $(\mathscr{S}, \mathscr{A})$ is subjective, and there is no notion of randomness involved.

2.3   Consonant Random Closed Sets

Let us assume that $\Omega = \mathbb{R}^d$ and $\mathscr{B} = 2^\Omega$. Let π be an upper semi-continuous map from $\mathbb{R}^d$ to $[0, 1]$, i.e., for any $s \in [0, 1]$, the set $^s\pi \stackrel{\text{def}}{=} \{x \in \mathbb{R}^d | \pi(x) \geq s\}$ is closed. Furthermore, assume that $\pi(x) = 1$ for some x. Let $S = [0, 1]$, $\mathscr{A}$ be the Borel σ-field on $[0, 1]$, μ the uniform measure, and Γ the mapping defined by $\Gamma(s) = {}^s\pi$. Then Γ is a random closed set [20]. We can observe that its focal sets are nested: it is said to be *consonant*. The plausibility function is then a possibility measure [25], and π is the corresponding possibility distribution. Function Pl can be computed as $Pl(B) = \sup_{x \in B} \pi(x)$, for any $B \subseteq \mathbb{R}^d$. In particular , $Pl\{x\} = \pi(x)$ for all $x \in \Omega$.

2.4   Dempster's Rule

Assume that we have two sources $(\mathscr{S}_i, \mathscr{A}_i, P_i, \Gamma_i)$ for $i = 1, 2$, where each Γ_i is a multi-valued mapping from $\mathscr{S}_i$ to 2^Ω, and each source induces a belief function Bel_i on $\mathscr{Y}$. Then, the orthogonal sum of Bel_1 and Bel_2, denoted as $Bel_1 \oplus Bel_2$ is induced by the source $(\mathscr{S}_1 \times \mathscr{S}_2, \mathscr{A}_1 \otimes \mathscr{A}_2, P_1 \otimes P_2, \Gamma_\cap)$, where $\mathscr{A}_1 \otimes \mathscr{A}_2$ is the tensor product algebra on the product space $\mathscr{S}_1 \times \mathscr{S}_2$, $P_1 \otimes P_2$ is the product measure, and $\Gamma_\cap(s_1, s_2) = \Gamma_1(s_1) \cap \Gamma_2(s_2)$. This operation is called Dempster's rule of combination [7]. It is the fundamental operation to combine belief functions induced by independent pieces of evidence in Dempster-Shafer theory.

3   Predictive Belief Functions

In this paper, we are concerned with the construction of predictive belief functions (PBF), i.e., belief functions that quantify the uncertain on future data Y, given the evidence of past data x. This problem can be illustrated by the following examples, which will be used throughout this paper.

Example 1. *We have observed the times between successive failures of an air-conditioning (AC) system, as shown in Table 1 [21]. We assume the time ξ between failures to have an exponential distribution $\mathscr{E}(\theta)$, with cdf*

$$F(\xi; \theta) = \left[1 - \exp(-\theta x)\right] I(\xi \geq 0),$$

where θ is the rate parameter. Here, the past data $x = (\xi_1, \ldots, \xi_n)$ is a realization of an iid sample $X = (\Xi_1, \ldots, \Xi_n)$, with $\Xi_i \sim \mathscr{E}(\theta)$, and Y is a random variable independent from X, also distributed as $\mathscr{E}(\theta)$. Based on these data and this model, what can we say about the time to the next failure of the system? □

Example 2. *The data shown in Fig. 1(a) are annual maximum sea-levels recorded at Port Pirie, a location just north of Adelaide, South Australia, over the period 1923–1987 [5]. The probability plot in Fig. 1(b) shows a good fit with the Gumbel distribution, with cdf*

$$F_X(\xi; \theta) = \exp\left(-\exp\left(-\frac{\xi - \mu}{\sigma}\right)\right), \tag{7}$$

where μ is the mode of the distribution, σ a scale parameter, and $\theta = (\mu, \sigma)$. Suppose that, based on these data, we want to predict the maximum sea level Y in the next $m = 10$ years. Assuming that the distribution of sea level will remain unchanged in the near future (i.e., neglecting, for instance, the effect of sea level rise due to climate change), the cdf of Y is

$$F_Y(y; \theta) = F_X(y; \theta)^m = \exp\left(-m\exp\left(-\frac{y - \mu}{\sigma}\right)\right). \tag{8}$$

The parameter θ is unknown, but the observed data provides information about it. How to represent this information, so as to quantify the uncertainty on Y? What can be, for instance, a sound definition for the degree of belief in the proposition $Y \geq 5$? □

Table 1. Times between successive failures of an air-conditioning system, from [21].

23	261	87	7	120	14	62	47	225	71
246	21	42	20	5	12	120	11	3	14
71	11	14	11	16	90	1	16	52	95

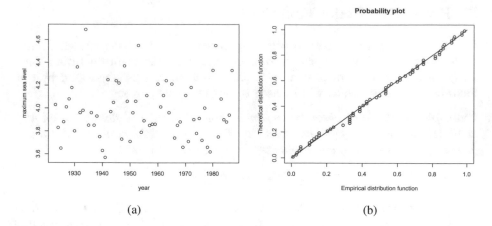

Fig. 1. Annual maximum sea-levels recorded at Port Pirie over the period 1923–1987 (a), and probability plot for the Gumbel fit to the data (c).

In general, the evidence on Y may consist in (1) the observed data x and (2) prior knowledge on θ, which can be assumed to be represented by a belief function Bel_θ^0. A predictive belief function on Y can thus be denoted as $Bel_Y(\cdot; x, Bel_\theta^0)$. If Bel_θ^0 is vacuous, we simple write $Bel_Y(\cdot; x)$. The following three requirements have been proposed for Bel_Y.

R0: Likelihood principle

As we assume X and Y to be independent, the observation of X provides information on Y only through the parameter θ. The likelihood principle [4,13] states that all relevant information about θ, after observing $X = x$, is contained in the likelihood function $L(\theta; x) = p(x; \theta)$. Formally, this principle means that two observations X and X' generated by two different random experiments, with probability distributions $p(x; \theta)$ and $p(x'; \theta)$, provide the same information about θ as long as $p(x; \theta)$ and $p(x'; \theta)$ are proportional, i.e., there is some constant $c = c(x, x')$ not depending on θ, such that $p(x; \theta) = c \cdot p(x'; \theta)$ for all $\theta \in \Theta$. Consequently, we should also have

$$\big(\forall \theta \in \Theta, p(x; \theta) = c \cdot p(x'; \theta)\big) \Rightarrow Bel_Y(\cdot; x) = Bel_Y(\cdot; x'). \tag{9}$$

The likelihood principle was shown by Birnbaum in [4] to follow from two principles generally accepted by most (but not all) statisticians: the conditionality principle (see also [3, p. 25]) and the sufficiency principle.

R1: Compatibility with Bayes

For some statisticians, Bayesian reasoning is a perfectly valid approach to statistical inference provided a prior probability distribution is available, but is questionable in the absence of such prior information. Many authors have attempted to generalize Bayesian inference to some "prior-free" method of inference. This was, in particular, Dempster's motivation in his early papers on belief functions [6,8]. If we adopt this point of view, then a predictive belief function should coincide with the Bayesian posterior predictive distribution if a probabilistic prior is available. Formally, if $Bel_\theta^0 = P_\theta^0$ is a probability measure, then the following equality should hold,

$$Bel_Y(A; x, P_\theta^0) = P_B(A|x) \tag{10}$$

for all measurable event $A \subseteq \mathscr{Y}$, where $P_B(\cdot|x)$ is the Bayesian posterior predictive probability measure corresponding to (1). This requirement ensures that the Bayesian and belief function approaches yield the same predictions when they are provided with exactly the same information.

A PBF verifying requirements (9) and (10) will be called a Type-I PBF. It can be seen as a representation of the evidence about Y from the observation of X, and possibly additional information on θ; it becomes the Bayesian predictive posterior distribution when combined with a probabilistic prior.

R2: Approximation of the true future data distribution

We may also consider that, if we knew the true value of parameter θ, then we would equate the predictive belief function with the true distribution $P_Y(\cdot; \theta)$ of Y. If we do not know θ, but we have only observed a sample x of X, then the predictive belief function should most of the time (i.e., for most of the observed samples) be less committed than $P_Y(\cdot; \theta)$ [1,10]. Formally, we may thus fix some $\alpha \in (0, 1)$, and require that, for any $\theta \in \Theta$,

$$P_X \left(Bel_Y(\cdot; X) \le P_Y(\cdot; \theta); \theta \right) \ge 1 - \alpha. \tag{11}$$

If $X = (\Xi_1, \ldots, \Xi_n)$ is a sequence of observations, a weaker requirement is to demand that (11) holds in the limit, as $n \to \infty$. A PBF verifying (11), at least asymptotyically, will be called a type-II PBF. For most of the samples, a type-II PBF is a lower approximation of the true probability distribution of Y. It can thus be compared to the plug-in distribution $P_Y(\cdot; \hat{\theta})$, which is also an approximation of $P_Y(\cdot; \theta)$. However, the PBF will generally be non-additive, as a consequence of accounting not only for random uncertainty, but also for estimation uncertainty.

R3: Calibration

Another line of reasoning, advocated by Martin and Liu [18], is to consider that plausibility values be *calibrated*, in the sense that the plausibility of the true value Y should be small with only a small probability [18, Chap. 9]. More precisely, for any $\theta \in \Theta$ and any $\alpha \in (0, 1)$, we my impose the following condition,

$$P_{X,Y}(pl_Y(Y; X) \le \alpha; \theta) \le \alpha, \tag{12}$$

or, equivalently,

$$P_{X,Y}(pl_Y(Y; X) > \alpha; \theta) \ge 1 - \alpha, \tag{13}$$

where $pl_Y(Y; X) = Pl_Y(\{Y\}; X)$ is the *contour function* evaluated at Y. Equations (12) and (13) may hold only asymptotically, as the sample size tends to infinity. It follows from (13) that the sets $\{y \in \mathscr{Y} | pl_Y(y; X) > \alpha\}$ are prediction sets at level $1 - \alpha$ (maybe, approximately). A PBF verifying (13) will be called a type-III PBF. It can be seen as encoding prediction sets at all levels; as such, it is somewhat similar to a frequentist predictive distribution; however, it is not required to be additive. Requirement (13) is very different from the previous two. In particular, a type-III PBF has no connection with the Bayesian predictive distribution, and it does not approximate the true distribution of Y. Rather, (12) establishes a correspondence between plausibilities and frequencies. A type III-PBF can be seen as a generalized prediction interval.

In the following section, we introduce a simple scheme that will allow us to construct PBF of each of the three kinds above, for any parametric model. We will also mention some alternative methods.

4   Construction of Predictive Belief Functions

In [14,15], the authors introduced a general method to construct PBFs, by writing the future data Y in the form

$$Y = \varphi(\theta, V), \tag{14}$$

where V is a pivotal variable with known distribution [6,15,18]. Equation (14) is called a φ-equation. It can be obtained by inverting the cdf of Y. More precisely, let us first assume that Y is continuous; we can then observe that $V = F_Y(Y; \theta)$ has a standard uniform distribution. Denoting by $F_Y^{-1}(\cdot; \theta)$ the inverse of the cdf $F_Y(\cdot; \theta)$, we get

$$Y = F_Y^{-1}(V; \theta), \tag{15}$$

with $V \sim \mathscr{U}([0, 1])$, which has the same form as (14). When Y is discrete, (15) is still valid if F_Y^{-1} now denotes the generalized inverse of F_Y,

$$F_Y^{-1}(V; \theta) = \inf\{y | F_Y(y; \theta) \geq V\}. \tag{16}$$

Example 3. *In the Air Conditioning example, it is assumed that $Y \sim \mathscr{E}(\theta)$, i.e., $F_Y(y; \theta) = 1 - \exp(-\theta y)$. From the equality $F_Y(Y; \theta) = V$, we get*

$$Y = -\frac{\log(1 - V)}{\theta}, \tag{17}$$

with $V \sim \mathscr{U}([0, 1])$.   □

Example 4. *Let Y be the maximum sea level in the next m years, with cdf given by (8). From the equality $F_Y(Y; \theta) = V$, we get $Y = \mu - \sigma \log \log(V^{-1/m})$, with $V \sim \mathscr{U}([0, 1])$.*   □

The plug-in prediction is obtained by plugging the MLE $\widehat{\theta}$ in (14),

$$\widehat{Y} = \varphi(\widehat{\theta}, V). \tag{18}$$

Now, the Bayesian posterior predictive distribution can be obtained by replacing the constant θ in (14) by a random variable θ_B with the posterior cdf $F_\theta(\cdot; x)$. We then get a random variable Y_B with cdf $F_B(y|x)$ given by (1). We can write

$$Y_B = \varphi(F_\theta^{-1}(U|x), V). \tag{19}$$

The three methods described in the sequel somehow generalize the above methods. They are based on (14), and on belief functions Bel_θ and Bel_V on θ and V induced, respectively, by random sets $\Gamma(U; x)$ and $\Lambda(W)$, where U and W are random variables. The predictive belief function on Y is then induced by the random set

$$\Pi(U, W; x) = \varphi(\Gamma(U; x), \Lambda(W)). \tag{20}$$

Assuming that $\Pi(u, w; x) \neq \emptyset$ for any u, v and x, we thus have

$$Bel_Y(A; x) = P_{U,W}\left\{\Pi(U, W; x) \subseteq A\right\}$$

and

$$Pl_Y(A; x) = P_{U,W}\left\{\Pi(U, W; x) \cap A \neq \emptyset\right\}$$

for all subset $A \subseteq \mathcal{Y}$ for which these expressions are well-defined.

The three methods described below differ in the choice of the random sets $\Gamma(U; x)$ and $\Lambda(W)$. As will we see, each of the three types of PBF described in Sect. 3 can be obtained by suitably choosing these two random sets.

4.1   Type-I Predictive Belief Functions

As shown in [11], Requirements R0 and R1 jointly imply that the contour function $pl(\theta, x)$ associated to $Bel_\theta(\cdot; x)$ should be proportional to the likelihood function $L(\cdot; x)$. The least committed belief function (in some sense, see [11]) that meets this constraint is the consonant belief function defined by the following contour function,

$$pl(\theta; x) = \frac{L(\theta; x)}{L(\widehat{\theta}; x)}, \tag{21}$$

where $\widehat{\theta}$ is a maximizer of $L(\theta; x)$, i.e., a maximum likelihood estimate (MLE) of θ, and it is assumed that $L(\widehat{\theta}; x) < +\infty$. As it is consonant, the plausibility of any hypothesis $H \subseteq \Theta$ is the supremum of the plausibilities of each individual values of θ inside H,

$$Pl_\theta(H; x) = \sup_{\theta \in H} pl(\theta; x). \tag{22}$$

The corresponding random set is defined by

$$\Gamma_\ell(U) = \{\theta \in \Theta | pl(\theta; x) \geq U\} \tag{23}$$

with $U \sim \mathscr{U}([0,1])$, i.e., it is the set of values of θ whose relative likelihood is larger than a uniformly distributed random variable U. This *likelihood-based belief function* was first introduced by Shafer [22], and it has been studied by Wasserman [23], among others.

The prediction method proposed in [14,15] consists in choosing Bel_θ defined by (21)–(22) as the belief function on θ, and P_V, the uniform probability distribution of V, as the belief function on V. The resulting PBF $Bel_{Y,\ell}(\cdot;x)$ is induced by the random set

$$\Pi_\ell(U,V;x) = \varphi(\Gamma_\ell(U;x),V), \qquad (24)$$

where (U,V) has a uniform distribution in $[0,1]^2$.

By construction, combining $Bel_\theta(\cdot;x)$ with a Bayesian prior P_θ^0 by Dempster's rule yields the Bayesian posterior $P_B(\cdot|x)$. The random set (24) then becomes

$$\Pi_B(U,V;x) = \varphi(F_B^{-1}(U|x),V), \qquad (25)$$

with (U,V) uniformly distribution in $[0,1]^2$. This random set is actually a random point, i.e., a random variable, and this rv is identical to (19): its distribution is the Bayesian posterior predictive distribution. Consequently, the PBF $Bel_{Y,\ell}$ constructed by this method meets requirements R0 and R1.

Example 5. *The contour function for the AC data of Example 1, assuming an exponential distribution, is shown in Fig. 2(a). As it is unimodal and continuous, the sets $\Gamma_\ell(u;x)$ are closed intervals $[\theta^-(u), \theta^+(u)]$, whose bounds can be approximated numerically as the roots of the equation $pl(\theta;x) = u$. From (17), the random set $\Pi_\ell(U,V;x)$ is then the random closed interval*

$$\Pi_\ell(U,V;x) = [Y^-(U,V;x), Y^+(U,V;x)],$$

with

$$Y^-(U,V;x) = -\frac{\log(1-V)}{\theta^+(U)}$$

and

$$Y^+(U,V;x) = -\frac{\log(1-V)}{\theta^-(U)}.$$

As shown by Dempster [9], the following equalities hold, for any $y \geq 0$,

$$Bel_Y((-\infty,y]) = P_{U,V}(Y^+(U,V;x) \leq y)$$
$$Pl_Y((-\infty,y]) = P_{U,V}(Y^-(U,V;x) \leq y),$$

i.e., they are the cdfs of, respectively, the upper and lower bounds of Π_ℓ. Functions $Bel_Y((-\infty,y])$ and $Pl_Y((-\infty,y])$ are called the lower and upper cdfs of the random set $\Pi_\ell(U,V;x)$. As explained in [15], they can be approximated by Monte Carlo simulation: let (u_i,v_i), $i = 1,\ldots,N$ be a pseudo-random sequence generated independently from the uniform distribution in $[0,1]^2$. Let $y_i^- = y^-(u_i,v_i;x)$ and $y_i^+ = y^+(u_i,v_i;x)$ be the corresponding realizations of the bounds of Π_ℓ.

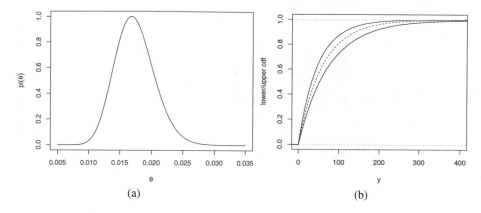

Fig. 2. AC data. (a): Contour function; (b): Lower and upper cdf (solid lines) and plug-in cdf (dotted line)

Then, the lower and upper cdfs can be approximated by the empirical cdfs of the y_i^+ and the y_i^-, respectively. These functions are plotted in Fig. 2(b), together with the plug-in cdf $F_Y(y; \widehat{\theta})$, with $\widehat{\theta} = 1/\bar{x}$. We can observe that the plug-in cdf is always included in the band defined by the lower and upper cdf, which is a consequence of the inequalities $\theta^-(u) \leq \widehat{\theta} \leq \theta^-(u)$ for any $u \in (0,1]$. We note that $\widehat{\theta} = \theta^-(1) = \theta^+(1)$. □

Example 6. *Let us now consider the Sea Level data of Example 2. The contour function (21) for these data is plotted in Fig. 3(a). As the level sets $\Gamma_\ell(u; x)$ of this function are closed and connected, the sets $\Pi_\ell(U, V; x)$ still are closed intervals in this case [15]. To find the bounds $Y^-(u, v; x)$ and $Y^+(u, v; x)$ for any pair (u, v), we now need to search for the minimum an the maximum of $\varphi(\theta, v)$, under the constraint $pl(\theta; x) \geq u$. This task can be performed by a nonlinear constrained optimization algorithm. The lower and upper cdfs computed using this method are shown in Fig. 3(b).* □

4.2 Type-II Predictive Belief Functions

The φ-equation (14) also allows us to construct a type-II PBF, such as defined in [10]. Let $C(X)$ be a confidence set for θ at level $1 - \alpha$, i.e.,

$$P_X(C(X) \ni \theta; \theta) = 1 - \alpha. \tag{26}$$

Consider the following random set,

$$\Pi_{Y,c}(V; x) = \varphi(C(x), V), \tag{27}$$

which is a special case of the general expression (20), with $\Gamma(U; x) = C(x)$ for all $U \in [0, 1]$, $W = V$ and $\Lambda(V) = V$. The following theorem states that the belief function induced by the random set (27) is a type-II PBF.

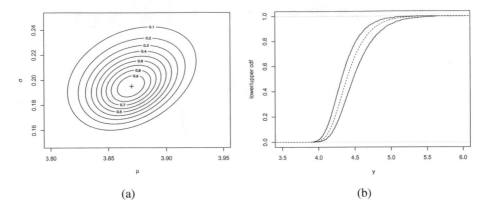

Fig. 3. Port Pirie sea-level data: (a): Contour plot of the relative likelihood function; (b): Lower and upper cdfs of the type-I PBF; the central broken line corresponds to the plug-in prediction.

Theorem 1. *Let* $Y = \varphi(\theta, V)$ *be a random variable, and* $C(X)$ *a confidence region for* θ *at level* $1 - \alpha$. *Then, the belief function* $Bel_{U,c}(\cdot; x)$ *induced by the random set* $\Pi_{Y,c}(V; x) = \varphi(C(x), V)$ *verifies*

$$P_X(Bel_{Y,c}(\cdot; X) \leq P_Y(\cdot; \theta); \theta) \geq 1 - \alpha, \qquad (28)$$

i.e., it is a type-II PBF.

Proof. If $\theta \in C(x)$, then $\varphi(\theta, V) \in \varphi(C(x), V)$ for any V. Consequently, the following implication holds for any measurable subset $A \subseteq \mathscr{Y}$, and any $x \in \mathscr{X}$,

$$\varphi(C(x), V) \subseteq A \Rightarrow \varphi(\theta, V) \in A.$$

Hence,

$$P_V(\varphi(C(x), V) \subseteq A) \leq P_V(\varphi(\theta, V) \subseteq A),$$

or, equivalently,

$$Bel_{Y,c}(A; x) \leq P_Y(A; \theta). \qquad (29)$$

As (29) holds whenever $\theta \in C(x)$, and $P_X(C(X) \ni \theta; \theta) = 1 - \alpha$, it follows that (29) holds for any measurable event A with probability at least $1 - \alpha$, i.e.,

$$P_X\left(Bel_{Y,c}(\cdot; X) \leq P_Y(\cdot; \theta); \theta\right) \geq 1 - \alpha.$$

$\square$

If $C(X)$ is an approximate confidence region, then obviously (28) will hold only approximately. In the case where $X = (X_1, \ldots, X_n)$ is iid, the likelihood function will often provide us with a means to obtain a confidence region on θ. From Wilks' theorem [24], we know that, under regularity conditions, $-2 \log pl(\theta; X)$ has approximately, for large n, a chi square distribution with p degrees of freedom, where p is the dimension of θ. Consequently, the sets

$$\Gamma_\ell(c; X) = \{\theta \in \Theta | pl(\theta; X) \geq c,$$

Table 2. Likelihood levels c defining approximate 95% confidence regions.

p	1	2	5	10	15
c	0.15	0.5	3.9e-03	1.1e-04	3.7e-06

with $c = \exp(-0.5\chi^2_{p;1-\alpha})$, are approximate confidence regions at level $1 - \alpha$. The corresponding predictive random set is

$$\Pi_{Y,c}(V;x) = \varphi(\Gamma_\ell(c;x), V). \tag{30}$$

We can see that this expression is similar to (24), except that, in (30), the relative likelihood function is cut at a fixed level c. A similar idea was explored in Ref. [26]. Table 2 gives values of c for different values of p and $\alpha = 0.05$. We can see that c decreases quickly with p, which means that the likelihood-based confidence regions and, consequently, the corresponding PBFs will become increasing imprecise as p increases. In particular, the likelihood-based type-II PBFs will typically be less committed than the type-I PBFs.

Example 7. *For the AC data, the likelihood-based confidence level at level $1 - \alpha = 0.95$ is*

$$[\theta^-(c), \theta^+(c)] = [0.01147, 0.02352],$$

with $c = 0.15$. It is very close to the exact confidence level at the same level,

$$\left[\frac{\widehat{\theta}\chi^2_{\alpha/2,2n}}{2n}, \frac{\widehat{\theta}\chi^2_{1-\alpha/2,2n}}{2n}\right] = [0.01132, 0.02329].$$

The corresponding Type-II PBF is induced by the random interval

$$\Pi_{Y,c}(V;x) = \left[-\frac{\log(1-V)}{\theta^+(c)}, -\frac{\log(1-V)}{\theta^-(c)}\right].$$

The lower and upper bounds of this interval have exponential distributions with rates $\theta^+(c)$ and $\theta^-(c)$, respectively. Figure 4 shows the corresponding lower and upper cdfs, together with those of the Type-I PBF computed in Example 5. We can see that the Type-II PBF at the 95% confidence level is less committed than the Type-I PBF.

Example 8. *Figure 5 shows the lower and upper cdfs of the type-II PBF constructed from the likelihood-based confidence region with $\alpha = 0.05$. The estimate of the true coverage probability, obtained using the parametric bootstrap method with $B = 5000$ bootstrap samples, was 0.94998, which is remarkably close to the nominal level. The simulation method to compute these functions is similar to that explained in Example 6, except that we now have $u_i = c = 0.05$ for $i = 1, \ldots, n$. The lower and upper cdfs form a confidence band on the true cdf of Y. Again, we observe that this band is larger than the one corresponding to the type-I PBF.*

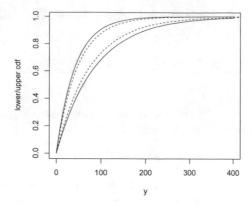

Fig. 4. Lower and upper cdfs of the type-II PBF for the AC data (solid lines). The type-I lower and upper cdf are shown as broken lines.

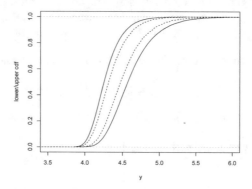

Fig. 5. Lower and upper cdfs of the type-II PBF for the sea-level example (solid lines). The type-I lower and upper cdf are shown as broken lines.

4.3   Type-III Predictive Belief Functions

The calibration condition (12) was introduced by Martin and Liu [17,18], in the context of their theory of Inferential Models (IMs). An equivalent formulation is to require that the random variable $pl_Y(Y; X)$ be stochastically not less than a random variable having a standard uniform distribution. In [18, Chap. 9], Martin and Liu propose a quite complex method for constructing PBFs verifying this requirements, based on IMs. It turns out that such Type-III PBFs (as we call them in this paper) can be generated by a simple construction procedure based on the φ-equation (14) and suitable belief functions on θ and V. Because the notion of Type-III PBFs is intimately related to prediction sets, and prediction sets at a given level can only be defined for continuous random variables, we will assume Y to be continuous in this section.

Let us first assume that θ is known. In that case, predicting $Y = \varphi(\theta, V)$ boils down to predicting $V = F(Y; \theta)$. Consider the random interval

$$\Lambda(W) = \left[\frac{W}{2}, 1 - \frac{W}{2} \right],$$

with $W \sim \mathscr{U}([0, 1])$. It is easy to check that the induced contour function is $pl(v) = 1 - |2v - 1|$ (it is a triangular possibility distribution with support $[0, 1]$ and mode 0.5), and $pl(V) \sim \mathscr{U}([0, 1])$. Consider the predictive random set

$$\Pi_Y(W) = \varphi(\theta, \Lambda(W)) \tag{31}$$

and the associated contour function

$$pl(y) = 1 - |1 - 2F(y; \theta)|. \tag{32}$$

It is clear that $pl(Y) = pl(V) \sim \mathscr{U}([0, 1])$, and the consonant belief function with contour function (32) verifies the calibration property (12). We can observe that the transformation (32) from the probability distribution of Y to this possibility distribution is an instance of the family of probability-possibility transformations studied in [12]. The mode of the possibility distribution is the median $y_{0.5} = \varphi(\theta, 0.5)$, and each α-cut $\Pi_Y(\alpha) = [y_{\alpha/2}, y_{1-\alpha/2}]$ with $\alpha \in (0, 1)$ is a prediction interval for Y, at level $1 - \alpha$.

Until now, we have assume θ to be known. When θ is unknown, we could think of replacing it by its MLE $\widehat{\theta}$, and proceed as above by applying the same probability-possibility distribution to the plug-in predictive distribution $F_Y(u; \widehat{\theta})$. As already mentioned, this approach would amount to neglecting the estimation uncertainty, and the α-cuts of the resulting possibility distribution could have a coverage probability significantly smaller than $1 - \alpha$. A better approach, following [16], is to consider the exact or approximate pivotal quantity $\widetilde{V} = F(Y; \widehat{\theta}(X))$. We assume that $\widehat{\theta}$ is a consistent estimator of θ as the information about θ increases, and $\widetilde{V}$ is asymptotically distributed as $\mathscr{U}([0, 1])$ [16]. However, for finite sample size, the distribution of $\widetilde{V}$ will generally not be uniform. Let G be the cdf of $\widetilde{V}$, assuming that it is pivotal, and let $\widetilde{\Lambda}(W)$ be the random interval

$$\widetilde{\Lambda}(W) = \left[G^{-1}(W/2), G^{-1}(1 - W/2) \right]$$

with $W \sim \mathscr{U}([0, 1])$ and corresponding contour function

$$pl(\widetilde{v}) = 1 - |1 - 2G(\widetilde{v})|.$$

The random set

$$\widetilde{\Pi}_Y(W; x) = \varphi(\widehat{\theta}(x), \widetilde{\Lambda}(W))$$

induces the contour function

$$pl(y; x) = 1 - \left| 1 - 2G\{F[y; \widehat{\theta}(x)]\} \right|. \tag{33}$$

As $G(F(Y; \widehat{\theta}(X))) \sim \mathscr{U}([0, 1])$, we have $pl(Y; X) \sim \mathscr{U}([0, 1])$, and the focal sets $\widetilde{\Pi}_Y(\alpha; X)$ are exact prediction intervals at level $1 - \alpha$. Consequently, the consonant belief function with contour function (33) is a type-III PBF. We can remark that it is obtained by applying the probability-possibility transformation (32) to the predictive confidence distribution $\widetilde{F}(y; x) = G\{F[y; \widehat{\theta}(x)]\}$.

When an analytical expression of the cdf G is not available, or $\widetilde{V}$ is only asymptotically pivotal, an approximate distribution $\widetilde{G}$ can be determined by a parametric bootstrap approach [16]. Specifically, let $x_1^*, \ldots, x_B^*$ be B and $y_1^*, \ldots, y_B^*$ be B bootstrap replicates of x and y, respectively. We can compute the corresponding values $\widetilde{v}_b^* = F(y_i^*; \widehat{\theta}(x_b^*))$, $b = 1, \ldots, B$, and the distribution of $\widetilde{V}$ can be approximated by the empirical cdf

$$\widetilde{G}(v) = \frac{1}{B} \sum_{b=1}^{B} I(\widetilde{v}_b^* \le v).$$

Example 9. *Consider again the AC example. For the exponential distribution, it has been shown [16] that the quantity*

$$\widetilde{V} = F(Y, \widehat{\theta}(X)) = 1 - \exp(-Y\widehat{\theta}(X))$$

is pivotal, and has the following cdf,

$$G(\widetilde{v}) = 1 - \left\{ 1 - \frac{1}{n} \log(1 - \widetilde{v}) \right\}^{-n}.$$

The predictive cdf is then

$$\widetilde{F}(y; x) = G\{F(y, \widehat{\theta}(x))\} = 1 - \left(1 + \frac{y\widehat{\theta}(x)}{n} \right)^{-n}$$

and the contour function of the type-III PBF is

$$pl(y; x) = 1 - \left| 2 \left(1 + \frac{y\widehat{\theta}(x)}{n} \right)^{-n} - 1 \right|. \tag{34}$$

Figure 6(a) shows the contour function (34) for the AC data (solid line), together with the contour function induced by the plug-in distribution (interrupted line). The two curves are quite close in this case: for $n = 30$, the distribution of $\widetilde{V}$ is already very close to the standard uniform distribution. Figure 6(b) shows the lower and upper cdfs of the PBF, together with the Type-I and Type-II $(1 - \alpha = 0.95)$ lower

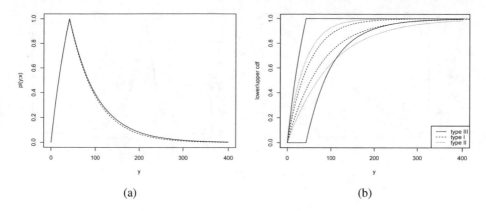

(a)   (b)

Fig. 6. AC example. (a): Contour function of the type-III PBF (solid line), and contour function induced by the plug-in distribution (interrupted line); (b): Lower and upper cdfs of the type-III PBF (solid lines). The type-I and type-II lower and upper cdf are shown, respectively, as interrupted and dotted lines.

and upper cdfs for the same data. As the Type-III PBF is consonant, the lower and upper cdfs can computed from the contour function as

$$Pl_Y((-\infty, y]) = \sup_{y' \le y} pl(y') = \begin{cases} pl(y; x) & \text{if } y \le \widetilde{F}^{-1}(0.5; x) \\ 1 & \text{otherwise,} \end{cases}$$

and

$$Bel_Y((-\infty, y]) = 1 - \sup_{y' > y} pl(y') = \left[1 - pl(y; x)\right] I\left(y > \widetilde{F}^{-1}(0.5; x)\right).$$

□

Example 10. *Let us now consider again the sea-level data. Here, the exact distribution of the quantity $\widetilde{V} = F_Y(Y; \hat{\theta}(X))$ is intractable, but it can be estimated by the parametric bootstrap technique. Figure 7(a) shows the bootstrap estimate of the distribution of $\widetilde{V}$, with $B = 10000$. There is clearly a small, but discernible departure from the uniform distribution. Figure 7(b) shows the contour function of the type-III PBF, together with that induced by the plug-in predictive distribution (corresponding to the approximation $G(\widetilde{v}) = \widetilde{v}$). Again, the two curves are close, but clearly discernible. With $n = 65$, the prediction intervals computed from the plug-in distribution have true coverage probabilities quite close to the stated ones. Finally, the lower and upper cdf of the type-III PBF for the Port-Pirie data are shown in Fig. 7(c), together with the corresponding functions for the type-I and type-II PBFs. Comparing Figs. 6(b) and 7(c), we can see that, in both cases, the type-I PBF is less committed than the type-III PBF. It is not clear, however, whether this result holds in general.* □

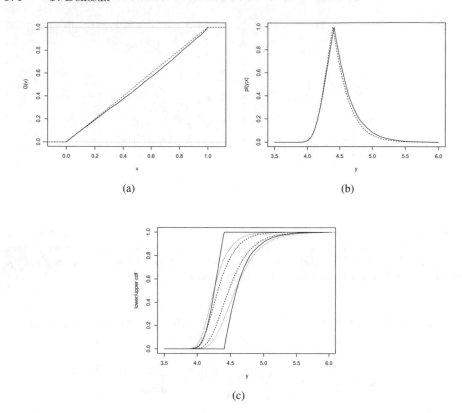

Fig. 7. Sea-level example. (a): Bootstrap estimate of the cdf G of $\widetilde{V} = F(Y, \widehat{\theta}(X))$; (b): Contour function of the type-III PBF (solid line), and contour function induced by the plug-in distribution (interrupted line); (c): Lower and upper cdfs of the type-III PBF (solid lines). The type-I and type-II lower and upper cdf are shown, respectively, as interrupted and dotted lines.

5   Conclusions

Being related to random sets, belief functions have greater expressivity than probability measures. In particular, the additional degrees of freedom of the belief function framework make it possible to distinguish between lack of information and randomness. In this paper, we have considered different ways of exploiting this high expressivity to quantify prediction uncertainty. Based on three distinct requirements, three different kinds of predictive belief functions have been distinguished, and construction procedures for each of them have been proposed. Type-I belief functions have a Bayesian flavor, and boil down to Bayesian posterior predictive belief functions when a prior probability distribution on the parameter is provided. In contrast, belief functions of the other types are frequentist in spirit. Type-II belief functions correspond to a family of probability measures, which contain the true distribution of the random variable of interest with some probability, in a repeated sampling setting. Type-III belief functions are "frequency-calibrated", in so far as the true value of the variable of interest rarely receives a small plausibility. It should be noticed by "frequentist" predictive belief functions (of types II and III) are not compatible with Bayesian inference, i.e., they do not allow us to recover the Bayesian posterior predictive distribution when combined with a Bayesian prior. It thus seems that the Bayesian and frequentist views cannot be easily reconciled, and different inference procedures have to coexist, just as frequentist and Bayesian procedures in mainstream statistics. Beyond philosophical arguments, the practicality of these construction procedures, as well as their interpretability and acceptability by decision-makers remain to be investigated.

References

1. Aregui, A., Denœux, T.: Constructing predictive belief functions from continuous sample data using confidence bands. In: G. De Cooman, J. Vejnarová, M. Zaffalon (eds.) Proceedings of the Fifth International Symposium on Imprecise Probability: Theories and Applications (ISIPTA 2007), pp. 11–20. Prague, Czech Republic (2007)
2. Barndorff-Nielsen, O.E., Cox, D.R.: Prediction and asymptotics. Bernoulli **2**(1), 319–340 (1996)
3. Berger, J.O., Wolpert, R.L.: The likelihood principle: a review, generalizations, and statistical implications. Lecture Notes-Monograph Series, vol. 6, 2nd edn. Institute of Mathematical Statistics, Hayward (1988)
4. Birnbaum, A.: On the foundations of statistical inference. J. Am. Stat. Assoc. **57**(298), 269–306 (1962)
5. Coles, S.G.: An Introduction to Statistical Modelling of Extreme Values. Springer, London (2001)
6. Dempster, A.P.: New methods for reasoning towards posterior distributions based on sample data. Ann. Math. Stat. **37**, 355–374 (1966)
7. Dempster, A.P.: Upper and lower probabilities induced by a multivalued mapping. Ann. Math. Stat. **38**, 325–339 (1967)
8. Dempster, A.P.: A generalization of Bayesian inference (with discussion). J. Roy. Stat. Soc. B **30**, 205–247 (1968)
9. Dempster, A.P.: Upper and lower probabilities generated by a random closed interval. Ann. Math. Stat. **39**(3), 957–966 (1968)
10. Denœux, T.: Constructing belief functions from sample data using multinomial confidence regions. Int. J. Approx. Reason. **42**(3), 228–252 (2006)
11. Denœux, T.: Likelihood-based belief function: justification and some extensions to low-quality data. Int. J. Approx. Reason. **55**(7), 1535–1547 (2014)
12. Dubois, D., Foulloy, L., Mauris, G., Prade, H.: Probability-possibility transformations, triangular fuzzy sets, and probabilistic inequalities. Reliab. Comput. **10**(4), 273–297 (2004)
13. Edwards, A.W.F.: Likelihood, expanded edn. The John Hopkins University Press, Baltimore (1992)
14. Kanjanatarakul, O., Sriboonchitta, S., Denœux, T.: Forecasting using belief functions: an application to marketing econometrics. Int. J. Approx. Reason. **55**(5), 1113–1128 (2014)

15. Kanjanatarakul, O., Sriboonchitta, S., Denœux, T.: Statistical estimation and prediction using belief functions: principles and application to some econometric models. Int. J. Approx. Reason. **72**, 71–94 (2016)
16. Lawless, J.F., Fredette, M.: Frequentist prediction intervals and predictive distribution. Biometrika **92**(3), 529–542 (2005)
17. Martin, R., Lingham, R.T.: Prior-free probabilistic prediction of future observations. Technometrics **58**(2), 225–235 (2016)
18. Martin, R., Liu, C.: Inferential Models: Reasoning with Uncertainty. CRC Press, Boca Raton (2016)
19. Nguyen, H.T.: On random sets and belief functions. J. Math. Anal. Appl. **65**, 531–542 (1978)
20. Nguyen, H.T.: An Introduction to Random Sets. Chapman and Hall/CRC Press, Boca Raton (2006)
21. Olkin, I., Gleser, L., Derman, C.: Probability Models and Applications. Macmillan, New York (1994)
22. Shafer, G.: A Mathematical Theory of Evidence. Princeton University Press, Princeton (1976)
23. Wasserman, L.A.: Belief functions and statistical evidence. Can. J. Stat. **18**(3), 183–196 (1990)
24. Wilks, S.S.: The large-sample distribution of the likelihood ratio for testing composite hypotheses. Ann. Math. Stat. **9**(1), 60–62 (1938)
25. Zadeh, L.A.: Fuzzy sets as a basis for a theory of possibility. Fuzzy Sets Syst. **1**, 3–28 (1978)
26. Zhu, K., Thianpaen, N., Kreinovich, V.: How to make plausibility-based forecasting more accurate. In: Kreinovich, V., Sriboonchitta, S., Huynh, V.N. (eds.) Robustness in Econometrics, pp. 99–110. Springer, Berlin, Cham, Switzerland (2017)

Kuznets Curve: A Simple Dynamical System-Based Explanation

Thongchai Dumrongpokaphan[1] and Vladik Kreinovich[2(✉)]

[1] Department of Mathematics, Faculty of Science, Chiang Mai University,
Chiang Mai, Thailand
tcd43@hotmail.com
[2] Department of Computer Science, University of Texas at El Paso,
500 W. University, El Paso, TX 79968, USA
vladik@utep.edu

Abstract. In the 1950s, a future Nobelist Simon Kuznets discovered the following phenomenon: as a country's economy improves, inequality first grows but then decreases. In this paper, we provide a simple dynamical system-based explanation for this empirical phenomenon.

1 Kuznets Curve: A Brief Reminder and Need for an Explanation

What is the Kuznets curve. In the 1950s, Simon Kuznets, an American economist of Russian origin, showed that as the country's Gross Domestic Product (GDP) increases, inequality first increases and then decreases again [1, 2, 4]. The resulting dependence on inequality on GDP looks like an inverted letter U and is thus called an *inverted U-shaped* dependence or the *Kunzets curve*. For this work, Professor Kuznets was awarded a Nobel Prize in Economics in 1971.

Kuznets curve: a controversy. The Kuznets curve is a purely empirical observation. Economists from different sides of the political spectrum have come up with different (and mutually exclusive) explanations for this empirical fact.

On the one hand, free-market champions use the Kuznets curve as an argument that the governments should not interfere with the free market: inequality will decrease by itself, as soon as the economy improves further. As Ronald Reagan used to say, the rising tide lifts all the boats. Based on this argument, these economists recommend that the best way to decrease inequality is to minimize the number of government regulations, and the free market will take care of it.

On the other hand, economists who support the need for government regulations note that while the decrease in inequality may indeed be an empirical fact, in all the developed countries, there was a lot of government intervention, and this intervention is what caused the inequality to decrease. Based on this argument, they recommend that the best way to decrease inequality is to continue with the government regulations.

© Springer International Publishing AG 2018
V. Kreinovich et al. (eds.), *Predictive Econometrics and Big Data*, Studies in Computational Intelligence 753, https://doi.org/10.1007/978-3-319-70942-0_9

An additional controversy. It should be mentioned that there is an additional controversy related to the Kuznets curve: namely, some researcher doubt that the Kunzets curve is indeed a universal phenomenon; see, e.g., [3].

What we do in this paper. In this paper, we show that the Kuznets curve phenomenon naturally follows from the general system-based analysis.

2   Analysis of the Problem

Let us describe the phenomenon in precise terms. We start in a situation when the overall economic output is small and therefore, most everyone is poor. In such situations, while there may be a small minority of relatively rich people, most people are poor. In this sense, there is not much inequality.

As the economy grows, people's incomes grow. For each person, his or her income grows until it reaches the level m_i expressing the capability of this person to earn money in the corresponding economy. People are different, so they have somewhat different rates v_i at which they move towards this larger income: some go faster, some go slower.

For simplicity, we can assume that for each person, the rate does not change with time, i.e., that the income of the i-th person income increases at this rate until it reaches the value m_i. At the rate v_i, this takes time $\dfrac{m_i}{v_i}$. So, under this assumption, at each moment of time t, the income $x_i(t)$ of the i-th person is equal to:

- $x_i(t) = v_i \cdot t$ when $t \leq \dfrac{m_i}{v_i}$, and
- $x_i(t) = m_i$ for $t \geq \dfrac{m_i}{v_i}$.

The values m_i are centered around the mean $\overline{m}$, with random deviations

$$\Delta m_i \overset{\text{def}}{=} m_i - \overline{m}.$$

Similarly, the rates v_i center around the mean $\overline{v}$, with random deviations $\Delta v_i \overset{\text{def}}{=} v_i - \overline{v}$.

Since there is no reason to believe that there is a correlation between m_i and v_i, we will assume these variables to be independent.

How can we describe inequality. Perfect equality means that everyone's income is the same. This is equivalent to saying that the standard deviation of income is 0. In general, if the standard deviation is equal to 10% of the average income, then it is reasonable to conclude that we have less inequality that when the standard deviation is equal to 20% of the average income. Thus, a natural measure of inequality is the ratio between the income's standard deviation and its mean value.

Now, we are ready to analyze how inequality changes when the economy improves. Kuznets curve considers three stages:

- the starting stage, when the inequality level is relatively low,
- the intermediate stage, when the level of inequality increases, and
- the final stage, when the level of inequality decreases.

We have already discussed that in the beginning, there is practically no inequality. So, to complete our analysis, we need to consider two other stages: the intermediate stage and the final stage.

What happens on the final stage. Let us start with the final stage, because, as we will see, this stage is easier to analyze. In this final stage, everyone reaches their potential m_i. Thus:

- the average income is equal to the average $\overline{m}$ of the values m_i, and
- the standard deviation is equal to the standard deviation σ_m of the differences Δm_i.

So, on the final stage, the inequality level is equal to the ratio

$$\frac{\sigma_m}{\overline{m}}. \tag{1}$$

What happens on the intermediate stage. In the beginning of the intermediate stage, when few people have reaches their potential m_i, the income of each person is equal to

$$x_i(t) = \frac{m_i}{v_i} \cdot t = \frac{\overline{m} + \Delta m_i}{\overline{v} + \Delta v_i} \cdot t.$$

Here, $\overline{m} + \Delta m_i$ can be represented as $\overline{m} \cdot \left(1 + \dfrac{\Delta m_i}{\overline{m}}\right)$ and similarly, $\overline{v} + \Delta v_i$ can be represented as $\overline{v} \cdot \left(1 + \dfrac{\Delta v_i}{\overline{v}}\right)$. Thus,

$$x_i(t) = \frac{\overline{m}}{\overline{v}} \cdot t \cdot \frac{1 + \dfrac{\Delta m_i}{\overline{m}}}{1 + \dfrac{\Delta v_i}{\overline{v}}}.$$

Differences between different people are, in most cases, not so large, so $|\Delta m_i| \ll \overline{m}$ and $\dfrac{\Delta m_i}{\overline{m}} \ll 1$. Similarly, we can conclude that $|\Delta v_i| \ll \overline{v}$ and thus, $\dfrac{\Delta v_i}{\overline{v}} \ll 1$. Thus, we can expand the above expression for $x_i(t)$ in terms of the small values $\dfrac{\Delta m_i}{\overline{m}}$ and $\dfrac{\Delta v_i}{\overline{v}}$ and keep only linear terms in this expansion. As a result, we get the following formula:

$$x_i(t) = \frac{\overline{m}}{\overline{v}} \cdot t \cdot \left(1 + \frac{\Delta m_i}{\overline{m}} - \frac{\Delta v_i}{\overline{v}}\right).$$

The mean value of Δm_i and Δv_i is 0, so the mean income is equal to

$$\overline{x}(t) = \frac{\overline{m}}{\overline{m}} \cdot t.$$

The standard deviation of Δm_i is equal to σ_m, so the standard deviation of the ratio $\dfrac{\Delta m_i}{\overline{m}}$ is equal to $\dfrac{\sigma_m}{\overline{m}}$. Similarly, the standard deviation of the ratio $\dfrac{\Delta v_i}{\overline{v}}$ is equal to $\dfrac{\sigma_v}{\overline{v}}$.

Since the quantities Δm_i and Δv_i are assumed to be independent, the variance of the expression

$$1 + \frac{\Delta m_i}{\overline{m}} - \frac{\Delta v_i}{\overline{v}} \tag{2}$$

is equal to the sum of the variances of $\dfrac{\Delta m_i}{\overline{m}}$ and $\dfrac{\Delta v_i}{\overline{v}}$. Thus, the corresponding standard deviation is equal to

$$\sqrt{\frac{\sigma_m^2}{(\overline{m})^2} + \frac{\sigma_v^2}{(\overline{v})^2}}.$$

The formula for $x_i(t)$ is obtained by multiplying this expression (2) by a constant $\dfrac{\overline{m}}{\overline{v}} \cdot t$. Thus, the standard deviation $\sigma_x(t)$ can be obtained by multiplying the standard deviation of the above expression (2) by the same constant:

$$\sigma_x(t) = \frac{\overline{m}}{\overline{v}} \cdot t \cdot \sqrt{\frac{\sigma_m^2}{(\overline{m})^2} + \frac{\sigma_v^2}{(\overline{v})^2}}.$$

Dividing this standard deviation y the mean $\overline{x}(t)$, we get the following formula for the inequality level at the intermediate stage:

$$\frac{\sigma_x(t)}{\overline{x}(t)} = \sqrt{\frac{\sigma_m^2}{(\overline{m})^2} + \frac{\sigma_v^2}{(\overline{v})^2}}. \tag{3}$$

Conclusion. By comparing the inequality level (3) at the intermediate stage and the inequality level (1) at the final stage, one can easily see that at the intermediate stage, the inequality is higher:

$$\sqrt{\frac{\sigma_m^2}{(\overline{m})^2} + \frac{\sigma_v^2}{(\overline{v})^2}} > \frac{\sigma_m}{\overline{m}}.$$

This is exactly the Kuznets curve phenomenon.

Thus, we have indeed arrived at a simple justification of the Kuznets curve phenomenon.

Acknowledgments. This work was supported by Chiang Mai University, Thailand. This work was also supported in part by the National Science Foundation grants HRD-0734825 and HRD-1242122 (Cyber-ShARE Center of Excellence) and DUE-0926721, and by an award "UTEP and Prudential Actuarial Science Academy and Pipeline Initiative" from Prudential Foundation.

References

1. Kuznets, S.: Economic growth and income inequality. Am. Econ. Rev. **45**, 1–28 (1955)
2. Maneejuk, P., Pastpipatkul, P., Sriboonchitta, S.: Economic growth and income inequality: evidence from Thailand. In: Huynh, V.-N., Inuiguchi, M., Le, B., Le, B.N., Denoeux, T. (eds.) Proceedings of the 5th International Symposium on Integrated Uncertainty in Knowledge Modeling and Decision Making IUKM 2016, pp. 649–663. Springer, Cham (2016)
3. Roberts, J.T., Thanos, N.D.: Trouble in Paradise: Globalization and Environmental Crises in Latin America. Routledge, London & New York (2003)
4. Yandle, B., Vijayaraghavan, M., Bhattarai, M.: The Environmental Kuznets Curve: A Primer. The Property and Environment Research Center (2000)

A Calibration-Based Method in Computing Bayesian Posterior Distributions with Applications in Stock Market

Dung Tien Nguyen, Son P. Nguyen, Uyen H. Pham[(✉)],
and Thien Dinh Nguyen

University of Economics and Law, VNU-HCM, Ho Chi Minh City, Vietnam
uyenph@uel.edu.vn

Abstract. Finding effective methods to compute or estimate posterior distributions of model parameters is of paramount importance in Bayesian statistics. In fact, Bayesian inference has only been extraordinarily popular in applications after the births of efficient algorithms like the Monte Carlo Markov Chain. Practicality of posterior distributions depends heavily on the combination of likelihood functions and prior distributions. In certain cases, closed-form formulas for posterior distributions can be attained; in this paper, based on the theory of distortion functions, a calibration-like method to calculate explicitly the posterior distributions for three crucial models, namely the normal, Poisson and Bernoulli is introduced. The paper ends with some applications in stock market.

Keywords: Calibration · Distortion function · Bayesian statistics
Posterior estimation · Stock market

1 Introduction

On the brink of the Fourth Industrial Revolution or Industry 4.0, data analytics has been playing a more and more vital role in every sector of any country's economy, especially in business and finance. Among the most effective tools in studying data is Bayesian statistics. In brief, Bayesian inference is the process of fitting a probability model to a set of data and summarizing the result by a probability distribution on the parameters of the model and on unobserved quantities such as predictions for new observations. According to [1], the Bayesian framework can be divided into the following three steps:

1. Setting up a full probability model - a joint probability distribution for all observable and unobservable quantities in a problem. The model should be consistent with knowledge about the underlying scientific problem and the data collection process.
2. Conditioning on observed data: calculating and interpreting the appropriate posterior distribution - the conditional probability distribution of the unobserved quantities of ultimate interest, given the observed data.

© Springer International Publishing AG 2018
V. Kreinovich et al. (eds.), *Predictive Econometrics and Big Data*, Studies in Computational Intelligence 753, https://doi.org/10.1007/978-3-319-70942-0_10

3. Evaluating the fit of the model and the implications of the resulting posterior distribution: does the model fit the data, are the substantive conclusions reasonable, and how sensitive are the results to the modeling assumptions in step 1? If necessary, one can alter or expand the model and repeat the three steps.

In short, we start with a likelihood function which describes how our phenomenon of interest generates data. Moreover, we are equipped with some prior information in the forms of expertise in the data domain or some observations on the data characteristics. This prior knowledge will also be described by a probability distribution called the prior distribution. In other words, the prior distribution summarizes things we know about the phenomenon before the data are analyzed. The main work of the whole process is step 2, where we combine the prior distribution with data (via the likelihood function) to create the posterior distribution. Hence, the posterior distribution summarizes what is known about the phenomenon after the data are taken into account.

The major contribution of this paper is another perspective on step 2. In our opinion, the posterior distribution is created by calibrating the prior distribution with data. The classical approach is via the likelihood function. We propose an alternative method for the calibration process using the theory of distortion functions.

2   A Review of Baysesian Statistics

In this section, four most fruitful likelihood functions in application are reviewed with some basic well-known facts that serves as a preliminary to the new approach in the next section.

In the following, let n and $\bar{x}$ be the size and the mean of the collected data x, respectively.

2.1   Normal distribution

Suppose, there is a normal-distributed population with known variance σ^2. We would like to infer the unknown mean μ. Let the likelihood function be $f(\theta \,|\, x, \sigma^2)$. In order to have a closed-form formula for the posterior distribution, we impose a normal prior distribution $\pi(\theta \,|\, \mu_0, \alpha^2)$ on the parameter space of μ. Here, both μ_0 and α^2 are known. Then, the posterior distribution for μ is also a normal distribution with mean

$$\mathrm{E}^{\text{posterior}}\left(\theta\right) = \frac{(\sigma^2/n)}{(\sigma^2/n) + \alpha^2}\mu_0 + \frac{\alpha^2}{(\sigma^2/n) + \alpha^2}\bar{x}$$

and variance

$$\mathrm{Var}^{\text{posterior}}\left(\theta\right) = \frac{1}{1/(\sigma^2/n) + 1/\alpha^2}$$

Thus, the posterior mean is a weighted sum of the prior mean and the sample mean. Similarly, the posterior variance is half the harmonic mean of the prior variance and the variance of the sample mean.

2.2 Poisson Distribution

In this case, $f(\theta \,|\, x)$ is the Poisson likelihood of with mean θ as the unknown parameter. The prior $\pi(\theta \,|\, \alpha, \beta)$ is a gamma distribution with constant parameters α and β. Then, the posterior is also a gamma distribution with parameters $\alpha + x_1 + \cdots + x_n$ and $\beta + n$, respectively. The mean is as follows

$$\mathrm{E}^{\text{posterior}}(\theta) = \frac{(\alpha + x_1 + \cdots + x_n)}{(\beta + n)} = \frac{(1/n \cdot \alpha/\beta + 1/\beta \cdot \bar{x})}{(1/n + 1/\beta)}$$

This is the Bayes estimator of the mean of the Poisson. It is a weighted average of the prior mean α/β and the sample mean $\bar{x}$ from the data.

2.3 Gamma Distribution

Consider the case when $f(\theta \,|\, \nu, x)$ is the Gamma likelihood with unknown parameter θ. (This includes the case $\nu = 1$ of an exponential distribution with parameter θ, and the case $\nu = 1/2$ of the squared normal distribution with mean zero and variance $1/2\theta$). The prior $\pi(\theta \,|\, \alpha, \beta)$ is chosen to be a gamma distribution with parameters α and β. Then, the posterior is also a gamma distribution with parameters $\alpha + n\nu$ and $\beta + x_1 + \cdots + x_n$, respectively. The posterior mean is

$$\mathrm{E}^{\text{posterior}}(\theta) = \frac{(\alpha + n\nu)}{(\beta + x_1 + \cdots + x_n)} = \frac{(1/(n\nu) + 1/\alpha}{(1/(n\nu) \cdot \beta/\alpha + 1/\alpha \cdot \bar{x}/\nu)}$$

This is the Bayes estimator of θ. It is a weighted harmonic average of the α/β and the estimator $\nu/\bar{x}$ from the sample.

2.4 Bernoulli Distribution

Consider the Bernoulli likelihood $f(\theta \,|\, x) = \theta^x (1-\theta)^{(1-x)}$ with unknown mean θ. The prior $\pi(\theta \,|\, \alpha, \beta)$ is selected to be a beta distribution with parameters α and β. Then, the posterior is also a beta distribution with parameters $\alpha + x_1 + \cdots + x_n$ and $\beta + n - (x_1 + \cdots + x_n)$ with mean

$$\begin{aligned}
\mathrm{E}^{\text{posterior}}(\theta) &= \frac{(\alpha + x_1 + \cdots + x_n)}{(\alpha + x_1 + \cdots + x_n + \beta + n - (x_1 + \cdots + x_n))} \\
&= \frac{(\alpha + x_1 + \cdots + x_n)}{(\alpha + \beta + n)} \\
&= \frac{(1/n \cdot \alpha/(\alpha + \beta) + 1/(\alpha + \beta) \cdot \bar{x})}{(1/n + 1/(\alpha + \beta))}
\end{aligned}$$

This is the Bayes estimator of the parameter θ. It is a weighted average of the prior probability $\alpha/(\alpha + \beta)$ and the sample proportion $\bar{x}$.

3   Distortion Function

Recall that distortion functions arose from the needs of a universal framework for pricing both financial and insurance risk in the 1990s.

In general, a function $g : [0, 1] \longrightarrow [0, 1]$ is a distortion function if g is non-decreasing and g satisfies $g(0) = 0$, $g(1) = 1$.

A distortion function allows us to adjust the probability measure defined by some distribution function $F(x)$. To be precise, let G be the function such that

$$1 - G(x) = g(1 - F(x))$$

Then, we have the following theorem:

Theorem 1. *G is also a cumulative distribution function if and only if g is left continuous at any $u = 1 - F(x)$, where $x \in \mathbf{R}^+$ and g is right-continuous at 0.*

Theorem 1 shows that any distortion function g satisfying the theorem's hypothesis transforms a distribution into a new distribution.

Wang, in his pioneer works, devised the following distortion function (see [4]),

Definition 1. *Let $\lambda \in R$ be a parameter. Define*

$$g_\lambda(p) = \Phi(\Phi^{-1}(p) + \lambda), 0 \le p \le 1$$

where Φ is the cumulative distribution function of the standard normal distribution.

Wang's function has been highly successful in capital asset pricing model (CAPM) and in evaluation of insurance risk. One highlight is the fact that it offers another way to construct the Black-Scholes equation. On the other hand, it has seen deep connections with expected utility theory and decision theory.

Following the successes of Wang's function, other authors such as Wirch (see [5]) have invented various other distortion functions and have also found a wide variety of applications.

Among diverse applications, the abilities of distortion functions to quantify certain subjective evaluations particularly interest us. That has led us to define two new families of distortion functions (see the next section). We have also found a novel method to update some important prior distributions to the corresponding posterior distributions in the bayesian framework.

Remark 1. All distortion functions so far satisfy Theorem 1. Our new families are of no exception.

4   Main Results

4.1   Two New Distortion Functions

In this section, we introduce two new families of distortion functions.

First is a distortion family to estimate the posterior distribution for the Bernoulli proportion θ in the next section. This family is derived from the beta distribution on two parameters.

Let I be the cumulative distribution function of the beta distribution on two parameters $\alpha > 0$ and $\beta > 0$, and I^{-1} be the corresponding quantile function.

Definition 2. *The dual beta distortion function* $g : [0,1] \longrightarrow [0,1]$ *has four parameters* α, β, α', β', *and is defined as follows*

$$g(p) = 1 - I\big[I^{-1}(1-p, \alpha, \beta), \alpha', \beta'\big]$$

Second is a distortion family built from the Gamma distribution which will be used to construct the posterior for the Poisson rate parameter.

Let Γ be the cumulative distribution function of the gamma distribution on two parameters $\alpha > 0$ and $\beta > 0$, and Γ^{-1} be the corresponding quantile function.

Definition 3. *The dual gamma distortion function* $g : [0,1] \longrightarrow [0,1]$ *has four parameters* α, β, α', β', *and is defined as follows*

$$g(p) = 1 - \Gamma\big[\Gamma^{-1}(1-p, \alpha, \beta), \alpha', \beta'\big]$$

4.2   Posterior Estimations

With a prior distribution and a functional form of the likelihood function, it is usually not easy to estimate the posterior distribution of the desired parameter. The reason is that numerical computations of integration are, in many cases, intractable. One popular way to avoid heavy calculations is to choose a prior distribution compatible with the likelihood function so that a closed-form formula for the posterior distribution exists, and can be computed theoretically.

In this work, when a close-formed formula exists, we propose a novel method to obtain the posterior distribution with no integration of the likelihood function. The gem of the method lies in the simplicity of the calculations. We can instantly write down the cumulative distribution function (cdf) for the posterior by calibrating the prior cdf with an appropriate distortion function.

The first theorem concerns the Wang's distortion function (see [2]). Here, the parameter of interest is the mean θ of a normally distributed population with known variance σ^2. We show that if the prior distribution for θ is chosen to be also a normal distribution, then the Wang's distortion function can always calibrate the prior distribution to achieve the correct formula for posterior distribution.

Theorem 2. *Let* $g(p) = \Phi\big[\lambda_1 \Phi^{-1}(p) + \lambda\big]$, $0 \le p \le 1$ *be the Wang's distortion function with two parameters* λ_1 *and* λ. *Suppose the prior distribution is the normal distribution* $N(\mu_0, \alpha^2)$. *Denote* $F(\theta)$ *to be the standardized cumulative distribution of the prior distribution*

$$F(\theta) = \Phi\left(\frac{\theta - \mu_0}{\alpha}\right)$$

Let

$$\lambda = \frac{|\alpha|}{\frac{\sigma}{\sqrt{n}} \sqrt{\frac{\sigma^2}{n} + \alpha^2}}$$

$$\lambda_1 = \alpha \sqrt{\frac{1}{\frac{\sigma^2}{n} + \frac{1}{\alpha^2}}}$$

Then, the standardized cumulative distribution function $G(\theta)$ of the posterior distribution is as follows

$$1 - G(\theta) = g(1 - F(\theta)) \tag{1}$$

Thus, the posterior cdf $G(\theta)$ is obtained via a calibration of the prior cdf $F(\theta)$.

Proof. Consider the right-hand side of Eq. (1)

$$g(1 - F(\theta)) = g\left[\Phi\left(\frac{\theta - \mu_0}{\alpha}\right)\right]$$

$$= \Phi\left[-\lambda_1 \cdot \frac{\theta - \mu_0}{\alpha} + \lambda\right]$$

$$= 1 - \Phi\left[\frac{\theta - (\mu_0 + \alpha\frac{\lambda}{\lambda_1})}{\frac{\alpha}{\lambda_1}}\right]$$

On the other hand, let

$$\mu_0' = \frac{\frac{\sigma^2}{n}}{\frac{\sigma^2}{n} + \alpha^2} \cdot \mu_0 + \frac{\alpha^2}{\frac{\sigma^2}{n} + \alpha^2} \cdot \bar{X}$$

$$\alpha'^2 = \frac{1}{\frac{1}{\sigma^2/n} + \frac{1}{\alpha^2}}$$

Then, by a direct calculation with the Bayes' formula, the cdf of the posterior is also a normal cdf with mean μ_0' and standard deviation α'. Moreover, $G(\theta)$ is exactly

$$G(\theta) = \Phi\left[\frac{\theta - (\mu_0 + \alpha\frac{\lambda}{\lambda_1})}{\frac{\alpha}{\lambda_1}}\right]$$

Therefore, $1 - G(\theta)$ is equal to the right-hand side of Eq. (1) which establishes the theorem. $\square$

The next theorem shows that our new dual beta distortion function with appropriate parameters recovers the posterior distribution for the Bernoulli proportion θ. As usual, x denotes the dataset collected and n denotes the size of x.

Theorem 3. *Let $g(p)$, $0 \le p \le 1$ be the dual beta distortion function with four parameters α, β, α', β'*

$$g(p) = 1 - I\big[I^{-1}(1 - p, \alpha, \beta), \alpha', \beta'\big]$$

where I is the beta distribution function with 2 parameters α, β. Suppose the prior distribution is $F(\theta) = I(\theta; \alpha, \beta)$
  Let

$$\alpha' = \alpha + x_1 + \cdots + x_n$$
$$\beta' = \beta + n - (x_1 + \cdots + x_n)$$

Then, the posterior distribution $G(\theta)$ for the proportion θ is exactly $I(\theta; \alpha', \beta')$. Moreover,

$$1 - G(\theta) = g(1 - F(\theta)) \tag{2}$$

Therefore, the posterior cdf $G(\theta)$ is obtained via a calibration of the prior cdf $F(\theta)$ by the dual beta distortion function g.

Proof. Consider the right-hand side

$$\begin{aligned}
g(1 - F(\theta)) &= g\left[1 - I(\theta; \alpha, \beta)\right] \\
&= 1 - I[I^{-1}(I(\theta; \alpha, \beta); \alpha, \beta); \alpha', \beta'] \\
&= 1 - I[\theta; \alpha', \beta']
\end{aligned}$$

On the other hand, by a direct calculation with the Bayes' formula, the posterior distribution for the Bernoulli proportion θ is exactly $I[\theta; \alpha', \beta']$. Therefore, $1 - G(\theta)$ is equal to the right-hand side of the Eq. (2)   □

Our third result is for the Poisson rate parameter θ with gamma prior distribution. Similar to all previous results, we prove that the posterior cdf for θ is obtained by calibrating the prior cdf using our new dual gamma distortion function.

Theorem 4. *Let $g(p)$ be the dual gamma distortion function with four parameters as follows*

$$g(p) = 1 - \Gamma\big[\Gamma^{-1}(1 - p; \alpha, \beta); \alpha_1, \beta_1\big]$$

where $\Gamma()$ is the gamma cumulative distribution function and Γ^{-1} is the corresponding quantile function.
  Suppose the prior cumulative distribution is $F(\theta) = \Gamma(\theta; \alpha, \beta)$
  Let

$$\alpha_1 = \alpha + x_1 + \cdots + x_n$$
$$\beta_1 = \beta + n$$

Then, the posterior cumulative distribution $G(\theta)$ is exactly $\Gamma(\theta; \alpha_1, \beta_1)$. Moreover,

$$1 - G(\theta) = g(1 - F(\theta)) \tag{3}$$

Therefore, the posterior cdf $G(\theta)$ is obtained via a calibration of the prior cdf $F(\theta)$ by the dual gamma distortion function g.

Proof. Consider the right-hand side

$$g(1 - F(\theta)) = g[1 - \Gamma(\theta; \alpha, \beta)]$$
$$= 1 - \Gamma[\Gamma^{-1}(\Gamma(\theta; \alpha, \beta); \alpha, \beta); \alpha_1, \beta_1]$$
$$= 1 - \Gamma[\theta; \alpha_1, \beta_1]$$

On the other hand, by a direct application of the Bayes' formula, the posterior cumulative distribution $G(\theta)$ is exactly $\Gamma(\theta; \alpha_1, \beta_1)$.

Therefore, $1 - G(\theta)$ is equal to the right-hand side of the Eq. (3). $\square$

5   Experimental Illustrations

We collected historical data of Vietnam stock, US stock, 5 popular foreign exchange rate and Gold price from Thomson Reuters as much as possible. We have 658 stocks in Vietnam, 53 stocks in US to experiment. See Table 1 for the duration of the time series data.

Table 1. Duration

	Quantity	From	To
Vietnamese stock	685	First trading day	Nov 16th 2015
Forex and Gold	6	Feb 6th, 1996	Nov 16th 2015
US stock	53	First trading day	Nov 16th 2015

Table 2 shows the exact rates of the data we had: (1) Exact rates of direction with Normal distribution and (2) Exact rates of confidence intervals with Gamma and Beta distributions.

Table 2. Rates

Exact rates (%)	Normal		Gamma	Beta
	Bayesian	Distortion		
0 − 50	0.1%	0.3%	6.7%	4.69%
50 − 60	0.3%	0.3%	12.2%	7.91%
60 − 70	2.9%	2.0%	24.3%	21.23%
70 − 80	55.8%	50.3%	39.7%	44.22%
80 − 90	39.5%	46.6%	16.0%	21.23%
90 − 100	1.3%	0.6%	1.2%	0.73%

Both estimations using the traditional Bayesian updating formula and the new distortion function formula show promising results in forecasting trends and predicting using credible intervals. In the case of normal distribution, about 95% of data gives the prediction accuracy from 70% to 90%. Meanwhile, in the case of beta and gamma distributions, more than 50% of data have the accuracy from 70% to 90% in credible interval prediction.

The results are good for forecasting risks in stock prices, hence, will be helpful for investors in portfolio management. The new method of Bayesian statistics in general, and distortion function in particular provides an alternative in time series prediction.

6   Conclusion

In this paper, we provide a new point of view towards computing the posterior distributions in bayesian statistics, namely via calibrations of the corresponding prior distributions. In our opinion, this perspective reflects the essence of bayesian inference where we start which some initial beliefs expressed through the prior distribution on the parameter space; then, as data arrive, we adjust our beliefs on our model parameters by mixing the information from the collected data with the initial beliefs to form the posterior distribution. The adjustments can be done by transforming the probability measure on the parameter space using a suitable distortion function. In the paper, we provide a few distortion functions to calibrate some popular prior distributions corresponding to highly-used likelihoods like the normal, Bernoulli and Poisson functions.

In applications, we also implement and compare this method with the traditional method using likelihood integration. The data in the study are real stock market data from the US and Vietnam; the predicting results are quite promising.

A lot of further research can be done: from estimating different parameters (apart from the mean and proportion) to extending the ideas of calibration using distortion to many other common distributions.

Regarding application, a number of questions in risk management and finance can be addressed by the new approach.

References

1. Gelman, A., Carlin, J.B., Stern, H.S., Dunson, D.B., Vehtari, A., Rubin, D.B.: Bayesian Data Analysis, 3rd edn. CRC Press (2013)
2. Wang, S.S.: A class of distortion operators for pricing financial and insurance risks. J. Risk Insur. **67**(1), 15–36 (2000)
3. Bolstad, W.M.: Introduction to Bayesian statistics. Wiley (2013)
4. Wang, S.S.: Premium calculation by transforming the layer premium density. In: Casualty Actuaries Reinsurance Semmar in New York, 26–27 June 1995
5. Wirch J.L.: Coherent beta risk measures for capital requirements. Ph.D. thesis, Ontario, Canada (1999)
6. Nguyen, H., Pham, U., Tran, H.: On some claims related to Choquet integral risk measures. Ann. Oper. Res. **195**, 5–31 (2012)

7. Acerbi, C.: Spectral measures of risk: a coherent representation of subjective risk aversion. J. Bank. Finance **26**(7), 1505–1518 (2002)
8. Artzner, P., Delbaen, F., Elber, J., Heath, D.: Coherent measures of risk. Mathe. Finance **9**, 203–228 (1999)
9. Artzner, P., Delbaen, F., Eber, J.M., Heath, D.: Coherent multiperiod risk adjusted values and Bellmans principle. Ann. Oper. Res. **152**, 5–22 (2007)
10. Bertoin, J.: Lévy processes. Cambridge University Press, Cambridge (1996)
11. Black, F., Scholes, M.: The pricing of options and corporate liabilities. J. Polit. Econ. **81**, 637–654 (1973)

How to Estimate Statistical Characteristics Based on a Sample: Nonparametric Maximum Likelihood Approach Leads to Sample Mean, Sample Variance, etc.

Vladik Kreinovich [1(✉)] and Thongchai Dumrongpokaphan[2]

[1] Department of Computer Science, University of Texas at El Paso,
500 W. University, El Paso, TX 79968, USA
vladik@utep.edu
[2] Department of Mathematics, Faculty of Science, Chiang Mai University,
Chiang Mai, Thailand
tcd43@hotmail.com

Abstract. In many practical situations, we need to estimate different statistical characteristics based on a sample. In some cases, we know that the corresponding probability distribution belongs to a known finite-parametric family of distributions. In such cases, a reasonable idea is to use the Maximum Likelihood method to estimate the corresponding parameters, and then to compute the value of the desired statistical characteristic for the distribution with these parameters.

In some practical situations, we do not know any family containing the unknown distribution. We show that in such nonparametric cases, the Maximum Likelihood approach leads to the use of sample mean, sample variance, etc.

1 Need to Estimate Statistical Characteristics Based on a Sample: Formulation of the Problem

Need to estimate statistical characteristics. In many practical situations, we need to estimate statistical characteristic of a certain random phenomenon based on a given sample.

For example, to check that for all the mass-produced gadgets from a given batch, the valued of the corresponding physical quantity are within the desired bounds, the ideal solution would be to measure the quantity for all the gadgets. This may be reasonable to do if these gadgets are intended for a spaceship, where a minor fault can lead to catastrophic results. However, in most applications, it is possible to save time and money by testing only a small sample, and by making statistical conclusions based on the results of this testing.

How do we estimate the statistical characteristics – finite-parametric case: main idea. In many situations, we know that the actual distribution belongs to a known finite-parametric family of distributions. For example, it is

V. Kreinovich et al. (eds.), *Predictive Econometrics and Big Data*, Studies in Computational Intelligence 753, https://doi.org/10.1007/978-3-319-70942-0_11

often known that the distribution is Gaussian (normal), for some (unknown) values of the mean μ and standard deviation σ. In general, we know that the corresponding probability density function (pdf) has the form $f(x \mid \theta)$ for some parameters $\theta = (\theta_1, \ldots, \theta_n)$.

In such situations, we first estimate the values of the parameters θ_i based on the sample, and then compute the values of the corresponding statistical characteristic (mean, standard deviation, kurtosis, etc.) corresponding to the estimates values θ_i.

How do we estimate the statistical characteristics – finite-parametric case: details. How do we estimate the values of the parameters θ_i based on the sample? A natural idea is to select the *most probable* values θ. How do we go from this idea to an algorithm?

To answer this question, let us first note that while theoretically, each of the parameters θ_i can take infinitely many values, in reality, for a given sample size, it is impossible to detect the difference between the nearby values θ_i and θ_i'. Thus, from the practical viewpoint, we have finitely many distinguishable cases.

In this description, we have finitely many possible combinations of parameters $\theta^{(1)}, \ldots, \theta^{(N)}$. We consider the case when all we know is that the actual pdf belongs to the family $f(x \mid \theta)$. There is no a priori reason to consider some of the possible values $\theta^{(k)}$ as more probable. Thus, before we start our observations, it is reasonable to consider these N hypotheses as equally probable: $P_0(\theta^{(k)}) = \dfrac{1}{N}$. This reasonable idea is known as the *Laplace Indeterminacy Principle*; see, e.g., [1].

We can now use the Bayes theorem to compute the probabilities $P(\theta^{(k)} \mid x)$ of different hypotheses $\theta^{(k)}$ after we have performed the observations, and these observations resulted in a sample $x = (x_1, \ldots, x_n)$:

$$P(\theta^{(k)} \mid x) = \frac{P(x \mid \theta^{(k)}) \cdot P_0(\theta^{(k)})}{\sum\limits_{i=1}^{N} P(x \mid \theta^{(i)}) \cdot P_0(\theta^{(i)})}.$$

Here, the probability $P(x \mid \theta^{(k)})$ is proportional to $f(x \mid \theta^{(k)})$. Dividing both numerator and denominator by $P_0 = \dfrac{1}{N}$, we thus conclude that

$$P(\theta^{(k)} \mid x) = c \cdot f(x \mid \theta^{(k)})$$

for some constant c.

Thus, selecting the most probable hypotheses $P(\theta^{(k)} \mid x) \to \max\limits_{k}$ is equivalent to finding the values θ for which, for the given sample x, the expression $f(x \mid \theta)$ attains its largest possible value. The expression $f(x \mid \theta)$ is known as *likelihood*, and the whole idea is known as the *Maximum Likelihood Method*; see, e.g., [2].

In particular, for Gaussian distribution, the Maximum Likelihood method leads to the sample mean

$$\widehat{\mu} \stackrel{\text{def}}{=} \frac{1}{n} \cdot \sum_{i=1}^{n} x_i$$

as the estimate for the mean, and to the sample variance

$$(\widehat{\sigma})^2 \stackrel{\text{def}}{=} \frac{1}{n} \cdot \sum_{i=1}^{n}(x_i - \widehat{\mu})^2$$

as the estimate for the variance.

What if we do not know the family? In some practical situations, we do not know a finite-parametric family of distributions that contains the actual one. In such situations, all we know is a sample. Based on this sample, how can we estimate the statistical characteristics of the corresponding distribution?

What we do in this paper. In this paper, we apply the Maximum Likelihood method to the above problem. It turns out that the resulting estimates are sample mean, sample variance, etc.

Thus, we get a justification for using these estimates beyond the case of the Gaussian distribution.

2   Nonparametric Maximum Likelihood Approach to Estimating Statistical Characteristics Based on a Sample: Continuous Case

Description of the case. Let us first consider the case when the random variable is continuous, i.e., when, in principle, it can take either all possible real values – or at least all possible real values from a certain interval.

Possibility to discretize. Similar to the above case, while theoretically, we can thus have infinitely many possible values of the random variable x, in reality, due to measurement uncertainty, very close values $x \approx x'$ are indistinguishable. Thus, in practice, we can safely assume that there are only finitely many distinguishable values $x^{(1)} < x^{(2)} < \ldots < x^{(M)}$.

Possible probability distributions. In the discretized representations, to describe the corresponding random variable, we need to describe the probabilities $p_i = p(x^{(i)})$ of each of M values $x^{(i)}$, $1 \leq i \leq M$. The only restriction on these probabilities is that they should be non-negative and add up to 1: $\sum_{i=1}^{M} p_i = 1$.

Let us apply the Maximum Likelihood Method: resulting formulation. According to the Maximum Likelihood Method, out of all possible probability distributions $\mathbf{p} = (p_1, \ldots, p_n)$, we should select a one for which the probability of observing a given sequence $x_1, \ldots, x_n$ is the largest.

The probability of observing each value x_i is equal to $p(x_i)$. It is usually assumed that different elements in the sample are independent, so the probability $p(x \mid \mathbf{p})$ of observing the whole sample $x = (x_1, \ldots, x_n)$ is equal to the product of these probabilities:

$$p(x \mid \mathbf{p}) = \prod_{i=1}^{n} p(x_i).$$

In the continuous case, the probability of observing the exact same number twice is zero, so we can safely assume that all the values x_i are different. In this case, the above product takes the form

$$p(x \,|\, \mathbf{p}) = \prod \{x_i : x_i \text{ has been observed}\}.$$

We need to find the values $p_1, \ldots, p_M$ that maximize this probability under the constraints that $p_i \geq 0$ and $\sum_{i=1}^{M} p_i = 1$.

Analysis of the problem. Let us explicitly describe the probability distribution that maximizes the corresponding likelihood.

First, let us notice that when the maximum is attained, the values p_i corresponding to un-observed values should be 0. Indeed, if $p_i > 0$ for one of the indices i corresponding to an un-observed value x_i, then we can, without changing the constraint $\sum_{i=1}^{M} p_i = 1$, decrease this value to 0 and instead increase one of the probabilities p_i corresponding to an observed value x_i.

Let I denote the set of all indices corresponding to observed values p_i. Then, in the optimal arrangement, we have $p_i = 0$ for $i \notin I$. So, the constraint $\sum_{i=1}^{M} p_i = 1$ takes the form $\sum_{i \in I} p_i = 1$, and the likelihood optimization problem takes the following form: $\prod_{i \in I} p_i \to \max$ under the constraint that $\sum_{i \in I} p_i = 1$.

This is a known and easy-to-solve optimization problem. The corresponding maximum is attained when all the probabilities p_i are equal to each other, i.e., when $p_i = \dfrac{1}{n}$. Thus, we arrive at the following conclusion.

Conclusion: we should use sample mean, sample variance, etc. In the non-parametric case, the maximum likelihood method implies that out of all possible probability distributions, we should select a distribution in which all sample values $x_1, \ldots, x_n$ appear with equal probability $p_i = \dfrac{1}{n}$, and no other values can appear.

So, as estimates of the desired statistical characteristics, we should select characteristics corresponding to this sample-based distribution. The mean of this distribution is equal to $\widehat{\mu} = \dfrac{1}{n} \cdot \sum_{i=1}^{n} x_i$, i.e., to the sample mean. The variance of this distribution is equal to $\dfrac{1}{n} \cdot \sum_{i=1}^{n} (x_i - \widehat{\mu})^2$, i.e., to the sample variance.

Thus, for the nonparametric case, the maximum likelihood method implies that we should use sample mean, sample variance, etc.

Discussion. Thus, we get a justification for using sample mean, sample variance, etc., in situations beyond their usual Gaussian-based justification.

3 Nonparametric Maximum Likelihood Approach to Estimating Statistical Characteristics Based on a Sample: Discrete Case

Description of the case. In the discrete case, we have a finite list of possible values $x^{(1)}, \ldots, x^{(M)}$. To describe a probability distribution, we need to describe the probabilities $p_i = p(x^{(i)})$ of these values.

Maximum Likelihood Approach: formulation of the optimization problem. For each sample $x_1, \ldots, x_n$, the corresponding likelihood $\prod\limits_{i=1}^{n} p(x_i)$ takes the form

$$p(x \mid \mathbf{p}) = \prod_{i=1}^{M} p_i^{n_i},$$

where n_i is the number of times the value $x^{(i)}$ appears in the sample.

We must find the probabilities p_i for which the likelihood attains its largest possible value under the constraint $\sum\limits_{i=1}^{n} p_i = 1$.

Optimizing the likelihood. To solve the above constraint optimization problem, we can use the Lagrange multiplier method that reduces it to the unconstrained optimization problem

$$\prod_{i=1}^{M} p_i^{n_i} + \lambda \cdot \left(\sum_{i=1}^{M} p_i - 1 \right) \to \max_{p}.$$

Differentiating this objective function with respect to p_i, taking into account that for $A \overset{\text{def}}{=} \prod\limits_{i=1}^{M} p_i^{n_i}$, we get

$$\frac{\partial A}{\partial p_i} = \prod_{j \neq i} p_j^{n_j} \cdot n_i \cdot p_i^{n_i - 1} = A \cdot \frac{n_i}{p_i},$$

and equating the derivative to 0, we conclude that

$$A \cdot \frac{n_i}{p_i} + \lambda = 0.$$

Thus, $p_i = \text{const} \cdot n_i$. The constraint that $\sum\limits_{i=1}^{M} p_i = 1$ implies that the constant is equal to 1 over the sum $\sum\limits_{i=1}^{n} n_i = n$. Thus, we get $p_i = \dfrac{n_i}{n}$. So, we arrive at the following conclusion.

Conclusion. In the discrete case, for each of the possible values $x^{(i)}$, we assign, as the probability p_i, the frequency $\dfrac{n_i}{n}$ with which this value appears in the observed sample.

This is the probability distribution that we should use to estimate different statistical characteristics. For this distribution, the mean is still equal to the sample mean, and the variance is still equal to the sample variance – same as for the continuous case.

However, e.g., for entropy, we get a value which is different from the continuous case: there, the entropy is always equal to

$$-\sum_{i\in I} p_i \cdot \ln(p_i) = -n \cdot \frac{1}{n} \cdot \ln\left(\frac{1}{n}\right) = \ln(n),$$

while in the discrete case, we have a different value

$$-\sum_{i\in I} p_i \cdot \ln(p_i) = -\sum_{i=1}^{M} \frac{n_i}{n} \cdot \ln\left(\frac{n_i}{n}\right).$$

Acknowledgments. This work was supported by Chiang Mai University, Thailand. This work was also supported in part by the National Science Foundation grants HRD-0734825 and HRD-1242122 (Cyber-ShARE Center of Excellence) and DUE-0926721, and by an award "UTEP and Prudential Actuarial Science Academy and Pipolino Initiative" from Prudential Foundation.

References

1. Jaynes, E.T., Bretthorst, G.L.: Probability Theory: The Logic of Science. Cambridge University Press, Cambridge, UK (2003)
2. Sheskin, D.J.: Handbook of Parametric and Nonparametric Statistical Procedures. Chapman and Hall/CRC, Boca Raton, Florida (2011)

How to Gauge Accuracy of Processing Big Data: Teaching Machine Learning Techniques to Gauge Their Own Accuracy

Vladik Kreinovich[1(✉)], Thongchai Dumrongpokaphan[2], Hung T. Nguyen[3,4], and Olga Kosheleva[1]

[1] Department of Computer Science, University of Texas at El Paso,
500 W. University, El Paso, TX 79968, USA
{vladik,olgak}@utep.edu
[2] Department of Mathematics, Faculty of Science, Chiang Mai University,
Chiang Mai, Thailand
tcd43@hotmail.com
[3] Department of Mathematical Sciences, New Mexico State University,
Las Cruces, NM 88003, USA
hunguyen@nmsu.edu
[4] Faculty of Economics, Chiang Mai University, Chiang Mai 50200, Thailand

Abstract. When the amount of data is reasonably small, we can usually fit this data to a simple model and use the traditional statistical methods both to estimate the parameters of this model and to gauge this model's accuracy. For big data, it is often no longer possible to fit them by a simple model. Thus, we need to use generic machine learning techniques to find the corresponding model. The current machine learning techniques estimate the values of the corresponding parameters, but they usually do not gauge the accuracy of the corresponding general nonlinear model. In this paper, we show how to modify the existing machine learning methodology so that it will not only estimate the parameters, but also estimate the accuracy of the resulting model.

1 Need to Gauge Accuracy of Big Data Processing

Need for data processing. In many practical situations, we are interested in the value of a quantity y which is difficult – or even impossible – to measure directly. For example, we may be interested:

- in tomorrow's temperature y, or
- in the distance y to a faraway star.

Since we cannot measure this quantity y directly, we have to measure it *indirectly*; namely:

- we measure easier-to-measure quantities $x_1, \ldots, x_n$ whose values determine y, i.e., for which $y = f(x_1, \ldots, x_n)$ for some function $f(x_1, \ldots, x_n)$, and then

V. Kreinovich et al. (eds.), *Predictive Econometrics and Big Data*, Studies in Computational Intelligence 753, https://doi.org/10.1007/978-3-319-70942-0_12

- we use the results $\widetilde{x}_i$ of measuring x_i and the known dependence $f(x_1, \ldots, x_n)$ to estimate y as $\widetilde{y} = f(\widetilde{x}_1, \ldots, \widetilde{x}_n)$.

The corresponding computations are known as *data processing*.

Need to find the corresponding dependence. To be able to perform data processing, we need to know the dependence $y = f(x_1, \ldots, x_n)$ between the corresponding quantities x_i and y.

- In some cases, this dependence can be determined based on the fundamental physical principles.
- However, in many other cases, this dependence needs to be determined experimentally, based on the known observation results.

Traditional approach to finding a dependence: a brief reminder. The traditional statistical approach to finding the desired dependence has been designed for situations in which the number of available observations is reasonably small; see, e.g., [4]. In such situations, we can usually fit the data with some simple model – e.g., with the linear regression model in which the dependence is linear: $f(x_1, \ldots, x_n) = a_0 + \sum_{i=1}^{n} a_i \cdot x_i$. For such models, the traditional statistical approach provides both:

- the estimates for the values of the corresponding parameters, and
- a good description of the accuracy of the corresponding estimated model $\widetilde{f}(x_1, \ldots, x_n)$, i.e., of the probability distribution of the approximation error $\Delta y \stackrel{\text{def}}{=} y - \widetilde{f}(x_1, \ldots, x_n)$.

The emergence of big data. In the last decades, the progress in computer and computer-based measurement technologies enabled us to get huge amounts of data, amounts far exceeding the sample sizes for which we can apply the traditional statistical techniques.

This phenomenon is known as *big data*; see, e.g., [3].

For big data, simple models are rarely possible. Real-life phenomena are usually very complex. When we have a reasonable small amount of data, we can still have simple approximate models that describe this data well. However, as we increase the amount of data, we can no longer use simple models.

This phenomenon is easy to explain. In general in statistics, based on a sample of size n, we can determine each parameter of the model with accuracy $\sim 1/\sqrt{n}$.

- When the sample size n is reasonably small, the resulting inaccuracy usually exceeds the size of the quadratic and higher order terms in the actual dependence. So, within this accuracy, we can safely assume that the dependence is linear.
- However, as the sample size increases, the accuracy $1/\sqrt{n}$ becomes smaller than the quadratic and higher order terms – and thus, these terms can no longer be ignored if we want to have a model that fits all the data.

This phenomenon is well known. For example:

- When a comet appears and we have only few of its observations, we can safely use simplified Newton's equations – that assumes that only the Sun and the Jupiter have to be taken into account – and get a good description of all the observed data.
- However, as the number of measurements increases, we have to take into account the gravitational influence of other planets to get a good fit with observations.

Machine learning techniques: the big-data analogs of the traditional statistical data processing. For large data sizes, we cannot use simple few-parametric models, we need complex models with large number of parameters. In most situations, it is not possible to guess the exact form of the corresponding non-linear dependence. So, we need to use techniques that do not assume any specific form, but try to extract the form of the dependence from the data itself.

Such techniques are known as *machine learning*; see, e.g., [1,2]. For most of machine learning approaches such as neural networks, there are *universal approximation* results that show that every possible non-linear function can be, with any given accuracy, approximated by the corresponding computational model.

Machine learning techniques: successes and limitations. Machine learning techniques have been very successful in many applications [1,2]. In many practical applications, they use the observations to come up with a dependence $\tilde{f}(x_1,\ldots,x_n)$ that provides a good fit for the observed data.

However, what is lacking in most machine learning techniques is a good understanding of how accurate is this model, i.e., what are the probabilities of different values of the approximation error $\Delta y = y - \tilde{f}(x_1,\ldots,x_n)$.

To be more precise, we can estimate overall characteristics of such an accuracy: e.g., we can take all the values of the approximation error corresponding to all the input observations $(x_1,\ldots,x_n,y)$ and find the standard deviation and the whole distribution. But this will be the overall distribution.

We know that in many cases, the approximation accuracy depends on the inputs $x_1,\ldots,x_n$:

- for some inputs, the model $\tilde{f}(x_1,\ldots,x_n)$ provides a more accurate approximation, while
- for other inputs, the model provides less accurate approximation.

Traditional statistical methods enable us to find the distribution of measurement errors corresponding to each individual tuple $x = (x_1,\ldots,x_n)$.

It is desirable to have something similar for machine learning techniques as well.

What we do in this paper. In this paper, we show how we can teach the existing machine learning techniques

- to not only the approximate dependence model, but
- also to automatically gauge the accuracy of the resulting model.

2   Our Main Idea

Analysis of the problem. We would like to be able, for each possible tuple $x = (x_1, \ldots, x_n)$, to generate not only a single numerical estimate for the corresponding value y, but also the whole conditional probability distribution describing possible values y corresponding to this tuple.

From the computational viewpoint, a natural way to simulate a probability distribution with a given cumulative distribution function (cdf) $F(Y) = \text{Prob}(y \leq Y)$ is:

- to start with the standard random number r which is uniformly distributed on the interval $[0,1]$ – and whose generation is supported by most programming languages and programming environments – and
- apply the inverse function $F^{-1}(p)$ to this random number.

The resulting values $y = F^{-1}(r)$ are indeed distributed according to the given probability distribution $F(Y)$.

The inverse function $F^{-1}(p)$ has a direct probabilistic meaning; namely,

- for each $p \in [0, 1]$,
- the value $r = F^{-1}(p)$ is the p-th *quantile*, i.e., the value x for which $F(x) = p$.

For $p = 0.5$, we get the median, for $p = 0.25$ and $p = 0.75$, we get the lower and upper quartiles, etc.

From this viewpoint, what we want is to be able,

- for every possible tuple $x = (x_1, \ldots, x_n)$ and
- for every possible value p,

to come up with the p-th quantile of the conditional y-distribution corresponding to this tuple. In precise terms, for the function $G_x(y) \stackrel{\text{def}}{=} F(y \mid x_1, \ldots, x_n)$, we want to be able to generate the quantile $q = G_x^{-1}(p)$ for which

$$\text{Prob}(y \leq q \mid x_1, \ldots, x_n) = p.$$

Let us denote this quantile q by $G(x_1, \ldots, x_n, p)$.

Thus, we want to generate a function $G(x_1, \ldots, x_n, p)$ of $n+1$ variables. Once we have this function, we will be able to find, for each tuple $x = (x_1, \ldots, x_n)$, the cdf corresponding y-distribution $F_x(y)$ – as the inverse function $F_x(p) = q_x^{-1}(p)$ to the function $q_x(p) \stackrel{\text{def}}{=} Q(x_1, \ldots, x_n, p)$.

Historical comment. The idea of combining stochastic processes with neural networks was originally proposed and actively promoted by Paul Werbos; see, e.g., [6,7] (see also [5]).

How can we use machine learning to find the desired function $Q(x_1, \ldots, x_n, p)$? In the ideal world, for each tuple $x = (x_1, \ldots, x_n)$, we would have several different observations in which we have

- these same values of x_i and
- different values of y.

In this case, from this sample of different values of y corresponding to the given tuple x, we will be able to determine the conditional probability distribution corresponding to this tuple x.

Specifically, once we have N such y-values, we can sort them into an increasing sequence $y_{(1)} \leq \ldots \leq y_{(N)}$. Crudely speaking, these values correspond to the quantiles $p = 1/N$, $p = 2/N$, etc. We therefore expect each predicted quantile $Q(x_1, \ldots, x_n, j/N)$ to be close to the corresponding value $y_{(j)}$.

In practice, we do not have such "same-x" observations: different observations correspond, in general, to different tuples x. Since we cannot have different observations corresponding to the *exact same* tuple x, a natural idea is to use observations corresponding to *nearby* tuples x.

Good news is that when we have big data, i.e., a *very large* amount of data, it is highly probable that a few of the observations will be close to x – even when the probability of being close to x is small.

Once we pick N such nearby tuples, we can sort the corresponding N y-values into an increasing sequence $y_{(1)} \leq \ldots \leq y_{(N)}$. Crudely speaking, these values correspond to the quantiles $p = 1/N$, $p = 2/N$, etc. In other words, each of these values $y_{(j)}$ correspond to $p = j/N$. We therefore expect each predicted quantile $Q(x_1, \ldots, x_n, j/N)$ to be close to the corresponding value $y_{(j)}$.

Thus, we arrive at the following algorithm for reconstructing the desired function $Q(x_1, \ldots, x_n, p)$.

Resulting algorithm. We start with the list of observations $\left(x_1^{(k)}, \ldots, x_n^{(k)}, y^{(k)} \right)$. Based on this list, we will form the extended tuples $\left(x_1^{(k)}, \ldots, x_n^{(k)}, p_j, y_j^{(k)} \right)$ as follows:

$1°$. We fix some number N – e.g., $N = 10$.

$2°$. Then, for each original observation $\left(x_1^{(k)}, \ldots, x_n^{(k)}, y^{(k)} \right)$, we do the following:

$2.1°$ We find $N - 1$ observations $\left(x_1^{(\ell)}, \ldots, x_n^{(\ell)}, y^{(\ell)} \right)$ in which the x-tuple

$$x^{(\ell)} = \left(x_1^{(\ell)}, \ldots, x_n^{(\ell)} \right) \text{ is the closest to the original } x\text{-tuple}$$

$$x^{(k)} = \left(x_1^{(k)}, \ldots, x_n^{(k)} \right).$$

$2.2°$ In the resulting $1 + (N - 1) = N$ observations, we have N different y-values:
- the original y-value $y^{(k)}$ and
- $N - 1$ values $y^{(\ell)}$ corresponding to $N - 1$ "nearest" observations.

Let us sort them into an increasing sequence

$$y_{(1)}^{(k)} \leq y_{(2)}^{(k)} \leq \ldots \leq y_{(N)}^{(k)}.$$

2.3° Then, we form N extended tuples

$$\left(x_1^{(k)}, \ldots, x_n^{(k)}, \frac{j}{N}, y_{(j)}^{(k)} \right).$$

3°. After that, we combine all the extended tuples corresponding to different original tuples into a single list.

4°. To the resulting single list, we apply a machine learning algorithm and thus, construct the desired function $Q(x_1, \ldots, x_n, p)$ that provides,

- for each tuple $x = (x_1, \ldots, x_n)$ and
- for each value $p \in [0, 1]$,

the p-th quantile of the y-distribution corresponding to this tuple x.

Which value N should we use: a comment

- The larger N, the more detailed is the resulting information about the probability distributions.
- On the other hand, if we take N to be too large, then some of the "closest" tuples $x^{(\ell)}$ may be too far away from the original tuple $x^{(k)}$ and thus, this description will not be very accurate.

So, here, we have a usual trade-off between accuracy and details – similar to what we have, e.g., when we divide the real line into bins to build a histogram approximating the actual probability density function $\rho(x)$:

- if we use smaller bins, we get more details, but
- these details are at the expense of accuracy of approximating $\rho(x)$: the smaller bins, the fewer observations are in each bin, and thus, the less accurately the corresponding frequencies represent the desired probabilities.

The actual value N should be determine empirically – or, if we have some prior information, by using this prior information. As a rule of thumb, we suggest using $N = 10$, since in our experience, this is where we get the largest number of reliable details.

Acknowledgments. This work was supported by Chiang Mai University, Thailand. This work was also supported in part by the National Science Foundation grants HRD-0734825 and HRD-1242122 (Cyber-ShARE Center of Excellence) and DUE-0926721, and by an award "UTEP and Prudential Actuarial Science Academy and Pipeline Initiative" from Prudential Foundation.
One of the authors (VK) is thankful to Paul Werbos for inspiring talks and discussions.

References

1. Bishop, C.M.: Pattern Recognition and Machine Learning. Springer, New York (2006)
2. Goodfellow, I., Bengio, Y., Courville, A.: Deep Learning. MIT Press, Cambridge (2016)
3. Mayer-Schönberger, V., Cukier, K.: Big Data: The Essential Guide to Work, Life, and Learning in the Age of Insight. John Murray Publishers, London (2017)
4. Sheskin, D.J.: Handbook of Parametric and Nonparametric Statistical Procedures. Chapman & Hall/CRC, Boca Raton, Florida (2011)
5. Turchetti, C.: Stochastic Models of Neural Networks. IOS Press, Amsterdam (2004)
6. Werbos, P.: A brain-like design to learn optimal decision strategies in complex environments. In: Karny, M., Warwick, K., Kurkova, V. (eds.) Dealing with Complexity: A Neural Networks Approach. Springer, London (1998)
7. Werbos, P.: Intelligence in the brain: a theory of how it works and how to build it. Neural Netw. **22**(3), 200–212 (2009)

How Better Are Predictive Models: Analysis on the Practically Important Example of Robust Interval Uncertainty

Vladik Kreinovich[1]([✉]), Hung T. Nguyen[2,3], Songsak Sriboonchitta[3], and Olga Kosheleva[1]

[1] Department of Computer Science, University of Texas at El Paso,
500 W. University, El Paso, TX 79968, USA
{vladik,olgak}@utep.edu
[2] Department of Mathematical Sciences, New Mexico State University,
Las Cruces, NM 88003, USA
hunguyen@nmsu.edu
[3] Faculty of Economics, Chiang Mai University, Chiang Mai 50200, Thailand
songsakecon@gmail.com

Abstract. One of the main applications of science and engineering is to predict future value of different quantities of interest. In the traditional statistical approach, we first use observations to estimate the parameters of an appropriate model, and then use the resulting estimates to make predictions. Recently, a relatively new *predictive* approach has been actively promoted, the approach where we make predictions directly from observations. It is known that in general, while the predictive approach requires more computations, it leads to more accurate predictions. In this paper, on the practically important example of robust interval uncertainty, we analyze how more accurate is the predictive approach. Our analysis shows that predictive models are indeed much more accurate: asymptotically, they lead to estimates which are $\sqrt{n}$ more accurate, where n is the number of estimated parameters.

1 Formulation of the Problem

Predictions Are Important. One of the main applications of science and engineering is to predict what will happen in the future:

- In science, we are most interesting in predicting what will happen "by itself" – e.g., where the Moon will be a year from now.
- In engineering, we are more interested in what will happen if we apply a certain control strategy – e.g., where a spaceship will be if we apply a certain trajectory correction.

In both science and engineering, prediction is one of the main objectives.

© Springer International Publishing AG 2018
V. Kreinovich et al. (eds.), *Predictive Econometrics and Big Data*, Studies in Computational Intelligence 753, https://doi.org/10.1007/978-3-319-70942-0_13

Traditional Statistics Approach to Prediction: Estimate then Predict.
The traditional statistical approach to prediction problems (see, e.g., [6]) is as
follows:

- First, we fix a statistical model with unknown parameters. For example,
  we can assume that the dependence of some quantity y on the quantities
  $x_1, \ldots, x_n$ is described by a linear dependence $y = a_0 + \sum_{i=1}^{n} a_i \cdot x_i + \varepsilon$, where ε
  is normally distributed with 0 mean and some standard deviation σ. In this
  case, the parameters are a_0, a_1, ..., a_n, and σ.
- Then, we use the observations to confirm this model and estimate the values
  of these parameters.
- After that, we use the model with the estimated values of the parameters to
  make the corresponding predictions.

Traditional Statistical Approach to Prediction: Advantages and Limitations. In the traditional approach, when we perform estimations, we do not
take into account what exactly characteristic we plan to predict. In the above
example, the same estimates for the parameters a_i and σ are used, whether we
are trying to predict the future value of the quantity y or whether we are trying to predict a different quantity z that depends on y and on several other
quantities.

A natural advantage of this approach is that a computationally intensive
parameter estimation part is performed only once, and the resulting estimates
can then be used to solve many different prediction problems. In the past, when
computations were much slower than now, this was a big advantage: by using pre-
computer estimates for the values of the corresponding parameters, we can perform many different predictions fast, without the need to re-do time-consuming
parameter estimation part.

With this advantages, come a potential limitation: hopefully, by tailoring
parameter estimation to a specific prediction problem, we may able to make
more accurate predictions.

Predictive Approach. In the past, because of the computer limitations, we
had to save on computations, and thus, the traditional approach was, in most
cases, all we could afford. However, now computers have become much faster. As
a result, in many practical situations, it has become possible to perform intensive
computations in a short period of time.

As a result, taking into account the above disadvantage of the traditional
approach, many researchers now advocate to use *predictive* approach to statistics,
in which we directly solve the prediction problem – i.e., in other words, on
the intermediate step of estimating the parameters, we take into account what
exactly quantities we need to predict; see, e.g., [1–3].

What We Do in This Paper. There are many examples of successful use of
the predictive approach. However, most of these examples remain anecdotal.

In this paper, on a practically important simple example of robust interval uncertainty, we prove a general result showing that predictive models indeed lead to more accurate predictions. Moreover, we provide a numerical measure of accuracy improvement.

2   Robust Interval Uncertainty: A Brief Reminder

Measurement Uncertainty. Data processing starts with values that come from measurement or from an expert estimate. Expert estimates are often important, but of course, a measuring instrument provides much more data than an expert. As a result, the overwhelming majority of data values come from measurements.

With the exception of simplest cases like counting number of people in a small group, measurement are not 100% accurate: the measurement result $\widetilde{x}$ is, in general, different from the actual (unknown) value of the corresponding quantity. In other words, in general, we have a non-zero *measurement error* $\Delta x \stackrel{\text{def}}{=} \widetilde{x} - x$.

What do We Know About Measurement Uncertainty: Case When We Know the Probability Distribution and Case of Robust Interval Uncertainty. In some situations, we know the probability distribution of the measurement error; for example, in many practical cases, we know that the measurement error is normally distributed, with 0 mean and known standard deviation σ.

However, in many practical situations, the only information that we have about the measurement error Δx is the upper bound Δ on its absolute value – the bound provided by the manufacturer of the measuring instrument; see, e.g., [5].

In other words, we only know that the probability distribution of the measurement error Δx is located on the interval $[-\Delta, \Delta]$, but we do not have any other information about the probability distribution. Such *interval uncertainty* is a particular case of the general *robust statistics*; see, e.g., [4].

Why cannot we always get this additional information? To get information about $\Delta x = \widetilde{x} - x$, we need to have information about the actual value x. In many practical situations, this is possible. Namely, in addition to the current measuring instrument (MI), we often also have a much more accurate ("standard") MI, so much more accurate that the corresponding measurement error can be safely ignored in comparison with the measurement error of our MI, and thus, the results of using the standard MI can be taken as the actual values.

In such a situation, we can find the probability distribution for the measurement error Δx if, for each of several quantities, we measure this quantity both by using the current MI and by using the standard MI. The difference between the two measurement results is a good approximation to the corresponding measurement error. Thus, the collection of such differences is a sample from the desired probability distribution for Δx. Based on this sample, we can find the corresponding probability distribution.

In many situations, however, our MI is already state-of-the-art, no more-accurate standard MI is possible. For example, in fundamental science, when we perform state-of-the-art measurements, we use state-of-the-art measuring instruments. For a billion-dollar project like space telescope or particle super-collider, the best MI are used. In this case, it is not possible to apply the above technique, so the best we can do is to use the bound Δ on the measurement error.

Another frequent case when we have to use Δ is the case of routine manufacturing. In this case, theoretically, we can calibrate every sensor, but sensors are cheap and calibrating them costs a lot – since it means using expensive state-of-the-art standard MIs. In routine manufacturing, such a calibration is just not financially possible – and not needed. For example, a simple thermometer for measuring a body temperature is reasonable cheap. If we had to calibrate each thermometer, it would become an order of magnitude more expensive – and what is the purpose? Honestly, all we need to know is whether a patient has a fever and, if yes, how severe, but the difference between, say 38.1 and 38.2 will not result in any changes in medical diagnosis or treatment.

Robust Interval Uncertainty Is What We Consider in This Paper. In view of the practical importance, in this paper, we consider the case of robust interval uncertainty.

3   Comparing Predictive and Traditional Statistics on the Example of Robust Interval Uncertainty: Analysis of the Problem

Let Us Describe the Traditional Approach in Precise Terms. Let y denote the quantity that we would like to predict.

To predict a quantity, we need to know the relation between this future quantity y and certain "estimate-able" quantities $x_1, \ldots, x_n$. Then, to predict y, we:

- estimate the quantities $x_1, \ldots, x_n$ based on the measurement results, and then
- use these estimates and the known relation between y and x_i to predict the desired future value y.

Let us denote the corresponding relation between y and x_i by $y = f(x_1, \ldots, x_n)$. Let us denote the measurement-based estimates for the quantities x_i by $\widetilde{x}_i$. In these terms, after generating these estimates, we get the following prediction for y:

$$\widetilde{y} \stackrel{\text{def}}{=} f(\widetilde{x}_1, \ldots, \widetilde{x}_n).$$

The quantities x_i are estimated based on measurement results. Let $v_1, \ldots, v_N$ denote all the quantities whose measurement results are used to estimate the quantities x_i. This estimation is based on the known relation between x_i and v_j.

Let us denote this relation by $x_i = g_i(v_1, \ldots, v_N)$, and let us denote the result of measuring each quantity v_j by $\widetilde{v}_j$. In these terms, the process of computing estimates $\widetilde{x}_i$ for the quantities x_i consists of the following two steps:

- first, we measure the quantities $v_1, \ldots, v_N$;
- then, the results $\widetilde{v}_1, \ldots, \widetilde{v}_N$ of measuring these quantities are used to produce the estimates $\widetilde{x}_i = g_i(\widetilde{v}_1, \ldots, \widetilde{v}_N)$.

Overall, the traditional approach takes the following form:

- first, we measure the quantities $v_1, \ldots, v_N$;
- then, the results $\widetilde{v}_1, \ldots, \widetilde{v}_N$ of measuring these quantities are used to produce the estimates $\widetilde{x}_i = g_i(\widetilde{v}_1, \ldots, \widetilde{v}_N)$;
- finally, we use the estimates $\widetilde{x}_i$ to compute the corresponding prediction

$$\widetilde{y} = f(\widetilde{x}_1, \ldots, \widetilde{x}_n).$$

How Will Predictive Approach Look in These Terms. The predictive approach means that, instead of first estimating the parameters x_i and then using these parameters to predict y, we predict y based directly on the measurement results v_j.

To make such a prediction, we need to know the relation between the predicted quantity y and the measurement results. Since we know that $y = f(x_1, \ldots, x_n)$ and that $x_i = g_i(v_1, \ldots, v_N)$, we thus conclude that $y = F(v_1, \ldots, v_N)$, where we denoted

$$F(v_1, \ldots, v_N) \overset{\text{def}}{=} f(g_1(v_1, \ldots, v_N), \ldots, g_n(v_1, \ldots, v_N)).$$

In these terms, the predictive approach to statistics takes the following form:

- first, we measure the quantities $v_1, \ldots, v_N$;
- then, the results $\widetilde{v}_1, \ldots, \widetilde{x}_N$ of measuring these quantities are used to produce the prediction $\widetilde{y} = F(\widetilde{v}_1, \ldots, \widetilde{v}_N)$.

How Accurate are These Estimates and Predictions? We are interested in the accuracy of the corresponding estimates and predictions.

For each estimated quantity x_i, the estimation error Δx_i is naturally defined as the difference $\widetilde{x}_i - x_i$ between the estimate $\widetilde{x}_i = g_i(\widetilde{v}_1, \ldots, \widetilde{v}_N)$ and the actual value $x_i = g_i(v_1, \ldots, v_N)$, i.e., the value that we would have got if we knew the exact values v_j of the measured quantities v_j. Similarly, for the prediction, the prediction error Δx_i is naturally defined as the difference $\widetilde{y} - y$ between the estimate $\widetilde{y} = f(\widetilde{x}_1, \ldots, \widetilde{x}_n)$ and the actual value $y = g_i(x_1, \ldots, x_n)$, i.e., the value that we would have got if we knew the exact values x_i of the estimated quantities x_i.

Measurements are usually reasonably accurate, which means that the measurement errors Δv_j are reasonably small. So, we can substitute the formula $v_j = \widetilde{v}_j - \Delta v_j$, expand the resulting expression for

$$\Delta x_i = g_i(\widetilde{v}_1, \ldots, \widetilde{v}_N) - g_i(v_1, \ldots, v_N)$$
$$= g_i(\widetilde{v}_1, \ldots, \widetilde{v}_N) - g_i(\widetilde{v}_1 - \Delta v_1, \ldots, \widetilde{v}_N - \Delta v_N)$$

in Taylor series, and keep only linear terms in this expansion. As a result, we get the following formula:

$$\Delta x_i = \sum_{i=1}^{N} g_{ij} \cdot \Delta v_j,$$

where we denoted $g_{ij} \overset{\text{def}}{=} \dfrac{\partial g_i}{\partial v_j}$.

What can we conclude about the value Δx_i? The only thing we know about each of the measurement errors Δv_j is that this measurement error can take any value from the interval $[-\Delta_j, \Delta_j]$. The above sum attains its largest possible value when each of the terms attains its largest value.

- when $g_{ij} \geq 0$, the term $g_{ij} \cdot \Delta v_j$ is an increasing function of Δv_j, so its maximum is attained when Δv_j attains its largest possible value $\Delta v_j = \Delta_j$; the resulting largest value of this term is $g_{ij} \cdot \Delta_j$;
- when $g_{ij} < 0$, the term $g_{ij} \cdot \Delta v_j$ is a decreasing function of Δv_j, so its maximum is attained when Δv_j attains its smallest possible value $\Delta v_j = -\Delta_j$; the resulting largest value of this term is $-g_{ij} \cdot \Delta_j$.

In both cases, the largest possible value of the term is equal to $|g_{ij}| \cdot \Delta_j$. Thus, the largest possible value Δ_i^x of Δx_i is equal to

$$\Delta_i^x = \sum_{j=1}^{N} |g_{ij}| \cdot \Delta_j. \tag{1}$$

One can easily check that the smallest possible value of Δx_i is equal to $-\Delta_i^x$. Thus, possible values of Δx_i form an interval $[-\Delta_i^x, \Delta_i^x]$.

Similarly, based on the estimates $\widetilde{x}_i$ and bounds Δ_i^x on the estimation errors, we can conclude that the possible values of the prediction error lie in the interval $[-\Delta, \Delta]$, where

$$\Delta = \sum_{i=1}^{n} |f_i| \cdot \Delta_i^x, \tag{2}$$

and we denoted $f_i \overset{\text{def}}{=} \dfrac{\partial f}{\partial x_i}$.

Alternatively, if we use the function $F(v_1, \ldots, v_N)$ to directly predict the value y from the measurement results, we conclude that the possible value of the prediction error lie in the interval $[-\delta, \delta]$, where

$$\delta = \sum_{j=1}^{N} |F_j| \cdot \Delta_j, \tag{3}$$

and we denoted $F_j \overset{\text{def}}{=} \dfrac{\partial F}{\partial v_j}$.

Preliminary Conclusion. Depending on whether we consider the traditional statistical approach or the predictive approach, we get the same estimate $\tilde{y}$ for the predicted quantity y. However, for the accuracy Δy, we have, in general, different bounds:

- if we use the traditional approach, then we get the bound Δ as described by the formulas (1) and (2);
- alternatively, if we use the predictive approach, we get the bound δ as described by the formula (3).

Comparing the Two Bounds. One can see that δ is the actual bound: in principle, the value δ can be attained if we take appropriate values of $\Delta v_j = \Delta_j \cdot \text{sign}(F_j)$.

Since all possible values of Δy also lie in the interval $[-\Delta, \Delta]$, the value δ also lies in this interval, thus the estimate δ coming from the predictive approach is smaller than or equal to the traditional estimate Δ. But is it better? Let us compare the two expressions.

If we substitute the expression (1) into the formula (2), we conclude that

$$\Delta - \sum_{i=1}^{n}\left(|f_i| \cdot \left(\sum_{j=1}^{N} |g_{ij}| \cdot \Delta_j\right)\right),$$

i.e., equivalently,

$$\Delta = \sum_{j=1}^{n} C_j, \tag{4}$$

where we denoted

$$C_j \overset{\text{def}}{=} \sum_{i=1}^{n} |f_i| \cdot |g_{ij}| \cdot \Delta_j.$$

Since $|a \cdot b| = |a| \cdot |b|$ and $\Delta_j > 0$, we can thus conclude that

$$C_j = \sum_{i=1}^{N} |c_{ij}|, \tag{5}$$

where we denoted $c_{ij} \overset{\text{def}}{=} f_i \cdot g_{ij} \cdot \Delta_j$.

On the other hand, the formula (3) takes the form

$$\delta = \sum_{j=1}^{n} c_j, \tag{6}$$

where $c_j \overset{\text{def}}{=} |F_j| \cdot \Delta_j$. By using the chain rule, we conclude that the derivative F_j of the composition function $F(v_1, \ldots, v_N)$ takes the form $F_j = \sum_{i=1}^{n} f_i \cdot g_{ij}$. Thus, the coefficient c_j in the formula (6) has the form

$$c_j = \left|\sum_{i=1}^{n} f_i \cdot g_{ij}\right| \cdot \Delta_j,$$

i.e., equivalently,

$$c_j = \left| \sum_{i=1}^{N} c_{ij} \right|. \tag{7}$$

By comparing formulas (4)–(5) with formulas (6)–(7), we can see that indeed $\delta \leq \Delta$: indeed, since $|a + b| \leq |a| + |b|$, we have

$$c_j = \left| \sum_{i=1}^{n} c_{ij} \right| \leq \sum_{i=1}^{n} |c_{ij}| = C_j$$

and thus, indeed

$$\delta = \sum_{j=1}^{N} c_j \leq \sum_{j=1}^{N} C_j = \Delta.$$

Is δ smaller? If yes, how smaller? To answer these equations, let us take into account that, in principle, each term $c_{ij} = f_i \cdot g_{ij} \cdot \Delta_j$ can take any real value, positive and negative. A priori, we do not have any reason to believe that positive values will be more frequent than negative ones, so it is reasonable to assume that the mean value of each such term is 0. Again, there is no reason to assume that the values c_{ij} are different, so it makes sense to assume that all these values are identically distributed. Finally, there is no reason to believe that there is correlation between different values, so its makes to consider them to be independent.

Under these assumptions, for large n, the sum $\sum_{i=1}^{n} c_{ij}$ is normally distributed, with 0 mean and variance which is n times larger than the variance σ^2 of the original distribution of c_{ij}. Thus, the means value of the absolute value c_j of this sum is proportional to its standard deviation $\sigma \cdot \sqrt{n}$.

On the other hand, the expected value μ of each term $|c_{ij}|$ is positive, thus, the expected value of the sum $C_j = \sum_{i=1}^{n} |c_{ij}|$ of n such independent terms is equal to $\mu \cdot n$.

For large n, $\mu \cdot n \gg \sigma \cdot \sqrt{n}$. Thus, we arrive at the following conclusion.

4   Conclusion

In this paper, we compare:

- the traditional statistical approach, in which we first use the observations to estimate the values of the parameters and then use these estimates for prediction, and
- the predictive approach to statistics, in which we make predictions directly from observations.

We make this comparison on the example of the practically important case of robust interval uncertainty, when the only information that we have about the corresponding measurement error is the upper bound provided by the manufacturer of the corresponding measurement instrument.

It turns out that while predictive techniques require more computations, they result in much more accurate estimates: asymptotically, $\sqrt{n}$ times more accurate, where n is the total number of parameters estimated in the traditional approach.

Acknowledgments. We acknowledge the partial support of the Center of Excellence in Econometrics, Faculty of Economics, Chiang Mai University, Thailand. This work was also supported in part by the National Science Foundation grants HRD-0734825 and HRD-1242122 (Cyber-ShARE Center of Excellence) and DUE-0926721, and by an award "UTEP and Prudential Actuarial Science Academy and Pipeline Initiative" from Prudential Foundation.

References

1. Briggs, W.: Uncertainty: The Soul of Modeling, Probability & Statistics. Springer, Cham (2016)
2. Dutta, J.: On predictive evaluation of econometric models. Int. Econ. Rev. **21**(2), 379–390 (1980)
3. Gneiting, T., Balabdaoui, F., Raftery, A.E.: Probabilsitic forecasts, calibration, and sharpness. J. R. Stat. Soc. Part B **69**(2), 243–268 (2007)
4. Huber, P.J., Ronchetti, E.M.: Robust Statistics. Wiley, Hoboken (2009)
5. Rabinovich, S.G.: Measurement Errors and Uncertainty: Theory and Practice. Springer, Berlin (2005)
6. Sheskin, D.J.: Handbook of Parametric and Nonparametric Statistical Procedures. Chapman & Hall/CRC, Boca Raton (2011)

Quantitative Justification for the Gravity Model in Economics

Vladik Kreinovich[1(✉)] and Songsak Sriboonchitta[2]

[1] Department of Computer Science, University of Texas at El Paso,
500 W. University, El Paso, TX 79968, USA
vladik@utep.edu
[2] Faculty of Economics, Chiang Mai University, Chiang Mai 50200, Thailand
songsakecon@gmail.com

Abstract. The gravity model in economics describes the trade flow between two countries as a function of their Gross Domestic Products (GDPs) and the distance between them. This model is motivated by the *qualitative* similarity between the desired dependence and the dependence of the gravity force (or potential energy) between the two bodies on their masses and on the distance between them. In this paper, we provide a *quantitative* justification for this economic formula.

1 Gravity Model in Economics: A Brief Introduction

What is gravity model. It is known that, in general:

- neighboring countries trade more than distant ones, and
- countries with larger Gross Domestic Product (GDP) g have a higher volume of trade than countries with smaller GDP.

Thus, in general, the trade flow t_{ij} between the two countries i and j:

- increases when the GDPs g_i and g_j increase and
- decreases with the distance r_{ij} increases.

A qualitatively similar phenomenon occurs in physics: the gravity force f_{ij} between the two bodies:

- increases when their masses m_i and m_j increase and
- decreases with the distance between then increases.

Similarly, the potential energy e_{ij} of the two bodies at distance r_{ij}:

- increases when the masses increase and
- decreases when the distance r_{ij} increases.

For the gravity force and for the potential energy, there are simple formulas:

$$f_{ij} = G \cdot \frac{m_i \cdot m_j}{r_{ij}^2}; \quad e_{ij} = G \cdot \frac{m_i \cdot m_j}{r_{ij}},$$

V. Kreinovich et al. (eds.), *Predictive Econometrics and Big Data*, Studies in Computational Intelligence 753, https://doi.org/10.1007/978-3-319-70942-0_14

for some constant G. Both these formulas are a particular case of a general formula

$$G \cdot \frac{m_i \cdot m_j}{r_{ij}^{\alpha}} :$$

for the force, we take $\alpha = 2$, and for the energy, we take $\alpha = 1$.

By using the analogy with the gravity formulas, researchers have proposed to use a similar formula to describe the dependence of the trade flow t_{ij} on the GDPs g_i and on the distance r_{ij}:

$$t_{ij} = G \cdot \frac{g_i \cdot g_j}{r_{ij}^{\alpha}}.$$

This formula – known as the *gravity model* in economics – has indeed been successfully used to describe the trade flows between different countries; see, e.g., [2–6].

Remaining problem and what we do in this paper. While an analogy with gravity provides a *qualitative* explanation for the gravity model, it is desirable to have a *quantitative* explanation as well. Such an explanation is provided in this paper.

2   Analysis of the Problem

What we want. We would like to have a formula that estimates the trade flow between the two countries t_{ij} as a function of their GDPs g_i and g_j and of the distance r_{ij} between the two countries. In other words, we would like to come up with a function $F(a, b, c)$ for which

$$t_{ij} = F(g_i, g_j, t_{ij}). \tag{1}$$

To describe the corresponding function $F(a, b, c)$, let us describe the natural properties of such a function.

First natural property: additivity. At first glance, the notion of a country seems to be very clear and well defined. However, there are many examples where this notion is not that clear. Sometimes, a country becomes a loose confederation of practically independent states. In other cases, several countries form such a close trade union – from Benelux to European Union – that most trade is regulated by the super-national organs and not by individual countries.

In all such cases, we have several different entities $i_1, \ldots, i_k, \ldots, i_\ell$ located nearby forming a single super-entity. If we apply the formula (1) to each individual entity i_k, we get the expression

$$t_{i_k j} = F(g_{i_k}, g_j, r_{i_k j}).$$

Since all the entities i_k are located close to each other, we can assume that the distances $r_{i_k j}$ are all the same: $r_{i_k j} = r_{ij}$. Thus, the above expression takes the form $t_{i_k j} = F(g_{i_k}, g_j, r_{ij})$.

By adding all these expressions, we can come up with the trade flow between the whole super-entity i and the country j:

$$t_{ij} = \sum_{k=1}^{\ell} t_{i_k j} = \sum_{k=1}^{\ell} F(g_{i_k}, g_j, r_{ij}). \tag{2}$$

Alternatively, we can treat the super-entity as a single country with the overall GDP $g_i = \sum_{k=1}^{\ell} g_{i_k}$. In this case, by applying the formula (1) to this super-entity, we get

$$t_{ij} = F(g_i, g_j, r_{ij}) = F\left(\sum_{k=1}^{\ell} g_{i_k}, g_j, r_{ij}\right). \tag{3}$$

It is reasonable to require that our estimate for the trade flow should not depend on whether we treat this loose confederation a single country or as several independent countries. By equating the estimates (2) and (3), we conclude that

$$F(g_{i_1}, g_j, r_{ij}) + \ldots + F(g_{i_\ell}, g_j, r_{ij}) = F(g_{i_1} + \ldots + g_{i_\ell}, g_j, r_{ij}).$$

In other words, we must have the following additivity property for all possible values a, ..., a', and b:

$$F(a, b, c) + \ldots + F(a', b, c) = F(a + \ldots + a', b, c). \tag{4}$$

A similar argument can be make if we consider the case when j is a loose confederation of states. In this case, the requirement that our estimate for the trade flow should not depend on whether we treat this loose confederation as a single country or as several independent countries leads to

$$F(g_i, g_{j_1}, r_{ij}) + \ldots + F(g_i, g_{j_\ell}, r_{ij}) = F(g_i, g_{j_1} + \ldots + g_{j_\ell}, r_{ij}),$$

i.e., to

$$F(a, b, c) + \ldots + F(a, b', c) = F(a, b + \ldots + b', c). \tag{5}$$

Second natural property: scale-invariance. The numerical value of the distance depends on what unit we use for measuring distance. For example, the distance in miles in different from the same distance in kilometers. If we replace the original unit with a one which is λ times smaller, all numerical values of the distance multiply by λ, i.e., each original numerical value r_{ij} is replaced by a new numerical value

$$r'_{ij} = \lambda \cdot r_{ij}.$$

It is reasonable to require that the estimates for the trade flow should not depend on what unit we use. Of course, we cannot simply require that $F(g_i, g_j, r_{ij}) = F(g_i, g_j, \lambda \cdot r_{ij})$ – this would mean that the trade flow does not depend on the distance at all. This is OK, since the numerical value of the trade flow also depends on what units we use: we get different numbers if we use US dollars or Thai Bahts. It is therefore reasonable to require that when we change the unit for measuring r_{ij}, then after an appropriate change $t_{ij} \to t'_{ij} = \mu \cdot t_{ij}$ in the measuring unit for trade flow we get the same formula. In other words, we require that for every $\lambda > 0$, there exists a $\mu > 0$ for which

$$F(g_i, g_j, \lambda \cdot r_{ij}) = \mu \cdot F(g_i, g_j, r_{ij}).$$

In other words, we require that

$$F(a, b, \lambda \cdot c) = \mu \cdot F(a, b, c). \tag{6}$$

Third natural property: monotonicity. The final natural property is that as the distance increases, the trade flow should decrease. In other words, the function $F(a, b, c)$ should be a decreasing function of c.

Now, we are ready to formulate our main result.

3   Definitions and the Main Result

Definition 1

- A non-negative function $F(a, b, c)$ of three non-negative variables is called additive *if the following two equalities hold for all possible values* $a, \ldots, a', b, \ldots, b'$, *and* c:

$$F(a, b, c) + \ldots + F(a', b, c) = F(a + \ldots + a', b, c);$$

$$F(a, b, c) + \ldots + F(a, b', c) = F(a, b + \ldots + b', c).$$

- A function $F(a, b, c)$ is called scale-invariant *if for every* λ, *there exists a* μ *for which, for all* a, b, *and* c, *we have*

$$F(a, b, \lambda \cdot c) = \mu \cdot F(a, b, c).$$

- A function $F(a, b, c)$ is called a trade function *if it is additive, scale-invariant, and increasing as a function of* c.

Proposition 1. *Every trade function has the form* $F(a, b, c) = G \cdot \dfrac{a \cdot b}{c^{\alpha}}$ *for some constants* G *and* α.

Discussion. Thus, we have indeed justified the gravity model.

Proof of Proposition 1

$1°$. Let us first use the additivity property.

For every b and c, we can consider an auxiliary function $f_{bc}(a) \stackrel{\text{def}}{=} F(a, b, c)$. In terms of this function, the first additivity property takes the form

$$f_{bc}(a + \ldots + a') = f_{bc}(a) + \ldots + f_{bc}(a').$$

Functions of one variable that satisfy this property are known as *additive*. It is known – see, e.g., [1] – that every non-negative additive function has the form $f(a) = k \cdot a$. Thus, $F(a, b, c) = f_{bc}(a)$ is equal to

$$F(a, b, c) = a \cdot k(b, c)$$

for some function $k(b, c)$.

Substituting this expression into the second additivity requirement, we conclude that

$$a \cdot k(b + \ldots + b', c) = a \cdot k(b, c) + \ldots + a \cdot k(b', c).$$

Dividing both sides of this equality by a, we conclude that

$$k(b + \ldots + b', c) = k(b, c) + \ldots + k(b', c).$$

Thus, the function $k_c(b) \stackrel{\text{def}}{=} k(b, c)$ is also additive. Hence, $k(b, c) = k_c(b) = b \cdot q(c)$ for some constant $q(c)$ depending on c. Substituting this expression for $k(b, c)$ into the formula describing $F(a, b, c)$ in terms of $k(b, c)$, we conclude that $F(a, b, c) = a \cdot b \cdot c(q)$.

Hence, to complete the proof, it is sufficient to find the function $q(c)$.

$2°$. For $a = b = 1$, we have $F(a, b, c) = q(c)$. Thus, for these a and b, the fact that $F(a, b, c)$ is a decreasing function of c implies that $q(c)$ is also an decreasing function of c.

$3°$. To find the function $q(c)$, let us now use scale invariance

$$F(a, b, \lambda \cdot c) = \mu(\lambda) \cdot F(a, b, c).$$

Substituting $F(a, b, c) = a \cdot b \cdot q(c)$ into this equality and dividing both sides by $a \cdot b$, we conclude that $q(\lambda \cdot c) = \mu(\lambda) \cdot q(c)$.

For every λ_1 and λ_2, we have

$$q((\lambda_1 \cdot \lambda_2) \cdot c) = \mu(\lambda_1 \cdot \lambda_2) \cdot q(c).$$

On the other hand, we also have $q(\lambda_2 \cdot c) = \mu(\lambda_2) \cdot q(c)$ and thus,

$$q(\lambda_1 \cdot (\lambda_2 \cdot c)) = \mu(\lambda_1) \cdot q(\lambda_2 \cdot c) = \mu(\lambda_1) \cdot \mu(\lambda_2) \cdot q(c).$$

By equating these two expressions for the same quantity $q(\lambda_1 \cdot \lambda_2 \cdot c)$, we conclude that

$$\mu(\lambda_1 \cdot \lambda_2) \cdot q(c) = \mu(\lambda_1) \cdot \mu(\lambda_2) \cdot q(c).$$

Dividing both sides by $q(c)$, we get

$$\mu(\lambda_1 \cdot \lambda_2) = \mu(\lambda_1) \cdot \mu(\lambda_2).$$

Functions $\mu(\lambda)$ with this property are known as *multiplicative*.

Here, for every c, we have $\mu(\lambda) = \dfrac{q(\lambda \cdot c)}{q(c)}$. In particular, for $c = 1$, we get

$\mu(\lambda) = \dfrac{q(\lambda)}{q(1)}$. Since $q(c)$ is an increasing function, we conclude that $\mu(\lambda)$ is also an increasing function.

It is known [1] that every monotonic multiplicative function has the form $\mu(\lambda) = \lambda^{-\alpha}$ for some $\alpha > 0$. From $q(\lambda) = \mu(\lambda) \cdot q(1)$, we can conclude that $q(c) = G \cdot c^{-\alpha}$, where we denoted $G \overset{\text{def}}{=} q(1)$.

The proposition is proven.

4   Where Do We Go from Here

Trade flow may depend on other characteristics. In the previous text, we assumed that the trade flow depends only on the GDPs and on the distance. In reality, the trade flow may also other depend on other characteristics, such as the country's population p_i. Indeed, intuitively, the larger the population, the more it consumes, so the larger its trade flow with other countries.

Similar to GDP, population is an additive property, in the sense that if two countries merge together, their population adds up. So, a natural question is: how can we describe the dependence of the trade flow on two or more additive characteristics?

Let us describe this problem in precise terms. Let us consider the case when each country is described by several additive characteristics, i.e., that g_i is now a vector consisting of several components $g_i = (g_{1i}, \ldots, g_{mi})$. We are interested in finding the dependence $t_{ij} = F(g_i, g_j, r_{ij})$.

Let us describe the reasonable properties of this dependence.

Additivity and monotonicity. Similarly to the GDP-only case, we can conclude that

$$F(g_{i_1} + \ldots + g_{i_\ell}, g_j, r_{ij}) = F(g_{i_1}, g_j, r_{ij}) + \ldots + F(g_{i_\ell}, g_j, r_{ij})$$

and

$$F(g_i, g_{j_1} + \ldots + g_{j_\ell}, r_{ij}) = F(g_i, g_{j_1}, r_{ij}) + \ldots + F(g_i, g_{j_\ell}, r_{ij}).$$

Also, similarly to the GDP-only case, it makes sense to require that the function $F(a, b, c)$ is a decreasing function of c.

Definition 2. Let $m > 1$.

- *A non-negative function $F(a, b, c)$ of three non-negative variables $a, b \in \mathrm{IR}^m$ and $c \in \mathrm{IR}$ is called* additive *if the following two equalities hold for all possible values $a, \ldots, a', b, \ldots, b'$, and c:*

$$F(a, b, c) + \ldots + F(a', b, c) = F(a + \ldots + a', b, c);$$

$$F(a, b, c) + \ldots + F(a, b', c) = F(a, b + \ldots + b', c).$$

- *A function $F(a, b, c)$ is called* scale-invariant *if for every λ, there exists a μ for which, for all a, b, and c, we have*

$$F(a, b, \lambda \cdot c) = \mu \cdot F(a, b, c).$$

- *A function $F(a, b, c)$ is called a* trade function *if it is additive, scale-invariant, and increasing as a function of c.*

Proposition 2. *Every trade function has the form*

$$F(g_i, g_j, r_{ij}) = \frac{\sum_\beta \sum_\gamma G_{\beta\gamma} \cdot g_{\beta i} \cdot g_{\gamma j}}{r_{ij}^\alpha}$$

for some constants $G_{\beta\gamma}$ and α.

Example. For the case of GDP g_i and population p_i, we have

$$t_{ij} = \frac{G_{gg} \cdot g_i \cdot g_j + G_{gp} \cdot g_i \cdot p_j + G_{pg} \cdot p_i \cdot g_j + G_{pp} \cdot p_i \cdot p_j}{r_{ij}^\alpha}.$$

An interesting property of this example is that, in contrast to the GDP-only case, when we always had $t_{ij} = t_{ji}$, we can have "asymmetric" trade flows for which $t_{ij} \neq t_{ji}$.

Proof of Proposition 2 is similar to the proof of Proposition 1: first additivity requirement implies that $F(a, b, c)$ is linear in a, second – that it is linear in b, so it is bilinear in a and b. Now, scale-invariance implies that all the coefficients of this bilinear dependence be proportional to $r_{ij}^{-\alpha}$ for some $\alpha > 0$.

Discussion. It would be nice to test these formulas on real data.

Acknowledgments. We acknowledge the partial support of the Center of Excellence in Econometrics, Faculty of Economics, Chiang Mai University, Thailand. This work was also supported in part by the National Science Foundation grants HRD-0734825 and HRD-1242122 (Cyber-ShARE Center of Excellence) and DUE-0926721, and by an award "UTEP and Prudential Actuarial Science Academy and Pipeline Initiative" from Prudential Foundation.

References

1. Aczel, J., Dhombres, J.: Functional Equations in Several Variables. Cambridge University Press, Cambridge (1989)
2. Anderson, J.E.: A theoretical foundation for the gravity equation. Am. Econ. Rev. **69**(1), 106–116 (1979)
3. Anderson, J.E., Van Wincoop, E.: Gravity with gravitas: a solution to the border puzzle. Am. Econ. Rev. **93**(1), 170–192 (2003)
4. Bergstrand, J.H.: The gravity equation in international trade: some microeconomic foundations and empirical evidence. Rev. Econ. Stat. **67**(3), 474–481 (1985)

5. Pastpipatkul, P., Boonyakunakorn, P., Sriboonchitta, S.: Thailand's export and ASEAN economic integration: a gravity model with state space approach. In: Huynh, V.-N., Inuigichi, M., Le, B., Le, B.N., Denoeux, T. (eds.) Proceedings of the 5th International Symposium on Integrated Uncertainty in Knowledge Modeling and Decision Making IUKM 2016, pp. 664–674. Springer, Cham (2016)
6. Tinbergen, J.: Shaping the World Economy: Suggestions for an International Economic Policy. The Twentieth Century Fund, New York (1962)

The Decomposition of Quadratic Forms Under Skew Normal Settings

Ziwei Ma[1], Weizhong Tian[2], Baokun Li[3], and Tonghui Wang[4(✉)]

[1] Department of Mathematical Sciences, New Mexico State University(USA)
and College of Science, Northwest A and F University, Xianyang, China
`ziweima@nmsu.edu`
[2] Department of Mathematical Sciences, Eastern New Mexico University (USA)
and School of Sciences, Xi'an University of Technology, Xi'an, China
`weizhong.tian@enmu.edu`
[3] School of Statistics, Southwestern University of Finance and Economics,
Chengdu, China
`bali@swufe.edu.cn`
[4] Department of Mathematical Sciences,
New Mexico State University, Las Cruces, USA
`twang@nmsu.edu`

Abstract. In this paper, the decomposition properties of noncentral skew chi-square distribution is studied. A given random variable U having a noncentral skew chi-square distribution with $k > 1$ degrees of freedom, can be partitioned into the sum of two independent random variables U_1 and U_2 such that U_1 has a noncentral skew chi-square distribution with 1 degree of freedom and U_2 has the noncentral chi-square distribution with $k-1$ degrees of freedom. Also if $k > 2$, this partition can be modified into $U = U_1 + U_2$, where U_1 has a noncentral skew chi-square distribution with 2 degrees of freedom and U_2 has a central chi-square distribution with $k - 2$ degrees of freedom. The densities of noncentral skew chi-square distributions with 1 degree of freedom, 2 degrees of freedom, and $k > 2$ degrees of freedom are derived, and their graphs are presented. For illustration of our main results, the linear regression model with skew normal errors is considered as an application.

Keywords: Skew normal distributions · Skew chi-square distributions
Quadratic form · Decomposition

1 Introduction

There are many real world problems in which we need to analyze the data which are asymmetrically distributed. We often assume that data are symmetrically or approximately normally distributed so that the classical statistical methods are applied. However, these data analysis results may have limitations such as a lack of robustness against departures from the normal distribution and invalid statistical inferences for skewed data. One of the solutions is to introduce an extra skewness (or shape) parameter to normal distributions, thus getting skew

© Springer International Publishing AG 2018
V. Kreinovich et al. (eds.), *Predictive Econometrics and Big Data*, Studies in Computational
Intelligence 753, https://doi.org/10.1007/978-3-319-70942-0_15

normal distributions which have location, scale and shape parameters, to fit various types of skewed data, see Azzalini and his collaborators' work [2–5] for univariate and multivariate cases, respectively (see more details in Azzalini and Capitanio [6]). In practice, the family of skew normal distributions is suitable for the analysis of skewed data which is unimodally distributed (see Hill, Arnold [1,11]).

For the situation when the location parameter is zero, the distribution of quadratic forms of skew normal vectors was discussed by Genton et al. [8] , Gupta and Huang [7] , and general skew-elliptical distribution was discussed by Fang [9] and Genton and Loperfido [10] . For the situation when location parameter is not zero, Wang et al. [12] defined the noncentral skew chi-square distribution, studied the quadratic forms of generalized skew normal vectors, and analyzed basic features such as moment generating function (mgf), independence and a version of Cochran's theorem. Later on, Ye and Wang [13] revisited the noncentral skew chi-square distribution, derived the probability density function (pdf) and applied this distribution to statistical inferences in linear mixed model with skew normal random effects and variance components models under skew normal setting (see Ye et al. [14]). In this paper, we study the decomposition of noncentral skew chi-square distributed random variable, and give two other representations of its pdf

This paper is organized as follows. Features of multivariate skew normal distribution and skew chi-square distribution such as their probability density function, moment generating function are discussed in Sect. 2. Decompositions of a noncentral skew chi-square distribution are studied in Sect. 3. For illustration, applications of main results are given in Sect. 4.

2   Skew-Normal and Skew Chi-Square Distributions

A random vector $Y \subset \mathbb{R}^k$ is said to have a *multivariate skew-normal distribution*, denoted by $Y \sim SN_k(\mu, \Sigma, \alpha)$, if its pdf is given by

$$f(y) = 2\phi_k(y; \mu, \Sigma) \Phi \left(\alpha' \Sigma^{-1/2} (y - \mu) \right),$$

where $\mu, \alpha \in \mathbb{R}^k$, Σ is a positive definite $k \times k$ matrix, $\phi_k(y; \mu, \Sigma)$ is the pdf of k-dimension normal distribution with mean μ and covariance matrix Σ, and $\Phi(\cdot)$ is the cumulative distribution function (cdf) of standard normal distribution $N(0, 1)$. The following lemmas will be used to prove our main results.

Lemma 2.1 *(Azzalini [6]). If $V \sim SN_k(\mu, \Sigma, \alpha)$, and A is an orthogonal $k \times k$ matrix, then*

$$A'V \sim SN_k(A'\mu, A'\Sigma A, A'\alpha).$$

Lemma 2.2 *(Wang et al. [12]). Let $Y \sim SN_n(\mu, I_n, \alpha)$. Then Y has the following properties.*

(a) *The moment generating function of* Y *is given by*

$$M_Y(t) = 2 \exp\left(t'\mu + \frac{t't}{2}\right) \Phi\left\{\frac{\alpha't}{(1 + \alpha'\alpha)^{1/2}}\right\},$$

for $t \in \mathbb{R}^n$ *and*

(b) *Two linear functions of* Y, $A'Y$ *and* $B'Y$ *are independent if and only if* (i) $A'B = 0$ *and* (ii) $A'\alpha = 0$ *or* $B'\alpha = 0$.

Lemma 2.3 *(Wang et al. [15]). Let* $Y \sim SN_n(\nu, I_n, \alpha_0)$, *and let* A *be a* $n \times k$ *matrix with full column rank, then the linear function of* Y, $A'Y \sim SN_k(\mu, \Sigma, \alpha)$, *where*

$$\mu = A'\nu, \qquad \Sigma = A'A, \qquad and \qquad \alpha = \frac{(A'A)^{-1/2}A'\alpha_0}{\sqrt{1 + \alpha_0'(I_n - A(A'A)^{-1}A')\alpha_0}}.$$

Recall that the noncentral chi-square distribution with k degrees of freedom and noncentrality parameter λ is defined as the distribution of $V'V$ where V follows k-dimensional normal distribution with location vector ν and scale matrix I_k, denoted as $V \sim N_k(\nu, I_k)$. The noncentral skew chi-square distribution was first defined in Wang et al. [12] and its modified version was given in Ye and Wang [13].

2.1   Noncentral Skew Chi-Square Distribution with Degrees of Freedom $k > 1$

Definition 1. *Let* $Y \sim SN_k(\mu, I_k, \alpha)$. *The distribution of* $U = Y'Y$ *is called as the* noncentral skew chi-square distribution *with degree of freedom* k, *the noncentrality parameter* $\lambda = \mu'\mu$, *and the skewness parameters* $\delta_1 = \mu'\alpha$, $\delta_2 = \alpha'\alpha$, *denoted by* $U \sim S\chi_k^2(\lambda, \delta_1, \delta_2)$.

The following Lemma lists the basic properties of noncentral skew chi-square distributions.

Lemma 2.4. *Let* $U \sim S\chi_k^2(\lambda, \delta_1, \delta_2)$.

(i) *(Ye and Wang [13]) The pdf of* U *is given by*

$$f_U(u; \lambda, \delta_1, \delta_2) = \frac{\exp\left\{-\frac{1}{2}(\lambda + u)\right\}}{\Gamma\left(\frac{1}{2}\right)\Gamma\left(\frac{k-1}{2}\right)2^{k/2-1}} h(u; \lambda, \delta_1, \delta_2), \quad u > 0, \qquad (1)$$

where

$$h(u; \lambda, \delta_1, \delta_2) = \int_{\sqrt{u}}^{\sqrt{u}} \exp\left(\lambda^{1/2}s\right)(u - s^2)^{\frac{k-3}{2}} \Phi\left\{\alpha_0\left(s - \lambda^{1/2}\right)\right\} ds$$

and $\alpha_0 = \lambda^{-1/2}\delta_1/(1 + \delta_2 - \delta_1^2/\lambda)^{1/2}$.

(ii) (Wang et al. [12]) The mgf of U is given by

$$M_U\left(t\right) = \frac{2\exp\left(t\left(1-2t\right)^{-1}\lambda\right)}{\left(1-2t\right)^{k/2}} \, \Phi\left\{\frac{2t\left(1-2t\right)^{-1}\delta_1}{\left[1+\delta_2\left(1-2t\right)^{-1}\right]^{1/2}}\right\} \quad (2)$$

for $0 < t < 1/2$.

Remark 2.1. Note that for $k = 1$,the integral of $h\left(x; \lambda, \delta_1, \delta_2\right)$ given in (1) is infinite so we need to discuss the noncentral skew chi-square distribution with 1 degree of freedom separately.

2.2 The Noncentral Skew Chi-Square Distribution with 1 Degree of Freedom

For $k = 1$, $Y^2 \sim S\chi_1^2\left(\lambda, \delta_1, \delta_2\right)$ with $Y \sim SN(\nu, 1, \alpha)$ where $\lambda = \nu^2, \delta_1 = \nu\alpha, \delta_2 = \alpha^2$ which implies these three parameters λ, δ_1, and δ_2 are determined by ν and α so that they are not independent parameters.

Theorem 1. *Suppose that $Y \sim SN\left(\nu, 1, \alpha\right)$ and $U = Y^2$, then the pdf of U is*

$$f_U\left(u\right) = u^{-1/2}\left\{\phi\left(\sqrt{u}; \nu, 1\right)\Phi\left(\alpha\left(\sqrt{u}-\nu\right)\right) + \phi\left(-\sqrt{u}; \nu, 1\right)\Phi\left(\alpha\left(-\sqrt{u}-\nu\right)\right)\right\} \quad (3)$$

for $u > 0$.

Proof. For any $u > 0$, consider the cdf of U

$$F_U\left(u\right) \equiv P\left(U \leq u\right) = P\left(Y^2 \leq u\right) = P\left(-\sqrt{u} \leq Y \leq \sqrt{u}\right)$$

$$= \int_{-\sqrt{u}}^{\sqrt{u}} 2\phi\left(s; \nu, 1\right)\Phi\left(\alpha\left(s-\nu\right)\right)\mathrm{d}s.$$

The desired results follows by taking the derivative of $F_U(u)$ with respect to u. $\qquad\square$

Remark 2.2. (i) When $\nu = 0$, the pdf of U given in (3) can be written as

$$f_U\left(u\right) = u^{-1/2}\phi\left(\sqrt{u}\right) = \frac{u^{-1/2}e^{-u/2}}{\sqrt{2\pi}} \quad \text{for} \quad u > 0$$

which is the pdf of $\chi_1^2(0)$ and is free of skewness parameter α.
(ii) When $\alpha = 0$, the pdf of U given in (3) can be written as

$$f_U(u, 1, \nu) = \frac{1}{2\sqrt{u}}\left(\phi(\sqrt{u}-\nu) + \phi(\sqrt{u}+\nu)\right) = \frac{1}{\sqrt{2\pi u}}e^{-(u+\nu^2)/2}\cosh(\nu\sqrt{u})$$

for $u > 0$ which is the pdf of $\chi_1^2(\lambda)$ with $\lambda = \nu^2$.

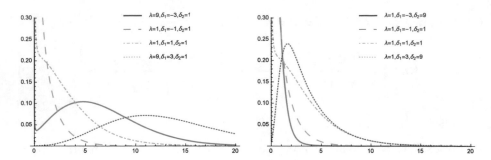

Fig. 1. Density curves of $S\chi_1^2$ with fixed δ_2 (left) and fixed λ (right)

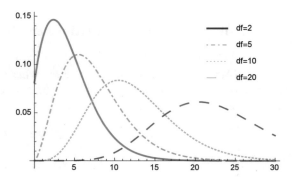

Fig. 2. Density curves of $S\chi_{df}^2(5, -4, 5)$ with degrees of freedom $df = 2, 5, 10, 20$

(iii) In general case, there is a relation among parameters λ, δ_1 and δ_2, which is $\delta_1^2 = \lambda\delta_2$. Therefore, these parameters are fully determined by ν and α. The density curves $S\chi_1^2(\lambda, \delta_1, \delta_2)$ are given Fig. 1. From Fig. 1, we can see that the changes in values of ν (taking values $-3, -1, 1, 3$) influence the shapes of the density curves for fixed $\alpha = 1$. Also, we can see that the changes in values of α (taking values $-3, -1, 1, 3$) influence the shape of density curves as well fixed location parameter $\nu = 1$.

Example 2.1. *Consider* $Y = (Y_1, \cdots, Y_k)' \sim SN_k(\mu\mathbf{1}_k, I_k, \alpha\mathbf{1}_k)$ *and let* $\overline{Y} = \frac{1}{k}\sum_{i=1}^{k} Y_i$. *Then it is easy to see that* $\sqrt{k}\,\overline{Y} \sim SN(\sqrt{k}\mu, 1, \sqrt{k}\alpha)$, *so by Definition 1, we have* $k\overline{Y}^2 \sim S\chi_1^2(\lambda, \delta_1, \delta_2)$ *with* $\lambda = k\mu^2, \delta_1 = k\mu\alpha$ *and* $\delta_2 = k\alpha^2$.

3   Decomposition of Noncentral Skew Chi-Square Distributions

First, we will prove a fundamental result of the decomposition of skew chi-square distributions.

Theorem 2. *Let $U \sim S\chi_k^2(\lambda, \delta_1, \delta_2)$ with $\delta_2 \neq 0$ and $k \geq 2$, then U can be partitioned into two independent random variables U_1 and U_2: $U = U_1 + U_2$, where $U_1 \sim S\chi_1^2(\lambda_1, \delta_1^*, \delta_2^*)$ and $U_2 \sim \chi_{k-1}^2(\lambda - \lambda_1)$ if and only if*
  (i) $\delta_i^* = \delta_i$ for $i = 1, 2$ and
  (ii) $\lambda_1 = \delta_1^2 / \delta_2$.

Proof. "If part" is trivial since the mgf of U can be rewritten as the product of the mgf's of U_1 and U_2. For the proof of "only if" part, let $Y_1 \sim SN(\mu_1, 1, \alpha_1)$ with $\alpha_1 = \sqrt{\delta_2}$ and $\mu_1 = \delta_1/\sqrt{\delta_2}$, and let $Y_2 \sim N_{k-1}(\mu_2, I_{k-1})$ with $\mu_2 = \left(\sqrt{\lambda - \delta_1^2/\delta_2}, 0, \cdots, 0\right)' \in \Re^{k-1}$. Assume that Y_1 and Y_2 are independent. Then the joint distribution of $Y = (Y_1, Y_2')'$ is $SN_k(\nu, I_k, \alpha)$, where $\nu = (\mu_1, \mu_2')'$ and $\alpha = (\alpha_1, 0, \cdots, 0)'$. Note that

$$\nu'\nu = \mu_1^2 + \mu_2'\mu_2 = \frac{\delta_1^2}{\delta_2} + \lambda - \frac{\delta_1^2}{\delta_2} = \lambda,$$

$$\nu'\alpha = \delta_1, \quad \text{and} \quad \alpha'\alpha = \delta_2,$$

which implies $Y'Y \sim S\chi_k^2(\lambda, \delta_1, \delta_2)$. On the other hand, $Y'Y = Y_1^2 + Y_2'Y_2$ where $Y_1^2 \sim S\chi_1^2(\lambda_1, \delta_1, \delta_2)$ with $\lambda_1 = \delta_1^2/\delta_2$ and $Y_2'Y_2 \sim \chi_{k-1}^2(\lambda - \lambda_1)$. Therefore, the desired decomposition is obtained with $U_1 = Y_1^2$ and $U_2 = Y_2'Y_2$.   □

Based on Theorem 2, the density of $U \sim S\chi_k^2(\lambda, \delta_1, \delta_2)$ can be derived.

Theorem 3. *The density of $U \sim S\chi_k^2(\lambda, \delta_1, \delta_2)$ with $k > 1$ is*

$$f_U(u) = \int_0^u g_1(u - v) h_1(v) \, dv \tag{4}$$

where $g_1(\cdot)$ is pdf of $S\chi_1^2(\lambda_1, \delta_1, \delta_2)$ and $h_1(\cdot)$ is the pdf of $\chi_{k-1}^2(\lambda_2)$ with $\lambda_1 = \frac{\delta_1^2}{\delta_2}$, and $\lambda_2 = \lambda - \lambda_1$.

Proof. By Definition 1, there exists $Y \sim SN_k(\mu, I_k, \alpha)$ such that $Y'Y \overset{d}{=} U$ where $\lambda = \mu'\mu$, $\delta_1 = \mu'\lambda$ and $\delta_2 = \alpha'\alpha$. From Theorem 2, we select $\alpha = (\alpha_1, 0, \cdots, 0)'$, in fact by Lemma 2.1, it is reasonable to assume that. It is easy to obtain $\alpha_1 = \delta_1/\sqrt{\delta_2}$, and $\lambda_1 = \delta_1^2/\delta_2$.   □

The following example shows the partition is not unique.

Example 3.1. *Suppose that $U \sim S\chi_3^2(4, 2, 4)$, then let $Y_i = (Y_{i1}, Y_{i2}, Y_{i3})' \sim SN_3(\mu_i, I_3, \alpha_i)$ for $i = 1, 2, 3$ with $\mu_1 = (1, \sqrt{3}, 0)'$, $\alpha_1 = (2, 0, 0)'$, $\mu_2 = (2, 0, 0)'$, $\alpha_2 = (1, \sqrt{3}, 0)'$, $\mu_3 = (\sqrt{2}, 0, \sqrt{2})'$, and $\alpha_3 = (\sqrt{2}, \sqrt{2}, 0)'$, then by definition of noncentral skew chi-square distributions, it is clear that $Y_i'Y_i \sim S\chi_3^2(4, 2, 4)$ for $i = 1, 2, 3$. However, the decompositions of these three quadratic forms are different.*
For $i = 1$, $\begin{pmatrix} Y_{11} \\ Y_{12} \end{pmatrix} \sim SN_2\left(\begin{pmatrix} 2 \\ 0 \end{pmatrix}, I_2, \begin{pmatrix} 1 \\ \sqrt{3} \end{pmatrix}\right)$ and $Y_{13} \sim N(0, 1)$ are independent which implies the decomposition of $Y_1'Y_1 = U_{11} + U_{12}$ where $U_{11} \sim S\chi_2^2(4, 2, 4)$ and $U_{12} \sim \chi_1^2(0)$ are independently distributed.

For $i = 2$, $Y_{21} \sim SN(1,2)$ and $\begin{pmatrix} Y_{21} \\ Y_{22} \end{pmatrix} \sim N_2 \left(\begin{pmatrix} \sqrt{3} \\ 0 \end{pmatrix}, I_2 \right)$ are independently distributed which implies the decomposition of $Y_2'Y_2 = U_{21} + U_{22}$ where $U_{21} \sim S\chi_1^2(1,2,4)$ and $U_{22} \sim \chi_2^2(3)$ are independently distributed.

For $i = 3$, $\begin{pmatrix} Y_{31} \\ Y_{32} \end{pmatrix} \sim SN_2 \left(\begin{pmatrix} \sqrt{2} \\ 0 \end{pmatrix}, I_2, \begin{pmatrix} \sqrt{2} \\ \sqrt{2} \end{pmatrix} \right)$ and $Y_{33} \sim N(\sqrt{2}, 1)$ are independently distributed as well, which implies the decomposition of $Y_3'Y_3 = U_{31} + U_{32}$ where $U_{31} \sim S\chi_2^2(2,2,4)$ and $U_{32} \sim \chi_1^2(2)$ are independently distributed.

To generalize above example, we obtain the following statement.

Corollary 3.1. *Let $U \sim S\chi_k^2(\lambda, \delta_1, \delta_2)$, $U_1 \sim S\chi_{k_1}^2(\lambda_1, \delta_1^*, \delta_2^*)$ with $\delta_2 \neq 0$ and $U_2 \sim \chi_{k-k_1}^2(\lambda - \lambda_1)$ for $k_1 \geq 2$ are independent. Then $U \overset{d}{=} U_1 + U_2$ if and only if*

(i) $\delta_1^2/\delta_2 \leq \lambda_1 \leq \lambda$;
(ii) $\delta_i^* = \delta_i$ for $i = 1, 2$.

Proof. "If part" is trivial based on the mgfs of U, U_1 and U_3. For "only if" Let us assume that $k_1 = 2$ the method used here for $k_1 = 2$ can be easily extend to $k_1 > 2$. Similarly to proof of Theorem 2, we construct a skew normal vector $Y = \begin{pmatrix} Y_1 \\ Y_2 \end{pmatrix}$ which can satisfy all conditions for the desired decomposition. Starting from Y_1, let $Y_1 \sim SN_2(\mu, I_2, \alpha)$ and $Y_2 \sim N_{k-2}(\nu, I_{k-2})$ with $\nu = (\sqrt{\lambda - \lambda_1}, 0, \cdots, 0)' \in \mathbb{R}^{k-2}$ independently such that for $\mu, \alpha \in \mathbb{R}^2$, $\mu'\mu = \lambda_1$, $\mu'\alpha = \delta_1$ and $\alpha'\alpha = \delta_2$ where $\delta_1^2/\delta_2 \leq \lambda_1 \leq \lambda$. Those equations system is always solvable, say $\alpha = (\sqrt{\delta_2}, 0)'$, $\mu = (\mu_1, \mu_2)'$ where $\mu_1 = \frac{\delta_1}{\sqrt{\delta_2}}$ and $\mu_2 = \sqrt{\lambda_1 - \delta_1^2/\delta_2}$. On one hand, we have

$$Y = \begin{pmatrix} Y_1 \\ Y_2 \end{pmatrix} \sim SN_k(\nu^*, I_k, \alpha^*)$$

where $\nu^* = \begin{pmatrix} \mu \\ \nu \end{pmatrix}$ and $\alpha^* = \begin{pmatrix} \alpha \\ 0 \end{pmatrix}$, and following simple algebra, $Y'Y \sim S\chi_k^2(\lambda, \delta_1, \delta_2)$. On the other hand, $Y'Y = Y_1'Y_1 + Y_2'Y_2$ where $Y_1'Y_1 \sim S\chi_2^2(\lambda_1, \delta_1, \delta_2)$ and $Y_2'Y_2 \sim \chi_{k-2}^2(\lambda - \lambda_1)$ independently. Combining both sides, the desired results follows.

For $k_1 > 2$ case, we just set $\mu, \alpha \in \mathbb{R}^{k_1}$ satisfying the same conditions, and adjust ν correspondingly. □

Consequently, we obtain another representation of pdf of U as follows.

Corollary 3.2. *The pdf of $U \sim S\chi_k^2(\lambda, \delta_1, \delta_2)$ is*

$$f_U(u) = \int_0^u g_2(u-v) h_2(v) \, dv \tag{5}$$

where $g_2(\cdot)$ is pdf of $S\chi_{k_1}^2(\lambda_1, \delta_1, \delta_2)$ and $h_2(\cdot)$ is the pdf of $\chi_{k-k_1}^2(\lambda - \lambda_1)$ for $k_1 \geq 2$ where $\delta_1^2/\delta_2 \leq \lambda_1 \leq \lambda$; .

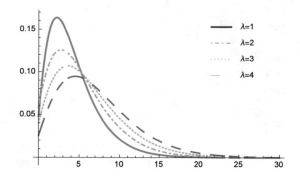

Fig. 3. Density curves of $S\chi_2^2(\lambda, 2, 5)$ with $\lambda = 1, 2, 3, 4$

Theorem 4. *For $U \sim S\chi_k^2(\lambda, \delta_1, \delta_2)$ where $k \geq 2$, the pdfs of U given by (1), (4) and (5) are identical.*

Proof. The pdfs given in (1), (4) and (5) have the same mgfs, therefore they are identical. □

The following graphs represent that the impact of parameters in noncentral skew chi-square distribution on the shape of curves of pdf of $S\chi_k^2(\lambda, \delta_1, \delta_2)$ for different values of degrees of freedom k and other parameters $\lambda, \delta_1, \delta_2$ (Figs. 2, 3, 4 and 5).

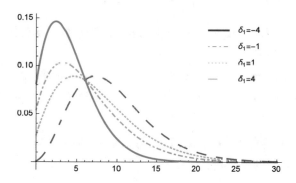

Fig. 4. Density curves of $S\chi_2^2(5, \delta_1, 5)$ with $\delta_1 = -4, -1, 1, 4$

Remark 3.1. Note that from Corollary 3.1 there could be more flexible decomposition on noncentrality if the degrees of freedom for skew chi-square is great than 1. So for the convenience, we choose $k_1 = 2$ and $\lambda_1 = \lambda$, then we have the following result.

Corollary 3.3. *Let $U \sim S\chi_k^2(\lambda, \delta_1, \delta_2)$ with $\delta_2 \neq 0$. Then U can be partitioned into the sum of two independent random variable U_1 and U_2 such that $U_1 \sim S\chi_2^2(\lambda, \delta_1, \delta_2)$ and $U_2 \sim \chi_{k-2}^2(0)$.*

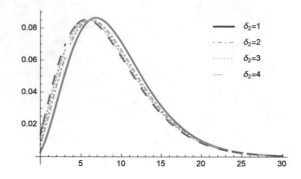

Fig. 5. Density curves of $S\chi_2^2(5, 2, \delta_2)$ with $\delta_2 = 1, 2, 3, 4$

Remark 3.2. (i) Notice above results, for all decompositions for $U \sim S\chi_k^2(\lambda, \delta_1, \delta_2)$, the noncentral skew chi-square parts have exactly the same skew parameters δ_1, δ_2 with original random variable U, which means "decompositions" is just decompositions of noncentral parameters λ and degrees of freedom k;

(ii) Further generalizing the decompositions is possible to represent a noncentral skew chi-square distribution $U \sim S\chi_k^2(\lambda, \delta_1, \delta_2)$ as the independent sum of more than two components, say $U \stackrel{d}{=} U_1 + \Sigma_{i=1}^k U_i$ where $U_1 \sim S\chi_{k_1}^2(\lambda_1, \delta_1, \delta_2)$, and $U_i \sim \chi_{k_i}^2(\lambda_i)$ for $i = 2, \cdots, k$ such that $\Sigma_{i=1}^k \lambda_i = \lambda$ and $\Sigma_{i=1}^k k_i = k$ where $|\delta_1| = \sqrt{\lambda_1 \delta_2}$ when $k_1 = 1$ or $|\delta_1| \leq \sqrt{\lambda_1 \delta_2}$ when $k_1 > 1$.

4   Partitions in Linear Regression Model with Skew Normal Errors

In this section, we will consider the regression model with skew normal errors , using decomposition results of skew chi-square distributions. Consider the regression model with skew normal errors:

$$y_i = \beta_0 + \beta_1 x_{i1} + \cdots + \beta_{p-1} x_{i,p-1} + \mathscr{E}_i \qquad \text{for} \qquad i = 1, \cdots, n$$

which is equivalent to the matrix form

$$Y = X\beta + \mathscr{E}, \tag{6}$$

where

$$Y = \begin{pmatrix} y_1 \\ \vdots \\ y_n \end{pmatrix}, \ X = \begin{pmatrix} 1 & x_{11} & \cdots & x_{1,p-1} \\ \vdots & \vdots & \vdots & \vdots \\ 1 & x_{n1} & \cdots & x_{n,p-1} \end{pmatrix}, \ \beta = \begin{pmatrix} \beta_0 \\ \vdots \\ \beta_{p-1} \end{pmatrix}, \ \mathscr{E} = \begin{pmatrix} \mathscr{E}_1 \\ \vdots \\ \mathscr{E}_n \end{pmatrix}.$$

Assume that the rank of design matrix X is p and $\mathscr{E} \sim SN_n(0, \sigma^2 I_n, \alpha)$ so that $Y \sim SN_n(X\beta, \sigma^2 I_n, \alpha)$. By Definition 1, we obtain

$$\frac{Y'Y}{\sigma^2} \sim S\chi_n^2(\lambda, \delta_1, \delta_2), \tag{7}$$

where $\lambda = \mu'\mu/\sigma^2$, $\delta_1 = \alpha'\mu/\sigma$ and $\delta_2 = \alpha'\alpha$ with $\mu = X\beta$. Consider the least squares estimator, $\hat{\beta}$, of β, and estimator, s^2, of σ^2, respectively, given by

$$\hat{\beta} = (X'X)^{-1}X'Y \quad \text{and} \quad s^2 = \frac{1}{n-p}Y'(I_n - P_X)Y,$$

where $P_X = X(X'X)^{-1}X'$. From Lemma 2.3, we obtain $\hat{\beta} \sim SN_p(\beta, \Sigma_1, \alpha_1)$, where

$$\Sigma_1 = \sigma^2(X'X)^{-1} \quad \text{and} \quad \alpha_1 = \frac{(X'X)^{1/2}X'\alpha}{\sqrt{1 + \alpha'(I_n - P_X)\alpha}}.$$

In many cases, random errors are assumed to be identically distributed so that $\alpha = \alpha_0 1_n$. Now we want to test the hypothesis $H_0 : \beta = \beta_0$ versus $H_1 : \beta \neq \beta_0$. We can use the partition

$$Y'Y = Y'P_X Y + Y'(I_n - P_X)Y \equiv U_1 + U_2. \tag{8}$$

Now it suffices to show that U_1 and U_2 have skew chi-square distributions and U_1 and U_2 are independent. We obtain by Lemma 2.2 that $P_X Y$ and $(I_n - P_X)Y$ are independent so that $\hat{\beta}$ and U_2 are independent. From Lemma 2.3, we know that

$$(X'X)^{\frac{1}{2}}\hat{\beta} = (X'X)^{-\frac{1}{2}}X'Y \sim SN_p\left((X'X)^{\frac{1}{2}}\beta, \sigma^2 I_p, \alpha_\beta\right),$$

where

$$\alpha_\beta = \frac{(X'X)^{-1/2}X'\alpha}{\sqrt{1 + \alpha'(I_n - P_X)\alpha}} = (X'X)^{-\frac{1}{2}}X'1_n\alpha_0.$$

Therefore by Definition 1, we obtain

$$\frac{(\hat{\beta} - \beta_0)'(X'X)(\hat{\beta} - \beta_0)}{\sigma^2} \sim S\chi_p^2(\lambda^*, \delta_1^*, \delta_2^*), \tag{9}$$

where

$$\lambda^* = \frac{(\beta - \beta_0)'(X'X)(\beta - \beta_0)}{\sigma^2}, \quad \delta_1^* = \frac{(\beta - \beta_0)'X'\alpha}{\sigma}, \quad \delta_2^* = n\alpha_0^2. \tag{10}$$

Note that $U_1 = Y'P_X Y = \hat{\beta}'(X'X)\hat{\beta}$ so that,

$$\frac{U_1}{\sigma^2} \sim S\chi_p^2\left(\beta'(X'X)\beta/\sigma^2, \alpha_0\beta'X'1_n/\sigma, n\alpha_0^2\right). \tag{11}$$

Also, it is easy to show that

$$\frac{U_2}{\sigma^2} = \frac{(n-p)S^2}{\sigma^2} \sim \chi_{n-p}^2(0). \tag{12}$$

Thus, we can define the test statistic for

$$H_0 : \quad \beta = \beta_0 \quad \text{v.s.} \quad H_a : \quad \beta \neq \beta_0$$

as

$$F = \frac{(\hat{\beta} - \beta_0)' X'X (\hat{\beta} - \beta_0)/p}{s^2} \sim SF_{p,n-p}(\lambda^*, \delta_1^*, \delta_2^*), \qquad (13)$$

the noncentral skew-F distribution with degrees of freedom p and $n - p$, noncentrality λ^* and skewness parameters δ_1^* and δ_2^* defined in (10). If H_0 is true, then $\lambda^* = 0$ and $\delta_1^* = 0$, $F \sim F_{p,n-p}(0)$. If H_0 is rejected, then $F \sim SF_{p,n-p}(\lambda^*, \delta_1^*, \delta_2^*)$.

Acknowledgments. Authors would like to thank referee's valuable comments which led to improvement of this paper.

References

1. Aronld, B.C., Beaver, R.J., Groenevld, R.A., Meeker, W.Q.: The nontruncated marginal of a truncated bivariate normal distribution. Psychometrica **58**(3), 471–488 (1993)
2. Azzalini, A.: A class of distributions which includes the normal ones. Scand. J. Statist. **12**(2), 171–178 (1985)
3. Azzalini, A.: Further results on a class of distributions which includes the normal ones. Statistica **46**(2), 199–208 (1986)
4. Azzalini, A., Dalla Valle, A.: The multivariate skew-normal distribution. Biometrika **83**(4), 715–726 (1996)
5. Azzalini, A., Capitanio, A.: Statistical applications of the multivariate skew normal distribution. J. R. Stat. Soc. **61**(3), 579–602 (1999)
6. Azzalini, A., Capitanio, A.: The Skew-Normal and Related Families, vol. 3. Cambridge University Press, Cambridge (2013)
7. Gupta, A.K., Huang, W.J.: Quadratic forms in skew normal variates. J. Math. Anal. Appl. **273**, 558–564 (2002)
8. Genton, M.G., He, L., Liu, X.: Moments of skew-normal random vectors and their quadratic forms. Stat. Probab. Lett. **51**, 319–325 (2001)
9. Fang, B.Q.: The skew elliptical distributions and their quadratic forms. J. Multivar. Anal. **87**, 298–314 (2003)
10. Genton, M.G., Loperfido, N.: Generalized skew-elliptical distributions and their quadratic forms. Ann. Inst. Stat. Math. **57**, 389–401 (2005)
11. Hill, M., Dixon, W.J.: Robustness in real life: a study of clinical laboratory data. Biometrics **38**, 377–396 (1982)
12. Wang, T., Li, B., Gupta, A.K.: Distribution of quadratic forms under skew normal settings. J. Multivar. Anal. **100**(3), 533–545 (2009)
13. Ye, R., Wang, T.: Inferences in linear mixed model with skew-normal random effects. Acta Math. Sin. Engl. Ser. **31**(4), 576–594 (2015)
14. Ye, R., Wang, T., Sukparungsee, S., Gupta, A.K.: Tests in variance components models under skew-normal settings. Metrika **78**(7), 885–904 (2015)
15. Wang, Z., Wang, C., Wang, T.: Estimation of location parameter in the skew normal setting with known coefficient of variation and skewness. Int. J. Intell. Technol. Appl. Stat. **9**(3), 191–208 (2016)

Joint Plausibility Regions for Parameters of Skew Normal Family

Ziwei Ma[1,2], Xiaonan Zhu[1], Tonghui Wang[1(✉)],
and Kittawit Autchariyapanitkul[3]

[1] Department of Mathematical Sciences, New Mexico State University,
Las Cruces, USA
{ziweima,xzhu,twang}@nmsu.edu
[2] College of Science, Northwest A&F University, Yangling, China
[3] Faculty of Economics, Maejo University, Chiang Mai, Thailand
kittawit_a@mju.ac.th

Abstract. The estimation of parameters is a challenge issue for skew normal family. Based on inferential models, the plausibility regions for two parameters of skew normal family are investigated in two cases, when either the scale parameter σ or the shape parameter δ is known. For illustration of our results, simulation studies are proceeded.

Keywords: Skew normal distribution · Inferential models
Plausibility regions

1 Introduction

Skew data sets occur in many diverse fields, such as economics, finance, biomedicine, environment, demography, and pharmacokinetics, just to name a few. In conventional procedure, practitioners assume that the data are normally distributed to proceed statistical analysis. This restrictive assumption, however, may result in not only a lack of robustness against departures from the normal distribution and but also in invalid statistical inferences, especially when data are skewed. One solution to analyze skewed data is to extend the normal family by introducing an extra parameter, thus getting skew normal distributions which have location, scale and shape parameters (see Azzalini and his collaborators' work [2,6] and reference therein). In many practical cases, skew normal distribution is suitable for the analysis of data which is unimodal empirical distributed but with some skewness (see Arnold [1], Hill [8]). In past three decades, the family of skew-normal distributions, including multivariate skew-normal distributions, has been studied by many authors, e.g. Azzalini [3–5], Wang et al. [17], Ye et al. [18].

However, estimating parameters of skew normal family is a challenge, especially in relation to estimating shape parameter (see Azzalini [4] and Pewsey [15]). Liseo and Loperfido [9] pointed out the likelihood function is

© Springer International Publishing AG 2018
V. Kreinovich et al. (eds.), *Predictive Econometrics and Big Data*, Studies in Computational Intelligence 753, https://doi.org/10.1007/978-3-319-70942-0_16

increasing with positive probability, which leads to an infinite maximum likelihood estimate for shape parameter, and the method of moments can give even
worse results. New methods are needed to solve this problem. Some methods
for estimation of parameters were studied by many authors, see Azzalini and
Capitanio [4], Sartori [16], Liseo and Loperfido [9], Debarshi [7] and Mameli et
al. [14]. In particular, Zhu et al. [19] applied inferential models (IMs) to construct
plausibility interval for shape parameter.

In this study, we construct plausibility regions for two parameters of skew
normal population in two cases, with either shape parameter or scale parameter
is known, by inferential models (IMs). IMs are new methods of statistical inference introduced by Martin and Liu [10,12]. Comparing with Fisher's fiducial
inference, Dempster-Shafer theory of belief functions and Bayesian inference,
IMs have several advantages: (i) IMs are free of prior distributions; (ii) IMs
depend only on the observed data. For more details of IMs, see Martin and his
collaborators' work [10–13].

This paper is organized as following. The basic concepts on skew-normal
distributions and IMs are introduced briefly in Sect. 2. Plausibility regions for
the parameters of skew normal population are obtained in two cases in Sect. 3.
Simulation studies are proceeded for illustration of our main result in Sect. 4.

2   Preliminaries

Throughout of this paper, we use $\phi(\cdot)$ and $\Phi(\cdot)$ to denote the probability density
function (pdf) and cumulative distribution function (cdf) of the standard normal
distribution, respectively. Let $F(\cdot)$ be the cdf of $\chi^2(0)$ distribution and $G(\cdot)$ be
the cdf of skew normal distribution, and $N(0,1)$ and Uniform$(0,1)$ represent the
standard normal and uniform distributions, respectively.

2.1   Brief Review of Skew-Normal Distributions

A random variable Z is said to be *skew-normal* distributed with the *shape parameter* λ, denoted by $Z \sim SN(\lambda)$, if its pdf is given by

$$f(z; \lambda) = 2\phi(z)\Phi(\lambda z), \qquad z, \lambda \in \mathbb{R}.$$

For any $\mu \in \mathbb{R}$ and $\sigma > 0$, the distribution of $X = \mu + \sigma Z$ is said to be *skew-
normal* distributed with the *location parameter* μ, the *scale parameter* σ and
the *shape parameter* λ, denoted by $X \sim SN(\mu, \sigma^2, \lambda)$ and the pdf of X is

$$f(x; \mu, \sigma^2, \lambda) = \frac{2}{\sigma}\phi\left(\frac{x-\mu}{\sigma}\right)\Phi\left(\frac{\lambda(x-\mu)}{\sigma}\right).$$

There is an alternative representation of $Z \sim SN(0,1,\lambda)$ given by

$$Z = \delta|Z_0| + \sqrt{1-\delta^2}Z_1, \qquad \delta \in (-1,1) \tag{1}$$

where Z_0 and Z_1 are independent $N(0,1)$ random variables and $\delta = \lambda/\sqrt{1+\lambda^2}$.
This stochastic representation plays a vital role in establishing our IMs. See
Azzalini and Capitanio's book [5] and reference therein for more details.

2.2   Inference Models

Let X be an observable random sample with a probability distribution $P_{X|\theta}$ on a sample space $\mathbb{X}$, where θ is an unknown parameter, $\theta \in \Theta$, a parameter space. Let U be an unobservable *auxiliary variable* on an auxiliary space $\mathbb{U}$, where although U is unobservable, we assume that U and $\mathbb{U}$ are known. An *association* is a map $a : \mathbb{U} \times \Theta \to \mathbb{X}$ such that

$$X = a(U, \theta).$$

For any given statistical assertion on parameters, an IM consists of following three steps.

Association Step (A-step). Suppose we have an association $X = a(U, \theta)$ and an observation $X = x$, where x could be a scalar or vector, then the unknown θ must satisfy

$$x = a(u^*, \theta)$$

for some unobserved u^* of U. So from the observation $X = x$, we have the set of solutions

$$\Theta_x(u) = \{\theta \in \Theta : x = a(u, \theta)\}, \qquad x \in \mathbb{X}, \qquad u \in \mathbb{U}.$$

Prediction Step (P-step). Since the true u^* is unobservable, to make a valid inference, the key point is to predicate u^*. Let $u \to S(u)$ be a set-value map from $\mathbb{U}$ to $\mathbb{S}$, a collection of P_U-measurable subsets of $\mathbb{U}$. Then the random set $S : \mathbb{U} \to \mathbb{S}$ is called a *predictive random set* of U with distribution $P_S = P_U \circ S^{-1}$. We will use S to predict u^*.

Combination Step (C-step). Define

$$\Theta_x(S) = \bigcup_{u \in S} \Theta_x(u).$$

For any assertion A of θ, i.e., $A \subseteq \Theta$, the *belief function* and *plausibility function* of A with respect to a predictive random set S are defined by,

$$\mathsf{bel}_x(A; S) = P_S\{\Theta_x(S) \subseteq A : \Theta_x(S) \neq \emptyset\};$$

$$\mathsf{pl}_x(A; S) = P_S\{\Theta_x(S) \not\subseteq A^c : \Theta_x(S) \neq \emptyset\}.$$

Note that

$$\mathsf{pl}_x(A; S) = 1 - \mathsf{bel}_x(A^c; S), \qquad \mathsf{bel}_x(A; S) + \mathsf{bel}_x(A^c; S) \leq 1, \qquad \text{for all} \quad A \subseteq \Theta.$$

Based on the plausibility function derived form an IM, the $100(1 - \alpha)\%$ level plausibility region for θ follows

$$\Pi_X(\alpha) = \{\theta : \mathsf{pl}_X(\theta; S) > \alpha\},$$

which is counter part of confidence regions in classical statistics.

3 Plausibility Regions for Parameters of the Skew Normal Family

Suppose that $X_1, \ldots, X_n$ are identical distributed random variables from the population $SN(\mu, \sigma^2, \lambda)$ with stochastic representations

$$X_i = \mu + \sigma(\delta|Z_0| + \sqrt{1 - \delta^2}Z_i), \qquad i = 1, \ldots, n, \qquad (2)$$

where $(Z_0, Z_1, \cdots, Z_n)' \sim N_{n+1}(0, I_{n+1})$ and $\delta = \lambda/\sqrt{1 + \lambda^2}$. See Azzalini and Capitanio [5] for details.

It is clear that $X_i \sim SN(\mu, \sigma^2, \lambda)$ for $i = 1, \cdots, n$, but they are dependent since they share the same component $|Z_0|$ when $\delta \neq 0$. The following result is needed for establishing our IMs.

Theorem 1. *Let $X_1, \ldots, X_n$ be identically distributed with stochastic representations given in (2). Let the sample mean and sample variance be*

$$\bar{X} = \frac{1}{n}\sum_{i=1}^{n}X_i \qquad and \qquad S^2 = \frac{1}{n-1}\sum_{i=1}^{n}\left(X_i - \bar{X}\right)^2,$$

respectively. Then $\bar{X}$ and S^2 are independent,

$$\bar{X} \sim SN\left(\mu, \frac{n\lambda^2 + 1}{n(1 + \lambda^2)}\sigma^2, \sqrt{n}\lambda\right) \qquad and \qquad \frac{(n-1)S^2}{\sigma^2(1 - \delta^2)} \sim \chi^2_{n-1}(0).$$

Proof. From the stochastic representations (2), it is easy to obtain

$$\bar{X} = \mu + \sigma\left(\delta|Z_0| + \sqrt{1 - \delta^2}\bar{Z}\right) \quad and \quad S^2 = \frac{\sigma^2(1 - \delta^2)}{n-1}\sum_{i=1}^{n}\left(Z_i - \bar{Z}\right)^2,$$

where $\bar{Z} = \frac{1}{n}\sum_{i=1}^{n}Z_i$. Note that $\bar{Z}$ and $(Z_1 - \bar{Z})$'s are independent so that $\bar{X}$ and S^2 are independent. For the distribution of $\bar{X}$, let $Z_* = \sqrt{n}\bar{Z}$, which is distributed as $N(0,1)$. Then

$$\bar{X} = \mu + \sigma\left(\delta|Z_0| + \sqrt{\frac{1 - \delta^2}{n}}Z_*\right)$$

$$= \mu + \sigma_*\left(\delta_*|Z_0| + \sqrt{1 - \delta_*^2}Z_*\right),$$

where $\delta_* = \frac{\sqrt{n}\delta}{\sqrt{1 + (n-1)\delta^2}}$ and $\sigma_* = \sigma\sqrt{\frac{1 + (n-1)\delta^2}{n}}$. Thus by definition, we have $\bar{X} \sim SN(\mu, \sigma_*^2, \lambda_*)$, where

$$\lambda^* = \frac{\delta_*}{\sqrt{1 - \delta_*^2}} = \frac{\sqrt{n}\delta}{\sqrt{1 - \delta^2}} = \sqrt{n}\lambda,$$

the distribution of $\bar{X}$ is obtained. The distribution of S^2 can be obtained directly from

$$\frac{(n-1)S^2}{\sigma^2(1 - \delta^2)} = \sum_{i=1}^{n}(Z_i - \bar{Z})^2 \sim \chi^2_{n-1}(0).$$

$\square$

3.1 Plausibility Function and Plausibility Region of (μ, σ) When δ Is Known

Assume that the skewness parameter λ (or δ) is known. We want to construct the plausibility region for unknown parameters (μ, σ) based on a sample $X_1, \ldots, X_n$ from a skew normal population.

A-step. From Theorem 1, we can have associations

$$\frac{(n-1)S^2}{\sigma^2(1-\delta^2)} = F_{n-1}^{-1}(U_1) \quad \text{and} \quad \bar{X} = \mu + G^{-1}(U_2), \tag{3}$$

where $F_{n-1}(\cdot)$ and $G(\cdot)$ are the cdf's of $\chi_{n-1}^2(0)$ and $SN\left(0, \frac{n\lambda^2+1}{n(1+\lambda^2)}\sigma^2, \sqrt{n}\lambda\right)$, respectively, and U_1, U_2 are independent uniformly distributed in interval $(0, 1)$. Thus for any observations $\bar{x}$ and s^2, and $u_1, u_2 \in (0, 1)$, we have the solution set

$$\Theta_{(\bar{x}, s^2)}(\mu, \sigma) = \left\{ (\mu, \sigma) : \bar{x} = \mu + G^{-1}(u_2), \frac{(n-1)s^2}{\sigma^2(1-\delta^2)} = F_{n-1}^{-1}(u_1) \right\}$$

$$- \left\{ (\mu, \sigma) : G(\bar{x} - \mu) = u_2, F_{n-1}\left(\frac{(n-1)s^2}{\sigma^2(1-\delta^2)}\right) = u_1 \right\}.$$

P-step. To predict auxiliary variables U_1 and U_2, we use the default predictive random set

$$S(U_1, U_2) = \{(u_1, u_2) : \max\{|u_1 - 0.5|, |u_2 - 0.5|\} \leq \max\{|U_1 - 0.5|, |U_2 - 0.5|\}\}.$$

C-step. By the P-step, we have the combined set

$$\Theta_{(\bar{x}, S^2)}(S) = \left\{ (\mu, \sigma) : \max\left\{ \left| G(\bar{x} - \mu) - 0.5 \right|, \left| F_{n-1}\left(\frac{(n-1)s^2}{\sigma^2(1-\delta^2)}\right) - 0.5 \right| \right\} \right.$$

$$\left. \leq \max\left\{ \left| U_1 - 0.5 \right|, \left| U_2 - 0.5 \right| \right\} \right\}.$$

Theorem 2. *For any singleton assertion* $A = \{(\mu, \sigma)\}$,

$$bel_{(\bar{x}, s^2)}(A; S) = 0,$$

$$pl_{(\bar{x}, s^2)}(A; S) = 1 - \max\left\{ \left| 2G(\bar{x} - \mu) - 1 \right|, \left| 2F_{n-1}\left(\frac{(n-1)s^2}{\sigma^2(1-\delta^2)}\right) - 1 \right| \right\}^2,$$

and the $100(1 - \alpha)\%$ *plausibility region*

$$\Pi_{\bar{X}, S^2}(\mu, \sigma) = \{(\mu, \sigma) : pl_{(\bar{x}, s^2)}(\mu, \sigma) \geq \alpha\}.$$

Proof. It is clear that $\{\Theta_{(\bar{x},s^2)}(S) \subseteq A\} = \emptyset$, so $\mathrm{bel}_{(\bar{x},s^2)}(A;S) = 0$.

$$\mathrm{pl}_{(\bar{x},s^2)}(A;S) = 1 - \mathrm{bel}_{(\bar{x},s^2)}(A^c;S) = 1 - P_S\left(\Theta_{(\bar{x},S^2)}(S) \subseteq A^c\right)$$
$$= 1 - \max\left\{\left|2G(\bar{x} - \mu) - 1\right|, \left|2F_{n-1}\left(\tfrac{(n-1)s^2}{\sigma^2(1-\delta^2)}\right) - 1\right|\right\}^2.$$

Thus we can obtain the $100(1 - \alpha)\%$ plausibility region by its definition.   □

The following example is used for the illustration of Theorem 2.

Example 3.1. *For a sample $X_1, \cdots, X_n$ from skew normal population as described above, graphs of plausibility function and the 95% plausibility region of (μ, σ) are listed in Figs. 1, 2 and 3.*

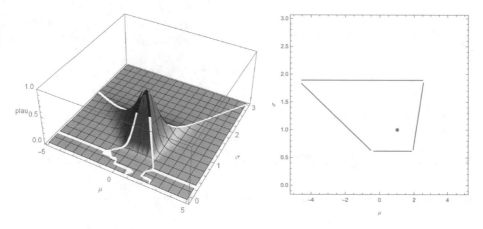

Fig. 1. Graphs of plausibility function and the 95% plausibility region of (μ, σ) based on simulated data with $\mu = 1, \sigma = 1, \delta = 1/\sqrt{2}$ and $n = 10$.

3.2   Plausibility Function and Plausibility Region for (μ, δ) When σ Is Known

Assume that the scale parameter σ is known. We want to construct the plausibility region for unknown parameters (μ, δ) based a sample $X_1, \ldots, X_n$ from a skew normal population.

A-step. From Theorem 1, we obtain the associations

$$\frac{(n-1)S^2}{\sigma^2(1-\delta^2)} = F_{n-1}^{-1}(U_1) \qquad \text{and} \qquad \bar{X} = \mu + G^{-1}(U_2), \tag{4}$$

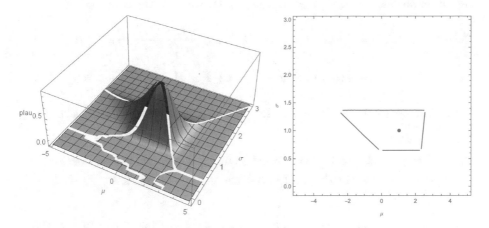

Fig. 2. Graphs of plausibility function and the 95% plausibility region of (μ, σ) based on simulated data with $\mu = 1, \sigma = 1, \delta = 1/\sqrt{2}$ and $n = 20$.

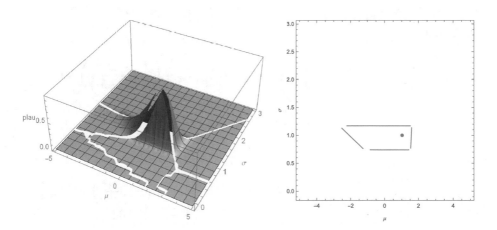

Fig. 3. Graphs of plausibility function and the 95% plausibility region of (μ, σ) based on simulated data with $\mu = 1, \sigma = 1, \delta = 1/\sqrt{2}$ and $n = 50$.

where $F_{n-1}(\cdot)$ and $G(\cdot)$ are cdf's of χ^2_{n-1} and $SN\left(0, \frac{n\lambda^2+1}{n(1+\lambda^2)}\sigma^2, \sqrt{n}\lambda\right)$ respectively, and U_1, U_2 are independent uniformly distributed in interval $(0,1)$. So for any observations $\bar{x}$ and s^2, and $u_1, u_2 \in (0,1)$, we have the solution set

$$\Theta_{(\bar{x},s^2)}(\mu,\delta) = \left\{(\mu,\delta) : \bar{x} = \mu + G^{-1}(u_2), \frac{(n-1)s^2}{\sigma^2(1-\delta^2)} = F^{-1}_{n-1}(u_1)\right\}$$

$$= \left\{(\mu,\delta) : G(\bar{x}-\mu) = u_2, F_{n-1}\left(\frac{(n-1)s^2}{\sigma^2(1-\delta^2)}\right) = u_1\right\}.$$

Note that to guarantee the solution set $\Theta_{(\bar{x},s^2)}(\mu,\delta) \neq \emptyset$, we need $u_1 \geq F\left(\frac{(n-1)s^2}{\sigma^2}\right)$.

P-step. To predict auxiliary variables U_1 and U_2, we should use an *elastic predictive random set* (see Martin and Liu [12] Chap. 5) as follow. If $F_{n-1}\left(\frac{(n-1)s^2}{\sigma^2}\right) \leq \frac{1}{2}$, we take

$$S(U_1, U_2) = \{(u_1, u_2) : \max\{|u_1 - 0.5|, |u_2 - 0.5|\} \leq \max\{|U_1 - 0.5|, |U_2 - 0.5|\}\},$$

otherwise, we take

$$S(U_1, U_2) = \left\{(u_1, u_2) : \max\left\{\left|F_{n-1}\left(\frac{(n-1)s^2}{\sigma^2}\right) - 0.5\right|, \left|u_1 - 0.5\right|, \left|u_2 - 0.5\right|\right\}\right.$$
$$\left. \leq \max\left\{\left|U_1 - 0.5\right|, |U_2 - 0.5|\right\}\right\}$$

C-step. By P-step, we have the combined set in two cases. If $F_{n-1}\left(\frac{(n-1)s^2}{\sigma^2}\right) \leq \frac{1}{2}$, then

$$\Theta_{(\bar{x},S^2)}(S) = \left\{(\mu,\delta) : \max\left\{\left|G(\bar{x}-\mu) - 0.5\right|, \left|F_{n-1}\left(\frac{(n-1)s^2}{\sigma^2(1-\delta^2)}\right) - 0.5\right|\right\}\right.$$
$$\left. \leq \max\left\{\left|U_1 - 0.5\right|, |U_2 - 0.5|\right\}\right\}.$$

Otherwise,

$$\Theta_{(\bar{x},S^2)}(S) = \left\{(\mu,\delta) : \max\left\{\begin{array}{c}\left|G(\bar{x}-\mu) - 0.5\right|, \left|F_{n-1}\left(\frac{(n-1)s^2}{\sigma^2(1-\delta^2)}\right) - 0.5\right|, \\ \left|F_{n-1}\left(\frac{(n-1)s^2}{\sigma^2}\right) - 0.5\right|\end{array}\right\}\right.$$
$$\left. \leq \max\left\{\left|U_1 - 0.5\right|, |U_2 - 0.5|\right\}\right\}.$$

Theorem 3. *For any singleton assertion* $A = \{(\mu,\delta)\}$,

$$bel_{(\bar{x},s^2)}(A; S) = 0,$$

If $F_{n-1}\left(\frac{(n-1)s^2}{\sigma^2}\right) \leq \frac{1}{2}$, then

$$pl_{(\bar{x},s^2)}(A; S) = 1 - \max\left\{\left|2G(\bar{x}-\mu)-1\right|, \left|2F_{n-1}\left(\frac{(n-1)s^2}{\sigma^2(1-\delta^2)}\right)-1\right|\right\}^2,$$

Otherwise,

$$pl_{(\bar{x},s^2)}(A; S) = 1 - \max\left\{\begin{array}{c}\left|2G(\bar{x}-\mu)-1\right|, \left|2F_{n-1}\left(\frac{(n-1)s^2}{\sigma^2(1-\delta^2)}\right)-1\right|, \\ \left|2F_{n-1}\left(\frac{(n-1)s^2}{\sigma^2}\right)-1\right|\end{array}\right\}^2,$$

and the $100(1-\alpha)\%$ plausibility region

$$\Pi_{\bar{X},S^2}(\mu,\delta) = \{(\mu,\delta) : pl_{(\bar{x},s^2)}(\mu,\delta) \geq \alpha\}.$$

Proof. It is clear that $\{\Theta_{(\bar{x},s^2)}(S) \subseteq A\} = \emptyset$, so $bel_{(\bar{x},s^2)}(A; S) = 0$. By definition of plausibility function, we can compute plausibility in two cases.

If $F_{n-1}\left(\frac{(n-1)s^2}{\sigma^2}\right) \leq \frac{1}{2}$, we have

$$pl_{(\bar{x},s^2)}(A; S) = 1 - bel_{(\bar{x},s^2)}(A^c; S) = 1 - P_S\left(\Theta_{(\bar{x},S^2)}(S) \subseteq A^c\right)$$
$$= 1 - \max\left\{\left|2G(\bar{x}-\mu)-1\right|, \left|2F_{n-1}\left(\frac{(n-1)s^2}{\sigma^2(1-\delta^2)}\right)-1\right|\right\}^2.$$

Otherwise,

$$pl_{(\bar{x},s^2)}(A; S) = 1 - bel_{(\bar{x},s^2)}(A^c; S) = 1 - P_S\left(\Theta_{(\bar{x},S^2)}(S) \subseteq A^c\right)$$
$$= 1 - \max\left\{\begin{array}{c}\left|2G(\bar{x}-\mu)-1\right|, \left|2F_{n-1}\left(\frac{(n-1)s^2}{\sigma^2(1-\delta^2)}\right)-1\right|, \\ \left|2F_{n-1}\left(\frac{(n-1)s^2}{\sigma^2}\right)-1\right|\end{array}\right\}^2. \quad \square$$

Remark. Note that plausibility functions in Theorems 2 and 3 are the same because we use the same association, but the plausibility regions are different, because in Theorem 2, we use this plausibility function to solve plausibility regions for (μ,σ), while in Theorem 3, we use this plausibility function to obtain plausibility regions for (μ,δ).

Similarly, the following example is used for the illustration of Theorem 3.

Example 3.2. *For a sample* $X_1,\cdots,X_n$ *from skew normal population as described above, graphs of plausibility function and the 95% plausibility region of* (μ,δ) *are given in Figs. 4, 5 and 6 for sample sizes* $n = 10, 20$ *and 50, respectively.*

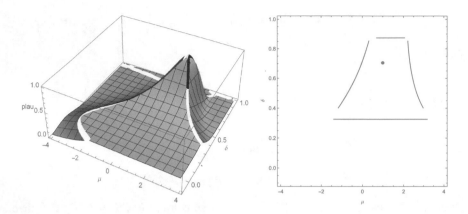

Fig. 4. Graphs of plausibility function and the 95% plausibility region of (μ, δ) based on simulated data with $\mu = 1, \sigma = 1, \delta = 1/\sqrt{2}$ and $n = 10$.

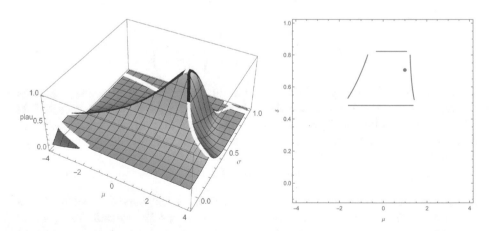

Fig. 5. Graphs of plausibility function and the 95% plausibility region of (μ, δ) based on simulated data with $\mu = 1, \sigma = 1, \delta = 1/\sqrt{2}$ and $n = 20$.

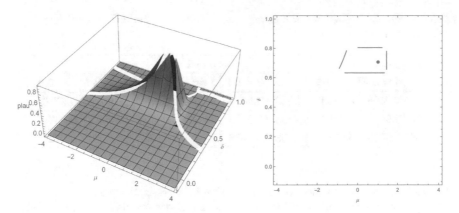

Fig. 6. Graphs of plausibility function and the 95% plausibility region of (μ, δ) based on simulated data with $\mu = 1, \sigma = 1, \delta = 1/\sqrt{2}$ and $n = 50$.

4   Simulation Studies

In this section, we perform two simulation studies on coverage probabilities of 95% plausibility regions for skew normal population of two cases discussed above.

4.1   Coverage Probabilities of the 95% Plausibility Regions for (μ, σ) When δ Is Known

When $\delta = 1/\sqrt{2}$, we choose sample sizes of $10, 20$ and 50 and simulate $10,000$ runs for different parameters, which is given in Table 1.

Table 1. Simulation results of coverage probabilities of the 95% plausibility regions for (μ, σ) when $\delta = 1/\sqrt{2}$.

n	$\mu = 0$				$\mu = 1$			
	$\sigma = 0.1$	$\sigma = 0.5$	$\sigma = 1$	$\sigma = 2$	$\sigma = 0.1$	$\sigma = 0.5$	$\sigma = 1$	$\sigma = 2$
10	0.9500	0.9449	0.9493	0.9503	0.9489	0.9540	0.9479	0.9513
20	0.9472	0.9524	0.9514	0.9493	0.9481	0.9502	0.9510	0.9463
50	0.9521	0.9537	0.9482	0.9499	0.9506	0.9485	0.9487	0.9507

4.2   Coverage Probabilities of 95% Plausibility Regions for (μ, δ) When σ Is Known

When $\sigma = 1$, we choose sample sizes of $10, 20$ and 50 and run the simulation $10,000$ times. Simulation results of coverage probabilities of the 95% plausibility regions for (μ, δ) is listed in Table 2.

Table 2. Simulation results of coverage probabilities of the 95% plausibility regions for (μ, δ) when $\sigma = 1$.

n	$\mu = 0$				$\mu = 1$			
	$\delta = 0.2$	$\delta = 0.4$	$\delta = 0.7$	$\delta = 0.9$	$\delta = 0.2$	$\delta = 0.4$	$\delta = 0.7$	$\delta = 0.9$
10	0.9495	0.9490	0.9518	0.9484	0.9502	0.9524	0.9513	0.9520
20	0.9487	0.9488	0.9493	0.9524	0.9515	0.9446	0.9481	0.9479
50	0.9513	0.9513	0.9508	0.9472	0.9489	0.9487	0.9540	0.9527

Acknowledgments. Authors would like to thank Professor Hung T. Nguyen for introducing this interesting and hot research topic to us. Also we would like to thank referee's valuable comments which led to improvement of this paper.

References

1. Aronld, B.C., Beaver, R.J., Groenevld, R.A., Meeker, W.Q.: The nontruncated marginal of a truncated bivariate normal distribution. Psychometrica **58**(3), 471–488 (1993)
2. Azzalini, A.: A class of distributions which includes the normal ones. Scand. J. Stat. **12**(2), 171–178 (1985)
3. Azzalini, A.: Further results on a class of distributions which includes the normal ones. Statistica **46**(2), 199–208 (1986)
4. Azzalini, A., Capitanio, A.: Statistical applications of the multivariate skew normal distribution. J. R. Stat. Soc. **61**(3), 579–602 (1999)
5. Azzalini, A., Capitanio, A.: The Skew-Normal and Related Families, vol. 3. Cambridge University Press, Cambridge (2013)
6. Azzalini, A., Dalla Valle, A.: The multivariate skew-normal distribution. Biometrika **83**(4), 715–726 (1996)
7. Dey, D. Estimation of the parameters of skew normal distribution by approximating the ratio of the normal density and distribution functions. Ph.D. thesis, University of California Riverside (2010)
8. Hill, M., Dixon, W.J.: Robustness in real life: a study of clinical laboratory data. Biometrics **38**, 377–396 (1982)
9. Liseo, B., Loperfido, N.: A note on reference priors for the scalar skew-normal distribution. J. Stat. Plan. Inference **136**(2), 373–389 (2006)
10. Martin, R., Liu, C.: Inferential models: a framework for prior-free posterior probabilistic inference. J. Am. Stat. Assoc. **108**(501), 301–313 (2013)
11. Martin, R.: Random sets and exact confidence regions. Sankhya A **76**(2), 288–304 (2014)
12. Martin, R., Liu, C.: Inferential models: reasoning with uncertainty. In: Monographs on statistics and Applied Probability, vol. 145. CRC Press (2015)
13. Martin, R., Lingham, R.T.: Prior-free probabilistic prediction of future observations. Technometrics **58**(2), 225–235 (2016)
14. Mameli, V., Musio, M., Sauleau, E., Biggeri, A.: Large sample confidence intervals for the skewness parameter of the skew-normal distribution based on Fisher's transformation. J. Appl. Stat. **39**(8), 1693–1702 (2012)

15. Pewsey, A.: Problems of inference for Azzalini's skewnormal distribution. J. Appl. Stat. **27**(7), 859–870 (2000)
16. Sartori, N.: Bias prevention of maximum likelihood estimates for scalar skew-normal and skew-t distributions. J. Stat. Plan. Inference **136**(12), 4259–4275 (2006)
17. Wang, T., Li, B., Gupta, A.K.: Distribution of quadratic forms under skew normal settings. J. Multivar. Anal. **100**(3), 533–545 (2009)
18. Ye, R., Wang, T., Gupta, A.K.: Distribution of matrix quadratic forms under skew-normal settings. J. Multivar. Anal. **131**, 229–239 (2014)
19. Zhu, X., Ma, Z., Wang, T., Teetranont, T.: Plausibility regions on the skewness parameter of skew normal distributions based on inferential models. In: Robustness in Econometrics, pp. 267–286. Springer (2017)

On Parameter Change Test for ARMA Models with Martingale Difference Errors

Haejune Oh and Sangyeol Lee[✉]

Department of Statistics, Seoul National University, Seoul 08826, Korea
sylee@stats.snu.ac.kr

Abstract. This study considers the CUSUM test for ARMA models with stationary martingale difference errors. CUSUM tests are widely used for detecting abrupt changes in time series models. Although they perform adequately in general, their performance is occasionally unsatisfactory in ARMA models. This motivates us to design a new test that can simultaneously detect the ARMA parameter and variance changes. Its null limiting distribution is derived under regularity conditions. Monte Carlo simulations confirm the validity of the proposed test.

Keywords: ARMA models · Parameter change test · CUSUM test Martingale difference errors

1 Introduction

This study considers the parameter change test for ARMA models with martingale difference errors. Since [11], the problem of testing for a parameter change has been an important issue in many applied fields such as economics, engineering and medicine. The change point problem has been a core issue among researchers since time series often suffer from structural changes owing to changes in policy and critical social events. It is widely known that detecting a change point is crucial for further inferences and ignoring it can lead to a false conclusion. The literature on the change point test for time series models is quite exhaustive. [5] consider to use the CUSUM of squares test to detect multiple changes of variance of independent samples. Since then, numerous studies have been devoted to the CUSUM tests for time series models: see [7] and the papers cited therein for earlier works.

Conventionally, the CUSUM test is constructed based on parameter estimates, score vectors, and residuals. Although the estimates-based test appeals much to practitioners, it often has severe size distortions as seen in [8,9]. The score vector-based test circumvents this phenomenon to a great extent in ARMA-GARCH models: see [1,3,10], who consider the score vector-based CUSUM test

This research is supported by Basic Science Research Program through the National Research Foundation of Korea (NRF) funded by the Ministry of Science, ICT and future Planning (No. 2015R1A2A2A010003894).

V. Kreinovich et al. (eds.), *Predictive Econometrics and Big Data*, Studies in Computational Intelligence 753, https://doi.org/10.1007/978-3-319-70942-0_17

in GARCH and AR models. However, this test cannot completely resolve the problem, and as an alternative, the residual-based test has been used as in [6]. Nevertheless, the test has a serious defect not to be able to detect a change of ARMA parameters (and location parameters within a more general framework). Hence, in this study, we intend to design a test that can simultaneously detect the ARMA parameter and variance changes in ARMA models with stationary martingale difference errors that accommodate a broad class of ARMA-GARCH type models. In particular, the variance change test turns out to be useful to detect a change in GARCH parameters when the errors follow a GARCH model: see Sect. 3.

The idea in our CUSUM set-up is to use the multiplications of the ARMA part and error terms in the construction of the CUSUM test. We do this because those terms form martingale differences under the null hypothesis, which, henceforth, leads to a desired Brownian bridge result, used for calculating critical values, whereas their values before and after the change point are significantly different and contribute to producing good powers. Compared with the score vector-based CUSUM test, this test has merit to be a lot simpler in implementation, has less size distortions, and produces good powers.

The organization of this paper is as follows. Section 2 introduces a new CUSUM test and investigates its limiting null distributions. Furthermore, the asymptotic behavior of the CUSUM test under alternatives is investigated. Section 3 conducts a simulation study and demonstrates the validity of the proposed test. Sections 4 and 5 provide the proofs and concluding remarks.

2   CUSUM Test for ARMA Models

Let us consider the ARMA(p,q) model:

$$y_t = \sum_{i=1}^{p} \alpha_i y_{t-1} + \epsilon_t + \sum_{j=1}^{q} \beta_j \epsilon_{t-j}, \tag{1}$$

where $\{\epsilon_t\}$ is either a sequence of i.i.d. random variables with zero mean and finite variance or strictly stationary martingale difference sequence (m.d.s.). We set $\theta = (\alpha_1, \ldots, \alpha_p, \beta_1, \ldots, \beta_q)^T \subset \Theta \in \mathbf{R}^{p+q}$.

In what follows, we assume the following conditions for the stationarity and invertibility of $\{y_t\}$:

(**A1**) For all $\theta \in \Theta$, $A(z)B(z) = 0$ implies $|z| > 1$,
    where $A(z) = 1 - \sum_{i=1}^{p} \alpha_i z^i$ and $B(z) = 1 + \sum_{j=1}^{q} \beta_j z^j$.
(**A2**) $A(z)$ and $B(z)$ have no common roots.

Assume that $y_{-p+1}, y_{-p+2}, \ldots, y_0, y_1, \ldots, y_n$ are observed and one wishes to test the following hypotheses:

$$H_0 : \text{The parameter } \theta \text{ does not change. vs.} \tag{2}$$
$$H_1 : \text{not } H_0.$$

For a test, we propose the CUSUM test based on $\{(y_t - \epsilon_t)\epsilon_t\}$ as follows:

$$T_n^\theta := \frac{1}{\sqrt{n}\kappa} \max_{1 \le k \le n} \left| \sum_{t=1}^{k}(y_t - \epsilon_t)\epsilon_t - \frac{k}{n}\sum_{t=1}^{n}(y_t - \epsilon_t)\epsilon_t \right|,$$

where $\kappa^2 = E\left((y_t-\epsilon_t)^2\epsilon_t^2\right)$. Notice that $\{(y_t-\epsilon_t)\epsilon_t\}$ forms a martingale difference sequence under the null hypothesis, which leads us to the weak convergence result as shown in Theorem 1. Under the alternative, T_n^θ is anticipated to have a large value owing to the variation change of terms $y_t - \epsilon_t$, rather than ϵ_t themselves.

Because ϵ_t are unobservable, we approximate them with observable $\{\hat{\epsilon}_t\}$ which are the processes defined recursively by

$$\hat{\epsilon}_t = y_t - \sum_{i=1}^{p}\hat{\alpha}_i y_{t-i} - \sum_{j=1}^{q}\hat{\beta}_j\hat{\epsilon}_{t-j},$$

where $\hat{\alpha}_1, \ldots, \hat{\alpha}_p$ and $\hat{\beta}_1, \ldots, \hat{\beta}_q$ are the estimators of $\alpha_1, \ldots, \alpha_p$ and $\beta_1, \ldots, \beta_q$, respectively. The initial values of $\hat{\epsilon}_{-q+1}, \ldots, \hat{\epsilon}_0$ are set to zero when $q > 0$.

Below we impose some regularity conditions:

(**A3**) Under H_0, $\sqrt{n}(\hat{\alpha}_i - \alpha_i) = O_P(1)$, $i = 1, \ldots, p$, and $\sqrt{n}(\hat{\beta}_j - \beta_j) = O_P(1)$, $j = 1, \ldots, q$.

Then, we have the following result.

Theorem 1. *Suppose that* (**A1**)–(**A3**) *and the following conditions hold:*

(**E1**) $\{\epsilon_t\}$ *is an i.i.d. sequence with zero mean, variance* σ^2 *and* $E\epsilon_t^4 < \infty$, *or*
(**E2**) $\{\epsilon_t\}$ *is a strongly mixing m.d.s. with* $(4 + \delta)$-*th moment for some* $\delta > 0$
and mixing-order $\alpha(\cdot)$ *satisfying*

$$E(\epsilon_t|\mathcal{F}_{t-1}) = 0, \quad E\epsilon_t^2 = \sigma^2, \quad \frac{1}{n}\sum_{t=1}^{n}E(\epsilon_t^2|\mathcal{F}_{t-1}) \to \sigma^2, \quad \frac{1}{n}E\left(\sum_{t=1}^{n}(\epsilon_t^2 - \sigma^2)\right)^2 \to \nu^2 > 0,$$

where $\mathcal{F}_t$ *is the* σ-*field generated by* ϵ_i, $i \le t$ *and* $\sum_k \alpha(k)^{\delta/(4+\delta)} < \infty$.

Then, under H_0, *as* $n \to \infty$,

$$\hat{T}_n^\theta := \frac{1}{\sqrt{n}\hat{\kappa}} \max_{1 \le k \le n} \left| \sum_{t=1}^{k}(y_t - \hat{\epsilon}_t)\hat{\epsilon}_t - \frac{k}{n}\sum_{t=1}^{n}(y_t - \hat{\epsilon}_t)\hat{\epsilon}_t \right| \xrightarrow{w} \sup_{0 \le s \le 1} |W_1^\circ(s)|,$$

where $\hat{\kappa}^2 = \frac{1}{n}\sum_{t=1}^{n}(y_t - \hat{\epsilon}_t)^2\hat{\epsilon}_t^2 - \left(\frac{1}{n}\sum_{t=1}^{n}(y_t - \hat{\epsilon}_t)\hat{\epsilon}_t\right)^2$ *and* $W_1^\circ(\cdot)$ *denotes a Brownian bridge: see [2] for its definition.*

We reject H_0 if $\hat{T}_n^\theta \ge C_\alpha$ at the nominal level α, where C_α is the $100(1-\alpha)$ quantile value of $\sup_{0 \le s \le 1}\|W_1^\circ(s)\|$. The critical values for $\alpha = 0.01, 0.05, 0.10$ are provided in Table 1 in [7].

Next, suppose that one wishes to test the following hypotheses:

$$H_0 : \sigma^2 \text{ does not change. vs.} \tag{3}$$
$$H_1 : \text{not } H_0.$$

In this case, the $\hat{T}_n^\theta$ cannot detect a variance change efficiently, and we consider the CUSUM test based on $\{\epsilon_t^2\}$:

$$T_n^\sigma := \frac{1}{\sqrt{n}\nu} \max_{1 \le k \le n} \left| \sum_{t=1}^k \epsilon_t^2 - \frac{k}{n} \sum_{t=1}^n \epsilon_t^2 \right|,$$

where $\nu^2 = E(\epsilon_0^2 - \sigma^2)^2 + 2 \sum_{t=1}^\infty E(\epsilon_0^2 - \sigma^2)(\epsilon_t^2 - \sigma^2)$.

Then, we have the following result.

Theorem 2. *Suppose that* **(A1)**–**(A3)** *hold and either* **(E1)** *or* **(E2)** *hold. Then, under H_0, as $n \to \infty$,*

$$\hat{T}_n^\sigma := \frac{1}{\sqrt{n}\hat{\nu}} \max_{1 \le k \le n} \left| \sum_{t=1}^k \hat{\epsilon}_t^2 - \frac{k}{n} \sum_{t=1}^n \hat{\epsilon}_t^2 \right| \xrightarrow{w} \sup_{0 \le s \le 1} |W_1^\circ(s)|,$$

where $\hat{\nu}^2 = \frac{1}{n} \sum_{t=1}^n \left(\hat{\epsilon}_t^2 - \frac{1}{n} \sum_{t=1}^n \hat{\epsilon}_t^2 \right)^2 + 2\frac{1}{n} \sum_{k=1}^l \sum_{t=k+1}^n \left(\hat{\epsilon}_t - \frac{1}{n} \sum_{t=1}^n \hat{\epsilon}_t^2 \right) \left(\hat{\epsilon}_{t-k} - \frac{1}{n} \sum_{t=1}^n \hat{\epsilon}_t^2 \right)$ *with* $l := l_n \to \infty$ *and* $l = o(n^{1/4})$ *as* $n \to \infty$ *(cf. [12]).*

Finally, suppose that one is interested in testing for a change in θ and σ simultaneously. Then, we set up the hypotheses:

$$H_0 : (\theta, \sigma^2) \text{ does not change. vs.} \tag{4}$$
$$H_1 : \text{not } H_0,$$

and consider the test:

$$\hat{T}_n = \max_{1 \le k \le n} \frac{1}{n} \left(\sum_{t=1}^k \hat{U}_t - \frac{k}{n} \sum_{t=1}^n \hat{U}_t \right)^T \hat{\Sigma}^{-1} \left(\sum_{t=1}^k \hat{U}_t - \frac{k}{n} \sum_{t=1}^n \hat{U}_t \right),$$

where $\hat{U}_t = \left((y_t - \hat{\epsilon}_t)\hat{\epsilon}_t, \hat{\epsilon}_t^2 \right)^T$ and $\hat{\Sigma} = \begin{pmatrix} \hat{\kappa} & 0 \\ 0 & \hat{\nu} \end{pmatrix}$.

The following theorem can be proven similarly to Theorems 1 and 2 and is omitted for brevity.

Theorem 3. *Suppose that* **(A1)**–**(A3)** *hold and either* **(E1)** *or* **(E2)** *hold. Then, under H_0, as $n \to \infty$,*

$$\hat{T}_n \xrightarrow{w} \sup_{0 \le s \le 1} \|W_2^\circ(s)\|^2,$$

where $W_2^\circ(\cdot)$ *denotes a 2-dimensional Brownian bridge.*

Compared to the score vector-based CUSUM test, $\hat{T}_n$ is easier to use in practice, because the dimension of the CUSUM test increases proportional to the number of unknown parameters, whereas it is fixed in our test. Our simulation study in the next section confirms the validity of the test, outperforming both $\hat{T}_n^\theta$ and $\hat{T}_n^\sigma$.

3   Simulation Study

In this section, we evaluate the performance of the proposed CUSUM tests for the ARMA(1,1) model with GARCH(1,1) errors:

$$y_t = \alpha y_{t-1} + \epsilon_t - \beta \epsilon_{t-1}, \tag{5}$$
$$\epsilon_t = \sqrt{h_t}\eta_t, \quad h_t = w + a\epsilon_{t-1}^2 + bh_{t-1},$$

where η_t are assumed to follow $N(0,1)$, $\sqrt{4/5}t(10)$, or $0.2N(1.6,1) + 0.8N(-0.4,0.2)$ (NM hereafter) distributions.

We consider the problem of testing the hypotheses:

H_0 : The true parameter $\theta = (\alpha, \beta)$ and σ^2 does not change over $y_1, \ldots, y_n$. vs.
H_1 : θ change to $\theta' = (\alpha', \beta')$ or σ^2 change to $\sigma^{2'}$ at $[n/2]$.

Note that $E\epsilon_t^2 = \sigma^2 = \frac{w}{1-a-b}$ for a stationary GARCH(1,1) model. For observations with $n = 300$ and 500 generated from models (5), we calculate the empirical sizes and powers for specific set-ups in (θ, σ^2) with 500 repetitions, summarized in Tables 1, 2 and 3.

The result shows that the estimate-based CUSUM test $\hat{T}_n^E$ appears to have severe size distortions even when the sample size is moderate, while the residual-based CUSUM tests have no severe size distortions. As anticipated, $\hat{T}_n^\sigma$ and $\hat{T}_n^\theta$ perform poorly for the change of ARMA parameters α and β and the variance σ^2, respectively, whereas $\hat{T}_n$ can effectively detect all changes. It can be also observed that the different types of innovation distributions have little influence on the result. Overall, our findings show that $\hat{T}_n$ makes a stable test and outperforms $\hat{T}_n^\theta$ and $\hat{T}_n^\sigma$ in terms of power.

4   Proofs

The lemma below can be easily proven in a manner similar to Theorem 2 of [13], and the proof is omitted for brevity.

Lemma 1. *Under the same conditions in Theorem 1, we have that for $j = 1, 2$,*

$$\max_{1 \le k \le n} \left| \frac{1}{\sqrt{n}} \sum_{t=1}^{k} \hat{\epsilon}_t^j - \frac{1}{\sqrt{n}} \sum_{t=1}^{k} \epsilon_t^j \right| = o_P(1).$$

Moreover, we have

$$\max_{1 \le k \le n} \left| \frac{1}{\sqrt{n}} \sum_{t=1}^{k} y_t(\hat{\epsilon}_t - \epsilon_t) - \frac{1}{\sqrt{n}} \sum_{t=1}^{k} y_t(\hat{\epsilon}_t - \epsilon_t) \right| = o_P(1).$$

Table 1. Empirical sizes and powers for the ARMA(1,1) model with GARCH(1,1) errors and $\eta_t \sim N(0,1)$

(α, β, w, a, b) nominal level		$n = 300$		$n = 500$	
		0.05	0.10	0.05	0.10
size (0.3, 0.2, 0.1, 0.1, 0.2)	$\hat{T}_n$	0.034	0.074	0.048	0.096
	$\hat{T}_n^\theta$	0.046	0.070	0.046	0.088
	$\hat{T}_n^\sigma$	0.032	0.062	0.058	0.118
	$\hat{T}_n^E$	0.204	0.292	0.146	0.210
α	$\hat{T}_n$	0.826	0.892	0.968	0.988
$0.3 \longrightarrow 0.7$	$\hat{T}_n^\theta$	0.882	0.936	0.982	0.990
	$\hat{T}_n^\sigma$	0.028	0.058	0.038	0.070
	$\hat{T}_n^E$	0.962	0.976	0.992	0.996
β	$\hat{T}_n$	0.494	0.614	0.770	0.852
$0.2 \longrightarrow 0.6$	$\hat{T}_n^\theta$	0.586	0.690	0.846	0.914
	$\hat{T}_n^\sigma$	0.042	0.074	0.038	0.082
	$\hat{T}_n^E$	0.896	0.934	0.976	0.982
w	T_n	1.000	1.000	1.000	1.000
$0.1 \longrightarrow 0.5$	$\hat{T}_n^\theta$	0.044	0.060	0.040	0.084
	$\hat{T}_n^\sigma$	1.000	1.000	1.000	1.000
a	$\hat{T}_n$	0.380	0.524	0.674	0.798
$0.1 \longrightarrow 0.5$	$\hat{T}_n^\theta$	0.022	0.050	0.040	0.072
	$\hat{T}_n^\sigma$	0.544	0.672	0.820	0.886
b	$\hat{T}_n$	0.932	0.976	1.000	1.000
$0.2 \longrightarrow 0.6$	$\hat{T}_n^\theta$	0.038	0.078	0.054	0.106
	$\hat{T}_n^\sigma$	0.976	0.990	1.000	1.000

Proof of Theorem 1. We express

$$(y_t - \hat{\epsilon}_t)\hat{\epsilon}_t = (y_t - \epsilon_t)\epsilon_t + \left(y_t(\hat{\epsilon}_t - \epsilon_t) - (\hat{\epsilon}_t^2 - \epsilon_t^2) \right).$$

Since $\{(y_t - \epsilon_t)\epsilon_t, \mathcal{F}_t\}$ is a square integrable martingale difference sequence, Theorem 23.1 of [2] implies that T_n^θ converges weakly to $\sup_{0 \leq s \leq 1} |W_1^\circ(s)|$. Hence, owing to Lemma 1,

$$\frac{1}{\sqrt{n}\kappa} \max_{1 \leq k \leq n} \left| \sum_{t=1}^{k} (y_t - \hat{\epsilon}_t)\hat{\epsilon}_t - \frac{k}{n} \sum_{t=1}^{n} (y_t - \hat{\epsilon}_t)\hat{\epsilon}_t \right| \xrightarrow{w} \sup_{0 \leq s \leq 1} |W_1^\circ(s)|.$$

Since $\hat{\kappa} \to \kappa$ in probability, the theorem is asserted. $\qquad\square$

Table 2. Empirical sizes and powers for the ARMA(1,1) model with GARCH(1,1) errors and $\eta_t \sim \sqrt{4/5}t(10)$

(α, β, w, a, b) nominal level		n = 300		n = 500	
		0.05	0.10	0.05	0.10
size (0.3, 0.2, 0.1, 0.1, 0.2)	$\hat{T}_n$	0.042	0.076	0.044	0.078
	$\hat{T}_n^\theta$	0.032	0.062	0.024	0.062
	$\hat{T}_n^\sigma$	0.048	0.094	0.044	0.084
	$\hat{T}_n^E$	0.224	0.288	0.216	0.278
α	$\hat{T}_n$	0.744	0.842	0.920	0.956
$0.3 \longrightarrow 0.7$	$\hat{T}_n^\theta$	0.828	0.878	0.956	0.964
	$\hat{T}_n^\sigma$	0.018	0.044	0.026	0.068
	$\hat{T}_n^E$	0.952	0.970	0.994	0.998
β	$\hat{T}_n$	0.398	0.520	0.666	0.776
$0.2 \longrightarrow 0.6$	$\hat{T}_n^\theta$	0.504	0.648	0.754	0.832
	$\hat{T}_n^\sigma$	0.016	0.044	0.028	0.084
	$\hat{T}_n^E$	0.852	0.896	0.968	0.974
w	$\hat{T}_n$	0.902	0.932	0.956	0.972
$0.1 \longrightarrow 0.5$	$\hat{T}_n^\theta$	0.026	0.054	0.040	0.064
	$\hat{T}_n^\sigma$	0.944	0.962	0.974	0.978
a	$\hat{T}_n$	0.134	0.224	0.272	0.422
$0.1 \longrightarrow 0.5$	$\hat{T}_n^\theta$	0.020	0.038	0.034	0.058
	$\hat{T}_n^\sigma$	0.236	0.366	0.388	0.536
b	$\hat{T}_n$	0.548	0.682	0.788	0.854
$0.2 \longrightarrow 0.6$	$\hat{T}_n^\theta$	0.034	0.074	0.038	0.082
	$\hat{T}_n^\sigma$	0.668	0.752	0.852	0.902

Proof of Theorem 2. Corollary 1 of [4] implies that T_n^σ converges weakly to $\sup_{0 \le s \le 1} |W_1^\circ(s)|$. Hence, owing to Lemma 1, we have

$$\frac{1}{\sqrt{n}\nu} \max_{1 \le k \le n} \left| \sum_{t=1}^k \hat{\epsilon}_t^2 - \frac{k}{n} \sum_{t=1}^n \hat{\epsilon}_t^2 \right| \xrightarrow{w} \sup_{0 \le s \le 1} |W_1^\circ(s)|.$$

Moreover, using Lemma 1 and Theorem 4.2 of [12], we can easily see that $\hat{\nu}$ is consistent. This asserts the theorem. □

Table 3. Empirical sizes and powers for the ARMA(1,1) model with GARCH(1,1) errors and $\eta_t \sim NM$

(α, β, w, a, b) nominal level		$n = 300$		$n = 500$	
		0.05	0.10	0.05	0.10
size (0.3, 0.2, 0.1, 0.1, 0.2)	$\hat{T}_n$	0.024	0.072	0.046	0.078
	$\hat{T}_n^{\theta}$	0.024	0.064	0.030	0.052
	$\hat{T}_n^{\sigma}$	0.030	0.088	0.054	0.088
	$\hat{T}_n^{E}$	0.212	0.266	0.188	0.260
α $0.3 \longrightarrow 0.7$	$\hat{T}_n$	0.944	0.964	0.944	0.964
	$\hat{T}_n^{\theta}$	0.954	0.974	0.954	0.974
	$\hat{T}_n^{\sigma}$	0.042	0.082	0.042	0.082
	$\hat{T}_n^{E}$	0.946	0.962	0.998	0.998
β $0.2 \longrightarrow 0.6$	$\hat{T}_n$	0.376	0.502	0.652	0.730
	$\hat{T}_n^{\theta}$	0.470	0.590	0.720	0.784
	$\hat{T}_n^{\sigma}$	0.038	0.064	0.032	0.072
	$\hat{T}_n^{E}$	0.814	0.864	0.940	0.970
w $0.1 \longrightarrow 0.5$	$\hat{T}_n$	0.978	1.000	1.000	1.000
	$\hat{T}_n^{\theta}$	0.024	0.056	0.018	0.048
	$\hat{T}_n^{\sigma}$	1.000	1.000	1.000	1.000
a $0.1 \longrightarrow 0.5$	$\hat{T}_n$	0.122	0.200	0.232	0.380
	$\hat{T}_n^{\theta}$	0.038	0.046	0.014	0.036
	$\hat{T}_n^{\sigma}$	0.192	0.314	0.368	0.514
b $0.2 \longrightarrow 0.6$	$\hat{T}_n$	0.560	0.694	0.820	0.900
	$\hat{T}_n^{\theta}$	0.024	0.060	0.038	0.056
	$\hat{T}_n^{\sigma}$	0.704	0.798	0.908	0.954

5 Concluding Remarks

In this study, we proposed a new CUSUM test which is handier to use in practice and performs adequately for ARMA models with martingale difference innovations. We derived its limiting null distribution and demonstrated its validity through Monte carlo simulations. In particular, the models under consideration accommodate ARMA-GARCH type models, and our test is applicable to detecting the change of GARCH parameters as well as ARMA parameters when dealing with ARMA-GARCH models. In fact, this problem can be handled in more general settings such as location-scale time series models, which is our on-going project.

Acknowledgements. We thank one anonymous referee for his/her careful reading and valuable comments.

References

1. Berkes, I., Horváth, L., Kokoszka, P.: Testing for parameter constancy in GARCH (p, q) models. Stat. Probab. Lett. **70**(4), 263–273 (2004)
2. Billingsley, P.: Convergence of Probability Measures. Wiley, New York (1968)
3. Gombay, E.: Change detection in autoregressive time series. J. Multivar. Anal. **99**(3), 451–464 (2008)
4. Herrndorf, N.: A functional central limit theorem for weakly dependent sequences of random variables. Ann. Probab. **12**, 141–153 (1984)
5. Inclán, C., Tiao, G.C.: Use of cumulative sums of squares for retrospective detection of changes of variance. J. Am. Stat. Assoc. **89**(427), 913–923 (1994)
6. Lee, J., Lee, S.: Parameter change test for nonlinear time series models with GARCH type errors. J. Korean Math. Soc. **52**, 503–522 (2015)
7. Lee, S., Ha, J., Na, O., Na, S.: The cusum test for parameter change in time series models. Scand. J. Stat. **30**(4), 781–796 (2003)
8. Lee, S., Oh, H.: Parameter change test for autoregressive conditional duration models. Ann. Inst. Stat. Math. **68**(3), 621–637 (2016)
9. Lee, S., Song, J.: Test for parameter change in ARMA models with GARCH innovations. Stat. Probab. Lett. **78**(13), 1990–1998 (2008)
10. Oh, H., Lee, S.: On score vector- and residual-based CUSUM tests in ARMA-GARCH models. Stat. Methods Appl. 1–22 (2017). https://doi.org/10.1007/s10260-017-0408-9
11. Page, E.S.: A test for a change in a parameter occurring at an unknown point. Biometrika **42**(3/4), 523–527 (1955)
12. Phillips, P.C.B.: Time series regression with a unit root. Econometrica **55**, 277–301 (1987)
13. Yu, H.: High moment partial sum processes of residuals in ARMA models and their applications. J. Time Ser. Anal. **28**(1), 72–91 (2007)

Agent-Based Modeling of Economic Instability

Akira Namatame[(⊠)]

National Defense Academy, Yokosuka, Japan
akiranamatame@gmail.com

Abstract. Networks increase interdependence, which creates challenges for managing economic risks. This is especially apparent in areas such as financial institutions and enterprise risk management, where the actions of a single agent (firm or bank) can impact all the other agents in interconnected networks. In this paper, we use agent-based modeling (ABM) in order to analyze how local defaults of supply chain participants propagate through the dynamic supply chain network and interbank networks and form avalanches of bankruptcy. We focus on the linkage dependence among agents at the micro-level and estimate the impact on the macro activities. Combining agent-based modeling with the network analysis can shed light on understanding the primary role of banks in lending to the wider real economy. Understanding the linkage dependency among firms and banks can help in the design of regulatory paradigms that rein in systemic risk while enhancing economic growth.

Keywords: Agent-based economics · Systemic risk
Evolving credit networks · Financial networks

1 Introduction

Macro economy has created well defined approaches and several tools that seemed to serve us for the last decades. However, recent economic fluctuations and financial crises emphasize the need of alternative frameworks and methodologies to be able to replicate such phenomena in order to a deeper understanding the mechanism of economic crisis and fluctuation. Financial markets are driven by the real economy and in turn they also have a profound effect on it. Understanding the feedback between these two sectors leads to a deeper understanding of the stability, robustness and efficiency of the economic system [8]. Agent-based approaches in Macroeconomics from bottom-up are getting more and more attention recently [13]. Research on this line has been initiated by the series of papers by Delli Gatti et al. [4,5]. Their model simulating the behavior of interacting heterogeneous agents (firms and banks) is able to generate a large number of stylized facts. To jointly account for an ensemble of the facts regarding both micro-macro properties together with macro aggregates including GDP growth rates, output volatility, business cycle phases, financial fragility, and bankruptcy cascades, agent-based approaches are getting more and more attention recently.

© Springer International Publishing AG 2018
V. Kreinovich et al. (eds.), *Predictive Econometrics and Big Data*, Studies in Computational Intelligence 753, https://doi.org/10.1007/978-3-319-70942-0_18

Historically financial markets were driven by the real economy and in turn they also had a profound effect on it. In recent decades, a massive transfer of resources from the productive sector to the financial sector has been one of the characteristics of global economic systems. This process is mainly responsible for the growing financial instability [9,14]. In production sectors, there has been dramatic increase in the output volatility and uncertainty. Financial inter-linkages play an important role in the emergence of financial instabilities. Recently many research have focused on the role of linkages along the two dimensions of contagion and liquidity, and they suggest that regulators have to look at the interplay of network topology, capital requirements, and market liquidity. In particular for contagion, the position of institutions in the network matters and their impact can be computed through stress tests even when there are no defaults in the system [2].

For the data-driven study using empirical data, many scholars use a collection of daily snapshots of the Italian interbank money market originally provided by the Italian electronic Market for Inter-bank, referred to e-MID in the text [6,11,18]. However, even central banks and regulators have only a dim view of the interconnections between banks at a moment in time, and thus the systemic risk in the financial networks, and each bank's contribution to this risk, are poorly known. A natural starting point is to utilize complementary approach to data-driven approach, basing their systemic risk measures on accessing and interpreting data on balance sheets and trading. As understanding of the most critical systemic attributes improves, this network description can be extended to wider jurisdictions and can record more detail: complex transactions such as credit risk derivatives, more complex institutional behavior such as internal risk limit systems and responses to counterparty risk changes.

In this paper, we investigate the effect of credit linkages on the macroeconomic activity by developing the network-based agent model. In particular, we study the linkage dependence among agents (firms and banks) at the micro-level and to estimate their impact on the macro activities such the GDP growth rate, the size and growth rate distributions of agents. We propose the model refinement strategy which validate through the some universal laws and properties based on empirical studies revealing statistical properties of macro-economic time series [19]. The purpose of the network-based model of systemic risks is to build up the dependence among agents (firms and banks) at the micro-level and to estimate their impact on the macro stability.

Phase structure of hypothetical financial systems: the relation between basic network parameters such as connectivity, homogeneity and uncertainty, and macroscopic systemic risk measures is non-linear, non-intuitive and difficult to predict. Such emergent features will reflect profound properties of real world financial networks that can be understood by first looking at deliberately simplified agent-based simulation models [16]. Simulation studies of complex hypothetical financial networks that map out these types of features will lead to improved understanding of the resilience of networks, and perhaps ultimately to pragmatic rules of thumb for network participants. This line of inquiry also links systemic network theory strongly to other areas of network science, from which we may draw additional ideas and intuition. We show that three stylized facts: a fat-tailed size distribution of the firm sizes, a tent- shaped growth rate distribution, the scaling relation of the growth rate variance with firm size. We then address the questions of validating and verifying simulations. We validate with the widely acknowledged "stylized facts" which describe the firm (and bank) growth rates of fat tails, tent distribution, volatility, etc., and recall that some of these properties are directly linked to the way time is taken into account. The growth of firm size, the distribution of firm sizes, the distribution of sizes of the new firms in each year and find it to be well approximated by a log-normal [19]. We validate the simulation results in terms of such as (i) the distribution of the logarithm of the growth rates, for a fixed growth period of one year, and for companies with approximately the same size S, displays an exponential form.

In the second part of our work, we investigate the effect of credit linkages on the firms activities to explain some key elements occurred during the recent economic and financial crisis. From this perspective, the network theory is a natural candidate for the analysis of interacting agent systems. The financial sector can be regarded as a set of agents (banks and firms) who interact with each other through financial transactions. These interactions are governed by a set of rules and regulations, and take place on an interaction graph of all connections between agents. The network of mutual credit relations between financial institutions and firms plays a key role in the risk for contagious defaults. In particular, we study the repercussions of inter-bank connectivity on agents' performances, bankruptcy waves and business cycle fluctuations. Our findings suggest that there are issues with the role that the bank system plays in the real economy and in pursuing economic growth. Indeed, our model shows that a heavily- interconnected inter-bank system increases financial fragility, leading to economic crises and distress contagion.

2   Economic Risks in a Connected World

There is empirical evidence that as the connectivity of a network increases, there is an increase in the network performance, but at the same time, there is an increase in the chance of risk contagion which is extremely large. If external shocks at some agents are propagated to the other connected agents due to failure, the domino effects often come with disastrous consequences. The network is only as strong as its weakest link, and trade-offs are most often connected to a function that models system performance management. The qualification of risks lies in their connections. An interdependent risk to one system may present an opportunity to other systems. Therefore a systemic risk impacts the integrity of the whole system as well as its components.

In a networked world, the risks faced by any one agent depend not only on that agents actions but also on those of others. The fact that the risk one actor faces is often determined in part by the activities of others gives a unique and complex structure to the incentives that agents face as they attempt to reduce their exposure to these interdependent risks. The concept of interdependent risks refers to situations in which multiple agents act separately generate common risk. Protective management can reduce the risk of a direct loss to each agent, but there is still some chance of suffering damage from others actions. The fact that the risk is often determined in part by the behavior of others imposes independent risk structures on the incentives that agents face for reducing risk or investing in risk mitigation measures. Kunreuther and Heal were initially led to analyze such situations by focusing on the interdependence of security problems [12]. An interdependent security setting is one in which each individual or firm that is part of an interconnected system must decide independently whether to adopt protective strategies that mitigate future losses. The analysis focused on protection against discrete, low-probability events in a variety of protective settings with somewhat different cost and benefit structures: airline security, computer security, fire protection, and vaccination. Under some circumstances, the interdependent security problem resembles the familiar prisoners dilemma in which the only equilibrium is the decision by all agents not to invest in protection even though everyone would be better off if they had decided to incur this cost. In other words, a protective strategy that would benefit all agents if widely adopted may not be worth its cost to any single agent and it is better off simply taking a free ride on the others' investments.

The financial crises triggered numerous studies on the systemic risks caused by contagion effects via interconnections in the modern banking networks. Systemic risks result in continuous large-scale defaults or systemic failure among the networked banks and financial institutions [3, 7]. The formulation of systemic risk can greatly benefit from a complex network approach. Allen and Gale introduced the use of network theories to enrich our understanding of financial systems and studied how the financial system responds to contagion when financial institutions are connected with different network topologies [1]. They how the banking network topology affect the stability of both finance market and product market by changing the density of connections among banks. While the risk of contagion may be expected to be larger in a highly interconnected banking system, we show that shocks may have complex effects on financial institutions as well as the firms.

Many studies analyze the financial systems such as financial stability and contagion using the network theory and other network analysis methods. In an interbank market, banks facing liquidity shortages may borrow liquidity from other banks that have liquidity surpluses. This system of liquidity swapping provides the interbank market with enhanced liquidity sharing. Furthermore, it also brings down the risk of contagion among the interconnected banks when unexpected problems arise. Solvency or liquidity problems of a single bank can travel through the interbank linkages to other banks and become a causality of systemic failure; this highlights the importance of interbank markets for financial stability [15, 17].

3   Agent Based Modeling

Our work is based on an existing agent-based model [10, 20]. Here, we consider multiple banks which can operate not only in the credit market but also in the inter-bank market. In our model, firms may ask for loans from banks to increase their production rate and profit. If contacted banks face liquidity shortage when trying to cover the firms' requirements, they may borrow from a surplus bank in the inter-bank network. In this market, therefore, lender banks share with borrowers bank the risk for the loan to the firm. We model the inter-bank network as preference attachment.

3.1   Firms Behaviour

The goods market is implemented following the model of Delli Gatti et al. [5] where output is supply-determined, that is firms sell all the output they optimally decide to produce. In the model of Delli Gatti, there are two types of firms: Downstream (D) firms that produce consumption goods using labor and intermediate goods; Upstream (U) firms that produce intermediate goods on demand from D firms.

The D firms demand both for labor $N_i(t), i = 1, 2, .., D$, and for intermediate goods $Q_i(t)$ depending on their financial conditions, that is captured by net worth $A_i(t)$. Respectively, the demands of the ith firm are given by

$$N_i(t) = c_1 A_i(t)^\beta, Q_i(t) = c_2 A_i(t)^\beta \qquad (1)$$

The consumption goods are sold at a stochastic price $u_i(t)$ that is a random variable extracted from a uniform distribution between $(0, 1)$. The U firms produce intermediate goods employing only labour, so that for the $j^{th}, j = 1, 2, .., U$, the demand is

$$Q_j(t) = c_3 N_j(t) \qquad (2)$$

Many D firms can be linked to a single U firm but each D firm has only one supplier for intermediate goods among the U firms. The price of intermediate goods is

$$p_j(t) = 1 + r_j(t) = \alpha A_j(t)^{-\alpha} \qquad (3)$$

where $r_j(t)$ is the interest on trade credit, which is assumed to be dependent only on the financial condition of the U firm. In particular, if the jth firm is not performing well, it will give credit with a less favourable interest rate. While the production of D firms is determined by their worth $A_i(t)$, the production of U firms is determined by the demand on the part of D firms.

$$Q_j(t) = c_2 \sum_{i=1}^{D} A_i(t)^\beta \qquad (4)$$

Analogously, the demand for labor will be

$$N_j(t) = c_1 \sum_{i=1}^{D} A_i(t)^\beta \qquad (5)$$

Each period a subset of firms enter in the credit market asking for credit. The amount of credit requested by companies is related to their investment expenditure, which is, therefore, dependent on interest rate and firm's economic situation. If the net worth of the firms is not sufficient to pay the wage bill, they will demand credit to a bank. For each firm the credit demand is

$$B(t) = c_4 N(t) - A(t) \tag{6}$$

where the functional form of $N(t)$ changes if we are considering U firm or D firm. We assume that many firms can be linked to a single bank but each firm has only one supplier of loans. Without entering in the details, we point out that the interest rate on loans is a decreasing function of the banks net worth and penalizes financially fragile firms.

The profits of the D and U firm are evaluated from the difference between their gains and the costs, and the profit of the jth U firm at time t is given by

$$\pi_j(t) = (1 + r_j(t))Q_j(t) - (1 + r_j(t))B_j(t) \tag{7}$$

At each time step the net worth of each firm is updated according to

$$A_j(t+1) = A_j(t) + \pi_j(t) - D_j(t) \tag{8}$$

Bankruptcy occurs if the net worth becomes negative. The bankrupt firm leaves the market. Therefore $D_j(t)$ in Eq. (8) is the "bad debt", that takes into account the possibility that a borrower cannot pay back the loan because it goes bankrupt (that is, $A_j(t) \leq 0$). In this framework the lenders are the U firms and the banks and both U and D firms can be borrowers. The total number of agents (U and D firms) is kept constant over time. Therefore when firms fail, they are replaced by new entrants which are on average smaller than incumbents. So, entrants' size is drawn from a uniform distribution centered around the mode of the size distribution of incumbent firms.

3.2   Bank Behavior

The primary purpose of banks is to channel their funds towards loans to companies. Consulted banks, having analyzed their own credit risk and the firm's risk, may grant the requested loan, when they have enough supply of liquidity. The supply of credit is a percentage of banks' equity because financial institutions adopt a system of risk management based upon an equity ratio. When consulted banks do not have liquidity to lend, they can enter in the interbank system, in order not to lose the opportunity of earning on investing firms.

Similar to firms, we have a constant population of competitive banks indexed by $j = 1, ..., B$. Each bank has a balance sheet structure defined as $S_j(t) = E_j(t) + D_j(t)$ with $S_j(t)$ being the credit supply, $E_j(t)$ the equity and $D_j(t)$ the debt. The primary function of banks activity is to lend their funds through loans to firms, as this is their way to make money via interest rates. Bank j offers its interest rate to the borrower firm i:

$$r_{j,i} = \sigma A_j(t)^{-\beta} + \theta l_i(t)^{-\theta} \tag{9}$$

Where $l_i(t)$ is the amount of lending. So the interest rate is decreasing with the borrower's financial robustness. In a sense, we adopt the principle according to which the interest rate charged by banks incorporates an external finance premium increasing with the leverage and, therefore, inversely related to the borrower's net worth.

When firm i needs loan, it contacts a number of randomly chosen banks. Credit linkages between firms and banks are defined by some bipartite graph. Contacted banks, checked the investment risk and their amount of liquidity, offer an interest rate in Eq. (9). After exploring the lending conditions of the contacted banks, each firm asks the consulted banks for credit starting with the one offering the lowest interest rate. If in the credit market, the contacted financial institutions do not have enough supply of liquidity to fully satisfy the firm's loan, then banks can use the inter-bank market. As in the credit market, the requiring bank (borrower) asks the lacking fraction of the loan requested by the firm from a number of randomly chosen banks (lenders). Among the contacted banks, the banks satisfying the risk threshold in Eq. (7) and having enough supply of liquidity offer the loan to the asking bank for an inter-bank interest rate, which equals the credit market interest rate in Eq. (8). Among this subset of offering banks, the borrower bank chooses the lender bank, starting with the one offering the lowest interest rate.

At the end of each period t, after trading has taken place, financial institutions update their profits. The bank's profit depends on interests on credit market (first term), on interests on inter-bank market (second term), which can be either positive or negative depending on bank j net position (lender or borrower), on interests paid on deposits and equity (third term). Bank net worth evolves according to:

$$E_j(t+1) = E_j(t) - 1 + \pi_j(t) - B_{ij} - B_{kj} \tag{10}$$

with the last two terms on the right side being firms and banks' bad debts respectively. Similar to firms, banks go bankrupt when their equity at time t becomes negative, and the failed bank leaves the market.

The total number of banks is kept constant over time. Therefore when banks fail, they are replaced by new entrants which are on average smaller than incumbents.

3.3   Credit Network Formation

We define the network formation dynamics, how D firms look for the linkage with U firm and how D and U firms look for a bank to ask their loans. In order to establish the product supply linkages, the firms take the partner choice rule, that is, they search for the minimum of the prices charged by a randomly selected set of possible suppliers. It can change supplier only if a better partner is found. Similarly in order to establish the credit linkages the firms take the partner choice rule: they search for the minimum of the interest rate charged among the loan offered banks. If contacted banks face liquidity shortage when trying to cover the firms' requirements, they may borrow from a surplus bank in the inter-bank system. In this market, therefore, lender banks share with borrower banks the risk for the loan to the firm.

3.4   Interbank Network Formation

We model the inter-bank network based on some connection rules: (case 01) random connection (random graph as a benchmark), (case 02) net-worth based connection, an agent with higher net worth is selected as a partner, (case 03) interest based connection, an agent offering a lower interest is selected as a partner.

Bankruptcies are determined as financially fragile firms fail, i.e. their net worth becomes negative. If one or more firms are not able to pay back their debts to the bank, the bank's balance sheet decreases and, consequently, the firms' bad debt, affecting the equity of banks, can also lead to bank failures. As banks, in case of shortage of liquidity, may enter the interbank market, the failure of borrower banks could lead to failures of lender banks. Agents' bad debt, thus, can bring about a cascade of bankruptcies among banks. The source of the domino effect may be due to indirect interaction between bankrupt firms and their lending banks through the credit market, on one side, and to direct interaction between lender and borrower banks through the inter-bank system, on the other side. Their findings suggest that there are issues with the role that the bank system plays in the real economy and in pursuing economic growth. Indeed, their model shows that a heavily-interconnected inter-bank system increases financial fragility, leading to economic crises and distress contagion. The process of contagion gains momentum and spreads quickly to a large number of banks once some critical banks fail.

4   Simulations and Results

The model is studied numerically for different values of the parameter p, which drives the inter-bank connectivity. We consider an economy consisting of $N = 1000$ firms and $B = 50$ banks and do simulation over the time period span of $T = 1000$. Each firm is initially given the same amount of capital $K_i(0) = 1$, net-worth $A_i(0) = 65$ and loan $L_i(0) = 35$. We also set other parameters as follows: $\tau = 4, \varphi = 0.8, c_i(i = 1 - 4) = 1, \lambda = 0.3, \alpha = 0.1, \chi = 0.8, \psi = 0.1, \theta = 0.05$. The probability of attachment between firms and banks in the credit market is $x = 0.05$. In this way the number of firm's out-going links is less than three. The reason for this is that in a highly connected random network, synchronization could be achieved via indirect links. The effects of direct contagion among financial institutions are easier to be tested in a network where indirect synchronization is less likely to arise. We repeat simulations 100 times with different random seeds. We start by analyzing the effect of inter-bank linkages on the systemic risk. Then we analyze the correlation between the financial and the real sector of the economy.

In our model, bankruptcies are determined as financially fragile firms fail, that is their net worth becomes negative. If one or more firms are not able to pay back their debts to the bank, the bank's balance sheet decreases and, consequently, the firms' bad debt, affecting the equity of banks, can also lead to bank failures. As banks, in case of shortage of liquidity, may enter the interbank market, the failure of borrower banks could lead to failures of lender banks. Agents' bad debt, thus, can bring about a cascade of bankruptcies among banks. The source of the domino effect may be due to indirect interaction between bankrupt firms and their lending banks through the credit market, on one side, and to direct interaction between lender and borrower banks through the inter-bank system, on the other side.

One of the goals of our work is the study of bankruptcy avalanches and their connection with the dynamics of the credit networks. Suppose that a random price fluctuation causes the bankruptcy of some U firms. Consequently, the loans they took will not be fulfilled and the worth of the lenders (banks and D firms) will decrease. Eventually, this will result in a bankruptcy of some of them and, more importantly, in an increase of the interest rates charged on their old and new borrowers. This, in turn, will increase the probability of a bankruptcy of a D firm, and so on. The credit network has a scale free structure and then the default of a highly connected agent may provoke an avalanche of bankruptcy.

The question we address here is whether phenomena of collective bankrupt-cies are related to the initial setting of parameters. Usually simulation starts with homogeneous firms and banks. Early stage of the simulation (before 200 period) many firms and banks default. However after that, especially banks and D-firms grow as extremely heterogeneous agents and the size distribution obeys power low, then a few banks and firms default occasionally. In order to answer this mysterious observation, we investigate the effects of inter-bank linkages on contagion phase in the financial market. In particular, we focus on one of the most extreme examples of systemic failure, namely bank bankruptcies. Figure 1

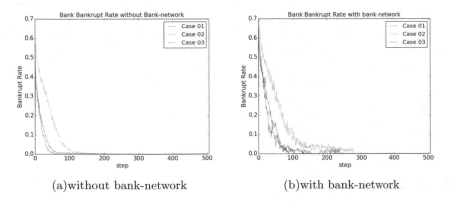

(a)without bank-network   (b)with bank-network

Fig. 1. The bankrupt rate of the banks. (a) The banks are not linked each other and operate independently. (b) The banks form financial networks.

shows the average number of failed banks. Collective bankruptcies arise from the complex nature of agent interaction. To better analyze this observation, we further need data-driven analysis of simulated data.

We are explicitly concerned with the potential of the inter-bank market to act as a contagion mechanism for liquidity crises and to determine the effect of the banks connectivity on macroeconomic outcomes such business cycle fluctuations and bankruptcies.

Our findings suggest that there are issues with the role of the central bank plays in the real economy and in pursuing economic growth. Indeed, our model shows that the central bank plays a great role in heavily shocked economy by clearing big debt. Without the central bank and it is the only the interbank network to absorb a big shock, then financial fragility increases, and leading to economic crises and distress contagion.

5   Conclusion

The main purpose of this paper was to study the contagious bank defaults, that is, the bank defaults that influence other banks through interconnectivity of the networked banking system, and not the defaults caused by the fundamental weakness of a given bank. Since failed banks are not able to honour their commitments in the interbank market, other banks may probably be influenced to default, which may affect more banks and cause further contagious defaults. By modeling a three sector economy with goods, credit and interbank market, we have been able to analyze the role of financial network on the agents performance and macro dynamics in terms of the growth and stability. Our results also support that the interaction among market participants (firms and banks) is a key element to reproduce important stylized facts about bankruptcy waves and business cycle fluctuations. In particular, we have shown that the existence the strong linkages among banks generates larger bankruptcy cascades due to the larger systemic risk. When our inter-bank network reaches the phase transition, the presence of many interconnected banks suggests that the credit network is more susceptible to the domino effect. In this case, in fact, when failures occur, many agents are potentially compromised. However, our model has shown that the relationship between risk propagation and risk sharing cannot be clearly defined. Our findings suggest that there are issues with the role that the bank system plays in the real economy and in pursuing economic growth. Our model also shows that a heavily- interconnected inter-bank system increases financial fragility, leading to economic crises and distress contagion if the role of the central bank is weak.

References

1. Allen, F., Gale, D.: Financial contagion. J. Polit. Econ. **108**(1), 1–33 (2000)
2. Battiston, S., et al.: DebtRank: too central to fail? Financial networks, the FED and systemic risk. Sci. Rep. **2**, 1–6 (2012). Nature Publishing

3. Bramoulle, Y., Kranton, R.: Risk-sharing networks. J. Econ. Behav. Organ. **64** (3–4), 275–294 (2007)
4. Delli Gatti, D., Di Guilmi, C., Gallegati, M., Palestrini, A.: A new approach to business fluctuations: heterogeneous interacting agents, scaling laws and financial fragility. J. Econ. Behav. Organ. **56**(4), 489–512 (2005)
5. Delli Gatti, D., Gallegati, M., Greenwal, B., Stiglitz, J.: The financial accelerator in an evolving credit network. J. Econ. Dyn. Control **34**, 1627–1650 (2010)
6. Fricke, D., Lux, T.: Core-Periphery Structure in the Overnight Money Market: Evidence from the E-mid trading platform. Kiel Working Paper, No. 1759 (2012)
7. Gai, P., Kapadia, S.: Contagion in financial networks. In: Proceedings of Royal Society Interface, pp. 466–480 (2010)
8. Gallegati, M., Greenwald, B., Richiardi, M.G., Stiglitz, J.E.: The asymmetric effect of diffusion processes: risk sharing and contagion. Glob. Econ. J. **8**(3), 1–20 (2008)
9. Haldane, A., May, R.: Systemic risk in banking ecosystems. Nature **469**, 351–355 (2011)
10. Ide, K., Namatame, A.: A mesoscopic approach to modeling and simulation of systemic risks. In: IEEE CIFEr (2014)
11. Iori, G., Jafarey, S., Padilla, F.G.: Systemic risk on the interbank market. J. Econ. Behav. Organ. **61**, 525–542 (2006)
12. Kunreuther, H., Heal, G.: Interdependent security. J. Risk Uncertain. **26**, 231–249 (2003)
13. Legendi, R., Gulyas, L.: Replication of the Macro ABM Model. CRISIS, working paper (2012)
14. May, R.M., Levin, S.A., Sugihara, G.: Ecology for bankers. Nature **451**, 893–895 (2008)
15. May, R.M., Arinaminpathy, N.: Systemic risk: the dynamics of model banking systems. J. R. Soc. **7**, 823–838 (2010)
16. Namatame, A., Chen, S.: Agent-Based Models and the Network Dynamics. Oxford University Press, Oxford (2015)
17. Nier, E., Yang, J., Yorulmazer, T., Alentorn, A.: Network models and financial stability. J. Econ. Dyn. Control **31**(6), 2033–2060 (2007)
18. Soramaki, K., Bech, M.L., Arnold, J., Glass, R.J., Beyeler, W.E.: The topology of interbank payment flows. Phys. A **379**, 317–333 (2007)
19. Stanley, M., Amaral, L., Sergey, V., Havlin, L., Leschhorn, H., Maass, P., Stanley, E.: Scaling behavior in the growth of companies. Nature **379**, 804–806 (1996)
20. Suzuki, Y., Namatame, A., Aruka, Y.: Agent-based modeling of economic volatility and risk propagation on evolving networks agent-based modeling of economic growth and crises. In: Handa, H., Ishibuchi, H., Ong, Y.S., Tan, K. (eds.) Proceedings in Adaptation, Learning and Optimization, vol. 1. Springer, Cham (2015)

A Bad Plan Is Better Than No Plan: A Theoretical Justification of an Empirical Observation

Songsak Sriboonchitta[1] and Vladik Kreinovich[2(✉)]

[1] Faculty of Economics, Chiang Mai University, Chiang Mai 50200, Thailand
songsakecon@gmail.com
[2] Department of Computer Science, University of Texas at El Paso,
500 W. University, El Paso, TX 79968, USA
vladik@utep.edu

Abstract. In his 2014 book "Zero to One", a software mogul Peter Thiel lists the lessons he learned from his business practice. Most of these lessons make intuitive sense, with one exception – his observation that "a bad plan is better than no plan" seems to be counterintuitive. In this paper, we provide a possible theoretical explanation for this somewhat counterintuitive empirical observation.

1 Formulation of the Problem

A bad plan is better than no plan: a counterintuitive empirical observation. In his 2014 book "Zero to One" [1], a software mogul Peter Thiel lists the lessons he learned from his business practice.

Most of these lessons make intuitive sense, with one exception – his observation that a bad plan is better than no plan. At first glance, this empirical observation seems to be counterintuitive.

What we do in this paper. In this paper, we provide a possible theoretical explanation for this empirical observation.

2 How to Describe This Problem in Precise Terms?

We need to describe actions. We decide between different plans of action. There may be many parameters that describe possible actions. For example, for the economy of a country, the central bank can set different borrowing rates, the government can set different values of the minimal wage and of unemployment benefits, etc. For a company, such parameters include percentage of income that goes into research and development, percentage of income that goes into advertisement, etc.

In general, let us denote the number of such parameters by n, and the parameters themselves by $x_1, \ldots, x_n$. From this viewpoint, selecting an action means

© Springer International Publishing AG 2018
V. Kreinovich et al. (eds.), *Predictive Econometrics and Big Data*, Studies in Computational Intelligence 753, https://doi.org/10.1007/978-3-319-70942-0_19

selecting the appropriate values of all these parameters – i.e., in mathematical terms, a point $x = (x_1, \ldots, x_n)$ in the corresponding n-dimensional space.

Initial state. Let $x_1^{(0)}, \ldots, x_n^{(0)}$ denote the values of the parameters corresponding to the current moment of time t_0. Our goal is to select parameters at future moments of time $t_1 = t_0 + h$, $t_2 = t_0 + 2h$, $\ldots$, $t_T = t_0 \cdot T \cdot h$, for some time quantum h.

Changes cannot be too radical. Whether we talk about the economy of a country or of a big company, it is very difficult to make fast drastic changes, there is a large amount of inertia in these economic systems.

Therefore, we will only consider possible actions $x_1, \ldots, x_n$ which are close to the initial state, i.e., which have the form $x_i = x_i^{(0)} + \Delta x_i$ for some small changes Δx_i.

Changes $x(t_{j+1}) - x(t_j)$ from one moment of time t_j to the next one t_{j+1} are even more limited. Let b be the upper bound on such changes:

$$\|x(t_{j+1}) - x(t_j)\| = \sqrt{\sum_{i=1}^{n} (r_i(t_{j+1}) - r_i(t_j))^2} \leq h$$

On the other hand, the very fact that we talk about changes means that we are not completely satisfied with the current situations. The more changes we undertake at each moment of time, the faster we will reach the desired state.

The size of each change is limited by the bound b. Within this limitation, the largest possible changes are changes of the largest possible size b. Thus, we assume that all the changes from one moment of time to the next one are of the same size b:

$$\|x(t_{j+1}) - x(t_j)\| = \sqrt{\sum_{i=1}^{n} (x_i(t_{j+1}) - x_i(t_j))^2} = b.$$

What is our objective. Since we talk about which plans are better, this assumes that we have agreed on how we gauge the effect of different plans, i.e., we have agreed on a numerical criterion y that describes, for each possible action, how good is the result of this action.

The value of this criterion depends on the action: $y = f(x_1, \ldots, x_n)$ for some function $f(x_1, \ldots, x_n)$. In some cases, we may know this function, but in general, we do not know the exact form of this function. In other words, we know what we want to optimize, but we do not necessarily know the exact consequences of each action.

Since changes are small, we can simplify the expression for the objective function. We are interested in the values $y = f(x_1, \ldots, x_n)$ of the agreed-upon objective function $f(x_1, \ldots, x_n)$ in the small vicinity of the original state

$x^{(0)} = (x_1^{(0)}, \ldots, x_n^{(0)})$. In other words, we are interested in the values

$$f(x_1, \ldots, x_n) = f(x_1^{(0)} + \Delta x_1, \ldots, x_n^{(0)} + \Delta x_n)$$

corresponding to small deviations Δx_i.

Since the deviations Δx_i are small, we can expand the objective function into Taylor series in Δx_i and keep only the main – linear – terms in this expansion. In other words, it makes sense to consider a linear approximation of the original objective function:

$$f(x_1, \ldots, x_n) = f(x_1^{(0)} + \Delta x_1, \ldots, x_n^{(0)} + \Delta x_n) = a_0 + \sum_{i=1}^{n} a_i \cdot \Delta x_i,$$

where $a_0 \stackrel{\text{def}}{=} f(x_1^{(0)}, \ldots, x_n^{(0)})$ and

$$a_i \stackrel{\text{def}}{=} \frac{\partial f}{\partial x_i}_{|x=x^{(0)}}.$$

Maximizing this expression is equivalent to maximizing the linear part

$$\sum_{i=1}^{n} a_i \cdot \Delta x_i.$$

Thus, if we denote the deviations Δx_i by u_i, we arrive at the following problem:

- we start with the values $u^{(0)} = (u_1, \ldots, u_n) = (0, \ldots, 0)$;
- at each moment of time, we change the action by a change of a given size b:

$$\|u(t_{j+1}) - u(t_j)\| = \sqrt{\sum_{i=1}^{n} (u_i(t_{j+1}) - u_i(t_j))^2} = b;$$

- we want to gradually change the values u_i so that the (unknown) objective function $\sum_{i=1}^{n} a_i \cdot u_i$ gets as large as possible.

What does "no plan" mean. An intuitive understanding of what "no plan" means is that at each moment of time, we undertake a random change, uncorrelated with all the previous changes.

In other words, at each moment of time, as the change vector $u(t_{j+1}) - u(t_j)$, we select a vector of length b with a random direction. Since we have no reason to select one of the possible directions, we thus consider all the directions to be equally probable.

In other words, we assume that the change vector uniformly distributed on the sphere of radius b, and that the changes corresponding to different moments of time are independent. The resulting trajectory $u(t)$ is thus an *n-dimensional random walk* [2] (or, equivalently, n-dimensional Brownian motion).

What we mean by a plan. Intuitively, a plan means that instead of going in different directions at different moments of time, we have a *systematic* change $u(t)$.

We consider local planning, for a few cycles $t_1, t_2, \ldots$, for which the difference $\Delta t \stackrel{\text{def}}{=} t - t_0$ is small. Thus, we can expand $u(t) = u(t_0 + \Delta t)$ into Taylor series and keep only linear terms in this expansion: $u(t) = u(t_0 + \Delta t) = u(t_0) + v \cdot \Delta t$, where

$$v \stackrel{\text{def}}{=} \frac{du}{dt}_{|t=t_0}.$$

By definition of the deviation $u(t)$, we have $u(t_0) = 0$, and thus, $u(t) = v \cdot \Delta t$. So, the change between each moment of time and the next one takes the form

$$u(t_{j+1}) - u(t_j) = v \cdot (t_{j+1} - t_j) = v \cdot h.$$

In other words, in contrast to the no-plan case, when changes at different moments of time are completely uncorrelated, here the changes at different moments of time are exactly the same.

Of course, from the practical viewpoint, they cannot always be the same: if a plan is bad, and we see that the desired objective functions decreases moment by moment, we will abandon this plan and select a new one.

Of course, it does not make sense to abandon the plan after a single decrease in the value of the objective function: it is known that even the best plans take some time to turn the economy around. Let us denote by m the reasonable number of decreases after which the plan will be abandoned.

From this viewpoint, selecting a plan means selecting a single change vector of length b, and following it for m steps, after which:

- if we had m decreases in the value of the objective function, we select a different plan,
- otherwise, we continue with the original plan for all T moments of time.

What we mean by a possible bad plan. We consider the situation in which we do not know the shape of the objective function. In this case, we do not know which change vector to select, so we select a random vector $w = (w_1, \ldots, w_n)$ of size a.

If $a \cdot w \stackrel{\text{def}}{=} \sum_{i=1}^{n} a_i \cdot w_i > 0$, the resulting plan will lead to a consistent improvement of an objective function, so we will have a *good plan*. Vice versa, if

$$a \cdot w = \sum_{i=1}^{n} a_i \cdot w_i < 0,$$

the resulting plan will lead to a consistent decrease of an objective function, so we will have a *bad plan*.

Resulting description of two strategies. In this paper, we compare two strategies:

- the no-plan strategy, and
- the possibly-bad-plan strategy.

In the no-plan strategy, we consider random walk with step size b.

In the possibly-bad-plan strategy, we select a random vector w of size b. If $a \cdot w < 0$, then after m moments of time, we select a new random vector, etc.

Which of the two strategies leads to better results? Both strategies rely on a random choice. So, for the same situation, the same strategy may lead to different results.

Our goal is to improve the value of the objective function. Each strategy sometimes improves this values, sometimes decreases it:

- For example, if every time we select a change vector which is improving, both strategies will improve the value of the objective function.
- On the other hand, if every time, we select a decreasing change vector, both strategies will decrease the value of the objective function.

Thus, for each of the two strategies, a reasonable performance measure is the probability that by the final time t_T, this strategy will increase the value of the objective function in comparison to its original value.

Let us compare these probabilities.

3   Comparing the Results of the No-Plan and the Possibly-Bad-Plan Strategies

Case of the no-plan strategy. Under this strategy, the vector $U \stackrel{\text{def}}{=} u(t_T)$ describing the difference between the final action and the original one is the sum of T independent change vectors, each of which has a random direction.

All original distributions are invariant with respect to rotations. Thus, for the sum of these change vectors, the distribution is still invariant with respect to rotations – and hence, the direction $e \stackrel{\text{def}}{=} \dfrac{U}{\|U\|}$ of the sum vector U is also random: uniform on the unit sphere.

This implies, in particular, that the distribution of e is the same as the distribution of the opposite vector $-e$.

The vector U leads to an improvement if $U \cdot a > 0$, i.e., equivalently, if $e \cdot a > 0$. Since e and $-e$ have the same distribution, the probability that $e \cdot a > 0$ is the same as the probability that $(-e) \cdot a > 0$, i.e., that $e \cdot a < 0$. The probability of a degenerate case $e \cdot a = 0$ is 0. Thus, with probability 1, we have two equally probable cases: improving and decreasing. Therefore, the probability of each of these cases is exactly $1/2$.

So, for the no-plan strategy, the probability of improvement is 0.5.

Case of the possibly-bad-plan strategy. In this case, similarly, with probability $1/2$, we select an improving change vector w, in which case the value of the objective function improves.

With the remaining probability $1/2$, we select a decreasing change vector, in which case the value of the objective function starts decreasing. However, it does not necessarily always decrease: after m steps, once we see that the value of the objective function decreases, we select a new change vector. In this case, with probability $1/2$, we will select an improving change vector – and with a positive probability, the resulting improvement in the remaining $T - m$ moments of time will compensate for the decrease in the first m steps.

Thus, in this case, the probability that this strategy will lead to an improvement is larger than $1/2$, since:

- in addition to the probability-$1/2$ situations in which we select an improving change vector from the very beginning,
- we also have situations in which we first decrease and then increase,
- and the probability of such additional situations is positive.

Conclusion: the possibly-bad-plan strategy is indeed better than the no-plan strategy

- For the no-plan-strategy, the probability of improvement is $1/2$, while
- for the possibly-bad-plan strategy, this probability is larger than $1/2$.

So, indeed, the possibly-bad-plan strategy is theoretically better than the no-plan strategy.

Thus, we indeed have a theoretical explanation for Thiel's empirical observation.

How better? How larger is the probability of success of the possibly-bad-plan strategy than $1/2$?

To answer this question, let us consider what happens when T is sufficiently large, i.e., in mathematical terms, when T tends to infinity. In this case, the probability that the no-plan method will succeed remains the same: $1/2$.

On the other hand, the probability that the possibly-bad strategy will succeed tends to 1. Indeed, if in this strategy, at some moment of time t, we select an improving change vector w, then with $T \to \infty$, the resulting increase $(T-t) \cdot (a \cdot w)$ will tend to infinity and thus, eventually overcome decreases that happened before moment t.

So, the only possibility not to improve is when we consistently select a decreasing vector w.

- The probability of selecting such a vector in the beginning is $1/2$.
- The probability of selecting it again after m iterations is also $1/2$, so the overall probability that we have a decrease for the first $2m$ moments of time is $(1/2)^2$.
- Similarly, the probability that for all T/m selections, we selected a decreasing vector, is equal to $(1/2)^{T/m}$.

When $T \to \infty$, this probability tends to 0 and thus, indeed, the probability that the possibly-bad-plan strategy will led to improvement tends to 1.

Acknowledgments. We acknowledge the partial support of the Center of Excellence in Econometrics, Faculty of Economics, Chiang Mai University, Thailand. This work was also supported in part by the National Science Foundation grants HRD-0734825 and HRD-1242122 (Cyber-ShARE Center of Excellence) and DUE-0926721, and by an award "UTEP and Prudential Actuarial Science Academy and Pipeline Initiative" from Prudential Foundation.

References

1. Thiel, P., Masters, B.: Zero to One: Notes on Startups, or How to Build the Future. Crown Business, New York (2014)
2. Voit, J.: The Statistical Mechanics of Financial Markets. Springer, Heidelberg (2010)

Shape Mixture Models Based on Multivariate Extended Skew Normal Distributions

Weizhong Tian[1,2], Tonghui Wang[3(✉)], Fengrong Wei[4], and Fang Dai[5]

[1] School of Science, Xian University of Technology, Xi'an 710054, China
weizhong.tian@enmu.edu
[2] Department of Mathematical Sciences, Eastern New Mexico University,
Portales, NM 88130, USA
[3] Department of Mathematical Sciences, New Mexico State University,
Las Cruces 88003, USA
twang@nmsu.edu
[4] Department of Mathematics, University of West Georgia,
Carrollton, GA 30018, USA
fwei@westga.edu
[5] School of Science, Xian University of Technology, Xi'an 710054, China
daifang@xaut.edu.cn

Abstract. In this paper, the class of the shape mixtures of extended skew normal distributions is introduced. The posterior distributions for the shaped parameters are obtained. The moment generating functions for the posterior distributions of the shaped parameters are discussed. Also Bayesian analysis for this shape mixture model is studied.

1 Introduction

In many real world problems, the assumptions of normality are violated as the data possess some level of skewness. The class of skew normal distributions is an extension of the normal distribution, allowing for the presence of skewness was introduced in the papers by Azzalini [3,4], according to which a random variable X is called a *skew normal* random variable if it has the probability density function

$$f(x) = 2\phi(x)\Phi(\alpha x),$$

where $\phi(\cdot)$ and $\Phi(\cdot)$ are the density function and the cumulative distribution function of the standard normal distribution, and α is called the shaped parameter.

Azzalini and Dalla Valle [4] defined the multivariate skew normal family. The random vector $\mathbf{V}$ is said to have a *multivariate skew normal* distribution with location parameter $\boldsymbol{\mu} \in \Re^n$, scale parameter Σ, and shape parameter $\boldsymbol{\alpha} \in \Re^n$, denoted by $\mathbf{V} \sim SN_n(\boldsymbol{\mu}, \Sigma, \boldsymbol{\alpha})$, if its probability density function is given by

$$f_{\mathbf{V}}(\mathbf{v}; \boldsymbol{\mu}, \Sigma, \boldsymbol{\alpha}) = 2\phi_n(\mathbf{v}; \boldsymbol{\mu}, \Sigma)\Phi\left(\boldsymbol{\alpha}'\Sigma^{-1/2}(\mathbf{v} - \boldsymbol{\mu})\right), \qquad \mathbf{v} \in \Re^n, \qquad (1)$$

© Springer International Publishing AG 2018
V. Kreinovich et al. (eds.), *Predictive Econometrics and Big Data*, Studies in Computational Intelligence 753, https://doi.org/10.1007/978-3-319-70942-0_20

where $\phi_n(\mathbf{v}; \boldsymbol{\mu}, \Sigma)$ is the n-dimensional normal density function with mean vector $\boldsymbol{\mu}$ and covariance matrix Σ. Since then, the class of multivariate skew normal distributions have been studied by many authors, see Azzalini [5], Gupta [8], and Szkely [11].

For the case where the location parameter is zero, the quadratic forms under skew normal settings were studied by Gupta and Huang [8], Genton et al. [7], Huang and Chen [9], and Loperfido [10]. For the general case where the location parameter is not zero, the quadratical forms of skew normal models based on stochastic representation were studied by Tian and Wang [13], and new versions of Cochran's theorem were obtained in Wang et al. [14] and Ye et al. [15].

The class of multivariate extended skew-normal distribution is an extension of the multivariate skew normal distribution. It was introduced by Tian et al. [12] and defined as follows. The random vector $\mathbf{Y} \in \Re^n$ is said to have a multivariate extended skew normal distribution with location parameter $\boldsymbol{\mu} \in \Re^n$, scale parameter $\Sigma \in M_{n \times n}$, shape parameter $\boldsymbol{\alpha} \in \Re^n$, and extended parameter $\gamma \geq -1$, denoted by $\mathbf{Y} \sim ESN_n(\boldsymbol{\mu}, \Sigma, \boldsymbol{\alpha}, \gamma)$ if its density function is given by

$$f_{\mathbf{Y}}(\mathbf{y}; \boldsymbol{\alpha}, \gamma) = \frac{2}{2 + \gamma} \phi_n(\mathbf{y}; \boldsymbol{\mu}, \Sigma) \left[1 + \gamma \Phi \left(\boldsymbol{\alpha}' \Sigma^{-\frac{1}{2}} (\mathbf{y} - \boldsymbol{\mu}) \right) \right], \qquad (2)$$

where $\phi_n(\mathbf{z})$ is the n-dimensional normal density function with mean vector $\mathbf{0}$ and covariance matrix I_n. For the case where $n = 1$, the distribution with the density function given in (2) is called the *extended skew normal distribution*, denoted by $Y \sim ESN(\mu, \sigma^2, \alpha, \gamma)$.

Recently, Arellano-Valle et al. [1,2] introduced the class of shape mixtures of skewed distributions by assuming that the shape vectors parameter follows a multivariate normal distribution or skew normal distribution. In this paper, a class of shape mixtures models was introduced by taking the shape parameter follow the skew normal mixing distribution, i.e., the conditional distribution of Y given $S = s$ is extended skew normal and the prior distribution of S is skew normal:

$$[Y|S = s] \sim ESN(\mu, \sigma^2, s, \gamma) \quad and \quad S \sim SN(\eta, \omega^2, \alpha). \qquad (3)$$

In (3), we assume the location-scale parameters (μ, σ^2), (η, ω^2), and extended parameter γ are known. Based on properties of the extended skew normal distribution given in Tian et al. [12], the class of shape mixture extended skew normal distribution is an extension of the one given in Arellano-Valle et al. [1,2].

In this paper, the class of shape mixtures of dependent and independent multivariate extended skew normal distributions with the Bayesian interpretations is analyzed. Moment generating functions for the posterior distribution of the shape parameters are discussed for different cases. In additional, some useful results based on the Bayesian analysis are obtained. The organization of this paper is as follows. The class of shape mixtures of dependent and independent multivariate extended skew normal distributions is studied in Sect. 2. The moment generating functions about the posterior distribution of shape parameters are obtained in Sect. 3. Bayesian analysis based on the shape mixtures models are discussed in Sect. 4, and some remarks on this model are given in Sect. 5.

2   Shape Mixture of Multivariate Extended Skew Normal Distribution

Let $\boldsymbol{\mu} = (\mu_1, \cdots, \mu_n)' \in \Re^n$ and $\boldsymbol{\eta} = (\eta_1, \cdots, \eta_n)' \in \Re^n$ be the location parameters, $\Sigma = (\sigma_{ij}) \in M_{n \times n}$, $\Omega = (\omega_{ij}) \in M_{n \times n}$ be positive definite scale matrices in which $\sigma_{ii} = \sigma_i^2, \omega_{ii} = \omega_i^2$. Let $D(\boldsymbol{\sigma})$, $D(\boldsymbol{\omega})$ be $n \times n$ diagonal matrices whose diagonal entries are the components of the vector $\boldsymbol{\sigma} = (\sigma_1^2, \cdots, \sigma_n^2)'$ and $\boldsymbol{\omega} = (\omega_1^2, \cdots, \omega_n^2)'$, and $\boldsymbol{\alpha} = (\alpha_1, \cdots, \alpha_n)' \in \Re^n$ be the vector of skewness parameters.

Consider an observable random vector $\mathbf{Y} = (Y_1, \cdots, Y_n)'$ whose distribution is from the multivariate extended skew normal family. The components $Y_1, \cdots, Y_n$ can be independent or dependent, conditionally on the shape parameter, which are different or common for the Y_i's. In this section, we will derive the shape mixture distributions for both independent and dependent components based on the extended skew normal distributions. The following Lemmas will be used to prove the properties of our main results.

Lemma 1 *(Azzalini and Capitanio [5]). If $\mathbf{V} \sim SN_n(\mathbf{0}, \Sigma, \boldsymbol{\alpha})$, and A is a nonsingular $n \times n$ matrix, then*

$$A'\mathbf{V} \sim SN_n(A'\Sigma A, A^{-1}\boldsymbol{\alpha}).$$

Lemma 2 *(Zacks [16] and Chen [6]). Let $\mathbf{U} \sim N_k(\mathbf{c}, C)$ be a non-siglar, mulivariate normal random vector, then for any fixed m-dimensional vector $\mathbf{a}$ and $m \times k$ matrix A, we have that,*

$$E[\Phi_m(\mathbf{a} + A\mathbf{U}|\mathbf{b}, B)] = \Phi_m\left(A\mathbf{c} + \mathbf{a}|\mathbf{b}, B + ACA'\right), \tag{4}$$

where $\Phi_m(\cdot|\mathbf{b}, B)$ denotes the cumulative density function of a multivariate normal distribution with mean vector $\mathbf{b}$ and covariance matrix B.

2.1   Mixtures on Different Shape Parameters

Assume that the vector of shape parameters $\mathbf{S} = (S_1, \cdots, S_n)' \in \Re^n$, with independent S_i's, and n observations $Y_1, \cdots, Y_n$ given $\mathbf{S} = \mathbf{s} = (s_1, \cdots, s_n)'$ are also independent. Then we have the following result.

Theorem 1. *Let $[Y_1|\mathbf{S} = \mathbf{s}], \cdots, [Y_n|\mathbf{S} = \mathbf{s}]$ be independent variables with $[Y_i|\mathbf{S} = \mathbf{s}] \sim ESN(\mu_i, \sigma_i^2, s_i, \gamma)$. Let $S_1, \cdots, S_n$ be independent random variables with $S_i \sim SN(\eta_i, \omega_i^2, \alpha_i)$, for $i = 1, \cdots, n$. Then the probability density function of $\mathbf{Y}$ is*

$$f(\mathbf{y}) = \left(\frac{4}{2 + \gamma}\right)^n |D(\boldsymbol{\sigma})|^{-\frac{1}{2}} \phi_n(\mathbf{z})$$

$$\times \prod_{i=1}^n \left\{ \frac{1}{2} + \gamma \Phi_2\left[\begin{pmatrix} \eta_i z_i \\ 0 \end{pmatrix}; \mathbf{0}, \begin{pmatrix} 1 + \omega_i^2 z_i^2 & \omega_i^2 z_i \alpha_i \\ \omega_i^2 z_i \alpha_i & 1 + \alpha_i^2 \end{pmatrix} \right) \right] \right\},$$

and the posterior density function of $\mathbf{S}|\mathbf{Y} = \mathbf{y}$ *is*

$$f_{\mathbf{S}|\mathbf{Y}=\mathbf{y}}(\mathbf{s}) = |D(\boldsymbol{\omega})|^{-\frac{1}{2}} \phi_n (\mathbf{u}) \prod_{i=1}^{n} \left[\Phi(\alpha_i u_i) \left(1 + \gamma \Phi (s_i z_i)\right) \right]$$

$$\times \prod_{i=1}^{n} \left\{ \frac{1}{2} + \gamma \Phi_2 \left[\begin{pmatrix} \eta_i z_i \\ 0 \end{pmatrix} ; \mathbf{0}, \begin{pmatrix} 1 + \omega_i^2 z_i^2 & \omega_i^2 z_i \alpha_i \\ \omega_i^2 z_i \alpha_i & 1 + \alpha_i^2 \end{pmatrix} \right] \right\}^{-1},$$

where $\mathbf{z} = D(\boldsymbol{\sigma})^{-\frac{1}{2}} (\mathbf{y} - \boldsymbol{\mu})$ *and* $\mathbf{u} = D(\boldsymbol{\omega})^{-\frac{1}{2}} (\mathbf{s} - \boldsymbol{\eta})$.

Proof. Let $\mathbf{z} = D(\boldsymbol{\sigma})^{-\frac{1}{2}} (\mathbf{y} - \boldsymbol{\mu})$ and $\mathbf{u} = D(\boldsymbol{\omega})^{-\frac{1}{2}} (\mathbf{s} - \boldsymbol{\eta})$. By (2), we know that the joint probability density function of $\mathbf{Y}|\mathbf{S} = \mathbf{s}$ and the joint probability density function of $\mathbf{S}$, are given, respectively, by

$$f_{\mathbf{Y}|\mathbf{S}=\mathbf{s}}(\mathbf{y}) = \left(\frac{2}{2+\gamma} \right)^n |D(\boldsymbol{\sigma})|^{-\frac{1}{2}} \prod_{i=1}^{n} \left\{ \phi(z_i)[1 + \gamma \Phi(s_i z_i)] \right\},$$

and

$$f(\mathbf{s}) = (2)^n |D(\boldsymbol{\omega})|^{-\frac{1}{2}} \prod_{i=1}^{n} \left\{ \phi(u_i) \Phi(\alpha_i u_i) \right\}.$$

Thus the joint probability density function of $\mathbf{Y}$ and $\mathbf{S}$ is

$$f(\mathbf{y}, \mathbf{s}) = f_{\mathbf{Y}|\mathbf{S}=\mathbf{s}}(\mathbf{y}) f(\mathbf{s}),$$

so that the marginal probability density function of $\mathbf{Y}$ can be obtained,

$$f(\mathbf{y}) = \left(\frac{4}{2+\gamma} \right)^n |D(\boldsymbol{\sigma})|^{-\frac{1}{2}} |D(\boldsymbol{\omega})|^{-\frac{1}{2}} \prod_{i=1}^{n} \int_{s_i} \left\{ \phi(z_i)\phi(t_i)\Phi(\alpha_i t_i)[1 + \gamma \Phi(s_i z_i)] \right\} ds_i$$

$$= \left(\frac{4}{2+\gamma} \right)^n |D(\boldsymbol{\sigma})|^{-\frac{1}{2}} \phi_n (\mathbf{z}) \prod_{i=1}^{n} \left\{ \frac{1}{2} + \gamma E \left[\Phi_2 \left[\begin{pmatrix} z_i \\ \alpha_i \end{pmatrix} t_i + \begin{pmatrix} \eta_i z_i \\ 0 \end{pmatrix} ; \mathbf{0}, I_2 \right] \right] \right\}.$$

Consequently, using Lemma 2 with $t_i \sim N(0, \omega_i^2)$ for $i = 1, \cdots, n$, the conditional distribution of $\mathbf{S}$ given $\mathbf{Y} = \mathbf{y}$ is obtained by a straightforward application of Bayes' Theorem. □

Note that if we assume that $\boldsymbol{\alpha} = 0$, then from Theorem 1, we can obtain the following result.

Corollary 2.1. *In Theorem 1, if we assume that* $\mathbf{S} \sim N_n(\boldsymbol{\eta}, D(\boldsymbol{\omega}))$, *then the probability density function of* $\mathbf{Y}$ *is*

$$f(\mathbf{y}) = \left(\frac{2}{2+\gamma} \right)^n |D(\boldsymbol{\sigma})|^{-1} \phi_n (\mathbf{z}) \prod_{i=1}^{n} \left\{ 1 + \gamma \Phi \left(\frac{\eta_i z_i}{\sqrt{1 + \omega_i^2 z_i^2}} \right) \right\},$$

and the posterior density function of $[\mathbf{S}|\mathbf{Y} = \mathbf{y}]$ *is*

$$f_{\mathbf{S}|\mathbf{Y}=\mathbf{y}}(\mathbf{s}) = 2^n |D(\boldsymbol{\omega})|^{-1} \phi_n(\mathbf{u}) \prod_{i=1}^{n} [(1 + \gamma \Phi(s_i z_i))] \tag{5}$$

$$\times \prod_{i=1}^{n} \left\{ 1 + \gamma \Phi\left(\frac{\eta_i z_i}{\sqrt{1 + \omega_i^2 z_i^2}} \right) \right\}^{-1},$$

where $\mathbf{z} = D(\boldsymbol{\sigma})^{-\frac{1}{2}}(\mathbf{y} - \boldsymbol{\mu})$, *and* $\mathbf{u} = D(\boldsymbol{\omega})^{-\frac{1}{2}}(\mathbf{s} - \boldsymbol{\eta})$.

Now we consider the case where the components of $\mathbf{S}$ are dependent so that $\mathbf{S}$ has a multivariate skew normal distribution, and $\mathbf{Y}$ given $\mathbf{S} = \mathbf{s}$ is sampled from a multivariate extended skew normal distribution.

Theorem 2. *Let* $[\mathbf{Y}|\mathbf{S} = \mathbf{s}] \sim ESN_n(\boldsymbol{\mu}, \Sigma, \mathbf{s}, \gamma)$ *and* $\mathbf{S} \sim SN_n(\boldsymbol{\eta}, \Omega, \boldsymbol{\alpha})$. *Then the probability density function of* $\mathbf{Y}$ *is*

$$f(\mathbf{y}) = \left(\frac{4}{2+\gamma} \right)^2 \phi_n(\mathbf{y}; \boldsymbol{\mu}, \Sigma)$$

$$\times \left\{ \frac{1}{2} + \gamma \Phi_2 \left[\begin{pmatrix} \boldsymbol{\eta}'\mathbf{a} \\ 0 \end{pmatrix}; \mathbf{0}, \begin{pmatrix} 1 \mid \boldsymbol{\alpha}'\Omega\mathbf{a} & \boldsymbol{\alpha}'\Omega^{\frac{1}{2}}\boldsymbol{\alpha} \\ \boldsymbol{\alpha}'\Omega^{\frac{1}{2}}\mathbf{z} & 1 + \boldsymbol{\alpha}'\boldsymbol{\alpha} \end{pmatrix} \right] \right\},$$

and the posterior density function of $[\mathbf{S}|\mathbf{Y} = \mathbf{y}]$ *is*

$$f_{\mathbf{S}|\mathbf{Y}=\mathbf{y}}(\mathbf{s}) = \phi_n(\mathbf{s}; \boldsymbol{\eta}, \Omega) [(1 + \gamma \Phi(\mathbf{s}'\mathbf{z})) \Phi(\boldsymbol{\alpha}'\mathbf{u})]$$

$$\times \left\{ \frac{1}{2} + \gamma \Phi_2 \left[\begin{pmatrix} \boldsymbol{\eta}'\mathbf{z} \\ 0 \end{pmatrix}; \mathbf{0}, \begin{pmatrix} 1 + \mathbf{z}'\Omega\mathbf{z} & \mathbf{z}'\Omega^{\frac{1}{2}}\boldsymbol{\alpha} \\ \boldsymbol{\alpha}'\Omega^{\frac{1}{2}}\mathbf{z} & 1 + \boldsymbol{\alpha}'\boldsymbol{\alpha} \end{pmatrix} \right] \right\}^{-1},$$

where $\mathbf{z} = \Sigma^{-\frac{1}{2}}(\mathbf{y} - \boldsymbol{\mu})$ *and* $\mathbf{u} = \Omega^{-\frac{1}{2}}(\mathbf{s} - \boldsymbol{\eta})$.

Proof. From (2), the conditional probability density function of $\mathbf{Y}|\mathbf{S} = \mathbf{s}$ is

$$f_{\mathbf{Y}|\mathbf{S}=\mathbf{s}}(\mathbf{y}) = \left(\frac{2}{2+\gamma} \right) \phi_n(\mathbf{y}; \boldsymbol{\mu}, \Sigma)[1 + \gamma \Phi(\mathbf{s}'\mathbf{z})],$$

and the probability density function of $\mathbf{S}$ is

$$f(\mathbf{s}) = 2\phi_n(\mathbf{s}; \boldsymbol{\eta}, \Omega)\Phi(\boldsymbol{\alpha}'\mathbf{t}),$$

where $\mathbf{z} = \Sigma^{-\frac{1}{2}}(\mathbf{y} - \boldsymbol{\mu})$ and $\mathbf{u} = \Omega^{-\frac{1}{2}}(\mathbf{s} - \boldsymbol{\eta})$. Similar to the proof of Theorem 1, we can obtain the probability density function of $\mathbf{Y}$

$$f(\mathbf{y}) = \left(\frac{4}{2+\gamma} \right) \phi_n(\mathbf{y}; \boldsymbol{\mu}, \Sigma) \int_{\mathbf{s}} \phi_n(\mathbf{s}; \boldsymbol{\eta}, \Omega)[1 + \gamma \Phi(\mathbf{s}'\mathbf{z})]\Phi(\boldsymbol{\alpha}'\mathbf{u})d\mathbf{s}$$

$$= \left(\frac{2}{2+\gamma} \right)^2 \phi_n(\mathbf{y}; \boldsymbol{\mu}, \Sigma)$$

$$\times \int_{\mathbf{s}} \phi_n(\mathbf{s}; \boldsymbol{\eta}, \Omega) \left\{ \Phi(\boldsymbol{\alpha}'\mathbf{u}) + \gamma \Phi_2 \left[\begin{pmatrix} \mathbf{z}' \\ \boldsymbol{\alpha}'\Omega^{-\frac{1}{2}} \end{pmatrix} \mathbf{s} - \begin{pmatrix} 0 \\ \boldsymbol{\alpha}'\Omega^{-\frac{1}{2}}\boldsymbol{\eta} \end{pmatrix}; \mathbf{0}, I_2 \right] \right\} d\mathbf{s}.$$

Consequently, the desired result follows from Lemma 2 with $\mathbf{s} \sim N_n(\boldsymbol{\eta}, \Omega)$ and Bayes' Theorem. □

Similarly with $\boldsymbol{\alpha} = 0$, we obtain the following result.

Corollary 2.2. *Let* $[\mathbf{Y}|\mathbf{S} = \mathbf{s}] \sim ESN_n(\boldsymbol{\mu}, \Sigma, \mathbf{s}, \gamma)$, *and* $\mathbf{S} \sim N_n(\boldsymbol{\eta}, \Omega)$. *Then the probability density function of* $\mathbf{Y}$ *is*

$$f(\mathbf{y}) = \left(\frac{2}{2+\gamma} \right) \phi_n (\mathbf{y}; \boldsymbol{\mu}, \Sigma) \left\{ 1 + \gamma \Phi \left(\frac{\boldsymbol{\eta}' \mathbf{z}}{\sqrt{1 + \mathbf{z}' \Omega \mathbf{z}}} \right) \right\}, \tag{6}$$

and the conditional probability density function of $[\mathbf{S}|\mathbf{Y} = \mathbf{y}]$ *is*

$$f_{\mathbf{S}|\mathbf{Y}=\mathbf{y}}(\mathbf{s}) = \frac{\phi_n(\mathbf{s}; \boldsymbol{\eta}, \Omega)\,(1 + \gamma \Phi(\mathbf{s}'\mathbf{z}))}{1 + \gamma \Phi \left(\dfrac{\boldsymbol{\eta}' \mathbf{z}}{\sqrt{1 + \mathbf{z}' \Omega \mathbf{z}}} \right)},$$

where $\mathbf{z} = \Sigma^{-\frac{1}{2}} (\mathbf{y} - \boldsymbol{\mu})$.

2.2 Mixtures on a common shape parameter

Assuming shape variable parameter $S \in \Re$, and we consider the situation where $\mathbf{Y}$ given S is sampled from a dependent multivariate extended skew normal distribution. Then we have the following result.

Theorem 3. *Let* $[\mathbf{Y}|S = s] \sim ESN_n(\boldsymbol{\mu}, \Sigma, s\mathbf{1}_n, \gamma)$, *and* $s \sim SN(\eta, \omega^2, \alpha)$. *Then the probability density function of* $\mathbf{Y}$ *is*

$$f(\mathbf{y}) = \left(\frac{4}{2+\gamma} \right) \phi_n (\mathbf{y}; \boldsymbol{\mu}, \Sigma) \tag{7}$$

$$\times \left\{ \frac{1}{2} + \gamma \Phi_2 \left[\begin{pmatrix} \eta \mathbf{1}'_n \mathbf{z} \\ 0 \end{pmatrix}; \mathbf{0}, \begin{pmatrix} 1 + n\omega^2 \mathbf{z}' \mathbf{z} & \omega^2 \alpha \mathbf{z}' \mathbf{1}_n \\ \alpha \omega^2 \mathbf{1}'_n \mathbf{z} & 1 + n\alpha^2 \omega^2 \end{pmatrix} \right] \right\},$$

and the posterior density function of $[\mathbf{S}|\mathbf{Y} = \mathbf{y}]$ *is*

$$f_{\mathbf{S}|\mathbf{Y}=\mathbf{y}}(\mathbf{s}) = \phi(s; \eta, \omega^2)\,(1 + \gamma \Phi(s\mathbf{1}'_n \mathbf{z}))\,\Phi(\alpha u) \tag{8}$$

$$\times \left\{ \frac{1}{2} + \gamma \Phi_2 \left[\begin{pmatrix} \eta \mathbf{1}'_n \mathbf{z} \\ 0 \end{pmatrix}; \mathbf{0}, \begin{pmatrix} 1 + n\omega^2 \mathbf{z}' \mathbf{z} & \omega^2 \alpha \mathbf{z}' \mathbf{1}_n \\ \alpha \omega^2 \mathbf{1}'_n \mathbf{z} & 1 + n\alpha^2 \omega^2 \end{pmatrix} \right] \right\}^{-1},$$

where $\mathbf{z} = \Sigma^{-\frac{1}{2}} (\mathbf{y} - \boldsymbol{\mu})$, *and* $u = \omega^{-1} (s - \eta)$.

Proof. The proof is similar to the proof given in Theorem 2 and thus it is omitted. □

Corollary 2.3. *Let* $[\mathbf{Y}|S = s] \sim ESN_n(\boldsymbol{\mu}, \Sigma, s\mathbf{1}_n, \gamma)$, *and* $s \sim SN(\eta, \omega^2)$. *Then the probability density function of* $\mathbf{Y}$ *is*

$$f(\mathbf{y}) = \left(\frac{2}{2+\gamma} \right) \phi_n (\mathbf{y}; \mu, \Sigma) \left\{ 1 + \gamma \Phi \left(\frac{\eta \mathbf{1}'_n \mathbf{z}}{\sqrt{1 + n\omega^2 \mathbf{z}' \mathbf{z}}} \right) \right\} \tag{9}$$

and the posterior density function of $[\mathbf{S}|\mathbf{Y} = \mathbf{y}]$ *is*

$$f_{\mathbf{S}|\mathbf{Y}=\mathbf{y}}(\mathbf{s}) = \frac{\phi(s;\eta,\omega^2)\left(1 + \gamma\Phi(s\mathbf{1}_n'\mathbf{z})\right)}{1 + \gamma\Phi\left(\dfrac{\eta\mathbf{1}_n'\mathbf{z}}{\sqrt{1 + n\omega^2\mathbf{z}'\mathbf{z}}}\right)}, \tag{10}$$

where $\mathbf{z} = \Sigma^{-\frac{1}{2}}(\mathbf{y} - \boldsymbol{\mu})$.

3   The Moment Generating Function for the Distribution of the Shape Parameters

In this section, the moment generating functions of different shape parameters for giving the known observed values under different cases are studied.

Theorem 4. *Let* $[Y_1|\mathbf{S} = \mathbf{s}], \cdots, [Y_n|\mathbf{S} = \mathbf{s}]$ *be independent random variables with* $[Y_i|\mathbf{S} = \mathbf{s}] \sim ESN(\mu_i, \sigma_i^2, s_i, \gamma)$. *Also let* $S_1, \cdots, S_n$ *be independent random variables with* $S_i \sim SN(\eta_i, \omega_i^2, \alpha_i)$. *Then the moment generating function of* $\mathbf{S}|\mathbf{Y} = \mathbf{y}$ *is*

$$M_{\mathbf{S}|\mathbf{Y}=\mathbf{y}}(\mathbf{t}) = C_0 \prod_{i=1}^{n} \exp\left\{t_i\eta_i + \frac{1}{2}t_i\omega_i^2\right\} \left\{\Phi(\alpha_i t_i\omega_i, 1 + \alpha_i^2)\right. \tag{11}$$
$$\left. + \gamma\Phi_2\left[\begin{pmatrix} \alpha_i t_i\omega_i \\ z_i\omega_i^2 t_i + z_i\eta_i \end{pmatrix}; \mathbf{0}, \begin{pmatrix} 1+\alpha_i^2 & \alpha_i z_i\omega_i \\ \alpha_i z_i\omega_i & 1+z_i^2\omega_i^2 \end{pmatrix}\right]\right\}$$

and the moment generating function of $\mathbf{S}'\mathbf{S}|\mathbf{Y} = \mathbf{y}$ *is*

$$M_{\mathbf{S}'\mathbf{S}|\mathbf{Y}=\mathbf{y}}(t)$$
$$= C_0 \prod_{i=1}^{n} \exp\left\{t\eta_i^2 + \frac{2t^2\omega_i^2\eta_i^2}{1 - 2\omega_i^2 t}\right\} \left\{\Phi\left(\frac{2\alpha_i\omega_i\eta_i t}{1 - 2\omega_i^2}, \frac{1 - 2\omega_i^2 + \alpha_i}{1 - 2\omega_i^2}\right)\right.$$
$$\left. + \gamma\Phi_2\left[\begin{pmatrix} \dfrac{\alpha_i}{1 - 2\omega_i^2} \\ \dfrac{z_i\omega_i}{1 - 2\omega_i^2} + z_i\eta_i \end{pmatrix}; \mathbf{0}, \begin{pmatrix} 1 + \dfrac{\alpha_i^2}{1 - 2\omega_i^2} & \dfrac{\alpha_i z_i\omega_i}{1 - 2\omega_i^2} \\ \dfrac{\alpha_i z_i\omega_i}{1 - 2\omega_i^2} & 1 + \dfrac{z_i^2\omega_i^2}{1 - 2\omega_i^2} \end{pmatrix}\right]\right\}, \tag{12}$$

where

$$C_0 = \prod_{i=1}^{n} \left\{\frac{1}{2} + \gamma\Phi_2\left[\begin{pmatrix} \eta_i z_i \\ 0 \end{pmatrix}; \mathbf{0}, \begin{pmatrix} 1 + \omega_i^2 z_i^2 & \omega_i^2 z_i\alpha_i \\ \omega_i^2 z_i\alpha_i & 1+\alpha_i^2 \end{pmatrix}\right]\right\}^{-1}$$

and $\mathbf{z} = D(\boldsymbol{\sigma})^{-\frac{1}{2}}(\mathbf{y} - \boldsymbol{\mu})$.

Proof. Let $\mathbf{z} = D(\boldsymbol{\sigma})^{-\frac{1}{2}}(\mathbf{y} - \boldsymbol{\mu})$ and $\mathbf{u} = D(\boldsymbol{\omega})^{-\frac{1}{2}}(\mathbf{s} - \boldsymbol{\eta})$. By Theorem 1, for $\mathbf{t} = (t_1, t_2, \cdots, t_n)' \in \Re^n$, the moment generating function of $\mathbf{S}|\mathbf{Y} = \mathbf{y}$ can be obtained as

$$M_{\mathbf{S}|\mathbf{Y}=\mathbf{y}}(\mathbf{t}) = C_0|D(\omega)|^{-1}\phi_n(\mathbf{u})\prod_{i=1}^{n}\int_{\Re}[\Phi(\alpha_i u_i)(1+\gamma\Phi(s_i z_i))\exp\{s_i t_i\}]\,ds_i$$

$$= C_0\prod_{i=1}^{n}\int_{\Re}\phi(u_i)\Phi(\alpha_i u_i)(1+\gamma\Phi[(\omega_i u_i+\eta_i)z_i])\exp\{t_i(\omega_i u_i+\eta_i)\}du_i$$

$$= C_0\prod_{i=1}^{n}\exp\left\{t_i\eta_i+\frac{1}{2}t_i\omega_i^2\right\}\int_{\Re}\phi(u_i;t_i\omega_i,1)\Phi(\alpha_i u_i)(1+\gamma\Phi[(\omega_i u_i+\eta_i)z_i])\,du_i.$$

Note that $\Phi(\alpha_i u_i)\Phi[(\omega_i u_i+\eta_i)z_i] = \Phi_2\left[\begin{pmatrix}\alpha_i\\z_i\omega_i\end{pmatrix}u_i+\begin{pmatrix}0\\z_i\eta_i\end{pmatrix};\mathbf{0},I_2\right]$ so that terms given in above expression can be simplified as follows.

$$\int_{\Re}\phi(u_i;t_i\omega_i,1)\Phi(\alpha_i u_i)du_i = E\left[\Phi(\alpha_i u_i)\right]$$

and

$$\int_{\Re}\phi(u_i;t_i\omega_i,1)\Phi_2\left[\begin{pmatrix}\alpha_i\\z_i\omega_i\end{pmatrix}u_i+\begin{pmatrix}0\\z_i\eta_i\end{pmatrix};\mathbf{0},I_2\right]du_i$$

$$= E\left\{\Phi_2\left[\begin{pmatrix}\alpha_i\\z_i\omega_i\end{pmatrix}u_i+\begin{pmatrix}0\\z_i\eta_i\end{pmatrix};\mathbf{0},I_2\right]\right\},$$

where $u_i \sim N(t_i\omega_i,1)$. Thus the desired result for $M_{\mathbf{S}|\mathbf{Y}=\mathbf{y}}(\mathbf{t})$ is obtained by Lemma 2.

Similarly, for $t \in \Re$, the moment generating function of $\mathbf{S'S}|\mathbf{Y}=\mathbf{y}$ is,

$$M_{\mathbf{S'S}|\mathbf{Y}=\mathbf{y}}(t) = C_0\prod_{i=1}^{n}\exp\left\{t\eta_i^2+\frac{2t^2\omega_i^2\eta_i^2}{1-2\omega_i^2 t}\right\}$$

$$\times\left\{E\left[\Phi(\alpha_i u_i)\right]+\gamma E\left[\Phi_2\left[\begin{pmatrix}\alpha_i\\z_i\omega_i\end{pmatrix}u_i+\begin{pmatrix}0\\z_i\eta_i\end{pmatrix};\mathbf{0},I_2\right]\right]\right\},$$

where $u_i \sim N\left(\frac{2\omega_i\eta_i t}{1-2\omega_i^2 t},(1-2\omega_i^2 t)^{-1}\right)$. Therefore by Lemma 2, the desired result is obtained. □

Theorem 5. Let $[\mathbf{Y}|\mathbf{S}=\mathbf{s}] \sim ESN_n(\boldsymbol{\mu},\Sigma,\mathbf{s},\gamma)$, and $\mathbf{S} \sim SN_n(\boldsymbol{\eta},\Omega,\boldsymbol{\alpha})$. Then

$$M_{\mathbf{S}|\mathbf{Y}=\mathbf{y}}(\mathbf{t}) = C_2\exp\left\{\boldsymbol{\eta'}\mathbf{t}+\frac{1}{2}\mathbf{t'}\Omega\mathbf{t}\right\}\left\{\Phi(\boldsymbol{\alpha'}\Omega^{\frac{1}{2}}\mathbf{t};0,1+\boldsymbol{\alpha'\alpha})\right. \tag{13}$$

$$\left.+\gamma\Phi_2\left[\begin{pmatrix}\boldsymbol{\alpha'}\Omega^{\frac{1}{2}}\mathbf{t}\\\mathbf{z}\Omega\mathbf{t}+\mathbf{z'}\boldsymbol{\eta}\end{pmatrix};\mathbf{0},\begin{pmatrix}1+\boldsymbol{\alpha'\alpha}&\boldsymbol{\alpha'}\Omega^{\frac{1}{2}}\mathbf{t}\\\boldsymbol{\alpha'}\Omega^{\frac{1}{2}}\mathbf{t}&1+\mathbf{z'}\Omega\mathbf{z}\end{pmatrix}\right]\right\},$$

$$M_{\mathbf{S'S}|\mathbf{Y}=\mathbf{y}}(t) = C_2\exp\left\{t\boldsymbol{\eta'\eta}+2t^2\boldsymbol{\eta'}\Omega^{\frac{1}{2}}(I_n-2\Omega t)^{-1}\Omega^{\frac{1}{2}}\boldsymbol{\eta}\right\}|I_n-2\omega t|^{-\frac{1}{2}} \tag{14}$$

$$\times\left\{\Phi\left(2t\boldsymbol{\alpha'}(I_n-2\Omega t)^{-1}\Omega^{\frac{1}{2}}\boldsymbol{\eta};0,1+\boldsymbol{\alpha'}(I_n-2\Omega t)^{-1}\boldsymbol{\alpha}\right)\right.$$

$$\left.+\gamma\Phi_2\left[\begin{pmatrix}2t\boldsymbol{\alpha'}(I_n-2\Omega t)^{-1}\Omega^{\frac{1}{2}}\boldsymbol{\eta}\\2t\mathbf{z'}\Omega^{\frac{1}{2}}(I_n-2\Omega t)^{-1}\Omega^{\frac{1}{2}}\boldsymbol{\eta}+\mathbf{z'}\boldsymbol{\eta}\end{pmatrix};\mathbf{0},\Psi\right]\right\},$$

where

$$C_2 = \left\{ \frac{1}{2} + \gamma \Phi_2 \left[\begin{pmatrix} \boldsymbol{\eta}'\mathbf{z} \\ 0 \end{pmatrix} ; \mathbf{0}, \begin{pmatrix} 1 + \mathbf{z}'\Omega\mathbf{z} & \mathbf{z}'\Omega^{\frac{1}{2}}\boldsymbol{\alpha} \\ \boldsymbol{\alpha}'\Omega^{\frac{1}{2}}\mathbf{z} & 1 + \boldsymbol{\alpha}'\boldsymbol{\alpha} \end{pmatrix} \right] \right\}^{-1},$$

$$\Psi = I_2 + \begin{pmatrix} \boldsymbol{\alpha}' \\ \mathbf{z}'\Omega^{\frac{1}{2}} \end{pmatrix}' (I_n - 2\Omega t)^{-1} \begin{pmatrix} \boldsymbol{\alpha}' \\ \mathbf{z}'\Omega^{\frac{1}{2}} \end{pmatrix}, \text{ and } \mathbf{z} = \Sigma^{-\frac{1}{2}}(\mathbf{y} - \boldsymbol{\mu}).$$

Proof. Let $\mathbf{z} = \Omega^{-\frac{1}{2}}(\mathbf{y} - \boldsymbol{\mu})$ and $\mathbf{u} = \Omega^{-\frac{1}{2}}(\mathbf{s} - \boldsymbol{\eta})$. By Theorem 2, for $\mathbf{t} \in \Re^n$, the moment generating function of $\mathbf{S}|\mathbf{Y} = \mathbf{y}$ is

$$M_{\mathbf{S}|\mathbf{Y}=\mathbf{y}}(\mathbf{t}) = C_2 \int_{\Re_n} \phi_n(\mathbf{s}; \boldsymbol{\eta}, \Omega) \left[(1 + \gamma\Phi(\mathbf{s}'\mathbf{z})) \Phi(\boldsymbol{\alpha}'\mathbf{u}) \right] d\mathbf{s}$$

$$= C_2 \int_{\Re_n} \phi_n(\mathbf{u}); \mathbf{0}, I_n) \exp\{ \left(\Omega^{\frac{1}{2}}\mathbf{u} + \boldsymbol{\eta} \right)' \mathbf{t} \}$$

$$\times \left[\left(1 + \gamma\Phi \left(\left(\Omega^{\frac{1}{2}}\mathbf{u} + \boldsymbol{\eta} \right)' \mathbf{z} \right) \right) \Phi(\boldsymbol{\alpha}'\mathbf{u}) \right] d\mathbf{u}.$$

By using following two facts,

$$\exp\left\{ -\frac{1}{2}\mathbf{u}'\mathbf{u} + \mathbf{u}'\Omega^{\frac{1}{2}}\mathbf{t} \right\} = \exp\left\{ -\frac{1}{2}\left(\mathbf{u} - \Omega^{\frac{1}{2}}\mathbf{t} \right)' \left(\mathbf{u} - \Omega^{\frac{1}{2}}\mathbf{t} \right) \right\} + \frac{1}{2}\mathbf{t}\Omega\mathbf{t}$$

and

$$\Phi(\boldsymbol{\alpha}'\mathbf{u})\Phi(\mathbf{z}'\Omega^{\frac{1}{2}}\mathbf{u} + \mathbf{z}'\boldsymbol{\eta}) = \Phi_2 \left[\begin{pmatrix} \boldsymbol{\alpha}' \\ \mathbf{z}'\Omega^{\frac{1}{2}} \end{pmatrix} \mathbf{u} + \begin{pmatrix} 0 \\ \mathbf{z}'\boldsymbol{\eta} \end{pmatrix} ; \mathbf{0}, I_2 \right].$$

the above expression can be simplified as

$$M_{\mathbf{S}|\mathbf{Y}=\mathbf{y}}(\mathbf{t}) = C_2 \exp\left\{ \boldsymbol{\eta}'\mathbf{t} + \frac{1}{2}\mathbf{t}'\Omega\mathbf{t} \right\}$$

$$\times \left\{ E\left[\Phi(\boldsymbol{\alpha}'\mathbf{U})\right] + \gamma E\left[\Phi_2 \left[\begin{pmatrix} \boldsymbol{\alpha}' \\ \mathbf{z}'\Omega^{\frac{1}{2}} \end{pmatrix} \mathbf{u} + \begin{pmatrix} 0 \\ \mathbf{z}'\boldsymbol{\eta} \end{pmatrix} ; \mathbf{0}, I_2 \right] \right] \right\},$$

where $\mathbf{u} \sim N_n \left(\Omega^{\frac{1}{2}}\mathbf{t}, I_n \right)$. Thus by Lemma 2, the desired result can be obtained.

Now for the moment generating function of $\mathbf{s}'\mathbf{s}|\mathbf{Y} = \mathbf{y}$, let $t \in \Re$. By using the argument,

$$\exp\left\{ -\frac{1}{2}\mathbf{u}'\mathbf{u} + (\Omega^{\frac{1}{2}}\mathbf{u} + \boldsymbol{\eta})'(\Omega^{\frac{1}{2}}\mathbf{u} + \boldsymbol{\eta})t \right\}$$

$$= \exp\{\boldsymbol{\eta}'\boldsymbol{\eta} t\} \exp\left\{ -\frac{1}{2}\left[\mathbf{u}'(I_n - 2\Omega t)\mathbf{u} - 4\mathbf{u}'\Omega^{\frac{1}{2}}\boldsymbol{\eta} t \right] \right\}$$

$$= \exp\left\{ \boldsymbol{\eta}'\boldsymbol{\eta} t + 2t^2\boldsymbol{\eta}'\Omega^{\frac{1}{2}}(I_n - 2\Omega t)^{-1}\Omega^{\frac{1}{2}}\boldsymbol{\eta} \right\} |I_n - 2\Omega t|^{-\frac{1}{2}}$$

$$\times \exp\left\{ -\frac{1}{2}\left[\mathbf{u} - 2t(I_n - 2\Omega t)^{-1}\Omega^{\frac{1}{2}}\boldsymbol{\eta} \right]' (I_n - 2\Omega t)\left[\mathbf{u} - 2t(I_n - 2\Omega t)^{-1}\Omega^{\frac{1}{2}}\boldsymbol{\eta} \right] \right\},$$

we obtain

$$M_{\mathbf{S'S}|\mathbf{Y}=\mathbf{y}}(t) = C_2 \exp\left\{t\boldsymbol{\eta}'\boldsymbol{\eta} + 2t^2\boldsymbol{\eta}'\Omega^{\frac{1}{2}}(I_n - 2\Omega t)^{-1}\Omega^{\frac{1}{2}}\boldsymbol{\eta}\right\} |I_n - 2\omega t|^{-\frac{1}{2}}$$

$$\times \left\{ E\left[\Phi(\boldsymbol{\alpha}'\mathbf{u})\right] + \gamma E\left[\Phi_2\left[\begin{pmatrix}\boldsymbol{\alpha}' \\ \mathbf{z}'\Omega^{\frac{1}{2}}\end{pmatrix}\mathbf{u} + \begin{pmatrix}0 \\ \mathbf{z}'\boldsymbol{\eta}\end{pmatrix}; \mathbf{0}, I_2\right]\right]\right\},$$

where $\mathbf{u} \sim N_n\left(2t(I_n - 2\Omega t)^{-1}\Omega^{\frac{1}{2}}\boldsymbol{\eta}, (I_n - 2\Omega t)^{-1}\right)$. By Lemma 2, the desired result follows. $\qquad\square$

Similarly we can prove the following result.

Theorem 6. *Let* $[\mathbf{Y}|S = s] \sim ESN_n(\boldsymbol{\mu}, \Sigma, s\mathbf{1}_n, \gamma)$, *and* $S \sim SN(\eta, \omega^2, \alpha)$. *Then the moment generating function of* $S|\mathbf{Y} = \mathbf{y}$ *is*

$$M_{S|\mathbf{Y}=\mathbf{y}}(t) = C_4 \exp\left\{\eta t + \frac{1}{2}t^2\omega\right\} \left\{\Phi(\alpha\omega; 0, 1+\alpha^2)\right. \tag{15}$$

$$\left. + \gamma\Phi_2\left[\begin{pmatrix}\alpha\omega \\ \mathbf{1_n}'\mathbf{z}\omega^2 + \mathbf{1_n}'\mathbf{z}\eta\end{pmatrix}; \mathbf{0}, \begin{pmatrix}1+\alpha^2 & \mathbf{1_n}'\mathbf{z}\omega\alpha \\ \mathbf{1_n}'\mathbf{z}\omega\alpha & 1+(\mathbf{1_n}'\mathbf{z}\omega)^2\end{pmatrix}\right]\right\}$$

and the moment generating function of $S^2|\mathbf{Y} = \mathbf{y}$ *is*

$$M_{s^2|\mathbf{Y}=\mathbf{y}}(t) = C_4 \exp\left\{\eta^2 t + \frac{2\omega^2\eta^2 t^2}{1-2\omega^2 t}\right\} \left\{\Phi\left(\frac{2\alpha\omega\eta t}{1-2\omega^2 t}; 0, 1+\alpha^2(1-2\omega^2 t)^{-1}\right)\right. \tag{16}$$

$$\left. + \gamma\Phi_2\left[\begin{pmatrix}\dfrac{2\alpha\omega\eta t}{1-2\omega^2 t} \\ \dfrac{2t\omega^2\eta\mathbf{1_n}'\mathbf{z}}{1-2\omega^2 t} + \mathbf{1_n}'\mathbf{z}\eta\end{pmatrix}; \mathbf{0}, \Psi_2\right]\right\},$$

where

$$C_4 = \left\{\frac{1}{2} + \gamma\Phi_2\left[\begin{pmatrix}\eta\mathbf{1_n}'\mathbf{z} \\ 0\end{pmatrix}; \mathbf{0}, \begin{pmatrix}1+n\omega^2\mathbf{z}'\mathbf{z} & \omega^2\alpha\mathbf{z}'\mathbf{1}_n \\ \alpha\omega^2\mathbf{1_n}'\mathbf{z} & 1+n\alpha^2\omega^2\end{pmatrix}\right]\right\}^{-1},$$

$$\Psi_2 = I_2 + \begin{pmatrix}\alpha \\ \mathbf{1_n}'\mathbf{z}'\omega\end{pmatrix}'(1-2\omega t)^{-1}\begin{pmatrix}\alpha \\ \mathbf{1_n}'\mathbf{z}\omega\end{pmatrix}, \text{ and } \mathbf{z} = \Sigma^{-\frac{1}{2}}(\mathbf{y}-\boldsymbol{\mu}).$$

4   Bayesian Analysis for the Shaped Mixtures Model

In this section, we will apply our main results to both linear regression model and Hierarchical model.

4.1   Robust Inference for Unknown Location and Scale Parameters

Consider the linear regression model $\mathbf{Y} = X\boldsymbol{\beta} + \boldsymbol{\mathcal{E}}$, where $Y = (Y_1, \ldots, Y_n)'$, $X\boldsymbol{\beta} = \boldsymbol{\mu} = (\mu_1, \cdots, \mu_n)'$ with $\mu_i = \boldsymbol{x}_i'\boldsymbol{\beta}$. Assume that location and scale parameters are unknown, denoted by $\boldsymbol{\theta} = (\boldsymbol{\mu}', \sigma^2)'$. We will use three examples to illustrate the robustness of $\boldsymbol{\theta}$. Note that the marginal likelihood functions of Y, $f(\mathbf{y}|\boldsymbol{\theta})$, can be obtain directly by setting up $\boldsymbol{\alpha} = 0$ in Theorems 1, 2 and 3. We will discuss the moment generating function of $f(\mathbf{Y}|\boldsymbol{\theta})$ under the assumption that the prior shape parameter follows the normal distribution. The following lemma will be used in our examples.

Lemma 3. *Let $W \sim N(a, b^2)$ and $Z \sim N(c, d^2)$ are independent. Then*

$$E\left[\Phi\left(\frac{cW}{\sqrt{1+d^2W^2}}\right)\right] = E[\Phi(WZ)].$$

Example 4.1. *Let $[Y_1|\mathbf{S} = \mathbf{s}], \ldots, [Y_n|\mathbf{S} = \mathbf{s}]$ be independent random variables with $[Y_i|\mathbf{S} = \mathbf{s}] \sim ESN(\mu_i, \sigma^2, s_i, \gamma)$ and let $\mathbf{S} \sim N_n(\boldsymbol{\eta}, D(\omega))$. Then the moment generating function of $\mathbf{Y}|\boldsymbol{\theta}$ is*

$$M_{\mathbf{Y}|\theta}(\mathbf{t}) = \left(\frac{2}{2+\gamma}\right)^n \exp\left\{\mathbf{t}'\boldsymbol{\mu} + \frac{1}{2}\mathbf{t}'D(\boldsymbol{\sigma})\mathbf{t}\right\} \prod_{i=1}^n \{1 + rE[\Phi(z_i u_i)]\}, \quad (17)$$

where $\mathbf{t} = (t_1, \cdots, t_n)' \in \Re^n$, $z_i \sim N(t_i\sigma^2, \sigma^2)$, $U_i \sim N(\eta_i, \omega_i^2)$, and u_i's are independent.

Proof. The desired result can be obtained by Corollary 2.1 and Lemma 2.   □

Example 4.2. *Let $[\mathbf{Y}|\mathbf{S} = \mathbf{s}] \sim ESN_n(\boldsymbol{\mu}, \sigma^2 I_n, \mathbf{s}, \gamma)$, $\mathbf{S} \sim N_n(\boldsymbol{\eta}, \Omega)$. Then the moment generating function of $\mathbf{Y}$ is*

$$M_{\mathbf{Y}}(\mathbf{t}) = \left(\frac{2}{2+\gamma}\right) \exp\left\{\mathbf{t}'\boldsymbol{\mu} + \frac{\sigma^2}{2}\mathbf{t}\mathbf{t}\right\}\left\{1 + \gamma E\left[\Phi\left(\frac{\boldsymbol{\eta}'\mathbf{z}}{\sqrt{1+\mathbf{z}'\Omega\mathbf{z}}}\right)\right]\right\}, \quad (18)$$

where $\mathbf{t} = (t_1, \cdots, t_n)' \in \Re^n$, and $\mathbf{Z} \sim N_n(\sigma\mathbf{t}, I_n)$.

Proof. The proof can be done directly by Corollary 2.2.   □

Similarly, by Corollary 2.3, we can obtain the following result.

Example 4.3. *Let $[\mathbf{Y}|S = \mathbf{s}] \sim ESN_n(\boldsymbol{\mu}, \sigma^2 I_n, s\mathbf{1}_n, \gamma)$, $S \sim SN(\eta, \omega^2)$. Then the moment generating function of $\mathbf{Y}'$ is*

$$M_{\mathbf{Y}|\theta}(\mathbf{t}) = \left(\frac{2}{2+\gamma}\right) \exp\left\{\mathbf{t}'\boldsymbol{\mu} + \frac{\sigma^2}{2}\mathbf{t}'\mathbf{t}\right\}\left\{1 + \gamma E\left[\Phi\left(\frac{\eta\mathbf{1}_n'\mathbf{z}}{\sqrt{1+\omega^2\mathbf{z}'\mathbf{1}_n\mathbf{1}_n'\mathbf{z}}}\right)\right]\right\}, (19)$$

where $\mathbf{t} \in \Re^n$, and $\mathbf{Z} \sim N_n(\sigma\mathbf{t}, I_n)$.

4.2   Hierarchical Bayesian Approach

In this section, we use the hierarchical Bayesian approach to construct a prior distribution on the shape parameter in the extended multivariate skew normal distribution. Suppose that we observed data from a random vector follows the multivariate extended skew normal distribution, where the location parameters and γ are assumed to be known, the prior distribution is the skew normal distribution conditional on $\boldsymbol{\theta}_0$, and $\boldsymbol{\theta}_0$ follows a normal distribution, we need to use our results to find the useful information for the parameter $\boldsymbol{\theta}$.

Example 4.4. *Let* $\mathbf{Y}|(\boldsymbol{\theta},\boldsymbol{\theta}_0) \sim ESN_n(\boldsymbol{\mu},\Sigma,\boldsymbol{\theta},\gamma)$, $\boldsymbol{\theta}|\boldsymbol{\theta}_0 \sim SN_n(0,\Omega,\boldsymbol{\theta}_0)$, and $\boldsymbol{\theta}_0 \sim N_n(\boldsymbol{\nu},\Psi)$. *Then the posterior density of* $\boldsymbol{\theta}$ *given* $\mathbf{Y} = \mathbf{y}$ *is*

$$f_{\boldsymbol{\theta}|\mathbf{y}}(\boldsymbol{\theta}) = \frac{4}{2+\gamma}\phi_n(\boldsymbol{\theta};0,\Omega)\left\{\Phi\left(\boldsymbol{\theta}'\Omega^{-\frac{1}{2}}\boldsymbol{\nu};0,\boldsymbol{\theta}'\Omega^{-\frac{1}{2}}\Psi\Omega^{-\frac{1}{2}}\boldsymbol{\theta}\right)\right.$$
$$\left. + \gamma\Phi_2\left[\begin{pmatrix}\boldsymbol{\theta}'\Omega^{-\frac{1}{2}}\boldsymbol{\nu}\\\boldsymbol{\theta}'\Sigma^{-\frac{1}{2}}(\mathbf{y}-\boldsymbol{\mu})\end{pmatrix};0,\begin{pmatrix}1+\boldsymbol{\theta}'\Omega^{-\frac{1}{2}}\Psi\Omega^{-\frac{1}{2}}\boldsymbol{\theta} & 0\\0 & 1\end{pmatrix}\right]\right\}.$$

Proof. By Theorem 2, we obtain the posterior density of $\boldsymbol{\theta}|(\mathbf{y},\boldsymbol{\theta}_0)$, given by

$$f_{\boldsymbol{\theta}|\mathbf{y},\boldsymbol{\theta}_0}(\boldsymbol{\theta}) = \frac{4}{2+\gamma}\phi_n(\boldsymbol{\theta};0,\Omega)\Phi(\boldsymbol{\theta}_0'\mathbf{u})[1+\gamma\Phi(\boldsymbol{\theta}'\mathbf{z})].$$

Note that $f_{\boldsymbol{\theta}|\mathbf{y}}(\boldsymbol{\theta},\boldsymbol{\theta}_0) = f_{\boldsymbol{\theta}|(\mathbf{y},\boldsymbol{\theta}_0)}(\boldsymbol{\theta})f(\boldsymbol{\theta}_0)$. Thus, by Bayesian calculation, we have

$$f_{\boldsymbol{\theta}|\mathbf{x}}(\boldsymbol{\theta}) = \int_{\Re^n} f_{\boldsymbol{\theta}|\mathbf{x},\boldsymbol{\theta}_0}(\boldsymbol{\theta})f(\boldsymbol{\theta}_0)d\boldsymbol{\theta}_0$$

and, by Lemma 2, the desired result follows after simplifications. $\square$

Remark 4.1. In Example 4.4, if we assume $\boldsymbol{\nu} = 0$, that is $\boldsymbol{\theta}_0 \sim N(0,\Psi)$, then the posterior density of $\boldsymbol{\theta}$ given $\mathbf{Y} = \mathbf{y}$ is,

$$f_{\boldsymbol{\theta}|\mathbf{x}}(\boldsymbol{\theta}) = 2\phi_n(\boldsymbol{\theta};0,\Omega)\Phi\left((\mathbf{x}-\boldsymbol{\mu})'\Sigma^{-\frac{1}{2}}\boldsymbol{\theta}\right),$$

which is free of parameters γ and Ψ. Also the moment generating function of $[\boldsymbol{\theta}|\mathbf{y}]$ and $[\boldsymbol{\theta}'\boldsymbol{\theta}|\mathbf{y}]$ are given by

$$M_{\boldsymbol{\theta}|\mathbf{y}}(\mathbf{t}) = 2\exp\left\{\frac{1}{2}\mathbf{t}'\Omega\mathbf{t}\right\}\Phi\left\{\frac{(\mathbf{y}-\boldsymbol{\mu})'\Sigma^{-\frac{1}{2}}\Omega^{\frac{3}{2}}\mathbf{t}}{\left[1+(\mathbf{y}-\boldsymbol{\mu})'\Sigma^{-\frac{1}{2}}\Omega^2\Sigma^{-\frac{1}{2}}(\mathbf{y}-\boldsymbol{\mu})\right]^{\frac{1}{2}}}\right\}, \quad \mathbf{t} \in \Re^n$$

and

$$M_{\boldsymbol{\theta}'\boldsymbol{\theta}|\mathbf{x}}(t) = |I_n - 2\Omega t|^{-\frac{1}{2}}, \quad t \in \Re,$$

respectively. Therefore the posterior distribution of $[\boldsymbol{\theta}'\boldsymbol{\theta}|\mathbf{y}]$ is the same as the one in the multivariate normal case.

5 Few Further Remarks

Continuing with the robust inference for the unknown location parameters and assume $\sigma^2 = 1$, we can obtain the moment generating function of $U = \mathbf{Y}'\mathbf{Y}$.

(i) Let $[\mathbf{Y}|\mathbf{S} = \mathbf{s}] \sim ESN_n(\boldsymbol{\mu},I_n,\mathbf{s},\gamma)$, and $\mathbf{S} \sim N_n(\boldsymbol{\eta},\Omega)$, then the moment generating function of U is

$$M_U(t) = \frac{2\exp\left\{\dfrac{t\boldsymbol{\mu}'\boldsymbol{\mu}}{1-2t}\right\}}{(2+\gamma)(1-2t)^{n/2}}\left\{1+\gamma E\left[\Phi\left(\frac{\boldsymbol{\eta}'\mathbf{z}}{\sqrt{1+\mathbf{z}'\Omega\mathbf{z}}}\right)\right]\right\}, \tag{20}$$

where $t \in \Re$, and $\mathbf{z} \sim N_n\left(\dfrac{2t\boldsymbol{\mu}}{1-2t},(1-2t)I_n\right)$.

(ii) Let $[\mathbf{Y}|S = s] \sim ESN_n(\boldsymbol{\mu}, I_n, s\mathbf{1}_n, \gamma)$, and $s \sim SN(\eta, \omega^2)$, then the moment generating function of U is

$$M_U(t) = \frac{2\exp\left\{\dfrac{t\boldsymbol{\mu}'\boldsymbol{\mu}}{1 - 2t}\right\}}{(2 + \gamma)(1 - 2t)^{n/2}}\left\{1 + \gamma E\left[\varPhi\left(\frac{\eta\mathbf{1}_n'\mathbf{z}}{\sqrt{1 + \omega^2(\mathbf{z}'\mathbf{1}_n)^2}}\right)\right]\right\}, \quad (21)$$

where $t \in \Re$, and $\mathbf{z} \sim N_n\left(\dfrac{2t\boldsymbol{\mu}}{1 - 2t}, (1 - 2t)I_n\right)$.

Remark 5.1. If we assume that $\boldsymbol{\eta} = 0$ in (20) or $\eta = 0$ in (21), then U in (1) and (ii) are both reduced into the noncentral chi-square distributed random variables. Also, if $\boldsymbol{\eta}'\mathbf{z} = 0$ in (20) or $\mathbf{1_n}'\mathbf{z} = 0$ in (21), then U in (1) and (ii) U are both reduced into the noncentral chi-square distribution.

Acknowledgement. The authors thank Professor Vladik Kreinovich for valuable suggestions that helped improve this article. The research of Weizhong Tian was partially supported by Internal Grants from Eastern New Mexico University and Funds from One Hundred Person Project of Shaanxi Province of China.

References

1. Arellano-Valle, R.B., Castro, L.M., Genton, M.G., Gmez, H.W.: Bayesian inference for shape mixtures of skewed distributions, with application to regression analysis. Bayesian Anal. **3**(3), 513–539 (2008)
2. Arellano-Valle, R.B., Genton, M.G., Loschi, R.H.: Shape mixtures of multivariate skew-normal distributions. J. Multivar. Anal. **100**(1), 91–101 (2009)
3. Azzalini, A.: A class of distributions which includes the normal ones. Scand. J. Stat. **12**(2), 171–178 (1985)
4. Azzalini, A., Dalla, A.: The multivariate skew-normal distribution. Biometrika **83**(4), 715–726 (1996)
5. Azzalini, A., Capitanio, A.: Statistical applications of the multivariate skew normal distribution. J. Royal Stat. Soc. Ser. B (Stat. Methodol.) **61**(3), 579–602 (1999)
6. Chen, J., Gupta, A.: Matrix variate skew normal distributions. Statistics **39**(3), 247–253 (2005)
7. Genton, M., He, L., Liu, X.: Moments of skew-normal random vectors and their quadratic forms. Stat. Probab. Lett. **51**(4), 319–325 (2001)
8. Gupta, A., Huang, W.: Quadratic forms in skew normal variates. J. Math. Anal. Appl. **273**(2), 558–564 (2002)
9. Huang, W., Chen, Y.: Quadratic forms of multivariate skew normal-symmetric distributions. Stat. Probab. Lett. **76**(9), 871–879 (2006)
10. Loperfido, N.: Quadratic forms of skew-normal random vectors. Stat. Probab. Lett. **54**(4), 381–387 (2001)
11. Szkely, G., Rizzo, M.: A new test for multivariate normality. J. Multivar. Anal. **93**(1), 58–80 (2005)
12. Tian, W., Wang, C., Wu, M., Wang, T.: The multivariate extended skew normal distribution and its quadratic forms. In: Causal Inference in Econometrics, pp. 153–169. Springer (2016)

13. Tian, W., Wang, T.: Quadratic forms of refined skew normal models based on stochastic representation. Random Operators Stoch. Equ. **24**(4), 225–234 (2016)
14. Wang, T., Li, B., Gupta, A.: Distribution of quadratic forms under skew normal settings. J. Multivar. Anal. **100**(3), 533–545 (2009)
15. Ye, R., Wang, T., Gupta, A.: Distribution of matrix quadratic forms under skew-normal settings. J. Multivar. Anal. **131**, 229–239 (2014)
16. Zacks, S.: Parametric Statistical Inference: Basic Theory and Modern Approaches. Elsevier, Philadelphia (2014)

Plausibility Regions on Parameters of the Skew Normal Distribution Based on Inferential Models

Xiaonan Zhu[1], Baokun Li[2], Mixia Wu[3], and Tonghui Wang[1(✉)]

[1] Department of Mathematical Sciences, New Mexico State University,
Las Cruces, USA
{xzhu,twang}@nmsu.edu
[2] School of Statistics, Southwestern University of Finance and Economy,
Chengdu, China
bali@swufe.edu.cn
[3] College of Applied Sciences, Beijing University of Technology, Beijing, China
wumixia@bjut.edu.cn

Abstract. In this paper, plausibility functions and $100(1-\alpha)\%$ plausibility regions on location parameter and scale parameter of skew normal distributions are obtained in several cases by using inferential models (IMs), which are new methods of statistical inference. Simulation studies and one real data example are given for illustration of our results.

Keywords: Skew normal distribution · Closed skew normal distribution · Noncentral closed skew chi-square distribution Plausibility function · Inferential model

1 Introduction

Skew normal distributions introduced by Azzalini in 1985 [1] are suitable for the analysis of data that is unimodal empirical distributed but with some skewness. Since then, skew normal distributions and their generalizations have been studied and used by many researchers, such as Azzalini and Dalla Valle [4], Azzalini and Capitanio [3], Wang et al. [11], Ye et al. [14], Ye and Wang [13] and Tian and Wang [10]. Although skew normal distributions have some properties similar to normal distributions, two important properties are absent: the closure for the joint distribution of independent skew normal variables and the closure under linear combinations. Therefore, to study statistical properties of a skew normal population by samples, we usually have to assume that the sample is identically distributed but not independent, e.g. Wang et al. [12] and Zhu et al. [15]. To overcome this issue, in this paper, we assume that components of the sample are independent and identically distributed (i.i.d.) from a skew normal population. Then based on results we obtained, plausibility functions and $100(1-\alpha)\%$ plausibility regions on location and scale parameters of skew normal distributions are studied by using inferential models (IMs). When location and scale parameters

© Springer International Publishing AG 2018
V. Kreinovich et al. (eds.), *Predictive Econometrics and Big Data*, Studies in Computational Intelligence 753, https://doi.org/10.1007/978-3-319-70942-0_21

are known, inferences for skewness parameter based on a dependent sample by using IMs were partially studied by Zhu et al. [15].

IMs are new methods of statistical inference introduced by Martin and Liu [8]. As Martin and Liu [9] note: *"Comparing with Fishers fiducial inference, Dempster-Shafer theory of belief functions and Bayesian inference, IMs not only provide data-dependent probabilistic measures of uncertainty about the unknown parameter, but does so with an automatic long-run frequency calibration property. Thus IMs produce exact prior-free and prior-less probabilistic inference."* See [6–9] for more details of IMs.

The paper is organized as follows. Skew normal distributions, their several extensions and IMs are briefly reviewed, and sampling distributions of skew normal populations are briefly studied in Sect. 2. Plausibility functions and $100(1-\alpha)\%$ plausibility regions about location parameter and scale parameter of skew normal distributions are obtained in several cases, and corresponding simulation studies are provided in Sect. 3. Plausibility functions and $100(1-\alpha)\%$ plausibility regions of difference between two location parameters are given and simulation studies are also provided in Sect. 4. One real data example is given to illustrate our results.

2   Preliminaries

2.1   Sampling Distributions from a Skew Normal Population

Throughout this paper, we use $M_{n\times m}$ to denote the set of all $n\times m$ matrices over the real field $\mathbb{R}$ and $\mathbb{R}^n = M_{n\times 1}$. For any $B \in M_{n\times m}$, B' is the transpose of B. $\mathbf{I}_n \in M_{n\times n}$ is the identity matrix. $\mathbf{1}_n$ is the column vector $(1, \cdots, 1)' \in \mathbb{R}^n$. $\mathbf{J}_n = \mathbf{1}_n\mathbf{1}_n'$ and $\bar{\mathbf{J}}_n = \frac{1}{n}\mathbf{J}_n$. Also $\phi_n(\cdot; \mu, \Sigma)$ and $\Phi_n(\cdot; \mu, \Sigma)$ are the probability density function (p.d.f.) and cumulative distribution function (c.d.f.), respectively, of an n-dimensional normal distribution with the mean vector $\mu \in \mathbb{R}^n$ and covariance matrix Σ. Specially, $\phi(\cdot)$ and $\Phi(\cdot)$ are the p.d.f. and c.d.f. of the univariate standard normal distribution.

The *skew normal* (SN) distribution is a generalization of the normal distribution introduced by Azzalini [1]. A random variable Z is said to be skew normal with *skewness parameter* λ, denoted by $Z \sim SN(\lambda)$, if its density function is

$$f(z; \lambda) = 2\phi(z)\Phi(\lambda z),$$

where $\lambda \in \mathbb{R}$. For any $\mu \in \mathbb{R}$ and $\sigma > 0$, let $X = \mu + \sigma Z$, then X is called a skew normal random variable with the *location parameter* μ, *scale parameter* σ and *skewness parameter* λ, denoted by $X \sim SN(\mu, \sigma^2, \lambda)$. The density function of X is

$$f(x) = \frac{2}{\sigma}\phi\left(\frac{x-\mu}{\sigma}\right)\Phi\left(\lambda\frac{x-\mu}{\sigma}\right).$$

Note that for any $Z \sim SN(\lambda)$, $Z^2 \sim \chi_1^2$ [2]. So if $X \sim SN(\mu, \sigma^2, \lambda)$, then $\frac{(X-\mu)^2}{\sigma^2} \sim \chi_1^2$.

Multivariate skew normal distributions were defined by Azzalini and Dalla Valle [4] and Azzalini and Capitanio [3]. An n-dimensional random vector $\mathbf{X}$ is said to have a multivariate skew normal distribution, denoted by $\mathbf{X} \sim SN_n(\Sigma, \lambda)$, if its density function is given by

$$f(\mathbf{x}; \Sigma, \lambda) = 2\phi_n(\mathbf{x}; \mathbf{0}_n, \Sigma)\Phi(\lambda'\mathbf{x}),$$

where parameters $\lambda \in \mathbb{R}^n$ and $\Sigma \in M_{n \times n}$ is positive definite. For more details of univariate and multivariate skew normal distributions, see [2].

An extension of multivariate skew normal distributions was defined by Wang et al. [11] as follows (which generalized Azzalini and Dalla Valle's definition). Let $\mathbf{Z} \sim SN_k(\mathbf{I}_k, \lambda)$. $\mathbf{Y} = \mu + B'\mathbf{Z}$ is called a multivariate skew normal random vector with location parameter $\mu \in \mathbb{R}^n$, scale parameter $B \in M_{k \times n}$ and shape parameter $\lambda \in \mathbb{R}^k$, and is denoted by $\mathbf{Y} \sim \mathbf{SN}_n(\mu, B, \lambda)$. The density of $\mathbf{Y}$, if exists, is given by

$$f(\mathbf{y}; \mu, B, \lambda) = \frac{2}{|\Sigma|^{\frac{1}{2}}} \phi_n\left(\Sigma^{-\frac{1}{2}}(\mathbf{y} - \mu)\right) \Phi\left[\frac{\lambda'B\Sigma^{-1}(\mathbf{y} - \mu)}{[1 + \lambda'(\mathbf{I}_k - B\Sigma^{-1}B')\lambda]^{\frac{1}{2}}}\right],$$

where $\Sigma = B'B$. Note that if $\mathbf{X} \sim SN_n(\Sigma, \lambda)$ or $\mathbf{SN}_n(\mu, B, \lambda)$ with $\lambda \neq \mathbf{0}_n$, then components of $\mathbf{X}$, $X_1, \cdots, X_n$, are not independent even if $\Sigma = \mathbf{I}_n$.

In order to include joint distributions and linear combinations of independent skew normal random variables, the *closed skew normal* distribution was defined by González-Farías et al. [5]. An n-dimensional random vector $\mathbf{X}$ is distributed according to a closed skew normal distribution with parameters m, μ, Σ, D, ν and Δ, denoted as $\mathbf{X} \sim CSN_{n,m}(\mu, \Sigma, D, \nu, \Delta)$, if its density is

$$f_{n,m}(\mathbf{x}; \mu, \Sigma, D, \Delta) = c\phi_n(\mathbf{x}; \mu, \Sigma)\Phi_m[D(\mathbf{x} - \mu); \nu, \Delta],$$

where $c^{-1} = \Phi_m(\mathbf{0}_m; \nu, \Delta + D\Sigma D')$, n and m are positive integers, $\mu \in \mathbb{R}^n$, $\nu \in \mathbb{R}^m$, $D \in M_{m \times n}$, and $\Sigma \in M_{n \times n}$ and $\Delta \in M_{m \times m}$ are positive definite. For simplicity, we use $CSN_{n,m}(\mu, \Sigma, D, \Delta)$ to denote $CSN_{n,m}(\mu, \Sigma, D, \mathbf{0}_m, \Delta)$ in this paper.

To study sampling distributions of $SN(\mu, \sigma^2, \lambda)$, we need the following definition.

Definition 1. *Let* $\mathbf{X} \sim CSN_{n,m}(\mu, \mathbf{I}_n, D, \Delta)$. *The distribution of* $\mathbf{X}'\mathbf{X}$ *is called a* **noncentral closed skew chi-square distribution** *with degrees of freedom* n, *the noncentrality parameter* $\lambda = \mu'\mu$, *skewness parameters* $\delta_1 = D\mu$ *and* $\delta_2 = DD'$, *and parameter* Δ, *denoted by* $\mathbf{X}'\mathbf{X} \sim CS\chi_n^2(\lambda, \delta_1, \delta_2, \Delta)$.

Sampling distributions of $SN(\mu, \sigma^2, \lambda)$ are given as follows.

Proposition 1. *Suppose that* $\mathbf{X} = (X_1, \cdots, X_n)'$ *is a random sample from* $SN(\mu, \sigma^2, \lambda)$, *i.e.,* $X_i \overset{i.i.d.}{\sim} SN(\mu, \sigma^2, \lambda)$, $i = 1, \cdots, n$. *Let* $\bar{X} = \frac{1}{n}\sum\limits_{i=1}^{n} X_i$ *and* $S^2 = \frac{1}{n-1}\sum\limits_{i=1}^{n}(X_i - \bar{X})^2$ *be the sample mean and sample variance respectively.* *Then*

$$\bar{X} \sim CSN_{1,n}\left(\mu, \frac{\sigma^2}{n}, \frac{\lambda}{\sigma}\mathbf{1}_n, (1 + \lambda^2)\mathbf{I}_n - \lambda^2\bar{\mathbf{J}}_\mathbf{n}\right),$$

and

$$\frac{(n-1)S^2}{\sigma^2} \sim CS\chi^2_{n-1}\left(0, \mathbf{0}_n, \lambda^2(\mathbf{I}_n - \bar{\mathbf{J}}_n), \mathbf{I}_n + \lambda^2\bar{\mathbf{J}}_n\right).$$

Proof. First, it can be shown that $\mathbf{X} \sim CSN_{n,n}\left(\mu\mathbf{1}_n, \sigma^2\mathbf{I}_n, \frac{\lambda}{\sigma}\mathbf{I}_n, \mathbf{I}_n\right)$. Thus, by Theorem 1 of [5],

$$\bar{X} = \frac{1}{n}\sum_{i=1}^n X_i \sim CSN_{1,n}\left(\mu, \frac{\sigma^2}{n}, \frac{\lambda}{\sigma}\mathbf{1}_n, (1+\lambda^2)\mathbf{I}_n - \lambda^2\bar{\mathbf{J}}_n\right).$$

Second, note that $\frac{(n-1)S^2}{\sigma^2} = \mathbf{X}'\left(\frac{1}{\sigma^2}(\mathbf{I}_n - \bar{\mathbf{J}}_n)\right)\mathbf{X}$. Since $\mathbf{I}_n - \bar{\mathbf{J}}_n$ is idempotent of rank $n-1$, there is an orthogonal matrix $P = [P_1, P_2] \in M_{n\times n}$ such that

$$\mathbf{I}_n - \bar{\mathbf{J}}_n = P\begin{pmatrix}\mathbf{I}_{n-1} & 0 \\ 0 & 0\end{pmatrix}P' = P_1P_1'.$$

Now let $\mathbf{Y} = \frac{1}{\sigma}P_1'\mathbf{X}$. Then $\mathbf{Y} \sim CSN_{n-1,n}\left(\frac{\mu}{\sigma}P_1'\mathbf{1}_n, \mathbf{I}_{n-1}, \lambda P_1, \mathbf{I}_n + \lambda^2\bar{\mathbf{J}}_n\right)$. Thus, by Definition 1, we have

$$\frac{(n-1)S^2}{\sigma^2} = \mathbf{Y}'\mathbf{Y} \sim CS\chi^2_{n-1}\left(0, \mathbf{0}_n, \lambda^2(\mathbf{I}_n - \bar{\mathbf{J}}_n), \mathbf{I}_n + \lambda^2\bar{\mathbf{J}}_n\right).$$

$\square$

Proposition 2. *Let* $\mathbf{X} = (X_1, \cdots, X_{n_1})'$ *and* $\mathbf{Y} = (Y_1, \cdots, Y_{n_2})'$ *be independent random samples from* $SN(\mu_1, \sigma_1^2, \lambda_1)$ *and* $SN(\mu_2, \sigma_2^2, \lambda_2)$, *respectively. Let* $\bar{X} = \frac{1}{n_1}\sum_{i=1}^{n_1} X_i$ *and* $\bar{Y} = \frac{1}{n_2}\sum_{j=1}^{n_2} Y_j$. *Then*

$$\bar{X} - \bar{Y} \sim CSN_{1,n_1+n_2}\left(\mu_1 - \mu_2, \frac{\sigma_1^2}{n_1} + \frac{\sigma_2^2}{n_2}, D_\star, \Delta_\star\right),$$

where

$$D_\star = \frac{1}{n_2\sigma_1^2 + n_1\sigma_2^2}\begin{pmatrix}n_2\lambda_1\sigma_1\mathbf{1}_{n_1} \\ -n_1\lambda_2\sigma_2\mathbf{1}_{n_2}\end{pmatrix},$$

and

$$\Delta_\star = \begin{pmatrix}(1+\lambda_1^2)\mathbf{I}_{n_1} - \frac{n_2\lambda_1^2\sigma_1^2}{n_2\sigma_1^2+n_1\sigma_2^2}\bar{\mathbf{J}}_{n_1} & \frac{\lambda_1\lambda_2\sigma_1\sigma_2}{n_2\sigma_1^2+n_1\sigma_2^2}\mathbf{1}_{n_1}\mathbf{1}'_{n_2} \\ \frac{\lambda_1\lambda_2\sigma_1\sigma_2}{n_2\sigma_1^2+n_1\sigma_2^2}\mathbf{1}_{n_2}\mathbf{1}'_{n_1} & (1+\lambda_2^2)\mathbf{I}_{n_2} - \frac{n_1\lambda_2^2\sigma_2^2}{n_2\sigma_1^2+n_1\sigma_2^2}\bar{\mathbf{J}}_{n_2}\end{pmatrix}.$$

Proof. Similar to the proof of Proposition 1, we can obtain

$$\bar{X} \sim CSN_{1,n_1}\left(\mu_1, \frac{\sigma_1^2}{n_1}, \frac{\lambda_1}{\sigma_1}\mathbf{1}_{n_1}, (1+\lambda_1^2)\mathbf{I}_{n_1} - \lambda_1^2\bar{\mathbf{J}}_{n_1}\right),$$

and

$$\bar{Y} \sim CSN_{1,n_2}\left(\mu_2, \frac{\sigma_2^2}{n_2}, \frac{\lambda_2}{\sigma_2}\mathbf{1}_{n_2}, (1+\lambda_2^2)\mathbf{I}_{n_2} - \lambda_2^2\bar{\mathbf{J}}_{n_2}\right).$$

Since $\bar{X}$ and $\bar{Y}$ are independent, by Theorem 3 of [5],

$$(\bar{X}, \bar{Y})' \sim CSN_{2,n_1+n_2}(\mu_*, \Sigma_*, D_*, \Delta_*),$$

where $\mu_* = (\mu_1, \mu_2)'$,

$$\Sigma_* = \begin{pmatrix} \frac{\sigma_1^2}{n_1} & 0 \\ 0 & \frac{\sigma_2^2}{n_2} \end{pmatrix}, \quad D_* = \begin{pmatrix} \frac{\lambda_1}{\sigma_1}\mathbf{1}_{n_1} & 0 \\ 0 & \frac{\lambda_2}{\sigma_2}\mathbf{1}_{n_2} \end{pmatrix},$$

and

$$\Delta_* = \begin{pmatrix} (1+\lambda_1^2)\mathbf{I}_{n_1} - \lambda_1^2\bar{\mathbf{J}}_{n_1} & 0 \\ 0 & (1+\lambda_2^2)\mathbf{I}_{n_2} - \lambda_2^2\bar{\mathbf{J}}_{n_2} \end{pmatrix}.$$

Note that $\bar{X} - \bar{Y} = (1, -1)(\bar{X}, \bar{Y})'$, so the desired result follows by Theorem 1 of [5]. $\qquad\square$

2.2   Inferential Models

In this subsection, let's briefly review IMs, which were introduced by Martin and Liu [8,9].

Let X be an observable random sample with a probability distribution $P_{X|\theta}$ on a sample space $\mathbb{X}$, where θ is an unknown parameter, $\theta \in \Theta$, a parameter space. Let U be a well-known but unobservable *auxiliary variable* on an auxiliary space $\mathbb{U}$. An *association* is a map $a : \mathbb{U} \times \Theta \to \mathbb{X}$ such that $X = a(U, \theta)$.

An IM consists of three steps based on a fixed association.

Association Step (A-step). Suppose we have an association $X = a(U, \theta)$ and an observation $X = x$, where x could be a scalar or vector, then the unknown θ must satisfy $x = a(u^*, \theta)$, for some unobserved u^* of U. So from the observation $X = x$, we can construct sets of solutions

$$\Theta_x(u) = \{\theta \in \Theta : x = a(u, \theta)\}, \qquad x \in \mathbb{X}, \qquad u \in \mathbb{U}.$$

Prediction Step (P-step). To predict the unknown u^*, let $u \to S(u)$ be a set-value map from $\mathbb{U}$ to $\mathbb{S}$, where $\mathbb{S}$ is a collection of P_U-measurable subsets of $\mathbb{U}$. Then the random set $S : \mathbb{U} \to \mathbb{S}$ is called a *predictive random set* of U with distribution $P_S = P_U \circ S^{-1}$. We will use S to predict u^*.

Combination Step (C-step). Define

$$\Theta_x(S) = \bigcup_{u \in S} \Theta_x(u).$$

For any assertion A of θ, i.e., $A \subseteq \Theta$, the *belief function* and *plausibility function* of A with respect to a predictive random set S are defined by,

$$\mathrm{bel}_x(A; S) = P_S\{\Theta_x(S) \subseteq A : \Theta_x(S) \neq \emptyset\};$$

$$\mathrm{pl}_x(A; S) = P_S\{\Theta_x(S) \not\subseteq A^c : \Theta_x(S) \neq \emptyset\}.$$

Note that

$$\mathsf{pl}_x(A; S) = 1 - \mathsf{bel}_x(A^c; S), \qquad \mathsf{bel}_x(A; S) + \mathsf{bel}_x(A^c; S) \leq 1, \qquad \text{for all } A \subseteq \Theta.$$

To make a good inference for assertions, we need some concepts of *validity*.

Let X and Y be two random variables. We say that X is *stochastically no smaller than* Y, denoted by $X \geq_{st} Y$, if $P(X > a) \geq P(Y > a)$, for all $a \in \mathbb{R}$. A predictive random set S is *valid* for predicting the unobserved auxiliary variable U if $\gamma_S(U)$, as a function of $U \sim P_U$, is stochastically no smaller than Uniform$(0,1)$, where γ_S is called the *contour function* of S defined by

$$\gamma_S(u) = P_S(S \ni u), \qquad u \in \mathbb{U}.$$

If $\gamma_S(U) \sim \text{Uniform}(0,1)$ then S is *efficient*.

Remark 1. *By Theorem 4.1 of [9], there is a simple way to construct a valid predictive random set as follows. Let $\mathbb{S}$ be a collection of subsets of $\mathbb{U}$. If $\mathbb{S}$ and P_U satisfy following conditions,*

(i) $\mathbb{S}$ is nested, i.e., all elements of $\mathbb{S}$ can be ordered by inclusions,
(ii) There is some $F \in \mathbb{S}$ such that $P_U(F) > 0$,
(iii) All closed subsets of $\mathbb{U}$ are P_U-measurable,
(iv) $\mathbb{S}$ contains $\emptyset, \mathbb{U}$, and all of the other elements are closed subsets,
and define a predictive random set S, with distribution P_S, supported on $\mathbb{S}$, such that

$$P_S\{F \subseteq K\} = \sup_{F \in \mathbb{S}; F \subseteq K} P_U(F), \qquad K \subseteq \mathbb{U},$$

then S is valid.

Suppose $X \sim P_{X|\theta}$ and let A be an assertion of interest. Then the IM with a belief function $\mathsf{bel}_x(\cdot \ ; S)$ is *valid* for A if

$$\sup_{\theta \notin A} P_{X|\theta}\{\mathsf{bel}_X(A; S) \geq 1 - \alpha\} \leq \alpha, \qquad \text{for all } \alpha \in (0,1).$$

The IM is said to be *valid* if it is valid for all A.

Remark 2. *By Theorem 4.2 of [9], if the predictive random set S is valid and $\Theta_x(S) \neq \emptyset$ with P_S-probability 1 for all x, then the IM is valid.*

Given an IM, a $100(1 - \alpha)\%$ *plausibility region* is defined by

$$\Pi_x(\alpha) = \{\theta \in \Theta : \mathsf{pl}_x(\theta; S) > \alpha\},$$

which is an IM-based counterpart of classical confidence intervals.

3   One-Sample Cases

Let $\mathbf{X} = (X_1, \cdots, X_n)'$ be a random sample from $SN(\mu, \sigma^2, \lambda)$. In this section, we are going to use $\mathbf{X}$ and IMs to make inferences about location parameter μ and scale parameter σ of $SN(\mu, \sigma^2, \lambda)$ in three different cases, (i) μ is unknown but σ and λ are known, (ii) σ is unknown but μ and λ are known, and (iii) σ and μ are both unknown but λ is known and we are interested in σ.

3.1   The Plausibility Function and Plausibility Region for μ

Let $\mathbf{X} \sim (X_1, \cdots, X_n)'$ be a random sample from $SN(\mu, \sigma^2, \lambda)$ with unknown μ and known σ and λ.

A-step: From Proposition 1, we know that

$$\bar{X} = \frac{1}{n}\sum_{i=1}^{n} X_i \sim CSN_{1,n}\left(\mu, \frac{\sigma^2}{n}, \frac{\lambda}{\sigma}\mathbf{1}_n, (1+\lambda^2)\mathbf{I}_n - \lambda^2\bar{\mathbf{J}}_n\right).$$

So by the property of linear transformations of CSN distributions [5], we can obtain an association

$$\bar{X} = \mu + F_1^{-1}(U),$$

where $U \sim \text{Uniform}(0,1)$ and F_1 is the c.d.f. of $CSN_{1,n}\left(0, \frac{\sigma^2}{n}, \frac{\lambda}{\sigma}\mathbf{1}_n, (1+\lambda^2)\right.$ $\left.\mathbf{I}_n - \lambda^2\bar{\mathbf{J}}_n\right)$. For any observation $\bar{x} \in \mathbb{R}$ and $u \in [0,1]$, we have $\Theta_{\bar{x}}(\mu) = \{\bar{x} - F_1^{-1}(u)\}$.

P-step: To predict auxiliary variables U, we use the default predictive random set

$$S(U) = [0.5 - |U - 0.5|, 0.5 + |U - 0.5|].$$

C-step: By the P-step, we have

$$\Theta_{\bar{x}}(S) = \left[\bar{x} - F_1^{-1}(0.5 + |U - 0.5|), \bar{x} - F_1^{-1}(0.5 - |U - 0.5|)\right].$$

So we can use the above IM to get the following result.

Theorem 1. *For any assertion $A = \{\mu\}$,*

$$bel_{\bar{x}}(A; S) = 0, \qquad pl_{\bar{x}}(A; S) = 1 - |2F_1(\bar{x} - \mu) - 1|,$$

and the $100(1 - \alpha)\%$ plausibility region $\Pi_{\bar{x}}(\mu)$ is

$$\bar{x} - F_1^{-1}\left(1 - \frac{\alpha}{2}\right) < \mu < \bar{x} - F_1^{-1}\left(\frac{\alpha}{2}\right).$$

Proof. It is clear that $\{\Theta_{\bar{x}}(S) \subseteq A\} = \emptyset$, so $bel_{\bar{x}}(A; S) = 0$.

$$\begin{aligned}
pl_{\bar{x}}(A; S) &= 1 - bel_{\bar{x}}(A^c; S) \\
&= 1 - P_S(\Theta_{\bar{x}}(S) \subseteq A^c) \\
&= 1 - P_U(\bar{x} - F_1^{-1}(0.5 - |U - 0.5|) < \mu) \\
&\quad - P_U(\bar{x} - F_1^{-1}(0.5 + |U - 0.5|) > \mu) \\
&= 1 - P_U(|U - 0.5| < |F_1(\bar{x} - \mu) - 0.5|) \\
&= 1 - |2F_1(\bar{x} - \mu) - 1|.
\end{aligned}$$

By the definition, $\Pi_{\bar{x}}(\mu) = \{\mu : pl_{\bar{x}}(\mu; S) > \alpha\}$. Let $pl_{\bar{x}}(A; S) = 1 - |2F_1(\bar{x} - \mu) - 1| > \alpha$ and then solve it for μ. We have $\bar{x} - F_1^{-1}\left(1 - \frac{\alpha}{2}\right) < \mu < \bar{x} - F_1^{-1}\left(\frac{\alpha}{2}\right)$. $\square$

In Table 1 and Fig. 1, we provide a simulation study for this case with different μ, σ, λ and sample size n.

Table 1. Simulation results of AL, AU, ALength and CP of 95% plausibility regions of μ when σ and λ are known.

$\mu = 0, \sigma = 1, \lambda = 1$					$\mu = 2, \sigma = 2, \lambda = 5$				
n	AL	AU	ALength	CP	n	AL	AU	ALength	CP
10	-0.5164	0.5071	1.0235	0.9480	10	1.1699	2.7094	1.5394	0.9420
20	-0.3614	0.3623	0.7237	0.9492	20	1.4181	2.5089	1.0908	0.9550
40	-0.2571	0.2545	0.5117	0.9520	40	1.6013	2.3729	0.7715	0.9540

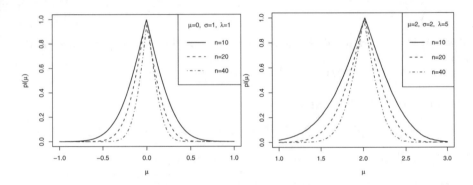

Fig. 1. Graphs of plausibility functions of μ based on simulated data when σ and λ are known.

Remark 3. *From the above proof, we can see that* $\text{bel}(A) = 0$ *for any single-point assertion A. It also holds in the following theorems.*

3.2   The Plausibility Function and Plausibility Region for σ

3.2.1   Location Parameter μ Is Known

Let $\mathbf{X} \sim (X_1, \cdots, X_n)'$ be a random sample from $SN(\mu, \sigma^2, \lambda)$ with unknown σ and known μ and λ.

A-step: From Sect. 2.1, we know that $\frac{X_i - \mu}{\sigma} \overset{i.i.d.}{\sim} SN(\lambda)$, $i = 1, \cdots, n$. So if we define a statistic $T = \sum\limits_{i=1}^{n} (X_i - \mu)^2$, then we have $\frac{T}{\sigma^2} \sim \chi_n^2$. It gives us an association

$$\frac{T}{\sigma^2} = F_2^{-1}(U),$$

where F_2 is the c.d.f. of χ_n^2 and $U \sim \text{Uniform}(0,1)$. For any observation $t > 0$ and $u \in [0,1]$, we have

$$\Theta_t(\sigma) = \left\{ \sqrt{\frac{t}{F_2^{-1}(u)}} \right\}.$$

P-step: To predict auxiliary variables U, we use the default predictive random set again

$$S(U) = [0.5 - |U - 0.5|, 0.5 + |U - 0.5|].$$

C-step: By the P-step, we have

$$\Theta_t(S) = \left[\sqrt{\frac{t}{F_2^{-1}(0.5 + |U - 0.5|)}}, \sqrt{\frac{t}{F_2^{-1}(0.5 - |U - 0.5|)}} \right].$$

Based on this IM, we have the below result.

Theorem 2. *For any assertion* $A = \{\sigma\}$,

$$bel_t(A; S) = 0, \qquad pl_t(A; S) = 1 - \left| 2F_2 \left(\frac{t}{\sigma^2} \right) - 1 \right|,$$

and the $100(1 - \alpha)\%$ *plausibility region* $\Pi_t(\sigma)$ *is*

$$\sqrt{\frac{t}{F_2^{-1}\left(1 - \frac{\alpha}{2}\right)}} < \sigma < \sqrt{\frac{t}{F_2^{-1}\left(\frac{\alpha}{2}\right)}}.$$

Proof.

$$\begin{aligned}
pl_t(A; S) &= 1 - bel_t(A^c; S) \\
&= 1 - P_S(\Theta_t(S) \subseteq A^c) \\
&= 1 - P_U \left(\sqrt{\frac{t}{F_2^{-1}(0.5 - |U - 0.5|)}} < \sigma \right) - P_U \left(\sqrt{\frac{t}{F_2^{-1}(0.5 + |U - 0.5|)}} > \sigma \right) \\
&= 1 - P_U \left(|U - 0.5| < \left| F_2 \left(\frac{t}{\sigma^2} \right) - 0.5 \right| \right) \\
&= 1 - \left| 2F_2 \left(\frac{t}{\sigma^2} \right) - 1 \right|.
\end{aligned}$$

By the definition, $\Pi_t(\sigma) = \{\sigma : pl_t(\sigma; S) > \alpha\}$. Let $pl_t(A; S) = 1 - \left| 2F_2 \left(\frac{t}{\sigma^2} \right) - 1 \right| > \alpha$ and then solve it for σ. We have $\sqrt{\frac{t}{F_2^{-1}\left(1 - \frac{\alpha}{2}\right)}} < \sigma < \sqrt{\frac{t}{F_2^{-1}\left(\frac{\alpha}{2}\right)}}$. $\qquad \square$

In Table 2 and Fig. 2, we provide a simulation study of this case with different σ, μ, λ and sample size n.

3.2.2 Location Parameter μ Is Unknown

Let $\mathbf{X} \sim (X_1, \cdots, X_n)'$ be a random sample from $SN(\mu, \sigma^2, \lambda)$ with unknown σ and μ and known λ.

Table 2. Simulation results of AL, AU, ALength and CP of 95% plausibility regions of σ when μ and λ are known.

$\sigma = 1, \mu = 0, \lambda = 1$					$\sigma = 2, \mu = 1, \lambda = 5$				
n	AL	AU	ALength	CP	n	AL	AU	ALength	CP
10	0.6803	1.7088	1.0285	0.9477	10	1.3597	3.4152	2.0555	0.9501
20	0.7562	1.4273	0.6711	0.9528	20	1.5086	2.8477	1.3390	0.9519
40	0.8168	1.2729	0.4561	0.9517	40	1.6343	2.5471	0.9127	0.9532

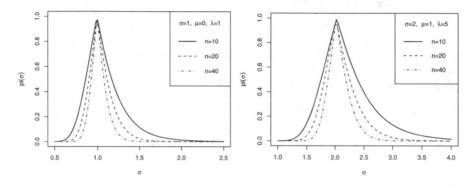

Fig. 2. Graphs of plausibility functions of σ based on simulated data when μ and λ are known.

A-step: In this case, let S^2 be the sample variance, i.e., $S^2 = \frac{1}{n-1} \sum_{i=1}^{n} (X_i - \bar{X})^2$. By Proposition 1, we know that

$$\frac{(n-1)S^2}{\sigma^2} \sim CS\chi_{n-1}^2(0, \mathbf{0}_n, \lambda^2(\mathbf{I}_n - \bar{\mathbf{J}}_n), \mathbf{I}_n + \lambda^2\bar{\mathbf{J}}_n).$$

So one association is

$$\frac{(n-1)S^2}{\sigma^2} = F_3^{-1}(U),$$

where F_3 is the c.d.f. of $CS\chi_{n-1}^2(0, \mathbf{0}_n, \lambda^2(\mathbf{I}_n - \bar{\mathbf{J}}_n), \mathbf{I}_n + \lambda^2\bar{\mathbf{J}}_n)$ and $U \sim$ Uniform$(0, 1)$. For any observation $s^2 > 0$ and $u \in [0, 1]$, we have

$$\Theta_{s^2}(\sigma) = \left\{ \sqrt{\frac{(n-1)s^2}{F_3^{-1}(u)}} \right\}.$$

P-step: To predict auxiliary variables U, let's still use the default predictive random set

$$S(U) = [0.5 - |U - 0.5|, 0.5 + |U - 0.5|].$$

C-step: By the P-step, we have

$$\Theta_{s^2}(S) = \left[\sqrt{\frac{(n-1)s^2}{F_3^{-1}(0.5 + |U - 0.5|)}}, \sqrt{\frac{(n-1)s^2}{F_3^{-1}(0.5 - |U - 0.5|)}} \right].$$

Thus, the following result can be obtained.

Theorem 3. *For any assertion* $A = \{\sigma\}$,

$$bel_{s^2}(A; S) = 0, \qquad pl_{s^2}(A; S) = 1 - \left| 2F_3\left(\frac{(n-1)s^2}{\sigma^2}\right) - 1 \right|,$$

and the $100(1 - \alpha)\%$ *plausibility region* $\Pi_{s^2}(\sigma)$ *is*

$$\sqrt{\frac{(n-1)s^2}{F_3^{-1}\left(1 - \frac{\alpha}{2}\right)}} < \sigma < \sqrt{\frac{(n-1)s^2}{F_3^{-1}\left(\frac{\alpha}{2}\right)}}.$$

Proof.

$$
\begin{aligned}
pl_{s^2}(A; S) &= 1 - bel_{s^2}(A^c; S) \\
&= 1 - P_S(\Theta_{s^2}(S) \subseteq A^c) \\
&= 1 - P_U\left(\sqrt{\frac{(n-1)s^2}{F_3^{-1}(0.5 - |U - 0.5|)}} < \sigma\right) - P_U\left(\sqrt{\frac{(n-1)s^2}{F_3^{-1}(0.5 + |U - 0.5|)}} > \sigma\right) \\
&= 1 - P_U\left(|U - 0.5| < \left|F_3\left(\frac{(n-1)s^2}{\sigma^2}\right) - 0.5\right|\right) \\
&= 1 - \left|2F_3\left(\frac{(n-1)s^2}{\sigma^2}\right) - 1\right|.
\end{aligned}
$$

By the definition, $\Pi_{s^2}(\sigma) = \{\sigma : pl_{s^2}(\sigma; S) > \alpha\}$. Let $pl_{s^2}(A; S) = 1 - \left|2F_3\left(\frac{s^2}{\sigma^2}\right) - 1\right| > \alpha$ and then solve it for σ. We have $\sqrt{\frac{(n-1)s^2}{F_3^{-1}\left(1 - \frac{\alpha}{2}\right)}} < \sigma < \sqrt{\frac{(n-1)s^2}{F_3^{-1}\left(\frac{\alpha}{2}\right)}}.$ $\qquad\square$

In Table 3 and Fig. 3, we provide a simulation study of this case with different σ, μ, λ and sample size n.

3.3 Simulation Study

In this subsection, we perform several simulation studies to compare average lower bounds (AL), average upper bounds (AU), average lengths (ALength) and coverage probabilities (CP) of 95% plausibility regions in three cases discussed above. We choose sample sizes of $10, 20$ and 40 for first two cases and sample sizes of $3, 5$ and 7 for the third case due to the difficulty of computations (See Remark 4 for more details). For each sample size, we simulated 10000 times for different parameters. We also provide graphs of plausibility functions based on simulated data with different parameters and different sample sizes.

Remark 4. *In this simulation study, we used n-dimensional spherical coordinates to derive the density function of closed skew chi-square distributions. However, our result is not efficient for computations. We are trying to derive a more concise expression of the density function.*

Table 3. Simulation results of AL, AU, ALength and CP of 95% plausibility regions of σ when μ is unknown but λ is known.

| $\sigma = 1, \mu = 0, \lambda = 1$ | | | | | $\sigma = 2, \mu = 1, \lambda = 5$ | | | | |
n	AL	AU	ALength	CP	n	AL	AU	ALength	CP
3	0.4607	5.6140	5.1533	0.9488	3	0.8574	11.8638	11.0063	0.9475
5	0.5614	2.7225	2.1611	0.9494	5	1.0595	5.8163	4.7568	0.9484
7	0.6246	2.1140	1.4893	0.9505	7	1.1744	4.5305	3.356	0.9509

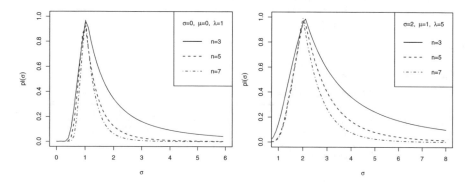

Fig. 3. Graphs of plausibility functions of σ based on simulated data when μ is unknown but λ is known.

4   Two-Sample Case

Let $\mathbf{X} = (X_1, \cdots, X_{n_1})'$ and $\mathbf{Y} = (Y_1, \cdots, Y_{n_2})'$ be independent random samples from $SN(\mu_1, \sigma_1^2, \lambda_1)$ and $SN(\mu_2, \sigma_2^2, \lambda_2)$, respectively. In this section, we are going to use $\mathbf{X}$ and $\mathbf{Y}$ to estimate the difference $\Delta\mu = \mu_1 - \mu_2$ when all other parameters are known.

A-step: By Proposition 2, we have

$$\bar{X} - \bar{Y} \sim CSN_{1,n_1+n_2}\left(\Delta\mu, \frac{\sigma_1^2}{n_1} + \frac{\sigma_2^2}{n_2}, D_\star, \Delta_\star\right),$$

where $D_\star$ and $\Delta_\star$ are given by Proposition 2. So we can obtain an association

$$\bar{X} - \bar{Y} = \Delta\mu + F_4^{-1}(U),$$

where $U \sim \text{Uniform}(0, 1)$ and F_4 is the c.d.f. of $CSN_{1,n_1+n_2}\left(0, \frac{\sigma_1^2}{n_1} + \frac{\sigma_2^2}{n_2}, D_\star, \Delta_\star\right)$. For any observation $\bar{x} - \bar{y} \in \mathbb{R}$ and $u \in [0, 1]$, we have $\Theta_{\bar{x}-\bar{y}}(\Delta\mu) = \{\bar{x} - \bar{y} - F_4^{-1}(u)\}$.

P-step: To predict auxiliary variables U, we use the default predictive random set

$$S(U) = [0.5 - |U - 0.5|, 0.5 + |U - 0.5|].$$

Table 4. Simulation results of AL, AU, ALength and CP of 95% plausibility regions for $\Delta\mu$ with different μ_i, σ_i, λ_i and sample sizes n_i, $i = 1, 2$.

$\mu_1 = 0, \sigma_1 = 1, \lambda_1 = 1$				$\mu_1 = 1, \sigma_1 = 2, \lambda_1 = 2$			
$\mu_2 = 0, \sigma_2 = 1, \lambda_2 = 1$				$\mu_2 = 0, \sigma_2 = 1, \lambda_2 = 1$			
$n_1 = n_2$	AL	AU	ALength	CP	$n_1 = n_2$	AL	AU	ALength	CP
10	-0.7265	0.7208	1.4473	0.9470	10	-0.0296	1.9862	2.0159	0.9532
15	-0.5892	0.5927	1.1819	0.9513	15	0.1725	1.8188	1.6463	0.9501
20	-0.5094	0.5139	1.0233	0.9550	20	0.2754	1.7008	1.4253	0.9509
$\mu_1 = 2, \sigma_1 = 2, \lambda_1 = 1$				$\mu_1 = 2, \sigma_1 = 2, \lambda_1 = 1$			
$\mu_2 = 1, \sigma_2 = 2, \lambda_2 = 2$				$\mu_2 = 0, \sigma_2 = 1, \lambda_2 = 1$			
$n_1 = n_2$	AL	AU	ALength	CP	$n_1 = n_2$	AL	AU	ALength	CP
10	-0.3435	2.3422	2.6857	0.9554	10	0.8559	3.1443	2.2887	0.9500
15	-0.0827	2.1100	2.1927	0.9482	15	1.0634	2.9320	1.8686	0.9512
20	0.0782	1.9583	1.8801	0.9478	20	1.1826	2.8011	1.6184	0.9510

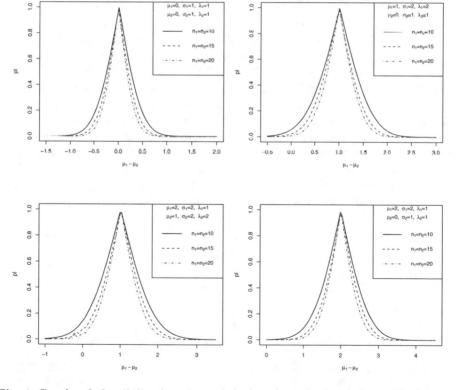

Fig. 4. Graphs of plausibility functions of $\Delta\mu$ based on simulated data with different μ_i, σ_i, λ_i and sample sizes n_i, $i = 1, 2$.

C-step: By the P-step, we have

$$\Theta_{\bar{x}-\bar{y}}(S) = \left[\bar{x} - \bar{y} - F_4^{-1}(0.5 + |U - 0.5|), \bar{x} - \bar{y} - F_4^{-1}(0.5 - |U - 0.5|) \right].$$

So we can use the above IM to get the following result.

Theorem 4. *For any assertion* $A = \{\Delta\mu\}$,

$$bel_{\bar{x}-\bar{y}}(A; S) = 0, \qquad pl_{\bar{x}-\bar{y}}(A; S) = 1 - |2F_4(\bar{x} - \bar{y} - \Delta\mu) - 1|,$$

and the $100(1 - \alpha)\%$ *plausibility region* $\Pi_{\bar{x}-\bar{y}}(\Delta\mu)$ *is*

$$\bar{x} - \bar{y} - F_4^{-1}\left(1 - \frac{\alpha}{2}\right) < \Delta\mu < \bar{x} - \bar{y} - F_4^{-1}\left(\frac{\alpha}{2}\right).$$

The proof of Theorem 4 is similar to the proof of Theorem 1. In Table 4 and Fig. 4, we provide a simulation study for this case with different μ_i, σ_i, λ_i and sample sizes n_i, $i = 1, 2$.

5   One Example

Lastly, we use one real data example to illustrate our results.

Example 5.1. *The data set was obtained from a study of leaf area index (LAI) of robinnia pseudoscacia in the Huaiping forest farm of Shannxi Province from June to October in 2010 (with permission of data owners). The LAI is given in Table 5 below. By [13], the LAI is approximately distributed as $SN(1.2585, 1.8332^2, 2.7966)$ via the method of moment estimation. We use Theorems 1 and 2 to explore the data again. For the location parameter μ, we assume MME $\hat{\sigma} = 1.8332$ and $\hat{\lambda} = 2.7966$ to be true parameters and then apply Theorem 1 to the data. The estimation of μ by IMs is 1.2602. Similarly, for the scale parameter σ, we assume MME $\hat{\mu} = 1.2585$ and $\hat{\lambda} = 2.7966$ to be true parameters and then apply Theorem 2 to the data. The estimation of σ by IMs is 1.8355. They are roughly equal to MME's. The corresponding graphs of plausibility functions of μ and σ are given in Fig. 5.*

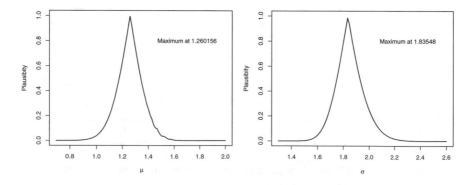

Fig. 5. Graphs of plausibility functions of μ and σ for Example 5.1.

Table 5. The observed values of LAI.

LAI			
June	July	September	October
4.87	3.32	2.05	1.50
5.00	3.02	2.12	1.46
4.72	3.28	2.24	1.55
5.16	3.63	2.56	1.27
5.11	3.68	2.67	1.26
5.03	3.79	2.61	1.37
5.36	3.68	2.42	1.87
5.17	4.06	2.58	1.75
5.56	4.13	2.56	1.81
4.48	2.92	1.84	1.98
4.55	3.05	1.94	1.89
4.69	3.02	1.95	1.71
2.54	2.78	2.29	1.29
3.09	2.35	1.94	1.34
2.79	2.40	2.20	1.29
3.80	3.28	1.56	1.10
3.61	3.45	1.40	1.04
3.53	2.85	1.36	1.08
2.51	3.05	1.60	0.86
2.41	2.78	1.50	0.70
2.80	2.72	1.88	0.82
3.23	2.64	1.63	1.19
3.46	2.88	1.66	1.24
3.12	3.00	1.62	1.14

Acknowledgments. We would like to thank Professor Hung T. Nguyen for introducing this interesting topic to us and referees for their valuable comments and suggestions which improve this paper.

References

1. Azzalini, A.: A class of distributions which includes the normal ones. Scand. J. Stat. **12**(2), 171–178 (1985)
2. Azzalini, A.: The Skew-Normal and Related Families, vol. 3. Cambridge University Press, Cambridge (2013)
3. Azzalini, A., Capitanio, A.: Statistical applications of the multivariate skew normal distribution. J. R. Stat. Soc. **61**(3), 579–602 (1999)

4. Azzalini, A., Dalla, V.A.: The multivariate skew-normal distribution. Biometrika **83**(4), 715–726 (1996)
5. González-Farías, G., Domínguez-Molina, A., Gupta, A.K.: Additive properties of skew normal random vectors. J. Stat. Plan. Infer. **126**(2), 521–534 (2004)
6. Martin, R.: Random sets and exact confidence regions. Sankhya A **76**(2), 288–304 (2014)
7. Martin, R., Lingham, R.T.: Prior-free probabilistic prediction of future observations. Technometrics **58**(2), 225–235 (2016)
8. Martin, R., Liu, C.: Inferential models: a framework for prior-free posterior probabilistic inference. J. Am. Stat. Assoc. **108**(501), 301–313 (2013)
9. Martin, R., Liu, C.: Inferential Models: Reasoning with Uncertainty, vol. 145. CRC Press, New York (2015)
10. Tian, W., Wang, T.: Quadratic forms of refined skew normal models based on stochastic representation. Random Oper. Stochast. Equ. **24**(4), 225–234 (2016)
11. Wang, T., Li, B., Gupta, A.K.: Distribution of quadratic forms under skew normal settings. J. Multivar. Anal. **100**(3), 533–545 (2009)
12. Wang, Z., Wang, C., Wang, T.: Estimation of location parameter in the skew normal setting with known coefficient of variation and skewness. Int. J. Intell. Technol. Appl. Stat. **9**(3), 191–208 (2016)
13. Ye, R., Wang, T.: Inferences in linear mixed models with skew-normal random effects. Acta Math. Sin. English Ser. **31**(4), 576–594 (2015)
14. Ye, R., Wang, T., Gupta, A.K.: Distribution of matrix quadratic forms under skew-normal settings. J. Multivar. Anal. **131**, 229–239 (2014)
15. Zhu, X., Ma, Z., Wang, T., Teetranont, T.: Plausibility regions on the skewness parameter of skew normal distributions based on inferential models. In: Robustness in Econometrics, pp. 267–286. Springer (2017)

Measures of Mutually Complete Dependence for Discrete Random Vectors

Xiaonan Zhu[1], Tonghui Wang[1(✉)], S. T. Boris Choy[2],
and Kittawit Autchariyapanitkul[3]

[1] Department of Mathematical Sciences, New Mexico State University,
Las Cruces, USA
{xzhu,twang}@nmsu.edu
[2] Discipline of Business Analytics, The University of Sydney, Sydney, Australia
boris.choy@sydney.edu.au
[3] Faculty of Economics, Maejo University, Chiang Mai, Thailand
Kittawit_a@mju.ac.th

Abstract. In this paper, a marginal-free measure of mutually complete dependence for discrete random vectors through subcopulas is defined, which generalizes the corresponding results for discrete random variables. Properties of the measure are studied and an estimator of the measure is introduced. Several examples are given for illustration of our results.

Keywords: Discrete random vector · Mutually complete dependence
Dependence measure · Subcopula

1 Introduction

Complete dependence (or functional dependence) is an important concept in many aspects of our life, such as econometrics, insurance, finance, etc. Recently, measures of (mutually) complete dependence have been defined and studied by many authors, e.g. Rényi [8], Schweizer and Wolff [9], Lancaster [5], Siburg and Stoimenov [12], Trutschnig [16], Dette et al. [3], Tasena and Dhompongsa [14], Shan et al. [11], Tasena and Dhompongsa [15] and Boonmee and Tasena [2]. However, measures in above papers have some drawbacks. Some measures only work for continuous random variables or vectors and some measures rely on marginal distributions (See Sect. 2 for a summary of several important measures). To the best of our knowledge, none of previously proposed measures are marginal-free and can describe (mutually) complete dependence for discrete random vectors. To overcome this issue, in this paper, we define a marginal-free measure of (mutually) complete dependence for discrete random vectors by using subcopluas, which extends the corresponding results of discrete random variables given in [11] to multivariate cases.

This paper is organized as follows. Some necessary concepts and definitions, and several measures of (mutually) complete dependence are reviewed briefly in Sect. 2. A marginal-free measure of (mutually) complete dependence for discrete random vectors is defined and properties of this measure are studied in Sect. 3. An estimator of the measure is introduced in Sect. 4.

© Springer International Publishing AG 2018
V. Kreinovich et al. (eds.), *Predictive Econometrics and Big Data*, Studies in Computational
Intelligence 753, https://doi.org/10.1007/978-3-319-70942-0_22

2 Preliminaries

Let $(\Omega, \mathscr{A}, P)$ be a probability space, where Ω is a sample space, $\mathscr{A}$ is a σ-algebra of Ω and P is a probability measure on $\mathscr{A}$. A *random variable* is a measurable function from Ω to the real line $\mathbb{R}$, and for any integer $n \geq 2$, an *n-dimensional random vector* is a measurable function from Ω to $\mathbb{R}^n$. For any $a = (a_1, \cdots, a_n)$ and $b = (b_1, \cdots, b_n) \in \mathbb{R}^n$, we say $a \leq b$ if and only if $a_i \leq b_i$ for all $i = 1, \cdots, n$. Let X and Y be random vectors defined on the same probability space. X and Y are said to be *independent* if and only if $P(X \leq x, Y \leq y) = P(X \leq x)P(Y \leq y)$ for all x and y. Y is *completely dependent* (CD) on X if Y is a measurable function of X almost surely, i.e., there is a measurable function ϕ such that $P(Y = \phi(X)) = 1$. X and Y are said to be *mutually completely dependent* (MCD) if X and Y are completely dependent on each other.

Let $E_1, \cdots, E_n$ be nonempty subsets of $\mathbb{R}$ and Q a real-valued function with the domain $Dom(Q) = E_1 \times \cdots \times E_n$. Let $[a, b] = [a_1, b_1] \times \cdots \times [a_n, b_n]$ such that all vertices of $[a, b]$ belong to $Dom(Q)$. The *Q-volume of* $[a, b]$ is defined by

$$\mathscr{V}_Q([a, b]) = \sum sgn(c)Q(c),$$

where the sum is taken over all vertices $c = (c_1, \cdots, c_n)$ of $[a, b]$, and

$$sgn(c) = \begin{cases} 1, & \text{if } c_i = a_i \text{ for an even number of } i's, \\ -1, & \text{if } c_i = a_i \text{ for an odd number of } i's. \end{cases}$$

An *n-dimensional subcopula* (or *n-subcopula* for short) is a function C with the following properties [7].

(i) The domain of C is $Dom(C) = D_1 \times \cdots \times D_n$, where $D_1, \cdots, D_n$ are nonempty subsets of the unit interval $I = [0, 1]$ containing 0 and 1;
(ii) C is *grounded*, i.e., for any $u = (u_1, \cdots, u_n) \in Dom(C)$, $C(u) = 0$ if at least one $u_i = 0$;
(iii) For any $u_i \in D_i$, $C(1, \cdots, 1, u_i, 1, \cdots, 1) = u_i$, $i = 1, \cdots, n$;
(iv) C is *n-increasing*, i.e., for any $u, v \in Dom(C)$ such that $u \leq v$, $\mathscr{V}_C([u, v]) \geq 0$.

For any n random variables $X_1, \cdots, X_n$, by Sklar's Theorem [13], there is a unique n-subcopula such that

$$H(x_1, \cdots, x_n) = C(F_1(x_1), \cdots, F_n(x_n)), \quad \text{for all } (x_1, \cdots, x_n) \in \overline{\mathbb{R}}^n,$$

where $\overline{\mathbb{R}} = \mathbb{R} \cup \{-\infty, \infty\}$, H is the joint cumulative distribution function (c.d.f.) of $X_1, \cdots, X_n$, and F_i is the marginal c.d.f. of X_i, $i = 1, \cdots, n$. In addition, if $X_1, \cdots, X_n$ are continuous, then $Dom(C) = I^n$ and the unique C is called the *n-copula* of $X_1, \cdots, X_n$. For more details about the copula theory, see [7].

Next, we are going to recall some measures of MCD and CD, which are equal to 0 if and only if two random variables (or vectors) are independent, and equal to 1 if and only if they are MCD or CD. In 2010, Siburg and Stoimenov [12] defined an MCD measure for continuous random variables as

$$\omega(X,Y) = \left(3\|C\|^2 - 2\right)^{\frac{1}{2}}, \tag{1}$$

where X and Y are continuous random variables with the copula C and $\|\cdot\|$ is the Sobolev norm of bivariate copulas given by

$$\|C\| = \left(\int\int |\nabla C(u,v)|^2 \, du \, dv\right)^{\frac{1}{2}},$$

where $\nabla C(u,v)$ is the gradient of $C(u,v)$.

In 2013, Tasena and Dhompongsa [14] generalized Siburg and Stoimenov's measure to multivariate cases as follows. Let $X_1, \cdots, X_n$ be continuous variables with the n-copula C. Define

$$\delta_i(X_1, \cdots, X_n) = \delta_i(C) = \frac{\int \cdots \int [\partial_i C(u_1, \cdots, u_n) - \pi_i C(u_1, \cdots, u_n)]^2 \, du_1 \cdots du_n}{\int \cdots \int \pi_i C(u_1, \cdots, u_n)(1 - \pi_i C(u_1, \cdots, u_n)) du_1 \cdots du_n},$$

where $\partial_i C$ is the partial derivative on the ith coordinate of C and $\pi_i C : I^{n-1} \to I$ is defined by $\pi_i C(u_1, \cdots, u_{n-1}) = C(u_1, \cdots, u_{i-1}, 1, u_i, \cdots, u_{n-1})$, $i = 1, 2, \cdots, n$. Let

$$\delta(X_1, \cdots, X_n) = \delta(C) = \frac{1}{n} \sum_{i=1}^n \delta_i(C). \tag{2}$$

Then δ is an MCD measure of $X_1, \cdots, X_n$.

In 2015, Shan et al. [11] considered discrete random variables. Let X and Y be two discrete random variables with the subcopula C. An MCD measure of X and Y is given by

$$\mu_t(X,Y) = \left(\frac{\|C\|_t^2 - L_t}{U_t - L_t}\right)^{\frac{1}{2}}, \tag{3}$$

where $t \in [0,1]$ and $\|C\|_t^2$ is the discrete norm of C defined by

$$\|C\|_t^2 = \sum_i \sum_j \left\{ \left(tC_{\Delta i,j}^2 + (1-t)C_{\Delta i,j+1}^2\right)\frac{\Delta v_j}{\Delta u_i} + \left(tC_{i,\Delta j}^2 + (1-t)C_{i+1,\Delta j}^2\right)\frac{\Delta u_i}{\Delta v_j} \right\},$$

$$C_{\Delta i,j} = C(u_{i+1}, v_j) - C(u_i, v_j), \quad C_{i,\Delta j} = C(u_i, v_{j+1}) - C(u_i, v_j),$$

$$\Delta u_i = u_{i+1} - u_i, \quad \Delta v_j = v_{j+1} - v_j,$$

$$L_t = \sum_i (tu_i^2 + (1-t)u_{i+1}^2)\Delta u_i + \sum_j (tv_j^2 + (1-t)v_{j+1}^2)\Delta v_j,$$

and

$$U_t = \sum_i (tu_i + (1-t)u_{i+1})\Delta u_i + \sum_j (tv_j + (1-t)v_{j+1})\Delta v_j.$$

In 2016 Tasena and Dhompongsa [15] defined a measure of CD for random vectors. Let X and Y be two random vectors. Define

$$\omega_k(Y|X) = \left[\int \int \left| F_{Y|X}(y|x) - \frac{1}{2} \right|^k dF_X(x)dF_Y(y) \right]^{\frac{1}{k}},$$

where $k \geq 1$. The measure of Y CD on X is given by

$$\overline{\omega}_k(Y|X) = \left[\frac{\omega_k^k(Y|X) - \omega_k^k(Y^\perp|X^\perp)}{\omega_k^k(Y|Y) - \omega_k^k(Y^\perp|X^\perp)} \right]^{\frac{1}{k}}, \tag{4}$$

where $X^\perp$ and $Y^\perp$ are independent random vectors with the same distributions as X and Y, respectively.

In the same period, Boonmee and Tasena [2] defined a measure of CD for continuous random vectors by using *linkages* which were introduced by Li et al. [6]. Let X and Y be two continuous random vectors with the linkage C. The measure of Y being completely dependent on X is defined by

$$\zeta_p(Y|X) = \left[\int \int \left| \frac{\partial}{\partial u} C(u,v) - \Pi(v) \right|^p dudv \right]^{\frac{1}{p}}, \tag{5}$$

where $\Pi(v) = \prod_{i=1}^{n} v_i$ for all $v = (v_1, \cdots, v_n) \in I^n$.

From above summaries we can see that measures given by (1), (2) and (5) only work for continuous random variables or vectors. The measure defined by (3) only works for bivariate discrete random variables. The measure given by (4) relies on marginal distributions of random vectors. Thus it is worth considering marginal-free measures of CD and MCD for discrete random vectors.

3   An MCD Measure for Discrete Random Vectors

In this section, we identify $\mathbb{R}^{n_1+\cdots+n_k}$ with $\mathbb{R}^{n_1} \times \cdots \times \mathbb{R}^{n_k}$, where $n_1, \cdots, n_k$ are positive integers. So any n-dimensional random vector X can be viewed as a tuple of n random variables, i.e., $X = (X_1, \cdots, X_n)$, where $X_1, \cdots, X_n$ are random variables. Also, if $Y = (Y_1, \cdots, Y_m)$ is an m-dimensional random vector, we use (X,Y) to denote the $(n+m)$-dimensional random vector $(X_1, \cdots, X_n, Y_1, \cdots, Y_m)$. Let $\psi = (\psi_1, \cdots, \psi_n) : \mathbb{R}^n \to \mathbb{R}^n$ be a function. ψ is said to be strictly increasing if and only if each component $\psi_i : \mathbb{R} \to \mathbb{R}$ is strictly increasing, i.e., for any a_i and $b_i \in \mathbb{R}$ such that $a_i < b_i$, we have $\psi_i(a_i) < \psi_i(b_i)$, $i = 1, \cdots, n$.

We will focus on *discrete random vectors* in this section, i.e., they can take on at most a countable number of possible values. Let $X = (X_1, \cdots, X_n) \in \mathscr{L}_1 \subseteq \mathbb{R}^n$ and $Y = (Y_1, \cdots, Y_m) \in \mathscr{L}_2 \subseteq \mathbb{R}^m$ be two discrete random vectors defined on the same probability space $(\Omega, \mathscr{A}, P)$. Their *joint* c.d.f. H, *marginal* c.d.f.'s F and G, and *marginal probability mass functions* (p.m.f.) f and g are defined, respectively, as follows.

$$H(x,y) = P(X \leq x, Y \leq y), \quad F(x) = P(X \leq x), \quad G(y) = P(Y \leq y),$$

$$f(x) = P(X = x), \quad \text{and} \quad g(y) = P(Y = y), \quad \text{for all} x \in \mathbb{R}^n \text{and} y \in \mathbb{R}^m.$$

Also, we use F_i and f_i to denote the marginal c.d.f. and p.m.f. of the ith component X_i of X, $i = 1, \cdots, n$, and use G_j and g_j to denote the marginal c.d.f. and p.m.f. of the jth component Y_j of Y, $j = 1, \cdots, m$, respectively. For simplicity, we assume that $f(x) \neq 0$ and $g(y) \neq 0$ for all $x \in \mathscr{L}_1$ and $y \in \mathscr{L}_2$. If C is the subcopula of the $(n+m)$-dimensional random vector (X,Y), i.e.,

$$H(x,y) = C(u(x), v(y)), \quad \text{for all} x \in \overline{\mathbb{R}}^n \text{and} y \in \overline{\mathbb{R}}^m,$$

where

$$u(x) = (u_1(x_1), \cdots, u_n(x_n)) = (F_1(x_1), \cdots, F_n(x_n)) \in I^n,$$

and

$$v(y) = (v_1(y_1), \cdots, v_m(y_m)) = (G_1(y_1), \cdots, G_m(y_m)) \in I^m,$$

for all $x \in \mathscr{L}_1$ and $y \in \mathscr{L}_2$, then C is said to be the *subcopula of X and Y*. In addition, for each vector $e \in E_1 \times \cdots \times E_n$, where E_i is a countable subset of $\overline{\mathbb{R}}$, let e_L be the *greatest lower bound* of e with respect to the coordinate wise order, i.e., $e_L = (e'_1, \cdots, e'_n)$ such that if there exists some element in E_i that is less than e_i, then e'_i is the greatest element in E_i so that $e'_i < e_i$, otherwise $e'_i = e_i$, $i = 1, \cdots, n$. We use 1_n and ∞_n to denote the n-dimensional constant vector $(1, \cdots, 1)$ and $(\infty, \cdots, \infty) \in \overline{\mathbb{R}}^n$.

To construct desired measures, a distance between two discrete random vectors is defined as follows.

Definition 1. Let X and Y be discrete random vectors. The *distance between the conditional distribution of Y given X and marginal distribution of Y* is defined by

$$\omega^2(Y|X) = \sum_{y \in \mathscr{L}_2, x \in \mathscr{L}_1} [P(Y \leq y|X = x) - G(y)]^2 f(x)g(y). \tag{6}$$

From the above definition, we can obtain the following two results.

Lemma 1. *For any discrete random vectors X and Y, we have $\omega^2(Y|X) \leq \omega^2_{max}(Y|X)$, where*

$$\omega^2_{max}(Y|X) = \sum_{y \in \mathscr{L}_2} \left[G(y) - (G(y))^2 \right] g(y).$$

Proof. By the definition,

$$\omega^2(Y|X) = \sum_{y \in \mathscr{L}_2, x \in \mathscr{L}_1} [P(Y \leq y|X = x) - G(y)]^2 f(x)g(y)$$

$$= \sum_{y \in \mathscr{L}_2, x \in \mathscr{L}_1} [P(Y \leq y|X = x)^2 - 2P(Y \leq y|X = x)G(y) + (G(y))^2] f(x)g(y)$$

$$\leq \sum_{y \in \mathscr{L}_2, x \in \mathscr{L}_1} [P(Y \leq y|X = x) - 2P(Y \leq y|X = x)G(y) + (G(y))^2] f(x)g(y)$$

$$= \sum_{y \in \mathscr{L}_2, x \in \mathscr{L}_1} [P(X = x, Y \le y) - 2P(X = x, Y \le y)G(y) + (G(y))^2 f(x)]g(y)$$

$$= \sum_{y \in \mathscr{L}_2} \left[G(y) - (G(y))^2 \right] g(y).$$

$\square$

Lemma 2. *Let X and Y be discrete random vectors. There is a function $\phi : \mathscr{L}_1 \to \mathscr{L}_2$ such that $\phi(X) = Y$ if and only if $\omega^2(Y|X) = \omega_{max}^2(Y|X)$, i.e., $P(Y \le y|X = x) = 0$ or 1 for all $(x, y) \in \mathscr{L}_1 \times \mathscr{L}_2$.*

Proof. For "if" part, suppose that $P(Y \le y|X = x) = 0$ or 1 for all $(x, y) \in \mathscr{L}_1 \times \mathscr{L}_2$. Then $\sum_{t \le y} P(Y = t|X = x) = 0$ or 1. So $\sum_{t \le y} P(X = x, Y = t) = 0$ or $P(X = x)$. Thus there exists a unique $y(x) \in \mathscr{L}_2$, which depends on x, such that $P(X = x, Y = y(x)) = P(X = x)$, and $P(X = x, Y = y) = 0$ for all $y \in \mathscr{L}_2$ with $y \ne y(x)$. Now if we define $\phi(x) = y(x)$ for all $x \in \mathscr{L}_1$, then $\phi(X) = Y$.

For "only if" part, suppose that $\phi(X) = Y$. Fix $x \in \mathscr{L}_1$. It is sufficient to show that $P(X = x, Y = y) = 0$ for all $y \ne \phi(x)$. Suppose that, on the contrary, there is $y' \in \mathscr{L}_2$ such that $y' \ne \phi(x)$ and $P(X = x, Y = y') \ne 0$, then there exists $\omega \in \Omega$ so that $X(\omega) = x$ and $Y(\omega) = y'$. So we have $\phi(X)(\omega) \ne Y(\omega)$. It's a contradiction. $\square$

Now we can define a measure of CD for two discrete random vectors as follows.

Definition 2. For any discrete random vectors X and Y, the *measure of Y being completely dependent on X* is given by

$$\mu(Y|X) = \left[\frac{\omega^2(Y|X)}{\omega_{max}^2(Y|X)} \right]^{\frac{1}{2}} = \left[\frac{\sum_{y \in \mathscr{L}_2, x \in \mathscr{L}_1} [P(Y \le y|X = x) - G(y)]^2 f(x)g(y)}{\sum_{y \in \mathscr{L}_2} \left[G(y) - (G(y))^2 \right] g(y)} \right]^{\frac{1}{2}}. \tag{7}$$

Properties of the measure $\mu(Y|X)$ are given as follows.

Theorem 1. *For any discrete random vectors X and Y, the measure $\mu(Y|X)$ has the following properties:*

(i) $0 \le \mu(Y|X) \le 1$;
(ii) $\mu(Y|X) = 0$ if and only if X and Y are independent;
(iii) $\mu(Y|X) = 1$ if and only if Y is a function of X;
(iv) $\mu(Y|X)$ is invariant under strictly increasing transformations of X and Y, i.e., if ψ_1 and ψ_2 are strictly increasing functions defined on $\mathscr{L}_1$ and $\mathscr{L}_2$, respectively, then $\mu(\psi_2(Y)|\psi_1(X)) = \mu(Y|X)$.

Proof. Property (i) is obvious by Lemma 1. For Property (ii), note that $\mu(Y|X) = 0$ if and only if $\omega^2(Y|X) = 0$. It is equivalent to $P(Y \le y|X = x) = G(y)$ for all $(x, y) \in \mathscr{L}_1 \times \mathscr{L}_2$, i.e., X and Y are independent. Property (iii) follows Lemma 2. Lastly, since ψ_1 and ψ_2 are strictly increasing, we have $P(\psi_1(X) \le \psi_1(x)) = P(X \le x)$ and $P(\psi_2(Y) \le \psi_2(y)) = P(Y \le y)$. Thus Property (iv) holds. $\square$

Remark 1. (i) It can be shown that the measure given by (7) is a discrete version of the measure $\overline{\omega}_k(Y|X)$ for random vectors given by (4) with $k = 2$. The difference here is that based on (7), we are going to define a marginal-free measure by using subcopulas for discrete random vectors.

(ii) The above measure may be simplified into

$$
\mu'(Y|X) = \left[\frac{\displaystyle\sum_{y \in \mathscr{L}_2, x \in \mathscr{L}_1} [P(Y \leq y | X = x) - G(y)]^2 f(x)}{\displaystyle\sum_{y \in \mathscr{L}_2} [G(y) - (G(y))^2]} \right]^{\frac{1}{2}}.
\tag{8}
$$

As indicated by Shan [10], $\mu'(Y|X)$ is well defined only if Y is a finite discrete random vector, i.e., if $\mathscr{L}_2$ is a finite set. Otherwise, $\sum_{y \in \mathscr{L}_2} [G(y) - (G(y))^2]$ may diverge. However, the measure $\mu(Y|X)$ given by (7) is well defined for all discrete random vectors.

(iii) It is easy to see that $\omega_{max}^2(Y|X) = 0$ if and only if Y is a constant random vector, i.e., if and only if there is $y \in \mathbb{R}^m$ such that $P(Y = y) = 1$. In this case, Y is clearly a function of X. Thus, without loss of generality, we assume that X and Y are not constant random vectors.

Since most multivariate dependence properties of random variables can be determined by their subcopula C, we are going to redefine the measure $\mu(Y|X)$ by using subcopulas such that $\mu(Y|X)$ is free of marginal distributions of X and Y. First, note that for any $x \in \mathscr{L}_1$ and $y \in \mathscr{L}_2$, we have

$$
G(y) = H(\infty_n, y) = C(1_n, v(y)),
\tag{9}
$$

$$
f(x) = \mathscr{V}_H([(x_L, \infty_m), (x, \infty_m)]) = \mathscr{V}_C([(u(x)_L, 1_m), (u(x), 1_m)]),
\tag{10}
$$

$$
g(y) = \mathscr{V}_H([(\infty_n, y_L), (\infty_n, y)]) = \mathscr{V}_C([(1_n, v(y)_L), (1_n, v(y))]),
\tag{11}
$$

and

$$
\begin{aligned}
P(Y \leq y | X = x) &= \frac{P(X = x, Y \leq y)}{P(X = x)} \\
&= \frac{\mathscr{V}_H([(x_L, y), (x, y)])}{\mathscr{V}_H([(x_L, \infty_m), (x, \infty_m)])} \\
&= \frac{\mathscr{V}_C([(u(x)_L, v(y)), (u(x), v(y))])}{\mathscr{V}_C([(u(x)_L, 1_m), (u(x), 1_m)])}
\end{aligned}
\tag{12}
$$

Thus, from Eqs. (9)–(12), we can redefine $\mu(Y|X)$ as follows.

Definition 3. Let X and Y be two discrete random vectors with the subcopula C. Suppose that the domain of C is $Dom(C) = \mathscr{L}_1' \times \mathscr{L}_2'$, where $\mathscr{L}_1' \subseteq I^n$ and $\mathscr{L}_2' \subseteq I^m$. The *measure of Y being completely dependent on X based on C* is given by

$$\mu_C(Y|X) = \left[\frac{\omega^2(Y|X)}{\omega_{max}^2(Y|X)} \right]^{\frac{1}{2}}$$

$$= \left[\frac{\sum\limits_{v \in \mathscr{L}_2'} \sum\limits_{u \in \mathscr{L}_1'} \left[\frac{\mathscr{V}_C([(u_L,v),(u,v)])}{\mathscr{V}_C([(u_L,1_m),(u,1_m)])} - C(1_n,v) \right]^2 \mathscr{V}_C([(u_L,1_m),(u,1_m)]) \mathscr{V}_C([(1_n,v_L),(1_n,v)])}{\sum\limits_{v \in \mathscr{L}_2'} [C(1_n,v) - (C(1_n,v))^2] \mathscr{V}_C([(1_n,v),(1_n,v_L)])} \right]^{\frac{1}{2}}. \tag{13}$$

Remark 2. Based on the same idea, if X and Y are continuous random vectors with the unique copula C, the measure $\mu_C(Y|X)$ given by (13) can be rewritten as

$$\mu_C(Y|X)) = \left[\frac{\int\int \left(\frac{\partial C}{\partial C_X} - C_Y \right)^2 \frac{\partial C_X}{\partial u} \frac{\partial C_Y}{\partial v} du dv}{\int C_Y (1 - C_Y) \frac{\partial C_Y}{\partial v} dv} \right]^{\frac{1}{2}},$$

where C_X and C_Y are copulas of X and Y. This is a marginal-free measure of CD for continuous random vectors.

By using $\mu_C(Y|X)$ defined in Definition 3, we can define a marginal-free measure of mutual complete dependence for two discrete random vectors as follows.

Definition 4. For any discrete random vectors X and Y with the subcopula C, the *MCD measure of X and Y* is defined by

$$\mu_C(X,Y) = \left[\frac{\omega^2(Y|X) + \omega^2(X|Y)}{\omega_{max}^2(Y|X) + \omega_{max}^2(X|Y)} \right]^{\frac{1}{2}}, \tag{14}$$

where $\omega^2(\cdot|\cdot)$ and $\omega_{max}^2(\cdot|\cdot)$ are the same as those given in Definition 3.

The properties of the measure $\mu_C(X,Y)$ are given in the following theorem. The proof is straightforward.

Theorem 2. *Let X and Y be two discrete random vectors with the subcopula C. The measure $\mu_C(X,Y)$ has following properties,*

(i) $\mu_C(X,Y) = \mu_C(Y,X)$;
(ii) $0 \le \mu_C(X,Y) \le 1$;
(iii) $\mu_C(X,Y) = 0$ *if and only if X and Y are independent;*
(iv) $\mu_C(X,Y) = 1$ *if and only if X and Y are MCD;*
(v) $\mu_C(X,Y)$ *is invariant under strictly increasing transformations of X and Y.*

Remark 3. (i) The insufficiency of 2-copulas to describe joint distributions with given multivariate marginal distributions was discussed by Genest et al. [4]. Let C be a 2-copula. They showed that

$$H(x_1, \cdots, x_{n_1}, x_{n_1+1}, \cdots, x_{n_1+n_2}) = C(H_1(x_1, \cdots, x_{n_1}), H_2(x_{n_1+1}, \cdots, x_{n_1+n_2}))$$

defines a $(n_1 + n_2)$ dimensional c.d.f., where $n_1 + n_2 \geq 3$, for all marginal c.d.f.'s H_1 and H_2 with dimensions n_1 and n_2, respectively, only if C is the bivariate independence copula, i.e.,

$$C(u,v) = uv, \quad \text{for all } (u,v) \in I^2.$$

Thus, in this work, we have to use the $(n+m)$-subcopula of (X,Y) to construct a marginal-free measure.

(ii) Both Shan et al. [11] and Tasena and Dhompongsa [15] tried to use copulas to construct measures of functional dependence for discrete random variables or vectors. However, we do not think that copulas should be used to construct measures for discrete random variables or vectors because, for fixed discrete random variables or vectors, the corresponding copulas may not be unique. Thus, as shown in their papers, if we have different copulas for two fixed discrete random variables, copula-based measures may give us different results.

(iii) Boonmee and Tasena [2] used linkages to construct a marginal-free measure of CD for continuous random vectors, but linkages have some defects. First, linkages are defined for continuous random vectors. Second, to find the linkage of two random vectors, they need to be transformed to uniform random vectors. It is not convenient in applications (See Li et al. [6] for more details of linkages). Thus, in this work, we prefer to use the subcopula of (X,Y) to construct marginal-free measures, since subcopulas are not only good for discrete random vectors but also more popular than linkages.

(iv) If both X and Y are discrete random variables with the 2-subcopula C, then we have

$$\omega^2(Y|X) = \sum_{v \in \mathscr{L}_2' u \in \mathscr{L}_1'} \left[\frac{C(u,v) - C(u_L,v)^2}{u - u_L} - v \right]^2 (u - u_L)(v - v_L),$$

$$\omega^2(X|Y) = \sum_{u \in \mathscr{L}_1' v \in \mathscr{L}_2'} \left[\frac{C(u,v) - C(u,v_L)^2}{v - v_L} - u \right]^2 (u - u_L)(v - v_L),$$

$$\omega_{max}^2(Y|X) = \sum_{v \in \mathscr{L}_2'} (v - v^2)(v - v_L) \quad \text{and} \quad \omega_{max}^2(X|Y) = \sum_{u \in \mathscr{L}_1'} (u - u^2)(u - u_L).$$

In this case, the measure $\mu_C(X,Y) = \left[\frac{\omega^2(Y|X) + \omega^2(X|Y)}{\omega_{max}^2(Y|X) + \omega_{max}^2(X|Y)} \right]^{\frac{1}{2}}$ is identical to the measure given by (3) with $t = 0$.

(v) If both X and Y are continuous random variables, i.e., $\max\{u - u_L, v - v_L\} \to 0$, then it can be show that

$$\mu_C(X,Y) = \left[\frac{\omega^2(Y|X) + \omega^2(X|Y)}{\omega_{max}^2(Y|X) + \omega_{max}^2(X|Y)} \right]^{\frac{1}{2}}$$

$$= \left\{ 3 \int \int \left[\left(\frac{\partial C}{\partial u} \right)^2 + \left(\frac{\partial C}{\partial v} \right)^2 \right] du dv - 2 \right\}^{\frac{1}{2}},$$

which is identical to the measure given by (1).

Next, we use two examples to illustrate our above results.

Example 1. Let $X = (X_1, X_2)$ be a random vector with the distribution given in Table 1. Let $Y = (Y_1, Y_2) = (X_1^2, X_2^2)$. Then the distribution of Y, the joint distribution of X and Y, and the corresponding subcopula are given in Tables 2, 3 and 4, respectively. It is easy to show that $\omega^2(Y|X) = \omega_{max}^2(Y|X) = 161/1458$, $\omega^2(X|Y) = 2699/38880$, $\omega_{max}^2(X|Y) = 469/2916$. So $\mu_C(Y|X) = 1$. $\mu_C(X|Y) = 0.6569$ and $\mu_C(X,Y) = 0.8142$.

Table 1. Distribution of X.

X_2	X_1			X_2
	-1	0	1	
-1	1/18	2/18	3/18	6/18
0	1/18	2/18	2/18	5/18
1	1/18	3/18	3/18	7/18
X_1	3/18	7/18	8/18	1

Table 2. Distribution of Y.

Y_2	Y_1		Y_2
	0	1	
0	2/18	3/18	5/18
1	5/18	8/18	13/18
Y_1	7/18	11/18	1

Table 3. Joint distribution of X and Y.

Y	X									Y
	$(-1,1)$	$(-1,0)$	$(-1,1)$	$(0,-1)$	$(0,0)$	$(0,1)$	$(1,-1)$	$(1,0)$	$(1,1)$	
$(0,0)$	0	0	0	0	$\frac{2}{18}$	0	0	0	0	$\frac{2}{18}$
$(0,1)$	0	0	0	$\frac{2}{18}$	0	$\frac{3}{18}$	0	0	0	$\frac{5}{18}$
$(1,0)$	0	$\frac{1}{18}$	0	0	0	0	0	$\frac{2}{18}$	0	$\frac{3}{18}$
$(1,1)$	$\frac{1}{18}$	0	$\frac{1}{18}$	0	0	0	$\frac{3}{18}$	0	$\frac{3}{18}$	$\frac{8}{18}$
X	$\frac{1}{18}$	$\frac{1}{18}$	$\frac{1}{18}$	$\frac{2}{18}$	$\frac{2}{18}$	$\frac{3}{18}$	$\frac{3}{18}$	$\frac{2}{18}$	$\frac{3}{18}$	1

Table 4. Subcopula of X and Y.

V	U								$(1,1)$
	$\left(\frac{3}{18},\frac{6}{18}\right)$	$\left(\frac{3}{18},\frac{11}{18}\right)$	$\left(\frac{3}{18},1\right)$	$\left(\frac{10}{18},\frac{6}{18}\right)$	$\left(\frac{10}{18},\frac{11}{18}\right)$	$\left(\frac{10}{18},1\right)$	$\left(1,\frac{6}{18}\right)$	$\left(1,\frac{11}{18}\right)$	
$\left(\frac{7}{18},\frac{5}{18}\right)$	0	0	0	0	$\frac{2}{18}$	$\frac{2}{18}$	0	$\frac{2}{18}$	$\frac{2}{18}$
$\left(\frac{7}{18},1\right)$	0	0	0	$\frac{2}{18}$	$\frac{4}{18}$	$\frac{7}{18}$	$\frac{2}{18}$	$\frac{4}{18}$	$\frac{7}{18}$
$\left(1,\frac{5}{18}\right)$	0	$\frac{1}{18}$	$\frac{1}{18}$	0	$\frac{3}{18}$	$\frac{3}{18}$	0	$\frac{5}{18}$	$\frac{5}{18}$
$(1,1)$	$\frac{1}{18}$	$\frac{2}{18}$	$\frac{3}{18}$	$\frac{3}{18}$	$\frac{6}{18}$	$\frac{10}{18}$	$\frac{6}{18}$	$\frac{11}{18}$	1

Example 2. Let $X = (X_1, X_2)$ be a discrete random vector, where X_1 is a geometric random variable with the success rate $p = \frac{1}{2}$, X_2 is a binomial random variable with the number of trials $n = 2$ and the success rate $p = \frac{1}{2}$, and X_1 and X_2 are independent. Let $Y = X_1 - X_2$. Then the joint distribution and subcopula of X and Y are given in Tables 5 and 6. By calculation, $\omega^2(Y|X) = \omega_{max}^2(Y|X) = 1223/7168$, $\omega^2(X|Y) = 3407543/30965760$ and $\omega_{max}^2(X|Y) = 1/3$. So $\mu_C(Y|X) = 1$, $\mu_C(X|Y) = 0.3301$ and $\mu_C(X,Y) = 0.5569$.

Table 5. Joint distribution of X and Y.

X	Y							$\cdots$	X
	-1	0	1	2	3	4	5		
$(1,0)$	0	0	$\frac{1}{2^3}$	0	0	0	0		$\frac{1}{2^3}$
$(1,1)$	0	$\frac{1}{2^2}$	0	0	0	0	0		$\frac{1}{2^2}$
$(1,2)$	$\frac{1}{2^3}$	0	0	0	0	0	0		$\frac{1}{2^3}$
$(2,0)$	0	0	0	$\frac{1}{2^4}$	0	0	0		$\frac{1}{2^4}$
$(2,1)$	0	0	$\frac{1}{2^3}$	0	0	0	0		$\frac{1}{2^3}$
$(2,2)$	0	$\frac{1}{2^4}$	0	0	0	0	0		$\frac{1}{2^4}$
$(3,0)$	0	0	0	0	$\frac{1}{2^5}$	0	0		$\frac{1}{2^5}$
$(3,1)$	0	0	0	$\frac{1}{2^4}$	0	0	0		$\frac{1}{2^4}$
$(3,2)$	0	0	$\frac{1}{2^5}$	0	0	0	0		$\frac{1}{2^5}$
$(4,0)$	0	0	0	0	0	$\frac{1}{2^6}$	0		$\frac{1}{2^6}$
$(4,1)$	0	0	0	0	$\frac{1}{2^5}$	0	0		$\frac{1}{2^5}$
$(4,2)$	0	0	0	$\frac{1}{2^6}$	0	0	0		$\frac{1}{2^6}$
$(5,0)$	0	0	0	0	0	0	$\frac{1}{2^7}$		$\frac{1}{2^7}$
$(5,1)$	0	0	0	0	0	$\frac{1}{2^6}$	0		$\frac{1}{2^6}$
$(5,2)$	0	0	0	0	$\frac{1}{2^7}$	0	0		$\frac{1}{2^7}$
$\vdots$									$\vdots$
Y	$\frac{1}{2^3}$	$\frac{2^2+1}{2^4}$	$\frac{2^3+1}{2^5}$	$\frac{2^3+1}{2^6}$	$\frac{2^3+1}{2^7}$	$\frac{2^3+1}{2^8}$	$\frac{2^3+1}{2^9}$	$\cdots$	1

Table 6. Subcopula of X and Y.

U	V					...
	$\frac{1}{2^3}$	$\frac{2^2+2+1}{2^4}$	$\frac{2^4+2^2+2+1}{2^5}$	$\frac{2^5+2^4+2^2+2+1}{2^6}$	$\frac{2^6+2^5+2^4+2^2+2+1}{2^7}$	
$(\frac{1}{2}, \frac{1}{2^2})$	0	0	$\frac{1}{2^3}$	$\frac{1}{2^3}$	$\frac{1}{2^3}$	
$(\frac{1}{2}, \frac{3}{2^2})$	0	$\frac{1}{2^2}$	$\frac{2+1}{2^3}$	$\frac{2+1}{2^3}$	$\frac{2+1}{2^3}$	
$(\frac{1}{2}, 1)$	$\frac{1}{2^3}$	$\frac{2+1}{2^3}$	$\frac{1}{2}$	$\frac{1}{2}$	$\frac{1}{2}$	
$(\frac{2+1}{2^2}, \frac{1}{2^2})$	0	0	$\frac{1}{2^3}$	$\frac{2+1}{2^4}$	$\frac{2+1}{2^4}$	
$(\frac{2+1}{2^2}, \frac{3}{2^2})$	0	$\frac{1}{2^2}$	$\frac{1}{2}$	$\frac{2^2+1}{2^4}$	$\frac{2^2+1}{2^4}$	
$(\frac{2+1}{2^2}, 1)$	$\frac{1}{2^3}$	$\frac{2^2+2+1}{2^4}$	$\frac{6+2^2+1}{2^4}$	$\frac{2+1}{2^2}$	$\frac{2+1}{2^2}$	
$(\frac{2^2+2+1}{2^2}, \frac{1}{2^2})$	0	0	$\frac{1}{2^3}$	$\frac{2+1}{2^4}$	$\frac{2^2+2+1}{2^5}$	
$(\frac{2^2+2+1}{2^2}, \frac{3}{2^2})$	0	$\frac{1}{2^2}$	$\frac{1}{2}$	$\frac{2^2+1}{2^3}$	$\frac{2^4+2^2 1}{2^5}$	
$(\frac{2^2+2+1}{2^2}, 1)$	$\frac{1}{2^3}$	$\frac{2^2+2+1}{2^4}$	$\frac{2^4+2^2+2+1}{2^5}$	$\frac{2^4+2^3+2+1}{2^5}$	$\frac{2^2+2+1}{2^3}$	
$\vdots$						

4   Estimators of $\mu(Y|X)$ and $\mu(X,Y)$

In this section, we are going to construct estimators of measures $\mu(Y|X)$ and $\mu(X,Y)$. Let $X \in \mathcal{L}_1$ and $Y \in \mathcal{L}_2$ be two discrete random vectors and $[n_{xy}]$ be their observed multi-way contingency table. Suppose that the total number of observation is n. For every $x \in \mathcal{L}_1$ and $y \in \mathcal{L}_2$, let n_{xy}, $n_{x\cdot}$ and $n_{\cdot y}$ be numbers of observations of (x,y), x and y, respectively, i.e., $n_{x\cdot} = \sum_{y \in \mathcal{L}_2} n_{xy}$ and $n_{\cdot y} = \sum_{x \in \mathcal{L}_1} n_{xy}$. If we define $\hat{p}_{xy} = n_{xy}/n$, $\hat{p}_{x\cdot} = n_{x\cdot}/n$, $\hat{p}_{\cdot y} = n_{\cdot y}/n$, $\hat{p}_{y|x} = \hat{p}_{xy}/\hat{p}_{x\cdot} = n_{xy}/n_{x\cdot}$ and $\hat{p}_{x|y} = \hat{p}_{xy}/\hat{p}_{\cdot y} = n_{xy}/n_{\cdot y}$, then estimators of measures $\mu(Y|X)$ and $\mu(X,Y)$ can be defined as follows.

Definition 5. Let $X \in \mathcal{L}_1$ and $Y \in \mathcal{L}_2$ be two discrete random vectors with a multi-way contingency table $[n_{xy}]$. *Estimators of* $\mu(Y|X)$ *and* $\mu(X,Y)$ *are given by*

$$\hat{\mu}(Y|X)\left[\frac{\hat{\omega}^2(Y|X)}{\hat{\omega}_{max}^2(Y|X)}\right]^{\frac{1}{2}},$$

and

$$\hat{\mu}(X,Y) = \left[\frac{\hat{\omega}^2(Y|X) + \hat{\omega}^2(X|Y)}{\hat{\omega}_{max}^2(Y|X) + \hat{\omega}_{max}^2(X|Y)}\right]^{\frac{1}{2}},$$

where

$$\hat{\omega}^2(Y|X) = \sum_{y \in \mathcal{L}_2, x \in \mathcal{L}_1}\left[\sum_{y' \le y,}\left(\hat{p}_{y'|x} - \hat{p}_{\cdot y'}\right)\right]^2 \hat{p}_{x\cdot}\hat{p}_{\cdot y},$$

$$\hat{\omega}_{max}^2(Y|X) = \sum_{y \in \mathcal{L}_2}\left[\sum_{y' \le y,}\hat{p}_{\cdot y'} - \left(\sum_{y' \le y,}\hat{p}_{\cdot y'}\right)^2\right]\hat{p}_{\cdot y},$$

and $\hat{\omega}^2(X|Y)$ and $\hat{\omega}^2_{max}(X|Y)$ are similarly defined as $\hat{\omega}^2(Y|X)$ and $\hat{\omega}^2_{max}(Y|X)$ by interchanging X and Y.

From the above definition, we have the following result. The proof is trivial.

Theorem 3. *Let X and Y be discrete random vectors. Estimators $\hat{\mu}(Y|X)$ and $\hat{\mu}(X,Y)$ have following properties,*

(i) $0 \leq \hat{\mu}(Y|X), \hat{\mu}(X,Y) \leq 1$;
(ii) *X and Y are empirically independent, i.e., $\hat{p}_{xy} = \hat{p}_{x\cdot}\hat{p}_{\cdot y}$ for all $(x,y) \in \mathscr{L}_1 \times \mathscr{L}_2$ if and only if $\hat{\mu}(Y|X) = \hat{\mu}(X,Y) = 0$;*
(iii) *$\hat{\mu}(Y|X) = 1$ if and only if Y is a function of X. And $\hat{\mu}(X,Y) = 1$ if and only if X and Y are functions of each other.*

Next, we use two example to illustrate our results.

Example 3. Suppose that we have the following multi-way contingency tables. Then from Table 7, we have $\hat{\mu}(Y|X) = 0.0516$, $\hat{\mu}(X|Y) = 0.0762$ and $\hat{\mu}(X,Y) = 0.0642$, so X and Y have very weak functional relations. However, from Table 8, we have $\hat{\mu}(Y|X) = 0.5746$, $\hat{\mu}(X|Y) = 0.0465$ and $\hat{\mu}(X,Y) = 0.3485$, so the functional dependence of Y on X is much stronger than the functional dependence of X on Y.

Example 4. The data given in Table 9 [1] is from a survey conducted by the Wright State University School of Medicine and the United Health Services in Dayton, Ohio. The survey asked students in their final year of a high school near Dayton, Ohio,

Table 7. Contingency table of X and Y.

Y	X				$n_{\cdot y}$
	$(1,1)$	$(1,2)$	$(2,1)$	$(2,2)$	
$(1,1)$	10	20	5	10	45
$(1,2)$	15	25	10	5	55
$(2,1)$	5	35	10	5	55
$(2,2)$	25	5	10	5	45
$n_{x\cdot}$	55	85	35	25	200

Table 8. Contingency table of X and Y.

Y	X				$n_{\cdot y}$
	$(1,1)$	$(1,2)$	$(2,1)$	$(2,2)$	
$(1,1)$	43	2	3	40	88
$(1,2)$	4	42	40	6	92
$(2,1)$	2	3	3	2	10
$(2,2)$	1	4	2	3	10
$n_{x\cdot}$	50	51	48	51	200

Table 9. Alcohol (A), Cigarette (C), and Marijuana (M) use for high school seniors.

Alcohol use	Cigarette use	Marijuana use	
		Yes	No
Yes	Yes	911	538
	No	44	456
No	Yes	3	43
	No	2	279

whether they had ever used alcohol, cigarettes, or marijuana. Denote the variables by A for alcohol use, C for cigarette use, and M for marijuana use. By Pearson's Chi-squared test (A,C) and M are not independent. The estimations of functional dependence between M and (A,C) are $\hat{\mu}(M|(A,C)) = 0.3097$, $\hat{\mu}((A,C)|M) = 0.2776$ and $\hat{\mu}((A,C),M) = 0.2893$.

Acknowledgments. We would like to thank Professor Hung T. Nguyen for introducing this interesting topic to us and referees for their valuable comments and suggestions which greatly improve this paper.

References

1. Agresti, A., Kateri, M.: Categorical Data Analysis, 2nd edn. Springer, Hoboken (2011)
2. Boonmee, T., Tasena, S.: Measure of complete dependence of random vectors. J. Math. Anal. Appl. **443**(1), 585–595 (2016)
3. Dette, H., Siburg, K.F., Stoimenov, P.A.: A copula-based non-parametric measure of regression dependence. Scand. J. Stat. **40**(1), 21–41 (2013)
4. Genest, C., Quesada Molina, J., Rodríguez Lallena, J.: De l'impossibilité de construire des lois à marges multidimensionnelles données à partir de copules. Comptes rendus de l'Académie des sciences. Série 1, Mathématique **320**(6), 723–726 (1995)
5. Lancaster, H.: Measures and indices of dependence. In: Kotz, S., Johnson, N.L. (eds.) Encyclopedia of Statistical Sciences, vol. 2, pp. 334–339. Wiley, New York (1982)
6. Li, H., Scarsini, M., Shaked, M.: Linkages: a tool for the construction of multivariate distributions with given nonoverlapping multivariate marginals. J. Multivar. Anal. **56**(1), 20–41 (1996)
7. Nelsen, R.B.: An Introduction to Copulas, 2nd edn. Springer, New York (2006)
8. Rényi, A.: On measures of dependence. Acta Math. Hung. **10**(3–4), 441–451 (1959)
9. Schweizer, B., Wolff, E.F.: On nonparametric measures of dependence for random variables. Ann. Stat. **9**(4), 879–885 (1981)
10. Shan, Q.: The measures of association and dependence through copulas. Ph.D. dissertation, New Mexico State University (2015)
11. Shan, Q., Wongyang, T., Wang, T., Tasena, S.: A measure of mutual complete dependence in discrete variables through subcopula. Int. J. Apporx. Reason. **65**, 11–23 (2015)
12. Siburg, K.F., Stoimenov, P.A.: A measure of mutual complete dependence. Metrika **71**(2), 239–251 (2010)

13. Sklar, A.: Fonctions de répartition á *n* dimensions et leurs marges. Publ. Inst. Stat. Univ. Paris **8**, 229–231 (1959)
14. Tasena, S., Dhompongsa, S.: A measure of multivariate mutual complete dependence. Int. J. Apporx. Reason. **54**(6), 748–761 (2013)
15. Tasena, S., Dhompongsa, S.: Measures of the functional dependence of random vectors. Int. J. Apporx. Reason. **68**, 15–26 (2016)
16. Trutschnig, W.: On a strong metric on the space of copulas and its induced dependence measure. J. Math. Anal. Appl. **384**(2), 690–705 (2011)

Applications

To Compare the Key Successful Factors When Choosing a Medical Institutions Among Taiwan, China, and Thailand

Tzong-Ru (Jiun-Shen) Lee[1], Yu-Ting Huang[2(✉)], Man-Yu Huang[3], and Huan-Yu Chen[4]

[1] Department of Marketing, National Chung Hsing University, No. 145 Xingda Rd., South Dist., Taichung City 402, Taiwan
trlee@dragon.nchu.edu.tw
[2] Department of Business Administration, National Taiwan University, Taipei, Taiwan
huangyt81@ntu.edu.tw
[3] National Cheng-Chi University, Taipei, Taiwan
lemonde6201@gmail.com
[4] National Chung Hsing University, Taichung, Taiwan
huanychen@gmail.com

Abstract. What are the most important factors that increases customers' (patients') satisfaction with a general clinic, dental clinic and cosmetic surgery clinic respectively in Asia? Our paper tries to answer the question by conducting survey in Taiwan, Thailand, and China (in the order of the time the survey was conducted), with Grey Relational Analysis (GRA) methodology applied to identify key successful factors in three regions. By the research, our paper found specific and interesting phenomenon of medical institutions in each regions. 'Doctor's Skill' is the general factor considered to be important across regions.

1 Introduction

Our paper aims to compare medical market among China, Thailand and Taiwan and to sort out the differences among the three markets. The ultimate goal of our paper is to identify the most vital factors, i.e. key successful factors, that influence customers' choice on medical institutions. The discoveries of this paper can be further applied in the medical industry.

Three types of medical institutions are chosen in this paper, which are general clinic, dental clinic and cosmetic surgery clinic. The former two are chosen as they account for over 50% of all types of medical institutions, while cosmetic surgery clinic is chosen as it is becoming a trend in these years. These medical institutions are among the scope of our study and are involved in the questionnaire.

The methodology applied in this paper is introduced in Literature Review in Sect. 2, while the details on the questionnaire design of this paper will be explained in Questionnaire Design in Sect. 3. Explanations on the process of

© Springer International Publishing AG 2018
V. Kreinovich et al. (eds.), *Predictive Econometrics and Big Data*, Studies in Computational Intelligence 753, https://doi.org/10.1007/978-3-319-70942-0_23

analysis will be in Quantitative Research in Sect. 4. Results and Discoveries are listed in Sects. 5 and 6 separately. Section 7 concludes the whole paper while, last but not least, a list of all the references that is mentioned in the paper. In this paper, we use the term "customer" instead of "patient" because people go to dental and cosmetic surgery clinic not only due to their illness, but also in the pursuit of a better appearance in most of the cases.

2 Literature Review

2.1 Critical Successful Factor (CSF)

Daniel D. Ronald of McKinsey Company proposed 'critical successful factor' (CSF) in 1961. Critical successful factor refers to the fundamental elements necessary for a company's or an institution's success. To ensure the representativeness of the result, two criteria are set for the critical successful factor: 1. The number of critical successful factors should be at least three and no more than six, 2. The number of critical successful factors should not exceed half the number of all the factors examined. Adopting Daniel's point of view, this paper will take the two criteria of the critical successful factor to be that of the 'key successful factors' of medical institutions sorted.

2.2 Grey Relational Analysis

Grey system theory is proposed by Deng (1989) to understand the degree of influence among a series of known information. If the information in the system can be completely analyzed and is distinctive, or the relations between input parameters and output parameters are clear, then we say the system is 'white system'. On the contrary, information that cannot be analyzed or relations that are completely unclear is 'black system'. Gray system is the one between black and white system, with vague construction, characteristics and parameters of information but not totally unclear.

Grey relational Analysis (GRA) is an impact evaluation model that measures the degree of similarity or difference between two sequences based on the grade of relation (Zhang and Liu 2011). There are three advantages that the methodology brings in the statistics perspective. First, the system can be established even with incomplete or vague information. For example, while many possible elements are involved in the reason why a president is elected (because the president is smart, honest, powerful, or shares the same hometown with the electors, etc.), GRA builds the system and sorts out the elements with highest relation to the event only. Second, while traditional regression statistics requires a large amounts of data to proceed to the analysis, GRA requires only four numbers of data to acquire a representative result. Last but not least, the calculation procedure is easy, and no complicated formulas are needed. Therefore, the Grey relational analysis (GRA) has been widely applied in various fields (Wei 2010).

Lin et al. (2002) tried to optimize EDM (electrical discharge machining) process with fuzzy logic and grey relational analysis (GRA). Chaang-Yung and Kun-Li Wen (2007) evaluated the relationship between company attributes and its financial performance by applying grey relational analysis (GRA). Kuo et al. (2008) applied grey relational analysis (GRA) to facilitate single alternative works that are best for all performance attributes. In optics and laser domain, Ulas and Ahmet (2008) presented an effective approach for the optimization of laser cutting process of St-37 steel with multiple performance characteristics based on the grey relational analysis (GRA). Noorul Haq et al. (2008) improved effectively the responses in drilling Al/ SiC process with GRA.

The calculation procedure of grey relational analysis (GRA) can be simplified into four major steps. To make the procedure more understandable, we will explain the steps with the data of Taiwanese clinics.

Step 1. Calculate the differences between referential sequences and comparative sequences

A Likert five-point scale is used to evaluate the criteria for the calculation of grey relational coefficients of all factors with the highest score being 5. Each factor is graded with a score of 5 (Very Important), 4 (Important), 3 (Average), 2 (Unimportant), or 1 (Very Unimportant). The evaluation point in the questionnaire is the so-called referential sequences, while the difference between the highest score (i.e. 5 in our paper) and the evaluation point is "comparative sequences". The less differences between referential sequences and comparatives sequences are, the more similar they are. The changes could be found from Tables 1 to 2.

Table 1. List of original scores of Taiwanese clinic (Comparative sequences)

Question items		Interviewees				
		1	2	$\cdots$	115	116
1	Clinic Division	*5	5	$\cdots$	4	4
2	Experts' recommendations	4	3	$\cdots$	2	3
$\cdots$	$\cdots$	$\cdots$	$\cdots$	$\cdots$	$\cdots$	$\cdots$
22	Payment Method: by Installments	4	1	$\cdots$	2	5
23	Clear and Specific Medical Instruction	5	5	$\cdots$	4	5

*Example: Interviewee of questionnaire No. 1 evaluated the score "5" to question ''.

Step 2. Calculate the grey relational coefficient with Eq. 1

Figure out the maximum (Δmax) and minimum (Δmin) of comparative sequences for each question items, and calculate the grey coefficient with Eq. (1).

$$\gamma X_{0k}, X_{ik} = \gamma 0ik = \Delta \min + \xi \Delta \max + \Delta X_{ik} + \xi \Delta \max \qquad (1)$$

Table 2. List of calculated differences between the highest score and comparative sequences of Taiwanese clinic (referential difference)

Question items		Interviewees				
		1	2	$\cdots$	115	116
1	Clinic Division	*0	0	$\cdots$	1	1
2	Experts' recommendations	1	2	$\cdots$	3	2
$\cdots$	$\cdots$	$\cdots$	$\cdots$	$\cdots$	$\cdots$	$\cdots$
22	Payment Method: by Installments	1	4	$\cdots$	3	0
23	Clear and Specific Medical Instruction	0	0	$\cdots$	1	0

*Example: Difference between original scores and referential alternative, i.e. referential difference sequences.

A. $i = 1, \cdots, m$; $k = 1, \cdots, m$
B. ξ is the identification coefficient and belongs to $[0, 1]$. ξ makes the best use when $\xi \leq 0.5463$, and our paper take it as 0.5
C. $X_0(k)$: Referential sequences
D. $X_i(k)$: Comparative sequences
E. $\gamma_{0i}(k)$: Grey relational coefficient
F. $|Dela\,$min and Δ max : A. The minimum and maximum of comparative sequences for each question items.

$X_i(k)$ is the score that the $k - th$ respondent answers factor i, where $i = 1, 2, \cdots, 23$; $k = 1, 2, \cdots, n$, and n is the amount of valid questionnaires.

Step 3. Calculate the grey relational grade with Eq. 2

$$\Gamma(x_0, x_i) = \Gamma_{0i} = \sum_{k=1}^{n} \beta_k \gamma[x_0(k), x_i(k)] \tag{2}$$

A. β_k is weight of each factor.

After step 1, with the sum of all the grey coefficients and the average of each question items, a system of grey relational grade will be established. The grey relational grade represents the relationship between sequence and its comparison sequence. If two factors in a system is changing toward the same tendency, then they have a high extent of synchronous change, as well as a high extent of the correlation. Then, grey relational grade, which is equal to grey relational coefficient under equal weighted index, is calculated (Table 3).

Step 4. Sort out key successful factors by raking grey relational grade
After calculating grey relational grade, we will rank all the factors through with the grade. The higher the grade is, the more important the factors is. All the question items by score are listed in Fig. 1. Finally, this paper obtains key successful factors with Ronald (1961)'s criteria.

Table 3. Grey coefficient of Taiwanese clinic

Question items		Interviewees				
		1	2	⋯	115	116
1	Clinic Division	1	0.6	⋯	0.67	1
2	Experts' recommendations	1	0.43	⋯	.33	1
⋯	⋯	⋯	⋯	⋯	⋯	⋯
22	Payment Method: by Installments	0.71	0.33	⋯	0.4	0.6
23	Clear and Specific Medical Instruction	0.71	0.43	⋯	1	1

Fig. 1. Taiwanese clinic grey relational grade in line chart

Here is an example by Taiwanese clinic data: (Due to limited space, this paper lists only the degree of key successful factors). According to Daniel's two critical successful factor criteria: 1. The amount of critical successful factors should be among three to six, and 2. The amount of critical successful factors should not exceed half the amount of all the factors examined, we would not involve factors after group 6 the amount of factors will exceed half of the total amount ($23/2 = 11.5 \approx 11$) and violate criterion 1. Moreover, we would not involve factors after group 3 as we should not include more than six factors to ensure the representativeness of our result (Table 4).

Therefore, our paper sorted out five key successful factors of Taiwanese clinic as showed in table below:

Table 4. Key successful factors of Taiwanese clinic

Group	No.	Key successful factors	Grey relational grade	Rank
1	15	Doctors Skill	0.885	1
	20	Treatment Effect	0.869	2
2	14	Doctor's Explanation before Doing Treatment	0.846	3
3	12	Nurse's Attitude	0.821	4
	11	Doctor's Attitude	0.807	5

Data of dental clinic and cosmetic surgery clinic in both Taiwan and Thailand are calculated with the same procedure.

3 Questionnaire Design

Three types of medical institutions are chosen in this paper, which are general clinic, dental clinic and cosmetic surgery clinic. The former two are chosen as they account for over 50% of all types of medical institutions, while cosmetic surgery clinic is chosen as it is becoming a trend in these years. These medical institutions are among the scope of our study and are involved in the questionnaire.

A five-point likert scale is the criteria of the questionnaire. Customers are asked to evaluate the importance of each factors, from very unimportant to very important, according to their own opinions.

The factors that influence customers' decisions are distributed into three stages in the questionnaire, which are before treatment, during treatment, and after treatment. All the factors in three stages are explained as follows.

3.1 Question Items (Factors) in the Questionnaire

In this section, 23 question items are classified three different periods, which are before treatment, during treatment, and after treatment. With this principle, ten question items (factors) were attributed to 'factors before treatment', eight question items (factors) were concluded in 'factors during treatment', and five question items (factors) were belonged to 'factors after treatment'.

Factors before Treatment

(1) Clinic Division
Patients select doctors according to their specialty before going to a medical institution (Boscarino and Steiber 1982; Lane and Linquist 1988). Therefore, clinic division is a question item in our paper.

(2) Experts' recommendations
Bayus (1985) claimed that positive or negative reputation is one of the key factors that influences customers' decision under the situation of insufficient information. The recommendations from other doctors is one of the most efficient ways to know the credibility of a doctor or a clinic/hospital (Wolinsky and Kurz (1984), Malhotra (1983)). Therefore, our paper includes 'experts' recommendations'.

(3) Social Media Recommendation
Gu et al. (2012) noticed that customers do more research before conducting 'high involvement treatment', such as dental correction and plastic surgery. So the questionnaire includes 'social recommendation on the website'.

(4) Transportation Convenience
The distant from the patients' location to the clinic is one of their concerns while choosing a medical institution. The closer one has a greater advantage as it reduces not only transportation costs but also time. Due to this reason, our paper will take this factor into consideration.

(5) Parking Space

More and more people go around with cars; therefore, a parking space is an essential requirement, especially for treatments in the dental clinic or the cosmetic surgery clinic that take a longer time.

(6) Waiting Time

'Waiting time' has been frequently mentioned since 1976. (Holtmann and Olsen 1976). Also, Otani et al. (2010a) included 'registration waiting time' into the questionnaire items. Thus, it will be included in this paper as well.

(7) Reservation Explanation

Huang et al. (2008) mentioned reservation explanation in their paper. Since reservation service is usual in dental clinic and cosmetic surgery clinic, a clear reservation explanation can be an influential element to enhance customers satisfaction. Therefore, 'reservation explanation' is one of the questionnaire items in our paper.

(8) Reputation of the Medical Institution

Trust and reputation are often taken into consideration while people are making decisions on whether or not to continue their interaction with a party in the future (Josang et al. 2007). They also serve as significant references for people to decide whether to start business (we mean whether to go to the clinic here) with the party. So 'reputation of the medical institution' will be included in our paper.

(9) Doctor's Specialty and Experience

Our paper included 'doctor's specialty and experience' since doctor's specialty and experience is the first thing that customers know about a doctor. Also, a doctor who is professional and well-experienced wins credit from customers.

(10) (Customer's) Service Experience

Beskind (1962) noted that customers make decision by their experience accumulated in the past. I.e. customers' high satisfaction keeps them tied with a company while unpleasant experiences force them to go for another.

Factors during Treatment

(11) Doctor's Attitude

Doctor is the person who interacts with customers the most during the treatment. Boudreaux (2004) also mentioned 'doctor's caring' in his research. Due to the reasons above, 'attitude: doctor's attitude' can be one of the questionnaire items in our paper.

(12) Nurse's Attitude

Wolinsky and Kurz (1984), Lane and Lindquist (1988), Boudreaux (2004), and Otani et al. (2010b) mentioned nurse's attitude to be an important issue in medical caring in their paper. Therefore, our paper included the factor as well.

(13) Other Staff's Attitude
'Other staff's' in this paper are the staffs except for doctors and nurses, including pharmacists, counter staff, etc. Doctors and nurses cannot finish the whole treatment process (from registration to distribute medicine) without the assistance from other staffs. Wolinsky and Kurz (1984), Lane and Lindquist (1988) even take 'pharmacists' attitude' as a questionnaire items.

(14) Doctor's Explanation before Doing Treatment
Boudreaux (2004) found that doctors' clear explanation about the treatment beforehand influences customer experience positively. Moreover, this clear explanation leads to a positive influence on customer satisfaction (Otani et al. 2005). Based on the studies, our paper put doctors' explanation before doing treatment into questionnaire.

(15) Doctors Skill
Cheng et al. (2003) found that 'doctor's skill' affected not only customers' satisfaction, but also decided if they will recommend to others. Otani et al. (2005) defines 'doctors skill' as careful treatment. Our paper adopted that 'doctors skill' is an important factor.

(16) Medical Environment (quiet, cleanness and comfort, etc.)
A quiet, clean and comfortable environment helps customers to rest and recover, so our paper put "medical environment (quiet, cleanness and comfort, etc.)" in, too.

(17) Privacy Reserve
Inhorn (2004) discussed the privacy issue during treatment. Although Inhorn's paper (2004) discussed cases of the Middle East, we consider the element to be worthwhile mentioning and include it in our paper.

(18) Medical Equipment
Whether the medical equipment is advanced relates to the comforts of customers and the effectiveness of the treatments. So this paper takes into consideration "medical equipment" as one of the factors.

Factors after Treatment

(19) Treatment Costs
Many papers studying the factors of patients' medical division choice mentioned that 'Medical Caring Cost' is one of the factors that patients concerned. (Boscarino and Steiber (1982), Malhotra (1983), Wolinsky and Kurz (1984), Lane and Lindquist (1988).

(20) Treatment Effect
Treatment effect is the direct proof or reference for customers to evaluate whether the money spent is worthy or not. "Treatment effect" is apparently an issue that customers care about, so our paper takes it in the questionnaire.

(21) Payment Method: Credit Card
Surgery and treatments like orthodontics and denture are expensive, which has made it inconvenient for patients to pay by cash at once. An alternative payment method should therefore be provided to enhance customer satisfaction. Therefore, "payment method: credit card" is included in our paper.

(22) Payment Method: by Installments
Similar reason with "Payment Method: Credit Card", paying in installments is an affordable alternative for customers. Thus, "Payment Method: by Installments" is included in this paper.

(23) Clear and Specific Medical Instruction
Clear and specific medical instruction can be taken as "after-sales services" in medical industry. It can be either beneficial or harmful to the impression of a medical institution. So "clear and specific medical instruction" is one of the factors in our paper.

4 Quantitative Research and Qualitative Research

Our paper conducts quantitative research in Taiwan and Thailand, and a qualitative research in China. The results of the researches are compared in the end of this paper. Although the two methodology applied seems to be so different at the first glance, the comparison is actually valid and meaningful. The key point that makes this possible is the 'confined inquiry range', i.e. the same question items.

During quantitative research, interviewees are asked to answer the importance of the 23 question items introduced in Sect. 3, and to sort out four to six key successful factors among them. While proceeding to the qualitative research in China, the sorted key successful factors in Taiwan and Thailand were main items in qualitative research interview process.

4.1 Quantitative Research

The most prominent advantage of quantitative research is that facts are demonstrated by numerical figures, which is more objective and can be assessed easily. We apply the quantitative method to analyze customers' needs in the cases of Taiwan and Thailand.

4.1.1 Subjects

The target subjects of this research in Thailand were experienced doctors and nurses, whereas in Taiwan they were patients in the general clinic, dental clinic and cosmetic surgery clinic. In Thailand, experienced doctors and nurses have enough knowledge regarding to their patients that they understand the needs of patients. Thus, the answers should be representative enough to reflect medical market in Thailand, and is adequate to be compared with the data of Taiwan.

4.1.2 Surveying Period

Thailand

The questionnaires were released in Thailand during 2014 November and December by paper. 90 questionnaires were collected from doctors and nurses in Chiang Mai (30 per type of medical institution). The overall response rate was 100%.

Taiwan

The questionnaire was released by both paper and online means from June 4th 2013 to July 2nd 2013. 116 questionnaires are collected in total, and 106 of them are valid. The overall response rate as 87.6%.

4.2 Qualitative Research

Our paper inquired patients in medical institution at the east coast of China instead of distributing questionnaires and conduct quantitative research for two reasons: first, it is time-consuming to collect all the questionnaires from 23 provinces in China; second, south east Chinese shares similar culture, language, belief, custom, etc. with that of the Taiwanese. Therefore, it is a reasonable assumption that the key successful factors in the two places will be similar. The final result of this research also supports this original assumption.

4.2.1 Subjects

This paper targets at 30 subjects that had experiences of going to different types of medical institutions. Most of the target subjects are from south east China since people here present the highest cultural similarity to those of Taiwan.

4.2.2 Surveying Period

The subjects were inquired from July 1st to August 31st in 2015. 30 subjects were inquired in total.

5   Results

5.1   Results

All the key successful factors of Taiwan and Thailand chosen follow Daniel's criteria mentioned in Literature Review in Sect. 2. Due to the space limit, this paper only displays key factors of each types of medical institutions. The analysis results in three places were listed.

The result of the inquiries in China is showed in Table 6 below:

Table 5. The key successful factors of Chinese customers in general clinic, dental clinic and cosmetic surgery clinic:

	General Clinic							Dental Clinic									Cosmetic surgery clinic							
Factors	*1B	D				A		B		D						A	B		D					A
	a	b	c	d	e	f	g	h	i	j	k	l	m	n	o	p	q	r	s	t	u	v	w	x
1									√								√		√	√				
2			√	√	√	√		√			√	√				√	√		√			√		√
3		√						√	√								√		√					
4					√												√							
5			√	√		√						√	√						√	√			√	
6		√	√			√	√	√	√		√					√	√	√	√	√	√			√
7			√			√			√	√	√					√			√	√	√			√
8		√		√	√			√				√	√			√	√	√	√	√		√		√
9		√		√	√	√		√	√	√						√	√	√	√					√
10		√		√	√	√		√		√		√			√	√	√	√	√			√	√	√
11			√			√		√			√					√	√	√	√			√		√
12			√							√		√	√	√		√			√		√	√		√
13					√	√	√	√	√	√						√	√					√		√
14		√				√	√	√	√	√	√					√	√	√	√			√		√
15		√		√		√			√	√	√					√			√	√	√		√	√
16						√			√	√	√			√		√			√		√	√		√
17		√		√				√	√							√			√		√	√		√
18		√				√		√	√	√	√		√			√	√							√
19			√	√	√			√	√		√	√				√	√	√			√	√		√
20		√		√				√	√	√						√	√	√	√			√		
21		√	√			√	√	√								√	√		√	√				√
22			√	√	√						√					√	√		√	√	√	√		√
23						√		√	√				√			√	√						√	√
24		√						√									√	√	√	√	√			
25						√		√	√	√		√		√		√	√							√
26		√	√					√				√							√		√			√
27						√		√	√							√	√	√						√
28		√					√	√									√	√						
29			√	√		√		√	√							√		√	√	√				√
30		√	√			√	√	√	√					√	√		√	√					√	√
31			√	√						√	√	√		√	√						√	√		√
32		√		√	√	√		√		√	√	√	√	√		√		√		√		√		√
33		√	√			√				√	√					√	√	√						√
34			√	√	√	√						√				√				√				√
35		√	√			√	√			√	√			√	√					√	√		√	√
Total		18	9	16	9	22	9	19	15	18	17	11	8	4	4	26	20	10	14	16	15	16	5	28

*Denotation: **B** denoted to 'Before', **D** denoted to 'During', and **A** denoted to 'After' in table 6

Table 6. *P.S. Letter denotation of 5.1.1.:

Letter	Denotation
Clinic	
a	N/A (There were no chosen factors in classification Before the treatment for Chinese clinic.)
b	Doctor's Skill
c	Doctor's Explanation before Doing Treatment
d	Doctor's Attitude
e	Nurse's Attitude
f	Treatment Effect
g	Clear and Specific Medical Instruction
Dental Clinic	
h	Doctor's Specialty and Experience
i	Customer's Experience
j	Doctor's Skill
k	Doctor's Explanation before Doing Treatment
l	Doctor's Attitude
m	Nurse's Attitude
n	Medical Environment (quiet, cleanness and comfort, etc.)
o	Privacy Reservation
p	Treatment Effect
Cosmetic surgery clinic	
q	Doctor's Specialty and Experience
r	Customer's Experience
s	Reputation of the Medical Institution
t	Doctor's Skill
u	Doctor's Explanation before Doing Treatment
v	Doctor's Attitude
w	Privacy Reservation
x	Treatment Effect

5.2   Results of the Survey

The horizontal title of Table 7 refers to period of treatment (i.e. before treatment, during treatment, and after treatment), while the vertical title of Table 7 refers to types of medical institutions. To understand the details of the key successful factors in three regions clearly, we add a second vertical subordinate title 'country' (Tables 5 and 6).

Factors listed in Table 7 are the key successful factors of each institution but not in the order of its priority. 'N/A' means no key successful factors are sorted out from the category. Some factors are mentioned repeatedly, while some are not. Factors mentioned repeatedly are the most significant ones. However, if a factor is mentioned only once in Table 7, the factor is especially important to the country in certain period.

Table 7. Overall result: key successful factors of three medical institution in three

Medical Institution	Country	Before Treatment	During Treatment	After Treatment
General Clinic	Taiwan	N/A	Doctor's Skill	Treatment Effect
			Doctor s Explanation before Doing Treatment	
			Nurse's Attitude	
			Doctor's Attitude	
	China	N/A	Doctor's Skill	Treatment Effect
			Doctor's Attitude	
	Thailand	N/A	Doctors Skill	Clear and Specific Medical Instruction
			Doctor s Explanation before Doing Treatment	
			Doctor's Attitude	
			Other Staff's Attitude	
Dental Clinic	Taiwan	Doctor's Specialty and Experience	Doctor's Skill	Treatment Effect
		Customer's Service Experience	Doctor's Attitude	
		-	Nurse's Attitude	
	China	Doctor's Specialty and Experience	Doctor's Skill	Treatment Effect
		Customer's Service Experience	Doctor s Explanation before Doing Treatment	
	Thailand	Customer's Service Experience	Doctor's Skill	N/A
			Doctor s Explanation before Doing Treatment	
			Doctor's Attitude	
			Medical Environment (quiet, cleanness and comfort, etc.)	
			Privacy Reservation	
Cosmetic surgery clinic	Taiwan	Doctor's Specialty and Experience	Doctor's Skill	Treatment Effect
		Reputation of the Medical Institution	Doctor's Explanation before Doing Treatment	
		-	Privacy Reservation	
	China	Doctor's Specialty and Experience	Doctor's Skill	Treatment Effect
		Reputation of the Medical Institution	Doctor's Explanation before Doing Treatment	
		-	Doctor's Attitude	
	Thailand	Doctor's Specialty and Experience	Doctor's Skill	N/A
		Customer's Service Experience	Doctor s Explanation before Doing Treatment	
		-	Doctor's Attitude	

6   Findings

This chapter will conclude and explain the results from the last chapter and reorganize the factors according to the frequency they show in three regions (Tables 8, 9 and 10). Since the results are from patients' intuition, and due to the page limit, we will focus on the factors regarded to be important only in a certain country and explore the reason behind the phenomenon. The explanation below is based on the interview with a doctor in Jen-Ai Hospital.

6.1   General Clinic

According to Table 7, 'doctor's skill' and 'doctor's attitude' are considered to be influential on customers' choice of clinics for all of the three regions. 'Doctor's explanation before doing treatment' affects Taiwanese and Thai, while 'treatment effect' is important to Taiwanese and Chinese. What we would like to discuss further in this paper is 'nurse's attitude' for Taiwanese, and 'other staff's attitude' and 'clear and specific medical instruction' for Thai. The frequency of the key successful factors is listed in Table 8.

Table 8. Frequency of clinic key successful factors in the three regions

Frequency	Key successful factor	Region
3	Doctor's skill	All
	Doctor's attitude	All
2	Doctor's explanation before doing treatment	TW, THI
	Treatment effect	TW, CN
1	Nurse's attitude	TW
	Other staff's attitude	THI
	Clear and specific medical instruction	THI

When choosing a general clinic,

(1) Nurse's attitude is an influential factor for Taiwanese customers
In Taiwanese clinics, 'nurse' is an influential role because nurse is in charge of most of the tasks in whole treatment procedure (e.g. registration, injection, medicine giving). In other words, customers spend more time with nurse than with doctor. Moreover, Taiwanese customers go to a clinic not only for physical treatment, but also for their mental comfort. Therefore nurse's attitude is especially important for the Taiwanese.

(2) 'Other staff's attitude' is an influential factor for Thai customers
Thai clinics pursue a high standardization process, which rely very much on the efforts of staffs. Most samples collected in this paper are from Maharaj Nakorn Chiang Mai Hospital, which is a famous and internationalized one. In high ranking hospitals, even registration time should not be wasted. In order to run a complex treatment procedure while at the same time to keep it at a good quality, the executers, i.e. other staffs, are doubtlessly important.

(3) 'Clear and specific medical instruction' is influential for Thai customers
Thai Government has been promoting 'medical tourism' since 2004. Such policy has pushed Thailand to be one of the medical leaders in East Asia, as well as the most famous 'medical tourism' destination for western people. To acquire customers' trusts, clear and specific medical instruction is a vital factor under the exotic medical environment.

6.2   Dental Clinic

After concluding key successful factors of dental clinics in the three regions, we found that 'customer's service experience' and 'doctor's skill' are vital factors to all. 'Doctors' specialty and experience' and 'treatment effect' are essential to Taiwanese and Chinese. 'Doctor's attitude' is shown as significant again to Taiwanese and Thai. 'Doctor's explanation before doing treatment' is important to Chinese and Thai. As for the factors that only influence one region, 'nurse's attitude' again has effect on the Taiwanese, while 'medical environment (quiet cleanness and comfort, etc.)' and 'privacy reservation' are fundamental to Thai in dental clinic. Table 9 shows the results mentioned. When choosing a dental clinic,

Table 9. Frequency of dental clinic key successful factor in the three regions

Frequency	Key successful factor	Region
3	Customer's service experience	All
	Doctors skill	All
2	Doctor's specialty and experience	TW, CN
	Treatment effect	TW, CN
	Doctor's attitude	TW, THI
	Doctor's explanation before doing treatment	CN, THI
1	Nurse's attitude	TW
	Medical environment (quiet cleanness and comfort, etc.)	THI
	Privacy reservation	THI

(1) 'Nurse's attitude' is an influential factor for Taiwanese customers
Dental treatment is not always a one-time treatment. Customers have to go back to the dental clinic for further treatment from time to time from a few weeks to several years (e.g. braces). As in the general clinics, nurses assist not only daily operations but also treatments, and they are the main receptors to customers. Therefore, 'nurse's attitude' is important to Taiwanese people in deciding a dental clinic.

(2) 'Medical environment (quiet, cleanness and comfort, etc.)' is an influential factor for Thai customers
Infection is a concern for customers who care about medical environment. The Thai people emphasize on 'medical environment' very much as infection is an issue in dental treatments. But this is an inference that has not yet been confirmed.

(3) 'Privacy reservation' to Thai
If you do not want to encounter anyone in a dental clinic, then privacy reservation must significant to you. Recent years, dental treatment includes not only therapies, but also teeth shaping, which is similar to orthopedics. Customers do not want their enhancement to be thought as 'artificial'. In contrast, they would rather to claim that they eat something special, massage cheek every day, etc.

to their delight. The reason why 'privacy reservation' is influential to Thai when choosing a dental clinic is similar to that of the Taiwanese. More explanation about the phenomena will be illustrated in Sect. 6.3 (1).

6.3   Cosmetic Surgery Clinic

After concluding cosmetic surgery clinic key successful factors for three regions, 'doctors specialty', 'doctors skill' and 'doctor's explanation before doing treatment' are fundamental to all of the three regions. 'Reputation of the medical institution' and 'treatment effect' is important to Taiwanese and Chinese; and 'doctor's attitude' is influential to Chinese and Thai customers. Last but not least, 'privacy reservation' is a major consideration for the Taiwanese, and 'customer's service experience is especially vital to the Thai while choosing an cosmetic surgery clinic. The result is presented in Table 10. When choosing an cosmetic surgery clinic:

Table 10. Frequency of cosmetic surgery clinic key successful factors in the three regions

Frequency	Key successful factor	Region
3	Doctor's specialty	All
	Doctor's skill	All
	Doctor's explanation	All
	Before doing treatment	
2	Reputation of the medical institution	TW, CN
	Treatment effect	TW, CN
	Doctor's attitude	CN, THI
1	Privacy reservation	TW
	Customer's service experience	THI

(1) Privacy reservation' to Taiwanese
Compared to 'privacy reservation' to Thai in dental clinic, Taiwanese not to be seen or known to have visited an cosmetic surgery clinic. Most Taiwanese go to cosmetic surgery clinics for a better appearance rather than accident recovery. They pursue not only 'beauty', but also 'nature beauty'. Although 'artificial beauty' is a kind of beauty, but somehow 'artificial' is related to 'fake'. People who go to cosmetic surgery clinic value their appearance, and hope to be considered as a nature beauty. If they are seen in an cosmetic surgery clinic, then the truth that they are not born to be a beauty will be known. That's why 'privacy reservation' is important to Taiwanese people.

(2) 'Customer's service experience' is an influential factor for Thai customers
The effect of an aesthetic surgery is shown directly on customers, from which the service experience is also direct and easy to tell. Also, aesthetic surgery and its

related treatment are usually not covered by Health Insurance and is at the customers' own expense. In a country with high quality medical environment where customers have more alternatives to choose from, it is to a clinic's advantage to possess the core competence that customers perceive to be important.

7   Conclusion

This paper sorts out the key successful factors for general clinic, dental clinic, and cosmetic surgery clinic in Taiwan, Thailand, and China. The methodology Grey Relational Analysis (GRA) is applied as it is effective in the case of incomplete data, which cannot be achieved by most of traditional statistical methods.

Generally speaking, Taiwanese and Chinese customers emphasize on "treatment effect" in the three kinds of medical institution. Also, 'doctors skill' is considered to be important across all the three medical institutions in the three regions. Throughout the findings, we conclude that doctor's skill is the core value and basic qualification for medical institutions in Taiwan and China. It is not only important but also is influential to other factors like 'reputation of the clinic', 'customer's experience', etc.

Although China and Thailand owns different medical conditions and resources, customers in both regions emphasize much on 'doctor's specialty and experience', 'doctor's explanation before doing treatment', and 'doctor's skill' in dental clinic and cosmetic surgery clinic. Chinese customers might require more on basic medical right, and Thai customers might pursue more on better quality.

Nowadays, it is noteworthy to understand the key successful factors of clinics since they are the indicators for clinics to meet customer's needs. The improvement of the whole industry will enhance its quality, and brings win-win business. We hope this paper can serve as the foundation for further studies on this matter.

References

Noorul Haq, A., Marimuthu, P., Jeyapaul, R.: Multi response optimization of machining parameters of drilling Al/SiC metal matrix composite using grey relational analysis in the Taguchi method. Int. J. Adv. Manufact. Technol. **37**(3–4), 250–255 (2008)

Boudreaux, E.D., O'Hea, E.L.: Patient satisfaction in the emergency department: a review of the literature and implications for practice. J. Emerg. Med. **26**(1), 13–26 (2004)

Boscarino, J., Steiber, S.R.: Hospital shopping and consumer choice. J. Health Care Mark. **2**(2), 15–23 (1982). Spring

Kung, C.-Y., Wen, K.-L.: Applying grey relational analysis and grey decision-making to evaluate the relationship between company attributes and its financial performance - a case study of venture capital enterprises in Taiwan. Decis. Support Syst. **43**(3), 842–852 (2007)

Cheng, S.-H., Yang, M.-C., Chiang, T.-L.: Patient satisfaction with and recommendation of a hospital: effects of interpersonal and technical aspects of hospital care. Int. J. Qual. Health Care **15**(4), 345–355 (2003)

Lin, C.L., Lin, J.L., Ko, T.C.: Optimisation of the EDM process based on the orthogo-nal array with fuzzy logic and grey relational analysis method. Int. J. Adv. Manufact. Technol. **19**(4), 271–277 (2002)

Daniel, R.D.: Management information crisis. Harvard Bus. Rev. **35**(5), 111–121 (1961)

Deng, J.L.: Introduction grey system. J. Grey Syst. **1**(1), 1–24 (1989)

Huang, F.-F., Chen, M.-S., Huang, L.-R.: The decision of patients and analysis of health check satisfaction. J. Orient. Inst. Technol. **28**, 37–46 (2008)

Marcia, I.C.: Privacy, privatization, and the politics of patronage: ethnographic chal-lenges to penetrating the secret world of middle eastern, hospital-based in vitro fertilization. Soc. Sci. Med. **59**(2004), 2095–2108 (2004)

Lin, J.L., Lin, C.L.: The use of the orthogonal array with grey relational analysis to optimize the electrical discharge machining process with multiple performance characteristics. Int. J. Mach. Tools Manufact. **42**(2), 237–244 (2002)

Audun, J., Roslan, I., Coin, B.: A survey of trust and reputation systems for online service provision. Decis. Support Syst. **43**, 618–644 (2007)

Otani, K., Waterman, B., Faulkner, K.M., Boslaughand, S., Dunagan, W.C.: How patient reactions to hospital care attributes affect the evaluation of overall quality of care, willingness to recommend and willingness to return. J. Health Care Manag. **55**(1), 25–57 (2010a)

Lane, P.M., Lindquist, J.D.: Hospital choice: a summary of the key empirical and hypothetical findings of the 1980s. J. Health Care Mark. **8**(4), 5–20 (1988)

Çaydaş, U., Hasslik, A.: Use of the grey relational analysis to determine optimum laser cutting parameters with multi-performance characteristics. Optics Laser Technol. **40**(7), 987–994 (2008)

Wei, G.W.: GRA method for multiple decision making the incomplete weight informa-tion in intuitionistic fuzzy setting. Knowl.-Based Syst. **23**(19), 243–247 (2010)

Wolinsky, F.D., Kurz, R.S.: How the public chooses and views hospital. Hosp. Health Serv. Adm. **29**, 58–67 (1984)

Kuo, Y., Yang, T., Huang, G.-W.: The use of grey relational analysis in solving multiple attribute decision-making problems. Comput. Indus. Eng. **55**(1), 80–93 (2008)

Zhang, S.F., Liu, S.Y.: A GRA-based intuitionistic fuzzy muilti-criteria group decision making method for personal selection. Expert Syst. Appl. **38**, 11401–11405 (2011)

Gu, B., Park, J., Konana, P.: Research note—the impact of external word-of-mouth sources on retailer sales of high-involvement products. Inf. Syst. Res. **23**(1), 182–196 (2012)

Beskind, H.: Psychiatric inpatient treatment of adolescents a review of clinical experi-ence. Compr. Psychiatry **3**(6), 354–369 (1962)

Otani, K., Herrmann, P.A., Kurz, R.S.: Patient satisfaction integration process: Are there any racial differences? Health Care Manage. Rev. **35**(2), 116–123 (2010b)

Bayus, B.L.: Word-of-mouth: the indirect effects of marketing efforts. J. Advertising Res. **25**, 31–9 (1985)

Holtmann, A.G., Olsen Jr., E.O.: The demand for dental care: a study of consumption and household production. J. Hum. Resour. 546–560 (1976)

Kuehn, A.: Consumer brand choice as a learning process. J. Advertising Res. **2**(March–April), 10–17 (1962)

Malhotra, N.K.: Stochastic modeling of consumer preferences for health care institu-tions. J. Health Care Market. **3**(4) (1983)

Otani, K., Kurz, R.S., Harris, L.E., Byrne, F.D.: Managing primary care using patient satisfaction measures/practitioner application. J. healthc. Manage. **50**(5), 311 (2005)

Forecasting Thailand's Exports to ASEAN with Non-linear Models

Petchaluck Boonyakunakorn[1]([⊠]), Pathairat Pastpipatkul[2],
and Songsak Sriboonchitta[2]

[1] Faculty of Economics, Chiang Mai University, Chiang Mai 52000, Thailand
petchaluckecon@gmail.com
[2] Center of Excellence in Econometrics, Chiang Mai University,
Chiang Mai Faculty of Economics, Chiang Mai University,
Chiang Mai 52000, Thailand
ppthairat@hotmail.com, songsakecon@gmail.com

Abstract. This work focuses on forecasting Thailand's exports to ASEAN. Thailand's exports to ASEAN reveal an overall increasing trend with a fluctuation since Thailand's exports are integrated in the global economy. However, the linear model might not be able to capture the behavior of Thailand's exports to ASEAN. Linear model cannot be applied in some phenomena such as fluctuation and structural breaks in time series data. In this study, we find that the Thailand's exports-to-ASEAN time series is non-linear via test of linearity, and find that there are two thresholds. Therefore, we forecast Thailand's exports to ASEAN with non-linear models. We employ four non-linear models, SETAR, LSTAR, MSAR, and Kink AR model. The simple linear AR model is also applied to compare with the non-linear models. To evaluate the forecasting performance of five different models, we use RMSE and MAE as criteria. The forecasting results indicate that the SETAR model is better than the other models. However, it is still not clear cut to conclude that the non-linear models outperform linear model. However, we can conclude that the SETAR is the most suitable for forecasting Thailand's exports to ASEAN compared with other non-linear models.

Keywords: Non-linear models · ASEAN · Forecasting
Thailand's exports

1 Introduction

Thailand's exports constitute a huge portion of Thai GDP, accounting for up to 70% of the country's economic output. Meanwhile, the world average value of exports of goods and services is only 30% of the world GDP. Clearly, economic growth relies heavily on trade. The biggest destination for Thai exports is the Association of Southeast Asian Nations (ASEAN) which accounts for approximately 25% of the total exports. Ten ASEAN members are Brunei, Cambodia,

© Springer International Publishing AG 2018
V. Kreinovich et al. (eds.), *Predictive Econometrics and Big Data*, Studies in Computational Intelligence 753, https://doi.org/10.1007/978-3-319-70942-0_24

340   P. Boonyakunakorn et al.

Indonesia, Laos, Malaysia, Myanmar, Philippines, Singapore, Vietnam and Thailand. The main objective of ASEAN is to promote economic and political cooperation among members. Becoming an ASEAN member provides a larger economy of scale, one of the advantages, which provides a significant support to Thai exports. However, with different sizes of markets and different geography it generates differences in the toughness of competition across export market destinations (Mayer et al. [11]).

ASEAN is recognized in recent years as the biggest export market for Thailand, and the above differences may influence stability in competing with other countries. Export policy becomes one of the most important contributing factors to successful exports. Thailand has been significantly focusing on maintaining its export capability to respond to changing environment. Therefore, economic development strategies should be in line with export policy. One of the key components supporting decision-making process is forecasting.

Forecasting is the heart of the strategic planning on exports. In addition, there is a notion that export policy decision often uses the historical record of exports. The data used in this study is also available for analysis with time lag. Therefore, a theoretical model that increases accuracy in export forecasting has gained more economic consideration. Figure 1 depicts Thailand's exports to ASEAN on monthly basis, revealing an overall increasing trend with a fluctuation. Due to global crisis, the value of Thailand's exports to ASEAN recorded the unusual lowest point in 2008, then it fluctuated significantly with an increasing trend during January 2009 and June 2011. After that, the export dropped substantially again in 2011 as the financial crisis began. These financial crises clearly pointed out that the global economy has an important role affecting Thailand's exports. Since 2012, it has remained the same with slightly downward trend.

Linear models have been widely used for forecasting with time series data. However, limitations caused by linear models have made practitioners encounter difficulties in dealing with certain situations, which cannot be explored under

Fig. 1. Thailand's exports to ASEAN during the period 2002–2016

linear model assumptions. Tong and Lim [16] has cautioned that there are phenomena that cannot be suitable to deal with using linear models, namely asymmetry, time-irreversibility, sudden bursts of very large amplitude at changeable time epoch, etc. Even though general linear models have been common in econometrics for most of the twentieth century, non-linear models have been introduced during the latter part, especially in econometrics for replicating key features of the business cycle (Ferrara et al. [7]). Also, some researchers (e.g. Brock and Potter [3]; Granger [8]) have supported that nonlinear behavior exists in financial time series and macroeconomic variables.

As non-linear models, Teräsvirta and Anderson [12] and Tong and Lim [16] proposed threshold models known as one of the regime-switching models. Among these regime-switching models, a model, known as self-exciting threshold AR (SETAR) model, is initially introduced. Regimes in non-linear models are typically defined by some lagged value of the time series. Then Hamilton [9] proposed Markov-switching model with the notion that the regimes are identified by some discrete Markov state variables.

Tiao and Tsay [14] stated that the threshold models provide better performance compared with other linear models with respect to forecasting US GDP. Djeddour and Doularouk [6], found an evidence that Threshold autoregressive model (TAR) outperforms ARMA model when forecasting US export crude oil in the mean square error. Bec et al. [1] proposed that an extension of threshold models can predict GDP growth adequately after a recession. Since the model allows for bounce-back effects. Camacho et al. [4] compared linear with Markov-switching models to investigate the forecasting performance of euro area GDP and found that Markov-switching models provide more information about the state of the economy. Ferrara et al. [7] found that the forecast accuracy of non-linear models is better in almost 40%–45% of studied cases.

Nevertheless, some researchers found that the non-linear models do not perform better than the linear models with respect to forecasting performance, when they applied linear and smooth-transition models to forecast industrial production and exchange rate series, see Boero and Marrocu [2], and Sarantis [17]. Their results show that Markov-switching forecasting models cannot operate as well as other linear models can. However, there is still no clear-cut evidence that non-linear models operate better than the other ones in terms of forecasting accuracy or reliability.

Figure 1 depicts that Thailand's exports to ASEAN gently fluctuate with nonlinearity during the period 2002–2016. Subsequently, we employ the non-linear model to forecast Thailand's exports. This paper will apply five models and one falls into the category of linear model and the rest fall into the category of non-linear model to forecast Thailand's exports to ASEAN. Autoregressive (AR) model, a linear model, followed by four non-linear models, Self-exciting autoregressive (SETAR), Logistic STAR (LSTAR), Markov-switching Autoregressive (MSAR) and Kink Autoregressive (AR) models. The five different models will be compared by means of Root Mean Squared Error (RMSE) and the Mean Absolute Error (MAE). This paper is divided into four sections. Section 2 describes methodology of the five different models employed in this paper. Section 3 presents the empirical results. Conclusion and some suggestion for further research are provided in the last section.

2   Methodology

2.1   The Linear Model

It is a simple Autoregressive (AR) model, a type of random process that current value depends on its lag values. It is commonly used to model and predict the future. For the sake of simplicity, we apply an autoregressive model of order 1. The AR (1) model can be expressed as

$$y_t = \phi_1 y_{t-1} + \sigma \varepsilon_t \tag{1}$$

where ϕ_1 are the AR coefficients at lag 1, $\varepsilon_t \sim WN(0,1)$, $\sigma > 0$ is the standard deviation of distribution term.

2.2   Non-linear Models

2.2.1   Threshold Autoregressive (TAR) Model

In general, Threshold Autoregressive model refers to piecewise linear models or regime switching models. Consequently, it is considered as one of non-linear models, which are firstly introduced by Tong [15]. A regime switch occurs when the threshold variable is held at the specific value of a variable. This is referred to as a threshold. The model parameters can change in TAR model to capture nonlinear dynamics, and they will change according to the value of a weakly exogenous threshold variable c_t.

In this paper, the threshold variable is the lagged dependent variable, which is y_t called Self-exciting autoregressive (SETAR) model. One of the advantages of SETAR is to allow greater flexibility in model parameters. During the SETAR process, the regime switching of a lag variable is based on the threshold values of the lag variable. Another advantage of SETAR model is the ability to capture some observed phenomena, which is in contrast to linear model, such as irreversibility or jumps.

The basic SETAR at lag 1 is written as;

$$y_t = \phi_1 y_{t-1} I(y_{t-1} \le c_i) + \phi_2 y_{t-1} I(y_{t-1} > c_i) + \varepsilon_t, \tag{2}$$

where y_t is the variable of interest, c is a threshold value. The assumption of ε_t is i.i d. mean zero sequence with a bound density function, and $E|u_t|^{2y} < \infty$ for some $\gamma > 2$. Another assumption is $|\phi_1| \le 1$ and $|\phi_2| \le 1$.

The main characteristic of TAR model is the discontinuous feature of the AR relationship. Smooth transition autoregressive (STAR) was proposed by Teräsvirta and Anderson [12] as its continuous feature is more natural. In STAR model, the indicator function is replaced by smooth function with sigmoid characteristics. The basic STAR takes the following form;

$$y_t = \phi_1 y_{t-1} \Phi_i(y_{t-1}; \psi_i \le c_i) + \phi_2 y_{t-1} \Phi_i(y_{t-1}; \psi_i > c_i) + \varepsilon_t \tag{3}$$

Now, the transition function $\Phi_i(y_{t-1}; \psi_i)$ is continuous function between 0 and 1, with parameter ψ_i. The option of transition function depends on its regime-switching behavior. The common use is logistic function, where f denotes the first-order logistic function with parameters $\psi = (\gamma_i, c_i)$ for regime $i : f(y_{t-1}; \psi_i) = (1 + exp(-\gamma_i(y_{t-1} - c_i)))^{-1}$. This resultant model is known as the Logistic STAR (LSTAR) model with its parameter c being threshold between two regimes.

2.2.2 Markov Switching Model

With this model, it is assumed that there tends to be a different regression model correlated with its regime. Regime-switching models build on the concept of a mixture of parametric distributions, whose mixture probabilities are subject to unobserved state variable. For Markov switching model, regime-switching models lie in the stochastic structure of the state variables which its state of unobserved process is modeled by discrete space Markov chain.

In this paper, we focus on a Markov switching Autoregressive (MSAR) model. A simple autoregressive model of order one in which the mean of the process switches between two regimes. In this paper, we consider two states; of high growth and of low growth according to a Markov chain process. It can be written as;

$$y_t = u_{s_t} + \phi_{s_t}(y_{t-1}) + \varepsilon_{S_t}, \varepsilon_{S_t} \sim (0, \sigma_{S_t}^2) \tag{4}$$

The regime transitions depend on the transition probabilities, which can be expressed as;

$$P(S_t = i | S_{t-1} = j) = \begin{bmatrix} P_{00} & P_{10} \\ P_{01} & P_{11} \end{bmatrix}, \tag{5}$$

where P_{ij} is the probability of transitioning from regime i, to regime j.

2.2.3 Kink Autoregression (AR) Model

Kink AR model comes from the combination of the simple AR model and the Kink model introduced by Hansen [10]. The advantage of Kink model is to have the continuous function in all variables. Meanwhile, the slope has a discontinuity at a threshold point. The lag data is separated into two (or more) groups based on its indicator function. For example, the Kink-AR (1) process of autoregressive order m can be expressed as;

$$y_t = u + \phi_{1i}y_{t-1}I(y_{t-d} \le r_i) + \phi_{2i}y_{t-i}I(y_{t-d} > r_i) + \varepsilon_t, \tag{6}$$

where y_t is interest variable, u is the mean of interest variable, ϕ_{i1} and is a lower regime autoregressive coefficients, meanwhile ϕ_{2i} is an upper regime one. I is an indicator variable by having y_{t-d} as a determinant of the switching point. r is the threshold parameter and it also represents a kink point value. It presents the regime of mean equations through its indicator function $I(y_{t-d} \le r_i)$ for a lower regime, meanwhile $I(y_{t-d} > r_i)$ for an upper one. ϵ_t is the innovation of the time series process.

2.3   Evaluation of Forecasting Model Performance

The accuracy of future forecast is the most concern. Therefore, it is essential for evaluating the forecasting performance with its forecasting reliability that it is based on out-of-sample performance. In this paper, only root mean square error (RMSE) and mean absolute error (MAE) are the selected methods used to measure forecasting accuracy and reliability.

2.3.1   Root Mean Square Error (RMSE)

RMSE benchmark evaluates the forecasting model performance by measuring the average magnitude of the error. It takes the square root of squared the difference between actual observations and forecasting values.

$$RMSE = \sqrt{\frac{1}{n} \sum_{i=1}^{n} (y_i - \hat{y}_i)^2} \tag{7}$$

2.3.2   Mean Absolute Error (MAE)

MAE is implied to measure the average magnitude of the errors, in the other word it considers only the magnitude, and ignore the error direction. Consequently, when all observation differences have the same weight, the average is over the test of the actual observations and forecasting values.

$$MAE = \frac{1}{n} \sum_{i-1}^{n} |y_t - \hat{y}_t| \tag{8}$$

The smallest values in each model generated by both RMSE and MAE are preferred and compared to each other since the smallest values denote the best forecasting ability of that model. Both values are in the range from zero to infinity to the error direction. The advantage of RMSE is to provide a relatively high weight to large errors. Therefore, it is more fitting to use when large errors are a concern. In contrast, Willmott and Matsuura [18] stated that the RMSE tends not to be an appropriate indicator of average model performance because it can lead to misleading measure of average error. Therefore, in this case MAE is more appropriate. Meanwhile Chai and Draxler [5] found that the RMSE is superior to MAE when model errors are expected to be a Gaussian distribution, and there are enough observations. Therefore, we apply both RMSE and MAE to access model performance.

3   Data

Thailand's exports to ASEAN data are monthly time series for the period from January 2002 to December 2016 consisting of 180 data. Thailand's exports to ASEAN values are calculated from the values of Thai exports to all ten ASEAN'S members; Brunei, Cambodia, Indonesia, Laos, Malaysia, Myanmar, Philippines,

Singapore and Vietnam. The export data is directly obtainable from Bank of Thailand (BOT) source. The application of the estimation method requires stationary data. We firstly use the log to transform Thailand's exports to ASEAN value, to reduce the asymmetry of time series data. Then we will apply unit roots test to check the stationarity.

In order to compare forecasting performance with respect to RMSE and MAE methods, the data is organized in the following terms. The data is firstly separated into two parts consisting of in sample and out of sample. The first 168 data contained in sample are for the period January 2001 to December 2015. The remaining data in out of sample are for the period January 2016 to December 2016. Secondly, the remaining data will be used to compare with the actual data via RMSE and MAE. The smallest RMSE and MAE is referred to better forecast performance.

4   Empirical Results

Unit roots test is initially applied to check the stationarity of the log of Thailand's exports to ASEAN value by using Augmented Dickey-Fuller (ADF) unit root test and Phillips-Perron (PP) unit root test. Table 1 shows that it is nonstationary for a unit root with drift in both ADF and PP unit root test, but it is stationary for a unit root with drift and trend. For the first difference, both unit root tests provide stationary results (Fig. 2).

Table 1. Unit root test

Test	Level		First difference	
	Constant	Constant and trend	Constant	Constant and trend
ADF	−2.158	−2.764	−19.393***	−19.461***
PP	−2.012	−3.414*	−20.572***	−21.651***

Note: *** significant at the 1% level, ** significant at the 5% level, * significant at the 10% level.

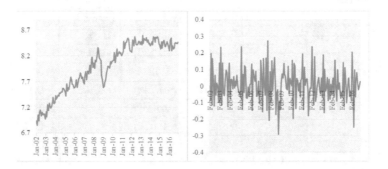

Fig. 2. Log of ASEAN series and ASEAN differentiated series

Table 2. Test of linearity

Hypothesis	Test
Test 1 vs 2	10.673*
Test 1 vs 3	13.897
Test 2 vs 3	3.041

Note: *** significant at the 1% level, ** significant at the 5% level, * significant at the 10% level.

In this paper, we employ test of linearity against bootstrap distribution from Hansen [10]. Both the linear AR versus one threshold TAR, and Linear AR versus two thresholds TAR are referred as the linearity test. If we reject the linearity, therefore there is presence of a threshold. The last test is one-threshold TAR versus two-thresholds TAR to choose whether there will be one or two thresholds. The result demonstrated in Table 2 indicates that two-threshold TAR is appropriate.

The order of the AR process is assumed as one lag for all the selected models. The fitted model criterion in this study is AIC. Table 3 shows that the SETAR

Table 3. Estimated parameters of the models

Parameter	Linear AR	SETAR	LSTAR	MSAR	Kink AR
Constant	0.011*	–	–	–	0.005
	(−0.007)	–	–	–	(−0.042)
Beta	−0.359***	–	–	–	–
	(−0.07)	–	–	–	–
Constant (L)	–	0.114***	0.274**	0.018	–
	–	(−0.04)	(−0.137)	(−0.008)	–
Beta (L)	–	0.316	1.17	−0.483***	−0.383*
	–	(−0.315)	(−0.726)	(−0.082)	(−0.196)
Constant (H)	–	0.00003	−0.272*	−0.018	–
	–	(−0.009)	(−0.139)	(−0.035)	–
Beta (H)	–	−0.249***	−1.437***	0.24	−0.336**
	–	(−0.104)	(−0.726)	(−0.301)	(−0.161)
Threshold	–	−0.066	−0.105***	–	0.013
	–	–	(−0.026)	–	(−0.138)
AIC	−797.484	−800.25	−798.815	−320.469	−311.828

Note: *** significant at the 1% level, ** significant at the 5% level, * significant at the 10% level.

displays the lowest AIC of –800.250 with its threshold value of –0.066; meanwhile the second lowest AIC of –798.815 belongs to LSTAR with its threshold value of –0.105 at the level of 1% significant, followed by Kink AR with the fifth lowest AIC of –311.828 and its threshold of 0.013.

For MSAR model, the time-varying probabilities in Table 4 show that the dependence relies on the transition probabilities with probability of remaining in the origin regime $P(S_t = 1|S_{t-1} = 1)$ being 0.9844 for the high output state and $P(S_t = 2|S_{t-1} = 2)$ being 0.8665 for the low output state.

Table 5 shows that the SETAR has the lowest errors in both RMSE and MAE criteria, which are 0.0928 and 0.0669 respectively, compared with the second lowest errors Linear AR model, 0.0989 and 0.0675 respectively. Whereas, the highest RMSE and MAE values are 0.1278 and 1.1065, which belong to Kink AR model (Figs. 3 and 4).

Table 4. Transition probability

Regime	1	2
1	0.9844	0.1334
2	0.0156	0.8665

Table 5. The forecasting performance of the model

Selected model	Linear AR	SETAR	LSTAR	MSAR	Kink AR
RMSE	0.0989**	0.0928***	0.1013	0.1102	0.1278
MAE	0.0675**	0.0669***	0.0747	0.0747	0.1065

Note: *** the first lowest values, ** the second lowest values.

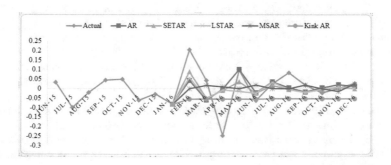

Fig. 3. Plotting actual values with predicted values of all the models

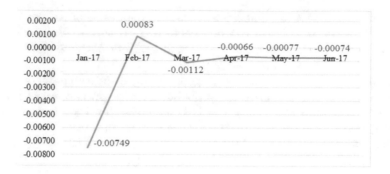

Fig. 4. Forecasting the growth of Thailand's exports to ASEAN for next 6 months

5   Conclusion

In this paper, we employ five different models namely Autoregressive (AR) model, a linear model, followed by four non-linear models, Self-exciting autoregressive (SETAR), Logistic STAR (LSTAR), Markov-switching Autoregressive (MSAR) and Kink Autoregressive (AR) models to forecast Thailand's exports to ASEAN. The results show that SETAR has the best forecasting performance in terms of lowest RMSE and MAE. The second best forecasting performance belongs to Linear AR model. Meanwhile, the worst forecasting performance is from Kink AR model.

This indicates that among different four non-linear models, the SETAR is the only method that outperforms Linear AR model. Therefore, it is not clear cut to conclude that all the non-linear models have outperformed linear forecasting model. However, we can conclude that the SETAR is the most suitable for forecasting Thailand's exports to ASEAN compared with other non-linear models.

References

1. Bec, F., Bouabdallah, O., Ferrara, L.: The way out of recessions: a forecasting analysis for some Euro area countries. Int. J. Forecast. **30**(3), 539–549 (2014)
2. Boero, G., Marrocu, E.: The performance of non-linear exchange rate models: a forecasting comparison. J. Forecast. **21**(7), 513–542 (2002)
3. Brock, W.A., Potter, S.M.: 8 Nonlinear time series and macroeconometrics. In: Maddala, G.S., Rao, C.R., Vinod, H.D. (Ed.) Handbook of statistics, North Holland, Amsterdam, vol. 11, pp. 195–229 (1993)
4. Camacho, M., Quiros, G.P., Poncela, P.: Green shoots and double dips in the euro area: a real time measure. Int. J. Forecast. **30**(3), 520–535 (2014)
5. Chai, T., Draxler, R.R.: Root mean square error (RMSE) or mean absolute error (MAE)? arguments against avoiding RMSE in the literature. Geoscientific Model Dev. **7**(3), 1247–1250 (2014)
6. Djeddour, K., Boularouk, Y.: Application of threshold autoregressive model: modeling and forecasting using US export crude oil data. Am. J. Oil Chem. Technol. **9**(1), 1–11 (2013)

7. Ferrara, L., Marcellino, M., Mogliani, M.: Macroeconomic forecasting during the great recession: the return of non-linearity? Int. J. Forecast. **31**(3), 664–679 (2015)
8. Granger, C.W.: Non-linear models: where do we go next-time varying parameter models. Stud. Nonlinear Dyn. Econometrics **12**(3), 1–9 (2008)
9. Hamilton, J.D.: A new approach to the economic analysis of nonstationary time series and the business cycle. Econometrica J. Econometric Soc. **57**, 357–384 (1989)
10. Hansen, B.: Testing for linearity. J. Econ. Surv. **13**(5), 551–576 (1999)
11. Mayer, T., Melitz, M.J., Ottaviano, G.I.: Market size, competition, and the product mix of exporters. Am. Econ. Rev. **104**(2), 495–536 (2014)
12. Teräsvirta, T., Anderson, H.M.: Characterizing nonlinearities in business cycles using smooth transition autoregressive models. J. Appl. Econometrics **7**(S1), 119–136 (1992)
13. Teräsvirta, T., Van Dijk, D., Medeiros, M.C.: Linear models, smooth transition autoregressions, and neural networks for forecasting macroeconomic time series: a re-examination. Int. J. Forecast. **21**(4), 755–774 (2005)
14. Tiao, G.C., Tsay, R.S.: Some advances in non-linear and adaptive modelling in time-series. J. Forecast. **13**(2), 109–131 (1994)
15. Tong, H.: On a threshold model. In: Chen, C. (ed.) Pattern Recognition and Signal Processing. NATO ASI Series E: Applied Sc. (29), pp. 575–586. Sijhoff and Noordhoff, Amsterdam (1978)
16. Tong, H., Lim, K.S.: Threshold autoregression, limit cycles and cyclical data. J. R. Stat. Soc. Ser. B (Methodological) **42**, 245 292 (1980)
17. Sarantis, N.: Modeling non-linearities in real effective exchange rates. J. Int. Money Finance **18**(1), 27–45 (1999)
18. Willmott, C.J., Matsuura, K.: Advantages of The mean absolute error (MAE) over The root mean square error (RMSE) in assessing average model performance. Clim. Res. **30**(1), 79–82 (2005)

Thailand in the Era of Digital Economy: How Does Digital Technology Promote Economic Growth?

Noppasit Chakpitak[1], Paravee Maneejuk[2,3]($\boxtimes$), Somsak Chanaim[2,3], and Songsak Sriboonchitta[2,3]

[1] International College, Chiang Mai University, Chiang Mai, Thailand
Nopasit@camt.info
[2] Center of Excellence in Econometrics, Chiang Mai University,
Chiang Mai 52000, Thailand
mparavee@gmail.com, somsak_ch@cmu.ac.th, songsakecon@gmail.com
[3] Faculty of Economics, Chiang Mai University, Chiang Mai 52000, Thailand

Abstract. As Thailand has undergone the reformation in both social and economic dimensions due to the digital economy, technologies are now becoming the new driving forces of economic growth. Therefore, an attempt of this study is to provide an empirical evidence on this issue, examining how increases in digital technologies impact the Thai economy. This study employs the stochastic frontier model estimated by entropy approach to model the production function. Because of a specific capability of this model, we are also able to find out how efficiently those technologies are utilized. The estimated results show that technologies can contribute positively to the Thai economy although the magnitudes are small. Moreover, our finding emphasizes that the digital technologies are not being used at the maximum capability, therefore, there is still a room for improvement in Thailand.

1 Introduction

Digital technology has profoundly become a player in our life, society as well as economy. As we can see, most people around the world are now online and spend a lot of time on the internet searching for contents, visiting social network sites, or making online purchases. Impacts of digital technology have also pervaded the ways entrepreneurs run their businesses, the ways markets are organized, and a lot more. These boundless connectivity and business transformations are examples of things that happen in the Digital Economy. But what does Digital Economy really mean? This term has become well known to people since Don Tapscott, a business executive, author, and business inspirator, introduced in his 1995 best-seller book, titled "The Digital Economy: Promise and Peril in the Age of Networked Intelligence." This book shows how internet and digital technologies impact businesses as well as our economy. Digital economy is simply an economy based on digital technologies by which economic activities are highly related to digital things. For example, customers can buy products and obtain

© Springer International Publishing AG 2018
V. Kreinovich et al. (eds.), *Predictive Econometrics and Big Data*, Studies in Computational Intelligence 753, https://doi.org/10.1007/978-3-319-70942-0_25

services through the digital transactions; businesses can use innovation and technology to improve their processes and outputs, increase their efficiency as well as profitability. Moreover, there are also many products have been invented by using advanced digital technologies, such as wireless networks and smartphone, which in turn enable network of economic activities to be more worldwide.

Thailand is one of the countries that considers the digital technologies as a new engine for driving economic growth. Let's get back to early stage of economic growth. We grew the economy by taking advantages of country's resources, in other words, the traditional drivers. But as the resources became limitedly available, the economy under traditional driving force went into a stage of maturity and eventually stagnation. As such, Thailand has shifted its attention to a new approach called "Creative Economy". This approach is employed prior to the digital economy; however, all are based upon the same thing that is knowledge (Powell and Snellman [10]). The Thai government at that time paid attention to the importance of knowledge and intellectual capabilities coupled with strong culture, in inventing new and unique products as happened in Japan's economy as well as in many developed economies to add greater products and services values. Despite the fact that this is the right way to propel the Thai economy forward, the development process under creativity could not reach the goal. This is due to many reasons: first, prices of creative products could change dramatically; and second, there were the disturbances and the unstable political situations during that time. Hence, this approach is then transferred to the present era of digitalization. The Thai government still considers the importance of knowledge and intellectual capabilities, but together with digital technologies. The government aims to grow the economy through the creation of digital technologies.

Many economists have spent their efforts examining the impacts of digitalization particularly in terms of information and communications technologies (ICTs) on the factor productivity and economic growth in empirical ways. The literature shows a significant role of digital technologies in stimulating developed economies, such as the US, Europe, Japan, and Australia (see, e.g., Qu et al. [11]; Gordon [12]; and Audretsch and Welfens [13]). However, less attention has been paid to the developing counties. In particular, based on our knowledge, the empirical study on this issue in the context of Thailand has not been conducted, so this gap becomes a great opportunity for us to fill up the gap by providing the empirical evidence on how rising in digital technologies impacts the economic growth in Thailand. In particular, we will estimate this impact using broad indicators, for example telephone and mobile cellular subscriptions, number of internet users, urban population growth, and gross capital formation. Moreover, we also aim at quantifying how efficiently those technologies are utilized. In particular, we would like to know whether Thailand can make full use of digital technology to stimulate its economy.

We will introduce a quantitative method used to accomplish these goals as well as the estimating technique and procedures through Sect. 3. But prior to that, we will describe all considered variables through a section of data description or Sect. 2. In Sect. 4, we will discuss about the estimated impacts of digital technologies on the Thai economy and also their estimated performances over time in terms of the efficiency scores. Finally, Sect. 5 concludes.

2   Data Description

As mentioned earlier, the expansion of digital economy can be seen through the continued growth of access to different technologies. However, this study is focusing on some specific indicators which can be described in the table below (Table 1).

All the variables of Thailand are collected annually from 1975 to 2015, from Thomson Reuters Datastream. In addition, we use Thailand's gross domestic product (GDP) measured in current US dollar as a dependent variable. The descriptive statistics of all considered variables are presented in Table 2, and the plots for each variable are shown in Fig. 1. This study relies on the framework of the neoclassical growth model in which the Cobb-Douglas production function is employed. But without a concern about the traditional economic driver, output is a function of only digital technologies: fixed telephone subscriptions, internet users, mobile cellular subscriptions, urban population growth, and gross capital formation.

3   Methodology

This study makes use of the stochastic frontier model (SFM) in examining how digital technologies impact the economic growth. The SFM was simultaneously proposed by Aigner et al. [1], Meeusen and van Den Broeck [3], and Battese and Coelli [2]. This model comprises two error components; one is statistical noise and the other is the technical inefficiency. The advantage of this model lies in measuring changes in output relative to changes in input variables, which are technologies. However, due to the limitation of our data, the conventional maximum likelihood for SFM is no longer appropriate. In addition, the assumption of normal distribution and half-normal distribution for the noise and technical inefficiency, respectively, becomes our concern. To improve the performance of this model, two error components should not be restricted to specific distributions, especially for a case that we only have a small sample size. To deal with this problem, this study considers the entropy estimation, introduced in Golan [9], which has been proved to be robust under the ill-posed and ill-conditioned problems. According to maximum entropy estimation, we establish a discrete set of support points for each parameter and then estimate the probability associated with each support point to arrive at the parameter estimate. For a stochastic production frontier estimated by the generalized maximum entropy (GME) approach, see, for examples, Macedo and Silva [15] and Campbell et al. [14].

Table 1. Data description

Variable	Label	Description
Fixed telephone subscriptions (per 100 people)	TEL_t	Fixed telephone subscriptions refers to the sum of active number of analoguefixed telephone lines, voice-over-IP (VoIP) subscriptions, fixed wireless local loop (WLL) subscriptions, ISDN voice-channel equivalents and fixed public payphones
Internet users (per 100 people)	INT_t	Internet users refers to the number of individuals who have used the internet in the last 12 months via computer, mobile phone, personal digital assistant, games machine, digital TV and other devices
Mobile cellular subscriptions (per 100 people)	MB_t	Mobile cellular subscriptions are subscriptions to a public mobile telephone service that provides access to the public switched telephone network (PSTN) using cellular technology. This variable includes the number of postpaid subscriptions and the number of active prepaid accounts (in the past 3 months). It excludes subscriptions via data cards or USB modems, subscriptions to public mobile data services, private trunked mobile radio, telepoint, radio paging and telemetry services
Urban population growth (annual %)	UP_t	Urban population refers to people living in urban areas as defined by national statistical offices. It is calculated using World Bank population estimates and urban ratios from the United Nations World Urbanization Prospects
Gross capital formation (Unit: current local currency)	GCF_t	Gross capital formation or gross domestic investment consists of outlays on additions to the fixed assets of the economy plus net changes in the level of inventories. Fixed assets include land improvements, plant, machinery and equipment, construction of roads, railways, and also buildings. Inventories are stocks of goods held by firms to meet temporary or unexpected fluctuations in production or sales, and work in progress. Moreover, net acquisitions of valuables are also considered capital formation

Table 2. Descriptive statistics

	$\log(GDP_t)$	$\log(TEL_t)$	$\log(INT_t)$	$\log(MB_t)$	$\log(UP_t)$	$\log(GCF_t)$
Mean	10.9906	0.539868	0.306654	0.483752	0.444884	11.90412
Median	11.09099	0.721735	0	0.221569	0.462357	12.05308
Maximum	11.62313	1.048138	1.542701	2.159684	0.697229	12.55023
Minimum	10.17268	−0.28623	−1.85387	−2.69897	0	10.90902
Std. Dev	0.414931	0.484124	0.815474	1.20145	0.18377	0.492912
Skewness	−0.27582	−0.42193	−0.40881	−0.40563	−0.50845	−0.55686
Kurtosis	2.01738	1.557321	2.9977	2.840715	2.370628	2.04439

Source: Calculation.

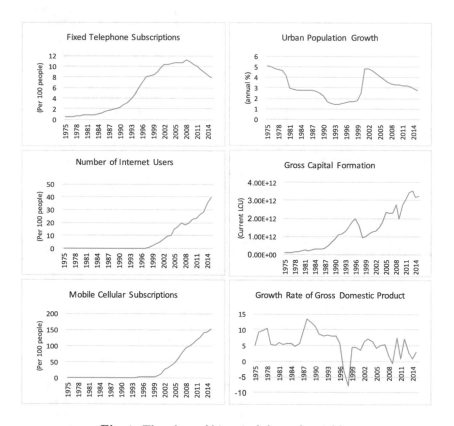

Fig. 1. The plots of historical data of variables

3.1   A Stochastic Frontier Model

The most general form of the stochastic frontier model (SFM) can be written as

$$Y_t = f(X'_{tk}\beta) \cdot TE \tag{1}$$

$$Y_t = X'_{tk}\beta + \varepsilon_t, \ t = 1, \cdots, T, \tag{2}$$

$$\varepsilon_t = U_t - V_t, \tag{3}$$

where Y_t is the output variable and X'_{tk} is $T \times K$ a matrix of K different input quantities. The term β represents $(T \times K)$ matrix of estimated parameters of the input variables. The function $f(\cdot)$ is the functional form of SFM which is imposed to be the Cobb–Douglas production function. The term TE denotes technical efficiency and ε_t is the composite error term which consists of the noise, U_t, and the inefficiency, V_t. In general, the distribution assumptions in Eq. (3) are normal distribution for U_t and half-normal for the random term, V_t. But in this study, we do not assume any distribution for U_t and V_t since the entropy estimation is proposed to estimate the model. However, a restriction on V_t is given as $V_t \in [0, \infty)$. These two error components are distributed identically and independently from each other and the regressor.

In the context of the SFM, the technical efficiency or TE can be defined as the ratio of the observed output to the corresponding frontier output, conditional on the levels of inputs. Therefore, the technical efficiency or TE is given by

$$TE_t = \frac{\exp[X_{tk} + V_t - U_t]}{\exp[X_{tk} + V_t]} \tag{4}$$

3.2   Estimating Technique: Generalized Maximum Entropy (GME) Approach

In this study, we propose to use maximum entropy estimator to estimate our unknown parameters in Eq. (2). Before, we have discussed briefly about the GME estimator. However, the advantage and the properties of this estimator are provided at length in Golan [9] and Mittelhammer et al. [7], and more recently in Tonini and Pede [5].

In brief, the core advantage of the GME estimator is that, first, it efficiently takes into account all the information contained in each data point. Second, it is less affected by outlier since we can give the probability weight between signal and noise in the objective function. Third, it is a robust estimator as it does not require any assumption regarding the error terms. And fourth, the GME estimator does not require strong behavioral assumptions on the underlying data generating process.

The maximum entropy concept consists of inferring the probability distribution that maximizes information entropy given a set of various constraints. Let p_k be a proper probability mass function on a finite set of β. Shannon [6] developed his information criteria and proposed a classical entropy, that is:

$$H(p) = -\sum_{k=1}^{K} p_k \log p_k \tag{5}$$

where $\sum_{k=1}^{K} p_k = 1$. The entropy measures the uncertainty of a distribution and reaches a maximum when p_k is uniformly distributed (Wu [8]). With the underlying model, the SFM, the objective entropy can be written as

$$H(p, w, v) = -\sum_{k=1}^{K}\sum_{m=1}^{M} p_{km} \log p_{km} - \sum_{t=1}^{T}\sum_{u=1}^{U} w_{tu} \log w_{w_{tu}} - \sum_{t=1}^{T}\sum_{n=1}^{N} v_{tn} \log v_{tn}$$

(6)

where p_{km}, w_{tu}, and v_{tn} are the probability of estimated parameters, noise U_t, and inefficiency V_t. The constraint of this objective function is given by

$$Y_t = \sum_{m=1}^{M} p_{km} z_{km} X_t^k + \sum_{m=1}^{M} w_{tm} h_{tm} - \sum_{m=1}^{M} v_{tm} g_{tm},$$

(7)

and

$$\sum_{m=1}^{M} p_{km} = 1, \ \sum_{m=1}^{M} w_{km} = 1, \ \sum_{m=1}^{M} v_{km} = 1.$$

(8)

The support values z_{km}, h_{tm}, and g_{tm}, are needed to estimate the unknown parameters, noise U_t, and inefficiency V_t in the SFM. Thus,

$$\beta_k = \sum_{m=1}^{M} p_{km} z_{km}$$

(9)

$$U_t = \sum_{m=1}^{M} w_{tm} h_{tm}$$

(10)

$$V_t = \sum_{m=1}^{M} v_{tm} g_{tm}$$

(11)

Suppose that SFM has one regressor $k = 1$; therefore, the optimization problems from Eqs. (7) and (9) can be solved by the Lagrangian method, which takes the form

$$L = H(p, w, v) + \lambda_1' \left(Y_t - \sum_{m=1}^{M} p_{km} z_{km} X_t^k - \sum_{m=1}^{M} w_{tm} h_{tm} + \sum_{m=1}^{M} v_{tm} g_{tm} \right)$$

$$+ \lambda_2'(1 - \sum_{m=1}^{M} p_m) + \lambda_3'(1 - \sum_{m=1}^{M} w_m) + \lambda_4'(1 - \sum_{m=1}^{M} v_m)$$

(12)

where λ_i', $i = 1, 2, 3, 4$ are the vectors of Lagrangian multiplier. The first-order conditions are:

$$\frac{\partial L}{\partial p_m} = -\log(p_m) - \sum_{m=1}^{M} \lambda_{1m} z_m X_t - \lambda_{2t} = 0 \tag{13}$$

$$\frac{\partial L}{\partial w_{tm}} = -\log(w_{tm}) - \sum_{m=1}^{M} \lambda_{1m} h_{tm} - \lambda_{3t} = 0 \tag{14}$$

$$\frac{\partial L}{\partial v_{tm}} = -\log(v_{tm}) - \sum_{m=1}^{M} \lambda_{1m} g_{tm} - \lambda_{4t} = 0 \tag{15}$$

$$\frac{\partial L}{\partial \lambda_1} = Y_t - \sum_{m=1}^{M} p_{km} z_{km} X_t^k - \sum_{m=1}^{M} w_{tm} h_{tm} + \sum_{m=1}^{M} v_{tm} g_{tm} = 0 \tag{16}$$

$$\frac{\partial L}{\partial \lambda_2} = 1 - \sum_{m=1}^{M} p_m \tag{17}$$

$$\frac{\partial L}{\partial \lambda_3} - 1 - \sum_{m=1}^{M} w_m \tag{18}$$

$$\frac{\partial L}{\partial \lambda_4} = 1 - \sum_{m=1}^{M} h_m \tag{19}$$

Solving the optimization of this problem, we yield the optimal and unique solution as in the following:

$$p_m = \frac{\exp[-z_m \sum_t \lambda_1 X_t]}{\sum_{m=1}^{M} \exp[-z_m \sum_t \lambda_1 X_t]} \tag{20}$$

$$w_{tm} = \frac{\exp[-\lambda_1 h_{tm}]}{\sum_{m=1}^{M} \exp[-\lambda_1 h_{tm}]} \tag{21}$$

$$v_{tm} = \frac{\exp[-\lambda_1 g_{tm}]}{\sum_{m=1}^{M} \exp[-\lambda_1 g_{tm}]} \tag{22}$$

Finally, this study employs a bootstrap approach as proposed by Efron [4] to estimate the standard error of each parameter. Bootstrapping is a general approach in statistical inference based on replacement of the true sampling distribution for a statistic by resampling from the actual data. In the bootstrap SFM procedure, the entropy method is used to estimate the parameters of the SFM. In this study, we estimate the bootstrap standard errors from 1,000 repetitions.

In this study, the supports for z_{km} are given as

$$[\widehat{\beta_k} - 3\widehat{\sigma}_{\beta_k}, \ \widehat{\beta_k} - 1.5\widehat{\sigma}_{\beta_k}, \ \widehat{\beta_k}, \ \widehat{\beta_k} + 1.5\widehat{\sigma}_{\beta_k}, \ \widehat{\beta_k} + 3\widehat{\sigma}_{\beta_k}],$$

the supports for h_{tm} as

$$[-\widehat{\sigma}_v, \ -.75\widehat{\sigma}_v, \ 0, \ .75\widehat{\sigma}_v, \ \widehat{\sigma}_v],$$

and the supports for g_{tm} as

$$\frac{\sigma_v}{6}[0, \ .75, \ 1.5, \ 2.25, \ 3].$$

We estimate β_k and σ_v by using package frontier in program R.

4   Empirical Results

This section is about the empirical results. We apply the previous algorithm to examine the impacts of digital technologies on the GDP. All considered variables within the framework of neoclassical economics, are modeled under the SFM, which is given by

$$\log(GDP)_t = \beta_0 + \beta_1 \log(TEL_t) + \beta_2 \log(INT_t) + \beta_3 \log(MB_t) + \beta_4 \log(UP_t) \\ + \ \beta_5 \log(GCF_t) + U_t + V_t.$$

Table 3 reports the bootstrap means, which are our estimated coefficients of the variables used in the growth equations above, and their 95% confident intervals from 1,000 repetitions of SFM. The result suggests that the effects of any changes in the given digital-technology variables including gross capital formation (GCF) will be translated into change in output of the country. However, the magnitude of the effects is surprisingly small. For example, the estimate for fixed telephone subscriptions (TEL) indicates that a 1% change in fixed telephone subscribers is associated with 0.0278% change in GDP. This effect size is quite similar to those of internet users (INT) and urban population growth (UP). The estimate for mobile cellular subscriptions (MB) is relatively lower in value. It can contribute just 0.0033% to the GDP. On the other hand, GCF just shows the highest effect size. We find that a 1% change in gross capital formation is associated with 0.7943% change in GDP.

The overall result indicates that technologies can contribute positively to Thailand's gross domestic product although the magnitudes are small. Figure 1 shows historical data for internet users and we can see that it increases gradually since 1996, when Thailand obtained internet access as the third country in Southeast Asia. During the second half of the 20th century, the number of internet users grew rapidly as broadband internet improved. Internet becomes readily available in most cities and towns, and expands to the countryside. Moreover, telephone service providers are also being nested within the internet, for example releasing 3G and 4G, which in turn enable people to access internet more easily. These rapid improvements make the number of internet hosts in Thailand the highest in Southeast Asia (World Factbook, [16]), and hence the growing digital economy. We believe that technologies especially internet can contribute to the economy much more than the value occurs in the empirical results. This is probably because productivity and efficiency of economic activities due to digital technologies are difficult to measure. So, we may think about a new measurement for this.

Table 3. Estimated results

Variable	Estimated coefficient	Bootstrap confidence interval 95%
Intercept	1.7303	$[1.6094, 1.8328]$
$\log(TEL_t)$	0.0278	$[-0.1612, 0.1860]$
$\log(INT_t)$	0.0101	$[-0.0202, 0.0400]$
$\log(MB_t)$	0.0033	$[-0.0114, 0.015]$
$\log(UP_t)$	0.0285	$[-0.1365, 0.2052]$
$\log(GCF_t)$	0.7943	$[0.7794, 0.8009]$

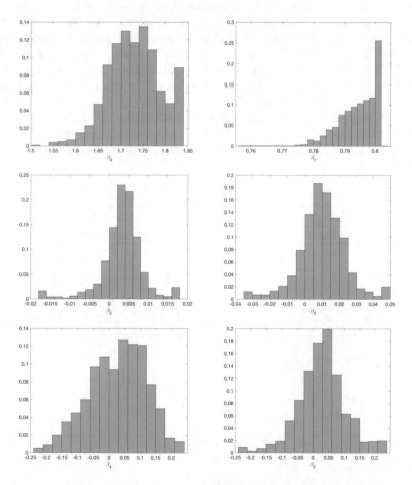

Fig. 2. Histogram plots for the bootstrap replications of parameters (β_0, β_1, β_2, β_3, β_4 and β_5) in the SFM

Figure 2 shows a graphical display of bootstrap parameters through histograms. All the bootstrap parameters in SFM are estimated by maximum entropy approach. We can see that β_0, β_2, β_3, β_4, and β_5 seem to have a reasonable shape. The bootstrap distribution of the intercept, and the coefficients of $\log(INT_t)$, $\log(MB_t)$, $\log(UP_t)$, and $\log(GCF_t)$ are reasonably symmetric.

4.1   Technical Efficiency

This section is conducted to answer the question of how efficiently those technologies are utilized. In fact, this is a contribution of the SFM allowing us to access the information about technical efficiency (TE). The technical efficiency is defined by the ratio of the observed output to the corresponding frontier output conditional on the levels of inputs used by a country. As such, the technologies are said to be used efficiently to produce output if they are used in minimum amount to produce maximum output, in other words, the highest TE score. Figure 3 displays the TE values of production under the digital economy varying between 0.765 and 0.776. The horizontal axis represents year, spanning from 1975 to 2015. This figure indicates that beyond the year 2005, Thailand can utilize the digital technologies to propel the economy in terms of GDP with higher efficiency levels than in the past. However, there is still a room for individuals, businesses, as well as the government to improve further in using the new technologies, since the TE score does not yet reach the maximum value, i.e. 1. This means the technologies and innovation are not being used at full efficiency at present.

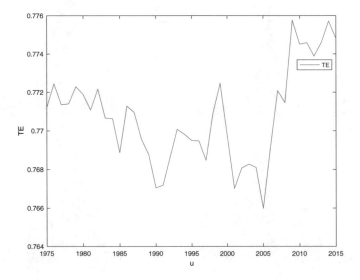

Fig. 3. Estimates of technical efficiency

5   Conclusion

Thailand nowadays undergoes the reformation in both social and economic dimensions due to the digital economy. The government, policy makers as well as academics realize the advantages of digital technologies in driving economic growth. This study attempts to provide an empirical evidence on this issue, examining how increases in digital technologies impact the Thai economy in terms of GDP. We estimate the impact using a set of variables, namely fixed telephone subscriptions, internet users, mobile cellular subscriptions, urban population growth, and gross capital formation. The stochastic frontier model estimated by entropy approach is used to model the production function. The specific capability of this model allows us to find out how efficiently those technologies are utilized. The estimated results show that technologies can contribute positively to the Thai economy although the magnitudes are small. The result on technical efficiency indicates that presently Thailand can make more use of the digital technologies than in the past. However, the result shows there is still a room for improvement in Thailand since the digital technologies are not being used at the maximum capability.

This is crucial because if we can make full use of digital technologies to produce output, to create new businesses, and to compete in a global market, and this will bring about a rise in the country's competitiveness. Apart from business, making full use of technologies will also bring about higher quality of life through an access to information and essential public services as well as entertainments. Moreover, Thai people will become digitally competent by international standards and be able to make full and creative use of technology for their jobs.

References

1. Aigner, D.J., Lovell, C.A.K., Schmidt, P.J.: Formulation and estimation of stochastic frontier production function models. J. Econ. **6**, 21–37 (1977)
2. Battese, G.E., Coelli, T.J.: Frontier production functions, technical efficiency and panel data: with application to paddy farmers in India. In: Gulledge, T.R., Lovell, C.A.K. (eds.) International Applications of Productivity and Efficiency Analysis, pp. 149–165. Springer, Dordrecht (1992)
3. Meeusen, W., van Den Broeck, J.: Efficiency estimation from Cobb-Douglas production functions with composed error. Int. Econ. Rev. **18**, 435–444 (1977)
4. Efron, B.: Bootstrap methods: another look at the jackknife. Ann. Stat. **7**, 1–26 (1979)
5. Tonini, A., Pede, V.: A generalized maximum entropy stochastic frontier measuring productivity accounting for spatial dependency. Entropy **13**(11), 1916–1927 (2011)
6. Shannon, C.E.: A mathematical theory of communication. ACM SIGMOBILE Mob. Comput. Commun. Rev. **5**(1), 3–55 (2001)
7. Mittelhammer, R.C., Judge, G.G., Miller, D.J.: Econometric Foundations Pack with CD-ROM, vol. 1. Cambridge University Press, Cambridge (2000)
8. Wu, X.: A weighted generalized maximum entropy estimator with a data-driven weight. Entropy **11**(4), 917–930 (2009)

9. Golan, A.: Information and entropy econometrics-editor's view. J. Econ. **107**(1), 1–15 (2002)
10. Powell, W.W., Snellman, K.: The knowledge economy. Annu. Rev. Sociol. **30**, 199–220 (2004)
11. Qu, J., Simes, R., O'Mahony, J.: How do digital technologies drive economic growth? Econ. Record **93**(S1), 57–69 (2016)
12. Gordon, R.J.: Secular stagnation on the supply side: US producivity growth in the long run. Commun. Strateg. **1**(100), 19–45 (2015)
13. Audretsch, D.B., Welfens, P.J.: The New Economy and Economic Growth in Europe and the US. Springer Science & Business Media, Heidelberg (2013)
14. Campbell, R., Rogers, K., Rezek, J.: Efficient frontier estimation: a maximum entropy approach. J. Prod. Anal. **30**(3), 213–221 (2008)
15. Macedo, P., Silva, E.: A stochastic production frontier model with a translog specification using the generalized maximum entropy estimator. Econ. Bull. **1**, 587–596 (2010)
16. United States. Central Intelligence Agency, Government Publications Office (eds.): The World Factbook 2014–15. Government Printing Office (2015)

Comparing Linear and Nonlinear Models in Forecasting Telephone Subscriptions Using Likelihood Based Belief Functions

Noppasit Chakpitak[1], Woraphon Yamaka[2(✉)], and Songsak Sriboonchitta[2]

[1] International College, Chiang Mai University, Chiang Mai, Thailand
[2] Faculty of Economics, Centre of Excellence in Econometrics,
Chiang Mai University, Chiang Mai, Thailand
woraphon.econ@gmail.com

Abstract. In this paper, we experiment with several different models with belief function to forecast Thai telephone subscribers. This approach will provide an uncertainty about predicted values and yield a predictive belief function that quantities the uncertainty about the future data. The proposed forecasting models include linear AR, Kink AR, Threshold AR, and Markov Switching AR models. Next, we compare the out-of-sample performance using RM3E and MAE. The results suggest that the out-of-sample belief function based KAR forecast is more accurate than other models. Finally, we find that the growth rate of Thai telephone subscription in 2016 will fall around 6.08%.

Keywords: Telephone subscriptions
Likelihood-based belief functions
Linear and nonlinear autoregressive model · Forecasting

1 Introduction

Over the last decade, the rapid development of mobile phone and the expansion of the Internet, information and communications technologies are decreasing the fixed telephone subscriptions. Telephone subscription here refers to the sum of active number of analogue fixed telephone lines, voice-over-IP (VoIP) subscriptions, fixed wireless local loop (WLL) subscriptions, ISDN voice-channel equivalents and fixed public payphones. Thus, forecasting the number of telephone subscribers is very important for resource allocation and network planning. Moreover, it can help network providers optimize their resources, obviate the waste of resources, and prepare the appropriate policy during this downturn of fixed telephone industry. So, a potent forecasting model is needed to obtain accurate predictions.

In the context of forecasting, we desire to have a prediction of the future growth of telephone subscription and assess uncertainty about them. Our study considers the work of Kanjanatarakul et al. [8] which proposed the use of theory of belief functions as an approach to achieve predicted intervals which is based on a mathematical theory of Shafer [13]. This approach involves two steps.

© Springer International Publishing AG 2018
V. Kreinovich et al. (eds.), *Predictive Econometrics and Big Data*, Studies in Computational Intelligence 753, https://doi.org/10.1007/978-3-319-70942-0_26

First step, a belief function on the parameter space in the forecasting model is estimated from the normalized likelihood given the observed data. Second, the prediction is made by writing the prediction value as a function of the parameter and an auxiliary random variable with known distribution. This approach is applied to the estimation by using the likelihood-based approach introduced in Shafer [13] and further justified in Denouex [5]. The application to the prediction problem can be found in the study of Kanjanatarakul et al. [8,9]. In this study, we have interest to employ this approach to make a forecast, however, we are worried about the selection of the appropriate forecasting models in the first step. As in the literature, the application of belief function based forecasting models have been found in many studies, such as Kanjanatarakul et al. [8,9], Autchariyapanitkul et al. [1], Thianpaen et al. [11], but these studies were conducted under a linear assumption. In economic data, there might exist both linear and nonlinear behaviours, and hence using a wrong forecasting model might bring about a large prediction bias [10].

The purpose of this study is to find the appropriate model to predict the future growth of telephone subscription in Thailand. Therefore we need to compare the forecasting performance of linear and nonlinear models based on belief function. In this study, we consider four competing Autoregressive (AR) models, that are linear model, Threshold model of Tong [12], Kink model of Chan and Tsay [2] and Hansen [7], and Markov Switching of Hamilton [6]. Our aim is to examine whether forecasts from a nonlinear model are preferable to those from a linear model in terms of forecast accuracy as well as forecast encompassing.

The remainder of this article is organized as follows. In the second section, we briefly review linear and nonlinear AR models. In the third section, the notion of likelihood based belief function will be recalled. In the fourth section, we explain the construction of a forecasting problem. In the fifth section, we forecast the growth of fixed telephone subscription. The final section provides a summary and concludes.

2   A Brief Review of Linear and Nonlinear Models

2.1   Linear AR Model

Let $\{y_t\}$, $t = 1, ..., T$ be the observed data which are expressed as a linear combination of past observations y_{t-p}. In the autoregressive model of order (p), we have

$$y_t = \alpha + \sum_{p=1}^{P} \beta_p y_{t-p} + \varepsilon_t \tag{1}$$

where β_p is the autoregressive coefficient at order (p). ε_t is the white noise which is assumed to be normally distributed with mean zero and variance σ^2.

2.2   Kink AR Model

The Kink model is proposed by Chan and Tsay [2] and Hansen [7] in the context of nonlinear regression, but in this study we use nonlinear autoregression context. The basic idea of kink autoregression is that a piecewise linear autoregression segment is split by threshold or kink parameter which focuses on lag variables (y_{t-p}). In this study we consider two regimes, hence the KAR model can be described as

$$y_t = \alpha_0 + \sum_{p=1}^{P} \beta_p^- (y_{t-p} \leq r_p) + \sum_{p=1}^{P} \beta_p^+ (y_{t-p} > r_p) + \varepsilon_t, \qquad (2)$$

where α_0 is the intercept term, β_p^- and β_p^+ are lower and upper regime autoregressive coefficients, respectively, and r_p is the threshold parameter or kink point value defining the regime through indicator function $(y_{t-p} \leq r_p)$ and $(y_{t-p} > r_p)$ for upper regime. The ε_t term is the innovations of the time series process, assumed to be normally distributed, $\varepsilon_t \sim N(0, \sigma^2)$.

2.3   Threshold AR Model

The Threshold model of Tong [12] is a nonlinear model which is decomposed into two regimes, inside which is autoregression, and the sudden transition across regimes is controlled by threshold variable r. This model is called "self-exciting threshold autoregression (SETAR)". It can be described as a piecewise linear approximation to the general univariate autoregressive model of order (p) such that

$$\begin{array}{ll} y_t = \alpha_0^{(1)} + \sum_{p=1}^{P} \beta_p^{(1)} y_{t-p} + \varepsilon_{1,t} & if \ y_{t-d} < r \\ y_t = \alpha_0^{(2)} + \sum_{p=1}^{P} \beta_p^{(2)} y_{t-p} + \varepsilon_{2,t} & if \ y_{t-d} \geq r \end{array} \qquad (3)$$

where $\alpha_0^{(j)}$ and $\beta_p^{(j)}$ are the estimated parameters of regime j in the model, r is threshold parameter and $\varepsilon_t \sim N(0, \sigma^2)$ is an $T \times 1$ vector of independent and identically distributed (iid) errors with normal distribution. Each regime is the function of the past realizations of y_t sequence itself. If $y_{t-d} < r$, the model is in lower regime at time t. In this model, we assume that the lag order (p) is the same in both regimes.

2.4   A Markov Switching AR Model

In the autoregression-type structure, we can establish the Markov switching autoregression model as

$$y_t = \beta_{0,S_t} + \sum_{p=1}^{p} \beta_{p,S_t} y_{t-p} + \varepsilon_{t,S_t} \qquad (4)$$

where $\varepsilon_{t,S_t} \sim i.i.d. \ N(0, \sigma_{S_t}^2)$, $y_{i,t}$ is dependent variable and y_{t-p} is lag dependent variable at time $t - p$. β_{0,S_t} denotes regime de-pendent intercept term β_{p,S_t} and

β_{p,S_t} denote the regime dependent autoregressive coefficients of AR(p). The state (or regime) is represented by S_t. Let $\{S_t\}_{t=1}^{T}$ be the finite state Markov chain with state or regime $\{1, ..., h\}$, thus the stationary transition probabilities

$$p_{ij} = P(S_{t+1} = i \,|\, S_t = j), \tag{5}$$

where p_{ij} is the probability of transition from regime i at time $t+1$ conditional on regime j at time t and $\sum_{j=1}^{h} p_{ij} = 1$. Thus, we can write the transition matrix as

$$P = \begin{bmatrix} p_{11} & p_{12} & \cdots & p_{1j} \\ p_{21} & \ddots & & p_{2j} \\ \vdots & & \ddots & \vdots \\ p_{i1} & p_{i2} & \cdots & p_{ij} \end{bmatrix} \tag{6}$$

3   Likelihood-Based Belief Function

Shafer proposed, on intuitive grounds, a more direct approach in which a belief function Bel_y^{Θ} on Θ is built from the likelihood function. In the recent writing with paper, Denoeux [5] justified three basic principles of likelihood based belief function, consisting the likelihood principle, compatibility with Bayesian inference and the least commitment principle. Let $y \in \mathbb{R}$ denote the observed data and $\theta \in \Theta$ is unknown parameter in the model. The likelihood function is a mapping L_y from Θ to $[0, \infty)$ which is defined by

$$L_y(\theta) = c f_\theta(y). \tag{7}$$

where $c > 0$ is an arbitrary multiplicative constant. When we rescale the likelihood into the interval [0,1], by transformation we obtain the relative likelihood

$$R_y(\theta) = \frac{L(\theta)}{L(\overline{\theta})}, \tag{8}$$

where $L(\theta)$ is a likelihood function and $L(\overline{\theta}) = \sup \theta \in \Theta L(\theta)$, i.e., maximum likelihood estimate (MLE) of θ. Shafer [13] mentioned that this relative likelihood refers to a "relative plausibility". Thus, it can be considered as the contour function $pl_y(\theta)$ of a belief function Bel_y^{Θ} on Θ

$$pl_y(\theta) = R_y(\theta) = \frac{L(\theta)}{L(\overline{\theta})} \tag{9}$$

If Bel_y^{Θ} is assumed to be consonant belief function defined from the normalized likelihood function given observed data, then the plausibility is given by

$$Pl_y^{\Theta}(H) = \sup_{\theta \in H} pl_y(\theta), \tag{10}$$

for any hypothesis $H \subseteq \Theta$. The focal sets of Bel_y^Θ, which are the plausibility regions, can be defined as the set of parameter values θ whose relative plausibility greater than some threshold

$$\Gamma_y(\omega) = \{\theta \in \Theta \,|pl_y(\theta) \geq \omega\}, \tag{11}$$

where ω is uniformly in $[0,1]$. This belief function Bel_y^Θ is equivalent to the random set induced by the Lebesgue measure λ on $[0,1]$ and the multi-valued mapping Γ_y from $[0,1] \to 2^\Theta$ [8].

4   Forecasting

In this study, we will propose a general solution to the forecasting problem in the context of linear and nonlinear models. The forecasting problem is the inverse of the previous one: given some knowledge about θ obtained by observing y, we now can make statements about some random quantity $y \in \mathbb{R}$ whose conditional distribution given y depends on θ.

4.1   Problem Solution

To forecast the future data $y_f = (y_{T+1}, ..., y_{T+h})$, the sampling model used by Dempster [3,4] is introduced here. In this model, the forecast data Y_f is expressed as a function of the proper probability mass function θ which is obtained by past observed data $y = (y_1, ..., y_T)$, and an unobserved auxiliary $s \in S$ variable with known probability measure μ not depending on θ

$$y_f = \varphi(\theta, s), \tag{12}$$

where φ is defined in such a way that the distribution of Y_f for fixed θ is $g_{z,\theta}(y_f)$. When y_f is a continuous random variable, Eq.(12) can be computed by

$$y_f = G_{y,\theta}^{-1}(s), \tag{13}$$

where $G_{y,\theta}^{-1}$ is the inverse conditional cumulative distribution function (cdf) of $Y_f \,|y$ and s is uniformly in $[0,1]$. Let Γ_y be the multi-valued mapping from $[0,1] \times U \to 2^Y$ with φ, we get a new multi-valued mapping Γ_y from $[0,1] \times S \to 2^Y$ defined as

$$\begin{aligned} \Gamma_y &: [0,1] \times U \to 2^Y \\ (\omega, s) &\to \varphi(\Gamma_y(\omega, s)). \end{aligned} \tag{14}$$

The predictive belief function Bel_y and plausibility Pl_y are then induced by the multi-valued mapping Γ_y and $\lambda \otimes \mu$ on $[0,1] \times U$ as follows:

$$Bel_y(H) = (\lambda \otimes \mu)\{(\omega_i, s_i)\,|\varphi_y(\Gamma_y(\omega_i, s_i) \subseteq H\}, \tag{15}$$

$$Pl_y(H) = Bel_y(H) = (\lambda \otimes \mu)\{(\omega_i, s_i)\,|\varphi_y(\Gamma_y(\omega_i, s_i) \cap H \neq \phi\}, \tag{16}$$

for all $H \subseteq Y$.

4.2   Example linear and nonlinear model: prediction

Easy to explain, here we rewrite our linear and three nonlinear models in Eqs.(1)–(4) as

$$y_t = f(\theta \,|y_{t-p}),\tag{17}$$

where $f(\cdot)$ is the distribution function of y_t for fixed θ. The data y_t is expressed as a function of the parameter θ and observed in the past y_{t-p}. Then, we can write, equivalently,

$$f_y(\theta, s) = f(\theta\,|y_{t-p}) + \sigma\phi^{-1}(s),\;\; U[0,1]\tag{18}$$

where $\phi(\cdot)$ is the normal quantile function and s has a uniform distribution in the interval [0,1].

To forecast $y_t \,|y_{t-1}$, the predictive belief function and plausibility function can then be approximated using Monte Carlo simulation as in the following:

(1) We draw ω_i and s_i independently from uniform [0,1] N draws. Then we calculate the focal sets, the interval defined by the following lower and upper bounds:

$$\begin{aligned}f_y(\Gamma_y(\omega, s)) &= \{f(\theta, s)\,|pl_y(\theta) \geq \omega\,\}\\&= [y^L(s_i, \omega_i), y^U(s_i, \omega_i)]\end{aligned}\tag{19}$$

where $y^L(s_i, \omega_i)$ and $y^U(s_i, \omega_i)$ is a lower bound and upper bound, respectively, which have normal likelihood $N(\min f_y(\omega_i, s_i))$and $N(max f_y(\omega_i, s_i))$, respectively. This optimization problem can be solved using the nonlinear optimization which takes the form as

$$y^L(s_i, \omega_i) = \min f_y(\omega_i, s_i)\tag{20}$$

subject to

$$pl_y(\theta) \geq \omega_i,\tag{21}$$

And

$$y^U(s_i, \omega_i) = \max f_y(\omega_i, s_i)\tag{22}$$

subject to

$$pl_y(\theta) \geq \omega_i,\tag{23}$$

(2) Thus, the predictive belief (Bel_y) and plausibility (Pl_y) functions that Y will be less than or equal to y_t are

$$Bel_y([0, y]) \approx \frac{1}{N}\sum_{i=1}^{N} F_{y_{t-p}, \min f_y(\omega_i, s_i)}(y),\tag{24}$$

$$Pl_y([0, y]) \approx \frac{1}{N}\sum_{i=1}^{N} F_{y_{t-p}, \max f_y(\omega_i, s_i)}(y),\tag{25}$$

Similarly, the lower and upper expectations of y^L and y^U with respect to Bel_y and Pl_y, respectively, can be approximated by

$$\bar{y}^L = \frac{1}{N} \sum_{i=1}^{N} \min f_y(\omega_i, s_i) \tag{26}$$

$$\bar{y}^U = \frac{1}{N} \sum_{i=1}^{N} \max f_y(\omega_i, s_i) \tag{27}$$

Finally, to obtain the predictive y_f, we take the mean of the sum of $\bar{y}^L$ and $\bar{y}^U$, that is

$$y_f = \frac{\bar{y}^L + \bar{y}^U}{2} \tag{28}$$

4.3   Constructing a Likelihood Function of Linear and Nonlinear Models

According to the definition of the likelihood function based belief function as described in Sect. 2, the likelihood $L(\theta \,|y)$ is very important and the approximation of Bel_y and Pl_y should only depend on the likelihood function. In this study, we consider 4 competing likelihood models: AR, Kink AR, Threshold AR, and Markov Switching AR where their functions can be written as follows:

(1) AR likelihood

$$L(\theta \,|y) = \frac{1}{\sqrt{2\pi\sigma_t^2}} \left(\frac{1}{2\sigma_j^2} (y_t - \alpha - \sum_{p=1}^{P} \beta_p y_{t-p})^2 \right) \tag{29}$$

(2) Kink likelihood

$$L(\theta \,|y) = \frac{1}{\sqrt{2\pi\sigma_t^2}} \left(\frac{1}{2\sigma_j^2} (y_t - \alpha_0 - \sum_{p=1}^{P} \beta_p^- (y_{t-p} \le r_p) - \sum_{p=1}^{P} \beta_p^+ (y_{t-p} > r_p))^2 \right) \tag{30}$$

(3) Threshold likelihood

$$L(\theta \,|y) = \prod_{j=1}^{2} \left\{ \frac{1}{\sqrt{2\pi\sigma_j^2}} \left(\frac{1}{2\sigma_j^2} (y_t - \alpha_0^{(1)} - \sum_{p=1}^{P} \beta_p^{(1)} y_{t-p})^2 \right) \right\} I(y_{t-d} < r) \cdot$$

$$\left\{ \frac{1}{\sqrt{2\pi\sigma_j^2}} \left(\frac{1}{2\sigma_j^2} (y_t - \alpha_0^{(2)} - \sum_{p=1}^{P} \beta_p^{(2)} y_{t-p})^2 \right) \right\} I(y_{t-d} \ge r) \tag{31}$$

(4) Markov Switching likelihood

$$L(\theta \,|y) = \sum_{j=1}^{2} \frac{1}{\sqrt{2\pi\sigma_{S_t=j}^2}} \left(\frac{1}{2\sigma_{S_t=j}^2}(y_t - \beta_{0,S_t=j} - \sum_{p=1}^{p} \beta_{p,S_t=j}y_{t-p})^2 \right)(S_t = j\,|\theta_{t-1})\,,$$

(32)

where $\Pr(S_t = j\,|\theta_{t-1})$ is filter probabilities obtained from Hamilton filter [6].

5 Forecasting the growth of fixed telephone subscription

Given the special characteristics of the growth of fixed telephone subscription, a number of alternative forecasting models have been proposed, including linear AR, KAR, TAR and MS-AR. The data set considered, derived from the Thomson Reuters Data stream, consists of yearly data, from the ending of 1975 to 2015, covering 40 observations. The data is plotted in Fig. 1. Then, we perform Augmented Dickey Fuller unit root test on fixed telephone subscription growth, and use model of constant term and trend term for testing, and determine lagging order number of AR model according to Akaiki Information Criterion (AIC). The ADF-statistic result is -1.4954, which is greater than the critical value at 1% significance level (-3.6104). This indicates that the growth of telephone subscription is stationary. Then, the lag selection for four competing models is shown in Table 1. The results show that lag 1 provides the lowest AIC for all models.

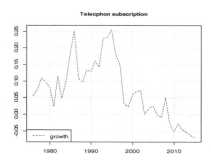

Fig. 1. Growth rate of telephone subscription in Thailand.

5.1 Assessing Forecasting Performance: Out of Sample Forecast

To compare the forecasting performance, our study focuses on the out-of-sample forecasting ability of the nonlinear and the linear models. It is important to note that we examine exclusively the out-of-sample forecasting ability of the four competing models. The out-of –sample forecasts refer to the period from

Table 1. Lag selection for linear and nonlinear models

AR	Lag(1)	Lag(2)	Lag(3)	Lag(4)
AIC	−237.0589	−235.3267	−234.066	−232.8945
KAR	Lag(1)	Lag(2)	Lag(3)	Lag(4)
AIC	−96.54266	−95.1245	−94.9974	−93.4575
TAR	Lag(1)	Lag(2)	Lag(3)	Lag(4)
AIC	−235.7989	−233.1144	−235.2477	NaN
MS-AR	Lag(1)	Lag(2)	Lag(3)	Lag(4)
AIC	−107.3489	−104.4907	NaN	NaN

Source: Calculation

Note : NaN refers to the estimation faced with the small degrees of freedom thus causing a maximum likelihood to fail to converge.

2011 to 2015, which is the testing period for our models. We perform a recursive forecasting to update telephone subscription growth year by year, totally 5 years, which give us enough data points to evaluate out-of sample forecast performance. We compare the out of sample forecasts using two approaches, namely the root mean square error (RMSE) and mean square error (MAE).We can compute the RMSE and MAE as in below:

$$MAE = \frac{1}{N} \sum_{j=1}^{N} |e_j|,$$

$$RMSE = \sqrt{\frac{1}{N} \sum_{j=1}^{N} e_j^2},$$

where $N=5$ and $e = y_{true} - y_f$.

The RMSE and MSE of the forecasts are reported in Table 2. The out of sample fore-casts made in the 2011-2015 are based on four different models. Here, the smaller the MAE and RMSE, the superior the corresponding model's performance. As indicated in this Table, the RMSE and MSE of the forecasts from the belief function based KAR model is lower than the other belief function based models. In addition, we also compare the performance of belief based model and non-belief based model. We can observe the better forecast when belief function is take into account. This implies that belief function based KAR outperforms, and in general more accurate than, the other forecasting models. We expect that if the model has a nonlinear characteristic and the assumption of the likelihood structure based belief function is not correctly specified, a less accurate prediction is obtained. Thus, we conclude that for telephone subscription growth, the belief function based KAR model is preferred to the other models on the basis of the forecast accuracy approach.

Table 2. Out of sample forecasting performance

Belief based model	Actual	AR(1)	KAR(1)	TAR(1)	MS-AR(1)
Forecast 2011	−0.028	−0.0234	−0.0307	−0.0247	0.045
Forecast 2012	−0.0456	−0.0388	−0.0392	−0.0128	−0.0566
Forecast 2013	−0.0535	0.0116	−0.0325	0.0019	−1.8343
Forecast 2014	−0.0634	0.0084	−0.0201	0.0212	0.7838
Forecast 2015	−0.0694	0.0092	−0.0552	0.0145	−1.2531
MAE		0.0454	0.0174	0.0606	0.7791
RMSE		0.0559	0.0226	0.052	1.0291
Non-Belief based model	Actual	AR(1)	KAR(1)	TAR(1)	MS-AR(1)
Forecast 2011	−0.028	−0.0259	−0.0106	−0.0297	−0.0226
Forecast 2012	−0.0456	−0.0043	−0.025	−0.0121	0.0086
Forecast 2013	−0.0535	0.0127	0.0052	0.0024	−0.0014
Forecast 2014	−0.0634	0.0263	−0.0052	0.0147	0.0204
Forecast 2015	−0.0694	0.0371	0.0187	0.0249	0.0141
MAE		0.0714	0.0553	0.0621	0.0627
RMSE		0.0611	0.0486	0.0527	0.0558

Source: Calculation

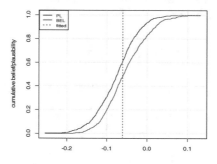

Fig. 2. Lower and upper cdf for one step ahead prediction: The forecasts made for 2016, with blue lines corresponding to the estimated expected values of forecast value..

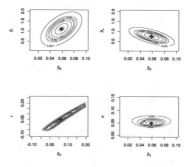

Fig. 3. Marginal contour functions $pl_y(\theta)$ in two-dimensional parameter sub-spaces: Kink AR model

5.2 One Step-Ahead Prediction for Likelihood Based Belief Function

The prediction is very important because of its relevance to the current policy debate on whether Thai telephone subscription growth will be up or down. The forecasts were made for year 2016. We then use belief function based KAR to predict telephone subscription growth in the future. Figure 2 shows the predictive belief function Bel^Y concerning telephone subscription growth in 2016. These plots pre-sent the lower and upper cumulative density functions (cdf) for the prediction problem obtained from $Pl_y(H)$ and $Bel_y(H)$. The expected prediction y_f in 2016 is plotted by the vertical red dotted line. We can observe that the value of growth rate is around -0.0608 in 2016.

Furthermore, Fig. 3 shows the accuracy of our estimation using the contour plots of plausibility $pl_y(\theta)$ in two-dimensional parameter subspaces of belief function based Kink AR model. These contour plots show two-dimensional slices of the contour function, where one of the five parameters is fixed to its maximum likelihood. The results of these contour plots show that the estimated parameters of the model can reach the maximum plausibility, as presented by the red cross. According to Fig. 3, we provide four sub-plots of plausibility function of two parameters. For example, consider the top left sub plot, we have a plausibility function $pl(\beta_0, \beta_1^-)$ of two parameters intercept β_0, and coefficient β_1^-. In the profile plausibility of the intercept parameter, we fix and just vary β_1^- and vice versa. We can observe that the estimated parameters of Kink likelihood can reach the maxi-mum plausibility. For other parameters, we also observe good optimization.

6 Concluding Remarks

The paper compared out-of-sample forecasts of yearly growth of Thai telephone subscription, generated by belief function based forecasting models. In this study, we proposed various linear and nonlinear AR models, namely linear AR, TAR, KAR, and MS-AR models, with belief function framework for the Thai telephone subscription. Prior to making a prediction value, Akaiki information criterion (AIC) is conducted to select the appropriate lag length for the four competing models. We found that lag 1 is appropriate since it showed the minimum value of AIC. The comparison of out-of-sample forecasts is carried out on the basis of two methods: RMSE and MAE. Over-all, the results indicate that the inclusion of nonlinear structure in the forecasting model is important in out-of-sample forecast. We find that belief function based KAR performs well in out-of-sample forecast relative to other models. Finally, we apply the best fit model to predict the growth rate of Thai telephone sub-scription in 2016 and find that the growth rate will fall around 6.08%.

Acknowledgements. The authors are grateful to Puay Ungphakorn Centre of Excellence in Econometrics, Faculty of Economics, Chiang Mai University for the financial support.

References

1. Autchariyapanitkul, K., Chanaim, S., Sriboonchitta, S., Denoeux, T.: Predicting stock returns in the capital asset pricing model using quantile regression and belief functions. In: Third International Conference Belief Functions: Theory and Applications, September 2014, Oxford, United-Kingdom. LNAI, vol. 8764, pp. 219–226. Springer (2014)
2. Chan, K., Tsay, R.S.: Limiting properties of the least squares estimator of a continuous threshold autoregressive model. Biometrika **45**, 413–426 (1998)
3. Dempster, A.P.: New methods for reasoning towards posterior distributions based on sample data. Ann. Math. Stat. **37**, 355–374 (1966)
4. Dempster, A.P.: The Dempster-Shafer calculus for statisticians. Int. J. Approx. Reason. **48**(2), 365–377 (2008)
5. Denouex, T.: Likelihood-based belief function: justification and some extensions to low-quality data. Int. J. Approx. Reason. **55**, 1535–1547 (2014)
6. Hamilton, J.D.: A new approach to the economic analysis of nonstationary time series and the business cycle. Econom. J. Econom. Soci. **57**, 357–384 (1989)
7. Hansen, B.E.: Regression kink with an unknown threshold. J. Bus. Econ. Stat. **35**, 228–240 (2017)
8. Kanjanatarakul, O., Sriboonchitta, S., Denoeux, T.: Forecasting using belief functions: an application to marketing econometrics. Int. J. Approx. Reason. **55**(5), 1113–1128 (2014)
9. Kanjanatarakul, O., Denoeux, T., Sriboonchitta, S.: Prediction of future observations using belief functions: a likelihood-based approach. Int. J. Approx. Reason. **72**, 71–94 (2016)
10. Khiewngamdee, C., Yamaka, W., Sriboonchitta, S. Forecasting asian credit default swap spreads: a comparison of multi-regime models. In: Robustness in Econometrics, pp. 471–489. Springer International Publishing (2017)
11. Thianpaen, N., Liu, J., Sriboonchitta, S.: Time series forecast using AR-belief approach. Thai J. Math. **14**, 527–541 (2016)
12. Tong, H.: Threshold Models in Non-linear Time Series Analysis. Lecture Notes in Statistics, vol. 21. Springer, Berlin (1983)
13. Shafer, G.A.: Mathematical Theory of Evidence. Princeton University Press, Princeton (1976)

Volatility in Thailand Stock Market Using High-Frequency Data

Saowaluk Duangin[1]([✉]), Jirakom Sirisrisakulchai[2], and Songsak Sriboonchitta[2]

[1] Faculty of Economics, Chiang Mai University, Chiang Mai 52000, Thailand
Saowaluk.econ@gmail.com
[2] Center of Excellence in Econometrics, Faculty of Economics,
Chiang Mai University, Chiang Mai 52000, Thailand
Sirisrisakulchai@hotmail.com, songsakecon@gmail.com

Abstract. The objective of this research is twofold: First, we aim to investigate the performance of conventional GARCH and GARCH-jump models when the data has high frequency. Second, the obtained conditional volatility from the best fit model is used to forecast and matched with the macroeconomic news announcement. We use GARCH and GARCH-jump models with high-frequency dataset of log return of Thailand stock market index (SET) from January, 2008 to December, 2015. We find that the volatility estimations by these two models have the same pattern but volatility estimation by GARCH-jump is higher than conventional GARCH model. However, the GARCH (1,1) and GARCH (1,1)-jump performances are non-stationary to estimate the volatility for 5 min interval return of SET but are stationary to estimate for 15 min, 30 min, 1 h, and 2 h returns of SET. Our results also show the matching jump point with macroeconomic news announcement. The empirical results support our assumption that macroeconomic news announcement may lead to volatility change in SET.

1 Introduction

There are many indicators can reflect the economy's health such as Gross Domestic Product (GDP), Consumer Price Index (CPI), unemployment rate, etc. However, the stock market can be used as an indicator of economic health. When an economy move to expansion period then the stock market expected to rising. There are many endogenous and exogenous factors affecting a stock market such as economic factors, world events, politics, natural disasters, etc. It is well known that new information can affect stock prices. The efficient market hypothesis (EMH) posits that "if markets are efficient and current, prices always reflect all available information". News is a major driving force of asset price volatility with implications on many segments of the economy both domestically and event globally. Investors need to know how news release affects stock return and volatility. When new information occurs, the stock market will respond to both bad and good news. Some traders and investors try to reason out or extract the information contained in news to re-allocate their portfolio and decide on risk management.

© Springer International Publishing AG 2018
V. Kreinovich et al. (eds.), *Predictive Econometrics and Big Data*, Studies in Computational Intelligence 753, https://doi.org/10.1007/978-3-319-70942-0_27

Macroeconomic reports such as gross domestic product, consumer price index, inflation rate, and unemployment rate are important sources of information transcribing the changes in economic fundamental indicators or policies into different metaphors of return and volatility of stock market. On August 1, 2008 consumer price index (CPI)[1] was announced to be 2.70%, which was well below the expected value of 3.85% reported by Bloomberg. The CPI was released at 10.30 am and the Stock Exchange of Thailand (SET) index went up sharply after the news release. The impact of macroeconomic news announcements on stock market has previously been focused on the link between macroeconomic news and developed markets such as U.S., U.K. and Germany. Andersen and Bollerslev [2] and Rangel [3] found significant impact of macroeconomic announcements on stock market.

Only few studies have investigated the impacts of macroeconomic news announcementa on Thailand stock market. Sidney [4] used low frequency data (daily and monthly data) and showed the impact of domestic and foreign macroeconomic announcements on SET by using GARCH type model estimation and found that SET responded to unexpected news but the impact of Thai news announcement on SET seemed higher than foreign news announcement.

However, stock will show quick reaction to news in the order of minutes or less. Currently, with the advanced computers and communications technology, the data are readily available in high frequency. The high-frequency data have a large sample size and complex structure. The high-frequency data is the data at frequencies higher than daily data. A recent research using high-frequency data at 5 min interval to investigate the impact of macroeconomic announcements on stock market and found that high-frequency data are significant for the identification of impact of news on market.

Hussain [5] investigated the response of European and U.S. equity markets to monetary policy. He found that the monetary policy decision had immediate influence on both European and U.S. stock markets. He also suggested that high-frequency data can separate the effect of monetary policy action from that of macroeconomic news announcement. Mastronardi et al. [6] estimated the impact of news announcements on Italian equity market. The results indicated that the response of the returns to the news is very quick, at least within ten minutes. Gurgul et al. [7] analyzed the impact of U.S. macroeconomic news announcements on Warsaw Stock Exchange and found that investors reacted to U.S. macroeconomic news immediately and quickly. Huang [8] studied the response of financial market to macroeconomic news announcements and found the significant jumps on news day more than no-news day and market response to news varied with economic situation.

For Thailand stock market, only Wongswan [9] used minute-by-minute data and found that the announcements from developed economies exerted influence on Thailand's stock market. Moreover, Wongswan [9] also suggested that the

---

[1] World Bank [1] "Consumer price index reflects changes in the cost to the average consumer of acquiring a basket of goods and services that may be fixed or changed at specified intervals, such as yearly."

important domestic macroeconomic announcements such as GDP and inflation had significant impact on SET as well.

The standard GARCH model has been applied and widely used to study the effect of macroeconomic news on stock market with low-frequency data. This standard GARCH model is appropriate when good and bad news have asymmetric shocks on return and volatility. Nikkinen et al. [10] used GARCH model for analyzing the behavior of volatilities for U.S. macroeconomic news on each of the 35 stock markets. Vrugt [11] used GARCH models that allow asymmetries, multiplicative and pre- and post- announcement effect for investigating the impact of U.S. and Japanese macroeconomic news announcements on each of the 4 stock markets.

Mastronardi et al. [6] estimated the impact of macroeconomic news on Italian equity market and focused on the overall impact of macroeconomic news using GARCH models to extract the conditional returns and variances. Ibrahim and Topuz [12] used GARCH model to examine the effect of news announcements on volatility behavior. Their results showed U.S. and Turkish GDP announcements had significant impact on volatility of Turkish stock market.

Previous studies used GARCH-type models that focused on conditional return and conditional variance of stock market returns. Hanousek et al. [13] estimated the impact of macroeconomic news announcements on European stock market using GARCH-M model which included conditional variance and conditional return. Brenner et al. [14] used the parsimonious multivariate GARCH-DCC model for study on volatility and co-movement of U.S. financial market upon macroeconomic news release. Caken et al. [15] assumed that a negative news will generate shock more than positive news. They also found that GJR-GARCH model was an appropriate estimation for analyzing impact of U.S. macroeconomic news announcements on stock markets. Wallenius et al. [16] applied EGARCH model with a Guassian normal distribution of error to study the effect of EU macroeconomic news on volatility of CIVETS stock markets.

However, GARCH-jump model has been applied in few studies. GARCH-Jump model is used to capture the volatility when jump occurs. Lu et al. [17] applied GARCH-jump model and bivariate GARCH model with correlated Poisson jump and compared these models with standard two dimensional GARCH models. Their result suggested that the performance of GARCH-jump model is better than GARCH model without jump. Rangel [3] presented Poisson-Gaussian-GARCH process with time-varying jump intensity for large market movement to examine the effect of macroeconomic announcements on stock market. Sidorov et al. [18] suggested GARCH-jump model augmented with news intensity to improve the power of GARCH-jump model. They compared these models and found that GARCH-jump model augmented with news intensity performs slightly better than GARCH model without jumps.

The objective of this research is twofold: First, we aim to investigate the performance of conventional GARCH and GARCH-jump models when the data has high frequency. Second, the obtained conditional volatility from the best fit model is forecasted and used to capture the effect of macroeconomic news announcements. This paper contributes to the existing literature in the way that econometric model contributes to the literature on GARCH-jump model with four distributions. We also contribute to the literature on estimating volatility using Thailand's stock market high-frequency data (5 min, 10 min, 1 h, and 2 h frequencies) from January 04, 2008 to December 29, 2015. Using this methodological framework, we analyze the following research question: "How Thailand's stock market reacts when macroeconomic news is released?"

The remainder of this study is organized as follows. Section 2 outlines the theoretical and assumption. Section 3 describes the model. Section 4 describes the data with summary statistics. Section 5 presents the empirical results. Finally, Sect. 6 contains conclusion.

2   Theoretical and Hypothesis

2.1   Theoretical

Macroeconomic news announcement influence stock market returns when news changes in the information set. The stock price will response to the surprise news. The stock price link with information set as Eq. (1) which stock price (P_t) equal the present discounted value of expected earning (d) given information set at time $t(I_t)$ with the discount rate (r_t).

$$P_t = E\left(\sum_{\tau=1}^{\infty} \frac{d_{t+\tau}}{1 + r_{t+\tau}} \middle| I_t\right) \tag{1}$$

Macroeconomic news announcements influence stock market returns when there is a change in the information set. The stock price will respond to the surprising news. The stock price links with information set as in Eq. (1) where stock price (P_t) equals the present discounted value of expected earning (d) given information set at time $t(I_t)$ with the discount rate (r_t).

The stock price is determined by many factors such as the risk-free rate of interest, growth expectation, and equity risk premium. Hence macroeconomic news has influence on stock price when the news leads to change in one or more of these factors. For example, an unexpected increase of GDP in the announcement may lead to the upward movement of stock price by decreasing the discount factor.

From the literature, Funke and Matsuda [19] suggested many hypotheses to explain the effect of macroeconomic news on return and volatility of stock market. In this study, we observe the following assumptions:

(i) Macroeconomic news has immediate effect on stock market

Many studies showed that stock market reacts within minutes after news release. Rangel [3] showed the results that the effect of shocks seems to have a short duration. Mastronardi et al. [6] suggested that the stock return response to the news is very quick at least within 10 min. Gurgul et al. [7] also suggested that investors immediately and quickly react to U.S. macroeconomic news.

(ii) Stock market will have different responses to different types of macroeconomic news

For example, interest rate that plays an important role in the economy may have a stronger effect on stock market than other news releases. Rangel [3] showed that only PPI and inflation shock have persistent effect on the market. Sidney [4] suggested that after financial crisis, oil price announcement has effect on Thai stock market (41%) more than the effect of GDP announcement (21%).

(iii) Macroeconomic news effect on stock market depends on the economic condition

Sometime, the stock market is a general measure of the state of economy. During economic expansion (recession), an economy grows (contracts) and generally leads to good (bad) stock market performance better than the time with lower (higher) rate of economic growth. For example, stock market performed badly during financial crisis in 2007–2008. De Goeij [20] suggested that the stocks respond differently to the same macroeconomic news in the different state of economy. They also presented the opposite effect of news as the good news during recession state is a bad news for stock market.

3   Model Description

The GARCH model proposed by Bollerslev [21] is an efficient way to deal with stock price data which usually observe the volatility clustering. From previous studies, GARCH model is most widely adopted to investigate the impact of news on stock price. However, we focus on the effect of macroeconomic news announcements on conditional means and variance of SET, and the conventional GARCH models might not be appropriate for evaluation in this study. To overcome this problem, we consider the GARCH-jump model proposed by Jorion [22] and developed by Maheuand McCurdy [23]. However, it is not clear whether news adds any value to a GARCH-jump model.

3.1   Univariate GARCH Model

The GARCH model is an important tool for analyzing the financial data especially to forecast volatility. In the application study, GRACH (1,1) model is sufficient to capture the volatility clustering in the data (Bollerslev, Chou and Kroner, [24]). Thus, the GARCH (1,1) specification for the volatility model following Bollerslev [21] can be expressed as

$$R_t = \alpha_0 + \varepsilon_t, \tag{2}$$

$$\varepsilon_t = \sigma_t z_t \tag{3}$$

$$\sigma_t^2 = \beta_0 + \beta_1 \varepsilon_{t-1}^2 + \beta_2 \sigma_{t-1}^2, \tag{4}$$

where R_t is log return of SET index from time $t-1$ to t, ε_t is the error term, z_t is standardize residual which is assumed to be normal (N), student-t (sT), skewed student-t (ssT), and skewed normal (sN) distributions. σ_t^2 is conditional variance as presented in Eq. (3). In this model, we need some restriction as the follows: $\beta_0, \beta_1, \beta_3 \geq 0$ and $\beta_2 + \beta_3 \leq 1$ which make the GARCH process stable.

3.2   GARCH-Jump Intensive Model

To estimate the impact of macroeconomic news on return and volatility of SET, GARCH-jump intensity model is used following Rangel [3]. We assume that investors know all information in time $t-1$, when they make investment decision at time t. In GARCH-jump model, it is assumed that the news process has two separate components which cause two types of error. The GARCH (1,1) with jump can be expressed as

$$R_t = \alpha_0 + (\varepsilon_{1,t} + \varepsilon_{2,t}), \tag{5}$$

$$\sigma_t^2 = \beta_0 + \beta_1 \left(\varepsilon_{1,t} + \varepsilon_{2,t} \right)^2 + \beta_2 \sigma_{t-1}^2. \tag{6}$$

Suppose the news process have two components are normal and unusual news. Thus, the error consists of two error components: $\varepsilon_{1,t}$ and $\varepsilon_{2,t}$ where $\varepsilon_{1,t}$ denotes normal news while $\varepsilon_{2,t}$ denotes unusual news announcements. In the first component errors reflects the impact of normal news on volatility such that:

$$\varepsilon_{1,t} = z_t \sigma_t, \tag{7}$$

where z_t assumed to be normal distribution, $z_t \sim N(0,1)$, student-t distribution, $z_t \sim sT(0,1,d.f.)$, skewed normal distribution, $ssN(0,1,\nu)$, skewed student-t distribution, $z_t \sim ssT(0,1,d.f.,\nu)$, $d.f.$ is degrees of freedom, and ν is skewness parameter.

Then, let J_t is cumulative jump size from $t-1$ to t, $J_t = \sum_{k=1}^{n_t} Y_{t,k}$ The second component $\varepsilon_{2,t}$ is defined by

$$\varepsilon_{2,t} = J_t - E(J_t | I_{t-1}), \tag{8}$$

where $Y_{t,k}$ is size of k-jump that occur from time $t-1$ to t, $1 \leq k \leq \eta_t$, and it is assumed to random form $Y_{t,k} \sim N(\theta, \delta^2)$. I_{t-1} denote the past information set containing the realized values of all relevant variables up to time $t-1$. η_t is number of jumps between time $t-1$ and t, which is random from Poisson distribution and its conditional density is

$$P(n_t = j | I_{t-1}) = \frac{\exp(-\lambda_t) \lambda_t^j}{j!}, \quad j = 0, 1, \cdots \tag{9}$$

where λ_t be intensity parameter of Poisson distribution and the conditional jump intensity has a form as an Autoregressive-moving-average (ARMA) process:

$$\lambda_t = a + b(\lambda_{t-1}) + c\xi_{t-1}, \tag{10}$$

where ξ_{t-1} is an intensity residual which is defined as

$$\xi_{t-1} = E\left(n_{t-1}|I_{t-1}\right) - \lambda_{t-1} = \sum_{j=0}^{\infty} j \cdot P(n_{t-1} = j|I_{t-1}) - \lambda_{t-1} \tag{11}$$

Therefore, following the derivation of Sidorov et al. [17] the error of unexpected news becomes

$$\varepsilon_{2.t} = \sum_{k=1}^{n_t} Y_{t,k} - \theta\lambda_t. \tag{12}$$

To estimate the unknown parameters in model, we conduct a maximum likelihood estimator (MLE) to estimate the unknown parameters and obtained the estimated parameters of our models. In addition, the Akaike information criterion (AIC) is conducted to compare the performance of standard GARCH and GARCH-jump models.

4   Data and Descriptive Statistics

4.1   The Stock Exchange of Thailand

We are interested in the short-lived effects of macroeconomic news announcements on the Stock Exchange of Thailand (SET). So, we choose high-frequency data set, which is obtained from Bloomberg database. The sample period is from the first trading day of 2008 to the last trading day of 2015 corresponding to the trading hours which give 116,806 return observations.

SET trading hours contain two sessions which are (i) Morning session: open at random between 9.55–10.00 am and close at 12.30 pm and (ii) Afternoon session: open at random between 2.25–2.30 am and close at random between 4.35–4.40 pm in Bangkok time (GMT+07:00). The data has been removed for the overnight and lunch break time. Moreover, from assumption (2) that assumes investors to react differently in different states of economy including financial crisis, we then separate the data set into two groups: crisis period (2008–2011) and non-crisis period (2012–2015). Table 1 present the list of the group of samples include 5 min, 10 min, 1 h, and 2 h data set. From the results of JB test that presented in Table 1 we can conclude to reject null hypothesis of normality test for all data sets. We use the return on SET ($R_{i,t}$) on the intervals i on trading day t define by price changes of the log of SET index;

$$R_{i,t} = \ln\left[\frac{SET_{i,t}}{SET_{i,t-1}}\right] \tag{13}$$

Table 1. Data description

	Thailand stock market index				
	5 min	15 min	30 min	1 h	2 h
Min	−0.08492	−0.085985	−0.08113	−0.08418	−0.08418
Max	0.073738	0.500774	0.078079	0.055464	0.055464
Mean	0.000003	0.000011	0.000017	0.000037	0.000062
SD	0.001415	0.003478	0.003139	0.004341	0.005506
variance	0.000002	0.000012	0.00001	0.000019	0.00003
Skewness	−1.112303	80.23905	−0.288609	−0.573509	19.24633
Kurtosis	309.421	11635.34	61.6277	29.7679	13.3329
Obs.	116806	37012	23497	11160	7007
JB test	−2.20E-16	−2.20E-16	−2.20E-16	−2.20E-16	−2.20E-16

4.2 Macroeconomic News

We examine the effect of macroeconomic news announcements on SET by focusing on different types of macroeconomic news for Thailand and U.S. The news related to three different areas are (i) news about the real sectors such as GDP and trade balance, (ii) news about the indicators such as consumer price index, and (iii) news about interest rate. Macroeconomic news is derived from Bloomberg database and in the case of the interest rate, it is derived from data provided by Bank of Thailand (BOT).

5 Empirical Result

As mentioned in the introduction the objective of this paper is to investigate the effect of macroeconomic news on Thai stock market. The GARCH models are used to estimate with different error distributions, and then the results are compared to choose the appropriate model for forecasting the conditional variance. There will be three main steps being studied. First the model selection will be investigated with Akaike Information Criterion (AIC) score. The second step is to show the performance of GARCH (1,1) and GARCH-jump models. The final step is matching jump with macroeconomic news announcement.

5.1 Model Selection

For GARCH and GARCH-jump models, four different types of error term distribution are considered in this paper. The error distribution is assumed to be a normal, Student's t-distribution, skewed normal and skewed Student's t-distribution. Therefore, AIC scores as presented in Table 2 are used to compare the fit of four distributions of GARCH (1,1) and GARCH (1,1) -jump models for each sample set for model selection. The model with the lowest AIC score indicates the best fit model with the given data.

According to AIC values in Table 2, sixteen models of GARCH (1,1) and GARCH-jump models are fit with skewed Student's t-distribution while seven models fit with Student's t-distribution.

5.2   GARCH and GARCH-Jump Model Performance

Based on the results of comparing the fit of error distributions, then we compare the different volatilities of GARCH (1,1) and GARCH-jump models. Table 3 shows the maximum likelihood estimates for conventional GARCH (1,1) and GARCH-jump model for log return of 5 min, 15 min, 30 min, 1 h, and 2 h interval of SET index. Moreover, the estimated variances of GARCH (1,1) and GARCH-jump models based on the fit distributions are shown in Figs. 1, 2, 3, 4 and 5.

The estimates of β_1 in GARCH (1,1) model are positive and slightly bigger than GARCH-jump model. All estimates of β_1 are significant at 1% level implying that volatility is governed by the error term of the mean equation in both GARCH (1,1) and GARCH-jump models. The estimates of β_2 in GARCH (1,1) and GARCH-jump models also positive, indicating that the part of variance in the previous period is carried over into the period.

For volatility persistence, $(\beta_1 + \beta_2)$ in GARCH model ranging from 0.9966 to 1.378 in average is more than one, suggesting that the results are non-stationary. This implies that the return in current period has no effect on the conditional variance in the future in GARCH model. While volatility persistence in GARCH-jump model ranging from 0.04490 to 1.41 with the average of 0.759821 is closer to one, thus suggesting that the results are stationary in GARCH-jump model. This implies that current period's return has a significant effect on the forecasted variance for many periods in the future. However, the volatility persistence of both GARCH and GARCH-jump models for 5 min interval calculated at 1.378 and 1.41 respectively is more than one, indicating that the 5 min return in current period has no effect on the conditional variance in the future.

The results presented in Table 3, also include the maximum likelihood estimate of jump and size of jump for GARCH-jump model. The estimates of parameter, θ, are positive implying that jumps are associated with positive movement in the price. Only the estimate of parameter, θ, of full sample 5 min interval for GARCH-jump model is negative. That means the jump is related to the negative moment in price.

The average size of jump, δ, for each sample group is different for different sample sets. For full sample of each return interval, jump size ranges from 0.18 to 0.90 with the average of 0.1177 with the full sample of 5 min interval having the largest jump size. For crisis period sample sets, jump size ranges from 0.091 to 0.12132 or averagely 0.1023. In addition, jump size of non-crisis period sample sets ranges from 0.101 to 0.143 or on average 0.12248.

To compare the fit of the two models for each sample group, AIC values can be used. For all sample sets except 5 min interval sample set, the AIC values of GARCH (1,1) model are lower than GARCH-jump model. Therefore, most of high-frequency data sample sets of SET are fit with GARCH (1,1) model.

Table 2. Akaike information criterion (AIC) of GARCH (1,1) and GARCH-jump for log return of SET index

Distribution	Normal distribution		Student's t distribution		Skewed normal distribution		Skewed student's t distribution	
AIC	Model 1	Model 2	Model 1	Model 2	Model 1	Model 2	Model 1	Model 2
5 min interval								
Full sample	−10.42926	−9.510514	−11.40829[a]	−11.05668[a]	−10.91726	−10.16307	−11.40828	−10.98551
Crisis	−8.721228	−9.508688	−9.669448	−10.57239	−10.49894	−9.598413	−10.96492[a]	−10.57329[a]
Non-crisis	−10.10628	−10.35973	−11.84327	−10.57239	−11.1569	−9.107664	−11.84351[a]	−11.57781[a]
15 min interval								
Full sample	−9.048074	−9.513328	−10.16938	−9.886377	−9.613663	−8.235634	−10.17038[a]	−9.900273[a]
Crisis	−8.721228	−8.569735[a]	−8.993172	−8.440301	−9.158108	−8.100525	−9.670103[a]	−8.440301
Non-crisis	−10.10628	−9.631970[a]	−10.51820[a]	−9.440204	−10.11536	−9.559096	NaN	−9.439454
30 min interval								
Full sample	−9.048385	−8.758237	−9.437911	−9.156204[a]	−9.048562	−8.731171	−9.438803[a]	−9.09839
Crisis	−8.531825	−8.331201[a]	−8.993172	−8.128483	−8.532273	−8.286697	−8.993449[a]	−6.378013
Non-crisis	−9.543667	−9.344096	−9.853482	−9.716459	−9.546483	−8.096991	−9.854886[a]	−9.728172[a]
1 h interval								
Full sample	−8.461647	−8.078114[a]	−8.685571	−7.600734	−8.42547	−6.252816	−8.685884[a]	−7.247264
Crisis	−7.851969	−7.582842[a]	−8.084771[a]	−6.76081	−7.851615	−7.57741	−8.084513	−6.765364
Non-crisis	−8.912634	−8.693096	−9.120711	−6.601504	−8.916469	−8.702813[a]	−9.121122[a]	−6.714225
2 h interval								
Full sample	−7.968553	−7.593852	−8.126073	−7.455873	−7.970744	−7.602265[a]	−8.126544[a]	−7.007179
Crisis	−7.437588	−7.145501	−7.599856[a]	−7.023626	−7.437127	−7.149645	−7.59956	−7.380108[a]
Non-crisis	−8.40484	−8.152931	−8.547664	−8.307640[a]	−8.412274	−8.165664	−8.548172[a]	−7.143676

Source: Calculation

Note: [a] is the smallest AIC value for each sample set.

Table 3. Maximum likelihood estimates of GARCH (1,1) and GARCH-jump for log return of SET index

	α_0	β_0	β_1	β_2	ν	d.f.	θ	δ	λ
5 min interval									
Full sample									
Model 1	1.84E-05	3.05E-07	1.00E+00	3.78E-01		2.53E+00			
	-1.70E-06	-1.13E-08	-3.66E-02	-1.07E-02		-2.08E-02			
Model 2	0.00001	0.0012	1	0.4145		2.46841	-0.00284	0.18611	0.18586
	-3.40E-06	-1.78E-06	-2.38E-01	-1.63E-03		-3.61E-03	7.30E-05	-2.03E-02	-1.43E-02
Crisis period									
Model 1	1.64E-05	5.13E-07	9.23E-01	2.91E-01	9.71E-01	2.72E+00			
	-4.92E-06	-2.06E-08	-3.78E-02	-1.24E-02	-6.35E-03	-3.35E-02			
Model 2	0.00001	0.00132	0.92238	0.25057	0.98791	2.7156	0.08952	0.09148	0.08834
	-8.31E-06	-1.84E-06	-2.67E-01	-0.00206	-0.01629	-0.01394	-0.00105	-0.01529	-0.00118
Non-crisis period									
Model 1	2.05E-05	2.88E-07	1.00E+00	3.20E-01	1.02E+00	2.46E+00			
	-2.95E-06	-1.48E-08	-5.28E-02	-1.74E-02	-5.13E-03	-2.50E-02			
Model 2	0.00001	0.00084	1	0.34864	0.97194	2.4345	0.0466	0.14316	0.09504
	-4.14E-06	-1.71E-06	-4.48E-01	-7.74E-04	-8.13E-03	-0.0067	-2.30E-04	-0.01543	-1.56E-03
15 min interval									
Full sample									
Model 1	3.64E-06	4.79E-08	3.96E-02	9.57E-01	9.59E-01	2.62E+00			
	-7.72E-06	-3.22E-09	-2.37E-03	-1.77E-03	-6.42E-03	-3.72E-02			
Model 2	0.00001	0.00008	0.03971	0.95734	0.95936	2.62612	0.09992	0.10007	0.10006
	-1.22E-05	1.81E-06	-0.0102	-4.60E-04	-7.92E-03	-1.53E-02	-0.00018	-0.02396	-1.19E-03
Crisis period									
Model 1	-1.50E-05	2.43E-07	5.90E-02	9.25E-01	9.62E-01	2.51E+00			
	-1.65E-05	-2.70E-08	-6.33E-03	-4.85E-03	-1.05E-02	-5.38E-02			
Model 2	0.00001	0.00085	0.10066	0.80009			0.10065	0.09993	0.10066
	-4.20E-05	-1.94E-06	-0.04726	-1.55E-03			-0.00076	-0.01751	-1.82E-03
Non-crisis period									
Model 1	3.79E-05	2.14E-08	0.02638	0.97242		2.65424			
	-6.60E-06	-2.57E-09	-0.00235	-0.00175		-0.05433			
Model 2	1.06E-05	1.30E-03	1.20E-02	4.60E-01			8.99E-02	1.01E-01	9.26E-02
	-2.06E-05	-1.90E-06	-4.62E-01	-3.31E-03			-0.0006	-0.01622	-1.47E-03
30 min interval									
Full sample									
Model 1	2.61E-05	4.54E-08	3.22E-02	9.69E-01	9.62E-01	2.75E+00			
	-1.37E-05	-5.11E-09	-2.52E-03	-1.92E-03	-8.17E-03	-5.36E-02			
Model 2	4.49E-05	1.04E-04	3.23E-02	9.69E-01		2.76E+00	9.99E-02	1.00E-01	1.00E-01
	-2.19E-05	-1.87E-06	-3.36E-02	-8.76E-05		-2.53E-02	-7.53E-04	-2.47E-02	-1.10E-03
Crisis period									
Model 1	1.63E-05	1.18E-07	3.44E-02	9.65E-01	9.72E-01	2.64E+00			
	-2.61E-05	-2.02E-08	-4.29E-03	-3.38E-03	-1.20E-02	-7.15E-02			
Model 2	1.31E-05	2.43E-03	1.83E-02	3.62E-01			9.53E-02	1.00E-01	9.72E-02
	-4.09E-05	-1.97E-06	-1.98E-01	-5.14E-03			-0.00177	-0.01924	-0.00351
Non-crisis period									
Model 1	2.60E-05	4.23E-08	3.12E-02	9.66E-01	9.48E-01	2.88E+00			
	-1.63E-05	-6.70E-09	-3.38E-03	-3.10E-03	-1.15E-02	-8.27E-02			
Model 2	0.00001	0.00001	0.08241	0.99393	0.89861	4.30075	0.0441	0.14712	0.14832
	-1.63E-05	-1.94E-06	-1.83E-02	-1.34E-04	-1.56E-02	-2.67E-01	-1.92E-03	-0.04281	-2.42E-02
1 h interval									
Full sample									
Model 1	8.66E-05	7.91E-08	4.48E-02	9.57E-01	9.71E-01	3.14E+00			
	-2.78E-05	-1.42E-08	-4.36E-03	-3.45E-02	-1.20E-02	-1.03E-01			
Model 2	0.00001	0.00291	0.03221	0.31299			0.0906	0.10065	0.09468
	-4.43E-05	-1.97E-06	-1.46E-01	-5.74E-03			-0.00225	-0.02428	-0.00526

(*continued*)

Table 3. (*continued*)

	α_0	β_0	β_1	β_2	ν	d.f.	θ	δ	λ
Crisis period									
Model 1	1.42E-04	1.69E-07	3.93E-02	9.59E-01		3.23E+00			
	−4.91E-05	−4.60E-08	−6.19E-03	−5.43E-03		−1.67E-01			
Model 2	3.24E-05	3.26E-03	2.42E-02	3.50E-01			9.60E-02	1.00E-01	9.77E-02
	−8.02E-05	−1.99E-06	−2.23E-01	−8.71E-03			−0.0083	−0.04236	−0.01979
Non-crisis period									
Model 1	7.74E-05	1.03E-07	4.12E-02	9.55E-01	9.65E-01	3.09E+00			
	−3.18E-05	−2.24E-08	−5.62E-03	−5.36E-03	−1.59E-02	−1.34E-01			
Model 2	0.00001	0.00271	0.03483	0.09899	0.98105		0.06597	0.10178	0.07876
	−4.06E-05	−1.98E-06	−1.66E-01	−1.21E-02	−1.72E-02		−0.00528	−0.03613	−8.30E-03
2 h interval									
Full sample									
Model 1	1.57E-04	1.78E-07	5.47E-02	9.43E-01	9.64E-01	3.77E+00			
	−4.58E-05	−3.66E-08	−5.91E-03	−5.38E-03	−1.51E-02	−1.80E-01			
Model 2	0.00001	0.005	0.04489	0.00001	0.98699		0.0935	0.1016	0.09227
	−6.75E-05	−9.87E-05	−1.82E-01	−2.42E-02	−1.99E-02		−0.00555	−0.03468	−3.03E-03
Crisis period									
Model 1	2.66E-04	3.26E-07	4.98E-02	9.46E-01		3.87E+00			
	−7.95E-05	−1.02E-07	−8.22E-03	−7.75E-03		−2.81E-01			
Model 2	0.00001	0.0008	0.07221	0.87993	0.95257	3.82591	0.12097	0.121323	0.12108
	−2.01E-04	−1.99E-06	−1.30E-01	−1.58E-03	−6.27E-02	−0.18319	−0.00057	−0.08332	−4.60E-03
Non-crisis period									
Model 1	1.37E-04	2.30E-07	4.95E-02	9.39E-01	9.58E-01	3.73E+00			
	−5.30E-05	−5.63E-08	−7.66E-03	−8.55E-03	−2.08E-02	−2.41E-01			
Model 2	0.00001	0.00029	0.06957	0.93256		3.66788	0.11757	0.11761	0.11759
	−7.53E-05	−1.98E-06	−1.63E-01	−7.08E-04		−0.16301	−0.00173	−0.06376	−0.00056

Source: Calculation.
Notes: Model 1 and Model 2 are GARCH (1,1) and GARCH-jump model respectively. Standard error (SE) are in parentheses.

Figure 1 top panel is volatility plot of 5 min interval of SET using the GARCH (1,1) model in Eqs. (2–4). Figure 1 bottom panel plots volatility of 5 min interval of SET using the GARCH-jump model in Eqs. (5–12). We can see that they have the same movement but volatility of GARCH-jump is much higher and changes more dramatically over time than GARCH (1,1) model. Figures 2, 3, 4 and 5 also show the higher volatility in GRACH-jump model, more than GARCH (1,1) model of 15 min, 30 min, 1 h, and 2 h interval sample sets which can be considered the same as the 5 min sample set.

5.3   Matching Jumps with Macroeconomic News Announcement

In this section, we study the role of macroeconomic news announcements on SET volatility during the 5 min time point of all 116,806 trading days. From Fig. 1 bottom panel, we focus some point of jumps to match with Thai and U.S. news announcements. The highest jump is on October 8, 2008 (11243th observation), corresponding to Bank of Thailand announcement on interest rate. Moreover, there appears volatility of SET jump in many points on the day of Band of Thailand announcement on interest rate such as on 11 March 2008 (expected = 2.30%,

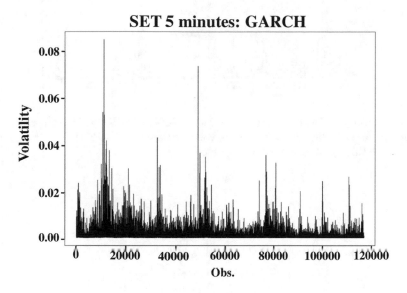

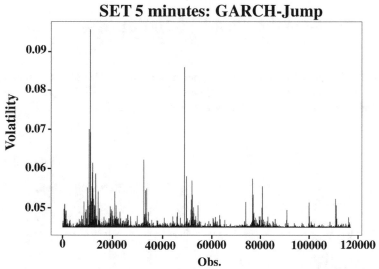

Fig. 1. Volatility of 5 min interval SET index: 2008:01–2015:12. Note: Vertical scales for each panel are different. Top panel is volatility plots of GARCH (1,1) model. Bottom panel is volatility plot of GARCH-jump model.

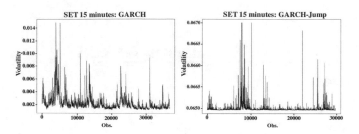

Fig. 2. Volatility of 15 min interval SET index: 2008:01–2015:12. Note: Vertical scales for each panel are different. Left panel is volatility plots of GARCH (1,1) model. Right panel is volatility plot of GARCH-jump model.

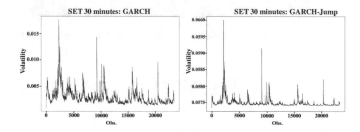

Fig. 3. Volatility of 30 min interval SET index: 2008:01–2015:12. Note: Vertical scales for each panel are different. Left panel is volatility plots of GARCH (1,1) model. Right panel is volatility plot of GARCH-jump model.

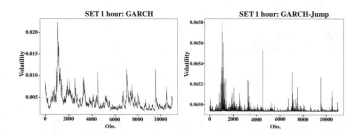

Fig. 4. Volatility of 1 h interval SET index: 2008:01–2015:12. Note: Vertical scales for each panel are different. Left panel is volatility plots of GARCH (1,1) model. Right panel is volatility plot of GARCH-jump model.

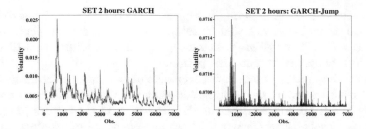

Fig. 5. Volatility of 2 h interval SET index: 2008:01–2015:12. Note: Vertical scales for each panel are different. Left panel is volatility plots of GARCH (1,1) model. Right panel is volatility plot of GARCH-jump model.

actual = 2.40%), 1 May 2009 (expected = 1.70%, actual = 1.80%) and 12 January 2011 (expected = 2.95%, actual = GARCH-jump2.90%).

The volatility of SET jump also took place on the day that Thailand foreign reserve announcement was made such as on 10 October 2008 (expected = $103.5b, actual = $102.1b), 4 April 2010 (expected = $144.8b, actual = $144.2b), and 12 December 2014 (expected − $157.1b, actual − $158.5b).

For U.S. macroeconomic news, we examine two types of news which are consumer price index (CPI) and gross domestic product (GDP) announcement. The CPI announcements match with volatility jump point on 30 September 2008 (expected = 5.60%, actual = 5.40%) and 15 May 2009 (expected = −0.60%, actual = −0.70%) etc. In addition, GDP announcement matches with jump point on 30 October 2008 (expected = −0.50%, actual = −0.30%). However, some points of jump have many news announcements such as on 16 January 2008, Thailand's interest rate was announced at 3.25% and U.S.'s CPI was announced at 4.10%.

6 Conclusions

This study examines the effect of macroeconomic news announcements on Thai stock market. The key assumption is that the news has a short-lived effect on return and volatility of SET. We study by using high-frequency data of SET return including 5 min, 10 min, 1 h, and 2 h data sets from January 04, 2008 to December 29, 2015.

The results show that the GARCH (1,1) and GARCH-jump model for most sample sets are fit with skewed Student's t-distribution and Student's t-distribution. Our results also show the estimated coefficients of GARCH (1,1) are positive and slightly bigger than those of GARCH-jump model. Moreover, the results indicate that the volatility estimation of two models have the same pattern but volatility of GARCH-jumps is higher than standard GARCH model. The finding also suggested that 5 min interval data set is not sufficient to estimate the volatility in GARCH (1,1) and GARCH-jump process, but these models are stationary for the lower data sets such as 15 min, 30 min, 1 h, and 2 h interval.

Jump size in non-crisis period is bigger than crisis-period. This result contradicts the general evidence that investors are more uncertain about economy during recession. Then the stock market should have a higher volatility during crisis period compared to non-crisis period.

By matching jump point with macroeconomic news announcements, our empirical results support our assumption that macroeconomic news announcement may lead to volatility change in SET. However, at the time of jump in SET return, many events also occurred. There emerge some suggestions for future work. Firstly, we need to test jump occurring to confirm that the surprising macroeconomic news announcements can cause jumps in the SET by following the jump detection of Huang [8] and Yao and Tian [25].

Secondly, macroeconomic news announcements can make jump of SET by increasing or decreasing volatility. Future works may incorporate Markov regime-switching in GARCH-jump model to estimate size and duration of the effect of macroeconomic news announcements on SET.

Acknowledgement. We gratefully acknowledge the support and generosity of Chiang Mai University and Centre of Excellence in Econometrics, Faculty of Economics, Chiang Mai University, without which the present study could not have been completed.

References

1. World Bank, Consumer Price Index for Thailand [DDOE02THA086NWDB], retrieved from FRED, Federal Reserve Bank of St. Louis, 6 August 2017. https://fred.stlouisfed.org/series/DDOE02THA086NWDB
2. Andersen, T.G., Bollerslev, T., Diebold, F.X., Vega, C.: Real-time price discovery in global stock, bond and foreign exchange markets. J. Int. Econ. **73**(2), 251–277 (2007)
3. Rangel, J.G.: Macroeconomic news, announcements, and stock market jump intensity dynamics. J. Bank. Finance **35**(5), 1263–1276 (2011)
4. Sidney, A.E.: The impact of information announcements on stock volatility. AU J. Manage. **10**(1) (2012)
5. Hussain, S.M.: Simultaneous monetary policy announcements and international stock markets response: an intraday analysis. J. Bank. Finance **35**(3), 752–764 (2011)
6. Mastronardi, R., Patané, M., Tucci, M.P.: Macroeconomic news and Italian equity market (2013)
7. Gurgul, H., Wójtowicz, T., Suliga, M.: The reaction of intraday WIG returns to the US macroeconomic news announcements. Metody Ilościowe w Badaniach Ekonomicznych **14**(1), 150–159 (2013)
8. Huang, X.: Macroeconomic news announcements, systemic risk, financial market volatility and jumps (2015)
9. Wongswan, J.: Transmission of information across international equity markets. Rev. Financ. Stud. **19**(4), 1157–1189 (2006)
10. Nikkinen, J., Omran, M., Sahlström, P., Äijö, J.: Global stock market reactions to scheduled US macroeconomic news announcements. Glob. Finance J. **17**(1), 92–104 (2006)

11. Vrugt, E.B.: US and Japanese macroeconomic news and stock market volatility in Asia-Pacific. Pac.-Basin Finance J. **17**(5), 611–627 (2009)
12. Gok, I.Y., Topuz, S.: The impact of the domestic and foreign macroeconomic news announcements on the Turkish stock market. Financ. Stud. **20**(3) (2016)
13. Hanousek, J., Kočenda, E., Kutan, A.M.: The reaction of asset prices to macroeconomic announcements in new EU markets: evidence from intraday data. J. Financ. Stab. **5**(2), 199–219 (2009)
14. Brenner, M., Pasquariello, P., Subrahmanyam, M.: On the volatility and comovement of US financial markets around macroeconomic news announcements. J. Financ. Quant. Anal. **44**(06), 1265–1289 (2009)
15. Cakan, E., Doytch, N., Upadhyaya, K.P.: Does US macroeconomic news make emerging financial markets riskier? Borsa Istanbul Rev. **15**(1), 37–43 (2015)
16. Wallenius, L., Fedorova, E., Ahmed, S., Collan, M.: Surprise effect of euro area macroeconomic announcements on CIVETS stock markets. Prague Econ. Pap. **26**(1) (2017)
17. Lu, X., Kawai, K.I., Maekawa, K.: Estimating bivariate GARCH-jump model based on high frequency data: the case of revaluation of the Chinese Yuan in July 2005. Asia-Pac. J. Oper. Res. **27**(02), 287–300 (2010)
18. Sidorov, S.P., Revutskiy, A., Faizliev, A., Korobov, E., Balash, V.: GARCH model with jumps: testing the impact of news intensity on stock volatility. In: Proceedings of the World Congress on Engineering, vol. 1 (2014)
19. Funke, N., Matsuda, A.: Macroeconomic news and stock returns in the United States and Germany. Ger. Econ. Rev. **7**(2), 189–210 (2006)
20. De Goeij, P., Hu, J., Werker, B.: Is macroeconomic announcement news priced? Bankers Mark. Investors **143**, 4–17 (2016)
21. Bollerslev, T.: Generalized autoregressive conditional heteroskedasticity. J. Econ. **31**(3), 307–327 (1986)
22. Jorion, P.: On jump processes in the foreign exchange and stock markets. Rev. Financ. Stud. **1**(4), 427–445 (1988)
23. Maheu, J.M., McCurdy, T.H.: News arrival, jump dynamics, and volatility components for individual stock returns. J. Finance **59**(2), 755–793 (2004)
24. Bollerslev, T., Chou, R.Y., Kroner, K.F.: ARCH modeling in finance: a review of the theory and empirical evidence. J. Econ. **52**(1–2), 5–59 (1992)
25. Yao, W., Tian, J.: The role of intra-day volatility pattern in jump detection: empirical evidence on how financial markets respond to macroeconomic news announcements (2015)

Technology Perception, Personality Traits and Online Purchase Intention of Taiwanese Consumers

Massoud Moslehpour[1,3](✉) ⓘ, Ha Le Thi Thanh[1,3],
and Pham Van Kien[2] ⓘ

[1] Department of Business Administration, Asia University, Taichung, Taiwan
mm@asia.edu.tw, writetodrm@gmail.com,
lethithanhha.cs2@ftu.edu.vn
[2] Business Administration Department, University of Economics and Finance,
Ho Chi Minh City, Vietnam
kienpv@buh.edu.vn
[3] College of Management, Business Administration, Asia University,
Taichung, Taiwan

Abstract. The aim of the study is to examine the influences of personality characteristics and perception of technology on e-purchase intention. This study uses a questionnaire survey in collecting relevant data. The target sample is Taiwanese consumers. Multi Regression Analysis is used to test the model and hypotheses. For the measurement model, descriptive analyses and factor analysis are assessed to verify the validity and reliability of the data. As results, the impact of perceived ease of use is the strongest influence on online buying intention. In addition, perceived usefulness, perceived ease of use, and openness to experience have significant impacts on online purchase intention, thereby mediating the relationship between consciousness and online purchase intention. Providing guidelines for strategic plan, technological project, marketing program decision, and website design for online suppliers. This study also has significant implications for personalization, e-commerce, and marketing in online stores. Due to the limited knowledge of the impact of personality traits and perception of technology on customers online purchase intention, the current study appears to be a newly emerging topic in the field of marketing research.

Keywords: Conscientiousness · Openness to experience
Perceived ease of use · Perceived usefulness · Online purchase intention

1 Introduction

Worldwide electronic commerce has significantly advanced due to the prominent development of the internet technology. It allows customers to seek for products' features, price, and functionality online. Consumers no longer need go to the actual store to browse, compare prices or shop for goods (Wind and Mahajan 2001). Accordingly, internet technology has reshaped the way customers buy merchandises.

© Springer International Publishing AG 2018
V. Kreinovich et al. (eds.), *Predictive Econometrics and Big Data*, Studies in Computational Intelligence 753, https://doi.org/10.1007/978-3-319-70942-0_28

Statista (2013) study of 4,887 participants indicates that a large number of online consumers earn $ 50,000 or more yearly and nearly 80% of them are with college degree academic level. The top categories of online sales relate to technological and electronic products, crafts, handmade products, accessories and clothes (Li 2013). E-commerce in Asia-Pacific region is becoming more and more popular. In 2013 Asia-Pacific reached to an increasing level of 30% which puts them to account for over one-third of total e-commerce sales all over the world. Regarding the online market in Taiwan, there were 17,656,414 Internet users in 2012 (International Telecommunication Union Statistics, 2012). An average Taiwanese spent annually NT$16,586 (US $568) in 2012 rising from NT$13,864 in 2010 on purchasing through the internet which accounts over 60% of online people for whole country. Since Taiwan's online shopping market is growing rapidly, it important to determine factors which drive purchase intention (Li 2013).

According to Tsai (2003), personality is one of the roots of understanding consumer purchase intention. Personality is the internal force that motivates customers to affect a particular behaviour, to unconsciously motivate customers; therefore, marketers have to understand the effect of personality as a direct link to the consumer's mind. Previous studies indicate that shopper's character is a substantial factor for the success of e-vendors (Barkhi and Wallace 2007). Online purchase intention has been described by different views, such as consuming in relative to demographical characteristics (Brown et al. 2003; Korgaonkar et al. 2004; Park et al. 2004; Stafford et al. 2004), emotional and psychological characteristics (Huang 2003; Xia 2002), realizing negative and positive aspects of virtual transactions (Bhatnagar and Ghose 2004; Featherman and Pavlou 2003; Garbarino and Strabilevitz 2004; Huang et al. 2004), shopping motivation (Novak et al. 2000; Wolfinbarger and Gilly 2001), and orientation of purchasing (Schiffman and Kanuk 2000). Nevertheless, a few of the previous research has investigated the influence of personality characteristics on customers purchase intention in online stores (Chen 2011; Zhou and Lu 2011). Furthermore, consuming behaviour is substantially influenced by individual factors of users (Mount et al. 2005). The authors of these previous studies conducted their research under the models of attitude which seem to be enough sustainable foundation. It is, therefore, undoubted that purchasing toward e-commerce is necessary to examine personality traits in the relationship with technology perception, and that makes the current study urgent and topical.

This paper attempts to provide unique insights into the online consumer's mindset by discussing the determinative factors affecting buying intention of participants. This study takes partially from the technology acceptance model and partially from the big five personality traits model. Individuals perceive technology differently. Furthermore, personality trait also makes a difference in the way consumers shop online. This study aims at discovering whether: (1) a client's perception about virtual vendors is adjusted by e-vendors' arrangements of their virtual environment based on consumer's personality trait, and (2) the adjustments would result in the ways which will cause clients to buy products easier (Barkhi and Wallace 2007). Taking into consideration the above-mentioned facts, this study touches on key characteristics to help sellers to comprehend online purchase intention and how to attract more customers in future.

Definition of Terms and Abbreviations
This section presents the definition of main concepts used in this study as well as the abbreviations used throughout this study. The key terms used throughout the research are defined as follows:

Online purchase intention (hereinafter referred to as INT). Online purchase intention is a plan to buy a particular food or service at a virtual store in the future in which buyers need to evaluate criterions toward quality of online shopping website, search information, and review product and so on (Hausman and Siekpe 2009; Poddar et al. 2009).

Perceived usefulness (hereinafter referred to as PU). Perceived usefulness demonstrates the level which a person believes that approaching a specific technology would enrich this individual's job performance (Ahmad and Barkhi 2011).

Perceived ease of use (hereinafter referred to as PEOU). Perceived ease of use is the ability to recognize and exploit the users' perception of usefulness of system related to its current and future usage as rated by users (Davis 1989).

Openness to experience (hereinafter referred to as OPE). "Openness to experience is described as being intellectually curious, open to new ideas, involves imaginative and creative cognition styles" (Migliore 2011, p. 39).

Conscientiousness (hereinafter referred to as CON). "Conscientiousness is described as the way individuals control, regulate, and direct their impulses, as related to decision-making and action-oriented behaviours" (Migliore 2011, p. 40).

2   Literature Review

2.1   Review on Influential Factors

Online purchase intention (INT)
E-purchase intention reflects the desire of clients to buy through the internet. It is believed that a shopper is more likely to buy from virtual stores when e-commerce sites provide satisfactory tools, including products or services catalogues, searching functions, pricing comparison sheets, buying carts, online payment systems, and outlining devices (Chen et al. 2010). Thus, considering the importance corresponding to each factor plays the important role for online vendors to draw and maintain consumers. The theory of reasoned action model was developed based on the foundation of Information Integration model. The direct factor of behaviour refers to result of a consumer's intention to decide a particular action. According to determinants of behavioural intention, there are two main features toward the aspects of personality which are shopper's attitude and the subjective criteria (Ajzen 1991). Thus, online shopping intention is most likely to consider as a long-term tendency of buyer to represent behaviour of buying in context of virtual stores (Chen 2011).

Perceived usefulness (PU)

"Usefulness of purchasing from a virtual store can be defined in terms of perceived benefits and the overall perceived advantages of shopping online. In addition to lower prices, the lower costs of information searching may contribute to the perceptions of the usefulness of making purchases from a virtual store when compared to making a purchase from a traditional store or even when compared to another virtual store" (Barkhi and Wallace 2007). As argued by Kim et al. (2003), the e-commerce websites supply functions and aid shoppers to perform exactly decisions of buying a product/service. Shambara (2013) observed that virtual stores can create helpful services for consumer's purchase. It is evident that the services are not as convenient as in traditional market (e.g. immediate comparison among a variety of products or services). Certainly, it will be counted as the convenience of shopping, and as a result in leading to the growth of advantageous attitudes toward web-based purchase.

Perceived ease of use (PEOU)

PEOU is easily to learn and become professional at accessing websites and internet functions (e.g. technology and web-interface) on e-commerce sites. It is about perceived necessary technological elements (Buton-Jones and Hubona 2005). More specifically, a technology is more favorable for using than another if it is most likely to be approved by online shoppers. Another words, the more complicated a technological application is perceived to be, the ratio of the website's users (Selamat et al. 2009).

Personality Traits

The researches of individual characteristic have used an essential analyzing tool for considering human behaviour. As the most common model of personality traits, the Big Five model evaluate the most noticeable sides of personality (Huang and Yang 2010). Human's traits are basic constructing slabs of personality, which is more likely to lead constant shapes of person's thinking, feeling, and manner (Pervin 1996). In addition, the prototype of Big Five can confirm impacts of time's flow (Hampson and Goldberg 2006) as well as cultural exchanges (Costa and McCrae 1995). This model includes elements of extraversion, conscientiousness, agreeableness, neuroticism (emotional stability) and openness to experience. In this paper, authors will limit the variables of personalities by using only conscientiousness as independent variable and openness to experiences as mediating variables. Both variables are related to the area of personal behaviour.

Conscientiousness (CON)

People with highly conscientious characteristic are often concentrated, careful, trustworthy, and well organized, whereas unconscientious persons are the most likely to express their distraction, disorganization and having flexibility (Migliore 2011). CON is identified by words of "precise," "efficient," "orderly," and "persistent". It is believed that CON persons normally concern about factor of effectiveness.

Openness to experience (OPE)

According to the study of Migliore (2011), open-minded persons to experience something new are individuals of intelligence, curiosity, free thought and flexible action which the measurement is based significantly on personality test (i.e. tendency to fantasize, emotional awareness, intellectual curiosity...) Conversely, others are referring to narrow and conservative thought. Therefore, open-mind is a noteworthy forecaster of the overall virtual use. Also, open-mind persons are more likely to experience an online shopping via the internet (McElroy et al. 2007). The result means that open-minded individual is more likely to approach e-purchase to confirm their inquisitiveness and find out freshly adventured practices (Tuten and Bosnjak 2001).

2.2   Interrelations Between Variables

The study of Devaraj et al. (2008) examined the usefulness of conscientiousness relating to two technological factors including perceived ease of use and intention to experience. It indicated that their relationship is more significant for persons referring to much more conscientiousness ($\beta = 0.16$). Punnoose (2012) ($\beta = 0.098$) and Sullivan (2012) ($\beta = 0.24$) also found that conscientiousness has an important influence on perceived usefulness.

As researched by Lim and Ting (2012), ($\beta = 0.410$) perceived usefulness is related to purchase attitude among online shoppers. If consumers find out the usefulness of an e-commerce website for shopping, they will have the better attitude which advantageously leads to intended purchase of a product or service. Other studies by Yoon and Barker Steege (2013) ($\beta = 0.346$), Punnoose (2012) ($\beta = 0.394$), (Aldás-Manzano et al. 2009) ($\beta = 0.164$), and Devaraj (2008) ($\beta = 0.19$) also confirm the relationship between PU and INT.

Several researchers have found a positive relation between the perceived ease of use and purchase intention. It is discussed that clear and understandable online shopping sites, which require for their users less mental efforts to make a purchase, are more attractive for potential customers, than more complicated ones (Childers et al. 2001). Lim and Ting (2012) considered that clients with "perceived ease of use" tend to have a higher intention of buying things from a virtual stores ($\beta = 0.395$). Thus, the ease of use relating to e-commerce website's functions and interfaces are urgent need of forecasting the user's intention towards e-purchase. Perceived ease of use is highly relevant to clients' manner in experience online shopping accordingly to applications of internet, which is strongly associated with intention to purchase ($\beta = 0.33$) (Yulihasri and Daud 2011).

Bosnjak et al. (2007) based on Mowen's hierarchical model of personality explain and predict people's disposition to make e-purchases (Mowen 2000). Almost all elements have positive relationship, including the relationship between conscientiousness and openness to experiences. This paper uses a hierarchy model with the following hierarchy: surface, situational, compound and elemental. CON and OPE are among the hierarchical elemental characteristic included in the Big Five dimensions of character (Mowen 2000). Figure 1 shows the meaningful interrelations among these three factors.

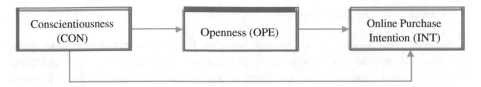

Fig. 1. The positive relationship between variables

2.3 Theoretical Framework and Hypotheses

The Framework guides the growth of the study hypotheses, which examines the interrelations among examination variables. The framework shows that personality traits can affect online purchase intention. The model, depicted on Fig. 2, suggests that e-purchase intention is a dependent variable (DV) which are influenced by the conscientiousness as independent variable (IV). Moreover, it is assumed that the effects of conscientiousness is fully mediated by perceived usefulness, perceived ease of use, and openness which in turn, influence on the dependent variable. The study hypotheses are:

H1: CON has positive association with PU
II2. CON has positive association with PEOU
H3: CON has positive association with OPE
H4: PU has positive association with INT
H5: PEOU has positive association with INT
H6: OPE has positive association with INT
H7: CON has positive association with INT
H8: PU, PEOU and OPE mediates the relationship between CON and INT.

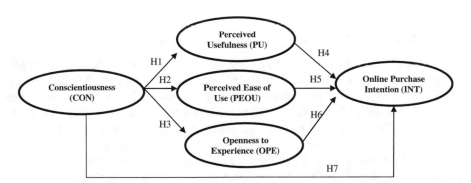

Fig. 2. Research model

3   Research Methodology

In the current study, we collected the data based on information gathered from previous literature review and then specific literature research area in Taiwan to evaluate the present situation about online shopping purchase of Taiwanese. Survey research instrument is used in order to study influential personality and technology on online purchase intention. IN this section we describe data analysis process by using quantitative methods. The results of statistical analysis will help to identify the results from the questionnaire and respondents to make the conclusion to obtain a general overview of the entire research.

3.1   Sample Selection

The population of this research is Taiwanese consumers. The use of technological applications to purchase products or services on virtual stores is a common way for Taiwanese. Consumers who are more comfortable to access technological applications can possess certain characteristics that will lead to their online shopping decision. Therefore, this situation can create a large number of online consumers in general and a potential e-commerce market in separate (Hayhoe et al. 2000). A total of 316 usable (out of 380) questionnaires were collected from Taiwanese consumers aged from 16 to 45 year-old.

The survey questionnaire of this paper is built fundamentally by the features which are chosen based on considerations for the research framework, definition of the variables and literature reviews. We use Likert scale for all of the research questions including five scores ranking from 1 to 5. In which, score 1 and score 5 are corresponding to "strongly disagree" and "strongly agree" perspectival.

The data collection committed through online questionnaire to let the respondents answer questions by using google drive and the paper-and-pencil survey. The time of collecting data is over a month period (1st of March, 2016 $\sim$ 1st April, 2016).

3.2   Data Analysis Procedure

At the end of the survey participation period, the results were exported to an excel file. The quantitative data was coded, tabulated and imported into the SPSS software. The SPSS software is used for this research to study factor analysis and descriptive data analysis to test the collected figures validly and reliably. Descriptive statistics shows the value of the personality measure for every personality dimensions. This study employs the multiple regression analysis to examine the hypotheses. The results are summarized by using descriptive statistics. The used tools are Cronbach's alpha, Kaiser-Meyer-Olkin, Bartlett's Test of Sphericity. To measure the factor loading the EFA is employed. Multi regression analysis is used to test the model and hypotheses.

4   Result and Findings

4.1   Demographic Characteristic

In this study 316 survey questionnaires are gathered from Taiwanese consumers. The demographic information including respondent's gender, age, education, income level and occupation are shown in Table 1 below.

Table 1. Demographic analysis

	Characteristics	Frequency	Proportion
Gender	Male	128	40.5%
	Female	188	59.5%
Age	16–25	171	54.1%
	26–35	129	40.8%
	35–45	16	5.0%
Education	High school or Below	29	9.8%
	Bachelor Degree	14	4.4%
	Master Degree	223	70.6%
	Doctor of Philosophy	50	15.8%
Income	5,000 NT or below	89	28.2%
	5,001–10,000 NT	30	9.5%
	10,001–20,000 NT	32	10.2%
	20,001–40,000 NT	132	41.8%
	40,001–60,000 NT	28	8.9%
	60,001 NT or above	5	1.6%
Occupation	Student	106	33.5%
	Employee	139	43.9%
	Self-employed	25	7.9%
	Unemployed	2	0.6%
	Other	44	13.9%
	Total	316	100%

4.2   Reliability Test

Reliable test, which should be carried out before an examination of validity, is regarded to reduce risks of random error and bring suitable outcomes not only over time but also across circumstances (Chen et al. 2015). The reliability coefficient of Cronbach's alpha conducted is obtained to evaluate internal consistency among the survey instruments (see Table 2). If the value of coefficient alpha is between 0.6 and 0.8 the instrument is considered reliable, and α value over 0.8 is considered highly reliable. The results of the reliability test indicate that CON and INT are reliable, and OPE, PU, PEOU are highly reliable as presented in Table 3. The whole instrument is highly reliable.

Table 2. Reliability statistics (N = 316)

Cronbach's alpha	Number of items
.855	21

Table 3. Cronbach's alpha

	Cronbach's Alpha	Number of items
CON	.792	5
OPE	.871	7
PU	.823	3
PEOU	.882	4
INT	.609	2

4.3 Exploratory Factor Analysis (EFA)

"Exploratory factor analysis is an important tool for researchers. It can be useful for refining measures, evaluating construct validity, and in some cases testing hypotheses" (Chen et al. 2015). This study conducts exploratory factor analysis, KMO and Bartlett's Test of Sphericity to test for the reliability of the instrument. The assumptions are that: (1) if the result of KMO is over 0.6 then the instrument passes the Bartlett's Test of Sphericity; (2) values of the anti-image correlation matrix and commonality are larger than 0.5; (3) Eigenvalues is over 1; (4) factor loadings should be 0.5 or higher and ideally 0.7 or higher (Coakes et al. 2009) then the instrument is valid. All factor loadings are higher than 0.5 which means this research is highly reliable. For the KMO measure of sampling adequacy, in this analysis the KMO is 0.835 which is higher than 0.6. The chi-square was $\chi2 = 3133.25$, it was significant $\rho < 001$. All factor loading are above 0.5 (Table 4).

4.4 Multi Regression Analysis

Multiple regression is a useful method that finds the independent factors' influence into dependent variable. Multiple regressions are largely conducted to discover the influences among variables relating to marketing investigation. The study of hypothesis is on the basic of the standardized path coefficient. Following to test of the hypothesis, the p-value of the r-path coefficient must be substantial at .05. This study intends to identify factors influencing conscientiousness towards purchase intention in online market. This research examines the variables effecting on online purchase intention. These factors, based on the research model, are Conscientiousness (CON), Perceived Usefulness (PU), Perceived Ease of Use (PEOU), and Openness to Experience (OPE). Table 5 indicates the results of multi regression analyses.

A multiple linear regression was calculated to predict online purchase intention based on conscientiousness, perceived usefulness, perceived ease of use, and openness to experience. A significant regression was found (F (4, 311) = 20.563, $p < .000$), with

Table 4. EFA table

Variables and items	Component loading
I see myself as someone who…	
CON1 (does a thorough job)	.806
CON2 (is a reliable worker)	.599
CON3 (perseveres until the task is finished)	.627
CON4 (does things efficiently)	.796
CON5 (makes plans and follows through with them)	.757
OPE1 (is original & comes up with new ideas)	.842
OPE2 (is curious about many different things)	.622
OPE3 (has an active imagination)	.839
OPE4 (is inventive)	.879
OPE5 (values artistic & aesthetic experiences)	.782
OPE6 (likes to reflect & plays with ideas)	.623
OPE7 (is sophisticated in art, music, or literature)	.595
Online shopping…	
PU1 (helps me to make purchases faster)	.844
PU2 (helps me to make cheaper purchases)	.801
PU3 (makes it easier for me to make purchases)	.869
PEOU1 (instructions are easy to follow)	.877
PEOU2 (is easy to learn how to use)	.893
PEOU3 (websites are easy to operate)	.866
PEOU4 (makes it easy to find what I want)	.645
INT1 (I am eager to learn about purchases online)	.790
INT2 (I intend to compare prices in an online store and in a traditional onc)	.764
KMO	**.835**
Cumulative %	65.187
P-value	.000

Table 5. Multi regression analysis CON, PU, PEOU, OPE and INT

R Square		Unstandardized Coefficients		Standardized Coefficients	T	Sig.
.209		B	Std. Error	Beta		
	(Constant)	1.305	.310		4.215	.000
	CON	.055	.071	.043	.770	.442
	OPE	.148	.059	.134*	2.518	.012
	PU	.106	.050	.112*	2.103	.036
	PEOU	.360	.055	.355***	6.527	.000

Dependent Variable: INT, *p < .05, **p < .01, ***p < .001

an R^2 of 0.209. Participants' predicted online purchase intention's equation is followed as: INT = 1.305 + 0.106 (PU) + 0.148 (OPE) + 0.360 (PEOU) (Fig. 3).

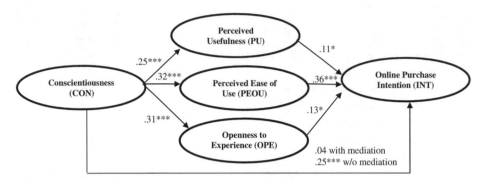

Fig. 3. Result of multiple regression analysis on proposed framework

5   Discussion and Conclusion

5.1   Discussion

For the overall study and from previous research, there are some similar topics that other researcher studied about personality traits and technology perception as they relate to online purchase intention. By combining personality traits and technology perception, this study offers eight hypotheses. These hypotheses are created and tested by using simple regression and multi regression analysis. The purpose of this research is to test the relationship between the independent variables (CON, PU, PEOU and OPE) and dependent variable (INT). Furthermore, the mediation by PU, PEOU, and OPE between CON and INT is tested.

CON has a significant influence on PU is supported ($\beta = 0.25***$). Devaraj et al. (2008), Punnoose (2012) and Sullivan (2012) observed support for this significant connection between CON and PU. Intentions to experience the technology is stronger for individuals with higher CON. This research supports previous research findings regarding traditional shopping for online shopping. Costa and McCare (1995) characterized CON personality trait as people who are naturally motivated and those who strive to aim at achievements at a high level and perform positive actions. The signs of conscientious person are self-control reflected in a need for performance, order, and persistence. Therefore, because CON represents an intrinsically driving force for improvement, conscientious people are cautious when reviewing whether technological applications will allow them to be more effective. On the other hand, if a CON person concludes that a technology is not beneficial, then he/she will not use it (Devaraj et al. 2008).

CON has a significant influence on PEOU ($\beta = 0.32***$). CON is one of the human characters of thinking inside out, persistence and cautioning. Is the characteristic of a person expecting that actions will end with highest possible results? Person of CON is often efficient and ordered as opposed to free and easy. It mean that this person who is generally organized and dependable will perform actions based on a clear and particular schedule instead of immediately thought. PEOU refers to technologies and interfaces on e-commerce websites which is more favorable to access than another is more likely to be accepted by the participants. Because they are very precise, perfectionist, and demand effectiveness, they prefer something that is very efficient to use. When they look for a website, they will eagerly compare for some features that make the website easier, and thus more efficient, to use. They are very aware of the usefulness of the feature in the website.

CON significantly influences OPE ($\beta = 0.31***$). Conscientious personalities usually are very concerned about effectiveness. OPE is implied as being intellectually curious, open to new ideas, involves imaginative and creative cognition styles (Migliore 2011). CON personality demands something that is efficient to do. Because of that reason, CON personality has to be more open with new idea which is developed in the environment which can make their life more efficient.

PU significantly influences INT is supported ($\beta = 0.11*$). Previous studies indicated that shoppers' PU of an online store positively influence their buying intentions and repurchase intention in the future. Online purchase intention is determined by perceived usefulness of technological innovation (Davis 1989, Tong 2010). The results of this study imply that Taiwanese customers that perceive technology as a useful tool for shopping online tend to buy more online. The result is consistent compared with studies of Aldás-Manzano et al. (2009) and Yoon and Steege (2013). They found that online shoppers in different countries have the alike expectation of looking for advantages of purchase via the internet by themselves. As a result of this, PU is the key driver of usage behaviour and intention to purchase.

PEOU has a significant influence on INT ($\beta = 0.36***$). This result agrees with similar previous research by Yulihasri and Daud (2011). Therefore, technologies which are perceived as easy to use will stimulate customers to purchase online. It was shown that clear and understandable online shopping sites, which require of their users less mental efforts to make a purchase, are more attractive for potential customers, than more complicated ones (Childers et al. 2001). This results are very important, since they revealed that convenient website, which is free from effort in order to make an order is one of the main factors of successful virtual store.

OPE is a sustainably determining factor in relating to INT, with ($\beta = 0.13*$). People with high OPE person, who is the most likely to explore something new, like to experience e-activities. As a given by Tuten and Bosnjak (2001), it is reasonable to satisfy this person's interest and find out new ways adventurously. This is also considered that OPE is the important forecasting element effecting on growth of using the internet's applications generally (McElroy et al. 2007). That means that people who exhibit this personality trait (openness) would better do purchase online than go to traditional store. The results are in accordance with Arnold and Reynolds (2003) who stated that a person with higher level of OPE is more likely to buy things in virtual stores.

CON significantly influence INT is supported ($\beta = 0.25***$). The awareness of high conscientious people is very high. They prefer to make themselves do something that is more efficient and well organized. They prefer to do something orderly and not something spontaneous. There is a possibility that conscientious people prefer to shop online rather than traditional shopping because shopping online is more organized, more efficient, and more specific to buy something, Moreover, going to the traditional stores, staying traffic, wandering around, wasting time with the possibility of buying nothing is not very attractive to conscientious people. Online store provides more detail information about prices, products, and even how to buy something until it arrives to the destination place. Thus, it will help users to compare the price or even the product from one website to another and help them to buy something that match with their need. Hence, it also will be more comfortable and more effective for those who seek for efficiency and precisely.

PU, PEOU, OPE mediate the relationship between CON and INT, is supported (β w/mediation = 0.04) and (β w/o mediation = 0.25***). Experiential shoppers and goal-oriented shoppers may be considered as two kinds of clients (Wolfinbarger and Gilly 2001; Barkhi and Wallace 2007). Conscientious people tend to be target-oriented clients, which have oriented tasks, efficiency, lucidity, and deliberateness and might enjoy the efficiency of virtual stores (Barkhi and Wallace 2007). CON influences INT. However, when mediated by three other variables (PU, PEOU and OPE) this influence is insignificant. CON's influence is diminished by the presence of mediating variables.

In summary, the study aims to explore whether CON have correlation with INT among Taiwanese people, examine the role of PU, PEOU, OPE as mediators in relationship between CON and INT. In particular, eight hypotheses were postulated. There are eight hypothesis and all hypothesis was accepted.

5.2   Implementations

This research shows evident that personal traits are an important determinant affecting e-shoppers' buying intention. As a result, it is a noteworthy background toward studying the field of online consuming behaviour. Actually, online vendors can use the result as fundamental reference material to plan an effective consumer-oriented strategy and attract many more consumers to buy things online.

The results also suggest that although consciousness people are attracted to shopping only, however, they are mainly influenced by the presence of other factors such as OPE, PEOU and PU. Their intention to purchase online will increase through PU, PEOU and their OPE of new technology. Online store owners and web designers can attract consciousness people by increasing PEOU and PU while considering the personality traits of OPE. Therefore, this research is more necessary to web designers, vendors supplying technological application and software and vendors selling products/services online particularly. For instant, the web-designer is needed to create eye-catching sales interface of websites, besides the technological applications of smart tools and functionality by a software company to help users easily manipulate even unprecedented access to technology previously. And duties of online providers is to find ways to encourage customer's interest in the products and services in their virtual stores by the ways of promotion, quality, new features and pricing competition.

In summary, this study indicated the significant roles of CON, PU, PEOU and OPE in predicting INT among Taiwanese consumers. Furthermore, the results showed that the role of PU, PEOU and OPE as mediators in the relationship between CON and INT is significant. In particular, eight hypotheses were postulated. There were eight hypotheses and all hypothesis were accepted.

References

Ajzen, I.: The theory of planned behavior. Organ. Behav. Hum. Decis. Process **50**(3), 179–211 (1991)

Ahmad, N., Barkhi, R.: The contextual and collaborative dimensions of avatar in real-time decision making. In: Supporting Real Time Decision-Making, pp. 133–156. Springer (2011)

Aldás-Manzano, J., Ruiz-Mafé, C., Sanz-Blas, S.: Exploring individual personality factors as drivers of M-shopping acceptance. Ind. Manage. Data Syst. **109**(6), 739–757 (2009). https://doi.org/10.1108/02635570910968018

Arnold, M.J., Reynolds, K.E.: Hedonic shopping motivation. J. Retail. **79**(2), 77–95 (2003)

Barkhi, R., Wallace, L.: The impact of personality type on purchasing decisions in virtual stores. Inf. Technol. Manage. **8**(4), 313–330 (2007)

Bakos, Y.: A strategic analysis of electronic marketplaces. MIS Q. **15**(3), 295–310 (1991)

Bhatnagar, A., Ghose, S.: A latent class segmentation analysis of E-shoppers. J. Bus. Res. **57**(5), 758–767 (2004)

Bosnjak, M., Galesic, M., Tuten, T.: Personality determinants of online shopping: Explaining online purchase intentions using a hierarchical approach. J. Bus. Res. **60**(6), 597–605 (2007)

Brown, M., Pope, N., Voges, K.: Buying or browsing? An exploration of shopping orientations and online purchase intention. Eur. J. Mark. **37**(11), 1666–1684 (2003)

Buton-Jones, A., Hubona, G.S.: Individual differences and usage behaviour: revisiting a technology acceptance model assumption. Date Base Adv. Inf. Syst. **36**(2), 58–77 (2005)

Chen, J.K., Batchuluun, A., Batnasan, J.: Services innovation impact to customer satisfaction and customer value enhancement in airport. Technol. Soc. **43**, 219–230 (2015)

Chen, Y.H., Hsu, I.C., Lin, C.C.: Website attributes that increase consumer purchase intention: A conjoint analysis. J. Bus. Res. **63**(9), 1007–1014 (2010)

Chen, T.: Personality traits hierarchy of online shoppers. Int. J. Mark. Stud. **3**(4), 23–39 (2011)

Childers, T.L., Carr, C.L., Peck, J., Carson, S.: Hedonic and utilitarian motivations for online retail shopping behavior. J. Retail. **77**(4), 511–535 (2001)

Coakes, S.J., Steed, L., Ong, C.: SPSS 16.0 for Windows: Analysis Without Anguish. Wiley, Australia (2009)

Davis, F.D.: Perceived usefulness, perceived ease of use, and user acceptance of information technology. MIS Q. **13**(3), 318–340 (1989). https://doi.org/10.2307/249008

Devaraj, S., Easley, R.F., Crant, J.M.: How does personality matter? Relating the five-factor model to technology acceptance and use. Inf. Syst. Res. **19**(1), 93–105 (2008)

Featherman, M., Pavlou, P.A.: Predicting E-services adoption: a perceived risk facets perspective. Int. J. Hum Comput. Stud. **59**(2), 451–474 (2003)

Garbarino, E., Strabilevitz, M.: Gender differences in the perceived risk of buying online and the effects of receiving a site recommendation. J. Bus. Res. **57**(5), 768–775 (2004)

Hampson, S.E., Goldberg, L.R.: A first large-cohort study of personality-trait stability over the 40 years between elementary school and midlife. J. Pers. Soc. Psychol. **91**(3), 763–779 (2006)

Hausman, A.V., Siekpe, J.S.: The effect of web interface features on consumer online purchase intentions. J. Bus. Res. **62**(1), 5–13 (2009)

Hayhoe, C., Leach, L., Turner, P., Bruin, M., Lawrence, F.: Differences in spending habits and credit use of college students. J. Consum. Aff. **34**(2), 113–133 (2000)

Huang, J.H., Yang, Y.C.: The relationship between personality traits and online shopping motivations. Social Behav. Pers. Int. J. **38**(5), 673–679 (2010)

Huang, M.H.: Modeling virtual exploratory and shopping dynamics: an environmental psychology approach. Inf. Manag. **41**(1), 39–47 (2003)

Huang, W.Y., Schrank, H., Dubinsky, A.J.: Effect of brand name on consumers' risk perceptions of online shopping. J. Consum. Behav. **4**(1), 40–50 (2004)

Kim, D., Ferrin, D., Rao, H.: Study of consumer trust on consumer expectations and satisfaction: The Korean experience. In: ACM International Conference Proceeding, vol. 50(1), pp. 310–315 (2003)

Kim, J.B.: An empirical study on consumer first purchase intention in online shopping: integrating initial trust and TAM. Bus. J. **12**(5), 125–150 (2012). https://doi.org/10.1007/s10660-012-9089-5

Korgaonkar, P., Silverblatt, R., Becerra, E.: Hispanics and patronage preferences for shopping from the internet. J. Comput. Mediated Commun. **9**(3), 31–44 (2004)

Li, J.: Study: online shopping behavior in the digital era. Mark. Res. **3**(5), 13–21 (2013)

Lim, W.M., Ting, D.H.: E-shopping: an analysis of the technology acceptance model. Modern Appl. Sci. **6**(4), 49–62 (2012). https://doi.org/10.5539/mas.v6n4p49

McElroy, J.C., Hendrickson, A.R., Townsend, A.M., DeMarie, S.M.: Dispositional factors in internet use: personality versus cognitive style. MIS Q. **31**(2), 809–820 (2007)

Migliore, L.A.: Relation between big five personality traits and Hofstede's cultural dimensions: Samples from the USA and India. Cross Cult. Manage. Int. J. **18**(1), 38–54 (2011)

Mount, M.K., Barrick, M.R., Scullen, S.M., Rounds, J.: Higher-order dimensions of the big five personality traits and the big six vocational interest types. Pers. Psychol. **58**(2), 447–478 (2005). https://doi.org/10.1111/j.1744-6570.2005.00468.x

Mowen, J.: The 3M Model of Motivation and Personality. Kluwer Academic Press, Norwell (2000)

Novak, T.P., Hoffman, D.L., Yung, Y.F.: Measuring the customer experience in online environments: a structural modeling approach. Mark. Sci. **19**(1), 22–42 (2000)

Park, J., Lee, D., Ahn, J.: Risk-focused e-commerce adoption model: a cross-country study. J. Global Inf. Manage. **5**(7), 6–30 (2004)

Pervin, L.A.: The Science of Personality. Wiley, New York (1996)

Poddar, A., Donthu, N., Wei, Y.: Web site customer orientation, web site quality, and purchase intentions: the role of web site personality. J. Bus. Res. **62**(3), 441–450 (2009)

Punnoose, A.C.: Determinants of intention to use e-learning based on the technology acceptance model. J. Inf. Technol. Educ. Res. **11**(3), 301–337 (2012)

Schiffman, L.G., Kanuk, L.L.: Consumer Behavior. Prentice-Hall, Upper Saddle River (2000)

Selamat, Z., Jaffar, N., Ong, B.H.: Technology acceptance in Malaysian banking industry. Eur. J. Econ. Finance Adm. Sci. **1**(17), 143–155 (2009)

Sosor: Factors influencing the Usage of Mobile Banking in a Mongolian Context, pp. 1–90. Asia University (2012)

Stafford, T.F., Turan, A., Raisinghani, M.S.: International and cross-cultural influences on online shopping behavior. J. Glob. Inf. Manage. **7**(2), 70–87 (2004)

Statista: Statistics and facts about global e-commerce (2013). http://www.statista.com/topics/871/online-shopping/

Tong, X.: A cross-national investigation of an extended technology acceptance model in the online shopping context. Int. J. Retail Distrib. Manage. **38**(10), 742–759 (2010)

Tsai, L.H.: Relationships between Personality Attributes and Internet Marketing (Doctoral dissertation). Retrieved from ProQuest Dissertation and Theses database. (UMI No. 3078519) (2003)

Tuten, T.L., Bosnjak, M.: Understanding differences in web usage: The role of need for cognition and the five factor model of personality. Soc. Behav. Pers. **29**(3), 391–398 (2001)

Venkatesh, V., Agarwal, R.: Turning visitors into customers: a usability-centric perspective on purchase behavior in electronic channels. Manage. Sci. **52**(3), 367–382 (2006)

Wind, J., Mahajan, V.: Digital Marketing: The Challenge of Digital Marketing. Wiley, New York (2001)

Wolfinbarger, M., Gilly, M.: Shopping online for freedom, control and fun. Calif. Manage. Rev. **43**(2), 34–56 (2001)

Xia, L.: Affect as information: the role of affect in consumer online behaviors. Adv. Consum. Res. **29**(1), 93–100 (2002)

Yoon, H.S., Steege, L.M.B.: Development of a quantitative model of the impact of customers' personality and perceptions on Internet banking use. J. Comput. Hum. Behav. **29**(3), 1113–1141 (2013)

Yuslihasri, I.A., Daud, A.K.: Factors that influence customers buying intention on shopping online. Int. J. Mark. Stud. **3**(1), 128–143 (2011)

Zhou, T., Lu, Y.: The effects of personality traits on user acceptance of mobile commerce. Int. J. Hum. Comput. Interact. **27**(6), 545–561 (2011)

Zikmund, W.G.: Business Research Methods. Dryden Press, Fort Worth (2000)

Analysis of Thailand's Foreign Direct Investment in CLMV Countries Using SUR Model with Missing Data

Chalerm Jaitang[1], Paravee Maneejuk[2,3]([✉]), Aree Wiboonpongse[1], and Songsak Sriboonchitta[2,3]

[1] Faculty of Economics, Prince of Songkla University, Song Khla 90110, Thailand
chalermjaitang@gmail.com, aree.w@psu.ac.th
[2] Center of Excellence in Econometrics, Chiang Mai University, Chiang Mai 52000, Thailand
mparavee@gmail.com, songsakecon@gmail.com
[3] Faculty of Economics, Chiang Mai University, Chiang Mai 52000, Thailand

Abstract. Thai enterprises and companies have turned their attentions to CLMV countries, since the establishment of ASEAN Economic Community (AEC) in 2015, due to market and production opportunities. This study is conducted with an attempt to provide useful information helping the Thai investors make investment decision. In particular, this study examines the determinants of outward Thailand's direct investment to the CLMV countries, and later estimates marginal effects. During the analysis, data unavailability in the CLMV and missing values in many available variables become destructive. This study handles this problem by using the bootstrap-based expectation maximization with bootstrapping (EMB). Once a complete data set is obtained, this study then employs the Seemingly Unrelated Regression (SUR) model to analyse the effects of the considered variables on Thailand's direct investment in the CLMV group. The estimated results show distinct determinants for the countries, which can be useful to investors.

1 Introduction

The eclectic paradigm provides that a firm engages in Foreign Direct Investment (FDI) if three main conditions are satisfied: an ownership advantage, a location advantage, and an internalization advantage [2]. Eclectic paradigm has remained the dominant analytical framework for accommodating a variety of operationally testable economic theories of FDI and foreign activities of multinational enterprises (MNEs) [4]. Dunning [3] proposed a four-stage investment development path to describe how countries inward and outward FDI positions evolve as local firms develop transitional corporations capacities. In stage one, there is little movement toward undertaking FDI. In stage two, there is still little movement toward outward investment. The outward investment that does occur is also most likely to support trade, but it is increasingly designed to support

© Springer International Publishing AG 2018
V. Kreinovich et al. (eds.), *Predictive Econometrics and Big Data*, Studies in Computational Intelligence 753, https://doi.org/10.1007/978-3-319-70942-0_29

products that require larger scale production and more capital. In stage three, as economy matures, companies seek to benefit from their distinctive capabilities and competencies. Outward investment may be driven by either resource or market-seeking motives. Finally, in stage four, the post-industrial or services stage, outward FDI depends more on capabilities through knowledge creation and the blurring of the distinction between products and services.

Four distinctive phases of Thai outward FDI can be discerned [11]. The first phase (early stage) before the first half of the 1980s saw a limited amount of Thailand's investment abroad. Much of the overseas investment during this phase went to a few key destinations such as the United States, Hong Kong (China), Singapore and Japan. These four economies accounted for over 85 % of the net Thai equity capital investment abroad. The second phase (take-off stage) took place between 1986 and 1996 when Thai outward FDI increased rapidly. Thai companies ventured further afield to such locations as Australia, Canada, as well as European countries. While the United States and Hong Kong (China) continued to be the principal host economies, other Asian countries, particularly ASEAN countries and China, have emerged as significant destinations because of the cost advantage, market size and business opportunities. The third phase (financial crisis impact stage), the period 1997 to 2002, saw a dramatic decline in Thai outward FDI due to the impact of the financial crisis, which significantly affected the ability of Thai enterprises to invest or maintain their investment abroad. Thai outward FDI to China, Europe, Hong Kong (China) and the United States fell considerably and flows to ASEAN during this period also fell by 36 % in absolute terms, compared with the period before the financial crisis (1989–1996). Despite the decline, Thai outward FDI at this stage was greater than in the earlier period. The fourth phase (recovering stage), which started in 2003, saw a recovery in outward FDI. ASEAN and China were the main recipients while FDI to Europe and the United States began to pick up but remained at the low level. Manufacturing was the most active sector for Thai outward FDI.

Thailand's integration into the ASEAN Economic Community (AEC) in 2015 is beneficial for Thai enterprises and companies investing in neighboring countries, especially Cambodia, Laos, Myanmar and Vietnam (henceforth CLMV) since they will bring about market and production opportunities for Thailand. The motive for firms to engage in foreign production and MNEs activity can be classified into four groups: natural resource seeking, market seeking, efficiency seeking, and strategic asset or capability seeking [5]. Thailand can use CLMV as a production base to solve its labor and energy problems. Moreover, Thailand, with its strategic geographical location, can position itself as a major service hub of the sub-region [10]. The development of transportation infrastructure in Thailand between 2013 and 2020, including the construction of new railway linking Bangkok with neighboring countries, is expected to reduce transportation costs within supply chain centered on Thailand [8]. The integrated mainland ASEAN will bring about larger market size which will allow Thai companies to invest more in CLMV countries. Thai companies can also relocate or expand their low value-added production activities to the neighbouring countries, especially

CLMV, while upgrading their domestic production to higher value-added activities. This study aims to analyze the opportunities exist in CLMV countries, and then provide this information to help Thai investors make investment decision. This study examines the relation between Thailand's direct investment in CLMV countries and four groups of the motive factors, and also estimates their marginal effects on Thailand's outward FDI.

The rest of the paper is organized as follows. Section 2 provides a brief definition of variables used in the study, which are classified into four different groups. The methodology is presented in Sect. 3. Section 4 provides the empirical finding. Finally, Sect. 5 contains conclusions.

2   Data Descriptions

This section presents the factors that may influence outward Thailand's direct investment (unit: million U.S. dollars). These factors relate not only to the country's economic performance, but also to policy decisions taken by its government. The factors are classified into four different groups namely natural resource seeking, market seeking, efficiency seeking, and strategic asset or capability seeking according to Dunning and Lundan [5] The first group is resource seeking containing minimum wage, labor force, corporate income tax rate, agriculture value added, industry value added, and service value added. The second group is market seeking including gross domestic product, gross domestic product growth rate, and gross domestic product per capita (in terms of constant 2010 US$). The third group is efficiency seeking containing openness rate, inflation rate, and exchange rate. The last group is capability seeking. The variables in this group consist of general government final consumption expenditure, gross capital formation, and patent. The definitions of variables in each group as well as their labels are illustrated in Table 1.

3   Methodology

Since the CLMV region, which is designated as least developed countries, is our area of interest, we are now facing a problem of data unavailability. During a process of data collection, we find that the data of CLMV countries are considerably unavailable. Some variables for some countries are unavailable and many of them are missing. This becomes a limitation of the analysis. So, we decided to eliminate the unavailable variables and considered only the variables that are officially provided. However, missing data occurred in the remaining variables is still an obstacle. In fact, missing data is ubiquitous problem for least developed and underdeveloped countries, and multiple imputation is a general approach to this problem. It creates a multiple filled in of the incomplete data set. In this study, we consider particularly the bootstrap-based EMB (Expectation Maximization with Bootstrapping) of Honaker et al. [6] for dealing with missing data occurred in CLMV countries. This algorithm uses the classical EM algorithm on the multiple bootstrapped data from incomplete data to draw values of the

Table 1. Data descriptions

Type of Motive	Variable	Label	Description
Resource seeking	Wage	W_{it}	Gross average nominal monthly wages (Unit: U.S. dollars per month)
	Labor force	L_{it}	Labor force comprises people ages 15 and older who supply labor for the production of goods and services during a specified period. It includes people who are currently employed and people who are unemployed but seeking work as well as first-time job-seekers. (Unit: absolute value in thousands)
	Corporate income tax rate	CIT_{it}	A direct tax levied on a juristic company or partnership carrying on business (Unit: %)
	Agriculture value added	A_{it}	Agriculture corresponds to ISIC divisions 1–5 and includes forestry, hunting, and fishing, as well as cultivation of crops and livestock production. (constant 2010 U.S. dollars) (Unit: million U.S. dollars)
	Industry value added	I_{it}	Industry corresponds to ISIC divisions 10–45 and includes manufacturing (ISIC divisions 15–37). It comprises value added in mining, manufacturing, construction, electricity, water, and gas. (constant 2010 U.S. dollars) (Unit: million U.S. dollars)
	Service value added	S_{it}	Services correspond to ISIC divisions 50–99. They include value added in wholesale and retail trade (including hotels and restaurants), transport, and government, financial, professional, and personal services such as education, health care, and real estate services. (constant 2010 U.S. dollars) (Unit: million U.S. dollars)
Market seeking	Gross domestic product growth rate	G_{it}	Annual percentage growth rate of GDP (Unit: %)
	Gross domestic product	GDP_{it}	GDP expressed in constant 2010 U.S. dollars (Unit: million U.S. dollars)
	Gross domestic product per capita	$GDPcapi_{it}$	GDP per capita is gross domestic product divided by midyear population (constant 2010 US$) (Unit: U.S. dollars)
Efficiency seeking	Inflation Rate	INF_{it}	CLMV countries' core inflation (Unit: $)
	Openness rate	$OPEN_{it}$	The ratio of trade (exports and imports) to GDP (Unit: %)
	Exchange rate	EX_{it}	Average exchange rate (Unit: local currency per U.S. dollar)
Capability seeking	Gross capital formation	Gcf_{it}	Consists of outlays on additions to the fixed assets of the economy plus net changes in the level of inventories (Unit: million U.S. dollars)
	General government final consumption expenditure	EXP_{it}	Includes all government current expenditures for purchases of goods and services (including compensation of employees), also includes most expenditures on national defense and security, but excludes government military expenditures that are part of government capital formation (Unit: million U.S. dollars)
	Patent	Pat_{it}	Patent applications are worldwide patent applications filed through the Patent Cooperation Treaty procedure or with a national patent office for exclusive rights for an invention—a product or process that provides a new way of doing something or offers a new technical solution to a problem.(Unit: pieces)

Note: i represents countries in CLMV and t corresponds to time.

complete data parameters, and then draw imputed values to fill in the missing values. After obtaining completed data set, we can use this data in the empirical analysis.

Empirically, this study employs a well-known Seemingly Unrelated Regression (SUR) model of Zellner [12] to analyze effects of the considered variables (as shown in Table 1) on Thailand's direct investment in CLMV countries, with regard to appropriateness of this research tool. A brief idea of this model is that it consists of several regression equations in which each equation can have its own dependent and independent variables [9]. However, all equations are related somehow by the error terms. When we consider economic condition of each of the four CLMV countries in detail, we will find that Cambodia, Laos, Myanmar, and Vietnam have both political and economic disparities. These in turn influence the direct investment from foreign countries, including Thailand. So, we may write regression equation for each of the CLMV countries separately, having its own FDI from Thailand and potentially different sets of explanatory variables.

3.1   Seemingly Unrelated Regression (SUR) Model

The seemingly unrelated regression (SUR) model is defined by the set of regressions. Suppose there are M regression equations,

$$y_{it} = x_{it}\beta_i + \varepsilon_{it} \qquad i = 1, ..., M; \ t = 1, ..., T, \tag{1}$$

which can be written in a vector form as

$$Y = X\beta + \varepsilon. \tag{2}$$

The subscript t corresponds to time and i represents equation number corresponding to the CLMV countries, where $C, L, M,$ and V stand for Cambodia, Laos, Myanmar, and Vietnam respectively. However, we are now showing a structure of the SUR model in general form with M regression equations, $i = 1, ..., M$. Y is a vector of dependent variables, X is a matrix of independent variables or so-called regressors, and β is a vector of unknown regression parameters. ε is a vector of the error terms which has a zero mean and variance-covariance matrix $\Gamma \otimes I$, where I is the unit matrix of order M and $\Gamma = \sigma_{ij}I$, $i \neq j$. Equation (2) can be written in the compact form as

$$X = \begin{bmatrix} x_{1t} & 0 & \cdots\cdots & 0 \\ 0 & x_{2t} & \cdots\cdots & 0 \\ \vdots & & \ddots & \vdots \\ \vdots & & & \ddots & \vdots \\ 0 & 0 & \cdots\cdots & x_{Mt} \end{bmatrix}, Y = \begin{bmatrix} y_{1t} \\ y_{2t} \\ \vdots \\ \vdots \\ y_{Mt} \end{bmatrix}, \beta = \begin{bmatrix} \beta_1 \\ \beta_2 \\ \vdots \\ \vdots \\ \beta_M \end{bmatrix}, \varepsilon = \begin{bmatrix} \varepsilon_{1t} \\ \varepsilon_{2t} \\ \vdots \\ \vdots \\ \varepsilon_{Mt} \end{bmatrix}. \tag{3}$$

In addition, this model assumes that the errors of each equation are related. Thus, under this assumption, the variance-covariance matrix is given by

$$\Gamma = \begin{bmatrix} \sigma_{11}I & \sigma_{12}I & \sigma_{1M}I \\ \sigma_{21}I & \sigma_{22}I & \sigma_{2M}I \\ \vdots & & \ddots & \vdots \\ \sigma_{M1}I & \cdots & \cdots & \sigma_{MM}I \end{bmatrix}. \tag{4}$$

The generalized least squares (GLS) estimator is conducted to estimate the unknown parameters β in the SUR model, which can be shown as

$$\beta = (X'\Gamma^{-1}X)^{-1}X'\Gamma^{-1}Y \tag{5}$$

3.2 Expectation Maximization with Bootstrapping (EMB) for Handling Missing Data

Under a missing at random (MAR) assumption of Honaker and King [7], the pattern of missing data D^{miss} only depends on the observed data D^{obs}. We can define the MAR assumption by

$$p(M\,|D) = p(M\,|D^{obs}), \tag{6}$$

where D denotes both data sets of D^{obs} and D^{miss}; and M is a missingness matrix with cell $m_{ij} = 1$ if $d_{ij} \in D^{mis}$ otherwise $m_{ij} = 0$. Here D is assumed to have a multivariate normal distribution with mean vector μ and covariance matrix Σ. Thus the likelihood of observed data is given by

$$p(D^{obs}, M\,|\Theta) - p(M\,|D^{obs})p(D^{obs}\,|\Theta) \tag{7}$$

where $\Theta = \{\mu, \Sigma\}$ is the complete-data parameters. Since we are interested in making inference about the complete data parameters, we can write the likelihood as

$$L(\Theta\,|D^{obs}) \propto p(D^{obs}\,|\Theta) \tag{8}$$

We can rewrite using the law of iterated expectations, and thereby

$$p(D^{obs}\,|\Theta) = \int p(D\,|\Theta)\,dD^{mis}. \tag{9}$$

In this study, we add Bayesian priors into individual cell values in order to improve the imputation models and obtain potentially valuable and extensive information. Thus, the posterior distribution can be done by combining flat priors with the likelihood function, Eq. (9), using Bayes theorem. Thus, the posterior distribution can be formed as follows

$$p(\Theta\,|D^{obs}) \propto p(D^{obs}\,|\Theta) = \int p(D\,|\Theta)\,dD^{mis} \tag{10}$$

Then, the EMB algorithm, which combines the classical EM algorithm with a bootstrap approach [1], takes draws from this posterior. For each draw, we bootstrap the data to simulate estimation uncertainty, and then run the EM algorithm to estimate the model of the posterior for the bootstrapped data. Once we have draws of the posterior of the completed data parameters, we create imputations by drawing values of D^{mis} from its distribution conditional on D^{obs} and the draws of Θ. In this study, the number of imputed data sets to create is $m = 10$, meaning that we have 10 imputed data sets for the SUR model. To combine the results, we estimate SUR model using these 10 data sets and obtain $(\beta^1, ..., \beta^{10})$. Then, the parameter coefficient of the multiple imputation point estimate is

$$\bar{\beta} = \frac{1}{m} \sum_{i=1}^{m} \beta_i. \tag{11}$$

The standard error of the parameter $\bar{\beta}$ is given by

$$SE(\beta)^2 = \frac{1}{10} \sum_{i=1}^{10} SE(\beta_i)^2 + \left(\sum_{i=1}^{10} (\beta_i - \bar{\beta})^2 / (10 - 1) \right)(1 + 1/10) \tag{12}$$

4   Empirical Analysis

This section consists of three parts. In the first subsection, we describe features of the data through descriptive statistics. Moreover, this subsection also presents a missingness map to illustrate where the missing value occurs in the data set. The missing values in data set are still of our concern, so that we go further and predict the missingnesses in the second subsection. Finally, the third subsection presents the main results of this study and some discussions on our findings.

4.1   Descriptive Statistics

This study considers the data of Cambodia, Laos, Myanmar, and Vietnam (CLMV countries) relevant to four groups of variables, as stated in Sect. 2, namely natural resource seeking, market seeking, efficiency seeking, and capability seeking. The variables contained in each group are described in Table 1. Moreover, we also use the data of Thailand's outward FDI to the CLMV countries as the dependent variable. This variable is denoted by FDIi, where subscript i represents countries in CLMV. All the data are collected annually from 1988 to 2015, covering 28 observations. We find that the data of corporate income tax rate and patent applications are not available for the case of CLMV countries, thus we have to eliminate these two terms. The descriptive statistics of the remaining variables are presented in Table 2.

To define more clearly the missing values, we then plot missingness map of the data for each country. The use of this map is to show which particular

Table 2. Descriptive statistics

	FDI	GDPcap	GDP	G	INF	Gcf	EX	L	EXP	OPEN	W	A	I	S
Cambodia	FDI_C													
Min	-3.44	314.9	1407	-0.091	-0.826	132.3	426.2	4072	1260	0.1171	20	788.5	202.3	658.5
Median	1.34	548.1	4477	0.09423	6.246	747.9	3944.8	6123	3521	1.116	46	1544.9	965	1609.5
Mean	2.009	598.8	7225	0.1058	26.097	1110.8	3374.3	6241	5979	0.9411	54.75	2233.3	1464.3	2501.9
Max	9.14	1020.9	19958	0.3242	191	3690.4	4184.9	8803	14977	1.4475	128	4822.3	4500	6706.8
SD	2.708	224.1	5561.21	0.0912	48.001	986.86	1101.01	1563.84	4652.6	0.4008	24.14	1407.3	1297.9	1870.62
Missing	9		1			1	2		8	1	4			
Laos	FDI_L													
Min	-4.41	402.8	1.26E+04	-2.01	-26.317	89.03	800.7	1833	2822	0.3391	8	302.6	202.3	320.6
Median	0.53	713.7	1.82E+09	7.011	7.742	326.66	8147.9	2547	7480	0.5466	12	746.1	965	741
Mean	2.583	810.1	3.79E+09	6.813	16.965	976.52	7498.6	2610	10932	0.5589	29.67	1127.8	1464.3	1483
Max	21.22	1531.2	1.24E+10	14.191	128.409	4131.73	10655.2	3572	26201	0.8005	110	2900	4500	4833.1
SD	6.013	326.1	3.66E+09	2.4853	29.497	1094.09	3151.2	5286	8358.1	0.1192	31.06	777.366	1230.3	1395.02
Missing	9		1			1	2		12	1	4			
Myanmar	FDI_M													
Min	-16.87	184.6	6.48E+09	-11.352	-1.723	449	5.44	19003	303	0.2901	20	2592	437.8	1488
Median	0.39	389.4	2.60E+10	8.564	10.82	1094	6.12	24715	3184	0.4014	30.97	5040	1138.7	3052
Mean	1.069	542.1	3.23E+10	8	16.612	4452	6.07	24414	5233	0.4133	33.28	7899	5384.4	7400
Max	21.47	1308.7	6.56E+10	13.844	58.104	23490	6.75	29918	18908	0.6896	61.06	18812	23000	25038
SD	6.506	360.4	2.36E+10	5.2727	17.105	6864.33	0.39	34426	5852.7	0.1171	13.21	5836.5	7751.5	8147.02
Missing	9		1			1	10		12	11	5			
Vietnam	FDI_V													
Min	-231.1	412.6	6.29E+09	4.688	-1.77	830.8	11032.6	29707	73419	0.1623	8	2507	1381	1842
Median	7.78	846.9	3.66E+10	5.734	8.23	10188.9	15858.9	43502	313627	1.1252	19	7862	12979	13060
Mean	99.7	925.5	6.35E+10	5.726	28.519	17737.6	16259.5	43407	474560	1.1174	25.39	12748	23078	25261
Max	937.3	1684.7	1.94E+11	9.54	374.354	49963.7	21697.6	55654	1242937	1.6982	54	35000	71694	90000
SD	251.72	383.5	5.84E+10	1.378	71.485	16435.6	3309.4	7817.45	398020.9	0.4075	16.32	10428.1	22805	25882.8
Missing	3		1			1	10		10	1	5			

Source: Calculation

observation in a series is not provided or missing. All data is illustrated with a grid in which its color refers to missingness status. The meaning of colors is that a blue grid represents missing value and a pink grid represents observed value. In each map, the horizontal axis shows variable labels and the vertical axis shows the number of observations, which is totally 28 observations. Figure 1 shows that Myanmar has maximum number of missing values at the border of these variables. This country does not provide long-term historical data of macroeconomic variables especially, for example, GDP, openness rate (OPEN), inflation (INF), and general government final consumption expenditure (EXP). Moreover, all the countries in CLMV do not provide the latest data on openness rate and gross capital formation (GCF). However, overall results suggest that FDIC, EXPL, GDPM, and EXPV are most likely to have missing data. (Note that the subscripts refer to countries in CLMV).

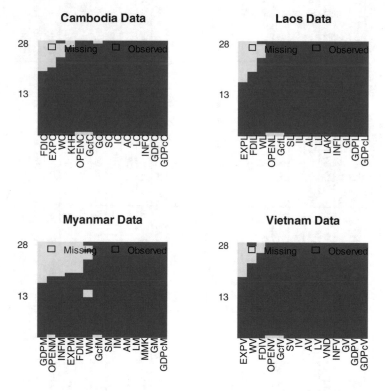

Fig. 1. Missingness map of the data for CLMV countries

4.2   Predicting Missing Values

Multiple imputation involves imputing values for each missing data in the matrix and creating a completed data set. Across these completed data sets, the observed values are the same, but the missing values are filled or replaced with a sample of values from the predictive distribution of missing data.

Figure 2 shows the prediction of missing values in the time series from 1988 to 2015. We illustrate the observed value and predicted value for gross capital formation of Cambodia (top left panel), foreign direct investment of Thailand to Cambodia (top right panel), inflation rate of Myanmar (bottom left panel), and government expenditure of Vietnam (bottom right panel). In these panels, the black points represent observed data while the red points are the mean imputation for each of the missing values, along with their 95% confidence bands. Here, we draw these bands by imputing each of missing values 100 times to obtain the imputation distribution for those observations. The imputation of each missing value is considerably reasonable as it is located within the line. After the imputation with EMB algorithm, we apply the SUR model to the completed data sets and use a generalized least squares (GLS) technique to estimate unknown parameters. Finally, we combine the estimated parameters obtained from the SUR model into one set of parameters, and then use the mean values of the new parameter set to make inferences and draw conclusions. Consequently, the estimated results are shown in Table 3, in the next subsection.

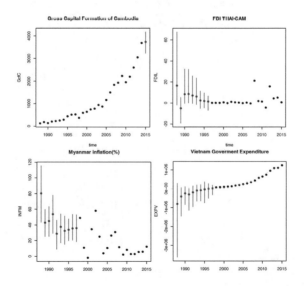

Fig. 2. Panels showing mean imputations with 95% bands (in red) and observed data point (in black)

4.3   Estimated Results

This section will analyze the effects of the considered variables (as shown in Table 3) on Thailand's direct investment in CLMV countries using Seemingly

Table 3. Estimated results for CLMV countries

Variable	FDI_C	FDI_L	FDI_M	FDI_V
Constant	-1.6738^a	9.4041	-11.4831	-2.1963
	(0.9105)	(8.5267)	(0.90637)	(2.8705)
GDPcap	0.0638	0.0546	0.0206^a	0.0065
	(0.0932)	(0.0452)	(0.0126)	(0.0563)
GDP	0.0350^c	0.0004^a	0.00001	-0.00001
	(0.0083)	(0.0003)	(0.00001)	(0.00001)
G	3.0190^c	0.3823	0.0022	-2.4471
	(1.0424)	(0.6091)	(0.0031)	(4.7427)
INF	-0.1406^b	-0.0416	0.0086	0.7371
	(0.0585)	(0.0522)	(0.0220)	(3.6177)
Gcf	0.0033	0.0097^c	0.0068^a	-0.0097^c
	(0.0057)	(-0.0010)	(0.0039)	(0.0027)
EX	-0.0030^a	-0.0009	-0.0085	0.0326
	(0.0019)	(0.0014)	(0.0547)	(0.0820)
L	0.0053^a	-0.0437	0.0041	0.0365
	(0.0029)	(0.0422)	(0.0052)	(0.0981)
EXP	0.0009	-0.0004	0.0006^b	0.0003
	(0.0004)	(0.0005)	(0.0003)	(0.0021)
OPEN	-4.2095	5.1161	1.9879	5.0173^*
	(11.064)	(8.3304)	(1.5165)	(3.0058)
W	-1.6738^a	0.3644	0.3033	0.7957
	(0.9105)	(0.3325)	(0.8297)	(3.1472)
A	0.0638	$-0.0291)$	-0.0183^c	0.1141^c
	(0.0932)	(0.0675)	(0.0019)	(0.0079)
I	0.0350^c	9.4041	-0.0026	-0.0642^b
	(0.0083)	(8.5267)	(0.0295)	(0.0328)
S	3.0190^c	0.0546	-0.0252	-2.4471
	(1.0424)	(0.0452)	(0.0296)	(4.7427)

Source: Calculation
Note: ^{a}, ^{b}, and ^{c} indicate significance at the 90%, 95% and
99% levels, respectively.
Standard errors are reported in parentheses.

Unrelated Regression model. The empirical results indicate that four motivations persuade Thailand's direct investment to the neighboring countries: Cambodia, Laos, Myanmar and Vietnam. Myanmar's GDP per capita (GDPcap), and market seeking motivation, attract Thai firms to invest in Myanmar. The implicit reason provides that Thai products are categorized to be higher brand premium in neighboring countries, especially Myanmar. That is the reason why

Myanmar's GDP per capita positively affects to Thai direct investment. Besides, gross domestic product (GDP) positively affects Thai companies to invest in Cambodia and Laos. For instance, in the agro-based industry, firms mainly expanded their business to Cambodia and Laos to seek new market opportunities. And GDP growth rate (G) also has favourable effect on Thailand's direct investment in Cambodia. The enlarged market will also allow Thai companies to invest more.

Meanwhile, inflation rate (INF) and exchange rate (EX), representing the efficiency seeking motivation, adversely affect Thai direct investment in Cambodia since the value of Cambodia currency fluctuated over time. Cambodia' average core inflation also was high, reflecting inefficiency of monetary measures. But Vietnam's openness rate (OPEN) has beneficial effect on Thai direct investment significantly.

On the other hand, the resource seeking motivation is interesting. First and foremost, average monthly wages have detrimental effect on Thailand's direct investment in Cambodia. A major incentive for a multinational enterprise to invest abroad is to outsource labor intensive production to countries with lower wages and Cambodia' average monthly wage is higher than others'. Furthermore, industry and service value added in Cambodia have satisfactory effect on Thai direct investment. Vietnam's agriculture value-added impacts positively on Thailand's direct investment but the Vietnam's industry value-added adversely affects Thailand's direct investment. Meanwhile Myanmar's agriculture value-added negatively affects Thailand's direct investment. With the increase of minimum wage in Thailand to 300 baht (roughly USD 10) per day, some Thai investors determine to invest more in CLMV especially labor intensive manufacturing sector in the high value-added. The number of labor force from where labor is supplied for the production of goods and services, has beneficial effect on Thailand's direct investment in Cambodia.

Consider the capability seeking motivation: general government final consumption expenditure (EXP) and gross capital formation (Gcf), they have advantageous effect on Thailand's direct investment in Myanmar. Besides, the outlays on gross capital formation (Gcf) also positively affect Thailand's direct investment in Laos but adversely affect that in Vietnam. Both factors work like pro-active measures of the recipient countries to promote the foreign direct investment.

5   Conclusions

The establishment of ASEAN Economic Community (AEC) in 2015, has offered Thai enterprises and companies an opportunity to invest in neighboring countries, especially Cambodia, Laos, Myanmar and Vietnam (henceforth CLMV). Investment in CLMV countries can bring about both market and production opportunities for Thailand's as the companies can relocate or expand their low value-added activities to those neighboring countries, while upgrading their domestic production to higher value-added activities. Accordingly, the key investment considerations in CLMV countries become our concern. The motives for firms to engage in foreign production and MNEs activities can be classified into four groups, namely natural resource seeking, market seeking, efficiency seeking, and capability seeking. And hence, this study examines the determinants of outward Thailand's direct investment to CLMV countries based on these motivations, and later estimate marginal effects.

But during the process of data collection, we faced a problem of data unavailability in the CLMV countries, and missing values in many available variables. So, we employ the bootstrap-based EMB (Expectation Maximization with Bootstrapping) for dealing with this missing data problem and eliminate variables that are unavailable. Once obtaining a completed data set, we proceed to the empirical analysis. The Seemingly Unrelated Regression (SUR) model is used to analyze the effects of the considered variables on Thailand's direct investment in CLMV. The estimated results suggest that among the specified variables, based on the four motivations, GDP per capita, gross capital formation, and government final consumption expenditure can attract Thai firms to invest in Myanmar. Gross domestic product as well as gross capital formation are considered advantageous compared to other variables for encouraging Thailand's direct investment in Laos, while gross capital formation and openness rate are encouraging the outward FDI of Thailand to Vietnam.

Cambodia seems to be the most interesting country for Thai investors to invest in. The investors are attracted by GDP and its growth rate, as well as labor force in Cambodia. However, our empirical result also shows that Cambodia does not succeed in persuading Thai firms with its lower wage due to the fact that its average monthly wage is higher than others'.

Acknowledgements. The authors are grateful to Centre of Excellence in Econometrics, Chiang Mai University for the financial support, and also Mr. Woraphon Yamaka who kindly helps ensure that our code is correct. Without their supports, this study could not be completed.

References

1. Dempster, A.P., Laird, N.M., Rubin, D.B.: Maximum likelihood from incomplete data via the EM algorithm. J. Roy. Stat. Soc. Ser. B (Methodol.) **39**, 1–38 (1977)
2. Dunning, J.H.: Explaining changing patterns of international production: in defence of the eclectic theory. Oxford Bull. Econ. Stat. **41**(4), 269–295 (1979)
3. Dunning, J. H. Multinational Enterprises and the Global Economy, pp. 272–274. Addison-Wesley, Wokingham (1993)
4. Dunning, J.H.: The eclectic paradigm as an envelope for economic and business theories of MNE activity. Int. Bus. Rev. **9**, 163–190 (2000)
5. Dunning, J.H., Lundan, S.M.: Multinational Enterprises and the Global Economy, 2nd edn. Edward Elgar, Cheltenham (2008)
6. Honaker, J., King, G.: What to do about missing values in time-series cross-section data. Am. J. Polit. Sci. **54**(2), 561–581 (2010)
7. Honaker, J., King, G., Blackwell, M.: Amelia II: a program for missing data. J. Stat. Softw. **45**(7), 1–47 (2011)

8. Oizumi, K.: The potential of the Thailand-Plus-One business model-a new fragmentation in east asia. RIM Pac. Bus. Ind. **13**, 2–20 (2013)
9. Pastpipatkul, P., Maneejuk, P., Wiboonpongse, A., Sriboonchitta, S.: Seemingly unrelated regression based copula: an application on Thai rice market. In: Causal Inference in Econometrics, pp. 437–450. Springer International Publishing (2016)
10. Rattanakhamfu, S., Tangkitvanich, S. Strategies and Challenges of Thai Companies to Invest CLMV. Thailand Development Research Institute (2015)
11. Wee, K.H.: Outward foreign direct investment by enterprise from Thailand. Transnational Corporations **16**(1), 89–116 (2007)
12. Zellner, A.: An efficient method of estimating seemingly unrelated regressions and tests for aggregation bias. J. Am. Stat. Assoc. **57**(298), 348–368 (1962)

The Role of Oil Price in the Forecasts of Agricultural Commodity Prices

Rossarin Osathanunkul[1(✉)], Chatchai Khiewngamdee[2], Woraphon Yamaka[3], and Songsak Sriboonchitta[3]

[1] Faculty of Economics, Chiang Mai University, Chiang Mai, Thailand
orossarin@gmail.com
[2] Department of Agricultural Economy and Development, Faculty of Agriculture, Chiang Mai University, Chiang Mai, Thailand
[3] Centre of Excellence in Econometrics, Faculty of Economics, Chiang Mai University, Chiang Mai, Thailand

Abstract. The objective of this paper is to examine whether including oil price to the agricultural prices forecasting model can improve the forecasting performance. We employ linear Bayesian vector autoregressive (BVAR) and Markov switching Bayesian vector autoregressive (MS-BVAR) as innovation tools to generate the out-of-sample forecast for the agricultural prices as well as compare the performance of these two forecasting models. The results show that the model which includes the information of oil price and its shock outperforms other models. More importantly, linear model performs well in one- to three-step-ahead forecasting, while Markov switching model presents greater forecasting accuracy in the longer time horizon.

1 Introduction

The relationship between agricultural commodity prices and crude oil price is of concern to both policy makers and academics. Increase in oil prices is considered to be a main driver for rising prices of agricultural crops since it is highly relevant in agricultural production. For instance, in practice, oil is required for running agricultural equipment, and used for transporting and distributing crops to retailers. In addition, crude oil prices also relate to the agricultural commodity prices in the context of substitution effect between fossil fuel and biofuel. Agricultural commodities such as corn, soybean, and oilseeds are alternatively used to produce fuel, called biofuel or alternative energy, in the light of unexpected increase in prices of fossil fuel like crude oil. This development of alternative energy, in turn, drives up the prices of these crops, and then brings about persistent increases in other agricultural commodities prices due to the limitation of planting area [1].

This situation has raised concerns among policy makers and academics about the transmission of oil price shocks to agricultural commodity prices. According to literature review, there are a vast number of studies on the price relations between crude oil and agricultural commodities, and the spillover effect of oil

© Springer International Publishing AG 2018
V. Kreinovich et al. (eds.), *Predictive Econometrics and Big Data*, Studies in Computational Intelligence 753, https://doi.org/10.1007/978-3-319-70942-0_30

price shocks on the commodity prices. For example, the studies of Saghaian [2] and Esmaeili and Shokoohi [3] have shown the empirical evidences supporting the relationship between oil prices and commodity prices, and also the influence of oil price shocks on prices of agricultural product. On the contrary, few studies could not find the significant effect of oil shocks on the variation in crop price, such as Zhang et al. [4] and Campiche et al. [5]. In other words, the unexpected increases in oil price could not explain all of commodity prices increase. The main reason is that other factors, in particular, government policies and world economic situation, can also cause the price variation, and often in emerging markets where the crop prices are more susceptible to domestic policies. Moreover, emerging markets do not have enough power to influence movements, so that they have to take the effects from external price shocks.

Recently, Gupta and Kotz [6] pointed out that the oil price shocks can produce the effect not only on agricultural commodity prices, but also on both macroeconomic and financial variables. These authors considered the use of unexpected oil price shocks in forecasting the nominal interest rate in South Africa, and disaggregated the shocks into positive and negative parts. They found that the forecasting with a special concern about oil price shocks can outperform and produce more accurate forecasts than the one excluding the shocks. This persuades us to revisit the issue of price volatility in agricultural commodity owing to unexpected change in oil price. To be more specific, this study focuses on the effects of oil price shocks on agricultural commodity prices, particularly on corn, wheat, and sugar, and aims to evaluate the role of oil price as well as its shocks in forecasting commodity price movements, as introduced by Gupta and Kotz [6].

In this study, we deal with the forecasting issue within the context of the vector autoregressive (VAR) and the Markov Switching vector autoregressive (MS VAR) models, focusing on the commodity forecasts and structural break in the mean growth rate of the commodity prices. To estimate the unknown parameters in the models, we employ Bayesian estimation. Pastpipatkul, et al. [7] and Gupta and Kotz [6] suggested that Bayesian estimation can avoid possible misspecification errors due to an identification of a stochastic or deterministic trend for the estimation of parameters.

The rest of the paper is organized as follows: Sect. 2 describes the methodology used in this study that is Markov-switching VAR model associated with Bayesian approach. Then data description and empirical results are provided in Sect. 3. Section 4 comprises the conclusion.

2 Methodology

2.1 Markov-Switching VAR

In this section, we briefly outline the structure and the estimation of the model that is used in the forecasting. The general specification of a Markov Switching

VAR model can be written as

$$Y_t' = C(s_t) + \sum_{p=1}^{P} A_p(s_t) Y_{t-p}' + \varepsilon_t' \Sigma^{-1}(s_t), \ t = 1, \cdots, T \tag{1}$$

where Y_t' is $n \times 1$ vector of endogenous variables, A_p is $n \times n$ matrix of regime dependent autoregressive coefficients order p, C is vector of regime dependent intercept terms, ε is the vector of n unobserved noise, and Σ is $n \times n$ positive matrix of variance. s_t is the time t realization of a discrete latent process that we call a "state" or "regime". To determine the number of lag p, we employ Bayesian information criterion (BIC) and the lowest BIC is preferred. Note that, in the case of linear VAR model, the state or regime latent variable, s_t, is not included in the model, and thus all unknown parameters are regime independent. Following Sim, Waggoner, and Zha [8], they provided a distributional assumption with densities of the MS-VAR disturbances as follows

$$P(\varepsilon_t | Y_{t-1}, (s_t), \Theta) = N(\varepsilon_t | 0_{n \times 1}, I_n), \tag{2}$$

and on the information set

$$P(Y_t | Y_{t-1}, s_t, \Theta) = N(Y_t | C(s_t), A_z(s_t), \Sigma_z(s_t)), \tag{3}$$

where Θ is all unknown parameters in the model. We also assume that the evolution of the latent variable driving the regime changes, (s_t), is governed by a first-order Markov chain with transition probabilities matrix. Suppose, we have h regimes, the transition matrix can take the form as:

$$Q = \begin{bmatrix} \rho_{11} & \rho_{12} & \cdots & \rho_{1h} \\ \rho_{21} & \rho_{22} & \cdots & \rho_{2h} \\ \vdots & & \ddots & \vdots \\ \rho_{h1} & \rho_{h2} & \cdots & \rho_{hh} \end{bmatrix} \tag{4}$$

In this study, we adopt a Bayesian approach to estimate all unknown parameters. This approach allows us to incorporate the prior distribution to the parametric likelihood function in order to construct the joint posterior distribution for the parameters and the latent variables (s_t). The likelihood of the MS-VAR model is built by Y_t.

$$\ln P(Y_t | Q, \Theta, s_t) = \sum_{t=1}^{T} \ln \left[\sum_{s_t \in h} P(Y_t | Q, \Theta) P(s_t | Q, \Theta) \right] \tag{5}$$

where $P(Y_t | Q, \Theta)$ is the density used to sample the probability that (s_t) is in regime i given $(s_{t-1}) = j$. To sample the parameter set in the model, we employ a Gibbs sampling algorithm as suggested by Sim, Waggoner, and Zha [8]. The estimation of the MS-VAR model in Eq. (1) depends on the joint posterior distribution of Θ and Q which is derived by Bayesrule, thus we have

$$P(Q, \Theta(s_t) | Y_T) \propto P(Y_t | Q, \Theta(s_t)) p(Q, \Theta(s_t)), \tag{6}$$

where $p(Q, \Theta(s_t))$ denotes the prior of Q and Θ. However, the estimation of this model is difficult to get the converge solution; the block EM algorithm of Nason and Tallman [9] is considered here and the blocks are Bayesian Vector Autoregressive (BVAR) regression coefficients for each regime (separating for intercepts, AR coefficient, and error covariance) and transition matrix.

For the prior of this model, we consider the prior distributions for the MS-VAR coefficients that belong to the following Normal-Inverse-Wishart prior. We assume $A(s_t)$, $N(n, \Sigma)$ and $\Sigma \sim IW(\Psi, d)$, where b, Σ, Ψ and d is a vector of hyperparameters. Then we employ the Markov chain Monte Carlo (MCMC) Gibbs sampler in order to estimate the marginal likelihoods and Bayes factor or marginal posterior distribution of interest for inference by running 2,000 steps of MCMC simulator as follows:

(1) Draw A^1 from $P(\Sigma^1 | \Sigma^0, C^0, A^0, Q^0, Y_t)$.
(2) Draw Σ^1 from $P(\Sigma^1 | \Sigma^0, C^0, A^1, Q^0, Y_t)$.
(3) Draw C^1 from $P(C^1 | \Sigma^1, C^0, A^1, Q^0, Y_t)$.
(4) Draw Q^1 from $P(C^1 | \Sigma^1, C^0, A^1, Q^0, Y_t)$.

This completes a Gibbs iteration and we obtain A_1, Γ_1, C^1, and Q^1. Then, using these new parameters as starting values in order to repeat the prior iteration of $A^{(i)}$, $\Gamma^{(i)}$, $C^{(i)}$, and $Q^{(i)}$ draws. Repeating the previous iterations for 1,000 times to obtain a sequence of random draws:

$$(A^1, \Sigma^1, C^1, Q^1), \cdots, (A^{10000}, \Sigma^{10000}, C^{10000}, Q^{10000}) \tag{7}$$

Finally, we can obtain the estimated parameters from the mean of each parameter draw.

3   Empirical Results

3.1   Data Analysis

We collected the data from Thomson Reuters database. The dataset used is weekly data which goes from April 1st, 2005 to March 31st, 2017 for a total of 627 observations and the data variables consist of CBoT Corn Futures, CBoT Wheat Futures, ICE-US Sugar Futures, and Crude Oil Futures. To compute oil price shock, we then define the oil price (OP) process as a random walk with the appropriate moving average representation:

$$OP_t = \alpha + \beta OP_{t-1} + \varepsilon_t$$

where ε_t is defined as a shock of oil price. Thus, this provides us with one additional measure of oil prices that represent oil price shock (Table 1).

Table 1. Data description

	Corn	Wheat	Sugar	Oil	shock
Mean	0.0831	0.0442	0.1086	−0.0184	0.0001
Median	0.2309	−0.0403	−0.2362	0.1722	0.1300
Maximum	20.2837	15.9499	20.0751	25.1791	23.5200
Minimum	−25.4272	−16.9865	−19.2067	−37.0059	−0.3673
Std. Dev	4.4984	4.6053	4.7304	5.3485	0.0534
Skewness	−42.3358	13.3674	4.2099	−46.63	−0.4909
Kurtosis	611.4265	387.9439	417.633	832.9197	8.0320
Jarque-Bera	27167.29	2203.749	3627.781	76346.15	684.5177
Probability	0	0.000016	0	0	0
ADF-test	−24.2844	−24.6926	−25.5893	−24.2499	−24.9737

3.2   Model Selection

In this study, we aim to compare the performance of oil price in forecasting agricultural prices including CBoT Corn Futures, CBoT Wheat Futures, ICE-US Sugar Futures, and Crude Oil Futures. Our benchmark case is the model that does not include any measure of oil prices, which we denote BVAR1 ($Y_t = \{Corn_t, Wheat_t, Sugar_t\}$). We then compare three commodity price models, the benchmark BVAR1 and the other two that include various features of oil price. The additional models are organized as follows:

(i) BVAR2 ($Y_t = \{Corn_t, Wheat_t, Sugar_t, Oil_t\}$) denotes the BVAR1 model that includes oil price data;

(ii) BVAR3 ($Y_t = \{Corn_t, Wheat_t, Sugar_t, Oil_t, shock_t\}$) denotes the BVAR1 model that includes oil price and oil price shocks. Hence, the BVAR1 model is nested within the respective BVAR2, and BVAR3 models. However, a non linear behavior might exist in the agriculture data. To investigate the nature of a potential nonlinear behavior in our data, we also consider various null and alternative hypotheses which result in the following two competing forecasting models: Model I: A BVAR model and Model II: A MS-BVAR model. Thus, we have six models for further investigation.

Prior to forecasting, we compare Bayesian inferences from the six alternative models considered in Subsect. 3.1. Throughout this section, all inferences are based on 10,000 Gibbs simulations, after discarding the initial 2,000 Gibbs simulations in order to mitigate the effects of initial conditions. Table 2 summarizes the marginal likelihoods of each of the models. Notice that the marginal likelihoods suggest that BVAR1, BVAR2 and BVAR3 are not dominated by MS-BVAR1, MS-BVAR2, and MS-BVAR3. Thus, we conclude that there exists no structural break in the agricultural prices.

Table 2. Value of log marginal likelihood

	Log marginal likelihood
BVAR1	3275.319
MS-BVAR1	2742.67
BVAR2	4313.178
MS-BVAR2	2910.925
BVAR3	6847.236
MS-BVAR3	5180.732

Source: Calculations.
Note: By using BIC selection, lag 1 is
an appropriate lag for all models.

3.3 Forecasting Performance over Time

In this section, we conduct an out-of-sample forecast in order to compare the forecasting performance among models. We firstly calculate the one- to eight-weekly-ahead forecasts and compare the forecast value with the real data. Here, this dataset is based on an out-of-sample period of week 8, 2017 to week 13, 2017, with an initial in-sample period that spans week 13, 2005 to week 7, 2017. Then, we analyze the forecasting performance among our proposed models with the BVAR1 model using root mean square error (RMSE). In doing so, we calculate the actual RMSE for BVAR1 model, while, for other models, we show the relative RMSE value. By providing the relative RMSE, we can easily compare the benchmark model with other models. In other words, a value of relative RMSE larger than one indicates that the benchmark BVAR1 model is superior to other models. Furthermore, we perform MSE-F test, proposed by McCracken [10] which is used to test the null hypothesis that the forecasting ability are equal between the restricted BVAR1 model and unrestricted MS-BVAR1, BVAR2, MS-BVAR2, BVAR3 and MS-BVAR3 models. The statistic is written as

$$\text{MSE-F} = \frac{(N - R - h + 1) \times \overline{d}}{MSE_1}$$

where N denotes the total samples, R represents the number of observations in an in-sample period, h is a forecasting horizon and $\overline{d} = MSE_0 - MSE_1$. In this case, $MSE_i = (N - R - h + 1)^{-1} \sum_{t=R}^{T-h} (u_{i,t+1})^2$, where i = 1,0 with u_i is the forecasting error. Since, the alternative hypothesis is that the MSE for the unrestricted model forecast, MSE_1, is less than the MSE for the restricted model forecast, MSE_0. Hence, a positive value of MSE-F statistic indicates that the restricted BVAR1 model forecasts are inferior to competing models.

Table 3 presents the RMSE for MS-BAVR1, BVAR2, MS-BVAR2, BVAR3 and MS-BVAR3 models relative to BVAR1 model which does not include oil price and oil price shock. The results show that, on average, the relative RMSE values of all models are less than one. This indicates that the forecasts from

Table 3. Out-of-sample root-mean square error (week 8, 2017 to week 13, 2017)

Model	1 steps	2 steps	3 steps	4 steps	5 steps	6 steps	7 steps	8 steps	Average
BVAR1	3.5386	3.2168	3.0654	2.9827	3.0344	2.9617	2.8894	2.7697	3.0573
MS-BVAR1	0.9996***	1.0166***	1.0039***	0.9869***	0.9913***	0.9971***	0.9857***	1.0004***	0.9977
BVAR2	0.9997***	0.9993***	0.9984***	1.0005***	0.9995***	1.0001***	1.0006***	1.0000***	0.9998
MS-BVAR2	1.0027***	1.0084***	1.0122***	0.9855***	0.9901***	0.9993***	0.9888***	0.9999***	0.9984
BVAR3	0.9996***	0.9995***	0.9983***	1.0004***	0.9994***	1.0001***	1.0004***	1.0001***	0.9997
MS-BVAR3	0.9995***	1.0040***	1.0043***	0.9899***	0.9885***	0.9869***	0.9852***	1.0040***	0.9953

BVAR1 model, which excludes oil price and oil price shock, are less precise than those from other models including Markov switching models. In addition, the model with the lowest value of RMSE is MS-BVAR3 model that includes oil price and oil price shock. It implies that including oil price and oil price shock to the model can improve the performance of agricultural prices forecasting. Moreover, based on MSE-F test, the improvements are statistically significant.

Interestingly, all nonlinear models, namely MS-BVAR1, MS-BVAR2, and MS-BVAR3, outperform their linear counterparts according to the lower value of relative RMSE despite the conclusion, in previous section, from the marginal likelihoods that there is no structural break in the data. One possible explanation is that nonlinear model performs better than linear model in longer time horizon forecasting and hence making the average value of relative RMSE of nonlinear model lower than linear model. As shown in Table 3, from four- to eight-step-ahead forecasting, the relative RMSE value of MS-BVAR1 is less than one, while those values are greater than one in the shorter steps. Likewise, the relative RMSE value of MS-BVAR2 and MS-BVAR3 from four- to eight-step-ahead forecasting are lower than BVAR2 and BVAR3, respectively. This finding is consistent with several researches, for instance, Altavilla and De Grauwe [11], Chen and Hong [12], and Marcellino [13]. Therefore, we can conclude that the Markov switching BVAR models are superior to their BVAR counterparts in longer time horizon forecasting.

4   Conclusions

This paper aims to investigate whether including oil price to the agricultural prices can improve the forecasting accuracy. We consider Bayesian vector autoregressive (BVAR) and Markov Switching Bayesian vector autoregressive (MS-BVAR) as a forecasting model in this study. We use weekly data to avoid the mismatch trading days across the markets. The data consist of CBoT Corn Futures, CBoT Wheat Futures, ICE-US Sugar Futures, and Crude Oil Futures. To compare the performance of the models, we compared out-of-sample forecasts of weekly agriculture returns. In this study, the dataset is based on an out-of-sample period of week 8, 2017 to week 13, 2017, with an initial in-sample period that spans week 13, 2005 to week 7, 2017.

We use BVAR models to generate the respective forecasts. We consider three BVAR models excluding oil price, including oil price, and including oil price and oil price shock denoted as BVAR1, BVAR2, and BVAR3, respectively. We also consider the Markov switching version of these three models. In this study, we further investigate the structural break in our forecasting model. We compare the performance of the BVAR1, BVAR2, and BVAR3 with a structural break in VAR. The results show that the logged marginal likelihoods for linear BVAR models are lower than MS-BVAR.

We conduct an out-of-sample forecast from one- to eight-step-ahead and compare the forecasting performance of our forecasting models using RMSE criterion. The results show that all of the competing models outperform BVAR1 model according to the average value of relative RMSE. Additionally, MS-BVAR3 model which includes oil price and oil price shock has the best forecasting performance in our study. It indicates that including oil price and oil price shock to the model can improve the forecasting performance of the agricultural prices. Furthermore, these improvements are statistically significant based on MSE-F test. More importantly, BVAR models perform well in one- to three-step-ahead forecasting. Markov switching BVAR models, however, present greater forecasting accuracy in the longer time horizon.

References

1. Wang, Y., Wu, C., Yang, L.: Oil price shocks and agricultural commodity prices. Energ. Econ. **44**, 22–35 (2014)
2. Saghaian, S.H.: The impact of the oil sector on commodity prices: correlation or causation? J. Agric. Appl. Econ. **42**(03), 477–485 (2010)
3. Esmaeili, A., Shokoohi, Z.: Assessing the effect of oil price on world food prices: application of principal component analysis. Energ. Policy **39**(2), 1022–1025 (2011)
4. Zhang, Z., Lohr, L., Escalante, C., Wetzstein, M.: Food versus fuel: what do prices tell us? Energ. Policy **38**(1), 445–451 (2010)
5. Campiche, J.L., Bryant, H.L., Richardson, J.W., Outlaw, J.L.: Examining the evolving correspondence between petroleum prices and agricultural commodity prices. In: The American Agricultural Economics Association Annual Meeting, Portland, OR (2007)
6. Gupta, R., Kotz, K.: The role of oil prices in the forecasts of South African interest rates: a Bayesian approach. Energ. Econ. **61**, 270–278 (2017)
7. Pastpipatkul, P., Yamaka, W., Wiboonpongse, A., Sriboonchitta, S.: Spillovers of quantitative easing on financial markets of Thailand, Indonesia, and the Philippines. In: International Symposium on Integrated Uncertainty in Knowledge Modelling and Decision Making, pp. 374–388. Springer (2015)
8. Sims, C.A., Waggoner, D.F., Zha, T.: Methods for inference in large multiple-equation Markov-switching models. J. Econ. **146**(2), 255–274 (2008)
9. Nason, J.M., Tallman, E.W.: Business cycles and financial crises: the roles of credit supply and demand shocks. Macroecon. Dyn. **19**(4), 836–882 (2015)
10. McCracken, M.W.: Asymptotics for out of sample tests of Granger causality. J. Econ. **140**(2), 719–752 (2007)
11. Altavilla, C., De Grauwe, P.: Forecasting and combining competing models of exchange rate determination. Appl. Econ. **42**(27), 3455–3480 (2010)
12. Chen, Q., Hong, Y.: Predictability of equity returns over different time horizons: a nonparametric approach. Manuscript, Cornell University (2010)
13. Marcellino, M.: A Comparison of Time Series Models for Forecasting GDP Growth and Inflation. Bocconi University, Italia (2007)

Does Forecasting Benefit from Mixed-Frequency Data Sampling Model: The Evidence from Forecasting GDP Growth Using Financial Factor in Thailand

Natthaphat Kingnetr[1(✉)], Tanaporn Tungtrakul[1], and Songsak Sriboonchitta[1,2]

[1] Faculty of Economics, Chiang Mai University, Chiang Mai 50200, Thailand
natthaphat.kingnetr@outlook.com
[2] Center of Excellence in Econometrics, Chiang Mai University, Chiang Mai 50200, Thailand

Abstract. It is common for macroeconomic data to be observed at different frequencies. This gives a challenge to analysts when forecasting with multivariate model is concerned. The mixed-frequency data sampling (MIDAS) model has been developed to deal with such problem. However, there are several MIDAS model specifications and they can affect forecasting outcomes. Thus, we investigate the forecasting performance of MIDAS model under different specifications. Using financial variable to forecast quarterly GDP growth in Thailand, our results suggest that U-MIDAS model significantly outperforms the traditional time-aggregate model and MIDAS models with weighting schemes. Additionally, MIDAS model with Beta weighting scheme exhibits greater forecasting precision than the time-aggregate model. This implies that MIDAS model may not be able to surpass the traditional time-aggregate model if inappropriate weighting scheme is used.

1 Introduction

Policy makers require reliable forecasting of economic growth. An accurately gross domestic product (GDP) measuring helps policy makers, economists, and investors determine appropriate policies and financial strategies. Forecast of real GDP growth depends on many economic variables, while the publication by statistical agencies of GDP data is generally delayed by one or two quarters. For forecasting GDP growth, Thailand's GDP is available only as quarterly data while other economic variables to be used as leading indicators may be available in monthly data. There is a huge literature including [6] for the United States of America and [3] for Euro area who employed financial variables as leading indicators of GDP growth. Ferrara and Marsilli [7] concluded that the stock index could improve forecasting accuracy on GDP growth.

Thus, involving data sampled at different frequencies in forecasting model seems be to beneficial. From the literature, a way of using high frequency indicators to forecast low frequency variable is the Mixed Data Sampling (MIDAS)

© Springer International Publishing AG 2018
V. Kreinovich et al. (eds.), *Predictive Econometrics and Big Data*, Studies in Computational Intelligence 753, https://doi.org/10.1007/978-3-319-70942-0_31

model proposed by Ghysels *et al.* [10]. It has been applied in various fields such as financial economics [13] and macroeconomics [4,5,15] to forecast GDP. Clements and Galvo [5] concluded that the predictive ability of the indicators in comparison with an autoregression is stronger. It also allows the regressed and the regressors to be sampled at different frequencies and is a parsimonious way of allowing lags of explanatory variables. MIDAS regression model combined with forecast combination schemes if large data sets are involved are computationally easy to implement and are less prone to specification errors. Based on the parsimony of representation argument[1], the higher frequency part of conditional expectation of MIDAS regression is often formulated in terms of aggregates which depend on a weighting function. However, there are many weighting schemes such as Step, Exponential Almon, and Beta (analogue of probability density function) [9].

The objective of this paper is to use such important financial leading indicator as Stock Exchange of Thailand (SET) index to forecast Thailands quarterly GDP growth by using the different weighted MIDAS models. In addition to MIDAS model with weighting schemes, we also consider the traditional time-aggregate model and the unrestricted MIDAS model. This will allow us to see whether high frequency data render any benefit in predicting lower frequency data, and if it does, which model specification performs the best in this setting of forecasting Thailands quarterly GDP growth. The result of study will be useful for government in imposing appropriate policies and strategies for stabilising countrys economy.

The organisation of this paper is as follows. Section 2 describes the scope of the data used in this study. Section 3 provides the methodology of this study and provides the estimation of this study. Section 4 discusses the empirical results. Conclusion of this study is drawn in Sect. 5.

2   Data

The data in this study consist of Thailand's quarterly gross domestic product (GDP) and monthly Stock Exchange of Thailand (SET) index. GDP is obtained from the Bank of Thailand while SET index is obtained from the Stock Exchange of Thailand. The series cover period of 2001Q1 to 2016Q4, while data during 2001Q1 to 2015Q4 are used for model estimation, the rest are left for out-of-sample forecast evaluation. All variables are transformed into year-to-year (Y-o-Y) growth rate to reduce the risk of having seasonality. Figures 1 and 2 provide the plot of GDP growth and SET index growth.

It can be seen from the figures that there is a huge drop in GDP growth around the end of 2008 and the beginning of 2009. The SET growth also changes in similar manner during the same period. This is believed to be the results of US financial crisis. Also, around the end of 2011, it can be seen that there is a drop in GDP growth

---

[1] Also called "The principle of parsimony", it states that the parsimonious model specification is the model that is optimally formed with the smallest numbers of parameters to be estimated [2].

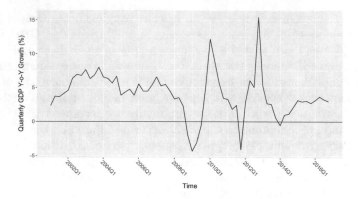

Fig. 1. Quarterly GDP growth 2001Q1 to 2016Q4

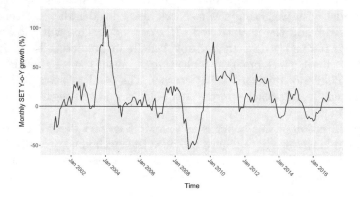

Fig. 2. Monthly SET index growth 2001Q1 to 2016Q4

which was caused by the great flood in Thailand. Again, the SET index growth follows in the same direction. These figures may suggest that financial variable such as the SET index is a potential predictor for GDP.

3   Methodology

Prior to model estimation and forecasting, it is recommended to check whether series in the study is stationary or not. Therefore, in this section, we begin with brief information regarding the unit root tests, followed by forecasting models employed in this study.

3.1   Unit Root Tests

We start with the Augmented Dickey-Fuller (ADF) test [18] which is very well-known and widely-used in empirical works. The test model can be specified as

$$\Delta y_t = \alpha_0 + \alpha_1 y_{t-1} + \sum_{i=1}^{p} \alpha_{2i} \Delta y_{t-i} + \varepsilon_t \tag{1}$$

where y_t is the time series being tested and ε_t is residual. The hypothesis testing can be specified as $H_0 : \alpha_1 = 0$ for non-stationary against $H_1 : \alpha_1 < 1$ for stationary.

Next, the Phillips-Perron (PP) test [17] has been frequently used as an alternative test to the ADF test. The test employs the same null hypothesis of non-stationary as in the ADF test. However, the advantage of this test is that the additional lagged dependent variable is not required in the presence of serial correlation. Additionally, it is robust to the functional form of the error term in the model since the test is non-parametric. However, the test requires large sample properties in order to perform well.

Unlike the ADF test and the PP test, the Kwiatkowski-Phillips-Schmidt-Shin (KPSS) test introduced by Kwiatkowski et al. [16] has the null hypothesis of stationary. With alternative way of interpreting the null hypothesis, the KPSS test complements other unit root tests.

By looking at the results from each test, we can have a better view before making a conclusion on whether the series is stationary, non-stationary, or inconclusive. This is important since the stationarity of the series is required for the forecasting models considered in the study. Now, we are going to describe the five approaches that incorporate higher frequency data in forecasting lower-frequency variables.

3.2   Time-Aggregate Model

Traditionally, when one working on forecasting that involves mixed frequency data, all series must be converted into the same frequency. That means all the series will be transformed into the frequency matching that of series which was observed at the lowest frequency. As pointed out by Armesto et al. [1], this can be easily done by taking an average of values from high frequency data within the time frame of low frequency data. For instance, we work on variable X which is measured monthly and Y being observed quarterly data. Then, X will be transformed to match the same frequency of Y by taking the average of X at each respective quarter. After transformation, we can now use the new Y to help predict X. This is so-called the time-aggregate model. Suppose that we are interested in one step forecast, the model can be mathematically specified as

$$Y_t = \alpha + \sum_{i=1}^{p} \beta_i L^i Y_t + \sum_{j=1}^{r} \gamma_j L^j \overline{X}_t + \varepsilon_t \tag{2}$$

with

$$\overline{X}_t = \frac{1}{m} \sum_{k=0}^{m-1} X_{t-(k/m)}^{(m)} \tag{3}$$

where Y_t is a lower-frequency variable; $X_{t-(k/m)}^{(m)}$ denotes the data from high frequency variable k periods prior to the low frequency period t; m is the frequency ratio between high and low frequency series (In the case of quarterly and monthly, $m = 3$ since the higher frequency monthly variable can be observed three times within each quarter); $\overline{X}_t$ is the average of $X_{t-(k/m)}^{(m)}$ at the low frequency period t. L is a lag operator such that $LY_t = Y_{t-1}$, $L^2 Y_t = Y_{t-2}$ and so on. i and j denote the selected lag lengths which are determined by Akaike Information Criterion (AIC). This approach is limited to the fact that it assumes coefficients of $X_{(t-k,m)}^H$ within each period t to be the same. In addition, there may be information loss due to the averaging [14].

3.3   MIDAS Regression Models

Ghysels *et al.* [10] proposed a Mixed Data Sampling (MIDAS) approach to deal with various frequencies in multivariate model. Particularly, a MIDAS regression tries to deal with a low-frequency variable by using higher frequency explanatory variables as a parsimonious distributed lag. It also does not use any aggregation procedure and can be modelled for the coefficients on the lagged explanatory variables as allowing long lags in distributed lag function with only small number of parameters that have to be estimated [5]. The general form of MIDAS model is given by

$$Y_t = \alpha + \gamma W\left(\theta\right) X_{t-h}^{(m)} + \varepsilon_t \tag{4}$$

where $X_{t-h}^{(m)}$ is an exogenous variable measured at higher frequency than Y_t. h is forecasting step. If $h = 1$, it means we are going to forecast the dependent variable by one period ahead using current and historical information of X. $W\left(\theta\right)$ smooths historical values of $X_{t-h}^{(m)}$. Unlike the time-aggregate model which simply takes the average, there are some weighting schemes here, controlled by estimated parameter θ that allows us to convert the variable more efficiently. It can be written as

$$W\left(\theta\right) = \sum_{k=1}^{K} \omega\left(k; \theta\right) L^{(k-1)/m} \tag{5}$$

where K is the optimal number of lagged high frequency variable to be employed in the model. L is the lag operator such that

$$L^{(k-1)/m} X_{t-h}^{(m)} = X_{t-h-\left(\frac{k-1}{m}\right)}^{(m)},$$

and $\omega\left(k; \theta\right)$ is the weighting function that can be in various forms. It can be noticed that it is possible to include the lagged dependent variable into the

MIDAS model. Tungtrakul *et al.* [19] found that it provides a better forecast accuracy. Hence, the general form of MIDAS model becomes

$$Y_t = \alpha + \sum_{i=h}^{p} \beta_i L^i Y_{t-h} + \gamma \left(\sum_{k=1}^{K} \omega(k;\theta) L^{(k-1)/m} X_{t-h}^{(m)} \right) + \varepsilon_t \qquad (6)$$

Now, we will discuss the different MIDAS weighting schemes employed in this study.

3.3.1   Step Weighting Scheme

Rather than transforming the high frequency variable to match the lower one, this approach directly includes all lags of high frequency variable into the model. It takes each lagged high frequency variable as an explanatory variable in the model. Thus, no information has been lost. The MIDAS model under step weighting scheme with step length of s can be specified as follows:

$$Y_t = \alpha + \sum_{i=h}^{p} \beta_i Y_{t-i} + \sum_{k=1}^{K} \gamma_{k,s} X_{t-h-(k-1)/m}^{(m)} + \varepsilon_t. \qquad (7)$$

However, this approach puts a restriction on the coefficient of lagged high frequency variable $(\gamma_{k,s})$, which is determined by the step parameter (s). For instance, if the step parameter is equal to three $(s=3)$, it means the first three lagged have the same coefficient, the next three lags will then employ another same coefficient. This pattern will continue to the last lag that is incorporated in the model.

For demonstration purpose, consider the case that $p=1$, $h=1$, $m=3$, $K=4$, and $s=2$, then the MIDAS model with step weighting scheme can be specified as

$$Y_t = \alpha + \beta_1 Y_{t-1} + \gamma_{1,2} X_{t-1}^{(3)} + \gamma_{2,2} X_{t-1-1/3}^{(3)} + \gamma_{3,2} X_{t-1-2/3}^{(3)} + \gamma_{4,2} X_{t-2}^{(3)} + \varepsilon_t. \qquad (8)$$

If Y_t is the GDP growth for the third quarter of 2017, then $X_{t-1}^{(3)}$ is a value of an indicator from June 2017, $X_{t-1-1/3}^{(3)}$ is from May 2017, $X_{t-1-2/3}^{(3)}$ is from April 2017, and $X_{t-2}^{(3)}$ is from March 2017. Also, the restriction on parameters are $\gamma_{1,2} = \gamma_{2,2}$, and $\gamma_{3,2} = \gamma_{4,2}$.

Another drawback of the step weighting scheme is that the model may suffer from large numbers of parameters due to high difference in frequency between high and low frequency series [1]. Suppose that we work on annual series and monthly, we can see that we have got at least 12 coefficients to be estimated. Thus, the estimation outcome may not be satisfactory.

3.3.2   Exponential Almon Weight

This weighting scheme has been employed in various empirical studies due to its flexibility despite involving a few parameters in estimation [8]. The weighing scheme can be specified as

$$\omega\left(k;\theta\right)=\frac{exp\left(k\theta_1+k^2\theta_2\right)}{\sum_{k=1}^{K}exp\left(j\theta_1+j^2\theta_2\right)}. \tag{9}$$

To have a better view how the MIDAS model with Exponential Almon weighting scheme is mathematically specified, let us consider the case that optimal lagged high frequency variable is 3 (or $K=3$), the ratio between high and low variables is 3 (or $m=3$), no lagged dependent variable, and forecasting for one step ahead ($h=1$). The model then can be written as follows

$$Y_t=\alpha+\gamma\left(\begin{array}{c}\frac{exp(\theta_1+\theta_2)}{\sum_{k=1}^{3}exp(k\theta_1+k^2\theta_2)}\left(X_{t-1}^{(3)}\right)\\+\frac{exp(2\theta_1+4\theta_2)}{\sum_{k=1}^{3}exp(k\theta_1+k^2\theta_2)}\left(X_{t-1-\left(\frac{1}{3}\right)}^{(3)}\right)\\+\frac{exp(3\theta_1+9\theta_2)}{\sum_{k=1}^{3}exp(k\theta_1+k^2\theta_2)}\left(X_{t-1-\left(\frac{2}{3}\right)}^{(3)}\right)\end{array}\right)+\varepsilon_t. \tag{10}$$

Suppose that Y_t is measured at the 4th quarter of 2017, then $X_{t-1}^{(3)}$ is from September 2017, $X_{t-1-\left(\frac{1}{3}\right)}^{(3)}$ is from August 2017 and $X_{t-1-\left(\frac{2}{3}\right)}^{(3)}$ is from July 2017. α, γ, θ_1, and θ_2 can be estimated by using either maximum likelihood approach or non-linear least squares (NLS) approach. Ghysels et al. [10] pointed out that the number of parameters in the MIDAS model with exponential Almon weight is not influenced by the number of lagged high frequency variables. This important feature of MIDAS regression model allows us to employ large lagged high frequency variables and, at the same time, maintain parsimonious parameter estimation [1,12].

3.3.3   Beta Weight

It is another weighting scheme, which is an analogue of probability density function. It has been considered in empirical works as alternative to the exponential Almon weight [11]. According to Armesto et al. [1], this weighting scheme can be specified as follows

$$\omega\left(k;\theta\right)=\frac{f\left(\frac{k}{K},\theta_1,\theta_2\right)}{\sum_{k=1}^{K}f\left(\frac{k}{K},\theta_1,\theta_2\right)}, \tag{11}$$

where

$$f\left(x,a,b\right)=\frac{x^{a-1}(1-x)^{b-1}\Gamma\left(a+b\right)}{\Gamma\left(a\right)\Gamma\left(b\right)}, \tag{12}$$

and

$$\Gamma\left(a\right)=\int_0^\infty e^{-x}x^{a-1}\ dx \tag{13}$$

is the gamma function. θ_1 and θ_2 are parameters that control the weighing value for each lagged high preference variable.

3.4   Unrestricted MIDAS (U-MIDAS) Model

It can be noticed MIDAS models with weighting schemes may not completely extract all information from high frequency variable [14] since it still involves a frequency transformation. Kingnetr *et al.* [14] further asserted that the forecasting outcome may be satisfactory in the MIDAS model with exponential Almon weight framework when the difference in sampling frequencies between variables in the study is relatively small. Additionally, the model requires assumption on weighting scheme which may or may not be appropriate for every series. The MIDAS with exponential Almon may work well with one series, but not another. Foroni and Marcellino [8] suggested an alternative approach, the unrestricted MIDAS (U-MIDAS) regression model, to deal with the issue.

The basic idea of U-MIDAS model is similar to the MIDAS with step weighting scheme, except that the coefficient of each lagged high frequency variable is allowed to differ. Suppose that low frequency data is measured quarterly, while the high frequency is measured monthly, the U-MIDAS model for h-step forecasting can be written as

$$Y_t = \alpha + \sum_{i=1}^{p} \beta_i Y_{t-h-i} + \sum_{k=1}^{K} \gamma_k X^{(m)}_{t-h-(k-1)/m} + \varepsilon_t. \tag{14}$$

Y_t is a quarterly variable at period t, $X^{(m)}_{t-h-(k-1)/m}$ is a monthly indicator measured at $k-1$ months prior to the last month of the quarter at period $t-h$, h is the forecasting step, m is a frequency ratio, K is a number of monthly data used to predict Y_t.

By taking each lagged high frequency variable as additional explanatory variable in the model, the parameters in U-MIDAS model can simply be estimated using OLS estimation [8]. However, the U-MIDAS will lose its parsimonious feature if the number of frequency ratio between high and low frequency variables is large. For instance, forecasting monthly series using daily series will involve more than 20 parameters to be estimated, which would lead to undesirable estimation and forecasting results.

4   Empirical Results

In this section, we begin with the results of unit root tests, followed by the results from each forecasting model considered in the study and discussion on their forecasting performances.

The results of unit root test are reported in Table 1. In the case of GDP growth, the null hypothesis of non-stationary cannot be rejected in the case of ADF test without intercept (specification C). However, the rest of the tests show that GDP growth is stationary. Similarly, all tests, except for ADF with trend and intercept, conclude that SET growth is stationary. Therefore, it is reasonable to conclude that both series are stationary and can be undergone model estimation and forecasting.

Table 1. Unit root tests

Test	Specification			Conclusion
	A: (Intercept)	B: (Trend and Intercept)	C: (None)	
Panel I: GDP growth				
ADF	0.000*	0.000*	0.286	Stationary
PP	0.003*	0.007*	0.028**	
KPSS	0.302	0.040	–	
Panel II: SET growth				
ADF	0.062***	0.166	0.044**	Stationary
PP	0.005*	0.023**	0.002*	
KPSS	0.087	0.058	–	

Note:
1. The null hypothesis of ADF and PP unit root tests is non-stationary, while the KPSS is stationary.
2. For ADF and PP tests, the number represents p-value.
3. For KPSS test, the number represents the test statistics.
4. The critical values for KPSS test for specification A (B) at 1%, 5%, and 10% are 0.739 (0.216), 0.463 (0.146), and 0.347 (0.119), respectively.
5. *, **, *** denote the rejection of null hypothesis at 1%, 5%, and 10% levels of statistical significance, respectively.

As far as model estimation is concerned, the data sample during the period of 2001Q1 to 2015Q4 is employed. The linear least squares estimation technique is used to estimate parameters for the time-aggregate model, while parameters in the MIDAS models are handled by the non-linear least squares. The optimal lag lengths for all models are chosen by Akaike information criterion (AIC) with the maximum of 24 lags. Then, we forecast the quarterly GDP growth rates for 2016. Since we are interested in comparing forecasting performance between models, it is advised to investigate how these models perform through figure. Figure 3 provides a plot of actual value of quarterly GDP growth rate and its forecasted values from different models.

It can be seen from Fig. 3 that the time-aggregate model, MIDAS model with Beta weighting, and U-MIDAS model seem to predict the GDP growth rate closer to the actual values than the MIDAS models with exponential Almon and step weighting schemes. However, as the forecasting horizon expands, the former three models seem to perform worse than the latter two. Table 2 provides forecasting results in details together with lag selection for each model.

It is possible to notice that it is still uncertain to see which model can generally perform better. Therefore, we now turn to the root mean square error (RMSE) for evaluation. Table 3 shows the RMSEs for each model at each forecasting horizon.

The results from Table 3 suggest that, overall, the unrestricted MIDAS (U-MIDAS) model exhibits higher forecasting accuracy than the rest of the models

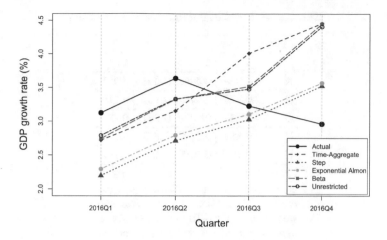

Fig. 3. Forecast and actual quarterly GDP growth in 2016

Table 2. Quarterly GDP growth forecast in 2016

Period	Model					Actual value
	A	B	C	D	E	
2016Q1	2.720	2.199	2.295	2.742	2.789	3.125
2016Q2	3.152	2.711	2.792	3.324	3.329	3.637
2016Q3	4.009	3.026	3.104	3.515	3.476	3.224
2016Q4	4.457	3.521	3.565	4.449	4.408	2.960
Lag selection	2	4	3	4	4	

Note: The description for each model is as follows, (A) Time-aggregate model, (B) MIDAS model with step weighting, (C) MIDAS model with exponential Almon weighting, and (D) MIDAS model with Beta weighting, and (E) U-MIDAS model. The number is rounded to nearest thousandth.

in this study. The conclusion here is also consistent with the recent empirical work by [14]. In addition, the superior in forecasting precision may due to the fact that, in U-MIDAS framework, information of high frequency variable is fully utilised. The results also suggest that the forecasting improvement is rather moderate, when it comes to the comparison between MIDAS models with weighting schemes and the traditional time-aggregate model. According to RMSEs, we can see that only the MIDAS model with beta weighting scheme can outperform the time-aggregate model in this study. This implies that using higher frequency will not improve the outcome after all if the inappropriate weighting scheme is chosen. Nevertheless, we can conclude that using high-frequency variable to predict the lower frequency one improves forecasting precision under U-MIDAS model, provided that the difference in frequency between series in a study is small.

Table 3. Forecast evaluation based on RMSEs

Model	Horizon			
	1	2	3	4
A	0.405	0.446	0.581	0.902
B	0.926	0.926	0.764	0.719
C	0.830	0.837	0.687	**0.668**
D	0.382	0.349	0.331	0.798
E	**0.335**	**0.322**	**0.300**	0.769

Note: The description for each model is as follows, (A) Time-aggregate model, (B) MIDAS model with step weighting, (C) MIDAS model with exponential Almon weighting, and (D) MIDAS model with Beta weighting, and (E) U-MIDAS model. The number is rounded to nearest thousandth. At each horizon, the lowest RMSEs are in **bold**.

5   Conclusion

In this paper, we investigate the forecasting performance of 5 different forecasting models, including the time-aggregate model, the MIDAS model with step weighting, exponential Almon weighting, and beta weighting, and the U-MIDAS model. Thailand's quarterly GDP growth was forecasted using a financial variable, SET index, as a predictor. Unlike the time-aggregate model, the MIDAS model with weighting scheme allows us to efficiently utilise the information of high frequency variable to forecast lower frequency variable. However, it still involves the concept of frequency conversion, as in the time-aggregate model, via weighting schemes.

On the other hand, the U-MIDAS model fully exhausts information of high frequency variable. The model directly incorporates high frequency variable into forecasting model without frequency conversion. The data in this study spans from 2001Q1 to 2016Q4 with 2016Q1 to 2016Q4 being left out for forecasting performance evaluation. Our results, based on RMSEs, show that the U-MIDAS model has greater forecasting precision than other models in this study. This implies that, under the U-MIDAS framework, using high frequency variable to predict lower frequency variable improves the forecasting accuracy.

In addition, we found that the improvement of using higher frequency variable to predict lower frequency variable is rather small when it comes to the MIDAS model with weighting scheme. The forecasting results may even be worst if the weighting scheme is not appropriately chosen. If one wishes to employ such MIDAS model, the results suggest that the MIDAS model with Beta weighting scheme performs best among other weighting schemes. Otherwise, the traditional time-aggregate seems to provide acceptable predicting accuracy for short-horizon.

Nevertheless, this study focused on four-period forecasting using single predictor and ignored the possibility of having structural break in time series due to the limitation of approaches. Therefore, the recommendation for future research would be the inclusion of additional predictors in the model, longer forecasting horizon and controlling for potential structural breaks.

Acknowledgements. The authors would like to thank the anonymous reviewer for useful suggestions which have greatly improved the quality of this paper. This research is supported by the Puay Ungphakorn Center of Excellence in Econometrics, Chiang Mai University.

References

1. Armesto, M.T., Engemann, K., Owyang, M.: Forecasting with mixed frequencies. Review **92**, 521–536 (2010)
2. Asteriou, D., Hall, S.G.: Applied Econometrics, 2nd edn. Palgrave Macmillan, Leicester (2011)
3. Bellégo, C., Ferrara, L.: Forecasting Euro-area recessions using time-varying binary response models for financial variables. Working papers 259, Banque de France (2009)
4. Clements, M.P., Galvão, A.B.: Macroeconomic forecasting with mixed-frequency data. J. Bus. Econ. Stat. **26**(4), 546–554 (2008)
5. Clements, M.P., Galvão, A.B.: Forecasting US output growth using leading indicators: an appraisal using MIDAS models. J. Appl. Econ. **24**(7), 1187–1206 (2009)
6. Estrella, A., Rodrigues, A.R., Schich, S.: How stable is the predictive power of the yield curve? Evidence from germany and the united states. Rev. Econ. Stat. **85**(3), 629–644 (2003)
7. Ferrara, L., Marsilli, C.: Financial variables as leading indicators of GDP growth: Evidence from a MIDAS approach during the Great Recession. Appl. Econ. Lett. **20**(3), 233–237 (2013)
8. Foroni, C., Marcellino, M.: A survey of econometric methods for mixed-frequency data. Working Paper 2013/06, Norges Bank (2013)
9. Ghysels, E., Kvedaras, V., Zemlys, V.: Mixed frequency data sampling regression models: the R package midasr. J. Stat. Softw. Art. **72**(4), 1–35 (2016)
10. Ghysels, E., Santa-Clara, P., Valkanov, R.: The MIDAS Touch: Mixed Data Sampling Regression Models. CIRANO Working Papers 2004s–20, CIRANO (2004)
11. Ghysels, E., Santa-Clara, P., Valkanov, R.: There is a risk-return trade-off after all. J. Financ. Econ. **76**(3), 509–548 (2005)
12. Ghysels, E., Santa-Clara, P., Valkanov, R.: Predicting volatility: getting the most out of return data sampled at different frequencies. J. Econ. **131**(1–2), 59–95 (2006)
13. Ghysels, E., Valkanov, R.I., Serrano, A.R.: Multi-period forecasts of volatility: direct, iterated, and mixed-data approaches. EFA 2009 Bergen Meetings Paper (2009)
14. Kingnetr, N., Tungtrakul, T., Sriboonchitta, S.: Forecasting GDP Growth in Thailand with Different Leading Indicators Using MIDAS Regression Models, pp. 511–521. Springer International Publishing, Cham (2017)

15. Kuzin, V., Marcellino, M., Schumacher, C.: MIDAS vs. mixed-frequency VAR: nowcasting GDP in the euro area. Int. J. Forecast. **27**(2), 529–542 (2011)
16. Kwiatkowski, D., Phillips, P.C., Schmidt, P., Shin, Y.: Testing the null hypothesis of stationarity against the alternative of a unit root. J. Econ. **54**(1), 159–178 (1992)
17. Phillips, P.C.B., Perron, P.: Testing for a unit root in time series regression. Biometrika **75**(2), 335 (1988)
18. Said, S.E., Dickey, D.A.: Testing for unit roots in autoregressive-moving average models of unknown order. Biometrika **71**(3), 599–607 (1984)
19. Tungtrakul, T., Kingnetr, N., Sriboonchitta, S.: An Empirical Confirmation of the Superior Performance of MIDAS over ARIMAX, pp. 601–611. Springer International Publishing, Cham (2016)

A Portfolio Optimization Between US Dollar Index and Some Asian Currencies with a Copula-EGARCH Approach

Ji Ma[1,2], Jianxu Liu[1,3(✉)], and Songsak Sriboonchitta[1,3]

[1] Faculty of Economics, Chiang Mai University, Chiang Mai 50200, Thailand
majiyn@hotmail.com, liujianxu1984@163.com, songsakecon@gmail.com
[2] Yunnan Academy of Social Sciences, Kunming 650031, China
[3] Center of Excellence in Econometrics, Chiang Mai University,
Chiang Mai 50200, Thailand

Abstract. There is a strong correlation between the value of the US dollar and the Asian currencies. EGARCH-copula model, with the skewed student-t distribution and the skewed general error distribution, can be used to capture the dependence correlation between US dollar and an Asian currency from those seven currencies in this paper. Building a bivariate portfolio based on the fitted EGARCH-copula models can be used to make portfolio optimization with the methods of max return, min risk and max sharpe ratio, to obtain a positive and reasonable return.

Keywords: Exchange rate · Copula-EGARCH · Portfolio optimization

1 Introduction

Markowitz (1952) propose the mean-variance model to construct the optimal portfolio which build up the foundation of the modern investment theory. At a given confidence level, Baumol (1963) create the concept of Value-at-Risk (VaR) to analyze the worst loss. Copula-GARCH approach can be used to analyze the conditional dependence structure for two correlated variables (Aloui et al. 2013; Sun et al. 2008; Huang et al. 2009; Patton 2012; Sriboonchitta et al. 2013; Wu and Lin 2014; Chen 2015). This paper uses a EGARCH-copulas to study the portfolio optimization between the US dollar (USD) index and seven major Asian currencies exchange rate. Aloui and Assa (2016) exchange rate corresponds to the trade weighted US dollar (TWEXB) index, measuring the movement of dollar against the currencies of a broad group of major U.S. trading partners. Therefore, in this paper we assume TWEXB as a tradable currencies, a financial asset, with its value as a price in US dollar to measure the value of Dollar. Daily logarithm returns for the eight assets have been used for EGARCH(1,1) with residuals of skewed student-t distribution or skewed general error distribution. Forty copula models has been estimated for model selection to conduct the parameter estimation and standard error with robustness of the most adequate model.

© Springer International Publishing AG 2018
V. Kreinovich et al. (eds.), *Predictive Econometrics and Big Data*, Studies in Computational Intelligence 753, https://doi.org/10.1007/978-3-319-70942-0_32

We used the daily data from January, 4, 2006 to December, 31, 2015 and the empirical results for daily data provided evidence of the dependence structure between US dollar and Asian currencies. Based on GARCH-copula model, VaR and optimization of portfolios can be estimated (Lee and Lin 2011; Wang et al. 2010). With the estimation and forecasting for eight assets and it's seven pair portfolio with EGARCH-copula models, the portfolio optimization of the seven portfolios made the positive and reasonable return.

We introduce the methodology in the following section. Then we describe the data and descriptive statistics. After showing the empirical results, the last section make the conclusion.

2   Methodology

We first fit skewed EGARCH models for univariate time series for obtaining the marginal distribution of the residuals. Then, we estimation the dependence structure for seven pairs assets with six copula functions (Gaussian, Student-t, Clayton, Frank, Gumbel and Joe copulas, see Necula 2010; McNeil 2015; Yan 2007; Kole et al. 2007; Wiboonpongse et al. 2015). Finally, portfolio optimization methods, such like the maximum mean, the minimum risk, the maximum Sharpe ratio, have been used to deal with the optimal portfolio allocation.

2.1   EGARCH with Skewed Distributions

An EGÁRCH(1,1) model was taken in this research We take the log return of assets as $\{X_t\}$, $t = 1, ..., T$ where μ is the expected return and ϵ_t is a zero-mean white noise,

$$X_t = \mu + \epsilon_t, \tag{1}$$

when the series ϵ_t is not serially independent, we can use a the exponential generalized autoregressive conditional heteroskedastic model (ϵ_t-EGARCH) by Nelson (1991) as:

$$\epsilon_t = \sigma_t \eta_t, \tag{2}$$

where η_t is standard skew general error distribution (*sged*) or standard skew student-t distribution (*sstd*) and

$$log(\sigma_t^2) = \omega + \alpha \left(|\eta_{t-1}| - E |\eta_{t-1}| \right) + \gamma \eta_{t-1} + \beta log(\sigma_{t-1}^2), \tag{3}$$

where $\eta_t \sim f_{sged}$ or $\eta_t \sim f_{sstd}$.

2.2   Copula and Skalar Theorem

Sklar's theorem (1959) which is the general idea of copula states that for a given joint multivariate distribution function there exist a function–known as copula function–such that

$$F_{XY}(x, y) = C(F_X(x), F_Y(y)), \tag{4}$$

where $F_{XY}(x, y)$ is the joint distribution of X and Y, $u = F_X(x)$ and $v = F_Y(y)$. If C is a copula function, then the function F_{XY} is a joint distribution function with margins F_X and F_Y. The conditional copula function (Patton 2006) can be expressed as

$$F_{XY|W}(x, y|w) = C(F_{X|W}(x|w), F_{Y|W}(y|w)|w), \qquad (5)$$

where W is the conditioning variable, $F_{X|W}(x|w)$ is the conditional distribution of $X|W = w$, $F_{Y|W}(y|w)$ is the conditional distribution of $Y|W = w$ and $F_{XY|W}(x, y|w)$ is the conditional distribution of $(X, Y)|W = w$.

2.3 Bivariate Copula Models

2.3.1 The Bivariate Gauss Copula
This bivariate Gaussian copula is given by

$$C^{Ga}(u_1, u_2, \rho) =$$
$$\int_{-\infty}^{\phi^{-1}(u_1)} \int_{-\infty}^{\phi^{-1}(u_2)} \frac{1}{2\pi\sqrt{(1-\rho^2)}} exp\left(-\frac{v_1^2 - 2\rho v_1 v_2 + v_2^2}{2(1-\rho^2)}\right) dv_2 dv_1, \quad (6)$$

ρ denotes the correlation of u_1 and u_2.

2.3.2 The Bivariate Student t-Copula
The bivariate student-t copula is the following function:

$$C^t(u_1, u_2, \rho, v) =$$
$$\int_{-\infty}^{t_v^{-1}(u_1)} \int_{-\infty}^{t_v^{-1}(u_2)} \frac{1}{2\pi\sqrt{(1-\rho^2)}} exp\left\{1 + \frac{r^2 - 2\rho rs + s^2}{v(1-\rho^2)}\right\}^{-\frac{v+2}{2}} dr\, ds, \quad (7)$$

where ρ is the linear correlation coefficient between the two random variables and t_v^{-1} denotes the inverse of the univariate Student-t distribution function with degrees of freedom.

2.3.3 The Bivariate Clayton Copula
The bivariate Clayton copula is given as following:

$$C_{Clayton}(u_1, u_2; \omega) = \left(u_1^{-\omega} + u_2^{-\omega} - 1\right)^{-\frac{1}{\omega}}, \qquad (8)$$

where $\omega \in [-1, \infty)$.

2.3.4   The Bivariate Frank Copula

The function of the bivariate frank copula is:

$$C_{Frank}(u_1, u_2; \lambda) = \frac{-1}{\lambda} log\left(\frac{\lambda\left(1 - e^{-\lambda}\right) - \left(1 - e^{-\lambda u_1}\right)\left(1 - e^{-\lambda u_2}\right)}{1 - e^{-\lambda}}\right), \quad (9)$$

where $\lambda \in (-\infty, 0) \cup (0, +\infty)$.

2.3.5   The Bivariate Gumbel Copula

The bivariate gumbel copula is given as below:

$$C_{Gumbel}(u_1, u_2; \delta) = exp\left(-\left(\left(-log_{u_1}\right)^{\delta} + \left(-log_{u_2}\right)^{\delta}\right)^{\frac{1}{\delta}}\right), \quad (10)$$

where $\delta \in [1, \infty)$.

2.3.6   The Bivariate Joe Copula

The function of the bivariate Joe copula is given as:

$$C_{Joe}(u, v; \theta) = 1 - \left[(1 - u)^{\theta} + (1 - v)^{\theta} - (1 - u)^{\theta}(1 - v)^{\theta}\right]^{\frac{1}{\theta}}, \quad (11)$$

where $\theta \in [1, \infty)$.

2.4   Portfolio Optimization

Suppose there are N risky assets whose returns are given by the random variables $R_1, ... R_n, R_N$. There are three methods based on portfolio optimization are imposed in terms of the maximum mean, the minimum risk, and the maximum Sharpe ratio.

2.4.1   Maximum Return

Then the maximize return mean-variance portfolio can be described as

$$max \quad R^T w, \quad (12)$$

$w_i \geq 0, i = 1, 2, ..., n$, subject to $\sum_{t=1}^{n} w_i = 1$, and $w^T \Sigma w \leq \sigma^2$. where w_i is defined as the vector of portfolio weight of asset i and restricted to be positive. $R^T w$ is the expected return of a portfolio. Σ is defined as the variance-covariance matrix. Therefore, the term $w^T \Sigma w$ represents the variance of the portfolio return which a variance of an efficient portfolio must not exceed the variance of an individual asset.

2.4.2 Minimum Risk

The method of the minimum risk, or variance, can be expressed as below:

$$min \quad \sigma_p^2 = w^T \Sigma w \quad s.t. \quad R^T w \geq \overline{R}, \tag{13}$$

where $\sum_{t=1}^{n} w_i = 1$, $(w_i \geq 0, i = 1, 2, ..., n)$ and Σ is the covariance matrix and the $\overline{R}$ is the given level of return.

2.4.3 Maximum Sharpe Ratio

The maximum Sharpe ratio of a given asset is defined by the expected excess return of an asset over the risk-free rate to its standard deviation as shown below:

$$max \quad S_a = \frac{E(R_a) - R_f}{\sigma_a}, \tag{14}$$

where S_a denotes the Sharpe ratio of portfolio (a) and $E(R_a) = R^T w$, and w is the weight of a portfolio. R_a and R_f are returns of portfolio (a) and risk-free asset respectively. In this study, R_f is assumed to be zero. Standard deviation of portfolio (a) defines by σ_a.

3 Data and Descriptive Statistics

This study uses eight currencies of US, China, Japan, Korea, India, Thailand, Malaysia, and Singapore. Those currencies are represented by TWEXB (Trade Weighted US Dollar Index), CNY, JPY, KRW, INR, THB, MYR, and SGD, respectively. Remarkably, TWEXB was used as an indicator to measure the value of US dollar in terms of a basket of major foreign currencies excluding US dollar. Therefore, in this paper we assume TWEXB as a tradable currency representing US dollar. We built seven pairs of portfolio for each Asian currencies with TWEXB. The data derives from the Federal Reserve Economic Data at the Federal Reserve Bank of S.T. Louis, the United States. The range of data are from January, 2, 2006 to December, 31, 2015 with a daily series. Log returns series has been calculated for fitting the model with totally 2512 daily returns for eight assets, which 2011 returns are for estimation and another 501 returns are for forecasting. The returns of eight assets are shown in Fig. 1.

The descriptive statistics for TWEXB and seven Asian currencies are reported in Table 1, which shows that the standard deviation of sex of seven Asian currencies, except CNY, are higher than that of TWEXB returns which shows that Asian currencies generally has higher volatilities. The skewness statistic for currencies of US, China, Japan, Korea, and Malaysia are negative, indicating the returns are significantly skewed to left, and for those of India, Thailand, Singapore are positive, indicating the returns skewed to right. Significant, thereby indicating that the oil returns are significantly skewed to the left.

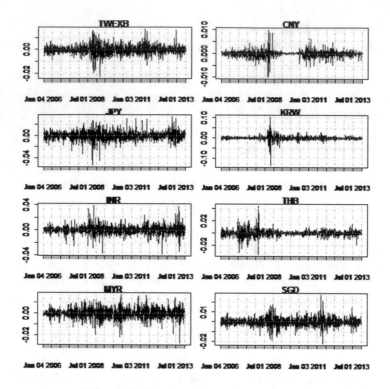

Fig. 1. Daily returns on eight currencies

Table 1. Descriptive statistics and stochastic properties of return series.

	Minimum	Maximum	Mean	Stdev.	Skewness	Kurtosis
TWEXB	−0.023	0.017	0.000	0.003	−0.067	4.61
CNY	−0.01	0.01	−0.0001	0.001	−0.063	15.813
JPY	−0.052	0.033	−0.0001	0.007	−0.304	4.928
KRW	−0.132	0.101	0.000	0.009	−0.663	44.649
INR	−0.038	0.039	0.0002	0.006	0.145	6.308
THB	−0.035	0.045	−0.0001	0.005	0.041	11.776
MYR	−0.027	0.017	−0.0001	0.004	−0.234	2.819
SGD	−0.022	0.027	−0.0001	0.004	0.265	5.095

With respect to the excess kurtosis statistics, the values all currencies, except MYR with slightly negative, are significantly positive (great than 3), thereby implying that the distribution of returns has larger, thicker tails than the normal distribution. Based on the statistics, we could assume the we should use skewed distribution on the residual for the single series to fit an EGARCH model.

4   Empirical Results

We use eight return series separately to fit the EGARCH(1,1) model with either SSTD and SGED distribution for the residuals and use AIC (Akaike information criterion) and BIC (Bayesian information criterion) to select the best model. The empirical result for both AIC and BIC tend to give the same suggestion on SSTD distribution for TWEXB, JPY, INR and SGD, also give the same advices on SGED distribution for CNY, KRW, THB, MYR. The detail showed in Table 2.

Using the fitted EGARCH(1,1) for each of the eight assets, the estimation for the parameters of the eight EGARCH models with robustness. We use the eight best fitted EGARCH models to figure out the residuals with dependence between TWEXB with each of the other seven Asian currencies results are listed in Table 3.

This research use 40 copulas to fit seven EGARCH-copula models, like Normal, Student-t, Clayton, Gumbel, Frank and Joe copulas, for seven pairs. Here we only show six copulas of the total 40. The estimated parameters of the correlation coefficient and its standard deviation for each adequate copula shows the statistically significant level for all parameters for the seven copula models. The details are showed in Table 4.

We use the goodness-of-fit test for the 40 copulas and the AIC shows that the pair of TWEXB/CNY can be best fitted by a Gaussian copula, the rest of the other six can be best fitted with a Student-t copula that shown in Table 5.

Seven pairs of the portfolio has been calculated for the optimization with three portfolio choice methods. The all three methods, including the max return, the min risk, the max Sharpe ratio, have been used for the portfolio optimization.

Table 2. Selecting eGARCH(1,1) model with SSTD and SGED by using AIC and BIC

Assets	AIC		BIC	
	SSTD	SGED	SSTD	SGED
TWEXB	-8.8177^a	-8.8158	-8.8010^a	-8.7991
CNY	-11.499	-11.499^a	-11.482	-11.483^a
JPY	-7.3071^a	-7.3014	-7.2903^a	-7.2847
KRW	-7.6434	-7.6485^a	-7.6266	-7.6318^a
INR	-7.7436^a	-7.7294	-7.7269^a	-7.7127
THB	-8.5665	-8.5675^a	-8.5498	-8.5508^a
MYR	-8.3783	-8.3856^a	-8.3616	-8.3688^a
SGD	-8.6568^a	-8.6483	-8.6401^a	-8.6316

Notes: EGARCH(1,1) with skewed student t distribution and SGED distribution. [a]Represents the chosen distribution.

Table 3. Parameter estimation with robustness for EGARCH(1,1) with skewed distributions

Pars	TWEXB	CNY	JPY	KRW
Dist	SSTD	SGED	SSTD	SGED
omega	−0.086***	−0.251***	−0.147***	−0.102***
	0.011	0.017	0.023	0.007
alpha	0.030**	0.013	−0.024	0.068***
	0.011	0.022	0.015	0.016
beta1	0.993***	0.982***	0.985***	0.990***
	0.001	0.001	0.002	0.0005
gamma1	0.107***	0.344***	0.11	0.156***
	0.022	0.011	0.087	0.006
skew	1.042***	1.000***	0.986***	1.053***
	0.03	0.001	0.026	0.024
shape	8.172	0.848***	5.781*	1.209***
	4.363	0.058	2.574	0.052
Pars	INR	THB	MYR	SGB
Dist	SSTD	SGED	SGED	SSTD
omega	−0.240***	−0.284***	−0.317***	−0.112***
	0.02	0.023	0.035	0.011
alpha	0.035***	0.013*	0.024	0.035**
	0.016	0.016	0.017	0.012
beta1	0.977***	0.975***	0.971***	0.990***
	0.002	0.002	0.003	0.001
gamma1	0.221***	0.299***	0.264***	0.127***
	0.01	0.023	0.027	0.03
skew	1.047***	1.000***	0.999995***	1.096***
	0.026	0.002	0.003	0.034
shape	4.233***	1.095***	1.151***	6.021***
	0.457	0.058	0.067	1.008

Notes: Significant level ***0.1%, **1%, *5%.

As we seen in Table 6, the results show that all three methods produce positive and reasonable return for two years. Among the three methods, the max Sharpe ratio make the best performance for US/CNY and US/JPY with 1.160 and 1.193, on another hand, max return works the best for other five pair of portfolios such like US/KRW, US/INR, US/THB, US/MYR and US/SGD with 1.210, 1.212, 1.158, 1.259 and 1.198.

Table 4. Estimates of the dependence parameters of different copula models

	Normal	Student-t			Clayton	Gumbel	Frank	Joe
TWEXB/CNY	**0.243*****	0.243***	0		0.219***	1.164***	1.441***	1.210***
	−0.021	−0.021	0		−0.029	−0.019	−0.135	−0.029
TWEXB/JPY	0.211***	**0.221*****	**4.546*****		0.258***	1.158***	1.350***	1.194***
	−0.021	**−0.024**	**−0.563**		−0.032	−0.019	−0.141	−0.027
TWEXB/KRW	0.501***	**0.503*****	**13.046*****		0.709***	1.443***	3.363***	1.551***
	−0.015	**−0.016**	**−4.049**		−0.039	−0.025	−0.147	−0.036
TWEXB/INR	0.462***	**0.466*****	**14.597*****		0.584***	1.399***	3.108***	1.512***
	−0.016	**−0.017**	**−4.874**		−0.037	−0.024	−0.145	−0.035
TWEXB/THB	0.496***	**0.498*****	**16.072*****		0.625***	1.457***	3.311***	1.605***
	−0.015	**−0.016**	**−5.829**		−0.037	−0.025	−0.146	−0.037
TWEXB/MYR	0.468***	0.479***	7.072***		0.640***	1.432***	3.238***	1.545***
	−0.016	−0.018	−1.294		−0.038	−0.025	−0.148	−0.037
TWEXB/SGD	0.777***	0.782***	7.776***		1.672***	2.215***	7.269***	2.576***
	−0.007	−0.008	−1.465		−0.056	−0.041	−0.191	−0.058

Notes: Significant level ***0.1%, **1%, *5%.

Table 5. Results for the goodness-of-fit test of different copula functions

	Normal	Std-t	Clayton	Gumbel	Frank	Joe
TWEXB/CNY	−119	–	−74	−93	−113	−62
TWEXB/JPY	−89	**−175**	−86	−110	−90	−84
TWEXB/KRW	−577	**−590**	−453	−519	−535	−384
TWEXB/INR	−478	**−490**	−336	−449	−465	−347
TWEXB/THB	−562	**−571**	−395	−533	−523	−419
TWEXB/MYR	−493	**−534**	−412	−468	−490	−337
TWEXB/SGD	−1856	**−1903**	−1450	−1775	−1744	−1401

Notes: t copula is a symmetrical and heavy tail distribution. All pairs, except TWEXB/CNY, has a tail dependence and symmetrical relation.

Table 6. Portfolio optimization via vary approaches

Portfolio	Max mean	Min risk	Sharpe ratio
US/CNY	1.077	1.136	**1.160**
US/JPY	1.180	1.171	**1.193**
US/KRW	**1.210**	1.152	1.150
US/INR	**1.212**	1.134	1.096
US/THB	**1.158**	1.153	1.125
US/MYR	**1.259**	1.253	1.166
US/SGD	**1.198**	1.161	1.128

5   Conclusion

In this paper, we introduced EGARCH model for skewed residuals for US dollar and seven Asian currencies. Then the bivariate EGARCH-copulas was used to make portfolios. To maximize the portfolio with four methods, the maximum portfolio found out a reasonable return from 1.16 to 1.259 for 512 trading days; The Gaussian copula and the student-t copula were selected by seven pairs of portfolio, respectively. Moreover, all four portfolio optimization methods figure out the positive returns. Among the three portfolio optimization methods, the max return and the max Sharpe ratio made better performance.

Acknowledgements. This work has been supported by the Faculty of Economics and the Puey Ungphakorn Centre of Excellence in Econometrics at Chiang Mai University.

References

Aloui, R., Ben Aissa, M.S., Nguyen, D.K.: Conditional dependence structure between oil prices and exchange rates: a copula-GARCH approach. J. Int. Money Fin. **32**, 719–738 (2013). https://doi.org/10.1016/j.jimonfin.2012.06.006

Aloui, R., Assa, M.S.B.: Relationship between oil, stock prices and exchange rates: a vine copula based GARCH method. North Am. J. Econ. Fin. **37**, 458–471 (2016)

Baumol, W.J.: An expected gain-confidence limit criterion for portfolio selection. Manag. Sci. **10**(1), 174–182 (1963)

Chen, Q.A., Wang, D., Pan, M.Y.: Multivariate time-varying G-H copula GARCH model and its application in the financial market risk measurement. Math. Prob. Eng. (2015). https://doi.org/10.1155/2015/286014

Huang, J.J., Lee, K.J., Liang, H.M., Lin, W.F.: Estimating value at risk of portfolio by conditional copula-GARCH method. Insur. Math. Econ. **45**(3), 315–324 (2009). https://doi.org/10.1016/j.insmatheco.2009.09.009

Jondeau, E., Rockinger, M.: The copula-GARCH model of conditional dependencies: an international stock market application. J. Int. Money Fin. **25**(5), 827–853 (2006). https://doi.org/10.1016/j.jimonfin.2006.04.007

Kole, E., Koedijk, K., Verbeek, M.: Selecting copulas for risk management. J. Bank. Fin. **31**(8), 2405–2423 (2007). https://doi.org/10.1016/j.jbankfin.2006.09.010

Lee, W.C., Lin, H.N.: Portfolio value at risk with Copula-ARMAX-GJR-GARCH model: evidence from the gold and silver futures. Afr. J. Bus. Manag. **5**(5), 1650–1662 (2011)

Markowitz, H.: Portfolio selection. J. Fin. **7**(1), 77–91 (1952)

McNeil, A.J.: Dependence modeling with copulas. J. Time Ser. Anal. **36**(4), 599–600 (2015). https://doi.org/10.1111/jtsa.12126

Necula, C.: A Copula-GARCH model. Ekonomska Istrazivanja-Economic Research **23**(2), 1–10 (2010)

Nelson, D.B.: Conditional heteroskedasticity in asset returns: a new approach. Econometrica **59**(2), 347–270 (1991)

Patton, A.J.: A review of copula models for economic time series. J. Multivar. Anal. **110**, 4–18 (2012). https://doi.org/10.1016/j.jmva.2012.02.021

Sriboonchitta, S., Nguyen, H.T., Wiboonpongse, A., Liu, J.X.: Modeling volatility and dependency of agricultural price and production indices of Thailand: static versus time-varying copulas. Int. J. Approximate Reasoning **54**(6), 793–808 (2013). https://doi.org/10.1016/j.ijar.2013.01.004

Sun, J.F., Frees, E.W., Rosenberg, M.A.: Heavy-tailed longitudinal data modeling using copulas. Insur. Math. Econ. **42**(2), 817–830 (2008). https://doi.org/10.1016/j.insmatheco.2007.09.009

Wang, Z.R., Chen, X.H., Jin, Y.B., Zhou, Y.J.: Estimating risk of foreign exchange portfolio: using VaR and CVaR based on GARCH-EVT-Copula model. Phys. A Stat. Mech. Appl. **389**(21), 4918–4928 (2010). https://doi.org/10.1016/j.physa.2010.07.012

Wiboonpongse, A., Liu, J.X., Sriboonchitta, S., Denoeux, T.: Modeling dependence between error components of the stochastic frontier model using copula: application to intercrop coffee production in Northern Thailand. Int. J. Approximate Reasoning **65**, 34–44 (2015). https://doi.org/10.1016/j.ijar.2015.04.001

Wu, C.C., Lin, Z.Y.: An economic evaluation of stock-bond return comovements with copula-based GARCH models. Quant. Fin. **14**(7), 1283–1296 (2014). https://doi.org/10.1080/14697688.2012.727213

Yan, J.: Enjoy the joy of copulas: with a package copula. J. Stat. Softw. **21**(4), 1–21 (2007)

Technical Efficiency Analysis of China's Agricultural Industry: A Stochastic Frontier Model with Panel Data

Ji Ma[1,2], Jianxu Liu[1,3(✉)], and Songsak Sriboonchitta[1,3]

[1] Faculty of Economics, Chiang Mai University, Chiang Mai 50200, Thailand
majiyn@hotmail.com, liujianxu1984@163.com, songsakecon@gmail.com
[2] Yunnan Academy of Social Sciences, Kunming 650031, China
[3] Center of Excellence in Econometrics, Chiang Mai University,
Chiang Mai 50200, Thailand

Abstract. This paper imposed the translog stochastic frontier production model to analyze the China's province-level agriculture productivity by using panel data during 2002–2012 on 31 provinces in China. The results show that China's province-level agriculture productivity has been improved for over 11 years. Hunan, Bejing and Shanghai approached the agriculture technical efficiency frontier. The agriculture technical efficiencies in underdeveloped area such like Guizhou, Yunnan and Anhui increased sharply and approached to the national province-level mean, 60%, in terms of the technical efficiencies over 11 years which, however, still have 40% space to be improved. We recommend that the provinces with lower technical efficiency, such as Anhui, Yunnan and Guizhou, should learn experiences from those provinces that have high technical efficiency so that improving the agricultural productivities of themselves.

Keywords: Stochastic frontier analysis · Translog production function
Technical efficiency · Agriculture productivity · China

1 Introduction

Under a national urbanization campaign, it is the first time that the proportion of rural population is less than 50% at the end of 2011 in China according to a government announcement in 2012. By going with a decrease in rural population, however, China's agriculture output has been increased for decade to feed a giant economic body with the most population in the world. It is the agriculture productivity that plays significant role. A large number of research papers focused on the development of China's agriculture productivity. Tong et al. (2012) point towards a rapid expansion of agricultural output and productivity during the 1980s and a slowdown during the 1990s, raising questions about the sustainability of these growth rates. Few studies cover the 2000s and most estimate productivity at the national rather than the provincial level. Therefore, this study aims to a province-level agriculture productivity during year 2002–2012.

V. Kreinovich et al. (eds.), *Predictive Econometrics and Big Data*, Studies in Computational Intelligence 753, https://doi.org/10.1007/978-3-319-70942-0_33

Liu et al. (2017) There are two primary methods of efficiency measures, namely stochastic frontiers and data envelopment analysis (DEA), which involve econometric methods and mathematical programming, respectively. Stochastic frontier models make assumptions about the functional form of production or cost functions, and can deal effectively with the presence of noise in the data, whereas DEA models make no assumptions about the functional forms, but cannot deal effectively with measurement error. Therefore, the stochastic frontier model has been maturely applied into analyzing the technical inefficiency or the technical efficiency. The first proposal of the stochastic frontier production function was token independently by Aigner et al. (1977) and Meeusen and van den Broeck (1977); Since after it had been extended by Forsund et al. (1980), Schmidt (1986), Battese (1992), and Greene (1993), Battese (1995). The stochastic frontier production function assumes there is existed a technical inefficiency for production activities. Based on an input factors analysis on production output, it took a two error component to analysis the production inefficiency. It will finally figure out the effect of the technical efficiency for both of technical change along with time and technical difference along with cross section. In this paper, we expand to a translog production function because the better adequate than Cobb-Douglas frontier model (Ngwenya et al. 2010) Furthermore, in order to explore the substitute elasticity effects, instead of Cobb-Douglas production function, the translog production frontier, Ngwenya et al. (1977), Pitt and Lee (1981), Coelli et al. (2005), is widely used to analyze technical efficiencies for developing countries.

Using the stochastic frontier analysis to study productivity has been imposed in China's relevant issues. Fan (1991) used a frontier production function to separate agricultural growth into input growth, technical change, institutional reform and efficiency change. Lin (1992) employed a fixed effects model on provincial data to evaluate the effects of decollectivization (HRS).

In this paper, we will firstly introduce the methodology, then describe the dataset applied in this paper, thirdly show and analyze the empirical results, make a conclusion at the last section.

2   Econometric Methodology

In this section, we summarize the method of the translog stochastic frontier production function analysis for the panel data. We first reviewed the translog stochastic frontier production function, then we use the model to estimate the technical inefficiencies and finally we estimate the technical efficiencies.

2.1   Translog Stochastic Frontier Production Function

In the paper, we imposed the translog stochastic frontier production model with panel data for China which include 31 provinces ($i = 1$ to 31). The time period covers eleven years ($t = 2002$ to 2012). Based on the production model we attempt to analyze the technical efficiency for the agriculture industry in China. The translog stochastic frontier production models listed as below

$$\ln Y_{it} = \beta_0 + \sum_{j=1}^{5} \beta_j \ln x_{j,it} + \frac{1}{2} \sum_{j=1}^{5} \sum_{k=1}^{5} \beta_{jk} \ln x_{j,it} \ln x_{k,it} + V_{it} - U_{it}, \quad (1)$$

and the symbols and unit of variables can be checked in Table 1.

Y, the output, denotes the nominal agricultural added value (including farming, forestry, husbandry and fishing);
x represents the factors of inputs on agriculture, including;
x_1, labor, represents the number of people employed in agricultural industry;
x_2, land, or the sown area of crops, refers to area of land sown or transplanted with crops regardless of being in cultivated area or non-cultivated area;
x_3, denotes mechanical power;
x_4, amount of fertilizer used;
x_5 denotes geomembrane, the plastic membrane for agriculture used;
β_j denotes the parameters of the input elasticity on output;
β_{jk} represents the alternative elasticity between input j and input k, where $\beta_{jk} = \beta_{kj}$;
V_{it}s are assumed to be independent and identically distributed as normal random variables with mean zero and variance σ_v^2 independent of the U_{it}s;
U_{it}s are non-negative random variables, associated with technical inefficiency of production, which are assumed to be independently distributed.

2.2 Technical Inefficiency

The technical inefficiency, U_{it}, is obtained by a truncation (at zero) of the normal distribution with variance σ_u^2 and mean $z_{it}\delta$ (Refer to Eq. 2).

$$U_{it} = \delta_0 + \Sigma_{m=1}^{4}\delta_m z_{m,it} + W_{it}, \quad (2)$$

$z_{m,it}$ denotes the explanatory variables which caused the technical inefficiencies, $m = 1, 2, 3, 4$ represents 4 explanatory variables, $i = 1, 2, ..., 31$ represents 31 provinces of the mainland China:
z_1 is population density;
z_2 is the non-agricultural GDP per capita;
z_3 is the available credit per capita;
z_4 is the number of teachers per hundred;
δ is the effects of the explanatory variables;
δ_0 is the constant term;
δ_m represents the effects of 4 explanatory variables, respectively;
W_{it} is defined by the truncation of the normal distribution with zero mean and variance σ_W^2.

2.3   Technical Efficiency

The technical efficiency of production for the i-th province at the t-th observation is defined by Eq. (3),

$$TE_{it} = exp(-U_{it}) = exp(-z_{it}\delta - W_{it}),\tag{3}$$

The parameters of the translog stochastic frontier production model, the technical inefficiency, and the technical efficiencies of china's agriculture are estimated and the variance parameters are expressed in terms of γ and σ^2 which are defined by $\gamma = \sigma_u^2/\sigma^2$ and $\sigma^2 = \sigma_u^2 + \sigma_v^2$. The estimation for the effects or parameters of the stochastic frontier model defined by Eqs. (1)–(3) are estimated by using the method of maximum likelihood.

3   Data

In this study, the dataset has 10 indicators which derive from the website of the National Bureau of Statistics of China (NBSC). The study period of data is from 2002 to 2012 because of the availability. It covers 31 provinces, autonomous regions and municipalities (using province instead in following content). Some variables directly obtained the data from NBSC website, such like the agriculture output, the employed labor in farming, forestry, husbandry, and fishery, the sown area, the mechanical power, the fertilizer and the geomembrane. However, some other variable has to be calculated by the author based on data from NBSC as following. The population density used the population divided the area of the province; the non-agricultural GDP per capita is the summation of the gross output value of the second and third industries divided population; the available credit per capita used the saving deposit in urban and rural households divided to the population; the number of teachers per hundred used the number of full-time Teachers divided population and times 100.

4   Empirical Results

A summary statistics of the variables in the translog stochastic frontier model is presented in Table 1. The descriptive province-level average values of 10 variables in the model along with 31 provinces in China includes: the agriculture output is about 99.6 billion Yuan; the amount of labor is somewhat 9.4 million persons; the land area is slightly more than 5000 thousand hectares; the mechanical power is a little more than 25 million Kilowatts; the fertilizer is about 1.64 million tons; the usage of geomembrane is 62.4 thousand tons; the population density is 405 persons per km^2 which the minimum 2.2 persons in the Tibet and 3754 persons in Shanghai; the non-agriculture GDP per capita is 21.6 thousand yuan; the available credit per capita is 16.9 thousand Yuan; and the number of teachers per hundred population is 0.7. Table 2 presents the parameter estimators for the translog stochastic frontier functions of the China's provincial level agriculture

Table 1. Summary of the China province-level agricultural production 2002–2012

Variable	Symbol	Unit	Mean	Min	Max
Output	y	10^8 Yuan	996.3	41.4	4281.7
Labor	x_1	10^4 Person	939.6	33.4	3393
Land	x_2	10^3 Hectare	5052.7	231.2	14262.2
Mechanical power	x_3	10^4 KW	2529.7	95.3	12419.9
Fertilizer	x_4	10^4 Ton	164	3	684.4
Geomembrane	x_5	Ton	62418.9	440.6	343524
Population density	z_1	person/km^2	405.2	2.2	3753.6
Non-agricultural GDP per capita	z_2	10^4 Yuan	2.1558	0.2508	9.0037
Available credit per capita	z_3	10^4 Yuan	1.6869	0.1977	10.4615
Number of teachers per hundred	z_4	Person	0.7033	0.3669	1.2669
Observations		341			

Sources: Author's tabulation based on data extracted from the Chinese National Statistical Bureau website.

productivity by using the maximum-likelihood method. The estimated parameters can be referred to the Eqs. (1) and (2) in the translog model. The estimation of the translog production function, Eq. (1), shows the positive effects of labor (β_1), fertilizer (β_4) and geomembrane (β_5) and the negative effects of mechanical power (β_3) on the agriculture output. However, the effect of land (β_2) is not statistical significant. The main two factors affect the output are fertilizer and labor with closely 0.66 and 0.62. It means 1% input increase on fertilizer and labor will cause 0.66% and 0.62% increase in agriculture output, respectively. Mechanical power has negative relation with agriculture output. It may because that it's applied on harvesting rather than sowing seeds and maintenance. If the price stays invariant, input on mechanic cant make effect on agriculture output, but it can increase the cost to cause a negative effect on output. Currently, China's agriculture value improvement doesn't benefit from increase on agricultural land supply, therefore the effect of land on agricultural output is not significant. The substitute elasticity of input β_{jk} is not the focus of the paper we only present it in the Table 2. In the inefficiency equation, there are a positive relation of the population density (δ_1) with 6.41 and a negative relation of the non-agriculture GDP per capita (δ_2) with -0.47 significantly on the inefficiency. Furthermore, it is not statistically significant for the available credit per capita (δ_3). However, it is somewhat statistically significant on 15% significant level for the number of teacher per hundred persons (δ_4) with -0.0074. We found that 1% population density increase will cause 6.41% increase in technical inefficiency. But 1% increase in the non-agriculture GDP per capita and the number of teacher per hundred persons will cause 0.47% and 0.0074 decrease in technical inefficiency. The detail is in Table 2.

The predicted mean of the technical efficiencies of the agriculture productivity for each province are presented in Table 3. The mean value of each province

Table 2. MLE estimates of the panel stochastic frontier model

	Parameter	Estimate	Std. error	z-value	$Pr(>|z|)$
(Intercept)	β_0	7.0871	0.0654	108.297	$<2.2e{-}16$***
ln(labor)	β_1	0.6189	0.0482	12.8282	$<2.2e{-}16$***
ln(land)	β_2	−0.0253	0.0886	−0.2862	0.7747
ln(mechanical)	β_3	−0.5048	0.0824	−6.1262	9.000e−10***
ln(fertilizer)	β_4	0.6582	0.0643	10.2356	$<2.2e{-}16$***
ln(geomembrane)	β_5	0.2382	0.0408	5.8428	5.133e−09***
$\frac{1}{2}[ln(labor)]^2$	β_{11}	−0.0918	0.1091	−0.8409	0.4004
$\frac{1}{2}[ln(landr)]^2$	β_{22}	−0.1046	0.1316	−0.7946	0.4269
$\frac{1}{2}[ln(machanical)]^2$	β_{33}	0.1719	0.0731	2.3509	0.0187*
$\frac{1}{2}[ln(fertilizer)]^2$	β_{44}	0.3487	0.105	3.3216	0.0009***
$\frac{1}{2}[ln(geomembrane)]^2$	β_{55}	−0.0611	0.0232	−2.6283	0.0086**
ln(labor) ln(land)	β_{12}	0.0000	0.1874	−0.0002	0.9998
ln(labor) ln(mechanical)	β_{13}	0.3247	0.1272	2.5526	0.0106909*
ln(labor) ln(fertilizer)	β_{14}	−0.1914	0.122	−1.5693	0.1166
ln(labor) ln(geomembrane)	β_{15}	−0.1135	0.0566	−2.0035	0.0451*
ln(land) ln(mechanical)	β_{23}	0.539	0.1634	3.2995	0.0010***
ln(land) ln(fertilizer)	β_{24}	0.1096	0.1631	0.6721	0.5015
ln(land) ln(geomembrane)	β_{25}	−0.1189	0.0684	−1.7371	0.0823
ln(mechanical) ln(fertilizer)	β_{34}	−1.1188	0.1875	−5.968	2.402e−09***
ln(mechanical) ln(geomembrane)	β_{35}	0.1123	0.0651	1.724	0.0847
ln(fertilizer) ln(geomembrane)	β_{45}	0.2475	0.0601	4.1151	3.869e−05***
Z_(Intercept)	δ_0	1.2805	0.0686	18.6782	$<2.2e{-}16$***
Z_(Population Density)	δ_1	6.4128	0.6469	9.9133	$<2.2e{-}16$***
Z_(Nonagricultural GDP per capita)	δ_2	−0.4767	0.0735	−6.4878	8.709e−11***
Z_(available credit per capita)	δ_3	0.0326	0.0765	0.4258	0.6703
Z_(teacher per hundred persons)	δ_4	0.0074	0.0047	−1.5764	0.1149
Sigma square	$\sigma^2 = \sigma_u^2 + \sigma_v^2$	0.0494	0.0049	10.0546	$<2.2e{-}16$***
Gamma	$\gamma = \sigma_u^2/\sigma^2$	0.4513	0.1466	3.0791	0.0021**

Notes: Significant codes 0 '***' 0.001 '**' 0.01 '*' 0.05 '.' 0.1 ' ' 1.

over 11 years, the predicted technical efficiencies, obtained from the translog model, range from 0.356 to 0.993. Hunan province has the highest agriculture technical efficiency 0.993 and the subsequent two provinces are Beijing and Tianjin with 0.906 and 0.845, respectively. On the contrary, Guizhou province has the lowest agriculture technical efficiency 0.356. The second and third last provinces are Yunnan and Anhui with the TEs 0.392 and 0.395. The results in Table 3 reveals there are huge technical efficiency gap in terms of the mean of the TEs between the top three and the bottom three provinces which shown in Fig. 1.

The variance parameter $\gamma = 0.45$ shows that the technical inefficiency take account for 45% of the total variance σ^2. It indicates that there exit significant technical inefficiencies in the agriculture of China during the period 2002–2012. The gap between the actual agriculture output and the frontier output were

Table 3. Summary statistics of TEs by province

Rank	Region	Mean	S.D	Rank	Region	Mean	S.D.
1	Hunan	0.993	0.002	17	Shanghai	0.530	0.260
2	Beijing	0.906	0.104	18	Hubei	0.522	0.243
3	Tianjin	0.845	0.171	19	Chongqing	0.511	0.206
4	Zhejiang	0.793	0.201	20	Sichuan	0.509	0.165
5	Inner Mongolia	0.728	0.237	21	Jiangxi	0.507	0.167
6	Liaoning	0.718	0.224	22	Hainan	0.489	0.171
7	Guangdong	0.711	0.222	23	Shaanxi	0.477	0.228
8	Jiangsu	0.683	0.268	24	Guangxi	0.472	0.168
9	Fujian	0.663	0.231	25	Tibet	0.460	0.084
10	Jilin	0.611	0.223	26	Shanxi	0.449	0.184
11	Heilongjiang	0.599	0.189	27	Gansu	0.409	0.117
12	Shandong	0.590	0.251	28	Henan	0.403	0.150
13	Qinghai	0.589	0.220	29	Anhui	0.395	0.149
14	Xinjiang	0.558	0.147	30	Yunnan	0.392	0.099
15	Ningxia	0.551	0.229	31	Guizhou	0.356	0.088
16	Hebei	0.548	0.205				

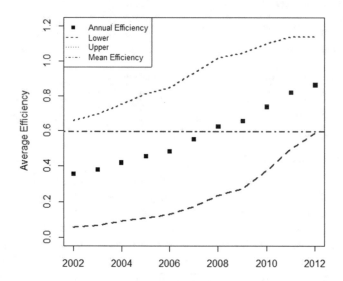

Fig. 1. Average efficiency of each year with 95% confidence interval

caused by the technical inefficiency. We choose Hunan, Beijing, Tianjing and Guizhou, Yunnan, Anhui as the top 3 and bottom 3 provinces in terms of the province-level TEs and the result shown in Fig. 2. Hunan has approached the

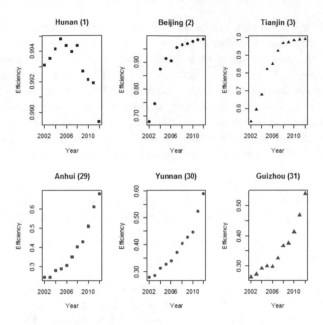

Fig. 2. Annual technical efficiency of Hunan, Beijing, Tianjin, Anhui, Yunnan, and Guizhou

frontier since 2002 until 2012 with fluctuation around 0.993. Beijing, Shanghai have been monotonic increased from 0.65, 0.5 in 2002 up to close to the production frontiers in 2012. For eleven years' development, the three lowest TEs provinces increased sharply from 0.24 in 2002 to, respectively, 0.65, 0.6 and 0.55 in 2012 which means all three provinces reached or close to the nation's average province-level TEs 0.6.

5   Conclusion

The study of the China's province-level agriculture output research was token by using the translog stochastic frontier production function analysis. The results point out (1) the technical efficiency of China's agriculture production has been increasing year after year. The agricultural advanced region has been approached to the technical efficiency frontier with a strong convergence trend suck like Beijing and Shanghai but a fluctuation in Hunan. The future research may find the reason of the variation. The technical efficiency in agriculture for the underdeveloped area increased sharply close to the average province level of China, however, still need to be improved for a big gap to the efficiency frontier; (2) High population density is the major reasons caused the technical inefficiencies. It may because the proportion of the agricultural population on total population has been continuously decreased to cause the insufficiency of the effective labor supply. The non-agriculture GDP per capita as a reason of reducing the technical inefficiency which are benefits from the development of the profitable non-agriculture industries. Education, in terms of number of teacher per hundred persons, has the positive effect on reducing the technical inefficiency. Therefore, the paper intends to give suggestions: In terms of improve the TEs, China still need to increase its input on labor, fertilizer and geomembrane especially for those provinces in the underdeveloped area to stimulate the agriculture output directly. Meanwhile, the country need to enlarge its investment in the second and third industries and to increase the number of teacher in underdeveloped area especially rural area to improve the agriculture system toward to the technical frontier.

Acknowledgements. This work has been supported by the Faculty of Economics and the Puey Ungphakorn Centre of Excellence in Econometrics at Chiang Mai University.

References

Aigner, D., Lovell, C.A.K., Schmidt, P.: Formulation and estimation of stochastic frontier production function models. J. Econom. **6**(1), 21–37 (1977)

Baten, M.A., Kamil, A.A., Haque, M.A.: Modeling technical inefficiencies effects in a stochastic frontier production function for panel data. Afr. J. Agric. Res. **4**(12), 1374–1382 (2009)

Battese, G.E.: Frontier production functions and technical efficiency: a survey of empirical applications in agricultural economics. Agric. Econ. **7**(3), 185–208 (1992)

Battese, G.E., Coelli, T.J.: Frontier Production Functions, Technical Efficiency and Panel Data: With Application to Paddy Farmers in India. In: Gulledge, T.R., Lovell, C.A.K. (eds.) International Applications of Productivity and Efficiency Analysis, pp. 149–165. Springer, Dordrecht (1992). A Special Issue of the Journal of Productivity Analysis

Battese, G.E., Coelli, T.J.: A model for technical inefficiency effects in a stochastic frontier production function for panel data. Empir. Econ. **20**(2), 325–332 (1995)

Battese, G.E., Rao, D.S.P., O'Donnell, C.J.: A metafrontier production function for estimation of technical efficiencies and technology gaps for firms operating under different technologies. J. Product. Anal. **21**(1), 91–103 (2004)

Bin, P., Vassallo, M.: The growth path of agricultural labor productivity in China: a latent growth curve model at the prefectural level. Economies **4**(3), 1–20 (2016)

Cao, K.H., Birchenall, J.A.: Agricultural productivity, structural change, and economic growth in post-reform China. J. Dev. Econ. **104**, 165 (2013)

Carter, C.A., Chen, J., Chu, B.J.: Agricultural productivity growth in China: farm level versus aggregate measurement. China Econ. Rev. **14**(1), 53–71 (2003)

Chen, P.C., Yu, M.M., Chang, C.C., Hsu, S.H.: Total factor productivity growth in China's agricultural sector. China Econ. Rev. **19**(4), 580–593 (2008)

Chen, Z., Song, S.: Efficiency and technology gap in China's agriculture: a regional meta-frontier analysis. China Econ. Rev. **19**(2), 287–296 (2008)

Coelli, T.J., Rao, D.S.P., O'Donnell, C.J., Battese, G.E.: An introduction to efficiency and productivity analysis. Springer (2005)

Fan, S.G., Zhang, X.B.: Production and productivity growth in Chinese agriculture: new national and regional measures. Econ. Dev. Cult. Change **50**(4), 819–838 (2002)

Foster, A.D., Rosenzweig, M.R.: Agricultural productivity growth, rural economic diversity, and economic reforms: India, 1970–2000. Econ. Dev. Cult. Change **52**(3), 509–542 (2004)

Gautam, M., Yu, B.X.: Agricultural productivity growth and drivers: a comparative study of China and India. China Agric. Econ. Rev. **7**(4), 573–600 (2015)

Hermann-Pillath, C., Kirchert, D., Pan, J.C.: Prefecture-level statistics as a source of data for research into China's regional development. China Q. **172**, 956–985 (2002)

Huang, C.J., Huang, T.H., Liu, N.H.: A new approach to estimating the metafrontier production function based on a stochastic frontier framework. J. Prod. Anal. **42**(3), 241–254 (2014)

Liu, J., Rahman, S., Sriboonchitta, S., Wiboonpongse, A.: Enhancing productivity and resource conservation by eliminating inefficiency of Thai rice farmers: a zero inefficiency stochastic frontier approach. Sustainability **9**(5), 770 (2017)

Meeusen, W., van Den Broeck, J.: Efficiency estimation from Cobb-Douglas production functions with composed error. Int. Econ. Rev. **18**(2), 435–444 (1977)

Ngwenya, S.A., Battese, G.E., Fleming, E.M.: The relationship between farm size and the technical inefficiency of production of wheat farmers in the eastern free state, province of South Africa. Agrekon **36**(3), 283–302 (1997). samevatting: die verhouding tussen plaasgrootte en die tegniese doeltreffendheid van koringboere in die oos vrystaat

Pitt, M.M., Lee, L.-F.: The measurement and sources of technical inefficiency in the Indonesian weaving industry. J. Dev. Econ. **9**(1), 43–64 (1981)

Schmidt, P.: Frontier production functions. Econom. Rev. **4**(2), 289–328 (1985)

Si, W., Wang, X.Q.: Productivity growth, technical efficiency, and technical change in China's soybean production. Afr. J. Agric. Res. **6**(25), 5606–5613 (2011)

Sriboonchitta, S., Liu, J., Wiboonpongse, A., Denoeux, T.: A double-copula stochastic frontier model with dependent error components and correction for sample selection. Int. J. Approx. Reason. **80**, 174–184 (2017)

Tong, H., Fulginiti, L.E., Sesmero, J.P.: Chinese Regional Agricultural Productivity: 1994–2005, p. 10. Lilyan E. Fulginiti Publications (2009)

Wiboonpongse, A., Liu, J., Sriboonchitta, S., Denoeux, T.: Modeling dependence between error components of the stochastic frontier model using copula: application to intercrop coffee production in Northern Thailand. Int. J. Approx. Reason. **65**, 34–44 (2015)

Zhang, Y.J., Brummer, B.: Productivity change and the effects of policy reform in China's agriculture since 1979. Asian Pac. Econ. Lit. **25**(2), 131–150 (2011)

Zhou, X.B., Li, K.W., Li, Q.: An analysis on technical efficiency in post-reform China. China Econ. Rev. **22**(3), 357–372 (2011)

Zhou, Y.H., Zhang, X.H., Tian, X., Geng, X.H., Zhang, P., Yan, B.J.: Technical and environmental efficiency of hog production in China - a stochastic frontier production function analysis. J. Integr. Agric. **14**(6), 1069–1080 (2015)

Empirical Models of Herding Behaviour for Asian Countries with Confucian Culture

Munkh-Ulzii[1] (ID), Massoud Moslehpour[2(✉)] (ID),
and Pham Van Kien[3] (ID)

[1] School of International, Relations and Public Administration,
National University of Mongolia, Ulan Bator, Mongolia
ulzii03@gmail.com
[2] Department of Business Administration, College of Management,
Asia University, Taichung, Taiwan
writetodrm@gmail.com, mm@asia.edu.tw
[3] International Business Faculty, Banking University,
Ho Chi Minh City, Vietnam
kienpv@buh.edu.vn

Abstract. The purpose of this study is to investigate the insights of herding behavior in the Confucian markets by conducting a set of empirical tests. More specifically this study investigates a sample of 7 countries and 13 markets to gain a deeper understanding of the causes of herding behavior and the potential factors that cause investors to behave in a group manner. Following a comprehensive review of the existing methodologies on herding behavior this study employs return dispersion approaches and herding tests developed by Chang et al. (2000) and Tan et al. (2008). This study investigates a sample of 13 stock markets of seven Asian economies with three different hypotheses. Those economies, which are considered to have Confucian culture, are mainland China, Hong Kong, Japan, South Korea, Taiwan, Singapore, and Vietnam. The hypotheses of this study aim to investigate formation of herding behavior in different market and economic circumstances. In testing the empirical models, this study uses OLS regression for the main test as well as regression with Newey and West (1987) for the robustness test of each result from OLS regression analysis. Data of this study consists of 13 index returns (Shanghai A and B share, Shenzhen A and B share, Hang Seng index, NIKKEI225, TOPIX, KOSDAQ, KOSPI, Straits Times Index, TAIEX, and indices of Hanoi and Hochiminh city Stock Exchanges) and returns of their constituent stocks. The time period of the sample data is from January 01, 1999 to December 31, 2014. All data were collected from the Thomson Reuters Datastream database. According to the empirical findings, all hypotheses are accepted. The sample markets demonstrate significant herding behavior in general and significant herding behavior in different markets conditions, such as in rising-falling markets and high-low market volatility states. This study has some major contributions to the literature of herding behavior and the link between herding behavior and cultural aspects. First, this study uses dataset of 13 Confucian stock markets of seven Asian economies, with time range from 1999 to 2014. Second, this research developed and tested three different hypotheses, and all of them are accepted. Third, this study adds the new dimension of the cultural aspects in order try to explain the root causes to herding behavior among

© Springer International Publishing AG 2018
V. Kreinovich et al. (eds.), *Predictive Econometrics and Big Data*, Studies in Computational Intelligence 753, https://doi.org/10.1007/978-3-319-70942-0_34

investors in the equity markets. Recognizing that the Confucian culture appears to be one of the most influential cultural aspects in management, this study examines herding behavior of Confucian culture in stock markets under the umbrella of one empirical study. According to the findings of this study, Confucian culture has a positive and significant effect on herding behavior among investors in equity markets.

Keywords: Herding behavior · Herding tendency · Emerging markets
Advanced markets · Confucian markets · Confucian culture

1 Introduction

It is difficult to make a precise definition about what is herding. Generally, it is a correlated behavioral pattern across individuals. Theoretically, herding behavior is a human behavior to mimic the behavior and actions of other people. Studies point out that herding behavior can be both rational and irrational (Chang et al. 2000). And, according to its motive, there are several types of herding behavior, such as information based herding, reputation based herding, and compensation based herding, as well as spurious herding. According to the sequential decision theory each trader observes the decisions of made by other investors in order to make his/her own decision. From the standpoint of the trader this is a rational act because decisions of other investors can include some important information for the trader. In this sense, herding behavior is rational.

1.1 Research Problem

Several studies have been held to catch up other factors that affect herding behavior on equity markets. Economou et al. (2011) tested Portuguese, Spanish, Italian and Greek markets to find herding effects by using daily stock return data during time period of 1998–2008. They found herding behavior from markets Greece and Italy, however they did not find evidence from Spanish and Portuguese markets. Mobarek et al. (2014) conducted another study on stock markets of eleven European countries using daily stock return data from 2001 to 2012. They divide sample countries into three categories – continental, PIIGS, and Nordic. They found that there is no herding during normal times, but the markets herd during crisis times and extreme market situation. Namely, Nordic markets were extensively impacted by the Eurozone crisis, while Continental and the PIIGS were more affected by the global financial crisis. However, these two studies found herding in cross-country level, the patterns and causes of herding were not the same from country to country, even though they all belong to the same general category – Europe. Otherwise, their outcomes suggest that there may exist some other reasons that make investors to herd, and that may be culture.

Therefore, above two studies not only found herding behavior from international markets, but also they found that stock markets of different countries react differently to the same events. It is thought that western markets are more mature, also the western personality is more individualistic; thus investors there would hardly display herding behavior. Unfortunately, these assumptions collapsed by their results. Therefore, now we know that even developed markets herd, and more interestingly cultural aspects may play an important role in herding besides information asymmetry, markets sentiments and fundamentals. However, findings of Economou et al. (2011) and Mobarek et al. (2014) suggest that cultural features matter in herding, the question of "why is it so" left inconclusive. Thus, now we have two studies support the idea of that culture may have an effect on formation of herding behavior. Are there other studies on how culture effects managerial decision making process?

Some more studies also discuss the potent impact of cultural factors on managerial decision making. Siegel et al. (2011) found the negative correlation between egalitarianism distance and international portfolio distribution. Aggarwal et al. (2012) found that the facets of culture have the potential to promote foreign portfolio management. Ferris et al. (2013) reported that managerial overconfidence tendency rises and falls with individualism and long-term orientation traits. Holderness (2009) showed that there exists a reciprocity among culture and ownership concentration of public corporations. Thus, these studies also found supportive results that national culture appears to be one of the main dominant factors in investment decision making.

Furthermore, culture might also emerge to be correlated with the aggregate economic activities of a country. Thus, studies also devoted attempts to check whether national culture has effects on the economic growth. Chow et al. (1991) found significant influence of individualism of employees on manufacturing performance. Li et al. (2013) found significant and positive relationship between GDP and individualism and uncertainty avoidance. Xue and Cheng (2013) discuss that culture and loan markets are correlative.

Thus, now we have more evidence on and confidence in that culture has a direct impact on behavioral decisions of people and has an indirect effect on aggregate economic activities. Therefore, we make the first assumption in our study that formation of herding tendency among investors might also be affected by cultural settings. And, formation herding tendency means herding behavior almost to happen. It is just a progression of a process. However, previous studies found out a clear evidence on how culture directly impacts on managerial decision making, they did not find out the actual reason of why culture does it. Yet next two studies did find a smoking gun.

Chang and Lin (2015) tested herding behavior among stock market investors at cross-country level. Their study sample, which is consisted of 50 stock markets, is made in a way to reflect as much geographical regions as possible. The specific feature of their study is that, besides using equity return data they employed Hofstede cultural index to check whether cultural aspects appear to be another reason why investor herd on the market. Interestingly, most of the herding behavior they found were mostly among Confucian markets. More specifically, they found that (1) markets under the influence of Confucian culture demonstrate a high power distance, low individualism, and high masculinity; (2) also, these dimensions found to have significantly positive effect on formation of herding tendency. The authors argue that people under the influence of Confucian culture, which focus on ethics, obedience, humanism, and collectivism, display the acceptance of unequal power distribution among the disadvantaged members of national institutions and organizations, close connection with the public, and the clear distinction between gender roles. In other words, investors in stock markets dominated by Confucian culture emphasize public morality and follow the behavior of the majority of people. It not only impacts investors' decision-making, but also frames their reaction to a certain informational shock. Otherwise, it effects on investors' behaviour in the stock markets. Therefore, Confucian philosophy may trigger investors to exhibit herding behavior.

Findings of Chang and Lin (2015) were in line with the findings of study conducted by Beckmann et al. (2008). Arguing that cultural differences translate into different behavior and that these differences are relevant for investment behavior, Beckmann et al. (2008) examined different cultured countries (Japan, Thailand, US, and Germany) by using cultural dimension approach of Hofstede. According to their findings, cultural dimensions of *individualism* and *power distance* were also found to be significantly correlated with herding behaviour. They found a strong significant evidence that asset managers in less individualistic countries exhibit more herding behaviour, and tend to follow market trends more closely. Moreover, societies with high power distance pay much attention on seniority. In societies, where power distance is high, age appears to be the strong factor in investment behavior, such as risk taking and portfolio allocation in a conservative way. Supporting this theory, Beckmann et al. (2008) also found that higher age positively influences on formation of herding tendency. Moreover, they conclude that societies with high power distance not only prefer older managers for promotion, but they also consider those who have less experience. Fifty percent of their sample was from Japan, which is also a Confucian country.

The common things of these two studies are, first, they both examined cultural effects on herding behavior; second, samples of either studies includes Confucian countries; third, they used same research instruments in their empirical analysis; and finally they found similar results. In summary, Chang and Lin (2015) found that Confucian markets tend to have high power distance, low individualism, and high masculinity, which appear to be the stepping stones to formation of herding behavior. While, Beckmann et al. (2008) found that markets under dominance of high power distance and low individualism more likely to exhibit herding behavior. And, the half of the sample of their study is data from Japan, which is a Confucian country.

Thus, these studies provide us with empirical evidences that encourage us to make the second assumption in our study that Confucian culture positively impacts on formation of herding tendency among investors in equity markets. What is the matter with Confucian culture, why it impacts on formation of herding tendency? This is a very interesting and big question. Thus, we will discuss about what Confucian culture is and the links between this culture and managerial decision making process in the next section.

1.2 Confucian Philosophy and Its Impacts on Managerial Decision-Making

The legacy of Confucian philosophy is a system that highlights the significance of hard work, loyalty, devotion, learning, and social hierarchy. What are the core teachings of Confucian? In Confucianism individuals are not regarded as important as a group. Thus, this philosophy teaches that a person should honor first his/her duty to family and society. In order words, personal needs to be sacrificed for the sake of group welfare. In this manner, it helped to develop a managerial mindset with greater emphasis on collectivism, teamwork, and harmony in relationships. Therefore, when

Chinese managers face difficult situations, they usually would not exhibit personal behavior to existing situation, rather they would follow the tendency of the group that he/she belong to. Chinese management culture differs from western culture in large extent. Confucian culture is based on collectivism, whereas the westerns tend to be very individualistic.

1.2.1 Confucian Management Theories "Guanxi" and "Mianzi"

There are two fundamental theories in Confucianism - "guanxi" and "mianzi", which have been powerful and essential among Chinese communities. Due to the strong foundation of these theories, every man in Chinese society is a businessman to expand his wealth and fame. And, the Confucian philosophy is mainly about how to make the nation flourish by enhancing individual wealth and fame via synergy of powerful network of collaboration and friendship.

Guanxi, which is the essence of Confucianism, is one of the most important factors to deal with when somebody enters into business and investment in Chinese society. Guanxi can be interpreted as "relationship" or "networking". Sometimes guanxi finds very complicated both for the members and non-members of the network. The problem is that it usually requires obligations or indebtedness, a mechanism of favors and debts among members in the network, which itself is based on relationships governed by Confucian teachings. Therefore, in order to establish and maintain a position among a guanxi network it requires some wisdom of manipulation of human feeling, reciprocity of favors and giving face (mianzi). Guanxi is very powerful among Chinese societies, so that, personal networks and loyalties are thought to be more important than organizational affiliations or legal standards (Luo 1997; Buttery and Leung 1998).

Even though the concept of "mianzi" has broad meaning, the degree of concern is much higher and its manifestations are somehow different in the Confucianism. In Confucian communities the notion of *mianzi* carries a meaning of showing respect for one's social status. To maintain face means to stay trustworthy and to honor obligations in one's guanxi. Sometimes, it is more important to give face to others than to protect own face (Buttery and Leung 1998). Giving favors to others shows that one has capacity beyond one's peers, and thus gains face or respect in return. In reciprocity, the person who have received a favor should return it whenever got to return it, even if this reciprocity may be harmful one's social reputation and lead to lose his face. Losing face is not only failure in reputation but also indication of lack of trust. Thus, subordinates are always advised to do their best to give superiors face by asking favors from them, but always to do it within the boundaries because of the code of reciprocity. In Confucian societies, guanxi and mianzi are two sides of the same thing, they are equally very important aspects (Luo 1997; Su and Littlefield 2001).

The core regions of the east Asian cultural sphere are *China, Taiwan, Hong Kong, South Korea, Japan*, and *Vietnam*, as well as territories settled by *Chinese people*, such as *Singapore*. We assume that the Confucian culture has a positive impact on formation of herding behaviour among investors in stock markets of Confucian countries. China, Korea, Japan, Taiwan, Singapore, and Vietnam are Confucian countries, thus investors of these countries are more likely to exhibit herding behavior.

2   Literature Review

Chan et al. (2000) examined the herding behavior among international markets, US, Hong Kong, Japan, South Korea, and Taiwan by employing daily stock price data with time range from January 1963 to December 1997. They used cross-sectional absolute deviation method, which is the revised version of the model that was developed by Christie and Huang (1995). They found no evidence of herding from markets of US and Hong Kong and partial evidence from Japan. However, they found significant herding from South Korea and Taiwan. They also found that macroeconomic information impacts heavily on formation of herding behavior than firm-specific information. Finally, their results conclude that stock return dispersion, as a function of the market return, is higher during up market than down market. They also tested Christie and Huang (1995) model, however, they did not find herding except market of Taiwan.

Following the Chan et al. (2000) model Tan et al. (2008) found herding behavior in Chinese A[1] and shares. Due to their hesitated in accuracy of estimation of beta, suggested by Chang et al. (2000), they used standard deviation in estimation of the return dispersion, which was adopted by Christie and Huang (1995). Moreover, they also tested the asymmetric impact of herding behavior by varying market return, trading volume, and volatility. They found herding behavior among investors of a share of Shanghai market under rising market circumstances with high volume of trading of stocks and volatility. However, there found no significant evidence of herding in B (see Footnote 1) shares. The main participants of A share market are local individual investors who are considered to be insufficient in skills and experience of investment. While investors of B shares market are mostly foreigners who possess with more skills and knowledge in investment rather investors of A share. In this fashion, they conclude that the discrepancies of investors of A and B shares may impact on variance of intensity of herding behavior in each market. This study also tested herding behaviour and cross-market information effect, however they did not find evidence for return dispersions of markets influence each other. According to Tan et al. (2008), the dissimilarity can be explained due to the difference of sample population.

Demirer et al. (2010) measures herding behavior by using daily data of stock returns for 689 stocks traded on the Taiwan Stock Exchange with time period from January 1995 to December 2006. In the analysis they used Christie and Huang (1995) model, Chan et al. (2000) model, and state space models. They found no evidence for herding when they used Christie and Huang (1995) model, however they did find significant herding when they used non-linear model of Chan et al. (2000) and the state space based models proposed by Hwang and Salmon (2004). They also found that the herding is stronger during periods of market losses. Thus, this paper suggest that investors need more diversified opportunities, specifically during periods of market losses. The authors pointed out the following three contributions of their study: First, this study contributes herding literature with empirical results of an emerging yet

---

[1] A shares are shares of the Renminbi currency that are purchased and traded on the Shanghai and Shenzhen stock exchanges. This is contrast to Renminbi B shares, which are owned by foreigners who cannot purchase A-shares due to Chinese government restrictions.

relatively sophisticated Taiwanese stock market at the sector level by using firm level data. Second, this paper used a set of different models. Third, this study discusses the practical usage of different herding measures for investors who face to systematic and unsystematic risks.

My and Truong (2011) examined existence of herding behavior in the Vietnamese stock market by using daily stock price data at the Ho Chi Minh Stock Exchange with time period from March 2002 to July 2007. In testing herding they used Christie and Huang (1995) model robustness test and Chan et al. (2000) model for the main tests. They also tested asymmetric effects of herding. This study found an evidence of herding behavior in the Vietnamese stock market. The authors discuss that this herding behavior can be explained due to the lack of transparency in information and financial management, the high magnitude of market volatility and thin trading. Moreover, the study found that return dispersion is less during upward markets than downward markets. It maybe be due to that Vietnamese investors behave more uniformly in rising markets than in declining markets. Based on the empirical findings, this study discusses that herding appears to be one of the reasons of market volatility, thus financial institutions may concern on policies adjustments to avoid from possible destabilization effects on the financial markets. With the Christie and Huang (1995) approach no herding was found.

Yao et al. (2014) measured existence of herding behavior among the Chinese A and B stock markets. To do so, they used daily and weekly firm level and market level data of equity prices for all firms and indices listed on the Shanghai Stock Exchange and the Shenzhen Stock Exchange with time period from January 1999 to December 2008. Additionally, monthly data on industry classification, market capitalization, earnings per share, and market-to-book values are also collected for all firms included in the sample. In data analysis, this study used modified version of Christie and Huang (1995) and Chan et al. (2000) models as herding test. They findings showed that the magnitude of herding behavior among investors was heterogeneous, specifically herding was stronger in the B-share markets. Also, they found that across markets herding was stronger at industry-level, was stronger for the largest and smallest stocks, and was stronger for growth stocks relative to value stocks. Results of their study show that herding was more prevalent during declining markets. Finally, they found that the magnitude of herding behavior was at tapering state. The authors explained it was maybe due to the effectiveness of regulatory reforms in China aimed at improving information efficiency and market integration.

Aiming to explore the determinants of investor decision-making in international stock markets Chang and Lin (2015) conducted a study on herding behaviour by using daily market return data and industrial index data for 50 stock markets to calculate the cross-sectional absolute deviation of returns. This study used extended version of Chan et al. (2000) model. Moreover, to investigate the influence of national culture on investment herding tendency, this study employs the national culture indexes proposed by Hofstede (2001) for empirical analysis. Finally, in order to test behavioral pitfalls on herding tendency, this study used daily data of price-to-book ratios as proxies for excessive optimism, while daily trading volumes data were used as proxies for overconfidence and the disposition effect. The time period of entire dataset of this study ranges from January 1965 to July 2011. The authors argue that unlike prior studies, their study examines the effects of national culture and behavioral pitfalls on investment decision-making process. Their empirical results show that herding behaviour exhibits in Confucian and less sophisticated stock markets. Moreover, this study found that some national culture indexes were closely correlated with herding behaviour. Finally, they found that behavioral pitfalls were dominant on formation of herding.

2.1 Hypothesis Development

Most of the studies on herding behavior used the return dispersion approach as the main research model. Thus, it is reasonable for us to follow the trend. Thus, we will generally refer to those studies in hypothesis development. Before we made a general proposition that Confucian culture has a positive impact on formation of herding behaviour among investors in stock markets of Confucian countries. To verify this assumption, we will go through the following three different hypotheses.

2.1.1 Development of Hypothesis 1

"Guanxi" and "Mianzi" are the core values and main management theories in Confucian philosophy. These theories have been the key factors in being China to be one of the strong markets in all times. In other words, a network of cooperation and friendship, which is powered by Confucian management theories, is simply a well-established market place. Thus so, China appears to be one of the places where market economy was born. With a such powerful foundation, it is apparent that Confucian philosophy is the most powerful management force in the world in terms of its influence on decision making process of managers. Therefore, the likelihood of influence of this culture on formation of herding behavior is much convincing.

Besides Confucian culture, there are also other factors may induce herding behavior. Moreover, there is a relatively high degree of government involvement in the operations of the equity markets, with the imposition of many restrictions, such as price limits, no short-selling and an intraday trading ban. The heavy intervention of the central bank in adjusting interest rate policy is thought to have some impact on investors in this market, especially in the case of upward adjustment. As, banks mobilize capital at higher costs, they are obliged to raise their lending interest rate on securities loans which might reduce investors' profitability. This potentially leads to a stronger tendency towards investor herding (Christie and Huang 1995; Chang et al. 2000; Tan et al. 2008; Economou et al. 2011; Lao and Singh 2011; Mobarek et al. 2014; Yao et al. 2014; Chang and Lin 2015). Thus, our study develops the following first hypothesis for empirical testing:

Hypothesis 1: Herding behaviour exists among the Confucian markets.

2.1.2 Development of Hypothesis 2 and 3

Presumably, investor behavior might be influenced by the direction of the market. In fact, some studies have found that the level of herding behavior changes under different market conditions, depending on whether the market is rising or declining. It is therefore hypothesized that investor have a greater tendency to herd in downward markets than in upward markets. It can be argued that the fear of the potential loss when the market is decreasing looms larger than the pleasure of potential gain when the market is increasing. As a result, this behavior is likely to induce investors to mimic the aggregate market when it is falling. A possible explanation for this behavior is that investors may experience less disappointment when others also make the same investment decision and this decision eventually turns out to be poor. Thus, herding behavior is expected to be more pronounced when a market is falling than when it is rising (Chang et al. 2000; Gleason et al. 2004; Tan et al. 2008; Chiang and Zheng 2010; Chiang et al. 2010; Lao and Singh 2011; My and Truong 2011; Bhaduri and Mahapatra 2013; Yao et al. 2014; Mobarek et al. 2014).

In addition, McQueen et al. (1996) found that while all stocks tend to respond quickly to negative macroeconomic news, small stocks tend to exhibit a delay in reacting to positive macroeconomic news. Since good macroeconomic news often entails an increase in stock prices, the slow reaction implies a postponement in the incorporation of good news into the prices of small stocks. This may therefore lead to an extra increase in market return dispersion in an increasing market but not in a declining market. This implies that herding is likely to be more pronounced in downward markets that are characterized by small stocks.

Christie and Huang (1995) and Chang et al. (2000) note that herding behaviour may be more pronounced during periods of market stress. Consequently, we investigate whether the herding behaviour documented above varies with market conditions. Specifically, we examine potential asymmetries in herding behaviour as the trading environment is characterized by different states of market returns and volatility. Return volatility and trading volume may help characterize such periods, so we use them to gain additional insight regarding the level of herding behaviour under different market conditions. Since the direction of the market return may affect investor behavior we develop the hypothesis two to examine possible asymmetries in herd behavior conditional on whether the market is rising or falling:

Hypothesis 2: Asymmetric herding behaviour exists among the Confucian markets when the market rises and falls.

Also, we develop hypothesis four to examine potential asymmetric effects of herding behavior with respect to market volatility:

Hypothesis 3: Asymmetric herding behaviour exists among the Confucian markets during high and low volatility states of markets.

3 Methodology and Data

3.1 Data Collection

This study uses daily returns of market indexes of Confucian stock markets and their constituent stocks. Data of this study consist from 13 indexes (Shanghai A and B share, Shenzhen A and B share, Hang Seng index, NIKKEI225, TOPIX, KOSDAQ, KOSPI, Straits Times Index, TAIEX, and indices of Hanoi and Hochimin city Stock Exchanges) of nine stock markets of six Confucian countries. Market return and individual stock returns calculated by Datastream by using discrete or simple return calculation method. All data were collected from the Thomson Reuters Datastream database. The time period of the sample data is from January 01, 1999 to December 31, 2014.

3.2   Methodology

This study conducts several tests and analysis, which can be called as data oriented and empirical model oriented in terms of their purpose. Normally, all tests and analysis use data, however one checks the reliability and validity the data, and makes sure it for an empirical analysis, while the another tests relationship between several variables by using the data. The data oriented tests in our study are test of descriptive statistics and unit root test as the Augmented Dickey-Fuller (ADF), while empirical model or empirical tests are correlation test and multiple regression analysis. Also, while we did empirical models estimation by using multiple regression analysis, we also conducted a robustness regression analysis by using regression with Newey-West standard errors (1987). In data analysis, we used softwares, such as Stata 13.1, Microsoft Excel, and MathType.

3.3   Empirical Models

3.3.1   Measuring Herding Behavior (H1)

A couple of approaches in detecting herding behaviour by using individual stock returns and market return, were developed by Christie and Huang (1995) and Chang et al. (2000). Christie and Huang (1995) discussed that the decision making process of investors depends upon overall market conditions. According to their view during normal time periods, rational asset pricing models anticipate that the return dispersion will increase with the absolute value of the market return, since individual investors are making decision based on their own information, which is diverse. However, during time periods of extreme market situations, investors tend to oppress their own beliefs, and mimic the collective actions in the market. Individual equity returns under these conditions tend to cluster around the overall market return. Therefore, they argue that herding will be more common during periods of occurrence of extreme returns on the market portfolio. They proposed the following equation in their empirical testing:

$$CSSD_t = \alpha + \beta^L D_t^L + \beta^U D_t^U + \varepsilon_t \qquad (1)$$

where $CSSD_t$ is cross-sectional standard deviation at time (t), α is an intercept, D_t^L is a dummy variable at time (t) taking value of 1 when market return at time (t) lies in the extreme lower tail of the return distribution, otherwise 0. D_t^U is a dummy variable at time (t) taking value of 1 when market return at time (t) lies in the extreme upper tail of the return distribution, otherwise 0. β^L is a coefficient at D_t^L. β^U is a coefficient at D_t^U. And ε_t is error term at time (t). To measure the return dispersion, Christie and Huang (1995) proposed the $CSSD$, which is expressed as:

$$CSSD_t = \sqrt{\frac{\sum_{i=1}^{N} \left(R_{i,t} - R_{m,t}\right)^2}{N - 1}} \tag{2}$$

where N is the number of firms in the portfolio. $R_{i,t}$ is individual stock returns of stock (i) at time (t). $R_{m,t}$ is market return at time (t). This model suggests that, if herding occurs, investors will make similar decisions, leading to lower return dispersions. Thus, statistically significant and negative coefficients β^L and β^U in Eq. (1) would indicate the presence of herding behaviour.

Demirer and Kutan (2006) used Christie and Huang (1995) method to test herding behaviour in Chinese stock markets. They employed daily stock returns of 375 firms but did not find evidence of herding. One of the drawbacks related with Christie and Huang (1995) approach is that it requires the definition of extreme returns. Christie and Huang (1995) discussed that this definition is arbitrary, and they employed values of one and five percent as the cut-off points to identify the upper and lower tails of the return distribution. In reality, traders not always find the same in their opinion about extreme return, and the characteristics of the return distribution may change over time. Moreover, herding behavior can happen to some extent over the entire return distribution, but become more noticeable during periods of market stress, and the Christie and Huang (1995) approach is designed to detect herding only during periods of extreme returns.

An optional method to the Christie and Huang (1995) approach for herding was proposed by Chang et al. (2000). They tested multiple international stock markets by using Christie and Huang (1995) method, and did not find evidence of herding behaviour in developed markets, such as the US and Hong Kong. However, they did find evidence of herding behaviour in the emerging stock markets of South Korea and Taiwan. Chang et al. (2000) discussed that the Christie and Huang (1995) method is a rigorous method, which requires a far greater degree of non-linearity to find herding behavior. While the method that they developed showed that when stock return dispersion is measured by the cross-sectional absolute deviation of returns, rational asset pricing models predict not only that dispersion of individual stocks is an increasing function of the market return, yet that the relation is linear as well. However, a heightened tendency of investors to herd around the market consensus during periods of large price movements is enough to convert the linear relation into a non-linear relation. Thus, to detect herding behavior, Chang et al. (2000) used a non-linear regression specification, which is alike in spirit to the market timing approach of Treynor and Mazuy (1966). The herding method of Chang et al. (2000) finds very helpful to herding over the entire distribution of market return with the following specification:

$$CSAD_t = \alpha + \gamma_1 |R_{m,t}| + \gamma_2 \left(R_{m,t}\right)^2 + \varepsilon_t \tag{3}$$

The $CSAD_t$ is cross-sectional absolute deviation at time (t), a measure of return dispersion, which is calculated as follows:

$$CSAD_t = \frac{1}{N}\sum_{i=1}^{N}|R_{i,t} - R_{m,t}| \tag{4}$$

where $|R_{m,t}|$ is absolute value of market return at time (t), $CSAD_t$ is a cross-sectional absolute deviation at time (t). Chang et al. (2000) discussed that rational asset pricing models suggest a linear relationship between the return dispersion in individual stock returns and market return. In order words, when the absolute value of the market return increases, return dispersion of individual stock also increases. During periods of comparatively large market price movements, traders may behave in an uniform manner, exhibiting herding behavior. Herding is likely to increase the correlation among stock returns, and the related dispersion among returns will decrease, or at least increase at a less-than-proportional rate with the portfolio average. Thus, Chang et al. (2000) included a nonlinear market return - $(R_{m,t})^2$ in their empirical model in Eq. (3), and a significantly negative coefficient γ_2, would indicate the occurrence of herding behaviour. As the market exhibits large price movements, investors more likely to degrade their own beliefs and follow the trend of the market. Return dispersion of individual stock return tend to either decrease or increase at a decreasing rate during this conditions.

The calculation approach of return dispersion measure $CSAD_t$ of this study follows Christie and Huang (1995), Gleason et al. (2004), and Tan et al. (2008), but not Chang et al. (2000). The reason of not following the method of Chang et al. (2000) is that their measure relies on the accuracy of the specification of a single market factor of the CAPM, which may be questionable (Tan et al. 2008). Studies, usually avoided their approach in measuring the $CSAD_t$ for the same reason.

Thus, in the empirical testing of the hypothesis 1–3 we will base on the Eq. (3). And, in testing hypothesis 7–9 we will use Eq. (4) for the calculation of the dependent variable, but independent variables will have different specifications. Finally, in examination of the hypothesis 10 we will use the Eqs. (1) and (2).

In the empirical examination of the hypothesis 1, the Eq. (5) will be utilized. The Eqs. (3) and (5) are same, except in the Eq. (5) the idea of multiple markets is infused.

$$CSAD_{i,t} = \alpha + \gamma_1|R_{i,m,t}| + \gamma_2(R_{i,m,t})^2 + \varepsilon_{i,t} \tag{5}$$

where $CSAD_{i,t}$ is return dispersion of market (i) at time (t), α is an intercept, $|R_{i,m,t}|$ is absolute value of market return of market (i) at time (t), $(R_{i,m,t})^2$ is squared value of market return of market (i) at time (t), $\varepsilon_{i,t}$ is error term of market (i) at time (t). The $CSAD_{i,t}$ for each market will be calculated according Eq. (4). A statistically significant and negative coefficient γ_2 would indicate the presence of herding behavior. In this analysis data of stock markets indices and their constituents of China, Hong Kong, Japan, South Korea, Singapore, and Taiwan, as well as Vietnam were used to analysis the empirical model.

3.3.2 Measuring Asymmetric Herding Behavior (H2 and 3)

Measuring Asymmetric Herding Behaviour During Up and Down Market (H2)
Several studies also examined the asymmetric herding behavior, however document different results. In all five markets (US, Hong Kong, Japan, South Korea and Taiwan) that Chang et al. (2000) examined, the return dispersion of individual stock increased as a function of the aggregate market return was higher during up market, comparatively to down market. Otherwise, investors exhibited herding behaviour in down market time. These results were consistent with the directional asymmetry documented by McQueen et al. (1996).

Using a dataset of NYSE (US) Gleason et al. (2004) found a weak presence of asymmetric herding behaviour during up and down markets. However, they discussed that investors may be more inclined to herd in down markets. Their findings indicated that return dispersion need not necessarily to be identical, during periods of market stress, in up and in down markets. According to McQueen et al. (1996) there is a delayed reaction to good news, whereas market participants react more quickly to bad news. In other word, it suggests that when traders react to bad news during down markets have a great tendency to follow the aggregate market. Otherwise, investors may afraid the potential loss during a down market time more than they enjoy the potential gain during an up market period of stress. This phenomenon is also described as "myopic loss aversion" (Benartzi and Thaler 1993). This type of trading behavior leads to lower dispersion and the possibility of herding in down markets. In overall, they concluded that their findings provided a weak support for hypothesis of myopic loss aversion.

Examining China, A and B share markets Tan et al. (2008) found that herding occurs in both rising and falling market conditions. Moreover, herding found to be stronger, especially for the Shanghai A-share market, during up markets, high trading volume states and high volatility states. However, there was a very weak asymmetry in the B-share markets. They documented that the obvious difference in investor behavior may be due to the different specifications of A and B share markets. A-share market is dominated by local investors, whereas the B share market is dominated by foreign investors.

Fu and Lin (2010) found an evidence of asymmetric reaction of herding that investors' tendency of the Shanghai and Shenzhen stock markets toward herding is significantly higher during down markets. My and Truong (2011) documented that in stock market of Vietnam, upward markets have less return dispersion than downward markets. In other word, investment behaviour in the stock market of Vietnam was more uniformly in up markets than in down markets.

By analyzing 18 markets, which is consist of markets from regions of Latin America, North America, Asia, Chiang and Zheng (2010) found that herding asymmetry is more profound in Asian markets (Hong Kong, Japan, China, South Korea, Indonesia, Malaysia, Singapore, Thailand, and Taiwan) during rising markets.

Lao and Singh (2011) found that herding behaviour is greater in Chinese stock markets when the market is falling and the trading volume is high. Moreover, Bhaduri and Mahapatra (2013) documented that herding behaviour is prevalent during up markets, which is in line with findings of Tan et al. (2008), but their study used data from Indian stock market. Finally, Yao et al. (2014) discussed that herding behaviour in Chinese stock markets were more pronounced under conditions of declining markets.

Mobarek et al. (2014) measured asymmetric herding behavior among 11 European markets by using dummy variable in detecting herding during up and down markets, high and low trading volume states, and high and low market volatility states. Their findings were somehow consistent with conclusion of McQueen et al. (1996). They found evidence of herding in down markets. Moreover, they found that some evidence of a significant herding effect during low volume trading periods. Finally, they detected a significant herding during high and low volatility periods.

In overall, most of the works, almost all of which are studies on stock markets of Confucian countries, found that herding behavior is more prevalent during down market than up market. Few tests were conducted by using dummy variable, however most of the studies employed the approaches specified in the Eqs. (6)–(9). Thus, we will also follow these models. Since the direction of the market return can influence on decisions of investors, we test possible asymmetric effects in herd behavior conditional on whether the market is rising or falling. Thus, the following empirical model will be estimated separately for positive and negative market returns:

$$CSAD_{i,t}^{UP} = \alpha + \gamma_1^{UP}\left|R_{i,m,t}^{UP}\right| + \gamma_2^{UP}\left(R_{i,m,t}^{UP}\right)^2 + \varepsilon_{i,t} \quad \text{If } R_{i,m,t} > 0 \qquad (6)$$

$$CSAD_{i,t}^{DOWN} = \alpha + \gamma_1^{DOWN}\left|R_{i,m,t}^{DOWN}\right| + \gamma_2^{DOWN}\left(R_{i,m,t}^{DOWN}\right)^2 + \varepsilon_{i,t} \quad \text{if } R_{i,m,t} < 0 \qquad (7)$$

where $CSAD_{i,t}^{UP}$ is return dispersion of market (i) at time (t) when the market rises, α is an intercept, $\left|R_{i,m,t}^{UP}\right|$ is absolute value of market return of market (i) at time (t) when the market rises, $\left(R_{i,m,t}^{UP}\right)^2$ is squared value of market return of market (i) at time (t) when the market rises, $\varepsilon_{i,t}$ is error term of market (i) at time (t). Similarly, the variables with superscript "DOWN" in the Eq. (7) refer to the scenario in which the market declines. The $CSAD_{i,t}$ for each market in the Eqs. (6) and (7) will be calculated according to the Eq. (4). A statistically significant and negative coefficients γ_2^{UP} or γ_2^{DOWN} would indicate the presence of herding behavior. In this analysis data of stock markets indices and their constituents of China, Hong Kong, Japan, South Korea, Singapore, and Taiwan, as well as Vietnam were used to analysis the empirical model.

Measuring Asymmetric Herding Behaviour During High and Low Volatility States (H3)

Furthermore, we examine asymmetric effects of herding behavior in regard to volatility of stock markets. As we did with trading volume, we characterize market volatility to be high when the observed volatility gets higher than the moving average of volatility over the previous 30 days. We define volatility as low when it does not exceed the moving average of volatility over the previous 30 days. We have also checked moving averages of 7 and 90 days, which is different from the settings of Tan et al. (2008), who used much longer time periods of 60, 90 and 120 day moving averages. We designed so, because studies discuss that sentiment of investors usually happens in short period of time (Chang et al. 2000). The asymmetric effects are examined using the following empirical models:

$$CSAD_{i,t}^{\delta^2-HIGH} = \alpha + \gamma_1^{\delta^2-HIGH}\left|R_{i,m,t}^{\delta^2-HIGH}\right| + \gamma_2^{\delta^2-HIGH}\left(R_{i,m,t}^{\delta^2-HIGH}\right)^2 + \varepsilon_{i,t} \qquad (8)$$

$$CSAD_{i,t}^{\delta^2-LOW} = \alpha + \gamma_1^{\delta^2-LOW}\left|R_{i,m,t}^{\delta^2-LOW}\right| + \gamma_2^{\delta^2-LOW}\left(R_{i,m,t}^{\delta^2-LOW}\right)^2 + \varepsilon_{i,t} \qquad (9)$$

where $CSAD_{i,t}^{\delta^2-HIGH}$ is return dispersion of market (i) at time (t) when market return volatility is high, α is an intercept, $\left|R_{i,m,t}^{\delta^2-HIGH}\right|$ is absolute value of market return of market (i) at time (t) when market return volatility is high, $\left(R_{i,m,t}^{\delta^2-HIGH}\right)^2$ is squared value of market return of market (i) at time (t) when market return volatility is high, $\varepsilon_{i,t}$ is error term of market (i) at time (t). Similarly, the variables with the superscript "$\delta^2 - LOW$" refer to the scenario in when market return volatility is low. The superscript "δ^2" refers to market return volatility, and it is calculated as the standard deviation of market return time with the square root of 252 trading days. The $CSAD_{i,t}$ for each market in the Eqs. (8) and (9) will be calculated according to the Eq. (4). A statistically significant and negative coefficients $\gamma_2^{\delta^2-HIGH}$ or $\gamma_2^{\delta^2-LOW}$ would indicate the presence of herding behavior. In this analysis data of stock markets indices and their constituents of China, Hong Kong, Japan, South Korea, Singapore, and Taiwan, as well as Vietnam were used to analysis the empirical model.

4   Data Analysis

4.1   Results of Descriptive Statistics

Table 1 analysis descriptive statistics of return dispersion measure of CSAD of sample markets. According to Table 1, stock markets of KOSDAQ (2.414891) and KOSPI (2.162403) of Korea Exchange and HNX (2.222067) of Hanoi city of Vietnam have the highest mean values, while they also have highest volatility as standard deviation of 0.8694664, 0.9576033, and 0.7862742 respectively. The lowest mean value got SHB (1.107081) of Shanghai B market of China, while lowest volatility resulted to NI225 (0.5594453) of Tokyo Exchange. According to the Eq. (4) CSAD cannot have negative values, thus we report only positive values. According to Table 4, the maximum value of CSAD resulted to VNINDEX (10.95739) market of Hochimin city exchange of Vietnam. Also, markets of KOSDAQ (8.036667) and KOSPI (10.22663) of Korea Exchange have higher maximum values.

Also, Chang et al. (2000) discussed that Asian equity market returns and returns of individual stocks are characterized by higher magnitudes of volatility with standard deviations, which is also supports our results, since our study concentrates on Asian stock markets, and they variability in terms of market return and return dispersion are a

Table 1. Descriptive statistics of return dispersion measure of CSAD

Variable	Obs	Mean	Std. Dev.	Min	Max
SHACSAD	3982	1.504224	.7679289	0	7.220784
SHBCSAD	4174	1.107081	.6864486	0	5.12972
SZACSAD	3877	1.453948	.6842355	0	6.678263
SZBCSAD	4174	1.272493	.693877	0	5.287333
HSICSAD	3610	1.204146	.5875868	0	7.300818
NI225CSAD	3587	1.252892	.5594453	0	4.767389
TOPIXCSAD	3913	1.455059	.6241186	0	6.133608
KOSDAQCSAD	3778	2.414891	.8694664	0	8.036667
KOSPICSAD	4168	2.162403	.9576033	0	10.22663
STICSAD	3679	1.175904	.5900925	0	5.554045
TAIEXCSAD	3779	1.549348	.6564749	0	6.014132
HNXINDEXCSAD	1899	2.222067	.7862742	0	5.575065
VNINDEXCSAD	1879	1.71228	.6571653	0	10.95739

quit high. Moreover, McQueen et al. (1996) and Gleason et al. (2004) discussed that investors may fear potential losses in the downward market price movements more than they enjoy the potential gains in the upward markets, which leads them to be likely to follow herds, the consequence being a reduction in return dispersion. Since the return dispersion (CSAD) tend to be low in most of the cases across our sample, except few markets discussed before, the conclusions of McQueen et al. (1996) and Gleason et al. (2004) also support our findings.

4.2 Results of ADF Unit Root Test

According to the studies on herding behavior, they checked the stationarity of their dependent variable, which is return dispersion (CSAD) (Chang et al. 2000; Lao and Singh 2011; Yao et al. 2014). Thus, this study has also checked the stationarity of the dependent variable. The dependent variables of the empirical models of our study are return dispersion measures of CSAD. Stationarity is checked by Dickey-Fuller test; however, it may create a problem of autocorrelation. Thus, Dickey-Fuller developed a test called Augmented Dickey-Fuller test. Thus, we used Augmented Dickey-Fuller (ADF) test. Results of ADF test rejected the null hypothesis, otherwise it means that the dependent variable of CSAD has no unit root and is stationary.

4.3 Results of the Empirical Models

Having studied about the background of herding behavior in stock markets we learnt that investors make decision differently according to many factors, such as experience, economic circumstances, or even weather situation and so on, which are all external factors. Moreover, several studies highlighted the impacts of some common force on herding behavior, which is identified as cultural aspects. If that is so, the one of the strongest cultures that impacts on not only in individual and social life, but also on business and economic world is the Confucian culture. Thus, we made an assumption that Confucian culture has a positive impact on formation of herding behaviour among investors in stock markets of Confucian countries. We did a comprehensive literature review on topics of herding behavior and Confucian aspects of business and economics, which led us to develop a consequent five hypotheses. These empirical models are designed to test whether the chosen sample of Confucian markets exhibit herding behavior in general and different market states, as well as whether investors of those markets herd during special events. Also, this study tested relationship between behavioral states and exhibition of herding behavior. In addition, as we mentioned before, a robustness regression analysis with Newey-West standard errors (1987) was conducted. Following studies on herding behavior, we present the key values and coefficients of our empirical analysis, such as the total number of observations in each model, adjusted R^2, coefficients of intercept, variables, and test statistics, as well as p-value. In overall, we have gotten quit good results. Thus, this section discusses about the results of our empirical tests through Tables 2, 3, 4, 5 and 6.

Table 2. Results of general herding test (H1)

Markets name (no. of obs.)	Panel A OLS regression				Panel B Newey-west regression		
	α	γ1	γ2	Adjusted R²	α	γ1	γ2
1. SHA (3982)	1.053 (62.56)***	0.564 (29.11)***	-0.053 (-14.91)***	0.249	1.053 (59.32)***	0.564 (27.33)***	-0.053 (-15.09)***
2. SHB (4174)	0.668 (47.19)***	0.479 (34.89)***	-.043 (-22.86)***	0.290	0.668 (48.15)***	0.479 (32.19)***	-0.043 (-22.42)***
3. SZA (3877)	1.008 (65.13)***	0.473 (27.91)***	-0.036 (-11.38)***	0.299	1.01 (56.62)***	0.473 (23.26)***	-0.035 (-8.55)***
4. SZB (4174)	0.839 (57.91)***	0.479 (33.96)***	-0.044 (-21.86)***	0.273	0.839 (55.36)***	0.479 (31.15)***	-0.044 (-21.24)***
5. HKG (3610)	0.873 (74.34)***	0.347 (28.94)***	-0.004 (-2.25)**	0.382	0.872 (58.41)***	0.347 (15.5)***	-0.004 (-0.81)
6. NI225 (3587)	0.953 (74.55)***	0.298 (23.25)***	-0.006 (-2.94)***	0.272	0.953 (66.45)***	0.298 (22.12)***	-0.006 (-3.18)***
7. TOPIX (3913)	1.070 (82.86)***	0.410 (28.94)***	-0.006 (-2.32)**	0.361	1.070 (71.08)***	0.410 (25.95)***	-0.006 (-2.03)**
8. KOSDAQ (3778)	1.845 (100.5)***	0.626 (32.34)***	-0.043 (-14.5)***	0.348	1.845 (75.49)***	0.626 (22.78)***	-0.043 (-8.52)***
9. KOSPI (4168)	1.545 (81.54)***	0.638 (32.36)***	-0.035 (-10.53)***	0.359	1.545 (74.29)***	0.638 (28.84)***	-0.035 (-9.54)***
10. STI (3679)	0.868 (63.5)***	0.404 (20.08)***	-0.010 (-2.17)**	0.278	0.868 (55.27)***	0.404 (14.4)***	-0.010 (-1.21)
11. TAIEX (3779)	1.126 (73.97)***	0.588 (28.6)***	-0.072 (-15.41)***	0.267	1.126 (47.78)***	0.588 (12.81)***	-0.071 (-5.47)***
12. HNX (1899)	1.686 (59.86)***	0.596 (21.41)***	-0.076 (-16.22)***	0.218	1.686 (45.43)***	0.596 (13.47)***	-0.075 (-8.65)***
13. VNINDEX (1879)	1.350 (51.16)***	0.692 (18.36)***	-0.174 (-18.25)***	0.155	1.351 (39.42)***	0.692 (12.54)***	-0.174 (-11.08)***

Notes: Numbers in parentheses are t-statistics, consistent standard errors. ***, ** and * represent statistical significance at the 1%, 5%, and 10% levels, respectively. This table reports results of estimation of the empirical model in the Eq. (5): $CSAD_{i,t} = \alpha + \gamma_1 |R_{i,m,t}| + \gamma_2 (R_{i,m,t})^2 + \varepsilon_{i,t}$.

4.3.1   Results of Measurement Herding Behavior (H1)

Hypothesis 1 assumes that herding behaviour exists among the Confucian markets. Table 2 reports results of the empirical testing the hypothesis 1, which was examined by the model in the Eq. (5). The independent variables of the model consist of market return, while the dependent variable is return dispersion measure of CSAD. According to the definition of the model in Eq. (5), a statistically significant and negative coefficient γ_2 would indicate herding behavior. All sample markets were examined by the model. The empirical model has been tested with OLS regression and regression with Newey-West standard errors (1987). The intercept α shows the location of return dispersion measure at 0 point. The coefficient γ_1 shows an increasing function of market return in absolute terms to the return dispersion measure.

OLS Regression: According to Panel A in Table 2, the coefficients of γ_2 are statistically significant at 1% and negative for all markets, except HSI, TOPIX, and STI. The

Table 3. Results of analysis of herding behaviour during UP markets (H2^{UP})

Markets name (no. of obs.)	Panel A OLS regression				Panel B Newey-west regression		
	α	γ1^{UP}	γ2^{UP}	Adjusted R²	α	γ1^{UP}	γ2^{UP}
1. SHA (2224)	0.916 (41.78)***	0.605 (23.86)***	-0.066 (-14.73)***	0.236	0.916 (39.79)***	0.605 (22.96)***	-0.066 (-13.97)***
2. SHB (2314)	0.560 (30.99)***	0.546 (30.77)***	-0.051 (-21.71)***	0.337	0.560 (31.99)***	0.546 (28.29)***	-0.051 (-21.42)***
3. SZA (2237)	0.862 (45.27)***	0.551 (25.69)***	-0.059 (-14.49)***	0.284	0.862 (38.75)***	0.551 (22.31)***	-0.059 (-11.78)***
4. SZB (2372)	0.700 (37.37)***	0.581 (31.39)***	-0.056 (-22.42)***	0.329	0.700 (36.27)***	0.581 (30.21)***	-0.056 (-24.63)***
5. HKG (1961)	0.774 (46.37)***	0.452 (26.31)***	-0.009 (-4.04)***	0.438	0.774 (38.63)***	0.452 (16.15)***	-0.009 (-1.6)
6. NI225 (1954)	0.838 (46.08)***	0.419 (22.73)***	-0.014 (-4.76)***	0.328	0.838 (40.55)***	0.419 (22.38)***	-0.014 (-6.26)***
7. TOPIX (2112)	0.924 (51.18)***	0.552 (28.04)***	-0.019 (-5.81)***	0.396	0.924 (43.27)***	0.552 (25.45)***	-0.019 (-5.61)***
8. KOSDAQ (2166)	1.673 (63)***	0.752 (24.17)***	-0.059 (-10.89)***	0.316	1.673 (46.14)***	0.752 (16.88)***	-0.059 (-6.03)***
9. KOSPI (2330)	1.396 (53.01)***	0.755 (25.13)***	-0.047 (-8.52)***	0.379	1.396 (46.47)***	0.755 (24.14)***	-0.047 (-8.75)***
10. STI (1992)	0.784 (40.97)***	0.526 (17.84)***	-0.020 (-2.88)***	0.339	0.784 (35.87)***	0.526 (13.17)***	-0.020 (-1.64)
11. TAIEX (2058)	0.960 (46.29)***	0.783 (25.87)***	-0.111 (-15.53)***	0.329	0.960 (39.23)***	0.783 (23.8)***	-0.111 (-12.78)***
12. HNX (962)	1.552 (36.94)***	0.682 (16.74)***	-0.085 (-12.91)***	0.240	1.552 (28.16)***	0.682 (10.35)***	-0.085 (-6.37)***
13. VNINDEX (1006)	1.139 (30.49)***	1.006 (17.91)***	-0.253 (-17.16)***	0.241	1.139 (25.43)***	1.006 (16.64)***	-0.253 (-18.88)***

Notes: Numbers in parentheses are t-statistics, consistent standard errors. ***, ** and * represent statistical significance at the 1%, 5%, and 10% levels, respectively. This table reports results of estimation of the empirical model in the Eq. (6): $CSAD_{i,t}^{UP} = \alpha + \gamma_1^{UP} \left| R_{i,m,t}^{UP} \right| + \gamma_2^{UP} \left(R_{i,m,t}^{UP} \right)^2 + \varepsilon_{i,t}$, when $R_{i,m,t} > 0$

Table 4. Results of analysis of herding behaviour during DOWN markets (H2^{DOWN})

Markets name (no. of obs.)	Panel A OLS regression				Panel B Newey-west regression		
	α	γ1^{DOWN}	γ2^{DOWN}	Adjusted R²	α	γ1^{DOWN}	γ2^{DOWN}
1. SHA (1758)	1.309 (52.62)***	0.391 (13.47)***	-0.015 (-2.81)***	0.272	1.309 (53.61)***	0.391 (11.65)***	-0.015 (-2.39)**
2. SHB (1860)	0.840 (37.72)***	0.356 (16.51)***	-0.027 (-8.68)***	0.227	0.840 (40.19)***	0.356 (15.46)***	-0.027 (-8.32)***
3. SZA (1640)	1.288 (54.13)***	0.29 (11.6)***	-0.000 (-0.01)	0.338	1.288 (58.35)***	0.291 (11.73)***	-0.000 (0.01)
4. SZB (1802)	1.082 (48.68)***	0.282 (12.83)***	-0.017 (-4.91)***	0.198	1.082 (50.06)***	0.282 (11.34)***	-0.017 (-4.39)***
5. HKG (1649)	0.993 (64.14)***	0.235 (14.53)***	0.001 (0.44)	0.322	0.993 (62.06)***	0.235 (10.38)***	0.001 (0.26)
6. NI225 (1633)	1.117 (68.01)***	0.132 (8.09)***	0.01 (3.83)***	0.232	1.117 (70.08)***	0.132 (7.3)***	0.009 (3.12)***
7. TOPIX (1801)	1.29 (74.98)***	0.186 (9.64)***	0.023 (6.74)***	0.360	1.791 (73.55)***	0.186 (7.56)***	0.023 (4.15)***
8. KOSDAQ (1612)	2.105 (92.88)***	0.462 (21.43)***	-0.025 (-8.31)***	0.415	2.105 (69.77)***	0.462 (10.75)***	-0.025 (-3.16)***
9. KOSPI (1838)	1.752 (65.92)***	0.490 (19.29)***	-0.020 (-5.2)***	0.335	1.75 (67.82)***	0.490 (14.51)***	-0.020 (-3.37)***
10. STI (1687)	0.975 (51.71)***	0.263 (9.93)***	0.002 (0.41)	0.210	0.975 (48.97)***	0.263 (7.95)***	0.002 (0.27)
11. TAIEX (1721)	1.332 (61.85)***	0.381 (14.14)***	-0.034 (-5.96)***	0.200	1.332 (50.91)***	0.381 (8.02)***	-0.034 (-2.62)***
12. HNX (937)	1.877 (51.88)***	0.450 (12.14)***	-0.055 (-8.65)***	0.179	1.877 (56.73)***	0.450 (11.03)***	-0.055 (-7.15)***
13. VNINDEX (873)	1.640 (48.09)***	0.312 (6.79)***	-0.087 (-7.72)***	0.064	1.640 (45.91)***	0.312 (4.64)***	-0.087 (-4.57)***

Notes: Numbers in parentheses are t-statistics, consistent standard errors. ***, ** and * represent statistical significance at the 1%, 5%, and 10% levels, respectively. This table reports results of estimation of the empirical model in the Eq. (7): $CSAD_{i,t}^{DOWN} = \alpha + \gamma_1^{DOWN} \left| R_{i,m,t}^{DOWN} \right| + \gamma_2^{DOWN} \left(R_{i,m,t}^{DOWN} \right)^2 + \varepsilon_{i,t}$, when $R_{i,m,t} < 0$.

Table 5. Results of analysis of herding behaviour during HIGH volatility states (H4$^{\text{HIGH VOL}}$)

Markets name (no. of obs.)	Panel A OLS regression				Panel B Newey-west regression		
	α	γ_1^{HIGH}	γ_2^{HIGH}	Adjusted R²	α	γ_1^{HIGH}	γ_2^{HIGH}
1. SHA (1864)	1.119 (40.95)***	0.532 (19.45)***	-0.049 (-11.00)***	0.230	1.119 (40.35)***	0.532 (18.61)***	-0.049 (-11.31)***
2. SHB (1958)	0.735 (31.59)***	0.4486 (22.43)***	-0.040 (-15.72)***	0.263	0.735 (32.16)***	0.448 (21.06)***	-0.040 (-15.55)***
3. SZA (1879)	1.046 (44.88)***	0.425 (18.80)***	-0.028 (-7.67)***	0.307	1.046 (39.25)***	0.425 (15.06)***	-0.028 (-5.39)***
4. SZB (2005)	0.931 (40.37)***	0.446 (22.78)***	-0.041 (-16.21)***	0.247	0.931 (38.77)***	0.446 (21.40)***	-0.041 (-16.14)***
5. HKG (1655)	0.929 (51.52)***	0.297 (18.65)***	0.001 (0.36)	0.429	0.929 (43.26)***	0.297 (10.83)***	0.001 (0.14)
6. NI225 (1600)	1.027 (49.74)***	0.256 (14.55)***	-0.001 (-0.70)	0.301	1.027 (44.58)***	0.256 (13.37)***	-0.002 (-0.75)
7. TOPIX (1787)	1.150 (59.79)***	0.346 (19.15)***	0.001 (0.42)	0.424	1.150 (49.95)***	0.346 (14.86)***	0.001 (0.28)
8. KOSDAQ (1688)	1.85 (65.63)***	0.573 (21.99)***	-0.036 (-9.90)***	0.376	1.852 (49.00)***	0.573 (14.80)***	-0.036 (-5.49)***
9. KOSPI (1921)	1.562 (52.11)***	0.612 (22.72)***	-0.031 (-7.89)***	0.376	1.562 (49.15)***	0.612 (19.67)***	-0.0318 (-6.80)***
10. STI (1717)	0.887 (41.84)***	0.385 (13.86)***	-0.007 (-1.34)	0.313	0.887 (37.66)***	0.385 (10.06)***	-0.008 (-0.77)
11. TAIEX (1720)	1.21 (53.13)***	0.477 (17.50)***	-0.055 (-9.91)***	0.229	1.215 (35.17)***	0.477 (8.14)***	-0.055 (-3.65)***
12. HNX (921)	1.742 (37.30)***	0.607 (14.65)***	-0.081 (-12.17)***	0.199	1.742 (28.52)***	0.607 (10.11)***	-0.081 (-7.48)***
13. VNINDEX (962)	1.379 (32.68)***	0.768 (13.97)***	-0.193 (-14.89)***	0.186	1.379 (25.05)***	0.768 (9.38)***	-0.193 (-9.16)***

Notes: Numbers in parentheses are *t*-statistics, consistent standard errors. ***, ** and * represent statistical significance at the 1%, 5%, and 10% levels, respectively. 30-day moving average of market volatility when volatility is high. This table reports results of estimation of the empirical model in the Eq. (8): $CSAD_{i,t}^{\delta^2-HIGH} = \alpha + \gamma_1^{\delta^2-HIGH}\left|R_{i,m,t}^{\delta^2-HIGH}\right| + \gamma_2^{\delta^2-HIGH}\left(R_{i,m,t}^{\delta^2-HIGH}\right)^2 + \varepsilon_{i,t}$.

Table 6. Results of analysis of herding behaviour during LOW volatility states (H4$^{\text{LOW VOL}}$)

Markets name (no. of obs.)	Panel A OLS regression				Panel B Newey-west regression		
	α	γ_1^{LOW}	γ_2^{LOW}	Adjusted R²	α	γ_1^{LOW}	γ_2^{LOW}
1. SHA (2047)	1.011 (45.57)***	0.603 (18.56)***	-0.0607 (-7.63)***	0.250	1.011 (40.92)***	0.604 (17.54)***	-0.060 (-7.54)***
2. SHB (2145)	0.619 (36.78)***	0.474 (24.05)***	-0.045 (-12.89)***	0.289	0.619 (36.22)***	0.474 (22.08)***	-0.045 (-12.09)***
3. SZA (1927)	0.952 (44.45)***	0.543 (18.60)***	-0.048 (-6.80)***	0.286	0.952 (37.90)***	0.543 (15.33)***	-0.048 (-5.26)***
4. SZB (2098)	0.777 (44.60)***	0.454 (21.84)***	-0.040 (10.75)***	0.266	0.777 (40.53)***	0.454 (19.10)***	-0.040 (-8.88)***
5. HKG (1884)	0.802 (47.28)***	0.436 (16.05)***	-0.024 (-3.30)***	0.303	0.802 (40.89)***	0.436 (12.80)***	-0.024 (-2.24)**
6. NI225 (1916)	0.861 (46.93)***	0.421 (13.65)***	-0.045 (-4.61)***	0.209	0.861 (38.56)***	0.421 (10.89)***	-0.045 (-3.43)***
7. TOPIX (2055)	0.949 (52.51)***	0.608 (17.41)***	-0.088 (-6.67)***	0.279	0.949 (38.65)***	0.608 (12.57)***	-0.088 (-4.56)***
8. KOSDAQ (2019)	1.81 (72.58)***	0.696 (19.11)***	-0.061 (-7.30)***	0.306	1.811 (53.08)***	0.696 (16.34)***	-0.061 (-6.74)***
9. KOSPI (2176)	1.506 (58.89)***	0.682 (18.96)***	-0.044 (-5.13)***	0.325	1.506 (48.48)***	0.682 (15.04)***	-0.044 (-3.79)***
10. STI (1891)	0.846 (45.40)***	0.400 (11.48)***	-0.004 (-0.40)	0.224	0.846 (39.41)***	0.400 (9.08)***	-0.004 (-0.26)
11. TAIEX (1988)	1.027 (48.84)***	0.752 (20.62)***	-0.111 (-10.30)***	0.293	1.027 (42.39)***	0.752 (20.40)***	-0.111 (-12.00)***
12. HNX (907)	1.617 (47.26)***	0.597 (13.73)***	-0.080 (-8.60)***	0.226	1.617 (35.72)***	0.597 (10.26)***	-0.080 (-6.37)***
13. VNINDEX (846)	1.361 (40.32)***	0.583 (19.86)***	-0.149 (-7.84)***	0.112	1.361 (30.51)***	0.583 (8.02)***	-0.149 (-7.42)***

Notes: Numbers in parentheses are *t*-statistics, consistent standard errors. ***, ** and * represent statistical significance at the 1%, 5%, and 10% levels, respectively. 30-day moving average of market volatility when volatility is low. This table reports results of estimation of the empirical model in the Eq. (9): $CSAD_{i,t}^{\delta^2-LOW} = \alpha + \gamma_1^{\delta^2-LOW}\left|R_{i,m,t}^{\delta^2-LOW}\right| + \gamma_2^{\delta^2-LOW}\left(R_{i,m,t}^{\delta^2-LOW}\right)^2 + \varepsilon_{i,t}$.

coefficient for the markets of HSI, TOPIX, and STI are negative and statistically significant at 5%.

The coefficients of t-statistics of γ_2 are in line with the results of p-values; the highest value in t-statistics has SHB market (−22.86), while lowest t-statistics value resulted to STI market (−2.17). The highest adjusted R^2 value resulted to the model of HSI market (.3815), while the model of VNINDEX market (.1548) appears to have the lowest adjusted R^2.

Robustness Regression: Results of the robustness regression with Newey-West standard errors (1987) in Panel B in Table 2 show that the coefficient γ_2 for all markets are negative, however, the p-values for the markets of HSI and STI are not statistically significant. P-values of rest of the markets are statistically significant at 1% and 5% (TOPIX). The coefficients γ_2 of t-statistics of the robustness test are in line with the results of p-values; the highest value in t-statistics has SHB market (−22.42), while lowest t-statistics value resulted to HSI market (−0.81).

Summary and Decision: In overall, according to the results of the model in Panel A and Panel B in Table 2, we accept the hypothesis 1.

Our findings from the testing of hypothesis 1 are consistent with the findings of similar studies, such as Tan et al. (2008), Chiang et al. (2010), Chiang and Zheng (2010), Demirer et al. (2010), Lao and Singh (2011), Bhaduri and Mahapatra (2013), Mobarek et al. (2014), Yao et al. (2014), and Chang and Lin (2015). These studies found some evidence of herding behavior among stock markets investors by using the methodology of Chang et al. (2000), which is shown in the Eq. (3). The half of these studies concentrated in Confucian markets, such as China, Taiwan, and South Korea. While there is another bunch of studies used the model of Chang et al. (2000) in the Eq. (3), however did not find evidence of herding. The half of those studies also targeted on Confucian markets, such as Vietnam and China.

By testing the hypothesis 1 with the empirical model in the Eq. (5), this study found an overwhelming evidence of herding behavior among the Confucian markets. However, previous studies tested herding among investors of Confucian markets, they used to concentrate only on a single country, such as China or Taiwan. However, our study brings all Confucian markets forward, which appears to be one of the contributions of this study to the existing literature of herding.

Discussion: The coefficient γ_2 is negative and statistically significant for all markets. Otherwise, the linear relation between $CSAD_t$ and $|R_{i,m,t}|$ obviously does not hold in our models. In order illustrate the relationship, we take the following general quadratic relationship between $CSAD_t$ and $|R_{i,m,t}|$:

$$CSAD_{i,t} = \alpha + \gamma_1 |R_{i,m,t}| + \gamma_2 (R_{i,m,t})^2 \tag{10}$$

According to the quadratic relationship, $CSAD_t$ reaches its maximum value when $|R_{i,m,t}|^* = -\left(\frac{\gamma_1}{2\gamma_2}\right)$. In other words, when $|R_{i,m,t}|$ increases over the range where realized average daily returns in absolute terms are less than $|R_{i,m,t}|^*$, then $CSAD_t$ is still in an increasing trend. However, as the value of $|R_{i,m,t}|$ exceeds the value of $|R_{i,m,t}|^*$, then return dispersion measure CSAD starts to increase at a decreasing rate. For instance, substituting the estimated coefficients for Taiwan Stock Exchange ($\gamma_1 = 0.5884697$ and $\gamma_2 = -0.0717432$) into the quadratic relation specified in the Eq. (10) indicates that $CSAD_t$ reaches a maximum when $|R_{i,m,t}| = |R_{i,m,t}|^* = 4.10\%$. This suggests that during large price movements in the market return that surpass the threshold level $|R_{i,m,t}|^*$, the $CSAD_t$ increases at decreasing rate as illustrated in the Fig. 1 (Chang et al. 2000).

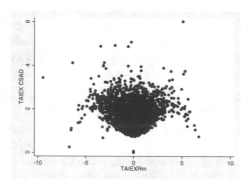

Fig. 1. Relationship between the daily return dispersion ($CSAD_{i,t}$) and the corresponding equally-weighted market return ($R_{i,m,t}$) for TAIEX of Taiwan Stock Exchange.

Furthermore, besides statistical significance and negativity of the coefficient γ_2, the size of the coefficient also shows magnitude of herding behavior in each market (Lao and Singh 2011). In this case, size of the coefficient γ_2 is larger in markets of TAIEX (−0.072), HNX (−0.076), and VNINDEX (−0.174). In contrast, size of the coefficient is smaller in markets of STI (−0.01), NI225/TOPIX (−0.006), and HSI (−0.004). The differences in size of the coefficient show that herding behavior is more prevalent in the markets of TAIEX, HNX, and VNINDEX, which are emerging markets. While, the advanced markets of Singapore, Japan, and Hong Kong display much less herding behavior.

4.3.2 Results of Measurement of Herding Behavior During Up and Down Markets (H2)

Hypothesis 2 assumes that asymmetric herding behaviour exists among the Confucian markets when the market rises and falls. Tables 3 and 4 reports results of the empirical testing the hypothesis 2, which was examined by the models in the Eq. (6), which is for up markets and Eq. (7), which is for down markets. According to the definition of the models in Eqs. (6) and (7), statistically significant and negative coefficients γ_2^{UP} and γ_2^{DOWN} would indicate herding behavior in up and down markets. All sample markets, which are totally 13 market, were examined by the models. The empirical models have been tested with OLS regression and regression with Newey-West standard errors

(1987). The intercept α shows the location of return dispersion measure at 0 point. The coefficient γ_1 shows an increasing function of market return in absolute terms to the return dispersion measure.

OLS Regression (UP MARKET): According to Panel A in Table 3, the coefficients of γ_2 for the whole sample during up markets are statistically significant at 1% and negative. The highest value of the adjusted R^2 during up markets is for HSI market (.4388), while the lowest value for it during up markets resulted to SHA market (.2356). The coefficients of t-statistics of γ_2 are in line with the results of p-values; the highest value in t-statistics during up markets resulted to SZB market (-22.42), while lowest value is for STI market (-2.88).

OLS Regression (DOWN MARKET): According to Panel A in Table 4, the coefficient γ_2 during down markets is statistically significant at 1% and negative for eight markets out of 13 markets. The coefficient γ_2 of markets of STI, TOPIX, NI225, and HSI and positive, as well in most of the cases statistically not significant. However the coefficient γ_2 for SZA market is negative, it is not statistically significant.

The highest value of the adjusted R^2 during down markets is for KOSDAQ market (.4146), while the lowest value for adjusted R^2 is for VNINDEX market (.0642). The coefficients of t-statistics of γ_2 are in line with the results of p-values, the highest value in t-statistics during down markets resulted to HNX market (-8.65).

Robustness Regression (UP MARKET): Results of robustness regression in Panel B in Table 6 show that coefficients γ_2 during up markets of 11 indices out of 13 markets are statistically significant at 1% and negative. The coefficients of markets of HSI and STI are negative but statistically not significant.

The highest value of t-statistics for the coefficient γ_2 during up markets has SZB (-24.63), while lowest value resulted to HSI market (-1.60).

Robustness Regression (DOWN MARKET): According to Panel B in Table 4, the coefficients γ_2 during down markets are statistically significant at 1% and negative for seven markets, while the coefficient for SHA is statistically significant at 5%. The coefficients of rest of the markets (SZA, HSI, NI225, TOPIX, STI) are either statistically not significant or positive.

The highest value of t-statistics for the coefficient γ_2 during down markets resulted to SHB market (-8.32), while lowest value resulted to SZA (-0.01).

Summary and Decision: In overall, according the results of the main regression and robustness regression analysis, investors more exhibit herding behavior during up markets than down markets. However, over 50% of down markets also displays herding behavior in both regression tests. Thus, we accept the hypothesis 2.

Our results contradict findings of others studies, which discussed that herding behavior is more prevalent during down markets (McQueen et al. 1996; Chang et al. 2000; Gleason et al. 2004; Fu and Lin 2010; Lao and Singh 2011; Yao et al. 2014; Mobarek et al. 2014). However, the results of hypothesis 2 examination are consistent with the findings of studies of Tan et al. (2008), Chiang and Zheng (2010), My and Truong (2011), and Bhaduri and Mahapatra (2013), which found that herding behavior more occurs during up market states. The studies, whose results are in line with our findings, used data from Asian markets.

4.3.3   Results of Measurement of Herding Behavior During High and Low Volatility States (H3)

Hypothesis 3 postulates that asymmetric herding behaviour exists among the Confucian markets during high and low volatility states of markets. Tables 5 and 6 report results of the empirical testing the hypothesis 3, which were examined by the empirical models in the Eq. (8), which is for high volatility states and Eq. (9), which is for low volatility states of sample markets. According to the definition of the empirical models in Eqs. (8) and (9), statistically significant and negative coefficients $\gamma_2^{\delta^2-HIGH}$ and $\gamma_2^{\delta^2-LOW}$ would indicate herding behavior during high and low volatility states. All sample markets were examined by these models. The empirical models have been tested with OLS regression and regression with Newey-West standard errors (1987). The intercept α shows the location of return dispersion measure at 0 point. The coefficient γ_1 shows an increasing function of market return in absolute terms to the return dispersion measure.

OLS Regression (HIGH VOLATILITY): According to Panel A in Table 5, the coefficients of $\gamma_2^{\delta^2-HIGH}$ for all markets, except the markets of HSI, NI225, TOPIX, and STI, are statistically significant at 1% and negative. Which means herding behavior exist during high volatility states in nine markets out of 13 markets. The coefficient for non-linearity for the markets of HSI, NI225, TOPIX, and STI are either statistically not significant or positive, which means they do not exhibit herding behavior during high volatility states.

The highest value of adjusted R^2 during the high volatility states resulted to the model of HSI market (0.4294), while lowest value got to the model of VNINDEX (0.1863) market. Moreover, the highest value in t-statistics during the high volatility states resulted to the model of SZB market (-16.21), while the lowest value got to the model of HSI market (0.36).

This study examined the Eq. (8) with moving averages of 7, 30, and 90 days, however Table 5 reports only the moving averages of 30 days. The results of 7 and 90 day moving averages are consistent with results of the coefficients of 30-day moving average.

OLS Regression (LOW VOLATILITY): According to Panel A in Table 6, the coefficient $\gamma_2^{\delta^2-LOW}$ of the empirical model in the Eq. (9) is statistically significant at 1% and negative for all sample markets, except market of STI. The coefficient for STI is negative but not statistically significant. This results means that herding behavior exist among the sample markets of this study during low volatility states of market.

The highest value in t-statistics during the low volatility states resulted to SHB market (-12.89), while the lowest value resulted to STI market (-0.40). The highest value in adjusted R^2 during the low volatility states resulted to the model of KOSPI market (0.3250), while the lowest value resulted to the model of VNINDEX market (0.1117).

This study examined the Eq. (9) with moving averages of 7, 30, and 90 days, however Table 6 reports only the moving averages of 30 days. The results of 7 and 90 day moving averages are consistent with results of the coefficients of 30-day moving average.

Robustness Regression (HIGH-VOLATILITY): According to Panel B in Table 5, the coefficient $\gamma_2^{\delta^2-HIGH}$ is statistically significant and negative for all markets, except markets of HSI, NI225, TOPIX, and STI. The coefficient $\gamma_2^{\delta^2-HIGH}$ for the markets of HSI, NI225, TOPIX, and STI are statistically not significant and positive.

The highest value in t-statistics resulted to SZB market (-16.14), while the lowest value is for HSI market (0.14). In overall, nine markets out of 13 markets exhibit herding behavior during high volatility states.

This study examined the Eq. (8) with moving averages of 7, 30, and 90 days, however Table 10 reports only the moving averages of 30 days. The results of 7 and 90 day moving averages are consistent with results of the coefficients of 30-day moving average.

Robustness Regression (LOW-VOLATILITY): According to Panel B in Table 6, the coefficient $\gamma_2^{\delta^2-LOW}$ of the empirical model in the Eq. (9) is statistically significant and negative for all sample markets, except market of STI. The only coefficient for non-linearity of HSI market is statistically significant at 5%, however the coefficients for rest of the markets are statistically significant at 1%. The coefficient for STI is negative but statistically not significant. The results means that herding behavior exist among the sample markets of this study during low volatility states of market.

The highest value in t-statistics during the low volatility states resulted to SHB market (-12.09), while the lowest value resulted to STI market (-0.26).

This study examined the Eq. (9) with moving averages of 7, 30, and 90 days, however Table 6 reports only the moving averages of 30 days. The results of 7 and 90 day moving averages are consistent with results of the coefficients of 30-day moving average.

Summary and Decision: According to the results in Tables 10 and 11, nine markets exhibit herding behavior out of 13 markets when the market volatility is high, while 12 markets display herding behavior out of 13 markets when market volatility is low. And, these outcomes are exactly in line with the results of robustness regression analysis. In orders words, more than 69.2% of the sample of this study exhibit herding behavior during periods of high and low volatility states of market, however herding is more prevalent when the market volatility is low. Thus, we accept the hypothesis 4.

Discussion on the Results of Asymmetric Herding Tests: This section has discussed about results of hypotheses 2 and 3, which were correspondingly tested by the empirical models in the Eqs. (6) and (7) and Eqs. (8) and (9). Generally, we found herding behavior overwhelmingly from all sample markets, in all specifications of our empirical models. However, if we look the results more in detail we can see that herding behavior is more profound: in up markets than down markets; herding behavior is more prevalent during low trading volume states than high trading volume states; and herding behavior is stronger during low volatility states than high volatility states.

These results in contradiction with findings of some studies, while they are in agreement of others studies as discussed below.

Some studies that examined asymmetric herding behavior found herding in both of rising and falling markets. However, they concluded that herding is more prevalent during declining markets, high trading volume states and high volatility states (McQueen et al. 1996; Chang et al. 2000; Gleason et al. 2004; Tan et al. 2008; Fu and Lin 2010; Lao and Singh 2011; Mobarek et al. 2014; Yao et al. 2014).

While, studies also found contradicting results. For example, while they found herding in falling markets Tan et al. (2008) also found herding from rising markets. Chiang and Zheng (2010) found that herding asymmetry is more profound in Asian markets (Hong Kong, Japan, China, South Korea, Indonesia, Malaysia, Singapore, Thailand, and Taiwan) during rising market states. My and Truong (2011) found that investment behaviour in the stock market of Vietnam was more uniformly in up markets. Bhaduri and Mahapatra (2013) found herding is more prevalent during up markets in Indian stock market. Mobarek et al. (2014) measured asymmetric herding behavior among 11 European markets by using dummy variable in detecting herding during up and down markets, high and low trading volume states, and high and low market volatility states. Moreover, they found that some evidence of a significant herding effect during low volume trading periods.

We also make a brief conclusion as follows. First, investors of sample markets found to exhibit more uniform behavior during up markets because up market is a bull market. During bull-market investors are more optimistic in trading. Bull market and investor optimism do not mean that investor make decision solely on they own. When investors too optimistic they also tend to exhibit risky behavior by doing risky investment decisions, such as buying too many stock either based on they own decision or based on decisions made by other traders (Kurov 2010). Thus, bull market states can also be a stepping-stone of possible herding behavior, which has been shown by empirical results of this study.

Second, investors of sample markets of this study found to display more herding behavior during low trading volume states than high trading volume states. According to studies in market sentiments, one of the common features of falling markets is low trading volume state. When the market is falling and the trading volume is low, there is a great uncertainty in the market. Because trading volume is falling the investors in the market would think that the market is also falling, thus they would probably decide to convert shares into cash by selling them out as soon as possible (Jansen and Tsai 2010). Therefore, this kind of investment sentiments make the fear on the market more contagious, finally end up with formation of herding behavior when the trading volume gets low.

Finally, investors of sample markets of this study exhibit more herding behavior during low volatility states of market. Low volatility state of a market is similar condition to up or rising markets. In order words, these features of a market are kind of positive and pleasant market condition for investors. As discussed before, when investors feel positive about the market they are more willing to invest more, even they are ready to make consensus decisions with others, in results of what herding behavior will happen.

5   Conclusion

The empirically tested hypothesis 1 assumed that a herding behaviour exists among the Confucian markets. The empirical test was conducted on all markets in our sample. According to our empirical findings, the hypothesis 1 was accepted. We found a significant herding behaviour from the entire sample population, no matter whether they are emerging or advance markets. Thus, our findings contradict the existing theory on herding, which discusses that herding behavior hardly exist among advanced markets, while support the rationale of this study that Confucian culture has a positive influence on formation of herding behavior.

The empirically tested hypotheses 2 and 3 correspondingly assumed that, first, an asymmetric herding behaviour exists among the Confucian markets during (H_2) the rising and falling markets and (H_3) during high and low volatility states of markets. The empirical tests have been conducted on whole sample markets. According to our empirical findings the hypotheses 2 and 3 are accepted. The details of our findings are broken down in the next paragraphs.

Generally, this study detected a significant herding behaviour from the entire sample in each state of market, where they are specified with high and low returns and volatility. However, detail results show that herding behaviour is more profound in (1) up markets than down markets; and (2) during low volatility states than high volatility states.

We try to deliberate the findings a bit more here. First, investors of sample markets display more uniform behaviour in up markets, because up market is a bull market. During bull markets, investors are more forward looking and optimistic in trading. Investors' optimism in bull markets does not mean that there is no herding behaviour. When investors get to be too optimistic, they also tend to make risky investment decisions, such as buying high, either based on their own decision or based on decisions made by other traders. Thus, bull market states have high probability to have a positive effect on formation of herding behaviour, which has been eventually demonstrated by the empirical findings of this study.

Second, we found that investors exhibit more herding behavior in low volatility states of market. When volatility is low, a market rather heads upward, which means it has similar features with a bull market. Otherwise, an up market exhibits positive and pleasant conditions for investors. Therefore, as investors feel positive about the market, they become more willing to make more investment, and even sometimes they join into a group decision, in results of what herding behavior happens.

References

Aggarwal, R., Kearney, C., Lucey, B.: Gravity and culture in foreign portfolio investment. J. Bank. Fin. **36**(2), 525–538 (2012)

Beckmann, D., Menkhoff, L., Suto, M.: Does culture influence asset managers' views and behavior? J. Econ. Behav. Organ. **67**(3), 624–643 (2008)

Bhaduri, S.N., Mahapatra, S.D.: Applying an alternative test of herding behavior: a case study of the Indian stock market. J. Asian Econ. **25**, 43–52 (2013)

Chang, E.C., Cheng, J.W., Khorana, A.: An examination of herd behavior in equity markets: an international perspective. J. Bank. Fin. **24**(10), 1651–1679 (2000)

Chang, C.H., Lin, S.J.: The effects of national culture and behavioral pitfalls on investors' decision-making: herding behavior in international stock markets. Int. Rev. Econ. Fin. **37**, 380–392 (2015)

Chiang, T.C., Zheng, D.: An empirical analysis of herd behavior in global stock markets. J. Bank. Fin. **34**(8), 1911–1921 (2010)

Chiang, T.C., Li, J., Tan, L.: Empirical investigation of herding behavior in Chinese stock markets: evidence from quantile regression analysis. Glob. Fin. J. **21**(1), 111–124 (2010)

Chow, C.W., Shields, M.D., Chan, Y.K.: The effects of management controls and national culture on manufacturing performance: an experimental investigation. Account. Organ. Soc. **16**(3), 209–226 (1991)

Christie, W.G., Huang, R.D.: Following the pied piper: do individual returns herd around the market? Finan. Anal. J. **51**(4), 31–37 (1995)

Demirer, R., Kutan, A.M.: Does herding behavior exist in Chinese stock markets? J. Int. Finan. Mark. Inst. Money **16**(2), 123–142 (2006)

Demirer, R., Kutan, A.M., Chen, C.D.: Do investors herd in emerging stock markets? Evidence from the Taiwanese market. J. Econ. Behav. Organ. **76**(2), 283–295 (2010)

Economou, F., Kostakis, A., Philippas, N.: Cross-country effects in herding behaviour: evidence from four south European markets. J. Int. Finan. Mark. Inst. Money **21**(3), 443–460 (2011)

Ferris, S.P., Jayaraman, N., Sabherwal, S.: CEO overconfidence and international merger and acquisition activity. J. Finan. Quant. Anal. **48**(01), 137–164 (2013)

Fu, T., Lin, M.: Herding in China equity market. Int. J. Econ. Finan. **2**(2), 148 (2010)

Gleason, K.C., Mathur, I., Peterson, M.A.: Analysis of intraday herding behavior among the sector ETFs. J. Empir. Finan. **11**(5), 681–694 (2004)

Holderness, C.G.: The myth of diffuse ownership in the United States. Rev. Finan. Stud. **22**(4), 1377–1408 (2009)

Hofstede, G.H.: Culture's Consequences: Comparing Values, Behaviors. Institutions and Organizations Across Nations. Sage, Thousand Oaks (2001)

Hwang, S., Salmon, M.: Market stress and herding. J. Empir. Finan. **11**(4), 585–616 (2004)

Jansen, D.W., Tsai, C.L.: Monetary policy and stock returns: financing constraints and asymmetries in bull and bear markets. J. Empir. Finan. **17**(5), 981–990 (2010)

Kurov, A.: Investor sentiment and the stock market's reaction to monetary policy. J. Bank. Finan. **34**(1), 139–149 (2010)

Lao, P., Singh, H.: Herding behaviour in the Chinese and Indian stock markets. J. Asian Econ. **22**(6), 495–506 (2011)

Li, K., Griffin, D., Yue, H., Zhao, L.: How does culture influence corporate risk-taking? J. Corp. Finan. **23**, 1–22 (2013)

McQueen, G., Pinegar, M., Thorley, S.: Delayed reaction to good news and the cross-autocorrelation of portfolio returns. J. Finan. **51**(3), 889–919 (1996)

Mobarek, A., Mollah, S., Keasey, K.: A cross-country analysis of herd behavior in Europe. J. Int. Finan. Mark. Inst. Money **32**, 107–127 (2014)

My, T.N., Truong, H.H.: Herding behavior in an emerging stock market: empirical evidence from Vietnam. Res. J. Bus. Manag. **5**(2), 51–76 (2011)

Newey, W.K., West, K.: A simple positive semi-definite, heteroskedasticity and autocorrelation consistent covariance matrix. Econometrica **55**, 703–708 (1987)

Siegel, J.I., Licht, A.N., Schwartz, S.H.: Egalitarianism and international investment. J. Finan. Econ. **102**(3), 621–642 (2011)

Tan, L., Chiang, T.C., Mason, J.R., Nelling, E.: Herding behavior in Chinese stock markets: an examination of A and B shares. Pacific-Basin Finan. J. **16**(1), 61–77 (2008)

Treynor, J., Mazuy, K.: Can mutual funds outguess the market? Harvard Bus. Rev. **44**(4), 131–136 (1966)

Xue, M., Cheng, W.: National culture, market condition and market share of foreign bank. Econ. Model. **33**, 991–997 (2013)

Yao, J., Ma, C., He, W.P.: Investor herding behaviour of Chinese stock market. Int. Rev. Econ. Finan. **29**, 12–29 (2014)

Forecasting the Growth of Total Debt Service Ratio with ARIMA and State Space Model

Kobpongkit Navapan[1], Petchaluck Boonyakunakorn[1(✉)],
and Songsak Sriboonchitta[1,2]

[1] Faculty of Economics, Chiang Mai University, Chiang Mai 52000, Thailand
kobpongkit.nav@gmail.com, petchaluckecon@gmail.com
[2] Center of Excellence in Econometrics, Chiang Mai University,
Chiang Mai 52000, Thailand
songsakecon@gmail.com

Abstract. Since the global financial crisis erupted in September 2008, many recent economists have been worried about the health of financial institutions. Consequently, many recent researches have put great emphasis on study of total debt service ratio (TDS) as one of the early warning indicators for financial crises. Accurate TDS forecasting can have a huge impact on effective financial management as a country can monitor the signal of financial crisis from a TDS's future trend. Therefore, the purpose of this paper is to find the modeling to forecast the growth of TDS. Autoregressive integrated moving average (ARIMA) models tends to be the most popular forecasting method with indispensable requirement of data stationarity. Meanwhile, State Space model (SSM) allows us to examine directly from original data without any data transformation for stationarity. Furthermore, it can model both structural changes or sudden jumps. The empirical result shows that the SSM expresses lower prediction errors with respect to RMSE and MAE in comparison with ARIMA.

Keywords: Total debt service ratio
Autoregressive integrated moving average model · State Space Model

1 Introduction

The global financial crisis in September 2008 is recognized as one of the worst financial crises since the Great Depression of the 1930s. It has pushed economies around the world into recession. Lots of concerns have been placed on studies of household debt level, as household debts can be related to loans that financial institutions lent to the individual in a country. The individual may use the loans with a variety of reasons, such as personal consumption or business purposes. Subsequently, an increased income would not only reflect to the individual's abilities of loan repayments, but also contribute to economic growth through loans. However, if excessive private sector debt remains for a long time, it may adversely affect economic stability.

© Springer International Publishing AG 2018
V. Kreinovich et al. (eds.), *Predictive Econometrics and Big Data*, Studies in Computational Intelligence 753, https://doi.org/10.1007/978-3-319-70942-0_35

One of the useful indicators for growing accretion of finance's vulnerabilities in recent economy is the total debt service ratio (TDS), which is fundamentally known as a key player of measurement of the financial limitations caused by private sector liability. Analyzing TDS time series intensively generates many advantages to both financial and economic forecasting, especially as an early warning indicator (EWI) for banking crisis. Recently there have been a lot of literature on EWIs of economic crisis, but there are a few of them focusing on accurate prediction ability. This paper is to fulfill the gap in this literature field.

Drehmann and Juselius [1] stated that the debt service ratio (DSR) can be used as a very useful EWI to forecast forthcoming systemic banking crises within one or two years in advance. Accordingly, the government agency or the central bank ideally uses total debt service ratio (TDS) to supervise financial institutions to enforce or seek cooperation in case not to let financial institutions to make loans to borrowers. An advantages of DSR in comparison with other leverage measures like debt-to-GDP ratio is to provide more accurate prediction, because it explicitly captures significant movements, specially interest rate movement of maturities that undermine repayment ability. Moreover, it was found that time series analysis of TDS prior to economic crashes can generate some information which is related to the size of the subsequent output losses and this characteristic can be used as an important indicator to warn for impending systemic banking crises. Drehmann and Juselius [2] showed that both the credit-to-GDP and DSR consistently outperform other measure, whereas the first one tends to be the best indicator for longer period whereas, the second dominates shorter period.

Furthermore Kida [9]'s study also showed that a decrease in TDS can be contributed by an increase in income. Relatively, if the TDS ratio is high, it can be referred to the state that a country's international finances are not healthy. The ratio should be at the range of 0 to 20%. One of the case study examples carried out by Faruqui [6] referred what happened in Canada. It was reported in the Bank of Canada's Financial System Review that the TDS continued rising but still below its historical average in the fourth quarter in 2007 suggesting that households' debt burden is still considered at manageable levels.

Moreover, a high level of TDS can cause a significantly negative impact on consumption and investment, proposed by Bank for International Settlements (BIS). There are many countries in Asia, for example Malaysia, and Hong Kong, where TDS is used as a financial tool to solve the debt problem. In Malaysia, it is used to deal with household or bubble in the economy with respect to vulnerable borrowers. Whereas it is used to deal with Hong Kong's bubble in real estate prices.

The measure of DSR is basically defined as interest payments and debt repayments divided by income. With this formula it provides advantageous ability to define factors such as changes in interest rate or maturities that affect borrowers' repayment capacity. Consequently, this advantage makes the DSR better than other established leverage measures, such as the debt-to-GDP ratio (Drehmann and Juselius [3]).

The DSR's properties have been explored to constructed TDS at the aggregate level with a basic idea of the measure that debt service cost—interest and amortizations—on the aggregate debt stock are repaid at a given lending rate in equal portions over the maturity of the loan (installment loans). The justification for the change defines that the differences between the repayment structures of individual loans will be canceled in the aggregate.

After exploring the basic formula for calculating the fixed debt service costs (DSC) of an installment loan and dividing it by income Y_{jt}, TDS formula is mathematically written as in the following equation. Furthermore, it can be interpreted into terms at aggregate level with the TDS for sector j at time t and it is expressed as

$$TDS_{jt} = \frac{DSC_t}{Y_t} = \frac{i_{jt}}{\left(1 - (1 + i_{jt})^{-s_{jt}}\right)} * \frac{D_{jt}}{Y_{jt}}, \tag{1}$$

where D_{jt} is the total stock of debt, i_{jt} indicates the average interest rate on the existing stock of debt per quarter, Y_{jt} indicates quarterly income, and s_{jt} indicates the average remaining maturity in quarters.

Accurate TDS ratio prediction tends to have a significant impact on effective financial management, which relatively contribute to economic stability. A signal of financial crisis from a TDS's future trend would be one of the supplementary information for impending financial crises. Therefore, the forecasting of the growth of DSR ratio should be paid more attention.

Among of forecasting models, we begins with an autoregressive moving average (ARMA) models for investigating the time series data, as firstly introduced by Box and Jenkins. ARMA only requires historical time series data to calculate along with an basic assumption of stationarility in data, which is that all mean, variance, and covariance should be constant over time.

Another point is that dynamic linear models (DLMs) as known as linear State Space Models (SSM) can be suitable for modeling with a wide range of sequence univariate or multivariate time series data. It provides more flexibilities than ARIMA in term of treating non-stationary data with instability in the mean and variance, model structural changes or sudden jumps. Additionally, SSM is easy to interpret and understand the results (Petris et al. [13]). SSM allows us to investigate directly from time series data. This is more natural to model time series data without its preliminary transformation.

As stated by Jacques and Siem [7], it was demonstrated that the ARIMA approach consists of a problem as the stationary of real time series data is hardly exist, especially in fields of economics, finance and social. Therefore, they suggest using SSM instead of ARIMA since SSM is never required stationarity time series data. SSM provides a powerful probabilistic structure, offering a flexible framework for many applications. Subsequently, the application of SSM in time series analysis has been widely increased, for instance, the studies of Durbin and Koopman [4] and Migon et al. [10].

ARMA can be represented as SSM. That may be usefully regarded the benefits of dynamic linear models. A recent study carried out by Omekara et al. [11] provides a support that SSM provides more adequate than ARIMA approaches. Therefore, this paper we will compare the predicted ability of ARIMA with SSM with respect to the growth of DSR. The following is divided into 5 sections. The second section focuses on methodology, meanwhile the third section describes the studied data. Empirical results are in the forth section and conclusion is in the last section.

2   Methodology

2.1   Autoregressive Integrated Moving Average (ARIMA) Model

It is common to find dependence in time series data. The primary objective of ARIMA model is to model this dependence and investigate univariate stochastic time series data. It is widely applied for forecasting since it offers great flexibility in forecasting of time series with many supportive evidences from academic studies. Moreover, it is a convenience methodology in terms of using only historical data.

ARMA model be composed of the Autoregressive (AR) and Moving average (MA) models. The AR refers to the level of its current observations rely on the level of its lagged observations. In time series data, the MA refers to the observations of a random variable at current time t can be influenced by both the shock at time t, and the shocks occurred before the time. The stationary time series Y_t is ARMA(p, q), which can be written as;

$$Y_t = u + \sum_{j=1}^{p} \phi_j Y_{t-j} + \sum_{j=1}^{q} \psi_j \varepsilon_{t-j} + \varepsilon_t \tag{2}$$

where ε_t is white noise process with variance σ^2 and the parameter $\phi_1, ..., \phi_p$ satisfy a stationary condition. ARMA model requires a stationary time series, which indicates that the mean, variance, and covariance should be constant over time. Otherwise it will result in spurious regression leading to biased results. The integrated autoregressive moving average (ARIMA) model will be applicable if the time series is non stationary. Therefore, the time series Y_t will be ARIMA (p, d, q) model. The time series is different stationary, where d the number of times is applied to difference a process before it is stationary. We use Autocorrelation function (ACF) and Partial Autocorrelation function (PACF) to guide us the lag of p and q. The best fitted model is confirmed AIC criterion.

2.2   The State Space Model (SSM)

Usually SSM is referred to probabilistic structure model describing the dependence relationship between the latent variable and the studied observations in terms of probabilities whether the data can be in either continuous or discrete forms (Kalman [8]). Kalman filter refers to an optimal algorithm for inferring linear Gaussian systems. Its primary objective is to draw inference from the observations to the relevant properties of unobserved series of vectors . Other objectives are to estimate of parameters and to forecast.

For a univariate time series y_t, we begin with a very simple model called random walk with noise model (or local level model) which is at this state expressed as

$$y_t = \theta_t + v_t, \qquad v_t \sim N(0, V), \tag{3}$$

$$\theta_{t+1} = \theta_t + w_t, \qquad w_t \sim N(0, W), \tag{4}$$

where θ_t denotes the unobserved level at time t, whereas v_t denotes the observation disturbance at time $t = (1, ..., n)$. The only parameters of the model are the observation and evolution variances V and W. Distributions v_t and w_t are assumed to be mutually independent. The Eq. (3) is called the observation equation or measurement equation. The Eq. (4) is called the state equation.

2.2.1   State Space Model Representation of ARIMA Models

The basic ARIMA (p, q) model as shown in (2) can be written the ARIMA models in state space form. Now we assume that u is typically denoted as zero, so the defining relation is expressed as

$$Y_t = \sum_{j=1}^{r} \phi_j Y_{t-j} + \sum_{j=1}^{r-1} \psi_j \varepsilon_{t-j} + \varepsilon_t, \tag{5}$$

where $r = max(p, q+1)$, $\phi_j = 0$ for $j > p$ and $\psi_j = 0$ for $j > q$. The ARIMA (p, q) models in state space form can be described by the matrices which can be showed as

$$F = \begin{bmatrix} 1\ 0 \dots 0 \end{bmatrix}, \ G = \begin{bmatrix} \phi_1 & 1 & 0 & \dots & 0 \\ \phi_2 & 0 & 1 & \dots & 0 \\ \vdots & \vdots & & \ddots & \\ \phi_{r-1} & 0 & \dots & 0 & 1 \\ \phi_r & 0 & \dots & 0 & 0 \end{bmatrix}, \ R = \begin{bmatrix} 1\ \psi_1 \dots \psi_{r-2}\ \psi_{r-1} \end{bmatrix} \tag{6}$$

After a r-dimensional state vector $\theta_t = (\theta_{1,t}, ..., \theta_{r,t})$ is defined, then the ARMA model can be referred to the following SSM representation:

$$Y_t = F\theta_t \tag{7}$$

$$\theta_{t+1} = G\theta_t + R\varepsilon_t \tag{8}$$

The two equations together in (7) and (8) are called dynamic linear models (DLMs) as known as linear SSM, where $V = 0$ and $W = RR'\sigma^2$ are contained. Whereas σ^2 represents the variance of the error sequence (ε_t). In this study, we use maximum likelihood to estimate dynamic system (Petris et al. [13]).

2.3 Evaluation of Forecasting Model Performance

To evaluate the forecasting model performance, we employ two benchmarks which are Root mean square error (RMSE) and Mean absolute error (MAE) for checking forecasting ability between the two models.

RMSE benchmark evaluates the forecasting model performance by measuring the average magnitude of the error. It takes the square root of squared the difference between actual observations and forecasting values. In other words, it is to measure the differences between actual values and predicted values. RMSE can be represented as;

$$RMSE = \sqrt{\frac{1}{n} \sum_{i=1}^{n} (y_j - \hat{y}_j)^2}, \tag{9}$$

where, y is the predicted value of time t, and $\hat{y}$ is the actual value.

MAE is also to measure the average magnitude of the errors. With this method, it principally reflects the magnitude and avoids the problem of positive and negative forecast errors offsetting one another. Consequently, MAE is computed as follows;

$$MAE = \frac{1}{n} \sum_{i=1}^{n} |y_i - \hat{y}_i| \tag{10}$$

The smallest values in each model generated by both RMSE and MAE are preferred since the smallest values denote the best forecasting ability of that model.

3 Data

The time series data sets of TDS from quarter 1 in 2001 to quarter 3 in 2016. There are 63 observations. The data is received from Bank for international settlements (BIS) database. The data is also divided into 2 periods due to a measure of predicted ability with RMSE and MAE. The first period known as in sample, covers a period from quarter 1 in 2001 to quarter 4 in 2014 with 56 observations in total and the second period known as out of sample covers a period from the latter two years with 7 observation in total.

4 Empirical Results

Using natural logarithms to summarize changes in terms of continuous compounding can be meaningful over looking a simple percent changes, however the graph on the left hand side in Fig. 1 indicates that the graph of log of TDS clearly is a non stationary during the studied period. Therefore, we transform the data to be the growth of TDS. The left hand graph in Fig. 1 shows the data with logarithm transformation and it clearly that the growth of TDS in the right hand side becomes more stationary compared to the graph of the log of TDS.

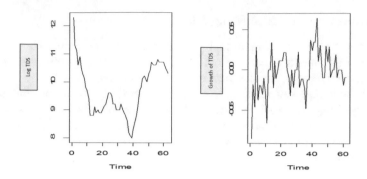

Fig. 1. Log of TDS and the growth of TDS

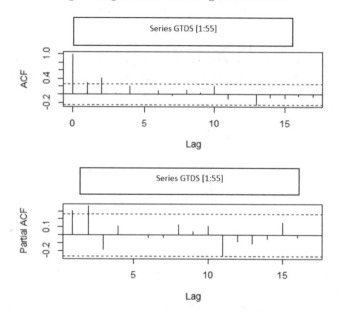

Fig. 2. Plot the ACF/PACF of the growth of TDS

Then we check the stationary of the growth of TDS by using the Augmented Dickey-Fuller (ADF), the result accepts hypothesis the time series are non stationary. Therefore, it becomes essential to first difference the growth of DSR. Then it becomes stationary time series data. Then we identify the numbers of p and q through plotting of ACF and PACF as shown in Fig. 2.

The Partial ACF plot shown in Fig. 2 has a significant cut off after lag 1 as pointed out by an AR (1) model. Therefore, the ARIMA (1,1,0) tends to be the first candidate models. Then ARIMA (1, 1, 0) is tested for fitting with variations including ARIMA (1, 1, 1), ARIMA (2, 1, 0), ARIMA (2, 1, 1) with respect to lowest numbers of lag p and lag q based on AIC criteria. The result show ARIMA (1, 1, 0) provides the lowest of AIC.

The SSM with representation of ARMA models, we also selected the best model ARMA (p, q) based on AIC criteria. ARMA (1, 1) in SS form provides the lowest value AIC.

In order to compare ARIMA (1, 1, 0) with ARMA(1, 1) in SS form, it is clearly that the lag of growth DSR are significant negatively related to present growth DSR as shown in Table 1.

Table 2 shows that the RMSE and MAE values of the two model are significantly different. The RMSE and MAE values of the ARIMA model are 00191 and 0.0181 respectively. Meanwhile, the smallest RMSE and MAE values, 0.0095 and 0.0099, belong to the SS model. It concludes that the SS model has the better forecasting performance with respect to the minimum values of RMSE and MAE.

Table 1. Estimated of ARIMA model and SS model

Model	ARIMA (1,1,0)		ARMA (1,1) in SS form	
	Coefficients	Std.Error	Coefficients	Std.Error
Intercept	–	–	−0.809	0.592
ar(1)	−0.597***	0.115	−7.536***	0.191
ma(1)	–	–	1.000	0.592

Note *** significant at 99 %

Table 2. Forecasting Criteria for the models.

Models	RMSE	MAE
ARIMA	0.0191	0.0181
SS model	0.0095*	0.0099*

Note *** lowest values compared to the other model

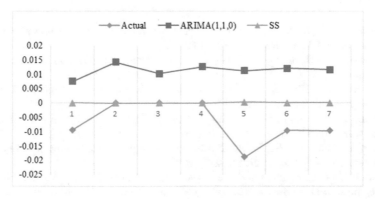

Fig. 3. Comparison of forecasting abilities between ARIMA and SS model

The previous findings show that the state space model has a better forecasting ability than ARIMA. Figure 3 shows three lines generated by actual data, and forecasting values generated by the ARIMA model and the state space model. Considering the actual values line as the center line, it is obvious that overall the ARIMA line is further than the state space line. This characteristic also supports the previous empirical results. Furthermore, the predicted values generated by the ARIMA model line are far higher than the actual values over the period.

5   Conclusion

The paper focuses on forecasting the growth of TDS in Thailand. The time series data sets of DSR from quarter 1 in 2001 to quarter 3 in 2016, totally are 63 observations. The data is divided into 2 periods. The first period known as in sample, covers a period from quarter 1 in 2001 to quarter 4 in 2014 with 56 observations in total and the second period known as out of sample covers a period from the latter two years with 7 observation in total. The studying of the ARIMA models, ARIMA $(1, 1, 0)$ provides the lowest AIC. Meanwhile, the SS model with representation of ARMA models, ARMA $(1, 1)$ in SS form provides the lowest AIC. To compare the forecasting performance of ARIMA model and SSM based on RMSE and MAE criteria, the result shows that SSM provides the better performance, as the predicted values generated by the ARIMA model are far higher than the actual values compared to the predicted values generated by the state space model over the period.

For further discussion as an informative EWI for forthcoming systemic banking crises, TDS's forecasting ability is accurate within first two years (Drehmann and Juselius [1]). Even though, the state space model tends to be the best model in this study, its predicted values generates almost a straight line. This characteristic could be another important issue in which needs a further exploration.

References

1. Drehmann, M., Juselius, M.: Do debt service costs affect macroeconomic and financial stability? BIS Q. Rev. pp. 21–34 (2012)
2. Drehmann, M., Juselius, M.: Evaluating early warning indicators of banking crises: satisfying policy requirements. Int. J. Forecast. **30**(3), 759–780 (2014)
3. Drehmann, M., Illes, A., Juselius, M., Santos, M.: How much income is used for debt payments? BIS Q. Rev. (2015)
4. Durbin, J., Koopman, S.J.: A simple and efficient simulation smoother for state space time series analysis. Biometrika **89**(3), 603–615 (2002)
5. Dynan, K., Johnson, K., Pence, K.: Recent changes to a measure of US household debt service. Fed. Res. Bull. **89**, 417 (2003)
6. Faruqui, U.: Indebtedness and the Household Financial Health: An Examination of the Canadian Debt Service Ratio Distribution. Bank of Canada, Ottawa, Canada (2008)
7. Jacques, J.F.C., Jan, K.S.: An Introduction to State Space Time Series Analysis. Oxford University Press inc, New York (2007). 174 pages

8. Kalman, R.E.: A new approach to linear filtering and prediction problems. J. Basic Eng. **82**(1), 35–45 (1960)
9. Kida, M.: Financial vulnerability of mortgage-indebted households in New Zealand-evidence from the Household Economic Survey. Reserve Bank N. Z. Bull. **72**(1), 5–12 (2009)
10. Migon, H.S., Gamerman, D., Lopes, H.F., Ferreira, M.A.: Dynamic models. Handb. Stat. **25**, 553–588 (2005)
11. Omekara, C.O., Okereke, O.E., Ehighibe, S.E.: Time series analysis of interest rate in Nigeria: a comparison of Arima and state space models. Int. J. Probab. Stat. **5**(2), 33–47 (2016)
12. Ramos, P., Santos, N., Rebelo, R.: Performance of state space and ARIMA models for consumer retail sales forecasting. Robot. Comput. Integr. Manuf. **34**, 151–163 (2015)
13. Petris, G., Petrone, S., Campagnoli, P.: Dynamic linear models. In: Dynamic Linear Models with R, pp. 31–84 (2009)

Effect of Macroeconomic Factors on Capital Structure of the Firms in Vietnam: Panel Vector Auto-regression Approach (PVAR)

Nguyen Ngoc Thach[1(✉)] and Tran Thi Kim Oanh[2]

[1] Head of Research Institute, Banking University of Ho Chi Minh City,
36 Ton That Dam Street, District 1, Ho Chi Minh City, Vietnam
thachnn@buh.edu.vn
[2] Banking University of Ho Chi Minh City, Ho Chi Minh City, Vietnam

Abstract. The article examines the impact of macroeconomicfactors on capital structure during the period of economic recession and economic recovery. The authors collected data from the financial statements of 82 firms listed in Vietnam stock market during the Quarter 1/2007-Quarter 2/2016 and using PVAR. The results demonstrate that during economic recession, the economic growth, the bond market, credit market positively impacted the capital structure whereas the stock market showed negative impacts on the capital structure. During economic recovery, economic growth positively impacted on the capital structure and the remaining macroeconomic variables negatively impacted on the capital structure. In addition, capital structure was affected bymicroeconomic variables such as profitability, asset structure, size, growth and liquidity.

Keywords: Capital structure · Economic recession · PVAR

1 Introduction

Maximizing profits and business value are important targets of enterprises. In order to achieve those targets, managers must employ right decisions in choosing investment opportunities as well as optimally organize and manger their businesses. Capital structure, one of financial tasks in corporate governance, plays a key role.

After witnessing a period of high economic growth in the first half of 2000s, the global economy experienced an economic recession from 2007 to 2010, which negatively affected the business activities of Vietnam's enterprises. The impact was demonstrated by a sharp increase in number of enterprises ceasing their operation during the period. According to the General Statistics Office of Vietnam (2015), in 2014, there were 58,322 enterprises the faced with difficulties and dissolved, 14.5% rising compared to that of the previous year. One of the main causes for the situation was the volatility in macroeconomic environment, which propelled businesses into financial difficulties. However, in the context

© Springer International Publishing AG 2018
V. Kreinovich et al. (eds.), *Predictive Econometrics and Big Data*, Studies in Computational Intelligence 753, https://doi.org/10.1007/978-3-319-70942-0_36

of such traumatized economy, most Vietnam's enterprises lacked a specific and long-term plan for capital restructure but still relied on subjective decisions regarding capital structure, ignoring the circumstances of the economy in each specific period.

Over the last several years, many studies on capital structure have been published. Most of them are about the impacts of micro-economic variables on the capital structure of businesses in different countries. These studies applied different approaches and methodologies, but mainly the Pooled OLS, the Fixed effect model (FEM), the Random effect model (REM) and the General method of moments (GMM). In this report, the authors use PVAR to analyze and compare the impact of those variables on the capital structure of Vietnam's businesses in the two periods of recession and recovery of the world economy.

2 Rationale and Empirical Studies

2.1 Rationale

Most studies regarding the capital structure focus on the following theories:

The MM theory

The theory was proposed by Modigliani and Miller (1958) based on the theory of perfect markets with the absence of taxes, concluding that the business value and the weighted average cost of capital (WACC) are independent of the capital structure. The theory was continued to be further studied in tax environment (1963), drawing a conclusion that the value of the business would increase if it utilizes debt from the benefits of tax shield. The weighted average cost of capital (WACC) of businesses utilizing debts is lower than that of debt-free businesses. However, the theory was based on unrealistic premises (perfect competitive market, absolute rationality, perfect information). Still, the theory serves as the basis for the emergence of more realistic theories later.

The pecking-order theory (POT)

The theory proposes a hierarchy of priorities in selecting funding options but does not address the existence or non-existence of an optimal capital structure for businesses. The POT theory states that the capital structure accords with the following funding order: internal capital from retained earnings, debts, last, new equity (Donaldson 1961).

Trade-off theory (TOT)

The static Trade-off theory was initiated by Kraus and Litzenberger (1973) and further developed into the dynamic TOT (Myers and Majluf 1984). According to the static TOT, enterprises can easily and quickly achieve their optimal capital structures, reflecting the tradeoff between debt's benefits from tax shield and the cost of capital exhaustion. Each enterprise has only one optimal capital structure. Conforming to the dynamic TOT, under the impacts of microeconomic and macroeconomic environment, enterprises cannot immediately reach the optimal capital structure without experiencing a gradual adjustment. Also, the optimal capital structure will vary in each specific period. Despite the difference in view, there is a common approach of the two theories based on the tradeoff between cost and benefit for business to obtain its optimal capital structure and maximize its value.

The theory of agency cost

The theory was proposed by Jensen and Meckling (1976). Agency cost incurs due to the asymmetry of information between enterprise's managers and owners. Therefore, enterprises tend to increase the use of debt in order to reduce agency cost because once doing so, managers must be more cautious in their financial decisions, business risks would be lessened and the efficiency of business activities would improve.

The signaling theory

The signaling theory was also developed based on the information asymmetry between enterprises and investors. Investors usually analyze enterprise's activities, speculate on the current situation and forecast the prospect of the enterprises. They believe issuing of debt is a positive signal about the business prospect, thus the stock price would go up. In contrast, that enterprises issue new equity indicates a negative signal, which would make stock price to fall. From the perspective of investors, only a business has not good prospects want to be funded by equity in order to share the risk of the business with new investors (Asquith and Mullins 1983).

The Market-timing theory

This theory was originated from the study of Baker and Wugler (2002), stating that the difference between market value and book value is the determining factor for enterprise's capital structure. In case of a high price-to-book ratio (P/B), enterprises will issue new equity to mobilize capital. Meanwhile, enterprises usually use debt when P/B is low.

The above mentioned theories have various views, however they do not conflict but rather complement each other in comprehensively explaining the manager's decisions of funding sources.

2.2   Other Related Studies

Although theoretical and empirical research on capital structure varies in perspectives and methodologies, they generally focus on the following aspects: capital structure is influenced by micro variables (Truong and Nguyen 2015; Vatavu 2015); the combined effect of micro and macro variables to capital structure (Jong et al. 2008; Nor et al. 2011; Khanna et al. 2015); capital structure impacts business value or determines the optimal capital structure threshold (Ahmad et al. 2012; Wang and Zhu 2014). However, the common limitation of these studies is that they merely focus on the capital structure of businesses in long term without analyzing specific economic contexts in each period (stage) of the economic cycle.

Since the 1970s, economic crises have occurred at high frequency, intensity and complexity, causing severe socio-economic consequences for many nations all over the world. This trend has resulted in studies on capital structure in combination with global and regional economic crises such as 1997–1998, 2007–2010. However, the number of these works remains modest. Other noticeable studies are Ariff et al. (2008); Fosberg (2013); Alves and Francisco (2015); Iqbal and Kume (2014). In Vietnam, currently only Truong and Nguyen (2015) refer to this issue. However, studies in Vietnam and abroad mainly focus on addressing and giving solutions to fix the impacts of economic downturns on capital structure without thoroughly analyzing the capital structure adjustment of businesses in response to the context of the economy during recession and recovery. This is a scientific gap in studying capital structure in Vietnam and abroad.

In terms of methodology, most of the published studies on this subject only use Pooled OLS, FEM, REM or GMM. The purpose of using these models is simply to verify the positive or negative impacts of macroeconomic variables on capital structure. However, they are unable to analyze the mechanism driving the effects of these variables to the capital structure decisions of enterprises as well as unable to explain how a macroeconomic shock impacts on the behavior of adjusting the capital structure of enterprises and how long this impact will last. Only Khanna and Associates (2015) used PVAR in panel data to study the impacts of macroeconomic variables such as economic growth, inflation and stock indexes on capital structure. In Vietnam, according to the authors, there has not been any studies applying PVAR to study this issue. This is a gap in study methodology because socio-economic characteristics in the context of unstable economic recession and recovery require the use of an appropriate method to study the impact of microeconomic and macroeconomic factors to the financial situation, especially the capital structure of the business. Therefore, in the present study, the authors analyze the impact of micro variables and the mechanism of the impact of macro variables on the capital structure of Vietnam's enterprises in the period of recession (from Q1/2007 to Q4/2010) and economic recovery (from Q1/2011 to Q2/2016) by using PVAR.

3 Data, Model and Study Methodology

3.1 Data

Based on grounded theories and empirical researchers, variables affecting the capital structure are selected to build the model (as shown in Table 1).

Table 1. Variables and measurement.

Symbol	Variables	Expectation	Theory	Measurement	Studies
Endogenous variables					
TDR	Capital structure	+		Total liabilities/Total assets	Vo et al. (2014); Vătavu (2015)
GDP	Economic growth	+ −	TOT, agency cost POT	Quarterly growth rates of real GDP	Ariff et al. (2008); Jong et al. (2008); Khanna et al. (2015)
RATE	Loan interest rate	+ −	MM POT, market-timing, TOT	Average loan interest	Nor et al. (2011); Allayannis et al. (2003); Zerriaa and Noubbigh (2015)
LNVNINDEX	Stock market	−	Market timing	Natural logarithm of VNINDEX	Jong et al. (2008); Alves and Francisco (2015); Khanna et al. (2015)
BOND	Bond market	+	TOT, agency cost	Market capitalization value/GDP	Jong et al. (2008); Nor et al. (2011)
Exogenous variables					
SIZE	Firm size	+ −	TOT, Agency cost, POT	Natural logarithm of total assets	Jong et al. (2008); Vo et al. (2014)
TANG	Asset structure	+ −	TOT, POT Agency cost	Fixed assets/Total assets	Vo et al. (2014); Vătavu (2015)
GRO	Firm growth	+ −	POT Agency cost	Quarterly growth rate of total assets	Jong et al. (2008); Vo et al. (2014)
LIQ	Short-term liquidity	+ −	TOT POT, Agency cost	Short-term assets/Short-term liabilities	Vo et al. (2014); Vătavu (2015)
VOL	Business risk	+ −	TOT POT	Standard deviation (EBIT/Total assets)	Vo et al. (2014); Vătavu (2015)
MTR	Coiporate income tax	+	TOT MM	Corporate income tax/Profit before tax	Jong et al. (2008); Vătavu (2015)

Note: (+) is positive relationship between capital structure and explanatory variables. (−) is negative relationship between capital structure and explanatory variables. Source: Author's compilation.

The authors used balance sheets extracted from financial statements of 82 randomly selected enterprises from those listed on the Vietnam Stock exchanges, which were continuously in operation from Q1/2007 to Q2/2016 ($82 \times 38 = 3,116$ observations). The samplcs were highly representative, provided by Ban Viet Capital Securities (VCSC). In addition, macroeconomic variables werecollected from IMF, ADB and AsianBondsOnline.

3.2 Models and Research Methodology

PVAR for economic recession period (Model 1) and economic recovery (Model 2) with latency k are described as follows:

$$Y_{it} = \mu_0 + A_1 Y_{it-1} + \ldots + A_k Y_{it-k} + \beta_x X_i t + e_{it}, \quad \forall i = 1, 2, \ldots, N, \ t = 1, 2, \ldots, T.$$

Where $Y_{it} = (TDR_{it}, GDP_{it}, RATE_{it}, LNVNINDEX_{it}, BOND_{it})$: is a random vector level of dependent variables; Y_{it-p}: vector level of dependent variables lentency; $A_1, A_2, \ldots, A_k$: matrices $k \times k$; X_{it}: exogenous vectors level $(1 \times k)$, including variables listed in Table 1; β_x: matrices $(l \times k)$ coefficient estimation; e_{it}: Fixed effects due to unobservable characteristics of enterprises and constant effects over time, $e_{it} | y_{it-1} \sim N(0; \sigma_e^2)$.

The classical VAR model is applied to the stationery and non-coherent time series, originated from Sims's study (1980) on the transmission mechanism of macro variables. Eakin et al. (1988) continued to propose VAR model to process panel data (PVAR). Since PVAR was proposed based on the classical VAR model, there were still some shortcomings such as the deviated estimated parameters or loss of observations when taking lags. To fix this disadvantage, Love and Zicchino (2006) introduced and used PVAR based on the application of GMM to ensure the uniformity of balance variances, preventing self-correlation and maintaining data conservation.

3.3 Basic Tests

3.3.1 Stationery Test

When estimating PVAR, the variables in use must be stationery. The authors used the Augmented Dickey-Fuller (ADF) to test the variables. Table 2 shows that all variables are stationery at 0.

Table 2. PVAR unit root test results.

VARIABLES	STATISTIC T	VARIABLES	STATISTIC T
TDR	526,1311***	TANG	291,4663***
ROE	2405,4509***	LIQ	702,1896***
VOL	1858,4298***	LNVNINDEX	381,9070***
SIZE	493,4786***	BOND	207,7378***
MTR	1638,3719***	RATE	207,7378***
GRO	1343,1092***	GDP	561,0923***

Note: ***corresponds to 1% of significance level. Source: Author's calculation.

Table 3. Lags criteria results of Model 1 and Model 1.

Latency	CD	Statistic J
Model 1		
1	0.9997	874.1953
2	0.9998	567.1583*
3	0.9999	867.7019
Model 2		
1	0.9979	1,286.025
2	0.9980	1,173.926
3	0.9986	510.0104*

Note: *represents the selected latency corresponding with criteria. Source: Author's calculation

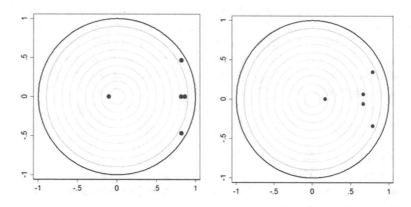

Fig. 1. AR root test results of Model 1 and Model 2. Source: Author's calculation.

3.3.2   Optimal Latency Test

Andrews and Lu (2001) proposed to use the Moment Model Selection Criteria (MMSC) with determination coefficient CD and J-Pvalue statistics to determine the optimal latency. The results shown in Table 3 indicate that PVAR optimal latency in model 1 is 2 and model 2 is 3.

3.3.3   Model Stability Test

Research conducted AR test. Figure 1 shows that all the solutions of Model 1 and Model 2 are in the unit circle. PVAR model ensures stability and sustainability.

4   Study Outcomes

Table 4 represent the regression results of Model 1 and Model 2.

Table 4. PVAR results of Model 1 and Model 2

Criteria	Model 1	Model 2
L.TDR	0.726***	0.722***
	[8.30]	[7.99]
L.LNVNINDEX	−0.178***	−0.0726***
	[−3.74]	[2.85]
L.BOND	3.597***	−0.495***
	[5.35]	[−5.03]
L.RATE	1.231^c	−0.0316
	[4.90]	[−0.56]
L.GDP	2.669***	0.105
	[3.53]	[0.84]
ROE	−0.251**	−0.0549*
	[−2.05]	[−1.50]
VOL	−0.26	0.219
	[−1.12]	[1.06]
SIZE	−0.328***	0.0369*
	[−3.09]	[0.62]
MTR	−0.131	−0.0356*
	[1.63]	[−1.91]
GRO	−0.00804**	0.00186*
	[−2.07]	[0.46]
TANG	0.299***	0.0810*
	[2.85]	[1.67]
LIQ	−0.00202*	0.000261*
	[1.53]	[0.93]
N	1230	1722

Note: *, **, *** correspond to the significance level of 10%, 5% and 1%; [] value of standard deviation. Source: Author's calculation.

4.1   Economic Recession

In order to analyze the mechanism and direction of the impact of macroeconomic variables on the capital structure of Vietnam's enterprises when shocks happen, the authors analyzed the push function (Fig. 2).

4.1.1   Impacts of Bond Market on the Capital Structure

As the bond market went up by one standard deviation, the bond market improved, enterprises increased 3.597% of debts in the first quarter and dampened in the fourth quarter. This result was consistent with the theory TOT and Jong et al. (2008); Nor et al. (2011).

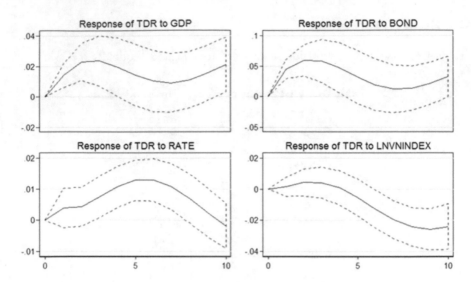

Fig. 2. Impulse response function Model 1. Source: Author's calculation.

4.1.2   Impacts of Economic Growth on the Capital Structure

Figure 2 shows that GDP has a positive impact on TDR. When GDP increased by one standard deviation, corporate debt use increased by 2.669% and the increase in debt use declined gradually as of Q3, in line with Ariff and Associates (2008).

4.1.3   Impacts of Credit Market on Capital Structure

With the shrinkage of credit market, credit balance declined, loan conditions became more difficult and RATE increased by one standard deviation but positively impacted TDR. TDR increased by 1.231% at 5% significance level. This increase lasted for nine quarters. The result is consistent with MM theory and Allayannis et al. (2003); Zerriaa and Noubbigh (2015).

4.1.4   Impacts of Stock Market on the Capital Structure

When the stock market increased by one standard deviation, TDR fell by 0.178% at 1% significance level. The decline of debt using lasted in 1 quarter, in agreement with Khanna et al. (2015) and Alves and Francisco (2015).

As analyzed above, there is a relation between TDR and macroeconomic variables in the context of economic downturn. However, TDR does not only depend on macroeconomic shocks but also under the effect of microeconomic variables.

The capital structure of the previous period positively influences the capital structure of the later one. This indicates that a rise in debt using in the previous period would make the debt using in the later period increase by 0.726%, harmonizing with Nor et al. (2011) and Khanna et al. (2015).

Profitability negatively correlated with the capital structure at 1% statistical significance level. The results was explained by POT, Nor et al. (2011) and Truong and Nguyen (2015). However, the results also demonstrated an inefficacy in using debts of Vietnam's enterprises, lowering businesse's profitability.

Firm size and scale negatively influences the capital structure. This indicates that during economic downturn, large-scale businesses usually have high profit, large equity, good reputation and financial capacity can easily issue new equity to the market. The result is analogous to Fosberg (2013) and Proenca et al. (2014).

Firm growth negatively affects the capital structure, which is similar to the Agency cost theory and Proenca et al. (2014) when studying the correlation in the economic recession period.

Asset structure has positive correlation with capital structure. Indeed, enterprises whose fixed assets are large when issuing secured debts or mortgaged debts are more likely to have access to loans and better policies. This results match with POT, TOT, Alves and Francisco (2015) and Iqbal and Kume (2014).

Solvency has negative impact on capital structure, consistent with POT, Nor et al. (2011) and Proenca et al. (2014) when studying this correlation in the economic recession period.

4.2   Economic Recovery

Similarly, Fig. 3 demonstrates the push function of Model 2.

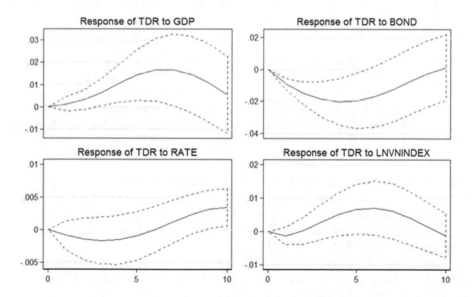

Fig. 3. Impulse response function Model 2. Source: Author's calculation.

4.2.1   Impacts of Bond Market on Capital Structure

Table 3 shows that when the bond market expanded by one standard deviation, TDR fell by 0.495% and this trend prolonged in 10 quarters. This result contradicts with the study on economic recession and TOT but consents to Vo et al. (2014). The reason for the result is the characteristics of bond markets of developing countries in general and Vietnam in particular, which is small in scale, undiversified products mainly comprising of Government bonds (which account for 95.4% of the market value). Besides, the market capitalization value of the government bond market increased from 13.68% GDP (economic recession) to 19.03% GDP during the observation period. In contrast, the scale of corporate bond market slightly decreased with market capitalization value fell from 1.01% GDP to 0.92% GDP. The cause of such development was the fact that credit market remained as a traditional mobilization source or Vietnam's enterprises, therefore when loan conditions loosened; there was a shift from issuing debts towards borrowing directly from financial institutions.

4.2.2   Impacts of Economic Growth on Capital Structure

Similar to the economic downturn period, Table 3 shows that when GDP increased by one standard deviation, TDR rose by 0.105% in Q1 and slowly diminished as of Q6. This implies a rise in consumption demand in that period, encouraging businesses to expand their production, which led to high capital demand. Furthermore, the low cost of capital exhaustion and reduction in bankruptcy risk made enterprises increase the use of debts to take advantages of the tax shields. This result is consistent to TOT, Jong et al. (2008); Nor et al. (2011) and Khanna et al. (2015).

4.2.3   Impacts of Credit Market on Capital Structure

In contrast with the study in economic recession, Table 3 shows that when credit market shrank, RATE increased by one standard deviation, TDR dropped by 0.312% at 5% of significance level. This trend lasted for 6 quarters. It indicates that the credit market positively correlates with TDR, consistent to POT and Nor et al. (2011)

4.2.4   Impacts of Stock Market on Capital Structure

Similar to the economic recession, Fig. 4.3 exhibits a decrease of TDR by 0.073% when the stock market went up by one standard deviation. The trend prolonged for 2 quarters since the economic shock. The result conforms with the market-timing theory and Vo et al. (2014).

The results of Model 2 also indicate that in economic recovery period, other variables such as solvency, scale and speed of growth positively influence the capital structure, contravening the results of the study in economic recession period. This result could be explained by Keynes (1936). That is, when the economy recovers, global and domestic demand for goods and services increases. Hence, in order to meet the demand of the market and the new business cycle,

Vietnam's firms expand their operations, increase their asset investment and actively seek new business opportunities. However, the result in this period also points out that Vietnam's businesses should prioritize their investment in short-term projects and/or assets with high profitability to increase their short-term solvency as well as minimize risks arising under the context of unstable and unsustainable economic recovery. Besides, businesses should also control and recalculate their taxable income in this period. This is also a significant source of capital enhancing capital resources for businesses.

5   Conclusion and Policy Recommendations

5.1   Conclusion

The PVAR regression results show that TDR is affected by macroeconomic variables such as bond market, economic growth, credit market and stock market. However, the direction and magnitude of the impact of macroeconomic variables on TDR varies. Vietnam economy under the global economic recession, macroeconomic variables had a strong impact on TDR, which suggests that managers were cautious during this period. However, as the global economy recovered and the process influenced the macroeconomic movements in the country, managers did not properly address the macroeconomic variables. Macroeconomic variables showed weak impacts on TDR, specifically:

The bond market has a positive impact on TDR during the economic downturn. A shock would increase the impact of the bond market, making businesses increase their debt use by 3.597%. Conversely, as the economy recovers, TDR fell by 0.495% under the impact of the bond market.

Economic growth has a positive impact on TDR in both economic recession and recovery. In particular, the strongest impact was during the economic downturn. Despite the difficulties of the period, economic growth still made TDR rise by 2.699%, in accordance with the characteristics of Vietnam economy during this period, whose growth rate was high due to recent WTO accession.

The credit markethas a negative impact on TDR during economic recession. Conversely, as the economy recovered, credit markets positively influenced TDR. Thereby we find that, despite the economic downturn, difficult operating conditions, financial exhaustion and bankruptcy risk, businesses still increased their use of debt. This demonstrates the fact that Vietnam's enterprises have low financial autonomy, depends heavily on loans as well as undiversified channels for capital mobilization.

The stock market has a negative impact on capital structure in both economic recession and recovery, in line with the Market-timing theory. However, the strongest impact was during the economic downturn, when a stock market shock made the TDR decrease by 0.178%.

In addition, the estimated PVAR model also indicates that during the economic downturn, the pace of TDR adjustment of Vietnam's enterprises (0.726%) is faster than that when the economy recovers (0.722%). This result reaffirms the content of the TOT theory that the capital structure of a business varies from time to time, however, this theory does not yet indicate the rate of firm's adjustment of target capital structure in each specific economic context. This study, therefore, provides additional evidences for the argument of the TOT theory. That is, during economic downturn, managers adjusts capital structure faster than in economic recovery to achieve the target capital structure, ensuring financial security and increasing the value of the company.

In addition, asset structure has a positive impact, while profitability has a negative impact on TDR. Other variables including solvency, firm growth and size have a negative impact on TDR in the context of economic recession. Conversely, as the economy recovers, these variables have a positive impact on corporate TDR. Furthermore, during the economic recovery period, businesses need to control expenses and especially recalculate their taxable income.

5.2   Policy Recommendations

For enterprises

First, enterprises should diversify their forms of capital mobilization to reduce debts, make the most of capital sources and increase the use of financial instruments.

Second, the capital restructure of businesses must be associated with each stages and target of development, specific financial situation, business size, domestic and foreign macroeconomic environment in line with stages of economic cycle- economic recession and recovery.

Third, restructure capital in the direction of increasing owner's equity and self-financing capacity of businesses. The study outcomes show that the profitability of enterprises negatively influences the capital structure. This proves that profitability is an important capital sources that can help firms actively meet their capital needs while still keeping control. Additionally, businesses should actively mobilize equity from outside such as issuing shares, doing joint ventures and associates.

Fourth, capital restructuring should be associated with the restructuring of investment portfolios, especially investment in fixed assets. However, enterprises should also note that the increase of fixed assets must be associated with capital structure adjustment in the direction of strengthening long-term source, avoiding financial imbalances. Besides, the study results also show that Vietnam's enterprises should prioritize investments in short-term projects or assets with high yields that can enhance their short-term liquidity and help avoid risks arising from the economic conditions of unstable and unsustainable recovery.

Fifth, capital restructure must be in line with the recalculation of taxable income and tax planning in order to optimize the amount of tax payable within the legal framework. It means that firms should recalculate their taxable income to achieve tax deduction that are higher than their reduction of business income.

For the Government and related agencies

First, the Government should simultaneously implement polices to stabilize the macro economy, curb inflation, ensure rational economic growth, create favorable business environment and assist enterprise to better access to the funds serving their purpose of restructuring.

Second, the Government should improve regulations and policies to promote and facilitate the development of financial markets, especially the bond market and the stock market. They are not only the channels that help mobilize capital for enterprises but also the channels for capital withdrawal under the market mechanism. Therefore, sound and developed stock and bond markets are significant conditions to ensure the success of the capital restructuring process of enterprises. Thus, the capital restructuring of enterprises must be associated with the development of these markets. The Government should employ policies to strengthen, stabilize and soundly develop credit markets.

Third, the corporate income tax is 22%, which is relatively higher than that of other countries in the region such as Singapore, Hong Kong (17%) and Taiwan (16.5%). The Government should adjust the corporate income tax to improve competitiveness and support enterprises in capital restructuring.

References

Ariff, M., Taufq, H., Shamsher, M.: How capital structure adjusts dynamically during fnancial crises. J. Faculty Bus. Bond Univ. **12**, 15–25 (2008)

Asquith, P., Mullins, D.W.: The impact of initiating dividend payments on shareholders. J. Bus. **56**(1), 77–96 (1983)

Ahmad, Z., Abdullah, N.M.H., Roslan, S.: Capital structure effect on firms performance: focusing on consumers and industrials sectors on Malaysian firms. Int. Rev. Bus. Res. Papers **8**(5), 137–155 (2012)

Alves, P., Francisco, P.: The impact of institutional environment in firms' capital structure during the recent financial crises. Q. Rev. Econ. Financ. **57**, 129–146 (2015)

Andrews, D.W.K., Lu, B.: Consistent model and moment selection procedures for GMM estimation with application to dynamic panel data models. J. Econometr. **101**, 123–164 (2001)

Allayannis, G., Brown, G.W., Klapper, L.F.: Capital structure and financial risk: evidence from foreign debt use in East Asia. J. Financ. **58**, 2667–2709 (2003)

Baker, M.P., Wurgler, J.: Market timing and capital structure'. J. Financ. **57**(1), 1–32 (2002)

Donaldson, G.: Corporate Debt Capacity: A Study of Corporate Debt Policy and the Determination of Corporate Debt Capacity. Harvard University, Massachusetts (1961)

Eakin, H.D., Newey, W., Rosen, H.S.: Estimating vector autoregressions with panel data. Econometrica **6**, 1371–1395 (1988)

Fosberg, R.H.: Short-term debt financing during the financial crisis. Int. J. Bus. Soc. Sci. **4**, 1–5 (2013)

Iqbal, A., Kume, O.: Impact of financial crisis on firms capital structure in UK, France, and Germany. Multinat. Financ. J. **18**, 249–280 (2014)

Jensen, M., Meckling, W.: Theory of the firm: managerial behaviour, agency costs ownership structure'. J. Financ. Econ. **3**(4), 305–360 (1976)

Jong, A.D., Kabir, R., Nguyen, T.T.: Capital structure around the world: the roles of of firm- and country-specific determinants. J. Banking Financ. **32**(9), 1954–1969 (2008)

Khanna, S., Srivastava, A., Medury, Y.: The effect of macroeconomic variables on the capital structure decisions of indian firms: a vector error correction model/vector autoregressive approach. Int. J. Econ. Financ. **5**(4), 968–978 (2015)

Keynes, J.M.: General Theory on Employment, Interest and Money. The University of Adelaide, Adelaide (1936)

Kraus, A., Litzenberger, R.H.: A state-preference model of optimal financial leverage. J. Financ. **33**, 911–922 (1973)

Love, I., Zicchino, L.: Financial development and dynamic investment behavior: evidence from panel VAR. Q. Rev. Econ. Financ. **46**, 190–210 (2006)

Modigliani, F., Miller, M.H.: The cost of capital, corporation finance and the theory of investment. Am. Econ. Rev. **48**, 261–297 (1958)

Myers, S.C., Majluf, N.S.: Corporate financing and investment decisions when firms have information that investors do not have. J. Financ. Econ. **13**, 187–221 (1984)

Nor, F.M., Haron, R., Ibrahim, K., Ibrahim, I., Alias, N.: Determinants of target capital structure evidence on south east Asia countries. J. Bus. Policy Res. **6**, 39–61 (2011)

Proenca, P., Laureano, R.M., Laureano, L.M.: Determinants of capital structure and the 2008 financial crisis: evidence from Portuguese SMEs. Procedia-Soc. Behav. Sci. **150**, 182–191 (2014)

Sims, C.: Macroeconomics and reality. Econometrica **48**, 1–47 (1980)

Genral Statistics Office (2015). Published information. http://www.vtc.vn/chuyen-gia-noi-gi-ve-trien-vong-kinhte-viet-nam-nam-2015-d188147.html. Accessed 10 Dec 2016

Truong, H.T., Nguyen, P.T.: Determinants of capital structure of a-reits and the global financial crisis. Pacific Rim Prop. Res. J. **18**, 3–19 (2015)

Vatavu, S.: The impact of capital structure on financial performance in Romanian listed companies. Procedia Econ. Financ. **32**, 1314–1322 (2015)

Vo, T.T.A., Tran, K.L., Le, T.N.A., Tran, T.D.: Study on the impact of macro factors on the capital structure of enterprises listed on Vietnam's stock exchanges. J. Econ. Dev. **207**, 19–27 (2014)

Wang, J., Zhu, W.: The impact of capital structure on corporate performance based on panel threshold model. Comput. Modell. New Technol. **18**(5), 162–167 (2014)

Zerriaa, M., Noubbigh, H.: Determinants of capital structure: evidence from Tunisian listed firms. Int. J. Bus. Manage. **10**, 121–135 (2015)

Emissions, Trade Openness, Urbanisation, and Income in Thailand: An Empirical Analysis

Rossarin Osathanunkul[1(✉)], Natthaphat Kingnetr[1], and Songsak Sriboonchitta[1,2]

[1] Faculty of Economics, Chiang Mai University, Chiang Mai 52000, Thailand
orossarin@gmail.com
[2] Center of Excellence in Econometrics, Chiang Mai University, Chiang Mai 52000, Thailand

Abstract. This study investigates the relationship between emissions, income, energy consumption, trade openness, and urbanisation in Thailand over the period of 1971 to 2014. The ARDL cointegration technique is employed and CUSUM and CUSUMSQ tests are used to ensure the stability of the estimated results. Our findings indicate there is a long run relationship among variables for the case of CO_2 emissions while there is none for the SO_2. The results indicate an increase in income can cause significantly more CO_2 emissions. Energy consumption also contributes to environmental degradation with slight impact, while there is no effect from trade openness. On the contrary, urbanisation greatly helps reduce CO_2 emissions in the long run.

Keywords: Emissions · Income · Thailand · EKC · ARDL

1 Introduction

Thailand is one of the emerging economies in Asia, and is expected to grow at a faster rate than high income economies. In 2015, the ASEAN Economic Community (AEC) was established. Thailand is one of the AEC members. The goal of the AEC is "to transform ASEAN into a region with free movement of goods, services, investment, skilled labour, and a freer flow of capital" [31]. One of the major priority actions for AEC is to develop and promote environmentally friendly industries. The rapid economic growth and the establishment of the AEC would impact Thailand environmentally. Increasing economic growth and trade openness may lead to an increase in demand for environmental inputs and consumptive goods, and a rise in production of goods and the subsequent pollution. Alternatively, although economic growth may lead to an initial increase in pollution, later as income growth, a nation can invest in environmentally friendly technologies that lead to a reduction in pollution.

In addition, the mitigation of carbon emissions through international and international policies will need to be agreed upon by the policy makers of the

© Springer International Publishing AG 2018
V. Kreinovich et al. (eds.), *Predictive Econometrics and Big Data*, Studies in Computational Intelligence 753, https://doi.org/10.1007/978-3-319-70942-0_37

major emitting countries that include China, the United States, India, Russia, the European Union, and Japan, that account for about seventy five percent of the global emissions. However, the cooperation of rapidly growing countries such as the Republic of Korea, Brazil, South Africa, Indonesia, Mexico, and Thailand is also needed to put constraints on carbon emissions [20]. Some of the policies include a local or global cap and trade system of carbon emission allowances, fuel efficiency and energy standards, emission credits, renewable energy portfolio standards, taxes, and subsidies. Therefore, it is important to empirically estimate the relationship between economic growth, carbon emissions, and energy consumption in Thailand.

The relationship between income and pollution has been well described as the environmental Kuznets curve (EKC) hypothesis introduced by Grossman and Krueger in 1991 [37]. The hypothesis indicates an inverted U-shaped relationship between emissions and income. It shows that countries initially face an increase in environmental degradation as their economy expands, but this will subsequently decrease when economic development reaches a certain level. Dinda [12] suggests that early in its development, a country would rely on agriculture and industry which draw heavily on natural resources in order to expand its economy, leading to a negative impact on environmental quality. This is referred to as the "scale effect" [5]. After a certain period, the structure of industry then changes into information-intensive industries and services, the population becomes more concerned about environmental quality leading to enforcement of environmental regulations, investments, protection, and advancement of new environmentally friendly technology results in a decrease in environmental damage. This is referred to as "the technique effect" [5].

An increase in trade openness due to a change in trade policy would cause a country to specialise in the production of goods that could have either a positive or negative impact on the environment depending on which is used, the pollution haven hypothesis (PHH) or the factor endowment hypothesis (FEH) [5,11,14]. Cole [10] argues that the pollution haven hypothesis (PHH) occurs when there is trade between a developed country with more stringent environmental regulations and a developing country with lax environmental regulations. The difference in regulations encourages pollution-intensive firms to move away from the developed country into the developing country. Hence, with the opening of trade, the level of pollution is expected to increase in countries with lax regulations while decreasing for countries with stringent regulations.

On the other hand, the factor endowment hypothesis (FEH) points to the difference in environmental regulations. It is argued that dirty goods are capital-intensive goods and more likely to be produced by a capital-abundant country than a capital-scarce country. By opening up to trade, dirty-intensive firms would move to a developed country from a developing country. Therefore, emissions in the capital-abundant country would rise, while the developing country benefits from the trade [38]. Brack [7] suggests that trade is good for the environment as it improves environmental standards in industry. When foreign environmental standards are high, exporting firms have to apply various standards when producing goods. This may reduce environmental damage and improve the firm's productivity.

In spite of many papers examining the EKC hypothesis, to our knowledge, only four studies investigating the applicability of EKC hypothesis in Thailand have been done so far. These studies are Arouri *et al.* [6], Saboori and Sulaiman [29], Bureecam [8], and Naito and Traesupap [21]. These studies employed empirical models which may result in multicollinearity from squared variables. It is the purpose of this research to avoid this potential problem. Therefore, this study will extend those by using the new empirical model for the EKC study suggested by Narayan and Narayan [23] using an autoregressive distributed lag (ARDL) approach, and contribute to the existing literature by also considering SO_2 emissions. The remainder of this paper is organized as follows. Review of literature is in Sect. 2. Section 3 gives the analytical methodology. Then, the results are presented and discussed in Sect. 4, and Sect. 5 provides conclusions and policy implications.

2   Literature Review

There have been numerous studies investigating the EKC hypothesis; however, only four studies consider Thailand. Though different emissions, economic indicators, and time scale were utilised, these studies concluded that the EKC hypothesis is applicable to Thailand.

Arouri et al. [6] studied the EKC hypothesis in the case of Thailand over the period of 1970 to 2010 using the ARDL approach. The study employed CO_2 emissions as the environmental degradation variable and real GDP per capita, energy consumption, trade openness and urbanisation as its determinants. Their study found that CO_2 emissions and real GDP per capita follow the pattern given by the EKC hypothesis. Energy consumption and trade openness increase the emissions while urbanisation causes a reduction.

Saboori and Sulaiman [29] investigated the relationship between CO_2 emissions, energy consumption, and economic growth in Association of Southeast Asian Nations (ASEAN) countries for the period of 1971 to 2009 using the ARDL approach. Their study found the applicability of the EKC hypothesis in Singapore and Thailand with the energy consumption promoting an environmental degradation.

Bureecam [8] investigated the relationship between municipal solid wastes generation, income, population and tax for the period of 1992 to 2006 using the ordinary least squares regression (OLS) method. The study found the applicability of the EKC hypothesis with the turning point of real GDP at 3.2 to 3.7 trillion Thai Baht approximately. Moreover, the result indicates that taxation and population contribute to an increase in solid wastes.

Naito and Traesupap [21] studied the effects of gross provincial product (GPP) and shrimp farming on mangrove deforestation in 23 provinces in Thailand during 1975 to 2004 using panel data technique. They found the applicability of the EKC hypothesis along an estimated value of GPP per capita at the turning point which has not been reached, and the shrimp farming has led to mangrove deforestation significantly. Although the development of semi-intensive and intensive aquaculture systems caused mangrove deforestation, the intensive shrimp farming, developed during the 1990s, improved the situation.

Surprisingly, other studies in the EKC literature found a controversy in the result. Studies which found support of the EKC hypothesis include those that; examined the relationship between energy consumption, CO_2 emissions, and economic growth in Europe [1], and heavy industries in Canada [15]; looked at CO_2 emissions and growth in developing economies [23]; looked at the relationship between CO_2 emissions, energy consumption, FDI, and GDP for Brazil, Russian Federation, India, and China [24]; and looked at relationship between CO_2 emissions, energy consumption, economic growth, trade openness, and urbanisation of newly industrialised countries [17]. Acaravci and Ozturk [1] and Hamit-Haggar [15] found the presence of a long-run relationship between carbon emissions, energy consumption, and real GDP. Within Europe, of which Acaravci and Ozturk [1] found the evidence of the EKC hypothesis only holds for Denmark and Italy. Additionally, Hamit-Haggar [15] showed that for the short run, economic growth has an impact on greenhouse gas emissions within industries. Their study found that energy consumption has an insignificant impact on economic growth. Hossain [17] found that over time higher energy consumption leads to an increase in CO_2 emissions. On the other hand, economic growth, trade openness, and urbanisation are good for the environment in the long run. Narayan and Narayan [23] suggested that if long run elasticity is smaller than short run elasticity then it implies that country is facing a decline in CO_2 emissions as the economy grows. Sharma [36] investigated the factors that affect CO_2 emissions. His results show that GDP per capita in middle and low income countries, energy consumption, and urbanisation have statistically significant positive influence on CO_2 emissions.

Studies that did not find evidence of the EKC included one investigating the relationship between CO_2 emissions, GDP per capita, capital, labour force, export, import, and energy consumption in Vietnam [4]. Jaunky [18] investigated the relationship between CO_2 emissions and income for high-income countries. His results show that there is a positive impact running from GDP per capita to CO_2 emissions. In terms of income elasticity, CO_2 emissions are inelastic in both periods with the long run being more inelastic. This reflects that over time the amount of CO_2 emissions will be stable and the EKC hypothesis will not hold. In addition, Al-Mulali et al. [4] found that fossil-fuel energy consumption increases the pollutants, while renewable energy consumption has no significant effect in reducing pollution. Furthermore, labour force helps reduce pollution. He and Richard [16] studied the relationship between per capital GDP and per capita CO_2 emissions in Canada. They used a semi-parametric and a non-flexible modelling methods and found little evidence in favour of the environmental Kuznets curve hypothesis. Wong and Lewis [40], instead of estimating the EKC for air pollutants, utilised water quality variables on the Lower Mekong Basin region in Asia. They did not find evidence of an EKC for water pollutants. They suggested that the results are entirely dependent on the model, error specification, and the type of pollutants as well.

Most of the previous studies above are based on the original empirical model specification for the EKC hypothesis which involves both GDP and squared of GDP variables in the model. Narayan and Narayan [23] pointed out that the results from the old model may have suffered from the problem of multicollinearity between GDP and its square.

3   Methodology and Data

3.1   Empirical Model

Following Saboori et al. [30], the general form of the EKC hypothesis can be formulated as

$$E = f(Y, Y^2, Z) \tag{1}$$

where E is a level of pollution, Y is the income, and Z are other variables which affect the emissions. The quadratic form of income (Y^2) will allow the model to capture the inverted U-shape relationship between emissions and income. However, Narayan and Narayan [23] pointed out that including the squared and cubed income may result in a biased result due to high correlation between them and suggested the alternative empirical model which excludes the squared term.

To empirically investigate the relationship between emissions, income, energy consumption, and trade openness in Thailand, the study will follow a similar methodology as in Al-Mulali *et al.* [4] which used the new model specification for the EKC hypothesis study proposed by Narayan and Narayan [23]. Two types of emissions are investigated in this study: CO_2 emissions (denoted as C) and SO_2 emissions (denoted as S). Therefore, the empirical model to be estimated can be specified as

$$\ln E_t = \alpha_0 + \alpha_1 \ln Y_t + \alpha_2 \ln EN_t + \alpha_3 \ln TR_t + \alpha_4 \ln UR_t + \varepsilon_t \tag{2}$$

where ln is natural logarithm[1]; E_t demotes emissions per capita; Y_t is real GDP per capita; EN_t is energy use per capita; TR_t is trade openness; UR_t is an urbanisation; ε_t is assumed to be the normal distributed error term with constant variance and zero mean.

However, a drawback of this model specification is that the turning point for income where emission begins to decline cannot be found, but the applicability of EKC hypothesis can still be investigated. The parameters in the model will be estimated under long-run and short-run frameworks (more on this later). By

---

[1] Using the logarithm transformation could reduce the risk of having heteroskedasticity and allows the estimated coefficients to be interpreted as elasticity, hence it reflects the impact from a percentage change in explanatory variable to the percentage change in dependent variable [41]. Also, Acaravci and Ozturk [1] pointed out that "the growth rate of the relevant variable will be obtained by their differenced logarithms.".

comparing the income elasticity of emissions (α_1) between the long run and the short run, one can conclude that EKC hypothesis is applicable if the long run is lower given that both values are positive [23]. The $\alpha_2 > 0$ indicates that energy consumption increases the emissions. Otherwise, it will decrease the emissions if $\alpha_2 < 0$. If an opening up to trade leads to the environmental improvement, then $\alpha_3 < 0$. Otherwise, $\alpha_3 > 0$ implies that trade contributes to environmental degradation. For the impact on emissions from urbanisation, it is expected to be negative ($\alpha_4 < 0$) as found by Arouri *et al.* [6].

3.2   Data Description

The period of study is limited by the availability of the data. The annual time series data within the period of 1971 to 2014 obtained from different sources will be used. Table 1 shows the descriptive statistics of the data.

Table 1. Descriptive statistics of variables, 1971 to 2014

Variable	Mean	Median	Maximum	Minimum	SD
CO_2 emissions per capita	2.059	2.061	4.099	0.473	1.246
SO_2 emissions per capita	0.010	0.010	0.018	0.004	0.004
Real GDP per capita	2911.747	2982.571	5635.643	946.849	1495.951
Energy use per capita	960.190	859.703	2012.058	360.578	526.189
Trade openness	84.540	78.213	140.437	34.802	36.363
Urbanisation	31.787	29.848	49.174	21.442	7.308

3.2.1   Data from the World Bank
Thailand's real GDP per capita is measured as 2010 US dollar, trade openness which is the sum of exports and imports of goods and services measured as a share of gross domestic product, and the urbanisation measured as the percentage of urban population to the total population.

3.2.2   Data from the International Energy Agency (IEA)
Thailand's energy use based on Total Primary Energy Supply (TPES) is measured as KTOE (thousand metric tons of oil equivalent). The data will later be transformed into per capita term and measured as KGOE (kilograms of oil equivalent).

3.2.3   Data from the Emissions Database for Global Atmospheric Research (EDGAR)
The data for CO_2 emissions and SO_2 emissions, measured as gigagram (Gg), are transformed into metric ton per capita. However, the data for SO_2 emissions is

only available up to 2010. The CO_2 emissions are evaluated based on fossil fuel use and industrial processes (cement production, carbonate use of limestone and dolomite, and non-energy use of fuels and other combustion). The data exclude short-cycle biomass burning (such as agricultural waste burning) and large-scale biomass burning (such as forest fires).

3.3   The ARDL Bounds Testing Approach to Cointegration

The autoregressive distributed lag (ARDL) bound testing approach to cointegration proposed by Pesaran and Shin [26] and developed to be used with a small sample size by Narayan [22] will be employed to investigate the relationship between the emissions, income, and energy consumption and trade openness. Faridi and Murtaza [13] suggest that the ARDL approach is far superior to the Engle-Granger and Johansen technique when working with a small sample size. Pesaran *et al.* [28] and Ahmed *et al.* [3] point out that the ARDL approach has a high flexibility. For instance, it can be used with a mix of I(0) and I(1) variables whereas the Engle and Granger method has to use I(1) variables. Thus, it is not necessary to conduct a unit roots test. Shahe Emran *et al.* [35] argue that the ARDL method could correct for an endogeneity of explanatory variables. Shahbaz *et al.* [33] assert that a dynamic unrestricted error correction model (UECM) used in the ARDL approach "integrate[s] the short run dynamics with the long-run equilibrium without losing any long-run information". Lastly, the estimates are unbiased and efficient [2]. The UECM specification in Arouri *et al.* [6] was different in terms of lag length specification compared to other papers in this literature. Thus, this study follows Shahbaz *et al.* [34] who have authored many papers in the EKC literature using the ARDL methodology. The UECMs for each case of emissions are expressed as follows:

$$\Delta \ln E_t = \alpha_0 + \sum_{i=1}^{p} \alpha_{1i} \Delta \ln E_{t-i} + \sum_{i=0}^{p} \alpha_{2i} \Delta \ln Y_{t-i} + \sum_{i=0}^{p} \alpha_{3i} \Delta \ln EN_{t-i}$$

$$+ \sum_{i=0}^{p} \alpha_{4i} \Delta \ln TR_{t-i} + \sum_{i=0}^{p} \alpha_{5i} \Delta \ln UR_{t-i} + \gamma_1 \ln E_{t-1}$$

$$+ \gamma_2 \ln Y_{t-1} + \gamma_3 \ln EN_{t-1} + \gamma_4 \ln TR_{t-1} + \gamma_5 \ln UR_{t-1} + \omega_t \quad (3)$$

where Δ is the first difference operator; ω_t is independently and identically distributed error term; p represents the lag length. The optimal lag lengths are decided using 2 different criteria that are the Akaike information criterion (AIC) and Schwarz criterion (SC). In the case that each criterion provides a different lag length, the SC will be used due to its success and popularity in time series modelling [9]. Pesaran *et al.* [27] argue that SC performs better than AIC in a small sample size and provides a consistent model selection. Nevertheless, Halicioglu [14] suggests that in a situation or case where SC gives the optimal lag length of zero, AIC should be used instead. Equation 3 will be estimated using the ordinary least squares (OLS) method. The conclusion for the existence of long-run relationships among the variables depends upon the F-test for the joint significance of the lagged levels of the variables. The null hypothesis of no cointegration can be specified as

$$H_0 : \gamma_1 = \gamma_2 = \gamma_3 = \gamma_4 = \gamma_5 = 0$$

Narayan [22] provides two sets of critical values for the test at different levels of statistical significance. The lower bound critical value is used if all variables are I(0) and the upper bound is used when all variables are I(1). If the F-test statistic lies above the upper bound critical value, the null hypothesis is rejected, concluding that variables are cointegrated. This can be applied even in the case of a mix of I(0) and I(1) variables. If the test statistic is lower than the lower bound critical value, the null hypothesis cannot be rejected. However, if the test statistic lies between the lower and upper bound critical values then the result would be inconclusive.

3.4   Long-Run Coefficient Estimation

If cointegration among the variables exists, the next step is to apply the ARDL model specification to estimate the long-run equilibrium model using the ordinary least squares (OLS) method. The ARDL model for this study can be written as

$$\ln E_t = \theta_0 + \sum_{i=1}^{k} \theta_{1i} \ln E_{t-i} + \sum_{i=0}^{m} \theta_{2i} \ln Y_{t-i} + \sum_{i=0}^{n} \theta_{3i} \ln EN_{t-i}$$
$$+ \sum_{i=0}^{p} \theta_{4i} \ln TR_{t-i} + \sum_{i=0}^{q} \theta_{5i} \ln UR_{t-i} + \mu_t \tag{4}$$

where $k, m, n, p,$ and q are the lag lengths. The lag lengths will be selected using AIC, SC and HQ. As in the cointegration test, the lag lengths from SC will be used in cases where other criteria provide a different lag length. θ are the estimated long-run multipliers. μ is the error term. The estimates obtained from the ARDL model will provide long-run coefficients for the cointegrating equations which can be specified as follows:

$$E_t = \alpha_0 + \alpha_1 \ln Y_t + \alpha_2 \ln EN_t + \alpha_3 \ln TR_t + \alpha_4 \ln UR_t + \varepsilon_t \tag{5}$$

where

$$\alpha_0 = \frac{\theta_0}{1 - \sum\limits_{i=1}^{k} \theta_{1i}} ; \alpha_1 = \frac{\sum\limits_{i=0}^{m} \theta_{2i}}{1 - \sum\limits_{i=1}^{k} \theta_{1i}} ; \alpha_2 = \frac{\sum\limits_{i=0}^{n} \theta_{3i}}{1 - \sum\limits_{i=1}^{k} \theta_{1i}}$$

$$\alpha_3 = \frac{\sum\limits_{i=0}^{p} \theta_{4i}}{1 - \sum\limits_{i=1}^{k} \theta_{1i}} ; \alpha_4 = \frac{\sum\limits_{i=0}^{q} \theta_{5i}}{1 - \sum\limits_{i=1}^{k} \theta_{1i}}$$

We can see that Eq. (5) is the same as Eq. (2), but the values of coefficients are now obtained through the ARDL modelling. Estimation of the long-run coefficients by imposing the OLS method directly on Eq. (2) may give a biased result since it does not take into account non-stationary variables, which is not the case for the ARDL approach [25]. In addition, the estimated coefficients are considered to be super consistent [39].

3.5   Short-Run Coefficient Estimation

After the ARDL model is estimated for the long-run coefficients, the next step is to estimate short-run dynamic coefficients by employing the error correction representation of the ARDL model. The general error correction model is specified as follows:

$$\Delta \ln E_t = \lambda_0 + \sum_{i=1}^{k} \lambda_{1i} \Delta \ln E_{t-i} + \sum_{i=0}^{m} \lambda_{2i} \Delta \ln Y_{t-i} + \sum_{i=0}^{n} \lambda_{3i} \Delta \ln EN_{t-i}$$

$$+ \sum_{i=0}^{p} \lambda_{4i} \Delta \ln TR_{t-i} + \sum_{i=0}^{q} \lambda_{5i} \Delta \ln UR_{t-i} + \psi ECT_{t-1} \qquad (6)$$

Here, ψ is the coefficient of the error correction term. It represents the speed of adjustment and is expected to have value between -1 and 0. This is the requirement to guarantee that the disequilibrium in dependent variables will eventually converge to an equilibrium level. ECT_{t-1} denotes the error correction term obtained from the cointegrating equation.

As already mentioned, the applicability of the EKC hypothesis can be seen from a comparison between long run elasticity and short run elasticity of emissions with respect to income. If they are statistically significant and the long run is lower than the short run, it can be concluded that there may be an inverted U-shape relationship between emissions and income, hence the EKC hypothesis [23].

3.6   Model Stability Tests

This study also applies the cumulative sum (CUSUM) and cumulative sum of squares (CUSUMSQ) tests on the estimated ARDL model to ensure the stability of the estimated coefficients. Halicioglu [14] argues that the results of cointegration do not imply that the estimated coefficients are stable. The tests focus on the recursive residuals which are standardised one-step-ahead prediction error [19]. In addition, the estimated model is likely to under-predict the period out of the sample range and result in recursive error. The CUSUM and CUSUMSQ tests are based on the plot of the sum of the recursive and test against the null hypothesis that it has an expected value of zero.

4   Empirical Results and Discussion

4.1   Unit Roots Tests

Although the unit roots test may be unnecessary as the ARDL approach can be used with variables being either I(0) or I(1), it is better to be cautious and apply the unit roots test to make sure that all variables are not I(2). We first employed two unit roots tests—the Augmented Dickey-Fuller (ADF) unit roots test and KPSS unit roots test—under two model specifications; with the intercept, and with the intercept and trend terms.

Table 2. Augmented Dickey-Fuller unit roots test

Variable	Level				1st Difference			
	Intercept		Intercept and trend		Intercept		Intercept and trend	
	Lag	Test statistic	Lag	Test statistic	Lag	Test statistic	Lag	Test statistic
$\ln C$	1	−1.044	1	−1.269	0	−4.017*	0	−4.065**
$\ln S$	1	−1.830	1	−1.873	0	−3.706*	0	−3.747**
$\ln Y$	1	−1.443	1	−1.629	0	−3.883*	0	−4.041**
$\ln EN$	1	−0.261	3	−2.684	0	−4.8923*	0	−4.815*
$\ln TR$	0	−1.295	0	−2.083	0	−6.964*	0	−6.983*
$\ln UR$	2	0.738	1	−2.831	1	−1.691	1	−2.026

Note: The null hypothesis of ADF test is non-stationary. * and ** denotes statistical significant at 1%, 5% level respectively

The results of the ADF tests as shown in Table 2 indicate that most of the variables are stationary at the 1st difference with a rejection of the null hypothesis of non-stationary at 5% level of statistical significance, while the urbanisation is not stationary. The findings from KPSS test as presented in Table 3 showed trade openness to be stationary at the level, while the rest are stationary at the 1st difference. Furthermore, we employ the ZA unit roots test proposed by Zivot and Andrews [42] since the ADF and KPSS unit roots test may give a biased result if the structural break is present in time series data. Following Arouri

Table 3. KPSS unit roots test

Variable	Level				1st Difference			
	Intercept		Intercept and trend		Intercept		Intercept and trend	
	NW	Test statistic	NW	Test statistic	NW	Test statistic	NW	Test statistic
$\ln C$	5	0.809*	1	0.145***	3	0.221	0	0.105
$\ln S$	5	0.524**	4	0.125***	3	0.201	3	0.066
$\ln Y$	5	0.828*	5	0.153**	3	0.169	3	0.073
$\ln EN$	5	0.826***	5	0.093***	4	0.098	4	0.091
$\ln TR$	5	0.822*	4	0.112	1	0.128	2	0.069
$\ln UR$	5	0.805*	5	0.156**	5	0.259	5	0.164**

Note: NW denotes Newey-West automatic bandwidth. The null hypothesis of KPSS test is stationary. *, **, and *** denotes the rejection of null hypothesis at 1%, 5%, and 10% level of statistical significance, respectively.

Table 4. ZA unit roots test

Variable	Level			1st Difference		
	Lag	Test statistic	Break year	Lag	Test statistic	Break year
$\ln C$	1	−3.489	1989	0	−5.707*	1997
$\ln S$	1	−3.933	1990	0	−4.361	1988
$\ln Y$	1	−3.957	1988	0	−4.958***	1996
$\ln EN$	3	−3.688	1988	0	−6.848*	1984
$\ln TR$	0	−3.479	1988	0	−7.817*	1987
$\ln UR$	1	−5.025***	2001	1	−5.075***	2001

Note: The critical values for 1%, 5%, and 10% level of significance are −5.57, −5.08, and −4.82, respectively. The null hypothesis of the test is non-stationary. *, **, and *** denotes the rejection of null hypothesis at 1%, 5%, and 10% level of statistical significance, respectively

et al. [6] and Shahbaz et al. [32], the intercept and trend model of the ZA unit roots test is employed in this study.

The ZA unit root test results from Table 4 show that urbanisation is stationary at the level and the rest are stationary at the 1st difference. Combining all the tests, it can be concluded that the data is a mixture of I(0) and I(1) variables. Therefore, the ARDL bounds test for cointegration is appropriate in this study.

4.2 Cointegration Test

After the unit roots tests indicating that all variables are not I(2), the next step is to find the existence of a long-run relationship between variables using the ARDL bounds test for cointegration.

The results from Table 5 indicate that there is a cointegrating equation in case of CO_2 emissions only. This means there is a long run relationship between CO_2

Table 5. ARDL cointegration test results

Case	Lag length	F-statistic	Conclusion
CO_2	1	10.525*	Cointegrated
SO_2	1	2.256	Not cointegrated
Critical value	Lower bound	Upper bound	
At 1% level	4.428 (4.590)	6.250 (6.368)	
At 5% level	3.202 (3.276)	4.544 (4.630)	
At 10% level	2.660 (2.696)	3.838 (3.898)	

Note:

1. *, ** and *** denotes statistically significant at 1%, 5% and 10% level respectively.

2. The critical values are from the Case III table provided by Narayan (2005) with $k = 4$, $n = 40$ for CO_2 emissions and $n = 35$ for SO_2 emissions (provided in parenthesis).

emission, income, energy consumption, and trade openness in Thailand and that the EKC hypothesis can be investigated. On the other hand, such relationship is not found for SO_2 emissions. Since the cointegration test indicates that variables are cointegrated, the next step is to estimate the relevant ARDL models.

Table 6 shows that the ARDL(1, 1, 1, 1, 1) is selected. The value of R^2 is 0.99 which is very high, indicating the strong correlation among variables. The diagnostic tests, under 1% level of statistical significance, indicate that the estimated ARDL model does not suffer from serial correlation, heteroskedasticity nor have model misspecification.

4.3   Long-Run and Short-Run Elasticities

Using the results from the selected ARDL model, the long-run elasticities can be obtained as shown in Table 7.

It can be seen that all variables, except for trade openness, are statistically significant at 1% level. The real GDP per capita has positive impact on CO_2 emissions. It seems CO_2 emissions is income elastic as an increase in real GDP per capita by 1% will increase the emissions by 1.43% in the long run. Energy consumption also has positive relationship with CO_2 emissions. It shows that 0.32% increase in CO_2 emissions is resulted from 1% increase in energy consumption in the long run; hence, CO_2 emissions is energy inelastic. Trade openness has negative impact on CO_2 emissions. The expansion of international trade by 1% is expected to decrease CO_2 emissions by approximately 0.15% but it is not statistically significant. Lastly, urbanisation greatly reduces CO_2 emissions at the rate of 1.08% for each 1% increase in the share of urban population to total population.

To find the applicability of EKC hypothesis, the estimation of short-run dynamic coefficients using error correction representation of the ARDL model is required [23]. The results for short-run elasticities are shown in Table 8. The

Table 6. ARDL Model estimation results

Panel I: ARDL(1, 1, 1, 1, 1) Dependent variable: $\ln C_t$

Regressor	Coefficient	Standard error	T - Ratio	Prob.
$\ln C_{t-1}$	0.292**	0.141	2.072	0.046
$\ln Y_t$	0.875*	0.182	4.817	0.000
$\ln Y_{t-1}$	0.138	0.263	0.524	0.604
$\ln EN_t$	0.328**	0.133	2.458	0.019
$\ln EN_{t-1}$	−0.102	0.131	−0.777	0.443
$\ln TR_t$	0.067	0.065	1.032	0.309
$\ln TR_{t-1}$	−0.175*	0.062	−2.829	0.008
$\ln UR_t$	0.784	0.719	1.091	0.283
$\ln UR_{t-1}$	−1.547**	−2.136	8.692	0.040
Constant	−6.033*	1.126	−5.356	0.000
R^2	0.999	Adjusted R^2	0.999	

Panel II: Diagnostic tests

A: Serial correlation	$F(1, 32) = 1.966 \ (0.171)$
B: Functional form	$F(1, 32) = 5.345 \ (0.027)$
C: Heteroskedasticity	$F(1, 40) = 0.788 \ (0.380)$

Note:
1. The null hypothesis for the diagnostic tests is as follows. (A): no serial correlation, (B): no functional form misspecification, and (C): no Heteroskedasticity. For information about the tests, see [25].
2. *, **, and *** indicate statistically significant at 1%, 5%, and 10% level, respectively.

coefficient of the error correction term has negative sign and is statistically significant at 1% level, confirming the existence of long-run relationship among variables and indicating that approximately 71% of disequilibrium in CO_2 emissions is adjusted within 1 year. The elasticity of CO_2 emissions with respect to real GDP per capita in the short run is 0.88 and statistically significant at 1% level, showing that a 1% increase in real GDP per capita will only increase CO_2 emissions by 0.88%, compared with the long run which is 1.43%. This means in the long run the impact from GDP toward CO_2 emissions increases as economy grows, rejecting an applicability of EKC hypothesis. The greater long run GDP elasticity of CO_2 emissions may imply that Thailand, as developing country, is in the beginning phase of EKC hypothesis where an increase in income will result in environmental degradation. Hence, the current data suggest that it has not reached the turning point. The result is contrast with the previous studies [6,29].

For energy consumption, it is statistically significant at 1% level. The elasticity of CO_2 emissions with respect to energy consumption per capita in the short run is about 0.33. This indicates that if energy consumption increases by 1%, CO_2 emissions is expected to increase by 0.33%, which is inelastic. Similarly, the

Table 7. Estimated long-run coefficients

Dependent variable: $\ln C_t$

Regressor	Coefficient	Standard error	T - Ratio	Prob.
$\ln Y_t$	1.430*	0.149	9.622	0.000
$\ln EN_t$	0.320*	0.095	3.357	0.002
$\ln TR_t$	−0.153	0.121	−1.261	0.216
$\ln UR_t$	−1.076*	0.097	−11.107	0.000

Note: * indicates statistically significant at 1% level.

Table 8. Estimated short-run coefficients

Dependent variable: $\Delta \ln C_t$

Regressor	Coefficient	Standard error	T - Ratio	Prob.
$\Delta \ln Y_t$	0.875*	0.138	6.348	0.000
$\Delta \ln EN_t$	0.328*	0.104	3.147	0.004
$\Delta \ln TR_t$	0.067	0.048	1.400	0.171
$\Delta \ln UR_t$	0.784**	0.376	2.086	0.045
ECT_{t-1}	−0.708*	0.092	−7.668	0.000
Constant	−6.033	0.787	−7.681	0.000

Note: *, **, and *** indicates statistically significant at 1%, 5% and 10% level, respectively.

energy elasticity of CO_2 emissions in the long run is about 0.32%. If nothing has been done, CO_2 emissions will increase further due to the use of energy in the long run though the impact seems to be slightly lower. Since the values are very close, it may be appropriate to conclude that the effect of energy use on CO_2 emissions is rather stable over time. The finding is contrast to Saboori and Sulaiman [29] where the elasticity is increasing over time and their study found that CO_2 emissions is energy elastic in the long run.

The elasticity of CO_2 emissions with respect to trade openness in the short run shows a positive sign. An increase in trade openness by 1%, CO_2 emissions is expected to increase 0.07% approximately, which is very small, suggesting that trade openness is inelastic. However, it is not statistically significant, in contrast to Arouri et al. [6]. The insignificant positive elasticity may reflect that the factor endowment hypothesis is not completely dominated by the pollution haven hypothesis.

For the case of urbanisation, a 1% increase is expected to increase CO_2 emissions by 0.78% which is contrast to that of Arouri et al. [6]. In addition, comparing with the result from the long run, it seems that the effect completely shifts to opposite direction where urbanisation reduces CO_2 emissions. This suggests that the CO_2 emissions will increase when the urban area begins to expand. With more migrations into cities and inefficient transportation management, more pollution is likely. As things start to settle down, a better public transportation and establishment is now well planned, people in urban area may tend to use less private transportation and experience a shorter commuting time reducing a chance of having higher CO_2 emissions later on.

4.4   Cumulative Sum (CUSUM) and Cumulative Sum of Squares (CUSUMSQ) Tests

The estimated ARDL model does not indicate that the estimated coefficients are stable over time. The conclusion for the stability of estimated coefficients depends upon the line plotted by the test. If the plotted lines are within 5% level of significance range, it is possible to say that the coefficient is stable over time [32].

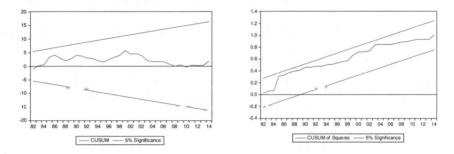

Fig. 1. CUSUM and CUSUMSQ plots

According to Fig. 1, the straight lines represent critical bounds at 5% level of significance. The coefficients estimated from the ARDL model are stable because the CUSUM and CUSUMSQ test lines are within the critical bounds.

5   Conclusions and Policy Implications

This study investigates the relationship between emissions, trade openness, urbanization, and income in Thailand over the period of 1971 to 2014 and examines the applicability of EKC hypothesis using a new model specification. CO_2 emissions and SO_2 emissions are investigated in this study. The ARDL cointegration technique is employed and CUSUM and CUSUMSQ tests are used to ensure the stability of the estimated results.

The results indicate that there is a long run relationship among variables for the case of CO_2 emissions while there is none for the SO_2. Therefore, only determinants of CO_2 emissions are further investigated in the study. Our results suggest that EKC hypothesis is not found in Thailand. An increase in income drives more CO_2 emissions at increasing rate. Energy consumption also contributes to environmental degradation with slight impact over time. In the case of trade openness, it does not exhibit significant impact on the emissions, suggesting the pollution haven hypothesis being offset by the factor endowment hypothesis. Lastly, we found that urbanisation contributes to CO_2 emissions in the short run, while it helps reduce CO_2 emissions in the long run.

In terms of policy implications, the study suggests that Thai policy makers should seriously focus on environmental regulations by setting a new emission standard or encouraging the use of new environmentally friendly equipment in industrial sector. In addition, the policy makers would also need to provide incentives to both the private producers and consumers of polluting goods, so there can be greater investment, adoption and widespread use of environmentally friendly goods. Incentives can be in the form of tax breaks, loan guarantees, and making technologies available to companies at low prices. In addition, taxes can be imposed on high polluting sources while subsidies are earmarked towards the low carbon technologies. This will significantly reduce CO_2 emissions in the future.

Even though energy consumption, trade openness and urbanisation still have moderate impacts on CO_2 emissions compared to the one from income. The study still would like to suggest that Thailand should put some efforts in slowing the growth of fossil based energy consumption. This can be achieved by investments in environmentally friendly and efficient technology, substituting for fossil based sources of energy with alternative sources such as wind and solar energy, and providing incentives to consumers and industry to encourage the use of low carbon alternatives. Turning to these renewable energy sources would help maintain a diversified future sustainable source of energy. In addition, it is believed to reduced pollution related mortality, reduced natural climate related disasters, improved health and productivity. As far as trade openness is concerned, the more stringent environmental regulation on establishing foreign firms. In case of urbanisation, the government should encourage the use of public transport with efficient provision of such services along with well-planned urbanisation.

To the extent how these variables are positively associated with carbon emissions, an understanding of the mechanism and channels through which these factors affect carbon emissions will be required. Efforts to identify these channels will be needed if meaningful carbon public policies and targets that yield reduction in carbon emissions are to be achieved. Given the current lack of understanding on these mechanisms, efforts to increase information in both the private and public sectors should be encouraged. This may require the policy makers to allocate resources to academic institutions and the private sector on researches seeking to determine the contribution of different production processes and mechanisms within the economy to aggregate emissions. A deeper understanding may help policy makers with tools to design and adopt trade and economic policies that help reduce CO_2 emissions.

Areas for further research include expanding and conducting the analysis on other types of emissions like particulate matters and considering the role of technological transfer and adoption through variables such as foreign direct investment in influencing income and emissions and the contribution of the value chain to emissions. Furthermore, the future work may thoroughly investigate the behaviour of emissions through the decomposition approach which concerns about scale, composition, and technical effects. This would provide a better view of how economic expansion contributes to the emissions.

Acknowledgements. The authors would like to thank the anonymous reviewer for useful suggestions which have greatly improved the quality of this paper. This research is supported by the Chiang Mai University Research Funding and the Puay Ungphakorn Centre of Excellence in Econometrics, Faculty of Economics, Chiang Mai University.

References

1. Acaravci, A., Ozturk, I.: On the relationship between energy consumption, CO_2 emissions and economic growth in Europe. Energy **35**(12), 5412–5420 (2010)
2. Afzal, M., Farooq, M.S., Ahmad, H.K., Begum, I., Quddus, M.A.: Relationship between school education and economic growth in Pakistan ardl bounds testing approach to cointegration. Pak. Econ. Soc. Rev. **48**(1), 39–60 (2010)
3. Ahmed, M.U., Muzib, M., Roy, A.: Price-wage spiral in Bangladesh: evidence from ardl bound testing approach. Int. J. Appl. Econ. **10**(2), 77–103 (2013)
4. Al-Mulali, U., Saboori, D., Ozturk, I.: Investigating the environmental Kuznets curve hypothesis in Vietnam. Energy Policy **76**, 123–131 (2015)
5. Antweiler, W., Copeland, B.R., Taylor, M.S.: Is free trade good for the environment? Am. Econ. Rev. **91**(4), 877–908 (2001)
6. Arouri, M., Shahbaz, M., Onchang, R., Islam, F., Teulon, F.: Environmental Kuznets curve in Thailand: cointegration and causality analysis (2014-204) (2014)
7. Brack, D.: Trade and Environment: Conflict or Compatibility? Royal Inst. of International Affairs (1998)
8. Bureecam, C.: An empirical analysis based on the environmental Kuznets curve hypothesis (ekc) in the relationship between Thailand's economic growth and environmental quality. In: 47th Kasetsart University Annual Conference, pp. 148–156. Bangkok (2009)
9. Cavanaugh, J.E., Neath, A.A.: Generalizing the derivation of the schwarz information criterion. Commun. Stat. Theory Methods **28**(1), 49–66 (1999)
10. Cole, M.A.: Trade, the pollution haven hypothesis and the environmental Kuznets curve: examining the linkages. Ecol. Econ. **48**(1), 71–81 (2004)
11. Cole, M.A., Elliott, R.J.R.: Determining the trade-environment composition effect: the role of capital, labor and environmental regulations. J. Environ. Econ. Manag. **46**(3), 363–383 (2003)
12. Dinda, S.: Environmental Kuznets curve hypothesis: a survey. Ecol. Econ. **49**(4), 431–455 (2004)
13. Faridi, M.Z., Murtaza, G.: Disaggregate energy consumption, agricultural output and economic growth in Pakistan. Pak. Dev. Rev. **52**(4), 493–516 (2013)
14. Halicioglu, F.: An econometric study of CO_2 emissions, energy consumption, income and foreign trade in Turkey. Energy Policy **37**(3), 1156–1164 (2009)
15. Hamit-Haggar, M.: Greenhouse gas emissions, energy consumption and economic growth: a panel cointegration analysis from Canadian industrial sector perspective. Energy Econ. **34**(1), 358–364 (2012)
16. He, J., Richard, P.: Environmental Kuznets curve for CO_2 in Canada. Ecol. Econ. **69**(5), 1083–1093 (2010)

17. Hossain, M.S.: Panel estimation for CO_2 emissions, energy consumption, economic growth, trade openness and urbanization of newly industrialized countries. Energy Policy **39**(11), 6991–6999 (2011)

18. Jaunky, V.C.: The CO_2 emissions-income nexus: evidence from rich countries. Energy Policy **39**(3), 1228–1240 (2011)

19. Kennedy, P.: A Guide to Econometrics, 5th edn. Blackwell, Malden (2003)

20. Libecap, G.D.: Addressing global environmental externalities: transaction costs considerations. J. Econ. Lit. **52**(2), 424–479 (2014)

21. Naito, T., Traesupap, S.: Is shrimp farming in Thailand ecologically sustainable? J. Fac. Econ. **16**, 55–75 (2006)

22. Narayan, P.K.: The saving and investment nexus for China: evidence from cointegration tests. Appl. Econ. **37**(17), 1979–1990 (2005)

23. Narayan, P.K., Narayan, S.: Carbon dioxide emissions and economic growth: panel data evidence from developing countries. Energy Policy **38**(1), 661–666 (2010)

24. Pao, H.T., Tsai, C.M.: Multivariate Granger causality between CO_2 emissions, energy consumption, FDI (foreign direct investment) and GDP (gross domestic product): evidence from a panel of BRIC (Brazil, Russian Federation, India, and China) countries. Energy **36**(1), 685–693 (2011)

25. Pesaran, B., Pesaran, M.H.: Time Series Econometrics Using Microfit 5.0. Oxford University Press, Oxford (2009)

26. Pesaran, M.H., Shin, Y.: An autoregressive distributed lag modelling approach to cointegration analysis, pp. 371–413. Cambridge University Press, Cambridge (1999)

27. Pesaran, M.H., Shin, Y., Smith, R.J.: Bounds testing approaches to the analysis of long-run relationships. Report, Faculty of Economics, University of Cambridge (1999)

28. Pesaran, M.H., Shin, Y., Smith, R.J.: Bounds testing approaches to the analysis of level relationships. J. Appl. Econ. **16**(3), 289–326 (2001)

29. Saboori, B., Sulaiman, J.: CO_2 emissions, energy consumption and economic growth in Association of Southeast Asian Nations (ASEAN) countries: a cointegration approach. Energy **55**, 813–822 (2013)

30. Saboori, B., Sulaiman, J., Mohd, S.: Economic growth and CO_2 emissions in Malaysia: a cointegration analysis of the environmental Kuznets curve. Energy Policy **51**, 184–191 (2012)

31. ASEAN Secretariat: ASEAN Economic Community Blueprint. ASEAN Secretariat, Jakarta, Indonesia (2008)

32. Shahbaz, M., Sbia, R., Hamdi, H., Ozturk, I.: Economic growth, electricity consumption, urbanization and environmental degradation relationship in United Arab Emirates. Ecol. Ind. **45**, 622–631 (2014)

33. Shahbaz, M., Shabbir, M.S., Butt, M.S.: Effect of financial development on agricultural growth in Pakistan. Int. J. Soc. Econ. **40**(8), 707–728 (2013)

34. Shahbaz, M., Uddin, G.S., Rehman, I.U., Imran, K.: Industrialization, electricity consumption and CO_2 emissions in Bangladesh. Renew. Sustain. Energy Rev. **31**, 575–586 (2014)

35. Shahe Emran, M., Shilpi, F., Alam, M.I.: Economic liberalization and price response of aggregate private investment: time series evidence from India. Can. J. Econ./Revue canadienne d'conomique **40**(3), 914–934 (2007)

36. Sharma, S.S.: Determinants of carbon dioxide emissions: empirical evidence from 69 countries. Appl. Energy **88**(1), 376–382 (2011)

37. Stern, D.I.: The environmental Kuznets curve: a primer. Report, Centre for Climate Economics & Policy, Crawford School of Public Policy, The Australian National University (2014)
38. Temurshoev, U.: Pollution haven hypothesis or factor endowment hypothesis: theory and empirical examination for the US and China. Report, The Center for Economic Research and Graduate Education - Economic Institute, Prague (2006)
39. Verbeek, M.: A Guide to Modern Econometrics, 4th edn. Wiley, Chichester (2012)
40. Wong, Y.L.A., Lewis, L.: The disappearing environmental Kuznets curve: a study of water quality in the lower Mekong basin (LMB). J. Environ. Manage. **131**, 415–425 (2013)
41. Wooldridge, J.M.: Introductory Econometrics: A Modern Approach, 6th edn. Cengage Learning, Boston (2015)
42. Zivot, E., Andrews, D.W.K.: Further evidence on the great crash, the oil-price shock, and the unit-root hypothesis. J. Bus. Econ. Stat. **10**(3), 251–270 (1992)

Analysis of Risk, Rate of Return and Dependency of REITs in ASIA with Capital Asset Pricing Model

Rungrapee Phadkantha[1], Woraphon Yamaka[1,2],
and Roengchai Tansuchat[1,2(✉)]

[1] Faculty of Economics, Chiang Mai University, Chiang Mai 50200, Thailand
rungrapee.ph@gmail.com, woraphon.econ@gmail.com
[2] Center of Excellence in Econometrics,
Chiang Mai University, Chiang Mai 50200, Thailand
roengchaitan@gmail.com

Abstract. This study introduces an approach to fitting a copula based seemingly unrelated regression to an interval-valued data set. This approach consists of fitting a model on the appropriate point of the interval values assumed by the variables in the learning set. To find the appropriate point of the interval values, we assign weights in calculating the appropriate value between intervals by using convex combination method. We apply this methodology to quantify the risk and dependence of Real Estate Investment Trust (REITs) in Asia. Our results suggest that Hong Kong and Japan markets have a positive sign of the beta and both markets have less volatility than the global REITs market. On the other hand, we find that the estimated beta for Singapore market shows a negative relationship with global REITs market. We conclude that Singapore market can be viewed as a hedge against higher risk in Asian REITs.

1 Introduction

This paper considers the relationship between returns and risk of Real Estate Investment Trust (REITs) in Asia. The steady growth in the Asian property market has been the major driver for the development of REITs in Asia over the last ten years. In November 2000, Japan was the first country in Asia to establish a REIT market. Since then, there have been seven countries, namely Japan, Singapore, South Korea, Thailand, Taiwan, Malaysia and Hong Kong, joined the REIT market. In addition, other Asian countries such as China, India, Pakistan and the Philippines are also in the process of implementing their own REIT regimes. In just over a decade, the number of Asian REITs has grown to 138 REITs across the seven Asian REIT markets contributing a market capitalization more than US \$118 billion. Asian REITs have become a major component of the global property portfolio, accounting for 12% of the global REIT market in 2012 [6].

© Springer International Publishing AG 2018
V. Kreinovich et al. (eds.), *Predictive Econometrics and Big Data*, Studies in Computational
Intelligence 753, https://doi.org/10.1007/978-3-319-70942-0_38

Asian REITs have also delivered strong performance over the last five years, significantly outperforming the major REIT markets of the US, UK and Australia. The development of Asian REITs is further supported by favorable changes in the regulatory structures in recent years. This has been received positively by both local and international investors [9]. Due to the ongoing demand for investment in the REITs, the price index tends to increase continuously over the last decade.

However, the decline of the price index due to the subprime mortgage crisis caused the investors to face huge risks, especially in Japan, Hong Kong and Singapore. These three countries have a high market capitalization in Asian, and the movements of their price indexes contribute a large effect to the global REITs markets [3]. Thus, this study would measure the risk of these three markets. To measure the risk, one of the most well-known approaches, Capital Asset Pricing Model (CAPM) of Sharpe [12] and Lintner [5] is considered in our study. This approach has an ability to explain the relationship between expected return and risk of each market. However, all these three markets belong to the Asian REITs market, any change in each market conditions or policy regulations may affect other markets. So we can expect that there might exist a correlation between each CAPM equation. Thus the model seems appropriate for this structural configuration is Seemingly Unrelated regression (SUR) model. After Zellner [17] introduced multivariate regressions, Seemingly Unrelated regression (SUR) model has become popular in both statistics and econometrics. In the recent years, due to the strong assumption in multivariate normal, a Copula approach of Sklar in 1959 has been applied to join the error term $\varepsilon_{i,t}$ of SUR model (see,Wichitaksorn [16], Pastpipatkul et al. [7]). Thus, this study employs a copula based SUR model as proposed in Pastpipatkul et al. [7] to quantify the risk of Japan, Hong Kong and Singapore markets. However, there has been other recent extension of the CAPM to the interval data, as seen in Piamsuwannakit et al. [11] and Phochanachan et al. [10]. Their work is about using the interval data is range of highest and lowest data to predict the return of the stock and applying CAPM to model this return. They found that using the interval data in CAPM can provide a better result than using the closing price for prediction. In this study, therefore, addresses a copula based SUR model for predicting interval data by using a convex combination method which was introduced in Chanaim et al. [2].

As a consequence, in this paper two contributions are made. First, copula based SUR model is adapted to interval data characterized by convex combination method and applied to the CAPM approach. Second, we will quantify the risk of Japan, Hong Kong and Singapore REIT markets and their dependency

The remainder of the paper is organized as follows. Section 2 briefly reviews a CAPM and a convex combination method, Sect. 3 presents our methodology, and Sect. 4 presents the simulation study. The application of our purposed model is reported in Sect. 5. Finally, Sect. 6 summarizes and presents the conclusions of this paper.

2 Review

2.1 Capital Asset Pricing Model

The basic idea behind Capital asset pricing model (CAPM) consists of two parts. The first part is the time value of money is the idea that money available at the present time is worth more than the same amount in the future due to its potential earning capacity. This core principle of finance holds that, provided money can earn interest, any amount of money is worth more the sooner it is received. TVM is also referred to as present discounted value which corresponds to the risk-free. The risk-free rate is the minimum return an investor expects for any investment with zero risk. In many application studies, the interest rate on a government bond and Treasury bill is often used as the risk-free rate for investors. The second part of the CAPM formula represents risk β and calculates the amount of compensation the investor needs for taking on additional risk. β reflects how risky an asset is compared to overall market risk and is a function of the volatility of the asset and the market as well as the correlation between the two.

$$\beta_i = \frac{cov(r_{it}, r_{Mt})}{\sigma_M^2} \tag{1}$$

where r_{it} is the return of the asset i, r_{Mt} is the return of the market, and σ_M^2 is the variance of the return of the market. According to the studies of Sharpe [12] and Lintner [5], CAPM is used to describe the relationship between beta of an asset and its corresponding return and returns can be explained as:

$$r_{it} - r_{ft} = \beta_0 + \beta_i, (r_{Mt} - r_{ft}) + \varepsilon_{it}, \tag{2}$$

where r_{ft} is risk free, ε_{it} is the random disturbance term in the CAPM at time t. The coefficient β_i refers to a relationship between the return of asset $i (r_{it})$ and the rate of return on market (r_{Mt}). Moreover, it is also viewed as a systematic risk. It represents a sensitivity of the asset i to the overall market. If $\beta_i < 1$, the asset is less volatile than the market. On the other hand, $\beta_i > 1$, the asset is more volatile than the market. In this study, we can view Hong Kong, Singapore, and Japan REITs as r_{it} while the global REITs can be viewed as r_{Mt}.

2.2 Center Method

The center method was first applied to time series model by Billard and Diday [1]. They proposed this method to the linear regression model. The main idea of this method is the estimated slope parameter is based on the center or mid-point of the interval data. Let $Y_t = [Y_t^U, Y_t^L]$, $i = 1, ..., T$ where Y_t^U denotes the upper bound value of observed interval Y_i while Y_t^L denotes the lower bound of observed interval Y_i. To obtain the mid-point value of Y_i, we can derive by

$$Y_t^C = \frac{Y_t^U + Y_t^L}{2} \tag{3}$$

2.3   Convex Combination Method

However, Chanaim et al. [2] suggested that the center method may lead to the misspecification problem since the midpoint of the intervals might not be the good representative of the intervals. To overcome this problem, Chanaim et al. [2] suggested a convex combination approach for interval data, which can be derived by

$$Y^{CC} = wY_t^U + (1-w)Y_t^L, \quad w \in [0,1] \tag{4}$$

where w is the weight parameter of the interval data with values [0,1]. The advantage of this method lies in the flexibility to assign weights in calculating the appropriate value between intervals.

3   Methodology

3.1   Copula Based Seemingly Unrelated Regression with Interval Valued Data

After Zellner [17] introduced multivariate regressions, Seemingly Unrelated regression (SUR) model has become popular in both statistics and economet-rics. In the recent years, due to the strong assumption in multivariate normal, a Copula approach of Sklar [13] in 1959 has been applied to join the error term of SUR model by Pastpipatkul et al. [7]. The present study addresses a copula based SUR model for predicting interval data. Our approach consists of fitting a model to the appropriate point of the interval values assumed by the interval variables in the learning set and applies this model to the lower and upper bounds of the interval values of the explanatory interval variables, $X_{i,t} = [X_{i,t}^U, X_{i,t}^L]$ to predict the lower and upper bounds of the interval value of the dependent variable, $Y_{it} = [Y_{it}^U, Y_{it}^L]$ respectively. Consider the structure of SUR model with n equations, the typical setup of SUR

$$Y_{i,t} = X'_{i,t}\beta_i + \varepsilon_{i,t} \quad , i = 1, ..., n; \ t = 1, ...T \tag{5}$$

where $Y_i = (T \times 1)$ vector of dependent variable in equation i^{th}. X_i is $(T \times K)$ matrix of K_i independent variables or explanatory variables in equation i^{th} and β_i is $(K \times 1)$ vector of an unknown parameters in each equation i^{th}. $\varepsilon_{i,t}$ is a vector of the error terms which are independent and identically distributed and also assumed to be correlated. Generally, the dependence between $\varepsilon_{i,t}$ are modeled through a multivariate distribution, especially the multivariate normal distribution. However, a limitation of multivariate distribution to the multivariate SUR model is the linear relationship. The correlation analysis can only capture linear relationship, but not many other types of dependence. Moreover, it cannot be used for heavy-tailed distributions. To relax this assumption, we can use the copulas to model the nonlinear dependence of error terms in the multivariate SUR model [7,8]. In this study, we address a copula based SUR model for predicting interval data, therefore a convex combination method is employed and we can rewrite Eq. 5 as follows:

$$
\begin{aligned}
w_y Y_{i,t} + (1 - w_y) Y_{i,t} &= (w_{i1} X_{1,t} + (1 - w_{i1}) X_{1,t}) \beta_1 + , ..., \\
&\quad + (w_{ik} X_{k,t} + (1 - w_{ik}) X_{k,t}) \beta_i + \varepsilon_{i,t}
\end{aligned}
\tag{6}
$$

3.2 Copulas

According to a general Sklar's theorem in 1959 [13], let H be an n-dimensional distribution with marginals F_i $i = 1, 2, ...n$. Then there exists an n-copula C such that for all $x_1, ..., x_n$ in $\bar{R}$

$$
H(x_1, ..., x_n) = C(F_1(x_1), ..., F_n(x_n))
\tag{7}
$$

where C is copula distribution function of a n-dimensional random variable. Furthermore, if the marginals are continuous, then the copula C is unique. Otherwise, C is uniquely determined on $\prod_{i=1}^{n} R(F_i)$ where $R(F_i)$ is denoted as the range of the marginal function F_i. Conversely, if C is an n-copula C and F_i $i = 1, 2, ...n$ are univariate marginal distributions of n variables, then the function $H : \bar{R}^n \rightarrow [0, 1]$ is also defined in Eq. (7). We can model the marginal distribution and joint dependence separately. If we have a continuous marginal distribution, the copula can be determined by

$$
C(u_1, ..., u_n) = C(F_1^{-1}(u_1), ..., F_n^{-1}(u_n))
\tag{8}
$$

where u is uniform [0,1]. There are two important classes of copula, namely Elliptical copulas and Archime-dean copulas. The symmetric Gaussian and t-copulas are the families of the copula in Elliptical class, which are simply the copulas of elliptical contoured distribution. For the asymmetric dependence cases, Clayton, Gumbel, Joe, and Frank are four important families in Archimedean class. In this study, we consider these six copulas to join the error terms of SUR model. For a brief review of the Copula approach, refer to Joe [4], Smith [15] and Smith, Gan and Kohn [14].

4 Experiment Study

4.1 Simulation Study

In the simulation study, we applied elliptical copulas (Gaussian and Student-t) and Archimedean copulas (Clayton, Gumbel, Joe and Frank) to model the dependence structure of the SUR. In this study, the simulation is the realization of SUR with two equations, thus we generated random data from the following specifications

$$
\begin{bmatrix} w_{y1} Y_{1,t}^U + (1 - w_{y1}) Y_{1,t}^L \\ w_{y2} Y_{2,t}^U + (1 - w_{y2}) Y_{2,t}^L \end{bmatrix} = \begin{cases} 2 + 1(w_{11} X_{11,t}^U + (1 - w_{11}) X_{11,t}^L) + \varepsilon_{1,t} \\ 0.5 - 2(w_{21} X_{21,t}^U + (1 - w_{21}) X_{21,t}^L) + \varepsilon_{2,t} \end{cases}
\tag{9}
$$

Table 1. True parameter value of simulation study when n=100

Parameter	Gaussian	Student-t	Clayton	Gumbel	Joe	Frank
β_{01}	2.0002	2.0003	2.0653	1.9974	1.9961	2.0241
	(0.0051)	(0.0038)	(0.0420)	(0.0489)	(0.0063)	(0.0285)
β_{11}	0.9250	1.0286	1.0775	0.9221	0.8283	1.0470
	(0.0335)	(0.0336)	(0.0976)	(0.0489)	(0.0415)	(0.0941)
σ_1	0.3283	0.3361	0.4461	0.4087	0.3797	0.4532
	(0.0164)	(0.0166)	(0.0250)	(0.0489)	(0.0184)	(0.0267)
β_{02}	0.4298	0.5257	0.5388	0.4101	0.3528	0.5067
	(0.0326)	(0.0331)	(0.0955)	(0.0468)	(0.0412)	(0.0935)
β_{21}	−2.0008	−2.0013	−1.9988	−1.9969	−2.0001	−1.9848
	(0.0045)	(0.0037)	(0.0266)	(0.0087)	(0.0071)	(0.0297)
σ_1	0.8299	0.8328	0.9479	0.9950	0.9584	0.9593
	(0.0031)	(0.0173)	(0.0285)	(0.0209)	(0.0186)	(0.0285)
ν_2	3.2548	0.4111	0.3598	3.5455	3.7454	3.8424
	(0.1588)	(0.0255)	(0.0157)	(0.5846)	(0.8657)	(0.8386)
w_{y1}	0.2984	0.3001	0.3011	0.3017	0.2977	0.3075
	(0.0026)	(0.0019)	(0.0147)	(0.0046)	(0.0043)	(0.0164)
w_{11}	0.3992	0.3993	0.3970	0.3950	0.4012	0.3958
	(0.0031)	(0.0022)	(0.0226)	(0.0209)	(0.0053)	(0.0217)
w_{y2}	0.5004	0.4989	0.4869	0.4933	0.4990	0.5010
	(0.0042)	(0.0030)	(0.0295)	(0.0084)	(0.0069)	(0.0935)
w_{21}	0.5996	0.5997	0.5823	0.5996	0.5964	0.5950
	(0.0046)	(0.0032)	(0.0273)	(0.0083)	(0.0076)	(0.0276)
θ	0.5001	0.5201	5.9945	5.0001	5.0021	4.1111
	(0.0255)	(0.0015)	(0.8569)	(0.4936)	(0.5302)	(0.9706)

Source: Calculation

Note: () is standard error.

where $w_{y1}, w_{11}, w_{y2}, w_{21}$ are set to be 0.3, 0.4, 0.5, and 0.6, respectively. The error terms are assumed to follow a normal distribution with $\varepsilon_{1,t} \sim N(0, 0.5)$ and $\varepsilon_{2,t} \sim t(0, 1, 4)$ for Eq. (9). For the dependence parameter for the copula function, we set the true value for the Gaussian and Student-t dependence coefficient at 0.5 with degree of freedom 4, the true value for the Clayton, Gumbel, Joe and Frank dependence coefficient is set at 5. The general simulation scheme goes as follows. We perform copula based SUR model using sample size n = 100 and n = 200 for all cases (Table 1).

The estimation results from Table 2 show the estimated parameters and we observed that our proposed model and method produced the unbiased parameter estimates when compared with true value. We find that the estimated parameters are close to the true value and the standard error is very low. In addition,

Table 2. True parameter value of simulation study when n=200

Parameter	Gaussian	Student-t	Clayton	Gumbel	Joe	Frank
β_{01}	1.9993	2.0009	2.0058	2.0005	2.0014	1.9908
	(0.0038)	(0.0025)	(0.0175)	(0.0045)	(0.0061)	(0.0207)
β_{11}	0.9758	1.0188	1.0766	0.9417	0.7936	0.0035
	(0.0255)	(0.0245)	(0.0535)	(0.0288)	(0.0305)	(0.0602)
σ_1	0.3612	0.3663	0.3597	0.3603	0.3957	0.4379
	(0.0128)	(0.0126)	(0.0144)	(0.0126)	(0.0134)	(0.0186)
β_{02}	0.4693	0.5154	0.6197	0.4491	0.3244	0.5352
	(0.0031)	(0.0242)	(0.0494)	(0.0287)	(0.0300)	(0.0595)
β_{21}	−2.0003	−2.0009	−1.9987	−2.0011	−1.9994	−1.9952
	(0.0036)	(0.0029)	(0.0169)	(0.0054)	(0.0057)	(0.0206)
σ_1	0.9553	0.9608	0.9509	0.9496	0.9775	0.9420
	(.01331)	(0.0131)	(0.0155)	(0.0130)	(0.0137)	(0.0213)
ν_2	.5789	4.4447	3.9248	3.9548	3.9248	4.5158
	(0.3090)	(0.4256)	(0.0102)	(0.0125)	(0.0671)	(0.7555)
w_{y1}	0.3002	0.3002	0.2942	0.3001	0.2979	0.2948
	(0.0018)	(0.0014)	(0.0084)	(0.0026)	(0.0027)	(0.0115
w_{11}	0.4001	0.3994	0.4115	0.4019	0.4002	0.3943
	(0.0023)	(0.0017)	(0.0118)	(0.0034)	(0.0040)	(0.0147)
w_{y2}	0.4983	0.4994	0.4941	0.4978	0.5023	0.5168
	(0.0031)	(0.0023)	(0.0154)	(0.0043)	(0.0051)	(0.0196)
w_{21}	0.5992	0.5991	0.6058	0.6008	0.5995	0.6028
	(0.0032)	(0.0023)	(0.0143)	(0.0025)	(0.0048)	(0.0188)
θ	0.4994	0.5158	5.0045	5.0001	4.9588	5.01255
	(0.1225)	(0.0010)	(0.5649)	(0.3442)	(0.3718)	(0.7124)

Source: Calculation

Note: () is standard error.

when the sample size increases, a more accurate estimation is obtained which is consistent with the $\sqrt{n}$-consistency of the coefficient estimators. Therefore, we can conclude that the proposed model has a good finite sample performance and is reasonable for CAPM application.

5   Application Study to CAPM Model with Interval REIT Returns

This section presents benchmark result. We analyze the returns and risks of ASIAN REITs markets.

5.1   Data Description

This paper considers three REITs in Asia. We use daily interval returns of three REITs in ASIA including Hong Kong, Japan and Singapore for the period of April 1, 2009 to February 24, 2017. The data is collected from Yahoo Finance and Thomson Reuters.

Global REITs (GL) is a free-float adjusted, market capitalization-weighted index designed to track the performance of listed real estate companies in both developed and emerging countries worldwide. Constituents of the Index are screened on liquidity, size and revenue.

Hong Kong REITs (HK) is real estate investment trusts listed in Hong Kong. The Hang Seng REIT Index will serve as the basis for index products, including funds and derivatives as well as a benchmark for investors in the REIT asset class.

Japan REITs (JP) is a capitalization-weighted index of all Real Estate Investment Trusts listed in the Tokyo Stock Exchange, and is calculated using the same methodology as the TOPIX. The index was developed with a base index value of 1000 as of March 31, 2003. This is a price return index.

Singapore REITs (SIN) is an index measures the performance of the 20 largest and most tradable trusts of the REIT sector listed on Singapore Exchange (SGX). The index provider adjusts for each Reits free float that is, the number of shares publicly available to investors and then attaches weights for each REIT constituent according to its free-float-adjusted market capitalization.

In this application study, we can write our model specifications as follows:

$$\begin{bmatrix} w_{y1}HK_{1,t}^U + (1-w_{y1})HK_{1,t}^L \\ w_{y2}JP_{2,t}^U + (1-w_{y2})JP_{2,t}^L \\ w_{y3}SIN_{3,t}^U + (1-w_{y3})SIN_{3,t}^L \end{bmatrix} = $$
$$\begin{bmatrix} \alpha_{HK} + \beta_{HK}(w_{11}GL_t^U + (1-w_{11})GL_t^L) + \varepsilon_{1,t} \\ \alpha_{JP} + \beta_{HK}(w_{21}GL_t^U + (1-w_{21})GL_t^L) + \varepsilon_{2,t} \\ \alpha_{SIN} + \beta_{HK}(w_{31}GL_t^U + (1-w_{31})GL_t^L) + \varepsilon_{3,t} \end{bmatrix}$$

Table 3. Summary statistics for interval valued data of three REITs in Asia.

	HK^U	HK^L	JP^U	JP^L	SIN^U	SIN^L	GL^U	GL^L
Mean	0.007448	−0.013814	0.01038	−0.00973	−0.004678	0.004135	0.006977	−0.005848
Median	0.005889	−0.010833	0.00869	−0.008451	−0.003701	0.003261	0.004826	−0.004203
Maximum	0.072484	0.016273	0.16169	0.068755	0.035914	0.063108	0.955659	0.885077
Minimum	−0.054514	−0.120214	−0.072972	−0.118814	−0.10479	−0.043515	−0.033312	−0.999721
Std. Dev	0.008914	0.011502	0.012763	0.012143	0.007764	0.007694	0.019843	0.026766
Skewness	1.268982	−2.819685	1.7256	−1.272618	−2.415195	1.525084	38.07373	−5.068424
Kurtosis	12.16828	16.07497	16.57828	11.55111	24.835	11.76051	1812.039	1085.047
Jarque-Bera	10882.47	24381.55	23602.73	9571.87	60137.04	10347.51	3.94E+08	1.41E+08
Probability	0.00000	0.00000	0.00000	0.00000	0.00000	0.00000	0.00000	0.00000
Obs	2886	2886	2886	2886	2886	2886	2886	2886
ADF-test	4.4805***	4.9929***	5.2992***	5.4118***	3.4803***	6.9890***	5.6122***	5.1792***

Source: Calculation

Note: *,**,*** denote significant at 90%, 95%, and 99%, respectively

The descriptive statistics are given in Table 3, which show that Japan REITs has the highest standard deviation, while Singapore REITs has the lowest. The Jarque-Bera calculated is compared to chi-square critical value at an independent degree 2. If the Jarque-Bera calculated is greater than that the data is not normally distributed. In this paper, the Jarque-Bera statistic indicates that all series are not normally distributed as it rejects the null hypothesis. The Augmented Dickey-Fuller (ADF) is applied to check unit roots in the series. Unit root test for each variable has a statistical significance level of 0.01. This means that all of REITs returns are stationary in characteristics. Therefore, these variables can be used to estimate the model in the next section.

5.2   Estimates Results

Table 3 shows AIC and BIC criteria for comparison of the copula families as well as the marginal distributions. To choose the best family and marginal distribution for our data, we will look at the minimum AIC and BIC criteria. As we can see, the minimum AIC and BIC are −62394.33 and −62286.91, respectively as bolded. So we should choose Gaussian as the appropriate copula for data and the proper distribution is student-t. We, thus, choose Gaussian as the copula function to link marginal distributions of REITs in Asia as joint distribution (Table 4).

Table 4. AIC and BIC criteria for model choice

Marginal distribution	Copula					
	Gaussian	Student-t	Frank	Joe	Clayton	Gumbel
Normal/Normal/ Normal	−53553.18 −53445.77	−55274.07 −55166.65	−55331.28 −55235.8	−51700.88 −51605.4	−55455.99 −55360.51	−52062.84 −51967.36
Normal/Normal/ Student-t	−43425.57 −43318.15	−57402.96 −57295.54	−47056.75 −46961.26	−45980.15 −45884.67	−56691.39 −56595.91	−46446.54 −46351.06
Normal/Student-t/ Normal	−49892.21 −49784.8	−55247.96 −55140.54	−48668.8 −48573.32	−47194.18 −47098.7	−56631.57 −56536.09	−47776.71 −47681.23
Normal/Student-t/ Student-t	−59988.07 −59880.65	−60494.55 −60387.14	−60620.71 −60525.23	−58069.73 −57974.25	−59211.22 −59115.74	−60442.27 −60346.79
Student-t/Normal/ Normal	−56464.46 −56357.05	−56519.5 −56412.08	−44745.65 −44650.17	−44883.61 −44788.13	−56527.03 −56431.55	−35962.46 −35866.98
Student-t/Normal/ Student-t	−57285.1 −57177.68	−58783.24 −58675.82	−54931.43 −54835.94	−51626.6 −51531.12	−57211.87 −57116.39	−58100.58 −58005.1
Student-t/Normal/ Normal	−55608.88 −55501.46	−58575.42 −58468	−59204.73 −59109.24	−58022.76 −57927.27	−56156.11 −56060.62	−57883.47 −57787.99
Student-t/ Student-t/ Student-t	**-62394.33** **−62286.91**	−61251.1 −61143.68	−61812.89 −61717.41	−62031.51 −61936.03	−62126.63 −62031.15	−61902.03 −61806.55

Source: Calculation

Tables 5 and 6 show the results of parameter estimation through the application of the CAPM model. Our main contribution is that we can estimate the weight parameters in order to find the appropriate value between intervals. These parameters seem to have econometric interpretation. If any weight is greater or lower than 0.5, it indicates that the market is likely to be in uptrend or downturn. Then, lets consider the estimated beta of each equation which indicates the sensitivity of securities to systematic risk. If the beta is equal to 1, it means that the individual market has changed in return equal to the global REITs return. If the beta is greater than 1, it means that the individual market is subject to a change in the Asian REITs return, which is higher than the global market risk. On the other hand, if the beta is less than zero, it indicates that the individual market has a change in the rate of return of the market where the risk is inversely proportional to the global REITs return. The results show that Hong Kong and Japan markets have a positive sign of the estimated beta, $\beta_{HK} = 0.30948$ and $\beta_{JP} = 0.17323$, hence this indicates that Hong Kong and Japan markets are less volatile than the global REITs. On the other hand, we find that the estimated beta for Singapore market shows a negative relationship with global REITs. Surprisingly, our beta has a negative sign. Can beta be negative? The answer is yes. The beta can be negative. A more intuitive way of thinking about this is that

Table 5. Estimated results

α_{HK}	Coefficient	S.E.
β_{HK}	-0.00076***	1.99E-04
σ_1	0.30948***	1.54E-02
w_{y1}	0.00565***	0.00010398
w_{y1}	0.61995***	0.00847734
α_{JP}	0.52438***	0.03678616
β_{JP}	-0.00066**	3.42E-04
σ_2	0.17323***	2.58E-02
w_{y2}	-0.00915***	1.66E-04
w_{21}	0.45527***	1.41E-02
α_{SP}	0.50092***	1.05E-01
β_{SP}	-0.00040**	1.75E-04
β_{SP}	-0.49251***	1.50E-02
σ_3	-0.00510***	8.62E-05
w_{y3}	0.55359***	2.03E-02
w_{31}	0.52375***	2.34E-02

Source: Calculation
Note: *,**,*** denote rejection of the null hypothesis at the 10%, 5% and 1% significance levels, respectively.

Table 6. Copula Parameter

	Dependence parameter from SUR Copula		
	Hong Kong	Japan	Singapore
Hong Kong	1.0000	0.1737***	−0.1079***
Japan	0.1737***	1.0000	−0.2445***
Singapore	−0.1079***	-0.2445***	1.0000

Source: Calculation

Note: *,**,*** denote rejection of the null hypothesis at the 10%, 5% and 1%, significance levels, respectively.

a negative beta investment represents insurance against some other risks that affect portfolio return. We can say that Singapore market can be viewed as a hedge against higher risk in Asian REITs.

Moreover, lets consider the copula dependence parameters in Table 6. The result depicts the dependency of each country on the basis of the dependence value. We find the significant positive dependence between Hong Kong and Japan REITs which is 0.1737 While Singapore and Hong Kong show a significant negative dependence which is −0.1079. Finally, the dependence coefficient of the Japan and Singapore REITs is −0.2445. These results confirm that Japan, Hong Kong, and Singapore RETIs are correlated to each other and our model is reasonable for use in this line of application

6   Conclusion

The high volatility of the REIT price indexes due to the hamburger crisis caused the investors to face huge risks, especially in Japan, Hong Kong and Singapore. These three countries have a high market capitalization in Asian, and the movements of their price indexes contribute a large effect to the global REITs market. Thus, in this study, it would be of great benefit to measure the risk of these three markets. To quantify the risk, we consider the CAPM approach. Moreover, we also expect a correlation between CAPM of each country. Thus, in this study, we employ a copula based SUR models as proposed in Pastpipatkul et al. [7] to quantify the risk of Japan, Hong Kong and Singapore markets. However, there has been other recent extension of the CAPM to the interval data, as seen in Piamsuwannakit et al. [11]. They found that using the interval data in CAPM can provide a better result than using the closing price for prediction. Thus, we propose copula based SUR model for analysis with interval data. To find the appropriate point data in the interval data we employ convex combination method which was introduced in Chanaim et al. (2016) [2].

Before we apply our model to the real data, we conduct a simulation study to confirm the accuracy of our model. The results confirm that our model is accurate and has a good finite sample performance and is reasonable for CAPM application. In the application study, we choose the best model specification using AIC and BIC. Thus Gaussian copula is chosen to be a linkage between Student-t of the Hong Kong, Japan, and Singapore CAPM equations.

Finally, the results of the best fit model specification show that Hong Kong and Japan markets have a positive sign of the beta and both markets are less volatile than the global REITs market. On the other hand, we find that the estimated beta for Singapore market shows a negative relationship with global REITs market. We conclude that Singapore market can be viewed as a hedge against higher risk in Asian REITs.

References

1. Billard, L., Diday, E.: Regression analysis for interval-valued data. In: Data Analysis, Classification, and Related Methods, pp. 369–374. Springer, Heidelberg (2000)
2. Chanaim, S., Sriboonchitta, S., Rungruang, C.: A convex combination method for linear regression with interval data. In: Proceedings of the 5th International Symposium on Integrated Uncertainty in Knowledge Modelling and Decision Making, IUKM 2016, Da Nang, Vietnam, 30 November– 2 December 2016, vol. 5, pp. 469–480. Springer (2016)
3. Fang, H., Chang, T.Y., Lee, Y.H., Chen, W.J.: The impact of macroeconomic factors on the real estate investment trust index return on Japan, Singapore and China. Investment Manage. Financ. Innov. 13(4-1) (2016)
4. Joe, H.: Asymptotic efficiency of the two-stage estimation method for copula-based models. J. Multivar. Anal. **94**(2), 401–419 (2005)
5. Lintner, J.: The valuation of risky assets and the selection of risky investments in stock portfolios and capital budgets. Rev. Econ. Stat. **47**(1), 13–37 (1965)
6. Newell, G.: The investment characteristics and benefits of Asian REITs for retail investors. Asia Pacific Real Estate Association (APREA) (2012)
7. Pastpipatkul, P., Maneejuk, P., Wiboonpongse, A., Sriboonchitta, S.: Seemingly unrelated regression based copula: an application on Thai rice market. In: Causal Inference in Econometrics, pp. 437–450. Springer (2016a)
8. Pastpipatkul, P., Panthamit, N., Yamaka, W., Sriboochitta, S.: A copula-based markov switching seemingly unrelated regression approach for analysis the demand and supply on sugar market. In: Proceedings of the 5th International Symposium on Integrated Uncertainty in Knowledge Modelling and Decision Making, IUKM 2016, Da Nang, Vietnam, 30 November–2 December 2016, vol. 5, pp. 481–492. Springer (2016b)
9. Pham, A.K.: An empirical analysis of real estate investment trusts in Asia: Structure, performance and strategic investment implications (2013)
10. Phochanachan, P., Pastpipatkul, P., Yamaka, W., Sriboonchitta, S.: Threshold regression for modeling symbolic interval data. Int. J. Appl. Bus. Econ. Res. **15**(7), 195–207 (2017)
11. Piamsuwannakit, S., Autchariyapanitkul, K., Sriboonchitta, S., Ouncharoen, R.: Capital asset pricing model with interval data. In: Integrated Uncertainty in Knowledge Modelling and Decision Making, pp. 163–170. Springer (2015)
12. Sharpe, W.F.: The Capital Asset Pricing Model: A "Multi-Beta" Interpretation. Stanford University, Graduate School of Business (1973)
13. Sklar, A.: Fonctions de de répartition n dimensions et leurs marges. Publications de l'Institut de Statistique de l'Universit de Paris (1959)

14. Smith, M.S., Gan, Q., Kohn, R.J.: Modelling dependence using skew t copulas: Bayesian inference and applications. J. Appl. Econometrics **27**(3), 500–522 (2012)
15. Smith, M.S., Khaled, M.A.: Estimation of copula models with discrete margins via Bayesian data augmentation. J. Am. Stat. Assoc. **107**(497), 290–303 (2012)
16. Wichitaksorn, N.: Estimation of bivariate copula-based seemingly unrelated Tobit models (2012)
17. Zellner, A.: An efficient method of estimating seemingly unrelated regressions and tests for aggregation bias. J. Am. Stat. Assoc. **57**(298), 348–368 (1962)

Risk Valuation of Precious Metal Returns by Histogram Valued Time Series

Pichayakone Rakpho[1], Woraphon Yamaka[1,2], and Roengchai Tansuchat[1,2(✉)]

[1] Faculty of Economics, Chiang Mai University, Chiang Mai 50200, Thailand
pichayakone@gmail.com, woraphon.econ@gmail.com, roengchaitan@gmail.com
[2] Center of Excellence in Econometrics, Chiang Mai University,
Chiang Mai 50200, Thailand

Abstract. The price of precious metals is highly volatile and it can bring both risk and fortune to traders and investors, and therefore should be examined. In this paper, we introduce an approach to fitting a Copula-GARCH to valued time series and apply this methodology to the daily histogram returns of precious metals consisting of gold, silver, and platinum. The study also conducts a simulation study to confirm the accuracy of the model and the result shows that our model performs well. In the empirical study, our results suggest investing on gold and platinum in high proportion while silver is not recommended for inclusion in the precious metal portfolio. Moreover, precious metal portfolio of the intraday 30-min returns gives lower risk when compared with portfolio of the intraday 60-min returns. Therefore, investors should not hold assets for long period of time because the long-term holding is likely to face a higher risk.

1 Introduction

Even though precious metals are recommended asset to hedge or to diversify in the portfolio in many poor situations such as a stock market crash, inflation or a declining dollar, precious metal price still has volatility under uncertainty situation particularly as occurred during the financial crisis of 2007–2009. The three precious metal prices from 2007 to 2017 are shown in Fig. 1. During this period with stock market collapse, credit crunch, and dollar's devaluation, the holding of stocks became a high-risk investment and thus investors preferred to hold precious metals as a safe haven in this troubled situation. However, after the economic crisis, investors are more interested in investing in precious metals. With more purchases, metal prices have continuously risen. In addition, precious metal price still had uncertainty situation in 2014. The main reason for this weakness in the second half of 2014 was US dollar appreciation [10]. As evident in Fig. 1, prices of all precious metals are volatile and move together. As mentioned above, the prices of precious metals are highly volatile and this bring both risk and fortune to traders and investors, and therefore should be examined. The Value at Risk (VaR) estimation is a tool to measure risk. It is a financial market instrument for measurement and evaluation of the portfolio

© Springer International Publishing AG 2018
V. Kreinovich et al. (eds.), *Predictive Econometrics and Big Data*, Studies in Computational Intelligence 753, https://doi.org/10.1007/978-3-319-70942-0_39

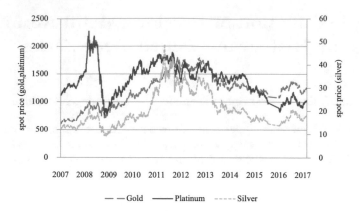

Fig. 1. Precious metal prices during 2007–2017 Source: Thomson Reuters, 2017

market risk associated with financial asset and commodity price movements. It represents in the form of the expected worst loss of a portfolio over a given time horizon at a given confidence level.

In the literature, VaR has been applied in metal markets to measure risk in several studies. Hammoudeh et al. [5] examined the volatility and correlation dynamics in price returns of gold, silver, platinum and palladium using copula model and VaR. The results are useful for participants in the global financial markets that needed to invest in precious metals in the light of high volatility. Demiralay and Ulusoy [2] predicted the Value at Risk of four major precious metals (gold, silver, platinum, and palladium) using FIGARCH, FIAPARCH and HYGARCH or long memory volatility models, under normal and student-t innovations distributions. The results showed that these models perform well in forecasting a one-day-ahead VaR and have potential implications for portfolio managers, producers, and policy makers. Khemawanit and Tansuchat [12] measured VaR by applying GARCH-EVT-Copula model and found that gold and silver should invest in a high investment proportion, whereas palladium and platinum have little investment proportion. However, most empirical researches about value at risk in precious metal returns which lower frequencies such as daily closing price weekly or monthly and used generally the GARCH models. In the analysis of the risks by the closing price still does not reflect the true volatility of precious metals. Due to the close prices is ignoring information that happened during the day. Therefore, In this paper purposes analysis of value at risk for precious metal returns using high frequency data. By using the full information contained in the histogram, we find that there are advantages in the estimation and prediction of a specific interval [4].

In financial markets, the price of an asset (stocks, bonds, exchange rates, etc.) during intraday trading is observed as high frequency data, i.e., tick by tick. However in a huge number of studies, many researchers prefer to analyze their works based on daily closing price or even at lower frequencies such as weekly or monthly. It may be claimed that tick-by-tick data or price will generate a huge amount of data from which it will be difficult to discriminate information from noise. But if we consider only the closing prices, we might miss valuable intraday information. Thus, in this study, we propose an alternative way to collect these information by constructing daily histogram-valued time series (HTS) from the intraday data [4]. The histograms, when indexed by time, will create a HTS. From a time series perspective, it is possible to define the HTS as a collection of histograms ordered over time.

In this context, an important issue is how volatility and dependence of precious metals can be measured, when the data are HTS. Here we consider Copula-GARCH model which proposed by Jondeau and Rockinger [10] as a tool to measure the risk and evaluate the dependence of precious metals. Hence, the study aim to fit the Copula-GARCH model to HTS data set. To deal with the HTS data set, we employ the method for histogram-valued variables which proposed by the study of Irpino and Verde [7]. They also presented an alternative definition of mean for histogram-valued variables, which produces a mean distribution, that they termed it barycentric histogram of the histograms. To the best of our knowledge, Copula-GARCH models for HTS has not been considered yet. Therefore, this fact becomes one of our motivations to work on this paper.

The remaining of the paper is organized as follows. Section 2 is the review of the methodology. Section 3 is estimation procedure. Section 4 shows the simulation study. Section 5 is on data description, and unit root tests. Section 6 describes the empirical results of estimates about the risk, VaR and ES, optimal portfolio, and optimal portfolio weights. Finally, Sect. 7 provides some concluding remarks.

2   Methodology

The Copula-GARCH model employed in this study was previously proposed by Patton [15], Rockinger and Jondeau [11] and it generally consists of two parts: the first part is to model the marginal distribution of returns by GARCH process and the second is a proper copula function to link the margins together. In this section, we briefly introduce Copula-GARCH with histogram valued data.

2.1   Histogram-Valued Variables

As we mention above, we consider a high frequency data, i.e., tick by tick. Although, tick-by-tick data can generate a huge amount of information, it will be difficult to deal with this huge data. Thus, we can transform this data into histogram data and find the appropreiate value between in the range of histogram. According to Dias and Brito [3], we can define histogram-valued variables as follows:

Definition 2.1. $Y_t = \{y(1), ..., y(T)\}$ is a histogram-valued variable at time t when to each time t corresponds an empirical distribution $y(t)$ that can be represented by a histogram

$$H_{y(1)} = \left\{ [\underline{I}_{y(t)1}, \overline{I}_{y(t)1}], p_{t1}; [\underline{I}_{y(t)2}, \overline{I}_{y(t)2}], p_{t2}; ... [\underline{I}_{y(t)n_t}, \overline{I}_{y(t)n_t}], p_{tn_t} \right\} \qquad (1)$$

where $\underline{I}_{y(t)i}$ and $\overline{I}_{y(t)i}$ represent the lower and upper bound of the $y(t)$ at each unit $i \in \{1, 2, ..., n_j\}$ which are uniformly distributed. p_{ti} is the frequency associated with the subinterval $[\underline{I}_{y(t)i}, \overline{I}_{y(t)i}]$ and $\sum_{i=1}^{n_j} p_{ij} = 1$, n_t is the number of subintervals for the empirical distribution $y(t)$, $t = 1, ..., T$.

To find the single value data in each $y(t)$, the quantile function, $\Psi_{Y_t}^{-1}$, which proposed in Irpino and Verde [7], is employed

$$\Psi_{Y_t}^{-1} =$$

$$\begin{cases} \underline{I}_{y(j)1} + \dfrac{t}{p_{t1}} a_{y(t)1} & \text{if } 0 \leq P < p_{j1}, \\[2mm] \underline{I}_{y(j)2} + \dfrac{t - p_{p1}}{p_{t2}} a_{y(t)2} & \text{if } p_{t1} \leq P < (p_{j1} + p_{j1}), \\[2mm] \vdots \\[2mm] \underline{I}_{y(j)n_t} + \dfrac{t - (p_{j2} + p_{t1} + \cdots + p_{tn_t})}{p_{tm}} a_{y(t)n_t} & \text{if } (p_{t2} + p_{t1} + \cdots + p_{tn_t}) \leq P < 1 \end{cases}$$

$$\qquad (2)$$

where $a_{y(t)i} = [\overline{I}_{y(t)i} - \underline{I}_{y(t)i}]$ with $\{i \in 1, \ldots n_t\}$.

Note that when we work with histogram-valued variables, the frequency associated with the subinterval p_{it} and the number of subintervals in the histogram n_t may be different; the subinterval of histogram $H_{y(t)}$ are considered ordered and disjoint, and if this is not the case, it must be possible to rewrite them in the required form Dias and Brito [3].

2.2   GARCH Model

Following Bollerslev [1], a GARCH model can be described by the following equations

$$y_{it} = c_i + h_{it} Z_{it} = c_i + \varepsilon_{it}, \qquad (3)$$

$$\eta_{it} = Z_{it}/h_{it}, \qquad (4)$$

$$h_{it}^2 = \alpha_0 + \sum_{m=1}^{M} \alpha_m \varepsilon_{it-m}^2 + \sum_{n=1}^{N} \beta_n h_{it-n}^2 \qquad (5)$$

where Eqs. 3 and 5 are, respectively, the conditional mean and variance equation, given past information. c_i is the intercept term of the mean equation, ε_{it} is the residual term of asset i at time t, h_{it} is the conditional variance and η_{it} is a sequence of i.i.d which is assumed to have a normal distribution, a student-t distribution, and a skewed-t distribution. $\alpha_m (m = 1, \ldots, M)$ and $\beta_n (n = 1, \ldots, N)$ are non negative parameters and $\alpha_m + \beta_n \leqslant 1$ to assure that $h_{it} > 0$ and covariance stationarity. As already mentioned, from Eq. 2 in this work we choose to represent the distributions by quantile functions. However, when we multiply a quantile function by a negative number we do not obtain a non-decreasing function. Therefore, it is necessary to impose positivity restrictions on the parameters of the model. Denoting $\Psi_{y_{it}}^{-1}$ the quantile function of the predicted distribution y_{it}; we rewrite the mean equation of GARCH regression model as follows:

$$\Psi_{y_{it}}^{-1} = c_i + \varepsilon_{it}, \tag{6}$$

2.3 Copula-GARCH Model

First of all, we briefly explain the definition of the copula which was introduced in Sklars theorem

Definition 2.3. A n-dimensional copula is a multivariate cumulative distribution function $c : [0,1]^d \rightarrow [0,1]$, whose margins have the uniform distribution on the interval $[0,1]$. Let F denote a d-dimensional distribution functions with marginal distribution function $F_{y_1}, \ldots, F_{y_d}$ then, there exists a copula C, such that

$$F(y_1, \ldots, y_d) = C(F_{y_1}(y_1), \ldots, F_{y_d}(y_d)), \forall (y_1, \ldots, y_d) \in R^d \tag{7}$$

where, the function F is the joint distribution function with marginal distribution functions $F_{y_1}, \ldots, F_{y_d}$ which are the marginal distributions specified as standard univariate GARCH processes. A copula is a cumulative distribution function with uniform marginal distributions. $u_i, C(u_i, \ldots, u_d) = Pr(U_1 \leqslant u_1), \ldots, Pr(U_d \leqslant u_d)$, where $u_i = F_{y_i}(y_i)$ is uniform $[0,1]$ interval.

In this study, we consider multivariate dimensional copulas with GARCH(1, 1) which is mostly employ in many financial data. We consider two families of copulas used in our paper, namely: the normal copula and the Student-t-copula, (see, Joe [9] and Hofert, Mchler, Mcneil [6]) to join the GARCH equations.

3 Estimation Procedure

To estimate the Copula-GARCH model with histogram value data, we begin to derive the log-likelihood for the model as follows:

$$\log l(\Theta, \theta \,|\Psi_{y_{it}}^{-1}) = \sum_{t=1}^{T} \log \left\{ [c_\theta(F_{y_1}(\Psi_{y_{1t}}^{-1}|\Theta_1), \ldots, F_{y_d}(\Psi_{y_{dt}}^{-1}|\Theta_d))] \prod_{i=1}^{d} f_i(\Psi_{y_{it}}^{-1}|\Theta_i) \right\} \tag{8}$$

where, Θ is the estimated parameter in GARCH equations and θ is estimated copula parameters. Note that the copula connects margins to a multivariate distribution function without any constrains on marginal distributions. We model the marginal distribution of the standardized innovations $\eta_{it}, i = 1, \ldots, d$, by cumulative distribution function (cdf). Then, we fit the six chosen families of the copulas to the vector of various marginal distributions namely: normal, student-t, and skewed student-t distributions. For each Copula-GARCH specification, we estimate the parameters by maximizing the log- likelihood function in Eq. (8). To select the best fit model, that with the lowest Akaike Information criteria and Bayesian Information criteria is preferred (see, pastpipatkul, [13,14]).

In addition, our study aims to measure the Value at Risk (VaR) of the portfolios of our histogram returns. In the next step, we generate 10,000 Monte Carlo histogram return of asset i, S_{it} from the best fit Copula-GARCH(1, 1). Then, the value of a portfolio at t is given by

$$V_t = \sum_{i=1}^{d} \omega_i S_{it},$$

where ω_i is set to be equally weighted. We define the profit and loss function from period $t = T$ to $t = T + 1$ as $L_{T+1} = V_{T+1} - V_T$. Thus, $VaR_{T+1}(\alpha)$ is the VaR at time $T + 1$, of a specific portfolio of assets over a time horizon from T to $T + 1$ with the confident level $1 - \alpha$, satisfies:

$$P(L_{T+1} > VaR_{T+1}(\alpha)) = 1 - \alpha$$

Thus,

$$P(L_{T+1} \leqslant VaR_{T+1}(\alpha)) = \alpha$$

and it is easily seen that $VaR_{T+1}(\alpha)$ is the α-th quantile from the distribution of L_{T+1}, i.e.

$$VaR_{T+1}(\alpha) = F^{-1}_{L_{T+1}(\alpha)}.$$

where $F_{L_{T+1}}$ stands for the distribution function of L_{T+1}.

In this computation, we compute all the risk measures at the 1%, 5%, and 10% levels. Additionally, We finally obtain expected return and risk to find the optimal weights of the portfolios.

4   Applied Examples

4.1   Simulation Example

In the simulation example, we applied Gaussian copulas to model the dependence structure of the GARCH(1, 1). The simulation is the realization of GARCH(1, 1) with three equations, thus we generated random histogram data from the following specifications:

$$Y_{it} = \Psi_{Y_{it}} \tag{9}$$

$$\Psi_{Y_{it}}^{-1} = c_i + h_{it}\varepsilon_{it}, \tag{10}$$

$$h_{it}^2 = \alpha_{i0} + \alpha_{i1}\varepsilon_{it-1}^2 + \beta_{i1}h_{it-1}^2, \ i = 1,\dots,3 \tag{11}$$

The error terms, ε_{it}, are assumed to follow $\varepsilon_{it} \sim N(0, 0.5)$ and $\varepsilon_{it} \sim t(0, 1, v = 4)$ and $\varepsilon_{3t} \sim std(0, 1, v = 4, skew = 1.5)$. The parameters of the model C, α_0, α_1 and β_1 are set to be $C = 0.2, 0.1, 0.3$, $\alpha_0 = 0.001, 0.001, 0.001$, $\alpha_1 = 0.2, 0.1, 0.3$ and $\beta_1 = 0.7, 0.8, 0.6$. For the dependence parameter for the copula function, we set the true value for the Gaussian dependence coefficient at 0.5. We run Copula-GARCH model with histogram valued data using sample size $n = 100$ and we then convert the single value data to histogram data by

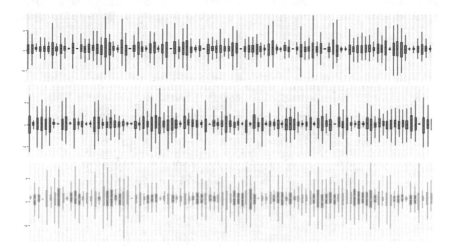

Fig. 2. Box plot of simulated histogram valued data Source: Calculations

Table 1. Simulation example results

Parameter	$i = 1$	$i = 2$	$i = 3$
c_1	0.1813 (0.0003)	0.1007 (0.0023)	0.2923 (0.0021)
α_{i0}	0.0004 (0.0001)	0.0004 (0.0020)	0.0004 (0.0015)
α_{i1}	0.2079 (0.0021)	0.0991 (0.0012)	0.2991 (0.0010)
β_{i1}	0.7929 (0.0001)	0.7929 (0.0561)	0.6929 (0.0497)
ν_1		3.5784 (0.0158)	3.7445 (0.1245)
$skew_i$			1.1154 (0.0024)
θ_1	1	0.4868 (0.1471)	0.4868 (0.1122)
θ_2	0.4868 (0.1471)	1	0.5043 (0.1312)
θ_3	0.4868 (0.1122)	0.5043 (0.1312)	1

Source: Calculation
Note: () is standard errors.

using a piecewise linear approximation of the empirical cumulative distribution function using the Ramer-Douglas-Peucker algorithm (see, Irpino [7,8]). We plot each histogram data for i equation in Fig. 2.

Table 1 shows the estimated parameters and we observed the model produced the unbiased parameter estimates when compared with the true values. We found that the estimated parameters are close to the true values and the standard error is reasonable.

5   Data

5.1   Descriptive Statistics

In the first step, we construct a daily histogram with the 30-min and 60-min form gold, silver, platinum returns during the period January 1, 2015 to December 8,

Table 2. Summary statistics for intraday precious metal

Intraday 30-min returns			
	Gold	Sliver	Platinum
Mean	−0.00000432	−0.00000695	−0.0000116
Std. dev	0.000364	0.000592	0.001047
Skewed	−1.062584	−3.323920	−135.3521
Kurtosis	289.0925	229.4621	22506.04
Min	−0.013834	−0.026695	−0.174192
Max	0.013763	0.014936	0.013285
Jarque-Bera	1.16E+08	72680126	7.17E+11
Probability	0.000000	0.000000	0.000000
ADF-test	(−183.13)***	(−126.23)***	(−183.52)***
Intraday 60-min returns			
	Gold	Sliver	Platinum
Mean	−0.00000662	−0.0000097	0.0000187
Std. dev	0.000365	0.000658	0.001421
Skewed	−0.700144	−4.807569	−108.3149
Kurtosis	153.2437	239.3651	13277.04
Min	−0.010918	−0.026695	−0.174192
Max	0.011210	0.014066	0.009315
Jarque-Bera	15982254	39618004	1.25E+11
Probability	0.000000	0.000000	0.000000
ADF-test	(−127.13)***	(−126.68)***	(−130.93)***

Source: Calculation
Note: ***, **, * are significant at 1%, 5%, and 10% level, respectively.

2016. All precious metals price information is derived from Bloomberg and precious metals are traded at London Metal Exchange. we use the precious metal prices (secondary data) to calculate the natural log returns which are defined as $r_{i,t} = \ln(P_{i,t}/P_{i,t-1})$ where $P_{i,t}$ is the ith metal price at time t, $r_{i,t}$ is the ith log return of metal price at time t, i and indicated the ith precious metal price. The descriptive statistics shows in Table 2. In log return of metal price every 30 and 60 min found that platinum has standard deviation higher than gold and silver. In additional, Table 2 shows that standard deviation of 60 min is higher than 30 min. Since the movement of stock prices need a time to be change. Thus, if we consider high frequency data i.e. 30 min, its prices might not change much when compare to low frequency data 60 min. Investors need time to make a buy or sell decided so the low frequency data should be less volatile than high frequency data. The Jarque-Bera statistic indicates that all series are not normally distributed because rejects the null hypothesis, thus the return series of precious metal price is non-normal distribution. The Augmented Dickey-Fuller (ADF) is applied to check unit roots in the series. Unit root test for each variable have a statistical significance level of 0.01. This means that all of precious metal returns is stationary characteristics. Therefore, these variables can be used to estimate Copula-GARCH model in the next step.

6 Result

Table 3 shows AIC and BIC for comparison of the copula families as well as the marginal distributions, when a daily histogram data is constructed from 30 and 60-min returns, respectively. To choose the best family and marginal distribution for our data, we will look at the minimum AIC and BIC. The results of these two criteria are the same. We can observe that student-t copula is chosen to be a linkage between skewed Student-t margins of gold, silver, and platinum equations. Thus, we will interpret our results from this model specification as presented in Table 4.

Figure 3 shows the results for estimated VaR and CVaR (ES). The calculated VaR and CVaR (ES) of the portfolio with an equally weighted portfolio of three precious metals (Gold, Silver and Platinum,) at level of 1%–10% under the equally weighted assumption of intraday 30-min returns and intraday 60-min returns. In period t + 1, the estimated CVaR (ES) is higher than VaR and the risk of intraday 60-min returns is more than intraday 30-min returns.

Figure 4 shows the results of the efficient frontiers of the portfolio under different expected returns, which come from the optimized portfolio based on mean-CVaR (ES) model. To get these results, we applied the Monte Carlo simulation to simulate a set of 10,000 samples and to estimate the expected shortfall of an optimal weighted portfolio.

For the discussion above, we focused on estimating the VaR and CVaR (ES) of an equally weighted portfolio. In Fig. 4, each plot here illustrates each portfolio.

Table 3. AIC and BIC criteria for model choice

Marginal distribution	30 min Copula				60 min Copula			
	Student-t		Normal		Student-t		Normal	
	BIC	AIC	BIC	AIC	BIC	AIC	BIC	AIC
N-N-N	−14369	−14440	−14296	−14366	−11671	−11742	−11647	−11717
N-N-T	−14445	−14515	−14387	−14458	−11699	−11769	−11678	−11748
N-N-sT	−14453	−14523	−14396	−14467	−11701	−11771	−11679	−11749
N-T-N	−14488	−14559	−14487	−14557	−11725	−11795	−11718	−11788
N-T-T	−14622	−14693	−14621	−14692	−11761	−11832	−11755	−11825
N-T-sT	−14633	−14703	−14632	−14702	−11764	−11834	−11757	−11827
N-sT-N	−14585	−14655	−14564	−14634	−11869	−11939	−11861	−11931
N-sT-T	−14680	−14750	−14664	−14734	−11902	−11972	−11895	−11965
N-sT-sT	−14690	−14760	−14674	−14744	−11906	−11977	−11898	−11968
T-N-N	−14430	−14500	−14357	−14428	−11773	−11844	−11747	−11818
T-N-T	−14515	−14586	−14456	−14527	−11805	−11876	−11779	−11850
T-N-sT	−14523	−14594	−14465	−14535	−11807	−11878	−11781	−11851
T-T-N	−14562	−14632	−14556	−14627	−11843	−11914	−11826	−11897
T-T-T	−14702	−14773	−14698	−14769	−11885	−11955	−11865	−11935
T-T-sT	−14713	−14783	−14708	−14779	−11887	−11958	−11866	−11937
T-sT-N	−14663	−14733	−14631	−14701	−11974	−12044	−11962	−12033
T-sT-T	−14768	−14839	−14738	−14808	−12013	−12083	−11997	−12067
T-sT-sT	−14778	−14848	−14748	−14818	−12018	−12088	−12000	−12071
sT-N-N	−14431	−14501	−14358	−14429	−11773	−11844	−11747	−11818
sT-N-T	−14516	−14587	−14457	−14527	−11805	−11876	−11779	−11850
sT-N-sT	−14524	−14595	−14466	−14536	−11807	−11878	−11781	−11851
sT-T-N	−14563	−14633	−14557	−14627	−11843	−11914	−11826	−11897
sT-T-T	−14703	−14773	−14699	−14769	−11885	−11956	−11865	−11935
sT-T-sT	−14713	−14784	−14709	−14780	−11887	−11958	−11866	−11937
sT-sT-N	−14664	−14734	−14632	−14702	−11974	−12044	−11962	−12033
sT-sT-T	−14769	−14839	−14739	−14809	−12013	−12083	−11997	−12067
sT-sT-sT	**−14779**	**−14849**	−14749	−14819	**−12018**	**−12088**	−12000	−12071

The first plot shows that low risk provides low return. It means that the greater return, the greater risky it will be. For the suggestion, for risk-averse investors, they would better choose the low return with low risk while risk-lover investors suit for high risk, high return.

Table 4. Estimation result of Copula-GARCH(1, 1)models

Intraday 30-min returns			
Variable	Gold	Silver	Platinum
C	−0.00099***	−0.00122***	−0.00220***
	(−6.119)	(−4.268)	(−1333.421)
α_0	0.00000	0.00006	0.00000
	(0.872)	(0.884)	(0.006)
α_1	0.02460***	0.35870	1.00000***
	(1.999)	(0.820)	(6.978)
β	0.96830***	0.45040*	0.52080***
	(70.398)	(2.292)	(17.973)
$skew$	0.97850***	0.97050***	0.91460***
	(20.002)	(21.643)	(27.293)
$shape$	3.69900***	2.22200***	3.00700***
	(5.916)	(8.309)	(13.493)
log-likelihood	2501.76	2237.488	2397.461
θ_1	1	0.572	0.5457
θ_2	0.572	1	0.6144
θ_3	0.5457	0.6144	1
Intraday 60-min returns			
Variable	Gold	Silver	Platinum
C	0.00115***	−0.00430***	−0.00167***
	(−3.49)	(−979.301)	(−3.467)
α_0	0.00000	0.00000	0.00001
	(0.971)	(0.005)	(1.671)
α_1	0.01029	1.00000***	0.05274*
	(0.905)	(6.668)	(2.349)
β	0.96080***	0.55560***	0.92980***
	(19.629)	(16.453)	(16.746)
$skew$	0.99860***	0.86380***	0.92980***
	(19.629)	(30.157)	(16.746)
$shape$	3.63800***	2.66900***	4.52000***
	(6.376)	(20.191)	(4.785)
log-likelihood	2087.514	1853.21	1876.643
θ_1	1	0.5042	0.5246
θ_2	0.5042	1	0.5112
θ_3	0.5246	0.5112	1

Source: Calculation
Note: ***, **, * are significant at 1%, 5%, and 10% level, respectively.

Table 5 shows the optimal investment proportion of precious metal portfolio at the efficient line. This result illustrates that most of the investment proportion is gold, followed by silver and platinum. We also observe that the risk of precious metal portfolio of the intraday 60-min returns is larger than portfolio

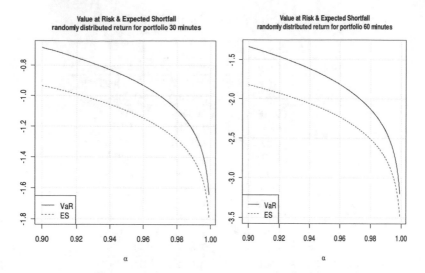

Fig. 3. Value at risk & expected shortfall at level of 1%–10% Source: Calculations

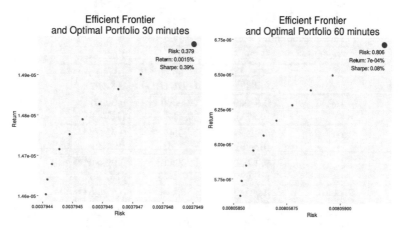

Fig. 4. The efficient frontiers of CVaR under mean Source: Calculations

of the intraday 30-min returns. This indicates that different frequency will give a different risk value. Therefore, we should consider the frequency of the data before measuring the risk of the portfolio. Then, lets consider the Sharpe ratio which is the average return earned in excess of the risk-free rate per unit of volatility or total risk [16]. Generally, the greater the value of the Sharpe ratio, the more attractive the risk-adjusted return. We find that the Sharpe ratio of precious metal portfolio, when portfolio of the intraday 30-min returns, is larger than the portfolio, when portfolio of the intraday 60-min returns.

Table 5. Optimal investment proportion of precious metal portfolio

Portfolio of the intraday 30-min returns

Portfolio	Gold	Silver	Platinum	Return	Risk	Sharpe
1	0.528679	0.000000	0.471321	0.001461%	0.379441%	0.003849
2	0.527224	0.000000	0.472776	0.001463%	0.379442%	0.003859
3	0.525769	0.000000	0.474232	0.001468%	0.379443%	0.003868
4	0.524313	0.000000	0.475687	0.001473%	0.379444%	0.003878
5	0.522858	0.000000	0.477142	0.001475%	0.379445%	0.003888
6	0.521403	0.000000	0.478597	0.001479%	0.379455%	0.003898
7	0.519948	0.000000	0.480052	0.001485%	0.379459%	0.003908
8	0.518492	0.000000	0.481508	0.001488%	0.379465%	0.003917
9	0.517037	0.000000	0.482963	0.00149%	0.379475%	0.003927
10	0.515582	0.000000	0.484418	0.0015%	0.3795%	0.003937

Portfolio of the intraday 60-min returns

Portfolio	Gold	Silver	Platinum	Return	Risk	Sharpe
1	0.811356	0.000000	0.188644	0.00057%	0.805855%	0.000699
2	0.810340	0.000000	0.189660	0.000574%	0.805857%	0.000712
3	0.809323	0.000000	0.190677	0.000581%	0.805859%	0.000725
4	0.808307	0.000000	0.191693	0.000595%	0.80586%	0.000739
5	0.807290	0.000000	0.192710	0.000611%	0.805872%	0.000752
6	0.806274	0.000000	0.193726	0.000624%	0.805876%	0.000766
7	0.805257	0.000000	0.194743	0.000627%	0.805884%	0.000779
8	0.804241	0.000000	0.195759	0.000645%	0.805886%	0.000793
9	0.803224	0.000000	0.196776	0.00065%	0.805888%	0.000806
10	0.802208	0.000000	0.197793	0.0007%	0.8059%	0.000819

Source: Calculation

Note: ***, **, * are significant at 1%, 5%, and 10% level, respectively.

7 Conclusion

In this paper, we focus on the risk that occurs from investing in precious metals traded at London Metal Exchange. We apply Copula-GARCH to fit daily histogram valued time series constructed from intraday 30-min returns and intraday 60-min returns of the precious metals (gold, silver, platinum). Then, we calculate the risk of our portfolios by applying Value at Risk (VaR) and Expected Shortfall (ES) concepts. The empirical results showed that student-t copula is the most appropriate copula function to link the GARCH(1, 1) with skewed student-t distribution of gold, silver, and platinum. The estimated VaR and ES are calculated based on 1%–10% levels. Finally, we obtain expected return and risk to find the optimal weights of the portfolios and the results suggest that gold and platinum should have a high investment proportion while silver should have a low proportion in the optimal portfolio. When, we compare the portfolios which were constructed from different frequencies. We find that a precious metal portfolio of the intraday 30-min returns gives lower risk when compared portfolio of the intraday 60-min returns. Therefore, investors should not hold assets for long period of time because it will lead to the higher risk.

References

1. Bollerslev, T.: Generalized autoregressive conditional heteroskedasticity. J. Econom. **31**(3), 307–327 (1986)
2. Demiralay, S., Ulusoy, V.: Non-linear volatility dynamics and risk management of precious metals. N. Am. J. Econ. Financ. **30**, 183–202 (2014)
3. Dias, S., Brito, P.: Distribution and symmetric distribution regression model for histogram-valued variables. arXiv preprint arXiv:1303.6199 (2013)
4. Gonzlez-Rivera, G., Arroyo, J.: Autocorrelation function of the daily histogram time series of SP500 intradaily returns (2009)
5. Hammoudeh, S., Malik, F., McAleer, M.: Risk management of precious metals. Q. Rev. Econ. Financ. **51**(4), 435–441 (2011)
6. Hofert, M., Mchler, M., Mcneil, A.J.: Likelihood inference for Archimedean copulas in high dimensions under known margins. J. Multivar. Anal. **110**, 133–150 (2012)
7. Irpino, A., Verde, R.: A new Wasserstein based distance for the hierarchical clustering of histogram symbolic data. In: Data Science and Classification, pp. 185–192. Springer, Heidelberg (2006)
8. Irpino, A.: R-Package HistDAWass (2016)
9. Joe, H.: Asymptotic efficiency of the two-stage estimation method for copula-based models. J. Multivar. Anal. **94**(2), 401–419 (2005)
10. Jondeau, E., Rockinger, M.: The Copula-GARCH model of conditional dependencies: an international stock market application. J. Int. Money Financ. **25**(5), 827–853 (2006)
11. Rockinger, M., Jondeau, E.: Conditional dependency of financial series: the copula-GARCH model. FAME research paper No. 69 (2002)
12. Khemawanit, K., Tansuchat, R.: The analysis of value at risk for precious metal returns by applying extreme value theory, Copula model and GARCH model. IJABER **14**(2), 1011–1025 (2016)
13. Pastpipatkul, P., Maneejuk, P., Wiboonpongse, A., Sriboonchitta, S.: Seemingly unrelated regression based copula: an application on Thai rice market. In: Causal Inference in Econometrics, pp. 437–450. Springer, Cham (2016)
14. Pastpipatkul, P., Panthamit, N., Yamaka, W., Sriboochitta, S.: A copula-based Markov switching seemingly unrelated regression approach for analysis the demand and supply on sugar market. In: Integrated Uncertainty in Knowledge Modelling and Decision Making: 5th International Symposium, IUKM 2016, Da Nang, Vietnam, 30 November–2 December 2016, pp. 481–492. Springer, Cham (2016)
15. Patton, A.J.: Applications of copula theory in financial econometrics. Unpublished Ph.D. dissertation, University of California, San Diego (2002)
16. Sharpe, W.F.: Mutual fund performance. J. Bus. **39**(S1), 119–138 (1966). https://doi.org/10.1086/294846
17. Verde, R., Irpino, A.: Ordinary least squares for histogram data based on Wasserstein distance. In: Proceedings of COMPSTAT 2010, pp. 581–588. Physica-Verlag HD (2010)

Factors Affecting Consumer Visiting Spa Shop: A Case in Taiwan

Meng-Chun Susan Shen[(✉)], I-Tien Chu, and Wan-Tran Huang

Department of Business Administration, Asia University, Taichung, Taiwan
kesalan@gmail.com, chuitien@gmail.com,
wthuangwantran7589@gmail.com

Abstract. The research aims to search for factors that influence the customer's decision-making process regarding spa services of a case study spa in Taiwan based on a Count Data Model. The estimation is via Poisson regression analysis and negative binomial regression analysis. 167 questionnaires were collected from Taiwanese customers. The results of both Poisson regression and negative binomial regression are statistically significant at the conventional levels, which provides the predictions of the consumer's decision-making. The study shows that the customer's demography and customer satisfaction towards the case study spa have an impact on a consumer's decision-making process when selecting spa services. Therefore, the spa can consider its marketing strategies based on the result.

Keywords: Spa · Consumer decision making · Count data analysis
Poisson regression · Negative binomial regression · STATA

1 Introduction

The healing power of spa, which has thousands of years of history, is believed to be a natural way for a person to recover from physical exhaustion and mental distress. With the aromatherapy and massage provided by spas, the customers are able to relax and find peace in mind. Hence, attracting people around the world to pay a visit, searching for balance of life, spa is currently a growing business and has become a competitive industry [1].

With the launch of new labor policy, "One Fixed Day Off and One Flexible Rest Day Policy", in Taiwan in 2017, the importance of resting and relaxing has become the center of Taiwanese's attention. For a long time, Taiwanese people have valued the importance of hard work and long working hours [2]. The labor policy is to assure that all employees are entitled to have sufficient rest periods to avoid any possible health issues in the future. The basic regulation is that a worker is entitled to have two regular days off every seven days. One day is a regular leave and the other one is a rest day [3]. Therefore, health services have expanded and grown in Taiwan to promote the well-being of people. Meanwhile, people are also searching for alternatives from health care facilities. An emphasis on prevention and health promotion makes health service business widely popular in Taiwan [4]. Among all the choices for people to relax when free,

© Springer International Publishing AG 2018
V. Kreinovich et al. (eds.), *Predictive Econometrics and Big Data*, Studies in Computational
Intelligence 753, https://doi.org/10.1007/978-3-319-70942-0_40

one of the popular options for Taiwanese is visit a spa. Spa entrepreneurs are trying to compete to be known and recognized from many aspects, such as service programs provided, interior decoration, and customer services. In other words, customers are the key factor that bring business success in the spa industry [5].

The purpose of the study is to find out the factors that affect the consumer's decision on visiting a spa shop. The factors may collectively influence the customer's decision-making process. Some of the factors are observed, but some are not. The observed are labeled as x while unobserved as y. The factors that affect the customer's decision through a function:

$$D = h(x, y). \tag{1}$$

We can look at the function as the decision-making process. In other words, given x and y, the customer's choice can be determined.

Since y is not observed, the customer's decision cannot yet be determined exactly. Hence, the unobserved terms are considered random with density $f(y)$. The probability that the customer chooses a particular spa and services from all possible outcomes is simply the probability that the unobserved factors are. Therefore, the behavioral process results in that outcome: $P(D|x) = \text{Prob}(y \text{ s.t. } h(x,y) = D)$.

We can define an indicator function $J[h(x,y) = D]$ that takes the value of 1 when the statement in brackets is true and 0 when the statement is false. In other words, $J[\cdot] = 1$ if the value of y, combined with x, induces the customer to choose outcome D, which is the going to a certain spa, and 0 if the value of y, combined with x, induces the customer to choose some different outcome. Then the probability that the customer chooses outcome D is the expected value of this indicator function. The expectation is over all possible values of the unobserved factors as follows:

$$
\begin{aligned}
P(D|x) &= \text{Prob}\left(J\big[h(x, y) = D\big] = 1\right) \\
&= \int J\big[h(x, y) = D\big]f(y)dy.
\end{aligned}
\tag{2}
$$

Stated in this form, the probability is an integral for the outcome of the behavioral process over all possible values of the unobserved factors. To calculate this probability, the integral must be evaluated [6].

The purpose of the study is to identify the unobserved factors that may influence the customer's decision making when choosing a spa. Then, the result can provide the case company the insight to improve their strategic plan and create marketing strategy by establishing the appropriate prices as well as promoting and maintaining customer loyalty. Moreover, the project can be used as a guideline for future research related to spa companies and help them understand the fundamental factors that make impact on customers choosing spa services.

2   Literature Review and Hypotheses

Nowadays, consumers demand a higher standard of service from the spa industry, which means the spa industry has become highly competitive. Hence, it is important for entrepreneur and staff of a spa to continuously improve and develop to increase their competitiveness to survive. Furthermore, the customer's satisfaction is the foundation of word of mouth marketing strategy. However, we must also search for other channels for positive reviews to be spread around.

Tabashi, searching for the most successful spas worldwide, suggests that a positive image created by having professional management has positive impact on customer's decision when choosing a spa [7]. In other words, a professional image can be created by having good organization structure, providing quality services to customers, and carrying good products. When a spa has healthy management in organization with strong marketing strategies, it can attract new customers to use the services and products, which can help develop the existing customer's loyalty to the spa.

Lu et al. claim that income is one of the important factors when customers choosing a spa after surveying Taiwanese lifestyle. Taiwanese long for good health, family time, and wealth [8]. In addition, according to a survey conducted by Taiwan Ministry of Economic Affairs in 2003, the percentage of Taiwanese people visiting hot springs increases every year. While the number of people traveling for health purposes continues to rise, hot spring tours are the result of consumers' needs in search for the balance of health, family time, and budget. While Taiwanese are willing to pay for health services, including spas, marketing strategies are needed for expanding the current number of customers.

Lu et al. also suggest that the perceived quality of service, price, and risk are highly related to the customer's decision making [9]. The findings help researchers and spas realize the factors that have impact on the consumer's decision making. Overall, good services and quality products with acceptable prices provided are more likely to satisfy customers. Therefore, the factors of administration and product will be used as a variable in this project of study.

Niratisayaputi claims that people usually visit the spas recommended by family, friends, and colleagues. While word of mouth plays an important role in decision making, customers also take location, promotion, and staff of a spa into consideration. However, the most important factor, based on the research, is the quality of the staff members whom the customers rely on for professional suggestions. They should demonstrate speaking politely, developing good relationship with customers, and offering quality treatments. In other words, putting the customers first is the quality that the spa goers are looking for. Moreover, the results between male and females are different. Male customers focus on services and the cleanness of the location. Female customers tend to check the prices each time they visit. Since the research implies that administration and staff are the main variables, the project will analyze the factors in this category in details [10].

Word of mouth, the information passing from one consumer to another, has impact on the customer's decision-making process when choosing a product or service [11]; however, it may not be the only channel that consumers obtain information. Therefore, the project also examines other channels of information, such as television shows and magazines, to find alternative accesses to promote services or products. Therefore, a spa

must create marketing strategies to reinforce quality consciousness, brand consciousness, fashion consciousness, recreation consciousness, impulse consciousness, price
consciousness, and brand loyalty.

Therefore, the hypothesis, the possible factors, that would influence the customer's
decision making are as follows.

Hypothesis 1: The customer chooses the case study spa based on the customer's
demography.

Hypotheses 2: The customer chooses the case study spa based on the customer's
satisfactory level towards the case spa's services, staff, and products.

Hypotheses 3: The customer chooses the case study spa based on the customer's
other concerns.

3   Methodology

To find out the factors that affect consumers visiting a spa shop, the project consists of
three stages. First, the initial interview to collect basic information regarding the care
study spa. Secondly, questionnaires are provided to the case study spa's customers to
collect data. The final stage is the count data analysis.

3.1   Case Study Spa Information

The case study spa chain was founded in 1997. Although it once expanded its branch to
Mainland China in 2006, but the services in China were terminated due to the difficulties
in human resources and financial support. Currently, the care study spa has 51 branch
stores with over 500 employees in Taiwan and has become one of the largest and famous
beauty spa in Taiwan. The case study company is still considering to branch out more
in the near future.

The case study spa focuses on serving female customers only. The philosophy of the
case study spa, "Beauty is a reflection of both inner and outer beauty," is to promote
outer beauty through both inner and outer treatments. Therefore, customers are provided
with various treatment programs, hoping that they will be able to find the ones that best
suit their needs.

While the case study spa is able to provide quality beauty products from domestic
and international brands, the greatest challenge for them is to find skillful professional
staff because the skills require long-term training and practice. Another issue for the
case study spa is the innovation of the business plan. The case study spa needs to
constantly renew and update their facilities and equipment; otherwise, they can lose their
current advantages and be out of the business.

3.2   Questionnaire and Data Collection

The questionnaire survey consists of three parts. Part 1 is the basic personal data.
Customers are asked to provide personal demographic data, such as gender, marital
status, age, occupation, income, education level, and sources of information (Table 1).

Table 1. Demographic variable names and description

Variable names	Variable description
Gender	Gender, but only female customers in this study
Status	Married or single
Age	Age
Occupation	Occupation
Income	Income per month
Education	Education level
Resources	How do you know about the spa case company?

Part 2 is the customer's satisfaction towards the case study spa in terms of quality of treatments, staff, products, and environment. This part provides more information regarding the factors that the customer may choose the case study spa. Part 3 is the survey of spa services used by the respondent within the past two months. This part shows the results of descriptive statistics as well as the reliability and validity analysis. The samples are analyzed with count data generated by STATA.

3.3 Count Data Analysis

Winkelman suggests that regression is a tool to apply for researchers who would like to explore data or weight the evidence in data for or against a specific hypothesis. Regression model is used in two cases. First, the function is to reduce the dimension of complex information. The second function is the test economic theories, quantify theoretical relations as well as predict and forecast [12].

Poisson Regression Model has two components: a distributional assumption and a specification of the mean parameter distribution. A model needs to be correctly specified to apply the results for maximum likelihood estimators. This requirement is more binding for count data models than for the normal linear models. The count model depends on the type of available data. Count data analysis in this project is estimated with STATA to provide support. Poisson process is a special case of count process which, in turn, is a special case of a stochastic process.

Cameron indicates that count data is concerned with models of event counts. An event count refers to the number of times an event occurs [13]. For example, the number of airline accidents or earthquakes. The main focus of count data is regression analysis of event counts. The benchmark model for count data is the Poisson distribution. It is useful at the outset to review fundamental properties and characterization results of the Poisson distribution [14].

4 Empirical Results

The respondents of the survey are Taiwanese customers who have experienced any forms of spa services in the case study spa within a duration of two months. The number of returned complete questionnaires is 167 units.

4.1 Personal Information Analysis

The result reveals that women aged between 25 to 54 are the primary users of the case study spa. The majority of the customers are office workers and self-employed with the average monthly income ranges between 20,000 NTD and 99,999 NTD. Approximately 80% of the participants have a college degree or higher. Therefore, the project suggests that women with higher level of education, stable jobs, and above-average income are most likely the target customers. They are most likely the ones who suffer from stress at work and are able to afford spa services. The primary information resource is still the word of mouth, while those customers received information from friends and colleges with similar background or working environment. However, approximately 40% of the participants obtained information from the Internet.

4.2 Customer Satisfaction and Other Factors

The second part of the questionnaire is to examine the customer's satisfaction. Consequently, the case study spa is able to improve current services and forming effective marketing strategies. On one hand, the spa should cultivate customer loyalty from current customers in order to attract new customers by sharing the reviews. On the other hand, the case study spa should carefully examine the not-so-popular programs and find ways to persuade potential customers to try the services.

Over half of the customers have been with the case study spa for 2 to 10 years. Approximately 80% of the customers spend from 1,000 NTD to 10,000 NTD at the spa every month. The spa provides various kinds of treatment, from facial treatments to body treatments and specialized treatments.

The satisfaction survey suggests that the top three popular treatments are "Aroma and hydraulic spa capsule," "Relaxing aesthetic muscle with ADS light," and "Lower body healthcare." Taiwanese female customers care much about the overall well-being and the fitness of lower body. However, with a total of 25 treatments provided by the case study spa, it is challenging to predict the possibility of purchase with such wide range of choices. Therefore, it relies on the customer's satisfactory level regarding the case study spa's staff and other possible factors.

The factors adopted in the questionnaire that may or may not influence the customer's decision-making process are location, spa administration, staff, and spa products. For location, 110 out of 167 customers consider location a very important factor. While a spa is located near their homes or offices, a greater possibility they would pay an initial visit. It would be even more attractive if the spa can provide free parking nearby. When it comes to administration and staff, customers expect to see enthusiastic and well-trained staff with proficient knowledge and skills. That is the initial trust between the spa and the customer. Finally, customers expect to have reliable spa products and varieties to choose from based on their needs and budget. Therefore, these are additional factors that the care study spa should consider about.

4.3 Reliability Test

To analyze the reliability of the measurement tools or the questionnaire survey, we put all the data and test Cronbach's alpha reliability coefficient before running other methods. To test construct reliability, values of Cronbach's alpha for all constructs are reported by using STATA. The value of Cronbach's alpha is estimated for each item

Table 2. Reliability assessment of variables

I tem	Obs	Sign	Item-test correlation	Item-test correlation	Average interitem covariance	Alpha
Status	167	+	0.3502	0.2823	.0284601	0.6316
Age	167	+	0.6698	0.5794	.0231476	0.5901
Occupation	167	+	0.4497	0.3304	.0262828	0.6209
Income	167	+	0.5397	0.3887	.024064	0.6099
Education	167	+	0.2510	0.1454	.0289568	0.6400
Resources	167	−	0.2300	0.1188	.0291762	0.6426
Knowcompany	167	+	0.3743	0.2548	.0273647	0.6296
Payment	167	+	0.5986	0.5114	.0247929	0.6036
Goodloc	167	−	0.2677	0.2005	.0291086	0.6367
Carpark	167	−	0.2445	0.2027	.0296087	0.6389
Surround	167	−	0.0747	0.0169	.0303724	0.6460
Space	167	−	0.1246	0.0756	.0301047	0.6433
Branches	167	−	0.0664	0.0018	.0304794	0.6477
Satisfied	167	+	0.1915	0.1191	.0296197	0.6413
Other	167	+	0.1156	0.0961	.0302824	0.6434
Goodknow	167	−	0.1014	0.0356	.0302392	0.6456
Stfavailable	167	+	0.2074	0.1585	.0296952	0.6400
Solveprob	167	−	0.0897	0.0486	.0302721	0.6441
Cantrust	167	−	0.1510	0.0831	.0299155	0.6432
Enthusiastic	167	+	0.2426	0.1713	.0292493	0.6382
Professional	167	−	0.1135	0.0477	.0301614	0.6450
Respondreq	167	+	0.1452	0.0784	.0299553	0.6434
Servicemind	167	+	0.2043	0.1385	.0295667	0.6403
Goodmang	167	+	0.1854	0.1237	.0297142	0.6411
Highquality	167	−	0.1691	0.1031	.0297994	0.6421
Wellbrand	167	+	0.3862	0.3274	.0283914	0.6303
Canchoose	167	+	0.1782	0.1082	.0297234	0.6419
Reliable	167	+	0.2148	0.1423	.0294467	0.6399
Chooseappate	167	−	0.2977	0.2393	.0290335	0.6354
Appateprice	167	−	0.1797	0.1116	.02972	0.6417
Satisfiedr~t	167	−	0.0264	−0.0453	.0307844	0.6504
Spavis	167	+	0.7555	0.4918	.0165355	0.6179
Test scale					.0285633	0.6439

and shown in Table 2. For all measures, if Cronbach's alpha value is greater than 0.60, it means the questionnaire survey has reasonable reliability.

4.4 Count Data Model Estimation

Count data model estimation is used to identify factors affecting customer's decision-making process. To analyze the process, all the questions from questionnaire survey are chosen to be independent variable that predict which factors have impact to the customer. The project survey chooses the count data analysis to analyze the data by using STATA.

The model suggests that there are 4 variables, marital status, age, occupation and information resources, having impact on spa visiting in case study spa because the p-value is less than 0.0, which means the result is significant. The negative binomial regression summary shows the relationship between spa visiting and customer demography. We find that the result of negative binomial regression and Poisson regression have some differences. Only two variables, age and occupation, are accepted because their p-value are less than 0.05.

The same procedure is repeated for spa location, spa administration and staff, and spa products. For spa location, good location is strongly significant while having branch spas can also be a selling point, according to the Poisson regression. With the negative binomial regression analysis, good location also shows strongly significant. For spa administration and staff, Poisson regression shows that 4 variables have significant results: staff are always available, service can be trusted, staff are enthusiastic to provide services, and staff have a service mind. However, only enthusiasm and service mind are accepted in the binomial regression model. Finally, for spa product, Poisson regression shows that good brands and choosing appropriate products are the key factors while binomial regression agrees with the result.

5   Conclusion

Based on the outcome of the study, there are quite a few unobserved factors that may influence the customer's decision-making process when choosing a spa. The case study spa attracts Taiwanese female customers between ages of 25 to 54, with sufficient monthly income ranging from NTD 20,000 to NTD 99,999. Those female customers are most likely well-educated and with stressful jobs. The customer's information resources are most likely word of mouth or the Internet. They care about the overall well-being and the fitness of the lower part of the body. When it comes to the customer's satisfaction, the customer may put emphasis on the case study spa's location, the staff's availability, the trustworthiness of the services as well as the enthusiasm of the staff. Finally, the customer may be influenced by the quality and the variety of the spa products.

In terms of the results of the study, we have a few recommendations for spas in Taiwan. First, the spa should develop a user-friendly website to promote services and products since the number of the Internet users is increasing. Inviting current customers and bloggers to write positive reviews may help. While word of mouth is still powerful, the spa company may encourage the current customers and develop their loyalty by providing them exclusive and extensive services if they are able to refer new customers. Future research can focus on the level of how the online reputation or online word of mouth influences the customer's decision-making process. Spa service is rather personal and intimate, which is different from buying clothes or accessories. How much can the customer trust a stranger on the Internet would be a question.

Moreover, the spa must provide staff training on a regular basis because they can directly influence the customer's satisfactory level. The staff must have workshops regarding how to take care of customer's individual needs, to improve professional skills and knowledge as well as to develop positive relation with customers. The case study spa provides a wide variety of treatments; therefore, the customer should depend on the staff's expertise to choose the appropriate treatment based on the customer's needs, budgets, and other concerns. A mentor-apprentice system can be formed to train the new staff while the workshop or training is not yet provided. The key is the develop the team spirit rather than having staff competing with each other.

In addition, as part of customer satisfaction, the spa should consider its pricing strategies and location. Since each customer has different buying power, depending on the customer's income, the price of treatments and products would have to be flexible enough to meet the customer's needs. When it comes to choosing a location, it depends on if the spa provides an easy access to the customer. A spa may choose to be located in the area where customers can get to by public transportation. However, again the spa is rather intimate and personal, some customers may prefer to drive for the sake of privacy. Therefore, having a few branches with parking space for easy access may play a role in customer satisfaction.

In the future, while the number of male customers using spa services increases, it is possible that the case study spa will expand their business to serve the male customers. Therefore, the case study spa should conduct a research on the male customer's needs before launching the services for male because male customers are physically and mentally different from the female ones. That is, treatments for both genders should be different if another set of research is created to survey on male customers' needs and satisfaction.

References

1. Dahlan, N., Yusoff, Y. M.: Service Innovation in the Business Models on the Spa and Med Beauty at the Saujana. Case Studies in Innovation, Centre for Service Research, Manchester Business School, The University of Manchester in Collaboration with SRII Service Innovation SIG, pp. 11–13 (2010)
2. Sui, C.: Deaths Spotlight Taiwan's 'Overwork' Culture. BBC News, Taipei (2012)
3. Ministry of Labor, Republic of China. https://laws.mol.gov.tw
4. Spas and the Global Wellness Market: Spas and the Global Wellness Market: Synergies and Opportunities, prepared by SRI International, May 2010
5. Lawrimore, E.W.: "Buck": The 5 Key Success Factors: A Powerful System for Total Business Success. Lawrimore Communications Inc. (2011)
6. Train, K.: Discrete Choice Methods with Simulation, pp. 3–6. Cambridge University Press (2002)

7. Tabashi, M.H.: Current research and events in the spa industry. Cornell Hosp. Q. **51**(1), 102–117 (2010)
8. Lu, I.Y., Shiu, J.Y.: Decision-making framework of customer perception of value in Taiwanese spa hotels. Soc. Behav. Pers. Int. J. **39**(9), 1183–1192 (2008)
9. Lu, I.Y., Shiu, J.Y.: Customers' behavioral intentions in the service industry: an empirical study of Taiwan spa hotels. Asian J. Qual. **10**(3), 73–85 (2012)
10. Niratisayaputi, N.: The Factor that Affect for Spa Business Entrepreneur, Faculty of Economics, Chiang Mai University (2010)
11. Ozcan, K.: Consumer-to-consumer interactions in a networked society: word-of-mouth theory, consumer experiences and network dynamics. Dissertation of University of Michigan (2004)
12. Winkelmann, R.: Econometric Analysis of Count Data. Springer, Heidelberg (2008)
13. Cameron, C., Trivedi, P.: Regression Analysis of Count Data. Cambridge University Press, Cambridge (1998)
14. Taylor, H.M., Karlin, S.: An Introduction to Stochastic Modelling. Academic Press, San Diego and New York (1994). Revised edition

The Understanding of Dependent Structure and Co-movement of World Stock Exchanges Under the Economic Cycle

Songsak Sriboonchitta, Chukiat Chaiboonsri[✉], and Jittima Singvejsakul

Faculty of Economics, Chiang Mai University, Chiang Mai, Thailand
songsakecon@gmail.com, chukiat1973@gmail.com, Jittimasvsk@gmail.com

Abstract. This study was to focus on the patterns of economic booms (bull markets) and recessions (bear markets) among world stock exchanges such as Europe (Euro Stoxx), USA (S&P 500), Asia (SSE composite index and Nikkei 225 index) and ASEAN (FTSE ASEAN). Monthly data was collected during 2000 to 2016. Econometrically, we employed Markov Switching Bayesian Vector Autoregressive model (MSBVAR) to determine regional switches within these financial data sets as well as CD-Vine copula approaches was used to explore the contagions and patterns of structural dependences. To clarify the connectional details in each type of switching regimes, the results presented the Elliptical copula was chosen and it indicated these monthly collected data contained symmetrical dynamics co-movements. In addition, it implied the stock markets were assumed to have small fluctuations since the governments had stable policies to control the risk and asymmetric information in financial markets efficiently. Base on CD-Vine copula trees, the results indicated Asia and European stock markets had a strongly dependence in economic booms and recessions during the pre-crisis period (2000 to 2008). Conversely, in the post crisis period, the US stock market and ASEAN stock market became the strong dependence with Europe. This meant that capital flows was mostly transferred between Europe and Asia financial markets during the pre-crisis periods (2009 to 2016). After that, the direction of capital flows were changed dramatically to the US stock market in the post-crisis periods. Predictively, this seems that the capital flows will return to European and US financial market, which these two continents have a strongly long-term financial dependence and deeply positive diplomacy.

Keywords: MSBVAR · CD-vine copula · Bull markets
Bear markets · Stock markets

1 Introduction

Because the financial crisis negatively affected the economic system in the United States during 2008, triggered by collapse in house prices, and caused

© Springer International Publishing AG 2018
V. Kreinovich et al. (eds.), *Predictive Econometrics and Big Data*, Studies in Computational Intelligence 753, https://doi.org/10.1007/978-3-319-70942-0_41

the Great Recession. This leaded the world economy to be suffered dramatically (Bloomberg 2009). This was the underline of the global financial crisis and caused European banks to enormously lost their liquidity in the ABS market. Moreover, the reliance on US currency for European banks had been sharply decreased (Lane 2013). Additionally, this can be seen from the low expansion rates of GDP in ASEAN, US, Europe, Japan and China during the period between 2000 and 2015, which were respectively represented in Figs. 1 and 2. For the pre-crisis (2000–2008), GDP in these five countries slightly grew up. In particularly, the economic expansion rate of japan did not change. In the post-crisis (2009–2015), this can be seen that the economic growth in many countries around the world continuously grew up. This is because the effect from the transferences of capital flows in the term of financial markets. Accordingly, this paper intensively explored a structural cycling pattern between them in the Worlds Stock

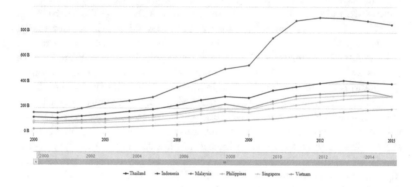

Fig. 1. Gross Domestic Product or GDP of ASEAN in current US dollar for during period of 2000–2015 Source: World development indicators.

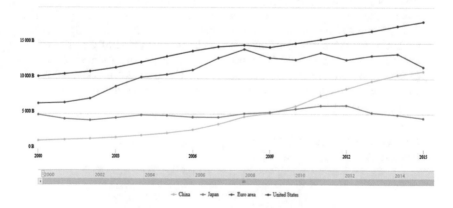

Fig. 2. Gross Domestic Product or GDP of China, Japan, Euro area and the United States in current US dollar for during period of 2000–2015 Source: World development indicators.

Exchanges as well as rare financial structural dependences, and these findings can be the solution to understand the deeply financial structures between major stock markets around the world that is useful information for supporting domestic and foreign investors to predict and plan their investments.

2   The Objective and Scope of Research

The objective of this research is to explore the pattern of structurally financial dependences in bull and bear markets among stock markets in US, Europe, Asia, and ASEAN during 2000 to 2016. The monthly time-series data such as in S&P 500 index (US), Europe (the Euro Stoxx), China (SSE composite index), Japan (Nikkei 225 index) and ASEAN (FTSE ASEAN) were collected to be considered, and they were divided into 2 periods: pre-crisis (2000 to 2008) and post-crisis (2009 to 2016).

3   Methodology

3.1   The Markovian Switching Bayesian VAR Model

This paper has two steps to determine the pattern of structural dependences among the capital markets. First, the Markov Switching Bayesian VAR model was employed to determine regime changes within data, and examine correlations among the European, US, Asia and ASEAN stock market. This found regimes for bull and bear markets.

The Markovian switching is constructed by combining two or more dynamic models via the Markovian switching mechanism (Hamilton 1994) and this can be shown in Eq. 1.

$$z_t = \alpha_0 + \beta z_{t-1} + \varepsilon_t, S_t = o, \tag{1}$$

$\varepsilon = $ i.i.d. random variables with zero means and variances σ_t^2
$|\beta| < 1$.

This is stationary AR (1) processed with mean $\alpha_0/(1 - \beta)$ when $S_t = 0$, and it switches to another stationary AR (1) process with mean $(\alpha_0 + \alpha_1)/(1 - \beta)$ when S_t is changed from zero to one. Then it provided that $\alpha_1 \neq 0$, this model admits two dynamic structures at different levels, depending on the value of state variables S_t.

The evaluation of the latent variable drives regime changes, S_t, is governed by the first-order Markov chain condition with constant transition probabilities expressed as the (SxS) transition probability matrix (P):

$$pr = (S_t = j | S_{t-1} = i) = p_{ij}, \tag{2}$$

$$p = \begin{pmatrix} p_{11} & p_{12} & \cdots & p_{1s} \\ p_{21} & p_{22} & \cdots & p_{2s} \\ \cdot & \cdot & \cdots & \cdot \\ p_{1s} & p_{2s} & \cdots & p_{ss} \end{pmatrix}. \tag{3}$$

Bayesian statistics was applied to do econometrical estimations, and this inference allows us to obtain a joint posterior distribution of parameters and latent variables. Bayesian simulated methods are well suited to estimate Markov Switching models (Kim and Nelson 1999). Conditionally, the value at risk (VaR) analysis allows parameters of the model to be considered as random variables. Generally, the typical VAR analysis is often constrained by the limited size of data sets, which are not compatible models with large numbers of parameters. The Bayesian method tackles this over-parameterisational problem by assigning initial probabilities into many parameters. Furthermore, the construction of a BVAR model will reduce the complexities involved future extensions (Canova 2007).

3.2   ARMA-GJR Model for Marginal Distributions

Technically, the CD-Vine copula was adopted to estimate the pattern of structural dependences among stock markets. We will find the major stock markets of bull and bear markets in pre-post crisis. ARMA-GJR model was used to conduct marginal distributions for the copula model. The form of the ARMA (P, Q)-GJR (K, L) model can be expressed as Eq. 4.

$$r_t = c + \Sigma_{i=1}^{p} \phi_i r_{t-i} + \Sigma_{i=1}^{p} \Psi_i \varepsilon_{t-i} + \varepsilon_i \tag{4}$$

$$\varepsilon_t = h_t \eta_t \tag{5}$$

$$h_t^2 = \omega + \Sigma_{i=1}^{k} \alpha_i \varepsilon_{t-i}^2 + \Sigma_{i=1}^{k} \gamma_i I[\varepsilon_{t-i} < o]\varepsilon_{t-i}^2 + \Sigma_{i=1}^{l} \beta_i h_{t-i}^2, \tag{6}$$

where $\Sigma_{i=1}^{p} \phi_i < 1, \omega > 0, \alpha_i > 0, \beta_i > 0, \alpha_i + \gamma_i > 0$ and $\Sigma_{i=1}^{k} \alpha_i + \Sigma_{i=1}^{l} \beta_i + \frac{1}{2}\Sigma_{i=1}^{k}\gamma_i < 1$. The formulas (4) and (6) are call mean equation and variance equation, respectively; the formula (5) describes the residual ε_t is consist of standard variance h_t and standardized residuals η_t; the leverage coefficient γ_i is applied to negative standardized residuals. In addition, the standardized residual are assumed to be the skewed student-t or skewed generalized error distribution and the cumulative distributions of standardized residuals are formed to plug into copula model.

3.3   Copula

The fundamental theorem is based on the concept of (Sklar 1959) and this can be shown in Eq. 7,

$$F(x_1, x_2, \ldots, x_n) = C(F_1(x_1), F_2(x_2), \ldots, F_n(x_n)). \tag{7}$$

F: n-dimensional distribution with marginal F_i , i $= 1$, 2, 3

$x_1, x_2, \ldots, x_n$:random vectors

C: n-copula for all $x_1, x_2, \ldots, x_n$.

The function C is a distribution function that has uniform margins between zero and one, and it is labelled as the copula function. It binds the univariate margins $F1$ and $F2$ to produce bivariate distribution F.

3.4   The C-D Vine Copulas Construction

Vine copula models are graphical representation to specify pair copula constructions (PCCs) introduced by (Joe 1996). These models are consequently developed by Bedford and Cook (2001, 2002). Basically, a principle for constructing multivariate copula generated from the product of bivariate pair copula was statistically described as canonical (C-) vines and (D-) vines by Aas et al. (2009). This contribution was a flexible model since bivariate copulas can easily accommodate complex structural dependences such as asymmetric dependences or strong joint tail behaviors (Joe et al. 2010). Based on previous reviews, this has been already pointed out the estimated patterns of relation among financial markets in world exchanges are defined as $X = x_1, x_2, x_3, x_4, x_5$, with marginal distribution function F_1, F_2, F_3, F_4, F_5, and corresponding densities. As a result, it can be written as Eq. 8.

$$f(x_1, x_2, x_3, x_4, x_5) = f(x_1)f(x_2|x_1)f(x_3|x_1, x_2)$$
$$f(x_4|x_1, x_2, x_3)f(x_5|x_1, x_2, x_3, x_4), \qquad (8)$$

where C is the copula associated with F via Sklar theorem. From Eq. 5, it can be determined the conditional density of x_2, and given x_1 as

$$f_{2|1}(x_2|x_1) = \frac{f(x_1, x_2)}{f_1(x_1)} = c_{1,2}(F_1(x_1), F_2(x_2))f_2(x_2), \qquad (9)$$

and

$$f_{2,3|1}(x_3|x_1, x_2) = \frac{f(x_2, x_3|x_1)}{f(x_2|x_1)}$$
$$= c_{2,3|1}(F(x_2|x_1), F(x_3, x_1))f(x_3|x_1)$$
$$= c_{2,3|1}(F(x_2|x_1), F(x_3, x_1))c_{1,3}(F_1(x_1), F_3(x_3))f_3(x_3), \qquad (10)$$

and

$$f_{3,4|1,2}(x_4|x_1, x_2, x_3) = \frac{f(x_3, x_4|x_1, x_2)}{f(x_2, x_3|x_1)}$$
$$= c_{3,4|1,2}(F(x_3|x_1, x_2), F(x_4|x_1, x_2))f_3(x_3)f(x_4|x_1, x_2)$$
$$= c_{3,4|1,2}(F(x_3|x_1, x_2), F(x_4|x_1, x_2))c_{1,4}$$
$$(F_1(x_1 F_4(x_4))f_4(x_4)c_{2,4}(F_2(x_2 F_4(x_4))f_4(x_4), \qquad (11)$$

and

$$f_{4,5|1,2,3}(x_5|x_1, x_2, x_3, x_4) = \frac{f(x_4, x_5|x_1, x_2, x_3)}{f(x_3, x_4|x_1, x_2)}$$

$$= c_{4,5|1,2,3}(F(x_4|x_1, x_2, x_3), F(x_5|x_1, x_2, x_3))f(x_5|x_1, x_2, x_3)$$
$$c_{1,5}(F_1(x_1)F_5(x_5))f_5(x_5)$$

$$= c_{4,5|1,2,3}(F(x_4|x_1, x_2, x_3), F(x_5|x_1, x_2, x_3))c_{2,5}$$

$$(F_1(x_1)F_5(x_5))f_5(x_5)c_{2,5}f_5(x_5)c_{3,5}(F_3(x_3)F_5(x_5))f_5(x_5). \qquad (12)$$

Therefore, the five-dimensional joint can be shown in terms of bivariate copula $c_{1,2}, c_{2,3|1}, c_{1,3}, c_{3,4|1,2}, c_{1,4}, c_{2,4}, c_{4,5|1,2,3}, c_{1,5}, c_{2,5}, c_{3,5}$ Based on graphical of canonical (C-) and D-vines copula was presented by Fig. 3.

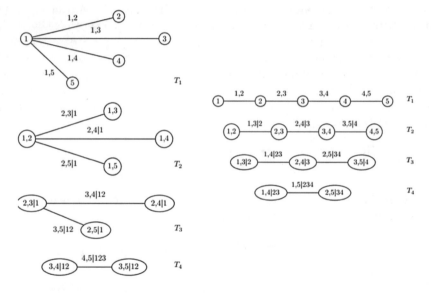

Fig. 3. Examples of five-dimensional C-vine tree (left panel) and D-vine tree (right panel) Source: Brechmann and Schepsmeier (2013)

Considering Fig. 3, on the left-panel trees represented the decomposition of a five-dimensional joint density function. The circled nodes are on the first-tree and it showed the four marginal density functions, f_1, f_2, f_3, f_4, f_5. The remaining nodes on the other trees are not used in the figure. Each edge corresponds to a pair-copula function.

On the other hand, on the right-panel trees represented the decomposition of five-dimensional joint density functions. The circles nodes showed the five marginal density functions written as f_1, f_2, f_3, f_4, f_5. Each edge is labeled with the pair-copula of the variables. The edges in level i become nodes for level $i+1$. The edges for the first tree are labeled as 1,2, 2,3, 3,4 and 4,5. The second tree has edges labeled as $1, 3|2, 2, 4|3$ and $3, 5|4$. The third tree's edges were labeled as $1, 4|23$ and $2, 5|34$. Finally, the tree number fourth has only one edge labeled as $1, 5|234$ (Durante and Sempi 2009).

3.5   Bivariate Copula Families

The package CD-Vine provides a wide range of bivariate copula families, which are divided into two major classes such as elliptical and Archimedean copulas (Joe 1997 and Nelsen 2006). Elliptical copulas are directly obtained by inverting Sklar Theorem (Eq. 7). Given a multivariate distribution function F with invertible margins F_1 and F_2, then

$$C(u_1, u_2) = F(F_1^{-1}(u_1), F(F_2^{-1}(u_2)), \tag{13}$$

C: F is elliptical
$u_1, u_2 \in [0, 1]$
F: distribution functions of invertible marginals F_1, F_2,

which are also implemented in CD-Vine, and they are the multivariate Student-t copula. Consequently, this type of copula models can be expressed in Eq. 20,

$$C(u_1, u_2, u_3, u_4, u_5) = t_{\rho,\nu}(t_\nu^{-1}(u_1), t_\nu^{-1}(u_2), t_\nu^{-1}(u_3), t_\nu^{-1}(u_4), t_\nu^{-1}(u_5)) \tag{14}$$

$\rho \in (-1,1)$ and is dependence parameter
ν: degree of freedom for student t copula $\nu > 2$.

Which $t_{\rho,\nu}$ is the multivariate Student-t distribution function contained correlation parameters, ρ and ν, t_ν^{-1} denotes the inverse univariate Student-t distribution function with ν degrees of freedom. Both copulas are obviously symmetric and have lower and upper tail dependence coefficients.

Multivariate Archimedean copulas, on the other hand, are defined as

$$C(u_1, u_2, u_3, u_4, u_5) = \Psi^{[-1]}(\Psi(u_1), \Psi(u_2), \Psi(u_3), \Psi(u_4), \Psi(u_5)), \tag{15}$$

where: $[0, 1] \cdots [0, \infty]$ is a continuous strictly decreasing convex function such that $\Psi(1) = 0$ and Ψ^{-1} is the pseudo-inverse,

$$\Psi^{-1}(t) = \Psi^{-1}(t), 0 \leq t \leq \Psi(0) \ or \ \Psi^{-1}(t) = 0, \Psi(0) \leq t \leq \infty, \tag{16}$$

Ψ is called the generator function of the copula C.

In addition, this paper implemented the common single parameter, which is in the Archimedean family (Clayton copula). This is a more flexible structure allows non-zero lower and upper tail to be the different dependent coefficient (Nelson 2006), then the Clayton are defined as

$$\Psi = \frac{1}{\theta}(t^{-\theta} - 1), \tag{17}$$

parameter range: $\theta > 0$

$Kendall's^{\tau}$: $\frac{\theta}{\theta+2}$,

Tail dependence (lower, upper): $(2^{\frac{-1}{\theta}}, 0)$.

4   Data Description

The world stock exchanges data considered in this study consisted five largest economics, for instances, the United States stock market (S&P 500 index), European stock markets (the Euro Stoxx), China stock market (SSE composite index), Japan stock market (Nikkei 225 index) and ASEAN stock markets (FTSE ASEAN). Basically, all of data was transformed to be standardized residuals of monthly log return observations (203 observations).

Considering Fig. 4, it provided the descriptive index returns of monthly data in world exchanges during 2000 to 2016. Furthermore, Table 1 presents the generally statistical data.

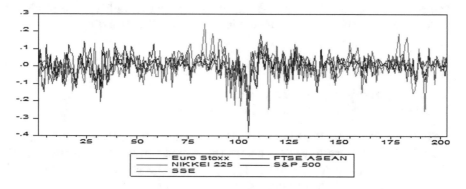

Fig. 4. The index return of monthly data in world exchange during period of 2000 to 2016. Source: Thomson Routers Corp database.

5   Empirical Results of Research

5.1   The Results of Marginal Testing for Copula Model Estimation

Based on the LM-test, this already confirmed that all of residual terms was satisfied for marginal models, which were employed to estimate the CD-Vine copula models. Additionally, the result of the KS testing already indicated that the marginal model is efficiently specified to estimate the CD-Vine copula model (Table 2).

Table 1. The descriptive statistics of the index return of monthly data in world exchanges during period of 2000 to 2016

Items	S&P 500	Euro Stoxx	SSE	NIKKEI 225	FTSE ASEAN
Mean	0.002332	−0.001768	0.003468	0.000108	0.010439
Median	0.007791	0.007097	0.007078	0.003953	0.017787
Maximum	0.102307	0.137046	0.242528	0.120888	0.18341
Minimum	−0.185636	−0.206236	−0.282783	−0.27216	−0.377193
Std. Dev	0.043343	0.054506	0.08062	0.05823	0.066147
Skewness	−0.728593	−0.638291	−0.543509	−0.735032	−1.124449
Kurtosis	4.492987	4.102723	4.554236	4.379683	8.028629
Jarque-Bera	36.81406	24.06951	30.42681	34.37986	256.6652
Probability	0	0.000006	0	0	0
Phillips-Perron test statistic	−12.76037	−13.01843	−12.65397	−12.37694	−11.36490
	(0.0000)	(0.0000)	(0.0000)	(0.0000)	(0.0000)
Sum	0.473446	−0.358927	0.704046	−0.022008	2.119088
Sum Sq. Dev	0.379481	−0.358927	1.312931	0.684933	0.883836
Observations	203	203	203	203	203

Table 2. Testing of the marginal distribution models based on LM-test (lag 2) and K-S test.

	S&P 500	Euro stoxx	SSE	Nikkei 225	FTSE ASEAN
L-M test	0.7985	0.4919	0.9574	0.2783	0.6446
K-S test	0.000	0.000	0.000	0.000	0.000

5.2 The Estimated Results of the Bull and Bear Markets in Pre-crisis and Post-crisis Periods Based on the Markovian Switching Bayesian VAR Model

Expressly, the results were represented in Table 3 showed that the Markovian Switching Bayesian VAR model computationally estimated the fluctuated regimes of five financial stock indexes. Econometrically, the regimes are defined as Bull and Bear market. First, the index of the S&P financial market contained boom periods rather than recessions, which were 113 months and 90 months, respectively. Second, Euro stock indexes had recession situations more than expansions, which were 94 months and 109 months, respectively. Third, the financial market in China (SSE) included expanding times more than recessions, which were 110 months and 93 months, respectively. Forth, Japanese financial equity (Nikkei 225) contained booming situations more than recessing times, which were 109 months and 94, respectively. Lastly, the financial market in South East Asia (FTSE ASEAN) had the fluctuated situations between bull and bear markets, which had 103 months for the booming periods and 100 months for recessions.

Table 3. Testing number of bull market and bear market based on MSBVAR

	S&P 500	Euro stoxx	SSE	Nikkei 225	FTSE ASEAN
Bull market (Months)	113	94	110	109	103
Bear market (Months)	90	109	93	94	100

5.3   The Estimation Results of the Contagion and Pattern of Structural Dependences Toward World Exchanges in Bull and Bear Markets Based on CD-Vine Copula Approach

There are two kinds of copula estimations. Elliptical and Archimedean copulas were used to estimate the pattern of dependences among world exchanges. The estimated result was investigated by CD-vine copula approach and it was represented in Appendix A. The best model based on AIC and BIC is Elliptical copula, which is the T-copula model. Accordingly, this result based on CD-vine indicated that there is a contagion among the two periods, which are the pre-crises periods (2000–2008) and post-crises periods (2009–2016).

5.4   The Results of Estimation in the Pattern of Structural Dependences Among Five Stock Markets of Economic Boom (Bull Market) and Economic Recession (Bear Market) Based on CD-Vine Trees from T-copula

5.4.1   Pre-crises Periods (2000–2008)
(a) The Elliptical t-copula of C-vine in Bull and Bear markets

As we see in Fig. 5, the financial market in Europe was assumed to be the central place that capital inflows and outflows were transferred during the post-crises periods. Obviously, in the Bull situation, the markets between Europe and Asia (ASEAN, Japan, and China) were the strongly structural dependence in terms of capital flows. Similarly, in the Bear market, the Asian financial market still strongly depended on the recessing time in the European market, but the US financial market had a weakly structural dependence with European in the post-crisis periods. As a result, this implied that the capital flows had been mostly transferred between Europe and Asia during 2000 to 2008.

(b) The Elliptical t-copula of D-vine in Bull and Bear markets

Considering Fig. 6, the D-vine copula model provided the different structural dependence from the C-vine model. In other words, the estimated result stated that the Asean stock market strongly depended on the Japanese financial market. This structural dependence was stronger than the pair of European and Asean. Accordingly, this can be indicated that most of capital inflows were transferred around Asia continent for bull situations during the pre-crises periods. On the other hand, for recessing times during pre-crises periods, the D-vine result

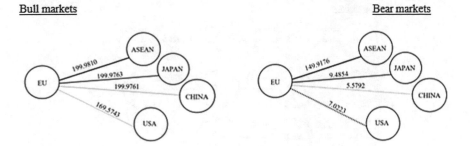

Fig. 5. The estimation results of the pattern of structural dependences among Bull and Bear markets in Elliptical (t-copula) from C-Vine during the pre-crises periods (2000–2008)

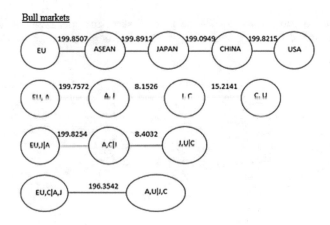

Fig. 6. The estimation results of the pattern of structural dependences among Bull markets in Elliptical (t-copula) from D-Vine during the pre-crises periods (2000–2008)

(as seen details in Fig. 7) showed that capital inflows were inversely moved from Asean to Europe, but the structural dependence between the US financial market (Euro stoxx) and European market are quite weak.

5.4.2   Post-crises Periods (2009–2016)
(a) The Elliptical t-copula of C-vine in Bull and Bear markets

Considering into C-vine's trees in Fig. 8, the European financial market and Asia stock indexes were a strong dependence during 2009 to 2016. In other words, capital inflows were still exchanged intensively between European and Asian stock markets after the economic crisis, especially the subprime crisis, had been passed. Conversely, US and Japanese stock markets became the strongly structural dependence with the European financial market in recessing periods.

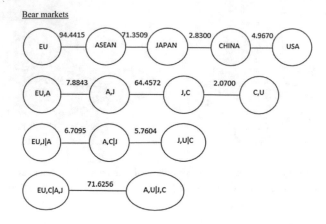

Fig. 7. The estimation results of the pattern of structural dependences among Bear markets in Elliptical (t-copula) from D-Vine during the pre-crises periods (2000–2008)

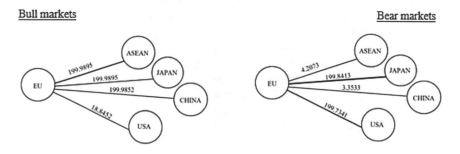

Fig. 8. The estimation results of the pattern of structural dependences among Bull and Bear markets in Elliptical (t-copula) from C-Vine during the post-crises periods (2009–2016)

(b) The Elliptical t-copula of D-vine in Bull and Bear markets

According to details of the D-vine copula in bull periods during the post-crises periods (as seen in Fig. 9), it is obvious that US and Asean stock markets strongly depended on the Euro financial market. This can be implied that capital inflows from Europe had been started to change the direction from Asian continent to North America. However, the structural dependences of financial markets between Asia, North America, and Europe were still strong in the post-crises periods. On the other hand, speaking to details of the D-vine copula in bear periods during the post-crises periods (as seen in Fig. 10), the result showed that capital inflows were transferred inside Asia continent rather than internationally moving to other continents in the recessing time during 2009 to 2016.

Bull markets

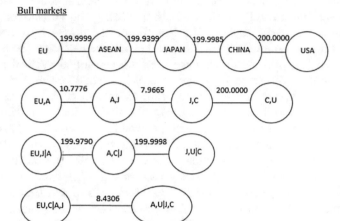

Fig. 9. The estimation results of the pattern of structural dependences among Bull markets in Elliptical (t-copula) from D-Vine during the post-crises periods (2009–2016)

Bear markets

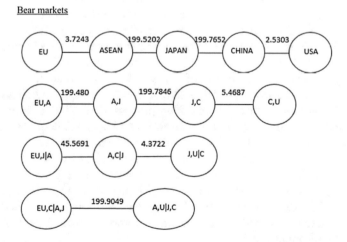

Fig. 10. The estimation results of the pattern of structural dependences among Bear markets in Elliptical (t-copula) from D-Vine during the post-crises periods (2009–2016)

6   Conclusion

For this paper, the patterns of structural dependences among world stock exchanges were successfully estimated. Empirically, the section of MSBVAR results were confirmedly divided the five financial indexes into two periods, including economic boom (bull markets) and economic recession (bear markets). This explained that all of five financial markets contained cyclical movements and fluctuated time-series trends, which cannot be directly estimated by assumptions of linearity. This study also found that there is a contagion among these financial indexes as well as two types of copula models, including Elliptical and Archimedean, were investigated. However, the Elliptical copula is chosen to estimate collected variables in this paper. The study on the structural cycling patterns clarified the Elliptical t-copula indicated the information is symmetric. This implied that investors could easily receive same information inside these five financial markets (Nermuth 1982). Therefore, this stated that governments have freedom choices to interfere the financial markets or let them adjust themselves to have an independently stable system for controlling risks and asymmetric information in their financial structures. Interestingly, the prior research of Lemmon and Ni (2008) found that speculative demands for equity options were positively related to most investor sentiments. Especially, if they have high leverage, they are also perfect vehicles for speculation. This empirical research confirmed that the Elliptical copula was suitable to estimate stock markets in this paper.

Specifically considering Elliptical CD-vine copula's results (t-copulas), in the pre-crisis (2000–2008), this seemed that capital flows were mostly transferred between Europe and Asia stocks in both bull and bear markets, but there was a small capital flow between US and European financial markets. In other words, there was the strongly structural dependence of European and Asia stocks since the financial crisis in US was starting, and this cause negatively impacted the confidence rate of financial sectors in US during that period. In the post-crisis (2009–2016), similar to the result of the pre-crisis periods, the capital flows between Europe and Asia were still a strong dependence in bull situations, and the financial markets between US and Europe were defined as a structural independence, meaning that capital flows from these two continents mostly moved out to other places rather than domestically transferring. Interestingly, in recessing time, the CD-vine copulas' results indicated that the direction of capital flows from Europe to US stock markets (North America) had been returned since US's economy was systematically recovered. Hence, this can be implied that the transference of funds, especially from Europe to US financial markets, would be predictively increased in the upcoming future, and financial investments in US can be positively mentioned.

Appendix A

See (Tables 4, 5, 6 and 7).

Table 4. C-vine copula testing in bull markets during pre-crisis and post-crisis periods

Canonicals (C-vine)	2000–2008		2009–2016	
Bull markets	Parameters	SE.	Parameters	SE.
	t-copula, clayton	t-copula, clayton	t-copula, clayton	t-copula, clayton
$\beta_{1,2}$	199.9810, 0.0604	0.448, 0.007	199.9895, 0.0000	0.0.35, 0.000
$\beta_{1,3}$	199.9763, 0.0000	2.398, 0.000	199.9895, 0.2128	0.012, 0.048
$\beta_{1,4}$	199.9761, 0.0000	0.020, 0.000	199.9852, 0.0000	0.015, 0.000
$\beta_{1,5}$	169.5743, 0.0000	948.211, 0.000	18.8452, 0.0001	0.002, 0.006
$\beta_{2,3|1}$	199.9466, 0.0000	0.292, 0.000	199.7074, 0.0000	0.217, 0.000
$\beta_{2,4|1}$	9.0388, 0.0994	16.042, 0.005	11.4471, 0.0925	28.950, 0.129
$\beta_{2,5|1}$	5.9405, 0.0603	7.870, 0.007	199.9219, 0.0001	0.125, 0.001
$\beta_{3,4|12}$	199.2320, 0.0000	1.474, 0.000	199.9749, 0.0302	0.023, 0.057
$\beta_{3,5|12}$	198.5553, 0.0000	3.183, 0.000	5.9234, 0.0566	4.964, 0.075
$\beta_{4,5|123}$	199.9657, 0.0000	0.022, 0.000	193.9490, 0.0003	373.888, 0.038
AIC	12.2180, 16.9958		5.3087, 11.6756	
BIC	30.7194, 35.4972		23.8102, 30.1771	
Log-likelihood	3.891, 1.502		7.346, 4.162	

Table 5. C-vine copula testing in bear markets during pre-crisis and post-crisis periods

Canonicals (C-vine)	2000–2008		2009–2016	
Bear markets	Parameters	SE.	Parameters	SE.
	t-copula, clayton	t-copula, clayton	t-copula, clayton	t-copula, clayton
$\beta_{1,2}$	149.9176, 0.0588	59.615, 0.013	4.2073, 0.1184	1.772, 0.112
$\beta_{1,3}$	9.4854, 0.0000	0.014, 0.000	199.8413, 0.0000	0.064, 0.000
$\beta_{1,4}$	5.5792, 0.0000	8.531, 0.000	3.3533, 0.1246	2.293, 0.070
$\beta_{1,5}$	7.0223, 0.0000	0.000, 0.000	199.7341, 0.0000	0.339, 0.000
$\beta_{2,3|1}$	47.7609, 0.1112	208.494, 0.065	199.4929, 0.0000	0.350, 0.000
$\beta_{2,4|1}$	41.2768, 0.0000	468.237, 0.000	199.9691, 0.0825	0.018, 0.070
$\beta_{2,5|1}$	61.3775, 0.0799	140.201, 0.096	4.6029, 0.0000	0.387, 0.000
$\beta_{3,4|12}$	4.9238, 0.1120	4.347, 0.102	199.4031, 0.0784	0.326, 0.067
$\beta_{3,5|12}$	2.0072, 0.2922	2.121, 0.119	6.5769, 0.0605	3.402, 0.091
$\beta_{4,5|123}$	5.2875, 0.0000	7.527, 0.000	4.7700, 0.1838	4.747, 0.102
AIC	2.0584, 10.8949		11.7961, 11.8260	
BIC	20.1250, 28.9615		29.8628, 29.8926	
Log-likelihood	8.971, 4.553		4.102, 4.087	

Table 6. D-vine copula testing in bull markets during pre-crisis and post-crisis periods

D-vine	2000–2008		2009–2016	
Bull market	Parameters	SE.	Parameters	SE.
	t-copula, clayton	t-copula, clayton	t-copula, clayton	t-copula, clayton
$\beta_{1,2}$	199.8507, 0.0322	0.257, 0.071	199.9999, 0.0000	0.000, 0.000
$\beta_{1,3}$	199.8912, 0.0000	0.164, 0.000	199.9399, 0.0000	0.331, 0.001
$\beta_{1,4}$	199.0949, 0.0000	1.233, 0.000	199.9985, 0.0000	0.000, 0.000
$\beta_{1,5}$	199.8215, 0.0002	0.286, 0.001	200.0000, 0.0008	0.001, 0.100
$\beta_{2,3|1}$	199.7572, 0.0000	0.411, 0.000	10.7776, 0.2105	0.000, 0.049
$\beta_{2,4|1}$	8.1526, 0.0915	0.002, 0.091	7.9665, 0.0900	15.599, 0.138
$\beta_{2,5|1}$	15.2141, 0.0000	44.325, 0.000	200.0000, 0.0355	17.621, 0.073
$\beta_{3,4|12}$	199.8254, 0.0000	0.202, 0.000	199.9790, 0.0000	0.000, 0.000
$\beta_{3,5|12}$	8.4032, 0.0445	14.787, 0.069	199.9998, 0.0000	0.003, 0.002
$\beta_{4,5|123}$	196.3542, 0.0000	5.720, 0.000	8.4306, 0.0002	18.963, 0.050
AIC	12.4287, 17.5261		5.2284, 12.2178	
BIC	30.9302, 36.0275		23.7298, 30.7193	
Log-likelihood	3.786, 1.237		7.386, 3.891	

Table 7. D-vine copula testing in bear markets during pre-crisis and post-crisis periods

D-vine	2000–2008		2009–2016	
Bear markets	Parameters	SE.	Parameters	SE.
	t-copula, clayton	t-copula, clayton	t-copula, clayton	t-copula, clayton
$\beta_{1,2}$	94.4415, 0.0480	293.303, 0.056	3.7243, 0.1396	0.132, 0.103
$\beta_{1,3}$	71.3509, 0.0955	172.282, 0.076	199.5202, 0.0000	0.860, 0.000
$\beta_{1,4}$	2.8300, 0.0001	0.000, 0.000	199.7652, 0.0688	0.941, 0.063
$\beta_{1,5}$	4.9670, 0.0000	0.001, 0.001	2.5303, 0.1818	2.523, 0.088
$\beta_{2,3|1}$	7.8843, 0.0000	17.143, 0.000	199.4808, 0.0000	0.499, 0.000
$\beta_{2,4|1}$	64.4572, 0.0000	482.304, 0.000	199.7846, 0.1164	7.407, 0.080
$\beta_{2,51}$	2.0700, 0.2827	0.215, 0.126	5.4687, 0.0444	5.356, 0.090
$\beta_{3,4|12}$	6.7095, 0.0000	12.632, 0.000	45.5691, 0.1238	4.600, 0.089
$\beta_{3,5|12}$	5.7604, 0.0334	5.838, 0.106	4.3722, 0.0000	502.558, 0.000
$\beta_{4,5|123}$	71.6256, 0.0000	225.931, 0.000	199.9049, 0.0001	0.508, 0.030
AIC	2.5875, 12.7461		10.7336, 11.9499	
BIC	20.6541, 30.8127		28.8002, 30.0165	
Log-likelihood	8.706, 3.627		4.633, 4.025	

References

1. Avdulai, K.: The Extreme Value Theory as a Tool to Measure Market Risk. Working paper 26/2011.IES FSV. Charles University (2011). http://ies.fsv.cuni.cz
2. Behrens, C.N., Lopes, H.F., Gamerman, D.: Bayesian analysis of extreme events with threshold estimation. Stat. Model. **4**, 227–244 (2004)
3. Chaithep, K., Sriboonchitta, S., Chaiboonsri, C., Pastpipatkul, P.: Value at risk analysis of gold price return using extreme value theory. EEQEL **1**(4), 151–168 (2012)
4. Christoffersen, P.: Evaluating interval forecasts. Int. Econ. Rev. **39**, 841–862 (1998)
5. Einmahl, J.H.J., Magnus, J.R.: Records in athletics through extreme-value theory. J. Am. Stat. Assoc. **103**, 1382–1391 (2008)

6. Embrechts, T., Resnick, S.T., Samorodnitsky, G.: Extreme value theory as a risk management tool. North Am. Actuar. J. **3**(2), 30–41 (1999)
7. Ernst, E., Stockhammer, E.: Macroeconomic Regimes: Business Cycle Theories Reconsidered. Working paper No. 99. Center for Empirical Macroeconomics, Department of Economics, University of Bielefeld (2003). http://www.wiwi. uni-bielefeld.de
8. Garrido, M.C., Lezaud, P.: Extreme value analysis : an introduction. Journal de la Socit Franaise de Statistique, 66–97 (2013). https://hal-enac.archives-ouvertes.fr/ hal-00917995
9. Hamilton, J.D.: Regime Switching Models. Palgrave Dictionary of Economics (2005)
10. Jang, J.B.: An extreme value theory approach for analyzing the extreme risk of the gold prices, 97–109 (2007)
11. King, R.G., Rebelo, S.T.: Resuscitating Real Business Cycles. Working paper 7534, National Bureau of Economic Research (2000). http://www.nber.org/papers/ w7534
12. Kisacik, A.: High volatility, heavy tails and extreme values in value at risk estimation. Institute of Applied Mathematics Financial Mathematics/Life Insurance Option Program Middle East Technical University, Term Project (2006)
13. Kupiec, P.: Techniques for verifying the accuracy of risk management models. J. Deriv. **3**, 73–84 (1995)
14. Manganelli, S., Engle, F.R.: Value at Risk Model in Finance. Working paper No.75. European Central Bank (2001)
15. Marimoutou, V., Raggad, B., Trabelsi, A.: Extreme value theory and value at risk: application to oil market (2006). https://halshs.archives-ouvertes.fr/ halshs-00410746
16. Mierlus-Mazilu, I.: On generalized Pareto distribution. Rom. J. Econ. Forecast. **1**, 107–117 (2010)
17. Mwamba, J.W.M., Hammoudeh, S., Gupta, R.: Financial Tail Risks and the Shapes of the Extreme Value Distribution: A Comparison between Conventional and Sharia-Compliant Stock Indexes. Working paper No. 80. Department of Economics Working Paper Series, University of Pretoria (2014)
18. Neves, C., Alves, M.I.F.: Testing extreme value conditions an overview and recent approaches. Stat. J. REVSTAT **6**, 83–100 (2008)
19. Perez, P.G., Murphy, D.: Filtered Historical Simulation Value-at-Risk Models and Their Competitors. Working paper No. 525. Bank of England (2015). http://www. bankofengland.co.uk/research/Pages/workingpapers/default.aspx
20. Perlin, M.: MS Regress - The MATLAB Package for Markov Regime Switching Models (2010). Available at SSRN: http://ssrn.com/abstract=1714016
21. Pickands, J.: Statistical inference using extreme order statistics. Ann. Stat. **3**, 110–131 (1975)
22. Rockafellar, R.T., Uryasev, S.: Conditional value-at-risk for general loss distributions. J. Bank. Financ. **26**, 1443–1471 (2002). http://www.elsevier.com/locate/ econbase
23. Sampara, J.B., Guillen, M., Santolino, M.: Beyond Value-at-Risk: Glue VaR Distortion Risk Measures. Working paper No. 2. Research Institute of Applied Economics, Department of Econometrics, Riskcenter - IREA University of Barcelona (2013)
24. Taghipour, A.: Banks, stock market and economic growth: the case of Iran. J. Iran. Econ. Rev. **14**(23), 19–40 (2009)

The Impacts of Macroeconomic Variables on Financials Sector and Property and Construction Sector Index Returns in Stock Exchange of Thailand Under Interdependence Scheme

Wilawan Srichaikul[1], Woraphon Yamaka[1,2], and Roengchai Tansuchat[1,2(✉)]

[1] Faculty of Economics, Chiang Mai University, Chiang Mai 50200, Thailand
srichaikul.w@gmail.com, woraphon.econ@gmail.com, roengchaitan@gmail.com
[2] Center of Excellence in Econometrics, Chiang Mai University,
Chiang Mai 50200, Thailand

Abstract. This paper investigates the impacts of macroeconomic variables, namely consumer price index, exchange rate, minimum loan rate and oil price movement, on the financials and the property & construction stock index return in the Stock Exchange of Thailand (SET). The monthly data is collected from January 2004 to November 2016, covering 155 observations. We employ a copula based SUR regression as a tool for this study. Ten copula functions are considered in this regression and the best copula function is selected based on Akaike's Information Criterion (AIC) and Bayesian Information Criterion (BIC). The estimated results show that Gumbel 270 copula is the most appropriate function for being the linkage between the marginal distributions of residuals of financials sector and property & construction sector equations. In addition, the marginal distribution is also tested, and the result shows that normal distribution is the best fit for the marginal distribution for both financials and property & construction equations. Our results suggest that the exchange rate can exert significant impact on both sectors. The dependency parameter also suggests that dependency between financials sector and property & construction sector is negative, and very low dependency, meaning when the impact of macroeconomic variables in one of these two sectors, it just has a little effect to another one sector.

1 Introduction

The Stock Exchange of Thailand (SET) is composed of eight main sectors, namely agro & food industry, consumer product, financials, industrials, property & construction, resources, services, and technology. Among these eight sectors, the financials sector and the property & construction sector account for the largest portion of the Stock Exchange of Thailand, with capital market share of 23.97 % and 12.08 % respectively [16]. The study of Abdelgalil [1] also suggested that

© Springer International Publishing AG 2018
V. Kreinovich et al. (eds.), *Predictive Econometrics and Big Data*, Studies in Computational Intelligence 753, https://doi.org/10.1007/978-3-319-70942-0_42

these two sectors have a strong relationship since the property & construction price can be used as an indicator of financial stability. Thus, these two sectors are highly relevant for monetary policy decision making.

Figure 1 shows the movements of financials sector and property & construction sector from 2004 to 2016. We observe that financials sector and property & construction sector have a high volatility over time. There are several factors affecting these movements of the two sectors; particularly, such macroeconomic variables as interest rate, exchange rate, oil price, and consumer price index.

From the literature, we found there are many macroeconomic factors that have impacts on stock price index returns. The study of Suriani et al. [14] explained that the market value of firms and the stock prices can be significantly affected by multiple factors where exchange rate fluctuation is an important one. According to the financial theory, the value of firm should be influenced by exchange rates and interest rates. The upward and downward exchange rate movements may determine the stock prices of the firms. Moreover, the study of Jamil and Ullah [8] suggested the foreign exchange rate as another factor that impacts the stock returns.

The interest rate is also called the cost of capital. Interest rate can be classified into two types: the savings interest rate and the borrowing interest rate. By that an increase in the interest rate affects the discount rate, which ultimately decreases the value of a stock. A related explanation is that an increase in the interest rate makes alternative investment opportunities more attractive. Specifically, when the interest rate rises, investors tend to invest less in stock and more in other investment assets, causing stock prices to fall [3]. Moreover, in the study of Talla [15] also suggested that inflation can affect stock market either positively or negatively. In addition, the unexpected and expected inflation are also determines the direction of the relationship between stock market and inflation. When demand exceeds supply, firms tend to increase their prices. This would increase their earnings, which would lead to an increase in dividends paid resulting in an increase in demand for the firms stock and eventually increasing its stocks value.

Fig. 1. Stock price index of financials sector and the property & construction sector in the Stock Exchange of Thailand Source: Thomson Reuters, 2017

Finally, another factor that affects the stock return is the movement of oil price. Many researchers have studied the relationship between the movement of oil price and stock market return such as Aydogan [2]; Ghosh and Kanjilal [5]; Jouini [9] and Pastpipatkul, et al. [12]. They also confirmed a relationship between stock and oil price. From these literature reviews, we found that stock market and macroeconomic factors are linked and there are various factors that affect stock market. Thus, in this study, we will introduce these factors and investigate the impact of these factors on SET market.

The main purpose of this paper is to examine the impact of macroeconomic variables on the financials sector and property & construction sector in SET markets. In addition, this paper studies to dependence between financials sector and property & construction sector. Therefore, this study applies copula based Seemingly Unrelated Regression (SUR) model which was introduced in Past-pipatkul, et al. [11] as a tool. By using this model, our study can gain more flexibility from its ability to link the different marginal distributions of residuals of each equation in the model. In particular it makes the model more realistic and far from the unrealistic assumption like a normal distribution. In the orig-inal SUR model of Zellner [18], Gaussian distributions were used, but the we do not want to make this strong assumption. Therefore, copula based model is consider in this study.

The remainder of this study is organized as follows. Section 2 briefly a methodologies used in this study, Sect. 3 presents estimation procedures, and Sect. 4 presents the data used in this study. In Sect. 5, we present the empirical results. Section 6 present our concludes.

2   Methodology

2.1   Seemingly Unrelated Regression (SUR) Model

The SUR model was introduced by Zellner [18]. It can be viewed as a system of linear regression equations or multivariate regression equations. The dependent variable, Y_i is different in each equation, and the error of equation is assumed to be correlated. Thus, SUR model can gain efficiency or improve estimation by combining information from different equations through the error terms. Con-sider the structure of SUR model, as it consists of several regression equations, let us say M equations, we can write the general from as in the following.

$$Y_i = X_i\beta_i + \varepsilon_i \qquad\qquad i = 1, \ldots, M \qquad\qquad (1)$$

where $Y_i = (T \times 1)$ vector with element y_{it}, X_i is $(T \times K_i)$ matrix of K_i inde-pendent variables or explanatory variables in equation $i = 1, \ldots, M$ and β_i is $(K \times 1)$ vector of an estimated parameter in each equation i^{th}. ε_i is a vector of the error terms errors in different moments of time are independence, but differ-ent error components at the same moment of time t are, in general, correlated and also assumed to be correlated across equations. Thus,

$$E[\varepsilon_{it}\varepsilon_{it} \mid X] = 0 \qquad\qquad (2)$$

whereas $E[\varepsilon_{it}\varepsilon_{jt} \mid X] = \sigma_{ij}$ and $\varepsilon \sim N(0, \Sigma)$. Σ is a matrix of non-negative variance-covariance for M equations, such that

$$\Sigma(\varepsilon_t \varepsilon_t^{'}) = \begin{bmatrix} \sigma_{11} & \sigma_{12} & & \sigma_{1M} \\ \sigma_{21} & \sigma_{22} & & \sigma_{2M} \\ \vdots & & \ddots & \vdots \\ \sigma_{M1} & \cdots & \cdots & \sigma_{MM} \end{bmatrix} \tag{3}$$

For example, consider SUR model with 2 equations, we can write matrix of the model as

$$\begin{bmatrix} y_{1t} \\ y_{2t} \end{bmatrix} = \begin{bmatrix} \beta_{10} \\ \beta_{20} \end{bmatrix} + \begin{bmatrix} \beta_{11} & \beta_{12} \\ \beta_{21} & \beta_{22} \end{bmatrix} \cdot \begin{bmatrix} x_{11t} & x_{12t} \\ x_{21t} & x_{22t} \end{bmatrix} + \begin{bmatrix} \varepsilon_{1t} \\ \varepsilon_{2t} \end{bmatrix}, \tag{4}$$

$$\Sigma(\varepsilon_t \varepsilon_t^{'}) = \begin{bmatrix} \varepsilon_{1t}^{'}\varepsilon_{1t} & \varepsilon_{1t}^{'}\varepsilon_{2t} \\ \varepsilon_{2t}^{'}\varepsilon_{1t} & \varepsilon_{2t}^{'}\varepsilon_{2t} \end{bmatrix} = \begin{bmatrix} \sigma_{11} & \sigma_{12} \\ \sigma_{21} & \sigma_{22} \end{bmatrix} \tag{5}$$

In the application study, x_t in each equation can be either different or the same. In this study, we consider using a maximum likelihood (ML) estimator to estimate all unknown parameters in the SUR model, thus the log-likelihood function of this model can be written as

$$L = \frac{1}{\sqrt{(2\pi)^{MT} |\Sigma|}} \exp\left(-\frac{1}{2} tr\left[(\varepsilon_{it})^{'}(\Sigma)^{-1}(\varepsilon_{it})\right]\right). \tag{6}$$

Taking logarithm, we obtain

$$\log L = -(\frac{MT}{2})\log(2\pi) + (\frac{T}{2})\log|\Sigma^{-1}| - \left(\frac{1}{2} tr\left[(\varepsilon_{it})^{'}(\Sigma^{-1})(\varepsilon_{it})\right]\right), \tag{7}$$

2.2 Basic concepts of copula

In this section, we review the theorem of the copula approach which is employed to improve the SUR model as mention before. The most fundamental theorem, which describes the dependence in copulas, is the Sklars theorem [13]. Sklar [13] has proposed the link between the marginal distributions which is possible to have different distributions with the same correlation, but different dependence structure. The linkage between the marginal distributions is called copula. Formally, let H be an M-dimensional joint distribution function of the random variables x_M with marginal distribution function F_M. They, there exists the n-copula C such that for all x_M.

$$H(x_1, \ldots, x_M) = C(F_1(x_1), \ldots, F_M(x_M)) \tag{8}$$

where C is copula distribution function of a M-dimensional random variable. If the marginals are continuous, C is unique. Equation 8 defines a multivariate distribution function F. Thus, we can model the marginal distribution and joint dependence separately. If we have a continuous marginal distribution, the copula can be determined by

$$C(u_1, \ldots, u_M) = C(F_1^{-1}(u_1), \ldots, F_M^{-1}(u_M)) \tag{9}$$

where u is uniform $[0, 1]$. In the copula approach, it proposes various families to join the marginal. In this study, we consider 10 classes of copulas namely, Gaussian, Student-t, Frank, Clayton, Gumbel, Joe, Clayton 90, Clayton 270, Gumbel 90 and Gumbel 270 copula in Hofert, Machler, and McNeil [6].

3   Estimation of Copula Based SUR Model

In this study, we consider bivariate copula based SUR model to derive the return in both financials and property & construction stock markets. First of all, the Augmented Dickey-Fuller (ADF) test is employed to check the stationary of our data series. Note that ML estimation is conducted in this model, therefore it is important to derive the complete likelihood of this model (see, Pastpipatkul, et al. [11]). The complete copula based SUR likelihood can derived by

$$\begin{aligned} \frac{\partial^2}{\partial u_1 \partial u_2} F(u_1, u_2) &= \frac{\partial^2}{\partial u_1 \partial u_2} C(F_1(u_1), F_1(u_1)) \\ &= f_1(u_1) f_2(u_2) c(F_1(u_1), F_2(u_2)) \end{aligned} \tag{10}$$

where u_1 and u_2 are the marginals of each equation which assumed to be either normal or student-t distribution, $f_1(u_1)$ and $f_2(u_2)$ are either normal or student-t density function, and $c(F_1(u_1), F_2(u_2))$ is a probability density function of bivariate copula. Then, we take a logarithm to transform Eq. 10, we obtain

$$\log L(\Theta) = \sum_{t=1}^{T} \log \left\{ (L(\Theta_1 \,|\, y_{1t}, X_1) + L(\Theta_2 \,|\, y_{2t}, X_2) + c(F_1(u_1), F_2(u_2)) \right\} \tag{11}$$

where $L(\Theta_1 \,|\, y_{1t}, X_1)$ and $L(\Theta_2 \,|\, y_{2t}, X_2)$ are the likelihood functions in Eq. 6 and $\log c(F_1(u_1), F_2(u_2))$ is a bivariate copula density assumed for Gaussian, Student-t, Frank, Clayton, Gumbel, Joe, Clayton 90, Clayton 270, Gumbel 90 and Gumbel 270 copulas. Finally, we use the ML estimator to maximize $\log L(\Theta)$ and obtain the estimated parameters of the model. However, in this study, we propose different marginal and copula families to SUR model, so we select the appropriate model specification using Akaiki Information Criteria (AIC) and Bayesian Information Criteria(BIC). The lowest AIC and BIC are preferred in this selection.

3.1   Macroeconomic Specification

$$P_{FINCIAL,t} = \beta_0 + \beta_1 CPI_t + \beta_2 EX_t + \beta_3 OIL\,PRICE_t + \beta_4 MLR_t + \varepsilon_t \tag{12}$$

$$P_{PROPCON,t} = \beta_0 + \beta_1 CPI_t + \beta_2 EX_t + \beta_3 OIL\,PRICE_t + \beta_4 MLR_t + \varepsilon_t \tag{13}$$

where $P_{FINCIAL,t}$ is the financials stock price index, $P_{PROPCON,t}$ is the property & construction stock price index, CPI_t is consumer price index, EX_t is the exchange rate per US. dollar, $OILPRICE_t$ is the crude oil price (WTI), MLR_t is the minimum loan rate.

4   Data

4.1   Descriptive Statistics

In this study is conducted using Financials stock price index, Property & Construction price index obtained from Thomson Reuters, Consumer price index (CPI), Oil price, Exchange rate (EX) and Minimum Loan Rate (MLR) obtained from Data Stream. The data is monthly from January 2004 to November 2016 covering 155 observations. The descriptive statistics are given in Table 1. It can be seen that oil price has the highest standard deviation, while CPI has the lowest. The Jarque-Bera statistic indicates that all series, except exchange rate, are not normally distributed because it rejects the null hypothesis. The Augmented Dickey-Fuller (ADF) test at level with none, intercept and trend and intercept suggest that the value of ADF test at this level of all variables are less than Mackinnon critical value at 1% level. Therefore, these variables are stationary.

Table 1. Summary statistics

	Financials	Property & construction	CPI	EX	Oil price	MLR
Mean	−0.0047	−0.0027	−0.002	0.0005	−0.0025	−0.0008
Median	−0.0153	−0.0095	−0.0018	0.0021	−0.015	0.0000
Max	0.3363	0.3742	0.0298	0.0524	0.3193	0.0714
Min	−0.1552	−0.1734	−0.0223	−0.0412	−0.238	−0.0715
Std. dev	0.0640	0.0685	0.0054	0.0166	0.1047	0.0170
Skewness	1.1168	1.0995	1.0841	−0.0745	0.4445	0.0027
Kurtosis	7.2433	8.0022	11.584	3.2139	3.1022	9.3472
Jarque-Bera	148.51***	192.83***	506.30***	0.4393	5.1736**	260.1884***
Augmented Dickey-Fuller						
None	(−9.73)***	(−9.02)***	(7.54)***	(−10.56)***	(−10.65)***	(−3.80)***
Intercept	(−9.75)***	(−9.01)***	(−8.18)***	(−10.54)***	(−10.63)***	(−3.80)***
Trend and intercept	(−9.72)***	(−9.06)***	(−8.34)***	(−10.66)***	(−10.69)***	(−3.88)***

Source: Calculation

Note: ***, **, * are significant at 1%, 5%, and 10% level, respectively.

5   Empirical Results

5.1   Estimation Results for the Financials Sector and Property and Construction Sector

First of all, we find the best fit model to explain the effects of macroeconomic variables on the financials and property & construction stock sectors. By comparison the minimum of AIC and BIC of various types of model specification. In this study, we consider ten copula functions, namely Gaussian, Student-t, Frank, Clayton, Gumbel, Joe, Clayton 90, Clayton 270, Gumbel 90 and Gumbel 270 copulas, to join the marginal distributions of financials and property & construction equations in SUR system. The assumption for each marginal distribution that we made here is either normal or student-t distribution. Therefore, we have 40 model specifications for the model selection. The results of AIC and BIC of model specifications are shown in Table 2.

From Table 2, we observe that among the trial runs of several alternative marginal distribution and copula functions for SUR model, it is evident that given normal distributions for both the financials and property & construction equations and joint distribution by rotated Gumbel 270-degree copula present the lowest AIC and BIC when compared with the other specifications. Rotated Gumbel 270-degree copula indicates that the dependence distribution between financials sector and property & construction sector is likely to have tail negative correlation rather than positive correlation, meaning that it may has a negative correlation between both sectors in the market downturn. In addition, we believe that the copulas could improve the efficiency of SUR, we also try to compare the results between the copula based SUR and the conventional SUR model. According to the results shown in Table 2, our result shows the superiority of copula based SUR model in this study. The estimated results from the best fit specification is shown in Table 3.

Table 3 shows the results of the best fit model. We find that the effect of exchange rate is statistically significant at 1% level and it has a negative impact on the financials and property & construction stocks. Meanwhile, other variables are not statistically significant to explain the financials and property & construction stock markets.

The exchange rate is significant for negative relationship with sector indices return. The rational reason to support the negative relationship between foreign exchange rate risk and sector indices return is demonstrated by the effect of the size of foreign currency in denominating asset and liabilities in balance sheet of companies listed in the same industry. The fluctuation in unanticipated movement of foreign currency directly affects the firm's balance sheet by creating transaction gain or losses based on the net foreign exposure.

We find that the minimum loan rate is insignificant. This is in consonance with the finding in a study by Wisudtitham [17] they also suggested that interest risk not the determinant on the sector indices return in Thailand market, because the big firms will have more borrowing alternatives than smaller firms and could seek out cheaper rates. Smaller firms will not have access to these alternatives,

Table 2. AIC and BIC criteria for model choice

Copula	Marginal distributions			
	Normal/ normal	Normal/ student-t	Student-t/ normal	Student-t/ student-t
Gaussian	1125.437 1165.085	1130.338 1173.036	1122.424 1165.122	1147.294 1193.042
Student-t	1132.871 1172.519	1127.494 1170.192	1120.480 1163.178	1157.121 1202.869
Clayton	1110.250 1149.898	1101.094 1143.792	1095.549 1138.247	1148.322 1194.070
Gumbel	1111.316 1150.964	1121.928 1164.626	1112.132 1154.830	1136.874 1182.622
Frank	1103.981 1143.629	1117.950 1160.648	1111.536 1154.234	1131.711 1177.459
Joe	1063.783 1103.432	1084.691 1127.389	1074.528 1117.226	1102.546 1148.294
clayton90	957.0425 006.6006	983.5725 1026.270	978.1121 1020.810	1004.682 1050.430
Clayton 270	957.0969 996.7450	983.5741 1026.272	978.230 1020.928	1004.682 1050.43
Gumbel90	933.4206 973.0687	958.4369 1001.135	953.861 996.559	978.4022 1024.150
Gumbel 270	**932.8928** **972.5409**	958.0118 1000.71	952.8069 995.5049	977.4244 1023.172
	SUR			
AIC	1082.238			
BIC	1033.662			

Source: Calculation

making them more vulnerable to interest rate changes. This means that a big firm will have less sensitive to change in the loan rate. Thus, we can conclude that the financials sector and property & construction sector are not be affected by minimum loan rate since these two sectors contain many big firms which are not sensitive to this loan rate.

The oil price is also not significant because Thai government has subsidized the energy consumption including liquefied petroleum gas (LPG), natural gas for vehicles (NGV), diesel, electricity and biofuel blends [7]. Hence, we expect that the effect of oil price is limit to both sectors. Then, consider the consumer price index, it is also insignificant. The same finding was found in the study of Limpanithiwat and Rungsombudpornkul [10] who also suggested that there is no apparent relationship between inflation and stock prices in Thailand. They mentioned that investors have different perspective toward the influence of consumer price index on stock prices in Thailand.

Table 3. Estimation of copula based SUR for the financials sector and property & construction sector

Financials	Copula based SUR	
	Coefficient	Std. error
Intercept	0.00516	0.00413
CPI	0.41847	0.93606
EX	−2.07767***	0.26916
OIL PRICE	0.003573	0.04928
MLR	−0.16595	0.26644
Property & construction		
Intercept	0.00083	0.00475
CPI	−0.37714	0.99191
EX	−2.38527***	0.27746
OIL PRICE	0.07363	0.05054
MLR	−0.40856	0.27262
Copula dependence	−1.10000***	0.09681

Source: Calculation
Note: ***, **, * are significant at 1%, 5%, and 10% level, respectively.

Finally, our model also provides a result of dependence between the financials and property & construction equations. The dependency between these two sectors is found to join by rotated Gumbel 270-degree copula. We can see that the estimated dependence is −1.1000, which corresponds to Kendalls tau coefficient (−0.0909), thus indicating a negative correlation between the financials sector and property & construction sector returns meaning when the impact of macroeconomic variables in one of these two sectors, it just has a little effect to another one sector.

6   Conclusion

In this paper, we employed copula based SUR to investigate the effect of macroeconomic variables, consisting of consumer price index, exchange rate, minimum loan rate and oil price movement, on the financials and the property & construction indices in the Stock Exchange of Thailand (SET). Among the trial runs, the results show that the model given normal distributions for both the financials and property & construction equations and joint distribution by Gumbel 270 copula is the most appropriate for investigating and explaining our problem. According to our estimated results, we find that consumer price index, minimum loan rate and the movement of oil price do not significantly affect both sectors. In the case of oil price, it is very important for investors to monitor since the Thai Government so far has subsidized the oil price in order to prevent any impacts from the shock of world oil price [7]. Moreover, Thai government should control exchange rate to have less fluctuate that will increase the confidence of investor. In addition, we also find very low dependency and likely to have tail negative correlation between financials and property & construction returns.

References

1. Abdelgalil, E.: Relationship between real estate and financial sectors in Dubai economy. J. Prop. Res. **18**, 1–18 (2005)
2. Aydogan, B.: Crude oil price shocks and stock returns: evidence from Turkish stock market under global liquidity conditions. Int. J. Energy Econ. Policy **5**(1), 54 (2015)
3. Forson, J.A., Janrattanagul, J.: Selected macroeconomic variables and stock market movements: empirical evidence from Thailand (2014)
4. Genest, C., MacKay, J.: The joy of copulas: bivariate distributions with uniform marginals. Am. Stat. **40**(4), 280–283 (1986)
5. Ghosh, S., Kanjilal, K.: Co-movement of international crude oil price and Indian stock market: evidences from nonlinear cointegration tests. Energy Econ. **53**, 111–117 (2014)
6. Hofert, M., Machler, M., McNeil, A.J.: Likelihood inference for Archimedean copulas in high dimensions under known margins. J. Multivar. Anal. **110**, 133–150 (2012)
7. IISD: A citizens' guide to energy subsidies in Thailand. The International Institute for Sustainable Development (2013)
8. Jamil, M., Ullah, N.: Impact of foreign exchange rate on stock prices. IOSR J. Bus. Manag. (IOSR-JBM) **7**(3), 45–51 (2013)
9. Jouini, J.: Return and volatility interaction between oil prices and stock markets in Saudi Arabia. J. Policy Model. **35**, 1124–1144 (2013)
10. Limpanithiwat, K., Rungsombudpornkul, L.: Relationship between inflation and stock prices in Thailand (2010)
11. Pastpipatkul, P., Maneejuk, P., Wiboonpongse, A., Sriboonchitta, S.: Seemingly unrelated regression based copula: an application on thai rice market. In: Causal Inference in Econometrics, pp. 437–450. Springer International Publishing (2016)
12. Pastpipatkul, P., Yamaka, W., Sriboonchitta, S.: Co-movement and dependency between New York stock exchange, London stock exchange, Tokyo stock exchange, oil price, and gold price. In: International Symposium on Integrated Uncertainty in Knowledge Modelling and Decision Making, pp. 362–373. Springer International Publishing, October 2015
13. Sklar, M.: Fonctions de répartition à n dimensions et leurs marges. Univ. Paris **8**, 229–231 (1959)
14. Suriani, S., Kumar, M.D., Jamil, F., Muneer, S.: Impact of exchange rate on stock market. Int. J. Econ. Financ. **5**(IS), 385–388 (2015)
15. Talla, J.T.: Impact of macroeconomic variables on the stock market prices of the Stockholm stock exchange (OMXS30). International Business School (2013)
16. The Stock Exchange of Thailand: SET Index Series (2017). https://marketdata.set.or.th/mkt/sectorialindices.do
17. Wisudtitham, K.: The Effect of Market, Interest Rate, and Exchange Rate Risks on Sector Indices Return: Evidence from the Stock Exchange of Thailand. Thammasat University, Bangkok (2013)
18. Zellner, A.: An efficient method of estimating seemingly unrelated regressions and tests for aggregation bias. J. Am. Stat. Assoc. **57**(298), 348–368 (1962)

Generalize Weighted in Interval Data for Fitting a Vector Autoregressive Model

Teerawut Teetranont[1(✉)], Woraphon Yamaka[1,2], and Songsak Sriboonchitta[1,2]

[1] Faculty of Economics, Chiang Mai University, Chiang Mai 52000, Thailand
teetranont@gmail.com, woraphon.econ@gmail.com, songsakecon@gmail.com
[2] Puey Ungphakorn Center of Excellence in Econometrics, Chiang Mai University, Chiang Mai 52000, Thailand

Abstract. This paper employ VAR model to analyse and investigate the relationship among oil, gold, and rubber prices. A convex combination approach is proposed to obtain appropriate value of the interval data in VAR model. The construction of interval VAR model based on the convex combination method for the analysis of their forecast performance are also introduced and discussed via the simulation study, as well as comparing the performance with conventional center method. To illustrate the usefulness of the proposed model, an empirical application on a weekly sample of commodity price is provided. The results show the performance of our proposed model and also provide some relationship between commodity prices.

1 Introduction

Since Billard and Diday [1,3] proposed a linear regression to interval-valued data,nmely, they propose to regress the centers of the intervals of the dependent variable on the centers of the intervals of the regressors. The model has been considered and applied in various studies during the last decade. The main advantage of interval data is that it can capture uncertain characteristics that cannot be fully described with a real number. In several situations, the available information is formalized in terms of intervals. Therefore, considering the minimum and maximum values of the interval-valued data may arise more complete insight about the phenomenon than considering the single-valued data [10,12]. Blanco-Fernndez, et al. [2] suggested that interval data are useful to model variables with uncertainty in their formalization, due to an imprecise observation or an inexact measurement, fluctuations, grouped data or censoring.

So far the literatures have been proposed to deal with interval data such as Center and MinMax methods of Billard and Diday [1], Center and Range method of Neto and Carvalho [6], and model M by Blanco-Fernndez, Corral, Gonzlez-Rodrguez [4]. These methods aim to construct the model without taking the interval as a whole. However, we expect that that these methods may not robust to explain the real behavior of the interval data. Especially, a center method, the mid-point value of data (weight $= 0.5$) is proposed to be a representative of the interval data and this may lead to the misspecification of the model.

© Springer International Publishing AG 2018
V. Kreinovich et al. (eds.), *Predictive Econometrics and Big Data*, Studies in Computational Intelligence 753, https://doi.org/10.1007/978-3-319-70942-0_43

Therefore, it will be of great benefit to relax this assumption of mid-point and assign appropriate weights between intervals. Thus in this study, a convex combination approach which proposed by Chanaim et al. [5] is employed to obtain the appropriate weights.

The main interest of this paper lies in the system equation models for the interval data. To our best knowledge, there have been no studies done on system equation with interval data before. Hence, we extend the linear regression of Billard and Diday [1], Neto and Carvalho [6], Fernndez, Corral, Gonzlez-Rodrguez [4] and Chanaim et al. [5] by proposing Vector Autoregressive(VAR) model for the Interval data context. Therefore, this become one of the main contribution of our study by proposing an Interval VAR with convex combination method. Furthermore, to show the performance of our model in the real application study, this is an additional study to the empirical literature, which focused on the relationships among gold, crude oil prices and rubber prices in commodity markets. Investigating these commodity prices has been attracting increasing interests among researchers, See Sang et al. [11], Lakshmi and Visalakshmi [8], Li and Yang [9] and Gupta et al. [7].

The commodities especially oil, gold and rubber have an important role in economics because of they are the main components of many common goods in the industry. The globalization causes these commodity prices become more integrated. It was believed that the performance of commodity prices would be affected by each others. Nevertheless, alongside this increased interest in futures trading in the commodity market, the data was normally collected and analyze d from single point type which is viewed as classical one. Due to an explosive growth in methods for dealing with data, the data are now not necessarily collected in the form of single point anymore. In contrast to earlier research, we thus replace a single value of commodity prices with the range of high and low historical weekly data into the VAR model. Specifically, using interval data instead of open, closing or average prices can improve the estimation. Incorporating interval regression into traditional financial econometrics improves efficiency in model estimations and forecasting.

As a consequent, in this paper two contributions are made. First, VAR model are adapted to Interval data characterized by convex combination method and Maximum likelihood techniques are used to estimate them. Second, we will try to identify if there is a causation relationship between the price of crude oil, gold, and rubber by using an Interval VAR model.

The paper is organized as follows. This paper is organized into five main sections including this introduction. Section 2 provides a literature review and research methodology. Section 3 is to put forward a simulation study. Section 4 is provide a data description and empirical results of the relationship among crude oil, gold, and rubber prices, while Sect. 5 summarizes and gives some concluding remarks.

2   Review and Research Methodology

In this section, we will briefly review a conventional center methods which were proposed Billard and Diday [1]. We then present the convex combination method to fitting a vector autoregressive model to interval-valued data.

2.1   Review of Center Methods and Convex Combination Method

Billard and Diday [1] proposed this method and applied in the linear regression model. The main idea of this method is the estimate slope parameter is based on the center of the interval data. Let $Y_i = [\underline{Y}_i, \overline{Y}_i]$, $i = 1, \cdots, T$ where $\overline{Y}_i$ is the maximum observed values of Y_i and $\underline{Y}_i$ the minimum observed values Y_i of $\underline{Y}_i$. Thus, we can compute the center of the intervals as

$$Y_c = \frac{\underline{Y}_i + \overline{Y}_i}{2} \tag{1}$$

However, the center method may lead to the misspecification problem since the midpoint of the intervals might not be an appropriate representative value of the intervals. To overcome the misspecification of the center method, Chanaim et al. [5] generalize the center method using convex combination approach, which can be rewritten in the simplest form as

$$Y^w = w\underline{Y} + (1 - w)\overline{Y}, \ w \in [0, 1] \tag{2}$$

where w is the weight parameter of the interval data with values $[0, 1]$. The advantage of this method lies in the flexibility to assign weights in calculating the appropriate value between intervals.

2.2   Vector Autoregressive (VAR) Model for Interval-Valued Data

Consider the Yth order autoregression for the k dimensional time series vector $(y_{1t}, \cdots, y_{kt})'$, $t = 1, \cdots, T$; this, for example, can be written in the general form as follows:

$$Y_t = A_0 + \sum_{i=1}^{p} A_i Y_{t-i} + u_t, \tag{3}$$

where y_t is the vector of endogenous variables A_0 is $k \times 1$ the vector of intercept term; A_i are $k \times k$ matrix of autoregressive coefficient and u_t is the error term with normal distribution $u_t \sim N(0, \Sigma)$. Thus, we can extend (3) as follows:

$$\begin{bmatrix} y_{1,t} \\ \vdots \\ y_{k,t} \end{bmatrix} = \begin{cases} A_{01} + A_{11}y_{1,t-1} + \cdots + A_{1p}y_{1,t-p} + u_{1,t} \\ \vdots \\ A_{0k} + A_{k1}y_{k,t-1} + \cdots + A_{kp}y_{k,t-p} + u_{k,t} \end{cases} \tag{4}$$

In this study, the convex combination method is applied to $VAR(p)$ model for interval-valued data, thus we can rewrite (3) as

$$Y_t^w = A_0 + \sum_{i=1}^{p} A_i Y_{t-i}^w + u_i, \tag{5}$$

where

$$Y_t^w = \begin{bmatrix} Y_{1,t}^w \\ \vdots \\ Y_{k,t}^w \end{bmatrix} = \begin{bmatrix} w_1 \overline{y}_{1,t} + (1 - w_1) \underline{y}_{1,t} \\ \vdots \\ w_k \overline{y}_{k,t} + (1 - w_k) \underline{y}_{k,t} \end{bmatrix} \tag{6}$$

Thus, for example, a bivariate $VAR(1)$ model equation by equation has the form

$$\begin{bmatrix} Y_{1,t}^w \\ Y_{2,t}^w \end{bmatrix} = \begin{cases} A_{01} + A_{11} Y_{1,t-1}^w + A_{11} Y_{2,t-1}^w + u_{1,t} \\ A_{02} + A_{21} Y_{1,t-1}^w + A_{22} Y_{2,t-1}^w + u_{2,t} \end{cases} \tag{7}$$

In the estimation technique, we employ the Maximum likelihood estimator to estimate all unknown parameters in the Interval VAR(p) model. Due disturbances are normal distributed, the conditional density is multivariate normal distributed:

$$Y_t^w \mid Y_{t-1}^w, \cdots, Y_{t-p}^w \sim N(A_0 + \sum_{i=1}^{p} A_i Y_{t-i}^w, \Sigma), \tag{8}$$

and the condition density of Interval VAR(p) model becomes:

$$f(A, \Sigma, c, w \mid Y) = (2\pi)^{(-k/2)} |\Sigma^{-1}|^{1/2} \left(- \left(\frac{1}{2\Sigma} (Y_t^w - c + \sum_{i=1}^{p} A_i Y_{t-i}^w)^2 \right) \right) \tag{9}$$

where W is the vector of weight w_k, see (2). Thus the likelihood function which is the product of each term of densities for $t = 1, \cdots, T$ can be derived as

$$L(\theta, Y) = \prod_{i=1}^{T} f(A, \Sigma, A_0, w \mid Y^w) \tag{10}$$

where $\theta = \{A, \Sigma, A_0, W\}$ is the parameter set of Interval VAR(p) model. Then, maximizing (11),

$$\theta = \arg\min L(\theta, Y^w) \tag{11}$$

We obtained the estimated parameter of this model, θ.

3   Simulation Result

In this section, a simulation study has been carried out in order to evaluate the performance of convex combination method in Interval VAR model. Specifically, the simulation study aims at offering a better insight into the efficiency of the method and its accuracy to capture the optimum weight in the intervals. For this purpose, we consider the following three cases of weight in the intervals:

Case 1: Center of interval data: $w_1 = 0.5$, $w_2 = 0.5$
Case 2: Deviate from center of interval data: $w_1 = 0.3$, $w_2 = 0.3$
Case 3: Extremely deviate from center of interval data: $w_1 = 0.1$, $w_2 = 0.9$

The general simulation scheme goes as follow. We perform 100 replications for bivariate Interval VAR(1) model using sample size $n = 50$ and $n = 100$ for all the three cases. In each simulation, we draw the initial values of the series from a uniform distribution with minimum value 0 and maximum value. Likewise, the error are drawn randomly from a multivariate normal distribution with mean 0 and covariance matrix,

$$\Sigma = \begin{bmatrix} 1 & 0.5 \\ 0.5 & 1 \end{bmatrix}.$$

Based on these initial values and innovations we simulate the data forward in time until we have 100 sample size. In this simulation study, we generate random data sets from these models specification:

$$\begin{bmatrix} w_1 y_{1,t} + (1 - w_1)y_{1,t} \\ w_2 y_{2,t} + (1 - w_2)y_{2,t} \end{bmatrix}$$
$$= \begin{cases} 2 + 1(w_1 y_{1,t-1} + (1 - w_1)y_{1,t-1}) - 1(w_2 y_{2,t-1} + (1 - w_2)y_{2,t-1}) + u_{1,t} \\ 0.4 + 0.2(w_2 y_{2,t-1} + (1 - w_2)y_{2,t-1}) + 0.7(w_1 y_{1,t-1} + (1 - w_1)y_{1,t-1}) + u_{2,t} \end{cases}$$

The resulting 100 independent sets of parameter estimates for each case is evaluated by their mean square error (MSE) and bias.

Next, we include a comparison in the simulation study in order to evaluate the performances of the convex combination method to center method in the Interval VAR context. The comparison between methods is accomplished by bias and MSE. The simulation results are summarized in Table 1.

According to Tables 1 and 2, we consider both small ($n = 50$) and large ($n = 100$) sample sizes. We can state that the Interval VAR parameters estimated using the convex combination method performs well in this simulation study and both Bias and MSE tend to go to zero when the sample size increases in all cases. We then compare the convex combination method to the center method and we observe that both case 1 and 2 have competitive performances. The Bias and MSE of the convex combination method exhibit a better performance in some estimated parameters thus we cannot conclude that the convex combination method outperforms center method with regard to the Bias and MSE. However, Tables 1 and 2 also displays some interesting differences. The Bias and MSE of the convex combination method are now less than that of center method. This prompts the questions: What has caused this relative change in Bias and MSE? To answer this question, Tables 1 and 2 provide a different cases of the weight of intervals and we can observe that when the true weight extremely deviate from center the Interval VAR parameters estimated using the convex combination method performs significantly better than center method. Thus, we conclude that the convex combination method outperforms the center method with regard to the Bias and MSE for the Monte Carlo experiments with the weight of intervals extremely deviate from the mid-point of intervals.

Table 1. Mean results of Bias and MSE on the validation set based on 100 independent repetitions.

Parameter		Convex combination method		Center method	
N = 100	TRUE	Bias	MSE	Bias	MSE
Case 1: center					
A_{01}	2	−0.1716	0.0579	0.6941	0.5329
A_{11}	1	−0.0038	0.0004	−0.0037	0.0004
A_{12}	−1	0.0043	0.0028	−0.0091	0.0027
A_{02}	0.4	−0.1605	0.0424	−0.0021	0.0501
A_{21}	0.2	−0.0008	0.0003	−0.0011	0.0003
A_{22}	0.7	−0.0102	0.002	−0.0215	0.0038
Σ_{11}	1	−0.0077	0.0235	1.3132	1.8594
Σ_{12}	0.5	−0.0047	0.0124	0.6189	0.4697
Σ_{22}	1	−0.0395	0.0233	1.2591	1.7124
w_1	0.5	−0.0922	0.0296		
w_2	0.5	0.0207	0.0073		
Case 2: deviate from center					
A_{01}	2	0.78275	0.6559	−0.0517	0.0409
A_{11}	1	0.0015	0.0005	−0.0043	0.0005
A_{12}	−1	−0.0154	0.0038	−0.0802	0.0715
A_{02}	0.4	−0.1194	0.0477	0.4556	0.2503
A_{21}	0.2	0.0039	0.0052	0.0018	0.0004
A_{22}	0.7	−0.0276	0.0037	−0.019	0.0026
Σ_{11}	1	0.0229	0.0224	1.4639	2.2718
Σ_{12}	0.5	−0.0265	0.01429	0.5731	0.3917
Σ_{22}	1	−0.0284	0.0236	1.3221	1.8626
w_1	0.3	0.2165	0.0524		
w_2	0.7	−0.1041	0.0203		
Case 3: extremely deviate from center					
A_{01}	2	−1.1631	1.4515	−3.8839	15.3848
A_{11}	1	−0.0001	0.0001	0.0035	0.0002
A_{12}	−1	−0.0263	0.001	−0.0319	0.0019
A_{02}	0.4	0.3685	1.3798	4.882	24.1735
A_{21}	0.2	−0.0007	0.0003	−0.0022	0.0002
A_{22}	0.7	0.006	0.0015	0.0201	0.0015
Σ_{11}	1	0.1209	0.0513	7.4393	56.4107
Σ_{12}	0.5	−0.1078	0.0708	−3.5829	13.2447
Σ_{22}	1	3.5238	1.3012	5.2646	28.5008
w_1	0.1	−0.4104	0.2185		
w_2	0.9	0.1429	0.0239		

Source: Calculation

Table 2. Mean results of Bias and MSE on the validation set based on 100 independent repetitions.

Parameter		Convex combination method		Center method	
$N = 100$	TRUE	Bias	MSE	Bias	MSE
Case 1: center					
A_{01}	2	−0.0072	0.0181	0.7798	0.6961
A_{11}	1	−0.0314	0.0022	−0.0009	0.0011
A_{12}	−1	0.0122	0.0038	−0.0104	0.0061
A_{02}	0.4	0.1359	0.0422	0.2341	0.1721
A_{21}	0.2	−0.0112	0.0013	0.0042	0.0016
A_{22}	0.7	−0.0287	0.0145	−0.0165	0.0059
Σ_{11}	1	−0.1043	0.0474	1.2754	1.8255
Σ_{12}	0.5	0.0723	0.1054	0.6103	0.4828
Σ_{22}	1	−0.1138	0.1012	1.3367	2.0085
w_1	0.5	0.0074	0.0668		
w_2	0.5	0.0085	0.0265		
Case 2: deviate from center					
A_{01}	2	0.8987	0.8492	2.1646	4.808
A_{11}	1	0.0298	0.0043	−0.0061	0.001
A_{12}	−1	−0.1083	0.0294	−0.0281	0.0068
A_{02}	0.4	−0.2053	0.1555	0.112	0.1291
A_{21}	0.2	0.0236	0.0025	−0.0002	0.001
A_{22}	0.7	−0.0574	0.0096	−0.0353	0.006
Σ_{11}	1	0.1737	0.1851	1.5744	2.7681
Σ_{12}	0.5	0.2809	0.3947	0.751	0.7344
Σ_{22}	1	−0.0621	0.0223	1.3744	2.1757
w_1	0.3	0.1713	0.0549		
w_2	0.7	−0.2013	0.0643		
Case 3: extremely deviate from center					
A_{01}	2	−1.1496	1.4172	−4.3242	19.6237
A_{11}	1	0.0118	0.0201	−0.0127	0.001
A_{12}	−1	0.0162	0.0569	−0.0241	0.0022
A_{02}	0.4	3.8648	15.8903	5.1518	27.414
A_{21}	0.2	−0.0026	0.0118	0.0071	0.0008
A_{22}	0.7	−0.0026	0.0076	0.0096	0.0022
Σ_{11}	1	0.0611	0.0832	7.1968	53.9514
Σ_{12}	0.5	−0.0824	0.1496	−3.3917	12.4925
Σ_{22}	1	3.1016	12.2641	5.2377	28.7605
w_1	0.1	−0.2347	0.2681		
w_2	0.9	0.1512	0.0368		

Source: Calculation

4   Empirical Results

4.1   Data Description

In this study, the data set consists of the Brent oil price and COMEX gold price, and Rubber sheet spread 3 (RSS3) rubber price for the period from 9 April 2013 to 6 October 2016, covering 785 observations. Interval data is the most important issue in examining the interaction among these three commodity prices. Therefore, we have considered the daily minimum and maximum of these prices and the data were collected form Thomson Reuters DataStream, Faculty of Economics, Chiang Mai University.

Due to, the prices are non-stationary, they exhibit an upward and downward trend. Therefore we need to transform our data to make them stationary. To preserve the interval format, we calculate daily returns with respect to the previous

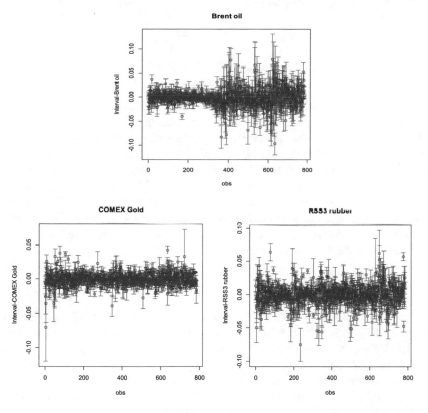

Fig. 1. Lower-upper limits of expanded and centered returns

Table 3. Summary statistics.

	rOil (Max)	rOil (Min)	rGold (Max)	rGold (Min)	rRubber (Max)	rRubber (Min)
Mean	0.0136	−0.0151	0.0058	−0.0062	0.0092	−0.0104
Median	0.0088	−0.0115	0.0045	−0.0053	0.0073	−0.0082
Maximum	0.1308	0.0529	0.0734	0.0366	0.0970	0.0511
Minimum	−0.0747	−0.1197	−0.0236	−0.1200	−0.0506	−0.1000
Std. dev.	0.0216	0.0194	0.0097	0.0109	0.0163	0.0161
Skewness	1.4952	−1.2394	1.1532	−2.1282	0.6956	−0.8222
Kurtosis	8.0826	6.5808	7.3102	20.7835	6.0319	6.0495
Jarque-Bera	1136.02	619.57	780.63	10922.73	363.50	392.13
ADF-test(prob.)	0	0	0	0	0	0

Source: Calculation

day weighted average price, that is,

$$r_{\max,t} = \frac{y_{\max,t} - y_{avg,t-1}}{y_{avg,t-1}}$$

$$r_{\min,t} = \frac{y_{\min,t} - y_{avg,t-1}}{y_{avg,t-1}}$$

where $y_{\min,t}$, $y_{\max,t}$ and $y_{avg,t-1}$ are minimum, maximum, and average price of our intervals. Figure 1 illustrates our interval returns.

The Table 3 shows the descriptive statistics of our intervals. The skewness of minimum values of rOil, rGold, and rRubber are negative, while a positive skewness are shown in their maximum values. For the excess kurtosis statistics, all of variables in this study are positive, thereby indicating that the distributions of returns have larger, thicker tails than the normal distribution. Similarly, the Jarque-Bera tests also confirm that all data series do not follow a normal distribution. In addition, the result of the Augmented-Dickey-Fuller test shows that t all series data are stationary.

4.2   Empirical Results

When conducting VAR model we need the optimal number of lags. To determine how many lags to use in the Interval VAR model, two selection criteria can be used. In this study, we employ Akaike information criterion (AIC) and Bayesian information criterion (BIC) to find an appropriate number of lag lengths and the minimum value of AIC and BIC is preferred. The result in Table 4 reveal that the AIC and BIC for lag = 1 are provided lowest values when compare to the others. Therefore, in this study, we choose the lag length p = 1 for Interval VAR model.

Next, the interval VAR parameters estimated using the convex combination and center methods are compared to show the performance of our model in real application study. In this study, we investigate the relationship among gold, oil and rubber prices and the results are shown in Table 5. The comparison of

Table 4. Interval VAR Lag length criteria

Lag	AIC	BIC
1	-36457.28^*	-36359.32^*
2	-38274.56	-37787.34
3	-38219.77	-37524.22
4	-38233.62	-37329.16
5	-38107.25	-36988.25

Source: Calculation

Table 5. Estimation results for Interval VAR(1)

		Convex combination method		Center method	
		Parameter	SE	Parameter	SE
rOil	A_{01}	-0.0283^{***}	0.0041	-0.0056	0.0006
	rOil(t-1)	0.5914	0.3546	0.2316^{***}	0.0348
	rGold(t-1)	0.517	0.5979	0.1325^*	0.08
	rRubber(t-1)	0.3825^*	0.1509	-0.1334	0.0899
rGold	A_{02}	-0.3318^{***}	0.0014	-0.0001	0.0003
	rOil(t-1)	2.5527^{***}	1.0241	-0.0058	0.0176
	rGold(t-1)	1.2752	2.9691	0.2843^{***}	0.0404
	rRubber(t-1)	0.9009	1.0028	-0.0748	0.0453
rRubber	A_{03}	3.3712^{***}	0.2665	-0.0001	0.0002
	rOil(t-1)	8.7365^{**}	3.4424	0.0425^{**}	0.0136
	rGold(t-1)	-1.4894	8.3092	0.0485^{***}	0.0313
	rRubber(t-1)	-1.0574	1.6341	-0.0216	0.0351
	w_1	0.8001^{***}	0.0024		
	w_2	0.1041^{***}	0.0003		
	w_3	0.4515^{***}	0.0036		
	LL	18249.4		7504.75	
	AIC	-36457.28		-14967.5	
	BIC	-36359.32		-14869.55	

methods is based on AIC and BIC which are the measure of model fit. We learned that Interval VAR with convex combination method has a lowest AIC and BIC which indicate that the results from the convex combination method is more prefer than the center method.

Consider, the estimated parameter result obtaining from Interval VAR with convex combination we can see that rOil is significantly driven by rRubber, an 1% change in the rRubber will change the rOil by 0.3825% in the same direction. Due to the fact that synthetic rubber is manufactured from the crude oil we expect that the decrease in natural rubber price could make synthetic rubber less attractive and thereby lowering the oil price. For the case of Gold price, the reaction of rGold is positive in first lag period of rOil. The main reason behind this relation is that prices of crude oil partly account for inflation. If gasoline, which is produced by oil, is more expensive, transport goods price will go up and thereby increasing an inflation. Then, gold price tend to appreciate with inflation rising. So, an increase in the price of crude oil can, eventually, translate into higher gold price. For rubber price, the reaction of the rRubber are positive in the first lag period of rOil, and the shock coefficients is statistically significant at 5% level. It is inconsistent with the expectation whereby the rubber price have been effected by the oil price due the fact that synthetic rubber is manufactured from the crude oil, thus crude prices drop, the rates of synthetic rubber will also drop and put high pressure to rubber price. As a result, natural rubber prices will drop in tandem with crude oil prices.

4.3   Impulse Response

The impulse response of Interval VAR model shows the response to one standard deviation shock in the error terms of other variables and the results are plot in Fig. 2. The X axis shows the time period and the Y shows the shock in the movement trend. In the left vertical panel, it displays the impulse response function for the changes in three commodity prices to a shock of oil price. The feedback of oil shock differs considerably among commodity prices. We observe that the shock in oil has a great and persistent positive effects on rubber and its price. It, then, falls sharply and reaches the steady state within 3 days. For

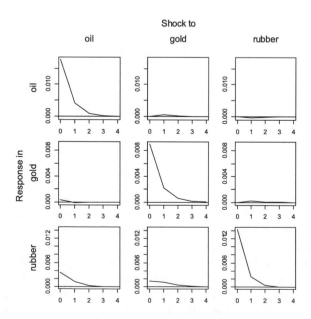

Fig. 2. Impulse response function

response in gold, it creates a small positive response to gold price. In the middle
vertical panel of Fig. 2, it displays the impulse response function for the changes
in three commodity prices to a shock of gold price. In this regime, a positive
response in oil, gold and gold are obtained whereas the response of its price
has larger response than the others. Lastly, the right vertical panel displays the
impulse response function for the changes in three commodity prices to a shock
of rubber price. The interesting results are obtained since the response of rubber
prices to the shock of its price is converge to their equilibrium within 2.5 days.
These results indicate that there is a long run relationship among rubber, gold
and oil prices; and their own prices.

5   Concluding Remarks

The main interest of this paper lies in the system equation models for the interval data. So far the literature had been focusing on the linear regression model
with interval data. To our best knowledge, there have been no studies done on
system equation with interval data before. Hence, we extend the linear regression to system equation VAR model. Thus, we propose a new estimation method
of Interval-valued data in the VAR model. We introduce a convex combination
method, which proposed in Chanaim et al. [5], to Interval VAR model. The
advantage of the convex combination method lies in the flexibility to assign
weights in calculating the appropriated point valued data in the intervals. To
show the performance of our purposed model, we have carried both simulation
study and application study. In the simulation study, our results show the superior of the convex combination method in Interval VAR model when the weight
of intervals is given to extremely deviate from the mid-point of intervals. In the
application study, we focus on the relationships among gold, crude oil prices and
rubber prices in commodity markets. The results show that there exhibit some
significant relationship among oil, gold and rubber prices in the interval context.
When we compare Interval VAR with convex combination method and center
method in the real application, we find that the convex combination method is
outperform the convention one.

References

1. Billard, L., Diday, E.: Regression analysis for interval-valued data. In: Data Analysis, Classification, and Related Methods, pp. 369–374. Springer International Publishing (2000)
2. Blanco-Fernndez, A., Garca-Brzana, M., Colubi, A., Kontoghiorghes, E.J.: Extensions of linear regression models based on set arithmetic for interval data. arXiv
   preprint arXiv:1210.5881 (2012)
3. Billard, L., Diday, E.: Symbolic regression analysis. In: Classification, Clustering,
   and Data Analysis, pp. 281–288(2002)
4. Blanco-Fernndez, A., Corral, N., Gonzlez-Rodrguez, G.: Estimation of a flexible
   simple linear model for interval data based on set arithmetic. Comput. Stat. Data
   Anal. **55**(9), 2568–2578 (2011)

5. Chanaim, S., Sriboonchitta, S., Rungruang, C.: A convex combination method for linear regression with interval data. In: Integrated Uncertainty in Knowledge Modelling and Decision Making: 5th International Symposium, IUKM 2016, Da Nang, Vietnam, 30 November–2 December 2016, Proceedings 5, pp. 469–480. Springer International Publishing (2016)
6. Neto, E.D.A.L., de Carvalho, F.D.A.: Constrained linear regression models for symbolic interval-valued variables. Comput. Stat. Data Anal. **54**(2), 333–347 (2010)
7. Gupta, R., Kean, G.J.S.E., Tsebe, M.A., Tsoanamatsie, N., Sato, J.R., et al.: Time-varyingcausality between oil and commodity prices in the presnce of structural breaks and nonlinearity. Economia Internazionale/Int. Econ. **68**(4), 469–491 (2015)
8. Lakshmi, P., Visalakshmi, S.: Reconnoitering the causal relationship in crude oil market during crisis. J. Bus. Manag. Sci. **1**(6), 128–132 (2013)
9. Li, M., Yang, L.: Modeling the volatility of futures return in rubber and oil–A Copula-based GARCH model approach. Econ. Model. **35**, 576–581 (2013)
10. Phochanachan, P., Pastpipatkul, P., Yamaka, W., Sriboonchitta, S.: Threshold regression for modeling symbolic interval data. Int. J. Appl. Bus. Econ. Res. **15**(7), 195–207 (2017)
11. Sang, W.C., Sriboonchitta, S., Rahman, S., Huang, W.T., Wiboonpongse, A.: Modeling volatility and interdependencies of Thai rubber spot price return with climatic factors, exchange rate and crude oil markets. J. Finan. Rev. **16**, 1–20 (2012)
12. Tibprasorn, P., Khiewngamdee, C., Yamaka, W., Sriboonchitta, S. Estimating efficiency of stock return with interval data. In: Robustness in Econometrics, pp. 667–678. Springer International Publishing (2017)

Asymmetric Effect with Quantile Regression for Interval-Valued Variables

Teerawut Teetranont[1(✉)], Woraphon Yamaka[1,2], and Songsak Sriboonchitta[1,2]

[1] Faculty of Economics, Chiang Mai University, Chiang Mai 52000, Thailand
teetranont@gmail.com, woraphon.econ@gmail.com, songsakecon@gmail.com
[2] Puey Ungphakorn Center of Excellence in Econometrics, Chiang Mai University, Chiang Mai 52000, Thailand

Abstract. In this paper, we propose a quantile regression with interval valued data using a convex combination method. The model we propose generalizes series of existing models, say typically with the center method. Three estimation techniques consisting EM algorithm, Least squares, Lasso penalty are presented to estimate the unknown parameters of our model. A series of Monte Carlo experiments are conducted to assess the performance of our proposed model. The results support our theoretical properties. Finally, we apply our model to empirical data in order to show the usefulness of the proposed model. The results imply that the EM algorithm provides a best fit estimation for our data set and captures the effect of oil differently across various quantile levels.

1 Introduction

In finance time series, such as stock, exchange rate, and asset prices, the data is mostly presented in single valued-data (closing price). Although it is useful in many situations, it fails in cases where a set of values is observed may times in each time period. For example, in the daily asset price, a sequence of prices is also available for each daily period. Therefore, if we consider either open price or close price and ignore the intraday variability of asset prices, this may bring about an inaccurate result and can not capture the fluctuation in the dynamics of the real economic and financial phenomena even in a single day [12]. In addition, if we use minimum and maximum values during the day and regress both dependent and independent variables with respect to the corresponding minimum and maximum values, we might not obtain a meaningful answers since a daily fluctuations are often very random. Thus, using values like daily minimum and daily maximum which are effected by these fluctuations does not lead to meaningful regression models.

To solve this problem, Arroyo et al. [1] suggested using interval valued data where the higher and lower prices of each sequence are considered. Giordani [6] also suggested that this kind of data offers a more complete insight about the phenomenon at hand than the average values Therefore, an interval-valued data y_t can be characterized by the pair of values y_t^U and y_t^L with $y_t^U > y_t^L$, where y_t^U and y_t^L denote as the upper and the lower bound of interval valued data, respectively.

© Springer International Publishing AG 2018
V. Kreinovich et al. (eds.), *Predictive Econometrics and Big Data*, Studies in Computational Intelligence 753, https://doi.org/10.1007/978-3-319-70942-0_44

Recently, interval valued data has been employed to deal with new data forms related to multivariate situation, pattern recognition, and artificial intelligence. It becomes important in the financial field and has also been applied in econometric studies especially in linear regression analysis. There are several proposals to fit a regression model to interval data. The topic has been extensively analyzed especially in the last decade, see Arroyo et al. [2], for review of the literatures. The regression model, which is a statistical tool for the investigation of relationships between dependent variables and explanatory variables, has received considerable attention and been proposed in the analysis of interval data in many studies. The linear regression with interval valued data was firstly introduced in Billard and Diday [3,4]. They introduced the simplest center method with linear regression where the representation of an interval can be done in terms of the midpoint, say $y_t^M = (y_t^U + y_t^L)/2$. Further approaches consider two separate regressions, namely lower bound and upper bound regression, and both regressions share the same regression coefficients. In a similar manner, Lima Neto and de Carvalho [10] extended this approach by running two different linear regression models with the first regression model on the midpoints of the intervals and the second one on the ranges which are obtained from $y_t^R = y_t^U - y_t^L$. Subsequently, Lima Neto and de Carvalho [10] imposed non-negative constraints on the regression coefficients of the model for the range and use an inequality constraint to ensure mathematical coherence between the predicted values of y_t^U and y_t^L.

Regression model for interval data has also increasingly become an area in economic and financial researches and can be found in many studies such as the works of Gonzalez and Lin [7], Rodrigues and Salish [14], Piamsuwannakit, et al. [11], and Tibprasorn, et al. [16]. However, these studies fit a regression line through the means of the covariates and do not focus on the information at the outlier. In general, stock prices and market data exhibit heavy tail, displaying the extreme price, thus the assumption of normality might lead to the wrong estimation result. In fact, economic factors are often notorious for containing extreme values due to erratic market reaction to news, thus presenting a non-Gaussian error distribution. To overcome these problems, Koenker and Bassett [8] extended the conventional Ordinary Least Squares (OLS) regression to quantile regression. The advantage of quantile regression is that it can provide estimates more robust against outliers in the response measurements and solve the estimation problems related to the impact of outliers and the fat-tailed error distribution. In addition, we can obtain a more comprehensive analysis of the relationship between response and covariate variables [17].

However, in this paper, we are going to extend the interval valued data to quantile regression. Moreover, we aim to construct the model without taking the midpoint of intervals. We expect that the center method may not be robust enough to explain the real behavior of the interval data nor able to solve the misspecification problem of the model. Therefore, it will be of great benefit to relax this assumption of mid-point and assign appropriate weights between intervals. Thus in this study, a convex combination approach which was proposed by

Chanaim et al. [5] is employed to obtain the appropriate weights. In addition, in the context of estimation technique, we introduce various estimation methods for estimating quantile regression with interval valued data. We first present an expected-maximum likelihood algorithm (EM-algorithm), least squares estimator (LSE) and lasso penalty of Tibshirani [15].

To the best of our knowledge, the estimation of quantile regression model with interval data using convex combination method has not been considered yet. This fact becomes one of our motivations to work on this paper. Furthermore, to show the performance of our model, we employ simulation study and real application study. This is an additional study to contribute to the empirical literature, with a focus on the impact of oil prices on the stock returns of Thailand. We focus on this market because it has an important role in the Thai economy and is attractive to foreign investors. The stock market may also reflect many scenarios in the domestic economy. When share price is high, the value of the firm relative to the replacement cost of its stock of capital is also high.

This paper is structured as follows: The next section introduces the quantile regression model with interval valued data and its estimation. Section 3 consists of simulations and demonstrates the estimation of parameters. Discussion is made in Sect. 4 regarding an empirical application to examine the impact of changes in real oil prices on the real stock returns of Thailand, Indonesia, and Philippines (TIP countries) while conclusions are given in Sect. 5.

2   Methodology

Koenker and Roger [9] introduced quantile regression as a variant of regression technique used in statistics and econometrics. It is in contrast to the linear regression model which results in estimates that are approximated at the conditional mean of the response variable given certain values of the predictor variables. In the classical quantile regression model, it was proposed to work with the time series where each period is represented by a single value data. Here, our study aims to go beyond that. The interval valued data is applied in quantile regression model using convex combination method.

2.1   Quantile Regression Model

2.1.1   Model Structure
The model is represented by

$$w_1(\tau)y_t^L + (1 - w(\tau))y_t^U = \alpha(\tau) + \sum_{i=1}^{k} \beta_i(\tau)[w_{i2}(\tau)x_{it}^L + (1 - w_{i2}(\tau))x_{it}^U] + \varepsilon_t(\tau) \ (1)$$

where y_t^L is the lower bound of dependent variable y_t and y_t^U the upper bound of dependent variable, $t = 1, \cdots, T$. x_{it}^L is matrix $T \times k$ lower bound of k independent variables and x_{it}^U is matrix $T \times k$ upper bound of k independent variables. w is the weight parameter of the interval data with values [0,1]. The advantage of this method lies in the flexibility to assign weights in calculating the appropriate value between intervals $\beta_i(\tau)$. is $1 \times k$ vector of coefficients and the error term $\varepsilon_t(\tau)$ has a distribution which depends on the quantile. Thus, $\tau^{th}(0 < \tau < 1)$ conditional quantile of y_t given x'_{it} is simply as

$$Q_{(w_1(\tau)y_t^L + (1-w_1(\tau))y_t^U)}(\tau \mid x^L, x^U) = [w_{i2}(\tau)x_{it}^L + (1 - w_{i2}(\tau))x_{it}^U]\beta_i(\tau) \quad (2)$$

In the simplest expression, we can rewrite (Eq. 1) in the reduced form equation as

$$[\overline{y}_t | w_1] = \alpha(\tau) + \sum_{i=1}^{k} [\overline{x}_{it} \mid w_{2i}]\beta_i(\tau) + \varepsilon_t(\tau) \quad (3)$$

where $\overline{y}_t | w_1$ is the vector of expected dependent variable conditional on w_1 and $\overline{x}_{it}$ is $T \times k$ matrix of k expected independent variables conditional on w_2.

2.1.2   Least Squares Estimation

In quantile regression with interval valued data, the estimation proceeds similarly to the conventional linear least squares estimator. To obtain an estimate of all conditional quantile function we simply minimize a weighted sum of absolute residuals (see, Koenker and Bassett [8]. The τ specific coefficient vector $\beta(\tau)$ can be estimated by minimizing the loss function:

$$\beta(\tau) = \arg\min_{\beta(\tau)} \sum_{j=1}^{n} \rho^\tau(\overline{y}_t - \alpha(\tau) - \overline{x}_{it}\beta_i(\tau)), \quad (4)$$

where the function $\rho^\tau(\cdot)$ is the check function, and this gives τ^{th} the sample quantile with its solution.

$$\rho^\tau(L) = \frac{(1-\tau)(L < 0) + (\tau)(L \geq 0)}{T} \quad (5)$$

where L is the loss function. The resulting minimization problem, when $\overline{x}_{it}\beta_i(\tau)$ is formulated as a linear function of parameters, can be solved by linear programming method.

2.1.3   An EM Algorithm

When the error is assumed to be asymmetrically Laplace distributed (ALD), quantile estimation is equivalent to the parametric case. The minimization of the objective function (4) is equivalent to the maximum likelihood theory. According to Yu and Moyeed [19], the ALD is a continuous probability distribution which is generalized from the Laplace distribution in which its probability density function is given by:

$$f(y \mid \mu, \sigma, \tau) = \frac{\tau(1-\tau)}{\sigma} \exp\left\{-\rho^\tau\left(\frac{y-\mu}{\sigma}\right)\right\} \quad (6)$$

where μ is location parameter, $\sigma > 0$ is scale parameter and $0 < \tau < 1$ is skew parameter. The $\varepsilon_t \sim ALD(0, \sigma, \tau)$ and i.i.d. in (2) is assumed. The function ρ^τ is check function defined by (5).

In this study, the EM algorithm is proposed to estimate our proposed model. In terms of (3), firstly, we utilize the same mixture representation of the ALD in Reed and Yu [13] as follows:

$$\varepsilon_t(\tau) = \theta z_t + \psi \sqrt{z_t e_t} \tag{7}$$

where $\theta = \dfrac{1 - 2\tau}{\tau(1 - \tau)}$, $\psi^2 = \dfrac{2}{\tau(1-]tau)\sigma}$, $z_t \sim \exp(1)$, $e_t \sim N(0, 1)$ and z_t is independent of e_t. Therefore, we can rewrite (3) as

$$[\overline{y}_t \mid w_1] = \alpha(\tau) + \overline{x}_{it} \mid w_{2i}\beta_i(\tau) + \psi \sqrt{z_t e_t}. \tag{8}$$

And the joint probability density function (6) becomes

$$f(y \mid \beta(\tau), \alpha(\tau), w(\tau), z) = \frac{1}{\psi\sqrt{2\pi z_t}} \exp\left\{-\rho^\tau\left(\frac{(\overline{y}_t \mid w_1 - \alpha(\tau) - \overline{x}_{2i}\beta_i(\tau))^2}{2\psi^2 z_t}\right)\right\} \tag{9}$$

According to Zhou et al. [20], in this EM algorithm, the estimation consists of two steps as in the following:

(1) Expectation likelihood (E-step): Given ith iteration with a current vector of $\Theta(\tau) = \{\beta(\tau), \alpha(\tau), w(\tau)\}$, we need to evaluate the expected complete-data log-likelihood function with respect to the conditional on z_t given $\overline{y}_t$ and $\Theta^i(\tau)$, that is given by

$$Q(\Theta(\tau)|\overline{y}, \Theta^{(i-1)}(\tau)) = E(L(\Theta(\tau)|\overline{y}, z)|\overline{y}, \Theta^{(i-1)}(\tau)), \tag{10}$$

where the log-likelihood function of complete data is

$$L(\Theta(\tau)|\overline{y}, z) = -T\log(2\sqrt{\pi}\psi) - \frac{1}{2}\sum_{t=1}^{T}\log z_t$$

$$-\frac{1}{2\psi^2}\sum_{t=1}^{T}\frac{(\overline{y}_t|w_1 - \alpha(\tau) - \overline{x}_{it}w_{2i}\beta_u(\tau) - \theta z_t)^2}{z_t} - \frac{1}{\sigma}\sum_{t=1}^{T}z_t \tag{11}$$

(2) Maximize Expectation likelihood (M-step): Update $\Theta^i(\tau)$ to $\Theta^{(i+1)}(\tau)$ by maximizing an expected likelihood function $Q(\Theta(\tau)|\overline{y}, \Theta^{(i-1)}(\tau))$. Then, let

$$\phi_t^i = E(z_t^{-1}|\overline{y}_i, \Theta) = \frac{1}{\tau(1-\tau)\cdot|\overline{y}_i|w_1 - \alpha(\tau) - \overline{x}_{it}|w_{2i}\beta_i(\tau)|}. \tag{12}$$

By solving

$$\frac{\partial Q(\Theta(\tau)|\overline{y}, \Theta^{(i-1)}}{\partial \beta} = 0, \tag{13}$$

$$\frac{\partial Q(\Theta(\tau)|\overline{y}, \Theta^{(i-1)}}{\partial w_i} = 0, \tag{14}$$

$$\frac{\partial Q(\Theta(\tau)|\overline{y}, \Theta^{(i-1)}}{\partial w_i} = 0. \tag{15}$$

Thus, we obtain $\{\beta^i(\tau), \alpha^i(\tau), w^i(\tau)\}$ and then we can repeat the above E step and M step until the largest change in the value of any parameter is convergence to minimum error. And the final convergence value of $\Theta(\tau) = \{\beta(\tau), \alpha\tau), w(\tau)\}$ is the MLE of unknown coefficient vector of our proposed model.

2.1.4   Lasso Penalty Method

Lasso Penalty is an extension of the lease squares method. The basic idea of this method is to penalize the coefficients of different covariates at a different level by using adaptive weights. Following the detail in Wu and Liu [18], the adaptive-lasso quantile regression estimated as weights was introduced and the loss function in (4) can be rewritten as follows

$$\beta(\tau) = \arg\min_{\beta(\tau)} \sum_{j=1}^{n} \rho^{\tau}(\overline{y}_t - \alpha(\tau) - \overline{x}_{it}\beta_i(\tau)) + \lambda \sum_{j=1}^{d} \overline{\omega}_j |\beta_i(\tau)_j| \qquad (16)$$

with respect to $\beta_i(\tau)$, where $\overline{\omega}_j$ is the weights which are obtained from $\overline{\omega}_j = \dfrac{1}{|\beta_i(\tau)_j|^{\gamma}}$ for $\gamma > 0$. λ is the value of the penalty parameter that determines how much shrinkage is done under the presumption that the first coefficient is an intercept parameter that should not be subject to the penalty. is a vector and it should have length equal to the number of the covariates. The adaptive-LASSO penalized quantile regression can also be solved using the linear programming method.

3   Experiment Study

3.1   Simulation Study

We perform Monte Carlo simulations to assess the finite sample performance of our proposed quantile regression with interval valued data using convex combination method. Three estimation methods are used consisting EM algorithm, Barrodale and Roberts least squares algorithm (LS), see detail in Koenker and Bassett [8] and Lasso penalty, see Wu and Liu [18]; and these three estimators are compared to determine the appropriate estimation technique for our proposed model.

The sampling experiments are based on the 3 model specifications

Model 1: Center of interval data: $w_1 = w_2 = 0.5$

$$0.5y_t^L + (1 - 0.5)y_t^U = 1 + 3(0.5x_t^L + (1 - 0.5)x_t^U) + \varepsilon_t$$

Model 2: Deviate from center of interval data: $w_1 = 0.3, \ w_2 = 0.7$

$$0.3y_t^L + (1 - 0.3)y_t^U = 1 + 3(0.7x_t^L + (1 - 0.7)x_t^U) + \varepsilon_t$$

Model 3: Extremely deviate from center of interval data: $w_1 = 0.2, \ w_2 = 0.9$

$$0.2y_t^L + (1 - 0.2)y_t^U = 1 + 3(0.9x_t^L + (1 - 0.9)x_t^U) + \varepsilon_t$$

In this section, we also consider different quantile levels of our proposed model; hence ε_t is assumed to be asymmetric Laplace distribution with skew parameter $\tau = (0.20, 0.5, 0.80)$. We simulate 50 and 100 samples for the Monte Carlo experiment using the specified parameters of each case. For each simulation case, we perform 100 replications for all the three cases. In each simulation, we proceed as follows.

(1) We draw $\varepsilon_t(\tau)$ from the ALD with mean 0, variance 1, and skew parameter $\tau = (0.20, 0.5, 0.80)$.
(2) Random variable x_t^U from normal distribution with mean 0, variance 1. Then, we generate x_t^U according to $x_t^L = x_t^U - U(0, 2)$. This guarantees that the bounds are not crossing each other.
(3) $\bar{x}_t$ of the intervals have been computed by $\bar{x}_t = w_1 x_t^U + (1 - w_1) x_t^L$ where w_1 is random $U(0, 1)$. And generate dependent variables $\bar{y}_t = \bar{x}_t \beta(\tau) + \varepsilon_t(\tau)$.
(4) Compute the upper and lower bounds of y_t by $y_t^L = \bar{y}_t - U(0, 2)$ and $y_t^U = \dfrac{\bar{y}_t - (1 - w_2) y_t^L}{w_2}$, where w_2 is random $U(0, 1)$.

To assess the performance of our proposed model and compare our results, the resulted $B = 100$ independent sets of parameter estimates for each case is evaluated by their Bias values:

$$Bias = \sum_{b=1}^{B} (\widetilde{\Theta}_b - \Theta_b^{true})/B,$$

where $\widetilde{\Theta}_b$ is the estimated parameter in each draw and Θ^{true} is the true parameter.

Then, we also include a comparison in the simulation study in order to evaluate the performances of the convex combination method and the center method in the quantile regression context. The comparison between methods is accomplished by bias and MSE. The simulation results are summarized in Tables 1, 2 and 3.

To investigate the performance of our model and compare different estimation algorithms, we repeat the process of data generation and parameter estimation 100 times independently using sample size 50 and 100. Tables 1, 2 and 3 summarize the results of 100 repetitions for the cases: model 1, model 2, and model 3. Bias values of the parameters, at each quantile level, are reported. We compare the estimated Bias values delivered by three cases of interval data: center, deviate from center and extremely deviate from center with those provided by Center and Convex combination methods. For case 1: when the appropriate weight is located at the center of interval data (Table 1), we find that center method completely outperforms the convex combination method at every quantile level and every sample size.

For cases 2 and 3, we present the simulation in Table 2 for the case of deviate from center and Table 3 for extremely deviate from center. Similar resuls are obtained. We find that with weight deviate from the center of interval, convex combination method has a lower Bias when compared with the center method at

Table 1. Simulation results for Model 1

Bias	Interval method	Center method			Convex combination method		
	Estimation	LS	EM	LASSO	LS	EM	LASSO
N=50	$\alpha(.2)$	0.0422	0.0627	0.0174	0.4916	0.2522	0.1644
	$\beta(.2)$	−0.0277	0.0063	−0.0203	−0.046	−0.0766	−0.0779
	$\sigma(.2)$	3.9801	−0.023	4.1726	3.9593	0.0006	3.8731
	$w_1(0.2)$	n.a	n.a	n.a	0.0638	0.1656	0.1917
	$w_2(0.2)$	n.a	n.a	n.a	−0.2136	−0.1916	−0.2566
N=100	$\alpha(.2)$	0.0077	0.0321	−0.0222	0.1685	0.2913	0.3048
	$\beta(.2)$	0.0042	−0.0294	−0.0346	−0.0235	0.1301	−0.0595
	$\sigma(.2)$	4.1274	−0.022	4.1004	4.2021	0.0317	4.1279
	$w_1(0.2)$	n.a	n.a	n.a	0.1011	0.1878	0.1229
	$w_2(0.2)$	n.a	n.a	n.a	−0.2716	−0.1508	−0.1969
N=50	$\alpha(.5)$	−0.0757	−0.0076	−0.0398	0.1424	0.2116	−0.0403
	$\beta(.5)$	0.0001	−0.0115	−0.0541	−0.0548	−0.1305	−0.0044
	$\sigma(.5)$	1.7882	−0.0185	1.7799	1.687	−0.0185	1.723
	$w_1(0.5)$	n.a	n.a	n.a	0.1742	0.1385	0.1529
	$w_2(0.5)$	n.a	n.a	n.a	−0.1642	−0.1481	−0.2231
N=100	$\alpha(.5)$	0.0045	−0.0359	−0.0106	−0.0953	0.1904	0.0661
	$\beta(.5)$	0.0032	−0.0325	−0.0318	−0.0003	−0.0003	−0.0192
	$\sigma(.5)$	1.849	−0.0109	1.8638	1.8165	−0.029	1.7444
	$w_1(0.5)$	n.a	n.a	n.a	0.2061	0.1074	0.1505
	$w_2(0.5)$	n.a	n.a	n.a	−0.2229	−0.1932	−0.2001
N=50	$\alpha(.8)$	−0.0369	−0.0146	−0.0785	−0.2996	−0.266	−0.4361
	$\beta(.8)$	−0.1281	0.014	−0.137	−0.01	0.0931	−0.1318
	$\sigma(.8)$	4.0642	−0.0003	3.8955	4.0966	−0.0419	4.0298
	$w_1(0.8)$	n.a	n.a	n.a	0.0914	0.2231	0.1271
	$w_2(0.8)$	n.a	n.a	n.a	−0.2323	−0.2655	−0.2607
N=100	$\alpha(.8)$	−0.0111	−0.0378	0.0042	−0.2111	−0.2462	−0.2155
	$\beta(.8)$	0.0345	0.0149	−0.0005	−0.0246	−0.1141	−0.0956
	$\sigma(.8)$	4.0923	−0.0054	4.0351	4.1604	−0.0247	4.1254
	$w_1(0.8)$	n.a	n.a	n.a	0.0961	0.1699	0.0805
	$w_2(0.8)$	n.a	n.a	n.a	−0.2457	−0.2177	−0.2626

Source: Calculation

Note: () is quantile level

every quantile level. Thus, we can indicate that the convex combination method outperforms the center method when the weight is deviated from the center of intervals for every quantile level.

This simulation study also compares the estimation algorithms to find the appropriate estimation technique that fits our proposed model. We find all estimation techniques to perform well in this simulation study since the estimated parameters values are close to the true values. However, it is not clear which estimator provides the best fit results. So, we can say that these estimation techniques can be used to estimate unknown parameters of our model. When we consider the number of observations, we find that given the large sample size (n = 100) the mean estimates have a smaller bias when compared with small sample size (n = 50).

Table 2. Simulation results for Model 2

Bias	Interval method	Center method			Convex combination method		
	Estimation	LS	EM	LASSO	LS	EM	LASSO
N=50	$\alpha(.2)$	−0.8605	−0.7763	−0.8007	−0.141	−0.6409	−0.5047
	$\beta(.2)$	0.0496	0.1197	0.0663	-0.0082	0.1374	−0.0588
	$\sigma(.2)$	4.0324	−0.0314	4.244	4.1122	0.035	3.9554
	$w_1(0.2)$	n.a	n.a	n.a	−0.0659	−0.0442	0.0919
	$w_2(0.2)$	n.a	n.a	n.a	−0.1382	−0.1264	−0.1871
N=100	$\alpha(.2)$	−0.8233	−0.8458	−0.8344	−0.1527	−0.1511	−0.3788
	$\beta(.2)$	0.0658	0.0865	0.0444	−0.0162	0.2004	−0.0892
	$\sigma(.2)$	4.1174	−0.0042	4.1936	4.2281	0.0076	4.156
	$w_1(0.2)$	n.a	n.a	n.a	−0.0105	0.0668	0.0388
	$w_2(0.2)$	n.a	n.a	n.a	−0.0982	−0.0864	−0.1299
N=50	$\alpha(.5)$	−0.8025	−0.7098	−0.8615	−0.1166	−0.0892	−0.0927
	$\beta(.5)$	0.0515	0.1365	0.0951	−0.0058	−0.0671	−0.0045
	$\sigma(.5)$	1.8398	0.0392	1.7864	1.8219	−0.0368	1.7802
	$w_1(0.5)$	n.a	n.a	n.a	0.0448	0.0792	0.0871
	$w_2(0.5)$	n.a	n.a	n.a	−0.0106	0.0059	0.0105
N=100	$\alpha(.5)$	−0.8551	−0.8809	−0.8477	0.0458	0.0158	−0.0513
	$\beta(.5)$	0.0813	0.1400	0.0730	0.0043	0.004	0.0737
	$\sigma(.5)$	1.7491	0.0207	1.8266	1.7524	0.0018	1.7801
	$w_1(0.5)$	n.a	n.a	n.a	0.0153	−0.0014	0.0457
	$w_2(0.5)$	n.a	n.a	n.a	0.0126	0.006	0.0134
N=50	$\alpha(.8)$	−0.7896	−0.671	−0.9064	−0.8724	−0.3592	−0.9064
	$\beta(.8)$	0.2235	−0.0318	0.0328	0.0376	−0.0266	−0.0749
	$\sigma(.8)$	4.1234	−0.0427	4.0622	3.8769	−0.0515	4.1015
	$w_1(0.8)$	n.a	n.a	n.a	0.1265	0.0919	0.0882
	$w_2(0.8)$	n.a	n.a	n.a	−0.1022	−0.012	−0.1304
N=100	$\alpha(.8)$	−0.8243	−0.8197	0.84	−0.5347	0.0477	−0.6141
	$\beta(.8)$	0.1257	0.0457	0.0516	0.0084	0.0035	−0.021
	$\sigma(.8)$	4.0812	0.0023	4.1731	4.1839	0.0011	4.189
	$w_1(0.8)$	n.a	n.a	n.a	0.0174	0.0087	0.0388
	$w_2(0.8)$	n.a	n.a	n.a	−0.1085	0.0165	−0.1276

Source: Calculation

Note: () is quantile level

In summary, from evaluating our model performance and its three estimation techniques' performances, we reach a similar conclusion to those obtained from evaluating their bias values. Our proposed model performs well in the simulation study and the convex combination method shows the superiority when the weight is deviated from the center. In addition, with a larger sample size, a lower bias estimate is obtained.

Table 3. Simulation results for Model 3

Bias	Interval method	Center method			Convex combination method		
	Estimation	LS	EM	LASSO	LS	EM	LASSO
N=50	$\alpha(.2)$	−1.3116	−1.2914	−1.2888	−0.4381	−0.2894	−0.6031
	$\beta(.2)$	0.0618	0.2148	0.1645	0.039	−0.1175	−0.0479
	$\sigma(.2)$	4.1467	0.0113	4.2156	3.9995	−0.0295	4.0379
	$w_1(0.2)$	n.a	n.a	n.a	0.1475	−0.1308	0.0131
	$w_2(0.2)$	n.a	n.a	n.a	−0.153	0.0017	−0.1038
N=100	$\alpha(.2)$	−1.3043	−1.3464	−1.3704	−0.4265	−0.1926	−0.5975
	$\beta(.2)$	0.1425	0.1118	0.05994	0.0565	−0.02	−0.0706
	$\sigma(.2)$	4.2235	0.0085	4.0601	4.0121	−0.2037	4.2214
	$w_1(0.2)$	n.a	n.a	n.a	0.0315	−0.0771	0.1114
	$w_2(0.2)$	n.a	n.a	n.a	−0.0007	0.0826	0.0021
N=50	$\alpha(.5)$	−1.1829	−1.3411	−1.3248	−0.3805	−0.1884	−0.5189
	$\beta(.5)$	0.2023	0.0466	0.0956	0.0115	0.0677	0.0043
	$\sigma(.5)$	1.93	0.0103	1.87	1.7785	−0.0673	1.754
	$w_1(0.5)$	n.a	n.a	n.a	0.0272	−0.0087	0.0912
	$w_2(0.5)$	n.a	n.a	n.a	0.0797	0.0962	0.0547
N=100	$\alpha(.5)$	−1.2798	−1.2935	−1.25	−0.0016	0.2913	−0.1409
	$\beta(.5)$	0.1334	0.1381	0.1118	−0.0036	0.13	−0.0416
	$\sigma(.5)$	1.8145	0.0301	1.8727	−0.0022	0.0317	1.7957
	$w_1(0.5)$	n.a	n.a	n.a	−0.1404	0.1878	−0.0988
	$w_2(0.5)$	n.a	n.a	n.a	0.1147	−0.1508	0.1029
N=50	$\alpha(.8)$	−1.2183	−1.2945	−1.3475	−1.2133	−0.6353	−1.1072
	$\beta(.8)$	0.1989	0.1292	0.1008	0.0584	−0.0901	0.0261
	$\sigma(.8)$	3.9875	0.0097	3.9895	4.0908	−0.1121	4.1459
	$w_1(0.8)$	n.a	n.a	n.a	0.0997	0.0198	0.0331
	$w_2(0.8)$	n.a	n.a	n.a	−0.0956	0.012	−0.0902
N=100	$\alpha(.8)$	−1.2986	−1.2752	−1.2095	−0.8271	−0.4554	−0.9515
	$\beta(.8)$	0.0986	0.1531	0.1485	−0.0377	−0.1329	−0.0235
	$\sigma(.8)$	4.1316	0.0186	4.1737	4.2708	−0.0284	4.2254
	$w_1(0.8)$	n.a	n.a	n.a	0.0072	−0.1407	0.0881
	$w_2(0.8)$	n.a	n.a	n.a	−0.0424	0.0852	−0.0449

Source: Calculation
Note: () is quantile level

4   Empirical Illustration: SET and Oil Low/High Return Interval

4.1   Data description

In this study, the daily data set consists of the Brent oil price and SET index of Thailand for the period from 4 January 2012 to 18 October 2016, covering 1209 observations. We have considered the daily minimum and maximum of these prices and the data were collected form Thomson Reuters DataStream, Faculty of Economics, Chiang Mai University. In addition, the prices are non-stationary; they exhibit an upward and downward trend. Therefore we need to transform our data to make them stationary. To preserve the interval format, we calculate daily returns with respect to the previous day weighted average price, that is,

$$ry_{\max,t} = \frac{y_{\max,t} - y_{avg,t}}{y_{avg,t-1}}$$

$$ry_{\min,t} = \frac{y_{\min,t} - y_{avg,t}}{y_{avg,t-1}},$$

where $y_{\min,t}$, $y_{\max,t}$ and $y_{avg,t-1}$ are minimum, maximum, and average price of our intervals. Figure 1 shows our interval returns, and the description of interval returns are shown in Table 4. We can observe that both lower and upper returns exhibit low volatilities where standard deviation of lower and upper oil returns, OILL and OILU), and standard deviation of lower and upper SET returns, SETL and SETU, are lower than 0.04, and they vary within range $[-0.1661 \; -0.1255]$; whereas the SET interval return varies within the wider range $[-0.0960 - 0.0691]$. Prior to estimating our proposed model, these interval returns have to be checked by the Augmented Dickey Fuller unit roots test and we find that all interval returns are stationary at the level with 1% significant.

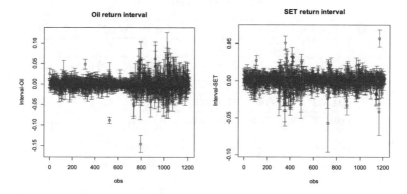

Fig. 1. High/Low returns of daily oil price and SET index

Table 4. Data description

	OILU	OILL	SETU	SETL
Mean	0.0109	−0.0134	0.0057	−0.005
Median	0.0086	−0.0097	0.0053	−0.0034
Maximum	0.1265	0.0365	0.0691	0.0458
Minimum	−0.1255	−0.1661	−0.0277	−0.096
Std. Dev	0.0339	0.0325	0.0082	0.0102
Skewness	−21.8241	−23.3027	0.9526	−1.5934
Kurtosis	658.2207	701.5841	9.2632	12.5008
JB-test	21704706	24672987	2157	5054
ADF-test	0	0	0	0

Source: Calculation

4.2   Empirical Results

In this study, we summarize and report the estimation results from our empirical data. We employ three estimation techniques and both convex combination and center methods as presented above which obtained the estimates $\alpha(\tau), \beta(\tau), \sigma(\tau), w_1(\tau), w_2(\tau)$ for $\tau = 0.1,\ 0.2,\ 0.5,\ 0.8,\ 0.9$. We run our proposed model using both convex combination and center methods delivered from LS, EM, and LASSO. We observe that the Akaki Information Criterion (AIC) of quantile regression using convex combination method shows a lower AIC when compared with quantile regression using center method. In addition, we compare the performance of the different estimation techniques by considering the AIC. We find that EM algorithm provides the smallest AIC at all quantile levels.

Moreover, our estimated results from EM-algorithm seem to have an economic interpretation. We found a significant effect of oil price on SET index. The empirical evidence shows the effect of oil to be different across various quantile levels (see, Fig. 2). The coefficient $\beta(\tau)$ of oil evolves from positive to

Table 5. Estimation results

Estimation	Center method			Convex combination method		
	LS	EM	LASSO	LS	EM	LASSO
$\alpha(0.1)$	−0.0046	−0.0051	−0.0093	−0.0447	−0.0311	−0.00311
	(0.0006)	(0.0001)	(0.0004)	(0.0005)	(0.0001)	(0.0004)
$\beta(0.1)$	0.0371	0.0389	0.00001	0.0341	0.0349	0.00001
	(0.0247)	(0.0098)	(0.0246)	(0.0298)	(0.0094)	(0.0219)
$\sigma(0.1)$		0.0014			0.0013	
		(0.0001)			(0.0001)	
$w_1(0.1)$				0.1142	0	0
				(0.0966)	(0.1463)	(0.1051)
$w_2(0.1)$				0.2976	0.7283	1
				(0.3725)	(0.4029)	(0.9495)
AIC	−7572.98	−7570.79	−7228.58	−7609.66	−7708.72	−7706.47
$\alpha(0.2)$	−0.0012	−0.0013	−0.0054	−0.0214	−0.0205	−0.0215
	(0.0003)	(0.0002)	(0.0003)	(0.0003)	(0.0001)	(0.0002)
$\beta(0.2)$	0.0236	0.0128	0.00001	0.012	0.0283	0.00001
	(0.0211)	(0.0102)	(0.0197)	(0.0179)	(0.0092)	(0.0193)
$\sigma(0.2)$		0.0021			0.002	
		(0.0001)			(0.0001)	
$w_1(0.2)$				0.1627	0	0.1627
				(0.0634)	(0.0395)	(0.0634)
$w_2(0.2)$				0.5166	1	0.5166
				(2.8889)	(0.5333)	(2.8889)

(*continued*)

Table 5. (*continued*)

Estimation	Center method			Convex combination method		
	LS	EM	LASSO	LS	EM	LASSO
AIC	−8004.4	−8001.42	−7726.51	−7998.8	−8100.85	−7998.8˙
$\alpha(0.5)$	0.0039	0.0039	0.001	0.0052	0.0053	0.0052
	(0.0002)	(0.0002)	(0.0003)	(0.0002)	(0.0002)	(0.0002)
$\beta(0.5)$	−0.0041	−0.0041	−0.00001	−0.001	−0.0035	−0.00001
	(0.0178)	(0.0114)	(0.0165)	(0.01558)	(0.0009)	(0.0159)
$\sigma(0.5)$		0.0029			0.0029	
		(0.0001)			(0.0001)	
$w_1(0.5)$				0.0085	0	0.0085
				(0.0001)	(0.0215)	(0.0001)
$w_2(0.5)$				0.5063	0	0.5063
				(8.7932)	(1.4043)	(8.7932)
AIC	−8312.58	−8310.59	−8175.71	−8347.02	−8350.29	−8349.01
$\alpha(0.8)$	0.0092	0.0095	0.0062	0.0113	0.0162	0.011
	(0.0003)	(0.0002)	(0.0003)	(0.0003)	(0.0002)	(0.0003)
$\beta(0.8)$	−0.0378	−0.027	−0.00001	−0.0171	−0.101	−0.00001
	(0.0216)	(0.0107)	(0.0198)	(0.0174)	(0.01)	(0.0162)
$\sigma(0.8)$		0.0022			0.0021	
		(0.0001)			(0.0001)	
$w_1(0.8)$				0	0.4843	0
				(3.2506)	(0.0014)	(0.3256)
$w_2(0.8)$				0	1	0
				(5.8502)	(0.1578)	(5.8502)
AIC	−7933.62	−7931.74	−7978.8	−7885.53	−7976.23	−7885.04
$\alpha(0.9)$	0.0126	0.013	0.0096	0.0525	0.0558	0.0531
	(0.0005)	(0.0001)	(0.0004)	(0.0005)	(0.0001)	(0.0004)
$\beta(0.9)$	−0.0272	−0.0213	−0.0001	−0.0279	−0.0164	−0.00001
	(0.0245)	(0.0099)	(0.024)	(0.0245)	(0.0083)	(0.02311)
$\sigma(0.9)$		0.0015			0.0014	
		(0.0001)			(0.0001)	
$w_1(0.9)$				0.1551	0.8791	0.1551
				(0.0034)	(0.0372)	(0.0034)
$w_2(0.9)$				0.8869	1	0.8869
				(2.3008)	(0.1478)	(2.3008)
AIC	−7466.86	−7464.8	−7601.266	−7466.85	−7984.54	−7466.19

Source: Calculation

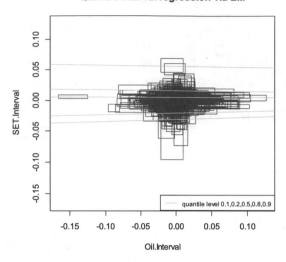

Fig. 2. Linear relationship between the interval Oil price and SET Index

negative as the quantile increases. Consider the $w_1(\tau)$ and $w_2(\tau)$, the different results are obtained. We observe that $w_1(\tau)$ of dependent variable SET interval return increases as the quantile increases while $w_2(\tau)$ of independent variable oil interval return is high at the extreme quantile level. This indicates that SET index tend to move to the upper bound when the market goes up. For oil interval returns, we can indicate that the returns tend to move to the upper bound when the market enters the extreme event such as market boom and market crash.

5   Conclusions

The analysis of interval-valued data has mainly focused on fitting classical regression models to the lower and upper bounds of the intervals; however, the optimal weight of intervals is computed at the center of the intervals. Moreover, the interval valued data mostly works on the linear regression model where the model is approximated at the conditional mean of the response variable. Thus, in this study, we aim to apply an interval valued data in quantile regression in order to estimate either the conditional median or other quantiles of the response variable. Moreover, to relax a strong assumption of the center method, we also proposed a convex combination method to find an appropriate weight between the intervals. Therefore, in this study, we proposed a quantile regression with interval valued data using convex combination method. Moreover, we also employ three estimation techniques, namely EM algorithm, Barrodale and Roberts least squares algorithm (LS), and Lasso penalty and then compare these estimators in order to determine the appropriate estimation technique for our proposed model. We have conducted a simulation study and shown the performance of our proposed model as well as compared the performance of center and convex combination methods. We find at any given weight deviated from center, convex combination appears superior to center method at every quantile level and estimation techniques. Furthermore, we have highlighted an empirical aspect of our model with interval time series of daily SET index and oil price to assess our model through real data set; and aimed at examining the effect of oil price on SET index. The results show that EM algorithm provides a best fit estimation for our data set and that the effect of oil varies across various quantile levels. The coefficient of $\beta(\tau)$ oil evolves from positive to negative as the quantile increases.

References

1. Arroyo, J., Espínola, R., Maté, C.: Different approaches to forecast interval time series: a comparison in finance. Comput. Econ. **37**(2), 169–191 (2011)
2. Arroyo, J., Gonzalez-Rivera, G., Mate, C.: Forecasting with interval and histogram data some financial applications. In: Ullahand, A., Giles, D. (eds.) Handbook of Empirical Economics and Finance, pp. 247–280. Chapman and Hall (2011)
3. Billard, L., Diday, E.: Regression analysis for interval-valued data. In: Data Analysis, Classification and Related Methods, Proceedings of the Seventh Conference of the International Federation of Classification Societies (IFCS 2000), pp. 369–374. Springer, Belgium (2000)
4. Billard, L., Diday, E.: Symbolic regression analysis. In: Proceedings of the Eighteenth Conference of the International Federation of Classification Societies (IFCS 2002), Classification, Clustering and Data Analysis, pp. 281–288. Springer, Poland (2002)
5. Chanaim, S., Sriboonchitta, S., Rungruang, C.: A convex combination method for linear regression with interval data. In: Proceedings of the 5th International Symposium on Integrated Uncertainty in Knowledge Modelling and Decision Making, IUKM 2016, Da Nang, Vietnam, 30 November–2 December 2016, pp. 469–480. Springer (2016)
6. Giordani, P.: Linear regression analysis for interval-valued data based on the Lasso technique. Technical report, 6 (2011)
7. Gonzalez-Rivera, G., Lin, W.: Interval-valued Time Series: Model Estimation based on Order Statistics (No. 201429) (2014)
8. Koenker, R.W., Bassett Jr., G.: Tests of linear hypotheses and I1 estimation. Econometrica Econometric Soc. **50**(1), 43–61 (1982)
9. Koenker, R.: Quantile Regression (No. 38). Cambridge University Press (2005)
10. Lima-Neto, E.A., De Carvalho, F.A.T.: Constrained linear regression models for symbolic interval-valued variables. Comput. Stat. Data Anal. **54**, 333–347 (2010)
11. Piamsuwannakit, S., Autchariyapanitkul, K., Sriboonchitta, S., Ouncharoen, R.: Capital asset pricing model with interval data. In: Integrated Uncertainty in Knowledge Modelling and Decision Making, pp. 163–170. Springer (2015)
12. Phochanachan, P., Pastpipatkul, P., Yamaka, W., Sriboonchitta, S.: Threshold regression for modeling symbolic interval data. Int. J. Appl. Bus. Econ. Res. **15**(7), 195–207 (2017)
13. Reed, C., Yu, K.: A Partially collapsed Gibbs sampler for Bayesian quantile regression (2009)
14. Rodrigues, P.M., Salish, N.: Modeling and forecasting interval time series with threshold models. Adv. Data Anal. Classif. **9**(1), 41–57 (2015)

15. Tibshirani, R.J.: Regression shrinkage and selection via the lasso. J. Roy. Statist. Soc. Ser. B **58**, 267–288 (1996)
16. Tibprasorn, P., Khiewngamdee, C., Yamaka, W., Sriboonchitta, S.: Estimating efficiency of stock return with interval data. In: Robustness in Econometrics, pp. 667–678. Springer (2017)
17. Waldmann, E., Kneib, T.: Bayesian bivariate quantile regression. Stat. Model. (2014). 1471082X14551247
18. Wu, Y., Liu, Y.: Variable selection in quantile regression. Stat. Sin. **19**, 801–817 (2009)
19. Yu, K., Moyeed, R.A.: Bayesian quantile regression. Stat. Probab. Lett. **54**(4), 437–447 (2001)
20. Zhou, Y.H., Ni, Z.X., Li, Y.: Quantile regression via the EM algorithm. Commun. Stat. Simul. Comput. **43**(10), 2162–2172 (2014)

The Future of Global Rice Consumption: Evidence from Dynamic Panel Data Approach

Duangthip Sirikanchanarak[1](✉), Tanaporn Tungtrakul[1],
and Songsak Sriboonchitta[1,2]

[1] Faculty of Economics, Chiang Mai University, Chiang Mai 50200, Thailand
doungtis@gmail.com, tanaporn.tung@gmail.com, songsakecon@gmail.com
[2] Center of Excellence in Econometrics, Chiang Mai University,
Chiang Mai 50200, Thailand

Abstract. This study investigates the future outlook of global rice consumption using dynamic panel data regression (DPD) with penalised fixed effect model. The three main factors affecting rice consumption include previous rice demand, GDP per capita, and world rice price. The data set covers 73 countries that is almost 80% of world rice consumption from 1960 to 2015. We separate these countries into 4 groups based on income levels classified by the World Bank including low income, lower middle-income, upper middle-income, and high income. The results show that, at the global scale, rice consumption is expected to be slightly higher. Such demand is driven by rising demand from the upper middle- and high income countries, while it is offset by the lower demand from lower middle- and low income countries.

Keywords: Dynamic panel data · Fixed effect · Forecasting
Rice consumption

1   Introduction

Rice is the main food of half of the world's population, and it supplies 20% of the calories consumed worldwide [10]. The decreasing growth of rice production is partly affected by climate change while there is increase in world population growth rate [19]. This may give rise to the need for new production approach to meet the competing demands for rice. There have been arguments that whether the genetically modified foods need or not. Global rice consumption per capita does not change much from year to year (0.27%). Therefore, if the global rice production increases, the excess of rice supply will be stored. On the other hand, if the global rice production decrease, the stocks will be used. However, many studies indicated that the food security issue is important for the policy makers to impose appropriate policy to achieve more food in the future. Moreover, the rice demand projections will help governments, donors, and businesses to allocate investment resources for increasing rice production and ensure that global rice

© Springer International Publishing AG 2018
V. Kreinovich et al. (eds.), *Predictive Econometrics and Big Data*, Studies in Computational
Intelligence 753, https://doi.org/10.1007/978-3-319-70942-0_45

supplies are enough and the producers also produce it as efficiently as possible [20]. The United Nations [21] predicted that the world's population will increase more than double by 2020, following the increase of population in China and India, the largest and second largest rice consumption countries in the world. The International Rice Research Institute (IRRI) [19] and the Food and Agriculture Organization (FAO) studied the future rice consumption by using time series data with linear regression model. They found that one of the key determinants of future rice consumption was the rate of population growth. Kubo and Purevdorj [10] conducted a research on the relationship between population and rice consumption. They concluded that whether rice would be enough for the increasing population depends on the rice production in Asia. Timmer et al. [20] discovered that the long-run dynamic of rice consumption may have four basic forces to consider or explain namely (1) population growth, (2) income growth, (3) declining real price for rice, and (4) the gradual shift of workers from rural to urban employment. They applied a step-wise regression model with time series to estimate the global rice consumption. Moreover, they suggested that panel estimation was the standard for estimating parameters when pooled cross-section and time-series data are available.

Apart from the changes in population that are usually studied in other literatures, the rice consumption per capita is also another important factor that affects the total consumption. In this case, if both factors rise, the overall rice demands will be accelerated. In contrast, even if the higher in population but partitioned with the lower in the consumption per capita, the total demand on rice will grow only in small amount. Abdullah et al. [1] predicted the rice consumption per capita would be 55.2-62.7 kg in 2015, 49.1–61.2 kg in 2025, and 42.4–58.1 kg in 2050. Alexandratos and Bruinsma [2] analyzed the world agriculture toward 2030/2050. They found that the world per capita rice consumption has levelled after the late 1980s, following mild declines in several countries of East and South Asia and small increases in other regions. These trends have projected to continue and the average of the developing countries may fall to 57 kg in 2050. FAO [6] reported the food outlook of global food markets that the rice consumption per capita in 2016 and 2017 were 54.5 kg and 54.6 kg, respectively.

This study employs a dynamic panel data (DPD) model with penalised fixed effects. The advantage of using the panel data approach is the increase in data points and therefore the power of statistical estimation [7]. The fixed effects model we have chosen is a common choice for economists, and is generally more appropriate than the random effects model [9]. The dynamic panel data approach has been widely used for analyzing various economic issues such as the relationship between energy consumption and GDP growth [7], the determinants of Foreign Direct Investment (FDI) into Central and Eastern European Countries [5], the influence of tourism and the composition of human capital on economic growth [16,23] and the effects of internet adoption on reducing corruption [13]. Their findings suggest that the dynamic panel data model is an accurate model for investigating the relationship between macroeconomic factors. Therefore, it is of practical interest to investigate the predictive performance of the dynamic panel data model. We do that by considering the forecasting of the global rice consumption based on key factors.

Our aim in this study is to forecast the global rice consumption per capita. We separate countries into 4 groups based on income levels including low income, lower middle-income, upper middle-income, and high income that are classified by the World Bank. We take three variables–previous rice demand, income, and world price of rice as the main drivers of change in rice consumption per capita to be estimated by the dynamic panel data regression model with fixed effect for forecasting. The main results of this research will be useful for the policy makers particularly in the middle- and low-income countries where rice consumption is the largest in the world. In addition, it is also appropriate for supporting the interests of producers, investors, and traders.

The remainder of this analytical task is organized into six parts. In Sect. 2, we brief the theoretical background. In Sect. 3, we describe the data set and provide methodology corresponding to the dynamic panel data regression model. The results on estimation and forecasting of global rice consumption are presented in Sect. 4. Policy implication is discussed in Sect. 5. And in the last section, we provide some concluding remarks.

2   Theoretical Background

The traditional theory of demand recognizes that the demand for a commodity is its quantity which consumers are able and willing to buy at various prices during a given period of time. So, for a commodity to have demand, the consumer must possess willingness to buy it, the ability or means to buy it, and it must be related to per unit of time i.e. per day, per week, per month or per year. The most important objective of demand analysis is forecasting demand. Forecasting refers to predicting the future level of demand on the basis of current and past trends, as the resource allocation relies on predicted results. There are two types of demand functions that are individual demand function and market demand function. The market demand function refers to the total demand for a commodity or service of all the buyers. In this research, we focus on individual demand function which is the basis of demand theory and it can be expressed mathematically as follows:

$$D_x = f(P_x, P_y, I, T, U) \tag{1}$$

where D_x is demand for a commodity x, P_x and P_y are the prices of commodities x and y, respectively. Commodities x and y are substitutes. I is income of a consumer and Engle curves [12], T is taste and preference of a consumer, and U is other variables such as quality of commodity, advertisement and publicity, fashion or demonstration effect, etc.

The individual demand function expressed above is a listing of variables that affect the individual demand. All the above factors play very important role in the determining demand for a commodity or service if all the above stated factors are taken as variable. In microeconomics [15], it is important to understand that *Law of Demand* assumes partial equilibrium which means that if other things remain constant then whenever the price of a commodity changes then the demand for that commodity changes in the opposite direction. It can be stated that, *"conditional on all else being equal, as the price of a good increases (↑), quantity demanded decreases (↓); conversely, as the price of a good decreases (↓), quantity demanded increases (↑)"*. If, on the other hand, general equilibrium analysis is used in explaining the demand then impact of some of these other factors can be explained as follows:

Price of a commodity. As the price of commodity falls a commodity becomes cheaper in a market and rational consumer will try to demand more units of the same to maximize his satisfaction and vice-versa when price rises. Therefore rise in price fall in demand and fall in price rise in demand. But if the good is inferior, the increase in purchasing power caused by the price decrease may cause less of the good to be bought.

In 1895, Robert Giffen observed Giffen's paradox. This paradox has been explained that the change in price and the resulting change in the quantity demanded could move in the same direction. A Giffen good is a staple food, such as bread or rice, which forms large percentage of the diet of the poorest sections of a society, and for which there are no close substitutes. From time to time the poor may supplement their diet with higher quality foods, and they may even consume the odd luxury, although their income will be such that they will not be able to save. A rise in the price of such a staple food will not result in a typical substitution effect, given there are no close substitutes. If the real incomes of the poor increase they would tend to reallocate some of this income to luxuries, and if real incomes decrease they would buy more of the staple good, meaning it is an inferior good. Assuming that the money incomes of the poor are constant in the short run, a rise in price of the staple food will reduce real income and lead to an inverse income effect.

Prices of substitute good. Demand for a commodity also depends upon the prices of its close substitutes. If price of close substitute falls then demand for that commodity also falls and vice-versa. Therefore demand also depends upon the number and degree of close substitutes available in market and the range of price change.

Income of a consumer. Consumer income is the basic determinant of the quantity demanded of the product. Generally the people with higher disposable income spend a larger amount of income than those with the lower income. Income demand relationship is more varied in nature than that between demand and its other determinants. To explain the varied relationship between income and demand we classify goods and services into four broad categories.

First, essential consumer good is necessary goods for the health, safety or welfare of consumers. The goods which people need no matter how high the price is are basic or necessary goods. The second is inferior good. It is a good which quantity demanded decreases when consumer income rises or quantity demanded rises when consumer income decrease. Inferior goods is often associated with lower socio-economic group. The third is normal good which consumer demand increase when their income increases. Last, prestige good or luxury good is often associated with wealth and the wealthy.

3   Data and Methodology

3.1   Data Set

The panel data contains 73 countries from 1960 to 2015. These countries include the major rice consuming countries covering 78% of the total world rice consumption. We divide them into four sub-panels based on the difference in income levels defined by the World Bank: low income group, lower middle-income group, upper middle-income group and high income group (Table 1). The variables used for the analysis include demand for rice per capita, rice price, and GDP per capita that is indicator of income per capita. This dataset is constructed from World Bank, United States Department of Agriculture (USDA), and United Nations Statistics Division (UN). We separated the data set into two sections. The actual data used for model selection and estimation are from 1960 to 2013, while 2014 and 2015 are left for forecast evaluation. All variables are transformed into growth rate.

Figure 1 presents information on rice consumption per capita and GDP per capita of aggregated 73 countries as a whole. This plot shows that since the early

Table 1. List of 73 sample countries, grouped by income level

Income level	Country
Low income countries	Afghanistan, Benin, Burkina Faso, Chad, Liberia, Madagascar, Nepal, Senegal, Sierra Leone, Togo
Lower middle income countries	Bangladesh, Bolivia, Cambodia, Cameroon, Cole d'Ivoire, Ghana, Guatemala, Honduras, India, Kenya, Mauritania, Morocco, Nicaragua, Nigeria, Pakistan, Papua, Philippines, Sri Lanka, Swaziland, Syrian, Zambia
Upper middle income countries	Algeria, China, Colombia, Costa Rica, Dominican, Ecuador, Guyana, Iran, Irag, Jamaica, Malaysia, Mexico, Panama, Peru, South Africa, Suriname, Thailand, Turkey, Venezuela
High income countries	Australia, Canada, Chile, Hong Kong, Israel, Japan, Korea, Singapore, Switzerland, Trinidad, United State, EU (12 countries)

Note that: Gross National Income (GNI) is the total incomes of citizens in the country living both domestically and internationally. In brief, we could say that GNI is the gross domestic product (GDP) plus the incomes earned from residents living overseas minus that earned from nonresident. World Bank classifies all countries into four income groups by using GNI per capita. Low income economies are defined as those with a GNI per capita of 1,025 USD or less; lower middle income economies are those with a GNI per capita between 1,026 USD and 4,035 USD; upper middle income economies are those with a GNI per capita between 4,036 USD and 12,475 USD; high-income economies are those with a GNI per capita of 12,476 USD or more.
Source: World Bank

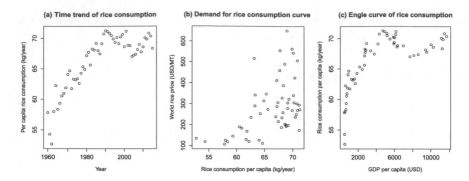

Fig. 1. Trend and Engel curve of global rice consumption per capita

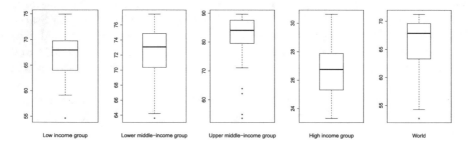

Fig. 2. Per capita rice consumption classified (kg) by income groups. The interquartile range box represents the middle 50% of the data set. The line in the rectangle shows the median and the horizontal lines on either side show the greatest value and least value, excluding outliers. Dots represent those who consumed a lot less than normal (outliers).

1990s, with strong economic growth in many countries particularly in major rice consuming countries such as China and India, the global rice consumption per capita had the upward trend from 58 kg per year in 1960's to 71 kg per year in 1990's (Fig. 1(a)). For the past two decades, global rice consumption per capita has been flat between 2001 and 2007 due to severe drought in China, India and Vietnam [14]. The resulted situation led India and Vietnam to announce the ban on non-Basmati rice export since 2007 and world rice prices so rapidly increased to record levels in 2008 that we called this situation as "the global rice crisis" [17]. After 2010, the trend of global rice consumption per capita kept growing. Figure 1(b) plots the demand for rice consumption per capita. It shows that change in price and that in quantity move in the same direction because rice is a staple food according to Giffen's paradox that we mentioned in the previous section. Furthermore, we found that a positive income elasticity of demand for rice is associated with normal and necessary goods; an increase in income will lead to an increase in rice consumption (Fig. 1(c)).

Table 2. Summary statistics of main variables

Variable	# of countries	Min	Max	Median	Mean	Std. dev.
(a) Rice consumption per capita growth rate						
Low income	10	−61.033	182.367	1.063	4.781	28.501
Lower middle-income	21	−80.544	348.303	0.378	5.217	30.614
Upper middle-income	19	−65.002	329.195	0.783	3.561	24.788
High income	23	−66.574	337.716	1.030	2.099	19.841
Total	4 (73)	−18.054	16.417	0.188	0.226	4.638
(b) GDP per capita growth rate						
Low income	10	−50.319	72.489	3.635	4.649	13.906
Lower middle-income	21	−62.313	115.362	5.806	6.128	14.023
Upper middle-income	19	−65.918	100.541	7.063	7.403	14.348
High income	23	−54.225	53.761	7.564	8.180	11.456
Total	4 (73)	−19.157	24.893	5.989	6.151	7.186
(c) Rice export price growth rate						
Thai WR5%		−40.634	123.702	2.993	6.121	28.816

Source: Calculation

Shown in Fig. 2 are rice consumption per capita classified by income level. We found that the upper middle-income countries have the highest amount of rice consumption per capita where rice is the main diet particularly China and India. Lower middle-income countries are the second largest rice consumer. The third largest demand for consumption is in low income countries because the population face with limited income and the high income countries is the minimal demand for rice consumption due to the different pattern of diet.

Table 2 presents descriptive statistics of variables by income levels to be used for estimation. We found that the rice consumption per capita growth in the low and lower middle-income groups have a larger average growth and higher volatility than the upper middle- and high income groups. It can be implied that the trend of rice consumption in the latter two groups may increase. The low and lower middle-income countries apparently have GDP per capita growth lower than the others which leads to the issue whether or not they will have sufficient food for consumption in the future. Therefore, this research aiming at forecasting the rice consumption per capita is important issue for dealing with this.

3.2   Methodology

In this section, we provide brief information regarding unit root tests and dynamic panel regression with penalised fixed effects model. The details of this study procedure are described as below.

3.2.1   Unit Root Tests

The panel unit root test should be conducted before using these data for estimation to avoid spurious regression. In this study, we use two unit root tests which are based on the cross-sectional independence hypothesis, and they include the LLC test by Levin et al. [11] and the IPS test by Im et al. [8].

3.2.2   Dynamic Panel

Static panel data model, random and fixed effects panel data models do not allow us to use observable information of previous periods in the model. However, many economic issues are dynamic by nature and use panel data structure to understand adjustment like in our study. The present rice demand depends on the past demand. Therefore, we conduct dynamic panel data models that use current and previous information and sensible modeling assumption. In the Arellano-Bond framework [3], the value of dependent variable in the previous period is used to predict the current value of the dependent variable. The relationship of interest can be written as:

$$y_{it} = x'_{it}\beta_1 + y_{i(t-1)}\beta_2 + \alpha_i + \epsilon_{it} \tag{2}$$

where y_{it} is the dependent variable for country i at time t, x_{it} are the set of explanatory variables, $y_{i(t-1)}$ is the previous period of dependent variable, α_i is unobserved country-specific and time invariant effect with $E(\alpha_i) = \alpha$ and $Var(\alpha_i) = \sigma^2_\alpha$, and ϵ_{it} are assumed to be independently distributed across countries with zero mean.

From the fixed effects framework, one assumption is the time invariant unobserved variable is related to the regressors. When unobservables and observables are correlated, the endogeneity problem occurs that yields inconsistent parameter estimates if we use a conventional linear panel data estimator. One solution is taking first difference of the relationship of interest in Eq. 2.

$$\Delta y_{it} = \Delta x'_{it}\beta_1 + \Delta y_{i(t-1)} + \Delta\epsilon_{it}$$
$$E(\Delta y_{i(t-1)}\Delta\epsilon_{it}) \neq 0 \tag{3}$$

where $\Delta y_{it} = y_{it} - y_{i(t-1)}$, $\Delta y_{i(t-1)} = y_{i(t-1)} - y_{i(t-2)}$, and $\Delta\epsilon_{it} = \epsilon_{it} - \epsilon_{i(t-1)}$. We take the first difference of Eq. 2 to get rid of α_i, which is correlated with our regressors. However, we created a new endogeneity problem. In Eq. 3, there is the correlation between the lagged dependent ($\Delta y_{i(t-1)}$) and error term ($\Delta\epsilon_{it}$). Thus, the estimation of Eq. 3 by Ordinary Least Squares (OLS) will be biased and provide inconsistent result. Arellano and Bond [3] suggest the second lags ($\Delta y_{i(t-s)}$ for $s \geq 2$) in level as instrumental variable (z) in Generalized Method of Moment (GMM) to solve this problem. It can generate the set of moment conditions which can be written as

$$E(\Delta y_{i(t-2)}\Delta\epsilon_{it}) = 0$$
$$E(\Delta y_{i(t-3)}\Delta\epsilon_{it}) = 0$$
$$...$$
$$E(\Delta y_{i(t-j)}\Delta\epsilon_{it}) = 0 \tag{4}$$

where $E[z_i \Delta \epsilon_i] = 0$ for $i = 1, 2, ...N$. As to the use of one-step or two-step estimator, Bond [4] suggested that these GMM estimators have focused on the results for the one-step estimator rather than the two step estimator. Huang et al. [7] also mentioned that the dependence of the two-step matrix on estimated parameters makes the usual asymptotic distribution approximations less reliable for two-step estimator. Thus, the robust one-step estimator is conducted.

Therefore, we forecast demand for rice per capita growth by using previous demand for rice, GDP per capita, and rice price as indicators. According to Eq. 4, we can specify the model as follows

$$PRC_{it} = \beta_1 PRC_{i(t-1)} + \beta_2 GDP_{it} + \beta_3 RP_{it} + \alpha_i + \epsilon_{it} \tag{5}$$

where PRC_{it} is per capita rice consumption of country i at time t, GDP_{it} is gross domestic product per capita of country i at time t, and RP_{it} is world rice price at time t.

The actual data used for model selection and estimation are from 1960 to 2013, while 2014 and 2015 are left for forecast evaluation. We calculate the out-of-sample root mean square error (RMSE) for each case. The calculation of RMSE can be written as:

$$RMSE = \sqrt{\frac{\sum_{t=1}^{h}(\widehat{PRC}_t - PRC_t)^2}{h}} \tag{6}$$

where h is the maximum number of forecasting period; $\widehat{PRC}_t$ and PRC_t are the forecast and actual value at period t respectively.

4   Empirical Results

In this section, we report the results of the unit root test and the discussion on the estimation and forecasting of the rice consumption per capita using DPD model. The LLC and IPS panel unit root tests are employed. The presence of the unit root cannot be found for all the variables of interest (Table 3). Therefore, we can conclude that all series are stationary and appropriate for further analysis.

The model specification are selected based on Akaike Information Criterion. The estimated results from the DPD regression are shown in Table 4. The demand for rice consumption depends on previous demand for rice consumption, GDP, and rice price. Further analysis reveals a negative feedback relationship between the current and the two last periods rice consumption per capita growth. It means that an increase in rice consumption per capita growth in the two last periods may bring about further decrease in rice consumption per capita growth in the current period. There is a negative relationship between rice consumption per capita growth and GDP growth in lower and upper middle-income groups, while the rest are positive relationship. In the case of rice consumption per capita growth and rice price growth, there is positive relationship in low and lower middle-income groups, while it is negative for the rest of groups.

Table 3. Unit root tests

Series	Levin, Lin & Chu t			Im, Pesaran and Shin W-stat		
	Statistic	Prob.	Inference	Statistic	Prob.	Inference
Low income						
$\%\Delta PRC_{i,t}$	−16.582	0.000	Stationary	−17.918	0.000	Stationary
$\%\Delta GDP_{i,t}$	−10.797	0.000	Stationary	−12.687	0.000	Stationary
$\%\Delta PR_{i,t}$	−18.864	0.000	Stationary	−16.597	0.000	Stationary
Lower middle-income						
$\%\Delta PRC_{i,t}$	−18.942	0.000	Stationary	−26.126	0.000	Stationary
$\%\Delta GDP_{i,t}$	−16.655	0.000	Stationary	−16.070	0.000	Stationary
$\%\Delta PR_{i,t}$	−27.337	0.000	Stationary	−24.052	0.000	Stationary
Upper middle-income						
$\%\Delta PRC_{i,t}$	−17.205	0.000	Stationary	−26.741	0.000	Stationary
$\%\Delta GDP_{i,t}$	−13.906	0.000	Stationary	−15.619	0.000	Stationary
$\%\Delta PR_{i,t}$	−26.003	0.000	Stationary	−22.878	0.000	Stationary
High income						
$\%\Delta PRC_{i,t}$	−18.267	0.000	Stationary	−19.333	0.000	Stationary
$\%\Delta GDP_{i,t}$	−9.472	0.000	Stationary	−10.890	0.000	Stationary
$\%\Delta PR_{i,t}$	−20.665	0.000	Stationary	−18.181	0.000	Stationary
World						
$\%\Delta PRC_{i,t}$	−10.139	0.000	Stationary	−12.859	0.000	Stationary
$\%\Delta GDP_{i,t}$	−6.658	0.000	Stationary	−5.609	0.000	Stationary
$\%\Delta PR_{i,t}$	−11.931	0.000	Stationary	−10.497	0.000	Stationary

Source: Calculation

Table 4. The estimated result from the dynamic panel

Independent variable	Low income (1)	Lower middle (2)	Upper middle (3)	High income (4)	World (5)
Constant	5.9297	6.8901	5.3313	3.0664	0.32574
$\%\Delta PRC_{i,t-2}$	−0.3556	−0.2608	−0.3283	−0.2814	−0.3143
$\%\Delta GDP_{i,t-2}$	0.1131	−0.1186	−0.0409	0.0081	0.0092
$\%\Delta PR_{i,t-2}$	0.0674	0.0712	−0.0028	−0.0599	−0.0019
Number of Obs	500	1,050	950	600	200
Number of group	10	21	19	12	4

Source: Calculation

Table 5. Out-of-sample forecast of the dynamic panel data

	2014			2015			Forecast	
	Actual	Forecast	RMSE	Actual	Forecast	RMSE	2016	2017
Low income	−0.156	−1.253	1.098	0.676	0.404	0.799	−0.953	−2.089
Lower middle	−0.684	−0.773	0.089	0.993	−2.539	2.499	−1.397	−0.473
Upper middle	0.598	−0.904	1.502	−1.374	0.336	1.609	−0.134	0.705
High income	−0.267	−0.664	0.397	−1.337	0.605	1.402	1.056	0.848
World	−0.913	0.204	1.117	−3.198	−0.269	2.217	0.346	0.982

Source: Calculation

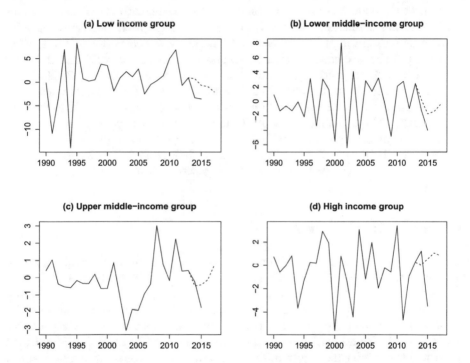

Fig. 3. Forecast and actual per capita rice consumption growth rate (— Actual, ⋯ Forecast)

Table 5 presents the performance of DPD model by considering root mean square error (RMSE) and we make prediction through 2017. We can see that the estimates from DPD model get wider forecasting error in the upper middle-income group than the rest of the groups. The changes in the rice consumption per capita growth in the upper middle-income group may be influenced by other factors which the model could not capture in this study. Hence, the RMSE in this case seems to become largest when considering 2 forecasting periods. Figure 3 plots the estimates and forecasts of rice consumption per capita growth based on DPD model compared with the actual series which are classified into four income groups. Moreover, the model performs relatively well for short period forecasting. Therefore, in general it can be concluded that the DVD model can be used to forecast rice consumption per capita growth quite well. However, for the rest of income groups, the predictive value is quite well based on RMSE. Thus, we can conclude that the DPD model is proper for estimating the per capita rice consumption growth.

Lastly, we use the model to forecast the rice consumption per capita growth in the two next years. Our projections indicate that the global rice consumption per capita growth of 73 countries has tendency to slowdown to 0.346% and 0.982% in 2016 and 2017, respectively due to decreasing rice consumption per capita growth in low income countries, lower middle-income countries, and upper middle-income countries. Conversely, the rice consumption per capita growth seem to increase in high income countries in the next two periods. While, FAO [6] predicted the rice consumption per capita growth rate in 2016 and 2017 are 0.0% and 0.18%, respectively. Abdullah et al. [1] predicted that 55.2–62.7 kg in 2015 and declined to 49.1–61.2 kg in 2025 and 42.4-58.1 kg in 2050. Timmer et al. [20] forecasted the total rice consumption trend to decrease since 2030 to 2050 while USDA [22] predicted that will slowdown in two next years. The results of all studies above revealed that the trend of world rice consumption is slowdown. However, we believe that our estimate model has good validation because we gain information from panel data and the performance of our model is more accurate.

5   Policy Implications

According to our findings, the factors of past rice consumption per capita growth, GDP growth, and world rice price growth can be used as leading indicators to forecast rice consumption per capita growth. Hence, policy makers including stakeholders in the rice related industries should attend to these factors. The negative relationship between the per capita rice consumption growth in this year ($\%\Delta PCR_t$) and that in the year before last year ($\%\Delta PCR_{t-2}$) indicates that the future of rice dietary trends is the decline in rice consumption per capita. These results are supported by the expert opinions from FAO and IRRI. The trend of health concern is rising leading to decreasing carbohydrate intakes. In addition, a report from World Bank [18] shows a sign of aging society because of the declining in population growth among rice-eating countries. It was argued that rapidly aging population will require less food per capita due to the natural decline in metabolic needs. Hence with these reasons, the development in the rice exporting countries should concentrate primarily on maintenance of existing yields including improving nutritional quality of rice in line with the trend of health concern and aging society such as organic rice. The agricultural policy should also encourage rice farmers to adjust their farming system to meet the needs of the global market. Lastly, the government should support the cross-disciplinary processing of rice for medical, industrial, energy uses, and etc.

6   Conclusions

This paper investigates the future of global rice consumption per capita growth using the dynamic panel data model. We separate countries into 4 groups based on income level classified by the World Bank. The panel data as a whole covers 73 countries from 1961 to 2015. We employ three key factors having effect on that–past rice consumption per capita growth, per capita GDP growth, and world rice price growth. The empirical results reveal that there is a negative relationship between current and previous rice consumption per capita growth. Trend of the demand for rice consumption may slowdown because of declining per capita rice consumption growth rate. The forecasting performance of the model performs quite well based on RMSE specially, in short period forecasting.

Lastly, we would recommend some policies for policy makers and stakeholders in rice related industries. The policy makers should develop agricultural strategies such as promoting the policies which support the changing of dietary patterns and the long run sustainable economic growth. For instance, the policy should concentrate on improving rice quality, encouraging alternative crops to rice farmers, and supporting the cross-disciplinary processing of rice. All of these suggestions would help increase the value addition and price of rice products. Furthermore, the benefits from these policies would help keep purchasing power of rice farmers when the demand for rice consumption tends to decrease.

Acknowledgements. The first author is grateful to the full scholarship from the Bank of Thailand. In addition, she would like to express much of her appreciations to Mr. Tanarat Rattanadamrongaksorn for his encouragement in bringing this interesting issue to our attention. Also, the second and third authors wish to thank the Puey Ungphakorn Centre of Excellence in Econometrics, Faculty of Economics, Chiang Mai University for giving them financial supports.

References

1. Abdullah, A.B., Ito, S., Adhana, K.: Estimate of rice consumption in Asian countries and the world towards 2050. In: Proceedings for Workshop and Conference on Rice in the World at Stake, vol. 2, pp. 28–43 (2006)
2. Alexandratos, N., Bruinsma, J., et al.: World agriculture towards 2030/2050: the 2012 revision. Technical report, ESA Working paper Rome, FAO (2012)
3. Arellano, M., Bond, S.: Some tests of specification for panel data: Monte Carlo evidence and an application to employment equations. Rev. Econ. Stud. **58**(2), 277–297 (1991)
4. Bond, S.R.: Dynamic panel data models: a guide to micro data methods and practice. Port. Econ. J. **1**(2), 141–162 (2002)
5. Carstensen, K., Toubal, F.: Foreign direct investment in Central and Eastern European countries: a dynamic panel analysis. J. Comp. Econ. **32**(1), 3–22 (2004)
6. Food and Agricultural Organization of The United Nation (FAO): Food outlook (2017). http://www.fao.org/3/a-i7343e.pdf
7. Huang, B.N., Hwang, M.J., Yang, C.W.: Causal relationship between energy consumption and GDP growth revisited: a dynamic panel data approach. Ecol. Econ. **67**(1), 41–54 (2008)

8. Im, K.S., Pesaran, M.H., Shin, Y.: Testing for unit roots in heterogeneous panels. J. Econom. **115**(1), 53–74 (2003)
9. Judson, R.A., Owen, A.L.: Estimating dynamic panel data models: a guide for macroeconomists. Econ. Lett. **65**(1), 9–15 (1999)
10. Kubo, M., Purevdorj, M., et al.: The future of rice production and consumption. J. Food Distrib. Res. **35**(1), 128–142 (2004)
11. Levin, A., Lin, C.F., Chu, C.S.J.: Unit root tests in panel data: asymptotic and finite-sample properties. J. Econom. **108**(1), 1–24 (2002)
12. Lewbel, A.: Engel curves. In: The New Palgrave Dictionary of Economics, 2 edn. (2008)
13. Lio, M.C., Liu, M.C., Ou, Y.P.: Can the internet reduce corruption? A cross-country study based on dynamic panel data models. Gov. Inf. Q. **28**(1), 47–53 (2011)
14. Mohanty, S.: Trends in global rice consumption. Rice Today **12**(1), 44–45 (2013)
15. Nicholson, W., Snyder, C.M.: Intermediate Microeconomics and Its Application. Cengage Learning (2014)
16. Sequeira, T.N., Maçãs Nunes, P.: Does tourism influence economic growth? A dynamic panel data approach. Appl. Econ. **40**(18), 2431–2441 (2008)
17. Sirikanchanarak, D., Liu, J., Sriboonchitta, S., Xie, J.: Analysis of transmission and co-movement of rice export prices between Thailand and Vietnam, pp. 333–346 (2016)
18. Smil, V., et al.: Feeding the world: how much more rice do we need. In: Rice is Life: Scientific Perspectives for the 21st Century, pp. 21–23 (2005)
19. The International Rice Research Institute (IRRI): Bigger harvest a cleaner planet (2000). http://www.irri.org/publications/annual/pdfs/ar2000/biggerharvests.pdf
20. Timmer, C.P., Block, S., Dawe, D.: Long-run dynamics of rice consumption, 1960–2050. In: Rice in the Global Economy: Strategic Research and Policy Issues for Food Security, pp. 139–174 (2010)
21. United Nations: World population to 2300. United Nations, New York (2004)
22. United States of Agricultural Department (USDA): Rice: world markets and trade (2017). https://apps.fas.usda.gov/psdonline/circulars/grain.pdf
23. Zhang, C., Zhuang, L.: The composition of human capital and economic growth: evidence from china using dynamic panel data analysis. China Econ. Rev. **22**(1), 165–171 (2011)

The Analysis of the Effect of Monetary Policy on Consumption and Investment in Thailand

Jirawan Suwannajak[1], Woraphon Yamaka[1,2], Songsak Sriboonchitta[1,2], and Roengchai Tansuchat[1,2(✉)]

[1] Faculty of Economics, Chiang Mai University, Chiang Mai 50200, Thailand
Suwannajak.j@gmail.com, woraphon.econ@gmail.com, roengchaitan@gmail.com
[2] Center of Excellence in Econometrics, Chiang Mai University, Chiang Mai 50200, Thailand

Abstract. This study highlights on the analysis of Thai monetary policy transmission channels, i.e. interest rate, credit, exchange rate, and asset price channels, to private consumption and private investment. The analytical methods are Time Varying Parameter Vector Autoregressive (TVP-VAR) with stochastic volatility, and its impulse response function. The results showed that the credit channel contribute the greatest impact on private consumption and investment. We also found that the effect of monetary policy to private consumption and investment are vary over time.

1 Introduction

Stabilizing and driving economic growth are targets of economic development in many countries. There are various factors affecting economic growth. Two of the important factors that drive economic growth are private consumption and investment [10] which are viewed as demand side factors [21]. Figure 1 shows the percentage of Thai private consumption which is about 11% of gross domestic product (GDP), while the percentage of Thai private investment is 45% of gross domestic product (GDP). Thus, it is important to stabilize these two factors. And, the most powerful tool of policy makers and government is monetary policy [2].

Monetary policy is implemented by central banks to stimulate or raise domestic private consumption and investment. To maintain the appropriate level of economic growth, production and inflation, the central bank needs to adjust its financial liquidity and interest rate through monetary instruments, such as reserve requirements and open market operations. The most important channels for adjusting the money supply are interest rate, bank deposit rate, long-term interest rates, credit expansion and exchange rate. Pootrakool and Subhaswadikul [22] suggested that these tools can stabilize price level and contribute to economic growth.

Rational Expectation is introduced in a standard New Keynesian model. Lucas [15] explained that the rational expectations hypothesis implies that every economic agent makes optimal use of information in forming expectations [8]. This hypothesis follows two assumptions: (i) the average estimate of the outcome

© Springer International Publishing AG 2018
V. Kreinovich et al. (eds.), *Predictive Econometrics and Big Data*, Studies in Computational Intelligence 753, https://doi.org/10.1007/978-3-319-70942-0_46

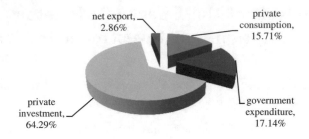

Fig. 1. Structure of Thai GDP Source: Office of the National Economic and Social Development Board, 2016

by participants in the market is correct and based on correct probability distributions, (ii) forecasting errors are uncorrelated in successive time periods [16]. While the assumption of rational expectations has little empirical implication, it has startling implications for effectiveness of monetary policy [17]. Firstly, Sargent and Wallace [23] showed that under the rational expectation the effect of a change in money on economic activity would only be temporary. In addition, if the rational expectations and not adaptive expectations are included in the analysis of Fisher nominal interest rate relation, the effect of the money has no lasting effect to the economy [4].

Secondly, Sargent and Wallaces [24] showed that changing the paths of output and employment are not affected by systematic monetary policy. However, the systematic monetary policy affects to expected inflation but has no affect to unemployment [17]. This is because rational economic agents expect the effect of the policy action and adopt it part of the information set on decision making [12]. Thirdly, Lucas [15] showed that in a business cycle framework of imperfect information, market clearing and rational expectations, monetary policy will only be effective as long as it distorts relative prices [16]. Therefore, by conducting of monetary policy must consider agent expectation. This is because result of rational expectation effect on effective of monetary policy and the impact to real economic sector [7].

There were numerous studies dealing with the transmission of monetary policy to economic growth, for examples, the studies by Kim [13], Angeloni and Kashyap [1], Benmank and Blinder [3], Petra [21], Fujiwara [9], and Pastpipatkul et al. [20]. Although these researches provided a better theoretical understanding, there is a scarcity in the number of empirical researches attempting to analyze how economic growth reacts to transmission of monetary policy over time.

To investigate the effective monetary policy transmission channels to private consumption and investment in Thailand, we need to understand the mechanisms of monetary policy transmission which can be distinguished into four channels i.e. credit channel, asset price channel, exchange rate channel, and interest rate channel (Fig. 2).

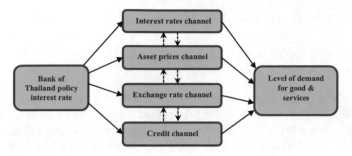

Fig. 2. The channel to transmission of monetary policy   Source: Bank of Thailand, 2012

The objectives of this study are investigate the effects of the four monetary policy transmission channels-credit, asset price, exchange rate, and interest rate-on private consumption and investment in Thailand over time, and conduct impulse response analysis for finding an explanation for the movement of the private consumption and investment when it comes as a shock to the system through different monetary policy transmission channels.

To achieve our objectives, we employ the Time Varying Parameter-Vector Autoregressive (TVP-VAR) with stochastic volatility model in order to capture the time-varying effects of monetary policy transmission on private consumption and private investment, and apply Bayesian estimation to estimate the unknown parameters in the model. Our concern is that the limited data sets in our study may lead a VAR model to face a degree of freedom problem. Van de Schoot et al. [26] suggested that Bayesian estimation can provide reliable results when sample size is small. It is more informative, flexible, and efficient than MLE and the implementation of Bayesian prior on the parameter of the model can reduce the estimation uncertainty and to obtain accurately the inference and the forecast [19].

The next section briefly summarizes some of the methodologies used in this study. In Sect. 3, we describe our data. The results are reported in Sect. 4. Finally, Sect. 5 provides the conclusion and policy suggestions.

2   Methodology

2.1   Bayesian VAR Models

To study the macroeconomic data, VAR model is often employed. However, the over-parameterization often exists in the estimation of VAR due to a large number of parameter estimates with too few observations. Doan et al. [6]; Sims and Zha [25] introduced estimating VAR model using Bayesian approach called BVAR model. Thus, BVAR has become popular in the macroeconomic studies.

Similar to VAR formula, the BVAR(p) model can be written as

$$Y_{i,t} = \sum_{p=1}^{p} \beta_p Y_{i,t-p} + \varepsilon_t \tag{1}$$

where $Y_{i,t}$ is a $k \times 1$ of $y_{i,t}$ which has a observation at time $t = (1, ..., T)$ for i^{th} variable. Y_{t-p} is a lag of Y_t this describes each variable with its own lag p variable and other variables based on lag p. β_j is $K \times K$ matrix of autoregressive coefficients, a_0 is the intercept term, while ε_t stands for the white noise processes of a zero-mean vector with positive definite contemporaneous covariance matrix Σ and zero covariance matrices at all the other lags.

2.2   Time Varying Parameter-Vector Autoregressive with Stochastic Volatility

The TVP-VAR with stochastic volatility model can be used to capture the potential time-varying nature of the underlying structure in any macroeconomic segment of interest in a flexible and robust manner. This model allows the shift in the parameters over time. Following [5, 11, 18] the Time-Varying VAR model can be written as

$$Y_{i,t} = \sum_{p=1}^{P} \beta_{i,t} Y_{i,t-p} + \varepsilon_t, \ \varepsilon_t \sim N(0, \Sigma_{t,1}) \tag{2}$$

Time varying equation is

$$\beta_{i,t} = F\beta_{i,t-1} + u_t, \ u_t \sim N(0, \Sigma_2) \tag{3}$$

Stochastic volatility equation is

$$\Sigma_{t,1} = \gamma(h_t), h_{t+1} = \phi h_{t+1} + \eta_t, \ \eta_t \sim N(0, \Sigma_3) \tag{4}$$

where $Y_{i,t}$ is a vector of endogenous variables, $\beta_{i,t}$ is a matrix of time varying parameters; $\Sigma_{t,1}$ is matrix of time varying standard deviation; ε_t is the error of the mean Eq. (2); F is a time varying coefficient of $\beta_{i,t-1}$; u_t is error of the time varying Eq. (3). h_t is stochastic volatility and η_t is error of the stochastic volatility Eq. (4). Here, we assume $\gamma > 0$ to guarantee that Σ_t is non-negative. To estimate these parameters, a Gibbs sampler is employed in this estimation. (See [5, 18]).

2.3   Prior, Likelihood, Posterior

In the Bayesian approach, to estimate the unknown parameters of the model, $\theta = \{\Sigma, F, \gamma, \phi, c\}$ we need to construct a posterior density of the model which is derived from

$$g(\theta \,|y) \propto f(y \,|\theta)p(\theta), \tag{5}$$

where, $f(y|\theta)$ is a density function or likelihood function and $p(\theta)$ is a prior information on θ which is assumed to be available in the form of density function after having updated by looking at the data. To estimate both BVAR and TVP-BVAR models, we conduct a Gibbs sampler where the prior is assumed to be normal-Wishart prior for BVAR model (see [14]). For TVP-BVAR model, the prior distributions proposed in this paper are chosen as follows: First of all, the priors for Σ is assumed to be inverse-Wishart. The priors for time varying F, γ, and ϕ are assumed to have normal distribution. Then, we generate the samples from the joint posterior distribution of the coefficients $\theta^n = \{\Sigma^n, F^n, \gamma^n, \phi^n\}$ by Gibbs sampling. Blocking Gibbs sampler is conducted here since it is easy to draw $g(\theta|y)$. Gibbs sampling is carried out in five steps, yielding the following algorithm (see [18]).

(1) Set the initial start for $\{\Sigma^0, F^0, \gamma^0, \phi^0\}$
(2) Draw Σ^n from $g(\Sigma^{n-1}, F^{n-1}, \gamma^{n-1}, \phi^{n-1}|y)$
(3) Draw F^n from $g(\Sigma^n, F^{n-1}, \gamma^{n-1}, \phi^{n-1}|y)$
(4) Draw γ^n and ϕ^n from $g(\Sigma^n, F^n, \gamma^{n-1}, \phi^{n-1}|y)$
(5) Repeat step (2)

These draws can be used as the basis for making inferences by appealing to suitable ergodic theorems for Markov chains. To derive the conditional posterior of each block, our study refers to Nakajima [18]. By using a standard Gibbs sampler algorithm for a given prior $p(\theta)$, we use the average of the Markov chain on θ as an estimate of θ. For the number of draws, we will repeat the algorithm until the posterior looks stationary and close to normal.

2.4 Macroeconomic Specification

From Fig. 2 and VAR Model, the relationship among 6 variables are specified as follow:

$$PCI_t = a_1 + \sum_{i=1}^{n} A_{1i}PCI_{t-i} + \sum_{i=1}^{n} B_{1i}PII_{t-i} + \sum_{i=1}^{n} C_{1i}IR_{t-i}$$
$$+ \sum_{i=1}^{n} D_{1i}CR_{t-i} + \sum_{i=1}^{n} E_{1i}SET_{t-i} + \sum_{i=1}^{n} F_{1i}REER_{t-i} + \varepsilon_{1t} \tag{6}$$

$$PII_t = a_2 + \sum_{i=1}^{n} A_{2i}PCI_{t-i} + \sum_{i=1}^{n} B_{2i}PII_{t-i} + \sum_{i=1}^{n} C_{2i}IR_{t-i}$$
$$+ \sum_{i=1}^{n} D_{2i}CR_{t-i} + \sum_{i=1}^{n} E_{2i}SET_{t-i} + \sum_{i=1}^{n} F_{2i}REER_{t-i} + \varepsilon_{2t} \tag{7}$$

$$IR_t = a_3 + \sum_{i=1}^{n} A_{3i} PCI_{t-i} + \sum_{i=1}^{n} B_{3i} PII_{t-i} + \sum_{i=1}^{n} C_{3i} IR_{t-i}$$
$$+ \sum_{i=1}^{n} D_{3i} CR_{t-i} + \sum_{i=1}^{n} E_{3i} SET_{t-i} + \sum_{i=1}^{n} F_{3i} REER_{t-i} + \varepsilon_{3t} \tag{8}$$

$$CR_t = a_4 + \sum_{i=1}^{n} A_{4i} PCI_{t-i} + \sum_{i=1}^{n} B_{4i} PII_{t-i} + \sum_{i=1}^{n} C_{4i} IR_{t-i}$$
$$+ \sum_{i=1}^{n} D_{4i} CR_{t-i} + \sum_{i=1}^{n} E_{4i} SET_{t-i} + \sum_{i=1}^{n} F_{4i} REER_{t-i} + \varepsilon_{4t} \tag{9}$$

$$SET_t = a_5 + \sum_{i=1}^{n} A_{5i} PCI_{t-i} + \sum_{i=1}^{n} B_{5i} PII_{t-i} + \sum_{i=1}^{n} C_{5i} IR_{t-i}$$
$$+ \sum_{i=1}^{n} D_{5i} CR_{t-i} + \sum_{i=1}^{n} E_{5i} SET_{t-i} + \sum_{i=1}^{n} F_{5i} REER_{t-i} + \varepsilon_{5t} \tag{10}$$

$$REER_t = a_6 + \sum_{i=1}^{n} A_{6i} PCI_{t-i} + \sum_{i=1}^{n} B_{6i} PII_{t-i} + \sum_{i=1}^{n} C_{6i} IR_{t-i}$$
$$+ \sum_{i=1}^{n} D_{6i} CR_{t-i} + \sum_{i=1}^{n} E_{6i} SET_{t-i} + \sum_{i=1}^{n} F_{6i} REER_{t-i} + \varepsilon_{6t} \tag{11}$$

where PCI_t is private consumption index, PII_t is private investment index, IR_t is interbank rate, CR_t is bank deposit rate, SET_t is Stock Price Index of Thailand, $REER_t$ is Real Effective Exchange Rate, which has a observation at time $t = (1, ..., T)$. a is constant. A B C D E and F is coefficients of lag. ε_t is stands for the white noise processes of a zero-mean vector with positive definite contemporaneous covariance matrix Σ and zero covariance matrices at all the other lags.

3   Data

3.1   Descriptive Statistic

In this study, we deal with growth rate of six variables which are private investment index (PII), private consumption index (PCI), commercial bank credit (CR), interbank rate (IR), exchange rate (REER), Stock Exchange of Thailand (SET). All variables are monthly data and collected from January, 2000 to December, 2016. The data are transform to the growth rate in order to avoid the nonstationary problem. The descriptive statistics are given in Table 1. The table shows that PCI has the highest standard deviation (3.8956), while CR has the lowest (0.0148). The Jarque-Bera statistic indicates that all series are not normally distributed became they reject the null hypothesis. The Augmented Dickey-Fuller (ADF) test at level with none, intercept, trend and intercept is also applied to check unit roots in the series. Unit root test for each variable has a statistical significance level of 0.01. Therefore, these variables are stationary.

Table 1. Lag length criteria BVAR

Variable		PII	PCI	IR	CR	REER	SET
Mean		0.0023	0.3716	0.0088	0.0049	−9.55E−05	0.0045
Max		0.0927	52.950	1.052	0.0685	0.0797	0.4059
Min		−0.468	−4.6969	−0.3397	−0.0594	−0.2335	−0.8175
Std. Dev.		0.0378	3.8956	0.1417	0.0148	0.0215	0.0963
Skewness		−9.5475	12.3456	3.8285	−0.0089	−5.9532	−2.5922
Kurtosis		119.746	165.523	27.1253	6.98066	70.7364	30.4060
Jarque-Bera		118368	228574.9	5418.94	134.031	40007.79	6580.348
Probability		0.00000	0.00000	0.00000	0.00000	0.00000	0.00000
ADF-test	None	$(-12.21)^{***}$	$(-14.26)^{***}$	$(-13.75)^{***}$	$(-3.31)^{***}$	$(-16.76)^{***}$	$(-13.88)^{***}$
	Intercept	$(-12.22)^{***}$	$(-14.36)^{***}$	$(-13.76)^{***}$	$(-5.26)^{***}$	$(-16.72)^{***}$	$(-13.89)^{***}$
	Trend and Intercept	$(-12.19)^{***}$	$(-14.63)^{***}$	$(-13.72)^{***}$	$(-5.40)^{***}$	$(-16.69)^{***}$	$(-13.94)^{***}$

Source: Calculation.
Note: *, **, and *** denote significant at 10%, 5%, and 1%, respectively

4 Results

Prior to analyzing the effect of monetary policy transmission through different channels on private consumption and investment in Thailand, we check the order of lag of our data from BVAR model. We conduct a Akaiki information criterion (AIC), Schwarz criterion (SQ), and Hannan–Quinn information criterion (HQ). The results indicate that lag 3 is better only for AIC, for other lag 1 is better (Table 2).

Table 2. Lag length criteria BVAR

Model	Lag	AIC	SC	HQ
BVAR.lag 1	1	−21.0853	−20.38034*	−20.79987*
BVAR.lag 2	2	−21.29097	−19.98177	−20.76089
BVAR.lag 3	3	−21.33768*	−19.42424	−20.56295
BVAR.lag 4	4	−21.16719	−18.64949	−20.1478

Source: Calculation.

4.1 Time Varying Effect of Monetary Policy

The results of the effect of monetary policy transmission through different channels on private investment and private consumption are illustrated in Figs. 3 and 4, respectively. The graph plot a matrix of Time Varying Parameters, $\beta_{i,t}$, from our TVP-VAR model. We found that the effect of (CR), which is denoted as credit channel, is positive to private investment and the range of the effect is around 0.0362 to 0.0368 (Fig. 3(a)). We observe that the effect increases since year 2011 to late 2012. After that the size of positive effect has gradually decreased. We expect that the European financial crisis in 2012 led the effect

(a) The effect of commercial bank credit (t-3) to private investment equation (t)

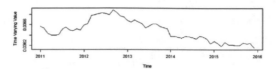

(b) The effect of interbank rate (t-3) to private investment equation (t)

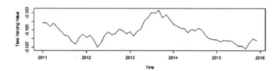

(c) The effect of effective exchange rate (t-3) to private investment equation (t)

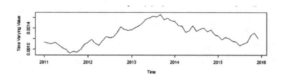

(d) The effect of SET index (t-3) to private investment equation (t)

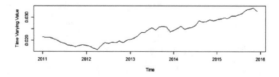

Fig. 3. Time varying effect of each channel on private investment index   Source: Calculations

of CR on private investment to become smaller. The reason is that the financial institutions and investors had less confidence in the economy and thereby retarding their investment decision.

For the interest rate (IR), we refer to it as expectation channel. The effect of IR on private investment is different from CR, see Fig. 3(b). We found that IR has a negative effect on private investment. We observed an interesting result in this effect. The negative effect of IR on investment is high during 2013–2014, which corresponds to the economic downturn in Europe. In times of economic downturn, the lower interest rates could encourage additional investment spending.

(a) The effect of commercial bank credit (t-3) to private consumption index equation (t)

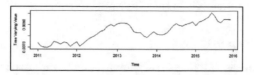

(b) The effect of interbank rate (t-3) to private consumption index equation (t)

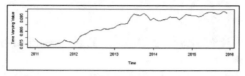

(c) The effect of effective exchange rate (t-3) to private consumption index equation (t)

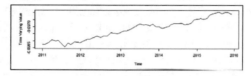

(d) The effect of set index (t-3) to private consumption index equation (t)

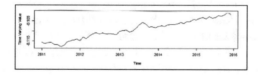

Fig. 4. Time varying effect of each channel on private consumption index   Source: Calculations

Consider the effect of the exchange rate (REER) and SET, which represent exchange rate and asset price channel, respectively. Both channels show the positive effect on private investment where the time varying effect of exchange rate channel has the range between 0.0010 and 0.0014 and the range between 0.107 and 0.103 for asset price channel (see Fig. 3(c) and (d)). In the case of exchange rate channel, we can observe a huge effect of exchange rate on private investment in late 2013.

This means that an increase in THB/USD (depreciation) would make exports more competitive and appear cheaper to foreigners. This would increase demand for exports and investment. We found that this period corresponds to the economic recovery in USA which is an important and large trading partner of Thailand. Therefore, the demand for Thai export products increased during this period.

For the effect of monetary policy transmission on private consumption, the results of Fig. 4 show that most monetary policy transmission channels generated positive effect except for asset price channel (SET). We also observe that the size of effect of these channels gradually increases overtime. Whereas, it is noteworthy that credit (CR) channel has higher fluctuation than the others.

4.2   Impulse Response

Finally, the responses of the private investment and private consumption to shocks in monetary policy transmission channels for over 5 months are illustrated in Figs. 5 and 6.

4.2.1   The Effect of Monetary Policy Transmission on Private Investment

Consider the interest rate channel, private investment index (PII) shows a small fluctuating response to interbank rate (IR) in the first four months and converged to equilibrium within 5 months. For credit channel, we found that PII responded to commercial bank rate (CR) shock with a small positive direction. We expect that firms can raise capital from sources other than commercial banks so that the shock of this channel did not affect investment (Fig. 5(b)).

In exchange rate channel, Fig. 5(c) shows that PII responded to the shock of real effective exchange rate (REER) in the positive rather than the negative direction. We observe that the shock of REER is not persistent. It has a positive response to the shock of REER and then falls sharply and reaches its minimum after about 3 months. However, the shock moved to equilibrium within 5 months.

For asset price channel in Fig. 5(d), it shows that SET index (SET) creates a positive sharp-shaped response in PII, following which it begins to decrease and eventually overshoots, thereby leading to a decrease in PII about 4–5 months later. Moreover, when we compare the effects of all channels on private investment, credit channel apparently had the most impact on private investment.

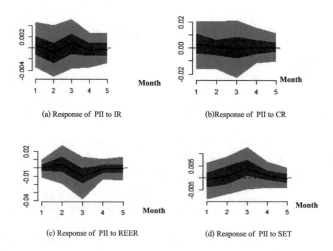

Fig. 5. Impulse response of private investment index to shock of monetary policy transmission channel Source: Calculations

4.2.2   The Effect of Monetary Policy Transmission on Private Consumption

Figure 6 also presents the impulse response function for the changes in private consumption to a shock of monetary policy transmission. The results illustrated that the shock of interbank rate (IR) causes the private consumption index (PCI) to fall initially before returning to the equilibrium within 4–5 months. For the shock of commercial bank rate (CR), the response of the private consumption is limited. This indicated that any shocks of CR will not lead to the private consumption change.

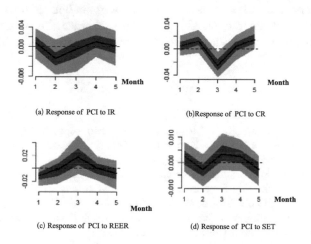

Fig. 6. Impulse response of private consumption index to shock of monetary policy transmission channel   Source: Calculations

Finally, consider the shock of real effective exchange rate (REER) and SET index (SET). The result depicted that the response of PCI is quite fluctuating since there are both positive and negative directions, see Fig. 6(c) and (d), respectively. In addition, when we compare the effects of all channels on private consumption, it is concluded that the most impact to private consumption is from credit channel.

5   Conclusion

Monetary policy is implemented to achieve macroeconomic target. It is necessary for policy makers to understand the effect of monetary policy transmitted through different channels on specific macroeconomic variables. The monetary policy transmission channels i.e. interest rate, exchange rate, asset price, and credit, can play an important role in monetary policy transmission. In this study, we employ a TVP-VAR model and estimate the unknown parameters using Bayesian approach. Our model has an ability to predict how possible changes in policy can affect the economy, and our result can help the government decide on the best monetary policy.

Our study points out that although credit channels it will have an impact on private consumption and investment faster than other channels, but it also affect volatile private consumption and investment. Therefore, the central bank should consider the transmission of monetary policy in other channels, such as the asset price channel, policy maker should develop a stock market to be more effective such as the dissemination of information on asset prices. To quality and transparent including money supply control in the system. To be according with the needs of the people. To reduce volatility of stock prices, which affect private consumption index. This will provide a great benefit for investment. In addition, the credit channel should be monitored since the private investment and consumption are very sensitive to the change in commercial banks' interest rates.

References

1. Angeloni, I., Kashyap, A.K., Mojon, B. (eds.): Monetary Policy Transmission in the Euro Area: A Study by the Eurosystem Monetary Transmission Network. Cambridge University Press, Cambridge (2003)
2. Banuso, F.B., Odior, E.S.: Macroeconomic volatility and government consumption expenditure: implication for public welfare: a dynamic macroeconometric stochastic model. Int. J. Econ. Fin. **4**(2), 140 (2012)
3. Bernanke, B.S., Blinder, A.S.: The federal funds and the channels of monetary transmission. Am. Econ. Rev. **82**, 901–921 (1992)
4. Blanchard, O.: What do we know about macroeconomics that Fisher and Wicksell did not? Q. J. Econ. **115**, 1375–1408 (2000)
5. Del Negro, M., Primiceri, G.E.: Time-varying structural vector autoregression and monetary policy: a corrigendum. Federal Reserve Bank of New York Staff Report 619 (2013)
6. Doan, T., Litterman, R., Sims, C.: Forecasting and conditional projection using realistic prior distributions. Econometric Rev. **3**, 1–144 (1984)
7. Eusepi, S., Giannoni, M.P., Preston, B.J.: Long-term debt pricing and monetary policy transmission under imperfect knowledge (2012)
8. Friedman, M.: Monetary policy: theory and practice. J. Money Credit Bank. **14**(1), 98–118 (1982)
9. Fujiwara, I.: Output composition of the monetary policy transmission mechanism in Japan. Topics in Macroecon. **4**(1) (2004)
10. Heim, J.J.: The impact of consumer confidence on consumption and investment spending. J. Appl. Bus. Econ. **11**(2), 37 (2010)
11. Kaewsompong, N., Sriboonchitta, S., Maneejuk, P., Yamaka, W.: Relationships among prices of rubber in ASEAN: Bayesian structural VAR model. Thai J. Math. 101–116 (2016)
12. Kantor, B.: Rational expectations and economic thought. J. Econ. Lit. **17**(4), 1422–1441 (1979)
13. Kim, H.E.: Was the Credit a Key Monetary Transmission Mechanism Following The Financial Crisis in The Republic of Korea? The World Bank Policy Working Paper No. 2013 (1999)

14. Koop, G., Korobilis, D.: Bayesian multivariate time series methods for empirical macroeconomics. Found. Trends Econometrics **3**(4), 267–358 (2010)
15. Lucas, R.E., Sargent, T.J. (eds.): Rational Expectations and Econometric Practice, vol. 2. University of Minnesota Press, Minneapolis (1981)
16. Macesich, G.: Monetary Policy and Rational Expectations. Praeger Publishers, New York (1987). 154 pages
17. Mankiw, N.G.: A quick refresher course in macroeconomics. J. Econ. Lit. **XXVIII**, 1645–1660 (1990)
18. Nakajima, J.: Time-Varying Parameter VAR model with stochastic volatility: An overview of methodology and empirical applications (No. 11-E-09). Institute for Monetary and Economic Studies, Bank of Japan (2011)
19. Pastpipatkul, P., Yamaka, W., Wiboonpongse, A., Sriboonchitta, S.: Spillovers of quantitative easing on financial markets of Thailand, Indonesia, and the Philippines. In: International Symposium on Integrated Uncertainty in Knowledge Modelling and Decision Making, pp. 374–388. Springer, Cham, October 2015
20. Pastpipatkul, P., Yamaka, W., Sriboonchitta, S.: Effect of quantitative easing on ASEAN-5 financial markets. In: Causal Inference in Econometrics, pp. 525–543. Springer International Publishing (2016)
21. Petra, G.K.: European Central Bank Working Paper Series: "Interest Rate Reaction Function and The Taylor Rule in The Euro Area" (2003). www.Sciendirect. com
22. Pooptakool, K., Subhaswadikul, M.: The monetary transmission mechanism. In: Bank of Thailand Research Symposium (2000)
23. Sargent, T.J., Wallace, N.: Rational expectations and the dynamics of hyperinflation. Int. Econ. Rev. **14**, 328–350 (1973)
24. Sargent, T.J., Wallace, N.: "Rational" expectations, the optimal monetary instrument, and the optimal money supply rule. J. Polit. Econ. **83**(2), 241–254 (1975)
25. Sims, C., Zha, T.: Bayesian methods for dynamic multivariate models. Int. Econ. Rev. **39**, 949–968 (1998)
26. Van De Schoot, R., Broere, J.J., Perryck, K.H., Zondervan-Zwijnenburg, M., Van Loey, N.E.: Analyzing small data sets using Bayesian estimation: the case of post-traumatic stress symptoms following mechanical ventilation in burn survivors. Eur. J. Psychotraumatol. **6**(1), 25216 (2015)

Investigating Relationship Between Gold Price and Crude Oil Price Using Interval Data with Copula Based GARCH

Teerawut Teetranont[1(✉)], Somsak Chanaim[1,2], Woraphon Yamaka[1,2], and Songsak Sriboonchitta[1,2]

[1] Faculty of Economics, Chiang Mai University, Chiang Mai 52000, Thailand
teetranont@gmail.com, somsak_ch@cmu.ac.th, woraphon.econ@gmail.com, songsakecon@gmail.com
[2] Center of Excellence in Econometrics, Chiang Mai University, Chiang Mai 52000, Thailand

Abstract. This study investigates and compares the performance of center method, equal weighted convex combination and unequal-weighted convex combination methods through various GARCH and copula-based approaches for the analysis of relationship between gold and crude oil prices using interval data in Comex and Nymex tradings. The results of this study confirm that unequal-weighted convex combination method improves the estimation and it tends to perform better than both the center method and its equal-weighted variant. In addition, the marginal from the best fit GARCH model is used to measure dependence via copula function in the form of Student-t copula as selected according to the lowest AIC among all candidates. Finally, we can conclude that there exists the dependence between Comex and Nymex not only in the normal event, but also in the extreme event.

1 Introduction

Commodities, especially gold and crude oil, and bond are important instruments to diversify the risk. In the calculation of optimum risky weights for portfolio, gold and crude oil are the most attractive commodities to be included for hedging risks in portfolio of investors. Gold and crude oil have played the vital role in economics. In the last decade, many economists paid much attention to investigating the volatility and the relationship between gold and crude oil prices. The most effective tool that is employed to measure this volatility and relation is Copula based GARCH model. The multivariate GARCH models have demonstrated to be useful and effective for analyzing the pattern of multivariate random series and estimating the conditional linear dependence of volatility or co-volatility in different markets.

The study of the volatility and the relationship between gold and crude oil prices, using Copula based GARCH, has been intensively conducted in the last decade. However, those studies investigated their works using a closing price

© Springer International Publishing AG 2018
V. Kreinovich et al. (eds.), *Predictive Econometrics and Big Data*, Studies in Computational Intelligence 753, https://doi.org/10.1007/978-3-319-70942-0_47

data series. Thus, if we consider only the closing prices, we might lack a valuable intraday information and the obtained results might not be reasonable [11]. Recently some studies have proposed to use interval data, e.g. the lowest and the highest price during each day or period of time, as an alternative to single value data. In the ideal world, we should be able to predict both the lowest daily price and the highest daily price. However, in practice, this is difficult, so we would like to predict at least some daily price between these bounds. In the past, researchers tried to predict the representative of the lowest and highest prices, for example, a Center and MinMax methods of Billard and Diday [3], Center and Range method of Neto and Carvalho [10], and model M by Blanco-Fernndez, Corral, Gonzlez-Rodrguez [4], to deal with the interval data. These methods aim to construct the model without taking the interval as a whole. However, we expect that these methods, especially center method, may not be robust enough to explain the real behavior of the interval data and this may lead to the misspecification of the model. Therefore, it will be of great benefit to relax this assumption of mid-point of center method and assign appropriate weights between intervals. Thus in this study, a convex combination method of Chanaim et al. [6] is employed to obtain the appropriate weights.

This study investigates and compares the dependence structure of crude oil and gold prices using different interval values for copula-based GARCH model estimation and prediction. The examined interval value methods include the center method, equal weighted, and unequal-weighted convex combination. The main findings will confirm the usefulness of the convex combination in copula-based GARCH approach for evaluating the relationship, joint distribution and co-movement between crude oil prices and gold prices for investors whose investment interest is in gold and crude oil.

The remainder of the paper is organized as follows: Sect. 2 provides methodology of study. Section 3 proposes the empirical results. Section 4 summarizes this paper.

2 Methodology

In this section, we brief the convex combination in GARCH, EGARCH, and GJR-GARCH models; and copula family for estimating joint density of the obtained marginal from the GARCH families.

2.1 Operation with Interval Arithmetic

Let $p_i = [\underline{p}_i, \overline{p}_i]$ be lower and upper interval data at time i. This data can be defined for arithmetic operations as in the following:

1. Addition

$$p_i + p_j = [\underline{p}_i + \underline{p}_j, \overline{p}_i + \overline{p}_j] \tag{1}$$

2. Subtraction

$$p_i - p_j = [\underline{p}_i - \underline{p}_j, \overline{p}_i - \underline{p}_j] \tag{2}$$

3. Multiplication

$$p_i.p_j = [\min A, \max A].A = \{\underline{p}_i\underline{p}_j, \underline{p}_i\overline{p}_j, \overline{p}_i\underline{p}_j, \overline{p}_i\overline{p}_j\} \tag{3}$$

4. Division , p > 0

$$\frac{1}{p_j} = [\frac{1}{\overline{p}_j}, \frac{1}{\underline{p}_j}] \tag{4}$$

5. Addition and Multiplication by scalar

$$p_i + a = \{\underline{p}_i + a, \overline{p}_i + a\} \tag{5}$$

$$a.p_i = \begin{cases} [a.\overline{p}, a.\underline{p}], & a < 0 \\ 0, & a = 0 \\ [a.\underline{p}, a.\overline{p}, & a > 0 \end{cases} \tag{6}$$

6. logarithm function, $p_i > 0$

$$\log p_i = \left| \log \underline{p}_i, \log \overline{p}_i \right| \tag{7}$$

2.2   Center Method

This method has been proposed by Billard and Diday [3]. The main idea is that it uses the center of the interval data p_t^c which is obtained from upper and lower values of interval, say $\underline{p}_t$ and $\overline{p}_t$, and can be derived by

$$p_t^c = \frac{\underline{p}_t + \overline{p}_t}{2} \tag{8}$$

2.3   Autoregressive Moving Average-GARCH Model

Many previous studies suggested that volatility of financial return data is not constant over time, but is rather clustered. This issue can be tackled using volatility modeling. Within a class of autoregressive processes with white noises having conditional heteroscedastic variances, this paper considers a GARCH(1,1) model to estimate the dynamic volatility. It is the workhorse model and mostly applied in many financial data. The model is able to reproduce the volatility dynamics of financial data. Thus, in this study, we consider ARMA(p,q)-GARCH(1,1) which can be written as

$$r_t = \phi_0 + \sum_{i=1}^{p} \phi_i r_{t-i} + \sum_{i=1}^{q} \varphi_i \varepsilon_{t-i} + \varepsilon_t \tag{9}$$

$$\varepsilon_t = \sigma_t \eta_t \tag{10}$$

$$\sigma_t^2 = \omega_0 + \omega_1\sigma_{t-1}^2 + \omega_2\varepsilon_{t-1}^2 \tag{11}$$

where η_t is a strong white noise which has normal distribution with mean zero and variance one. σ_t^2 is the conditional variance in GARCH process by Tim Bollerslev [5]. Some standard restrictions on the variance parameters are given.

$$\omega_1, \ \omega_2 > 0, \ \omega_1 + \omega_2 \ < 1 \tag{12}$$

Furthermore, Simon [12] presented a family of variance models in asymmetry EGARCH model and GJR-GARCH model.

2.4   EGARCH (Exponential GARCH)

From GARCH model, by introducing the parameters λ and ν, for $\lambda = \nu = 1$. Then we can rewrite GARCH(1,1), Eq. (11) as

$$\log \ \sigma_t^2 = \omega_0 + \omega_1 \log \ \sigma_{t-1}^2 + \omega_2 \left[\frac{|\varepsilon_{t-j}|}{\sigma_{t-j}} - E\left\{\frac{|\varepsilon_{t-j}|}{\sigma_{t-j}}\right\}\right] + \omega_3 \left(\frac{|\varepsilon_{t-j}|}{\sigma_{t-j}}\right) \tag{13}$$

The form of the expected value terms associated with ARCH coefficients in the EGARCH equation depends on the distribution of innovation. If the innovation distribution is Gaussian, then

$$E\left\{\frac{|\varepsilon_{t-j}|}{\sigma_{t-j}}\right\} = E\left\{|Z_{t-j}|\right\} = \sqrt{\frac{2}{\pi}} \tag{14}$$

If the innovation distribution is Student's t with $\nu > 2$ degrees of freedom, then

$$E\left\{\frac{|\varepsilon_{t-j}|}{\sigma_{t-j}}\right\} = E\left\{|Z_{t-j}|\right\} = \sqrt{\frac{v-2}{\pi}}\frac{\Gamma\left(\frac{v-1}{2}\right)}{\Gamma\left(\frac{v}{2}\right)} \tag{15}$$

2.5   GJR-GARCH

The GJR-GARCH model is a GARCH variant that includes leverage terms for modeling an asymmetric volatility clustering. In the GJR formulation, large negative changes are more likely to be clustered than positive changes. The GJR model is named for Glosten, Jagannathan, and Runkle [8]. The GJR-GARCH model is a recursive equation for the variance process, and the simple GJR-GARCH(1,1) can be written as

$$\sigma_t^2 = \omega_0 + \omega_1\sigma_{t-1}^2 + \omega_2\varepsilon_{t-1}^2 + \omega_3 I[\varepsilon_{t-1} < 0]\varepsilon_{t-1}^2 \tag{16}$$

The indicator function $I[\varepsilon_{t-1} < 0]$ equals 1 if $\varepsilon_{t-1} < 0$, and 0 otherwise. Thus, the leverage coefficients are applied to negative innovations, giving negative changes additional weight. For stationarity and positivity, the GJR model has the following constraints

$$\omega_0 > 0, \ \omega_1 \geqslant 0, \ \omega_2 \geqslant 0, \ \omega_2 + \omega_3 \geqslant 0, \ \omega_1 + \omega_2 + \omega_3 < 1$$

2.6   Convex Combination Method

The convex combination method is applied to deal with the interval return data, where the appropriate value over the range of interval can be computed by

$$r_t^{cc} = \alpha_0 \underline{r}_t + (1 - \alpha_1)\bar{r}_t \tag{17}$$

where α_0 *and* α_1 are the weighted parameters with value between 0 and 1. In this study, we consider both fixed weighted and unequal-weighted convex combination methods. Thus, we set $\alpha = 0.5$ for fixed weighted convex combination while $\alpha \ \varepsilon \ [0, 1]$ is set as the parameter to be estimated for unequal-weighted convex combination method. For example, in the case of ARMA(1,1)-GARCH(1,1), we can rewrite Eqs. (9)–(11) as

$$\alpha \underline{r}_t + (1 - \alpha)\bar{r}_t = \phi_0 + \phi_1(\alpha \underline{r}_{t-1} + (1 - \alpha)\bar{r}_{t-1}) + \varphi_1 \varepsilon_{t-1} + \varepsilon_t \tag{18}$$

$$r_t^{cc} = \phi_0 + \phi_i r_{t-1}^{cc} + \varphi_1 \varepsilon_{t-1} + \varepsilon_t$$

$$\varepsilon_t = \sigma_t \eta_t, \eta_t \sim N(0, 1) \tag{19}$$

$$\sigma_t^2 = \omega_0 + \omega_1 \sigma_{t-1}^2 + \omega_2 \varepsilon_{t-1}^2, \omega_0, \omega_1, \omega_2 > 0 \tag{20}$$

2.7   Model Selection by Akaike Information Criterion (AIC)

In this study, we compare our models using Akaike information criterion applied from Kullback Leibler Information. It is defined as:

$$AIC = -2\ln(\hat{L}) + 2K \tag{21}$$

where $\hat{L}$ is maximized value of likelihood function, K is the number of parameters in the model.

2.8   Bivariate Copula Approach

Let X, Y be random variables, the continuous marginal distributions are $F(x)$, $G(y)$ then $H(x, y)$ is a joint distribution, then 2-dimensional copulas $C : [0, 1]^2 \to [0, 1]$ can be defined by
Copula if property

$$F_i(x_i) = u, \ G(y_i) = v \ and \ H(x, y) = C(u, v)$$

so

$$C(0, v) = C(u, 0) = 0 \quad C(u, 1) = u, \ C(1, v) = v, u < u', v < v'$$
$$C(u', v') - C(u, v') - C(u', v) + C(u.v) \geq 0,$$

where C is copula function of marginal distribution random 2 variables. If marginal has continuous distribution, the copula function is

- Gaussian Copula

$$C(u,v) = \Phi(\Phi^{-1}(u), \Phi^{-1}(v)) \tag{22}$$

Lower and upper tail dependence or order parameters of Gaussian Copula is $k_L = k_u = \frac{2}{(1+\rho)}.\Phi_n^{-1}$ is quantile function for normal distribution function and $x = \Phi^{-1}(u), y = \Phi^{-1}(v)$ and $u, v \in [0,1]$.

- Student$-$t Copula

$$C(u,v) = \int_{-\infty}^{t_v^{-1}(u)} \int_{-\infty}^{t_v^{-1}(v)} f_{t_{1(v)}}(x,y) dx dy \tag{23}$$

where $t_v^{-1}(u)$ and $t_v^{-1}(v)$ are quantile functions with student$-$t distribution, where v is degree of freedom and $f_{t_{1(v)}}(x,y)$ is joint density function.

- Frank Copula

$$C(u,v) = -\frac{1}{\theta} \log[1 + (e^{-\theta u} - 1)\frac{(e^{-\theta v} - 1)}{(e^{-\theta} - 1)}], \theta \in R - \{0\} \tag{24}$$

- Clayton Copula

$$C(u,v) = (u^{-\theta} + v^{-\theta} - 1)^{-\frac{1}{\theta}}, \theta > 0 \tag{25}$$

- Gumbel Copula

$$C(u,v) = exp(-[(-\log u)^{\theta} + (-\log v)^{\theta}]^{\frac{1}{\theta}}), \theta \geqslant 1 \tag{26}$$

- Joe Copula

$$C(u,v) = 1 - (u^{\alpha} + v^{\alpha} - u^{\alpha}v^{\alpha})^{\frac{1}{\alpha}}, \theta \geq 1 \tag{27}$$

Furthermore,this study also uses the bivariate copula family, presented by Joe [9] for asymmetric lower and upper tail dependence including,
- BB1 coupla

$$C(u,v;\theta,\delta) = 1 + [(u^{-\theta} - 1)^{\delta}]^{-\frac{1}{\theta}}, \theta > 0, \delta \geqslant 0 \tag{28}$$

- BB2 coupla

$$C(u,v;\theta,\delta) = 1 + \delta^{-1} \log(e^{\delta(u^{-\theta}-1)} + e^{\delta(v^{-\theta}-1)} - 1)]^{-\frac{1}{\theta}}, \theta, \delta > 0 \tag{29}$$

- BB3 copula

$$C(u,v;\theta,\delta) = exp(-[\delta^{-1} \log(e^{\delta \tilde{u}^{-\theta}} + e^{\delta \tilde{v}^{-\theta}} - 1)]^{-\frac{1}{\theta}}), \theta \geqslant 0, \delta \geqslant 0 \tag{30}$$

- BB4 copula

$$C(u, v; \theta, \delta) = (u^{-\theta} + v^{-\theta} - 1 - [(u^{-\theta} - 1)^{-\delta} + v^{-\theta} - 1)^{-\delta}]^{-\frac{1}{\delta}})^{-\frac{1}{\theta}}, \theta \geq 1, \delta \geq 0 \quad (31)$$

- BB5 copula

$$C(u, v; \theta, \delta) = exp(-[x^{\theta} + y^{\theta} - (x^{-\theta\delta} + y^{-\theta\delta})^{-\frac{1}{\delta}}]^{\frac{1}{\theta}}), \theta \geqslant 1, \delta \geqslant 0 \quad (32)$$

- BB6 copula

$$C(u, v; \theta, \delta) = 1 - (1 - exp-[\log(1 - u^{-\theta}))^{\delta} + (-\log(1 - v^{-\theta}))^{\delta}]^{\frac{1}{\delta}})^{\frac{1}{\theta}}, \theta \geqslant 1, \delta \geqslant 0 \quad (33)$$

- BB7 copula

$$C(u, v; \theta, \delta) = 1 - (1 - [(1 - u^{-\theta})^{-\delta} + (1 - v^{-\theta})^{-\delta} - 1]^{-\frac{1}{\delta}})^{-\frac{1}{\theta}}, \theta \geqslant 1, \delta \geqslant 0 \quad (34)$$

- BB8 copula

$$C(u, v; \theta, \delta) = \delta^{-1}(1 - \{1 - \eta^{-1}[1 - (1 - \delta u)^{\vartheta}][1 - (1 - \delta v)^{\vartheta}]\}^{\frac{1}{\vartheta}} \quad (35)$$

$$\text{where } \vartheta \leqslant 1, 0 < \delta \leqslant 1, \eta = 1 - (1 - \delta)^{\vartheta}$$

3   Empirical Result

3.1   Data Description

The data set consists of the Comex and Nymex for the period from 8 May 2009 to 15 July 2016, covering 376 observations. Interval data is the most important issue in examining the interaction among these commodity prices. Therefore, we have considered the weekly minimum and maximum of these prices and they were collected from Thomson Reuters DataStream. The data description is shown in Table 1 and the interval return plot is show in Figs. 1 and 2.

3.2   Results of Optimal Weights for ARMA-GARCH, ARMA-EGARCH and ARMA-GJR GARCH Model Using Convex Combination Method

In this section, we use Comex and Nymex interval returns to estimate ARMA-GARCH(1,1), ARMA-EGARCH and ARMA-GJR-GARCH models with the convex combination method to find the appropriate weights in the model in the range of [0,1]. We conduct a grid search to find the best fit weight. Here, the AIC is used to determine the appropriate weight in the interval [0,1] and the results are shown in Table 2. Then, we compare three GARCH models

Table 1. Data description and statistics of Comex and Nymex using interval return data

Statistics	Comex		Nymex	
	Low	High	Low	High
Min	−0.1834	−0.0073	−0.276	0.0051
Max	0.013	0.1226	0.0012	0.2848
Mean	−0.0315	0.0337	−0.0673	0.0662
Variance	0.0006	0.0005	0.0024	0.002
Skewness	−1.9016	0.9806	−1.4245	1.5509
Kurtosis	10.1393	4.5066	5.4954	6.2506

source: calculation

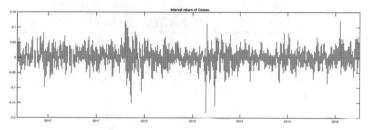

Fig. 1. Comex interval return

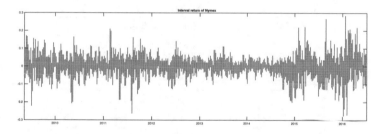

Fig. 2. Nymex interval return

using Akaike Information Criterion (AIC) and the lowest AIC is preferred. The results are also provided in Table 2 and we find that the GJR-GARCH model with Student−t distribution is appropriate for present volatility of Comex and EGARCH model with Student−t distribution is appropriate for present volatility of Nymex. Therefore, we use this GARCH specification to obtain our marginals.

Table 3 presents the results of Comex from the estimation by GJR-GARCH models. The results show that $\omega_1 + \omega_2 = 0.89$. This indicates that Comex exhibits a significantly high persistent volatility. Table 4 presents the results of Nymex from the estimation by EGARCH models. The results show that $\omega_1 + \omega_2 = 0.96$. This indicates that Nymex exhibits a significantly high persistent volatility. Moreover, we try to compare the results of the model with

Table 2. Results of AIC value in family of GARCH

Family of GARCH	ARMA(p,q)	Distribution	Comex		ARMA(p,q)	Nymex	
			AIC	α		AIC	α
GARCH(1,1)	p = 2, q = 0	Normal	−1975.88	0.41	p = 4, q = 3	−1534.17	0.4
	p = 4, q = 3	Student-t	−1981.73	0.51	p = 4, q = 3	−1532.77	0.43
EGARCH(1,1)	p = 4, q = 3	Normal	−1981.03	0.41	p = 2, q = 1	−1560.48	0.45
	p = 4, q = 3	Student-t	−1982.6	0.41	p = 4, q = 3	−1567.34*	0.40*
GJR GARCH(1,1)	p = 2, q = 0	Normal	−1973.88	0.41	p = 4, q = 3	−1555.03	0.42
	p = 4, q = 3	Student-t	−1983.73*	0.40*	p = 4, q = 3	−1555.57	0.4

source: calculation

Table 3. Estimated results of Comex by ARMA GJR-GARCH

ARIMA(4,0,3) with Student-t distribution		
Parameter	Value	Standard error
Constant	0.00761	0.0019
AR(1)	−0.06159	0.09549
AR(2)	−0.59763	0.03891
AR(3)	0.50141	0.06967
AR(4)	−0.21909	0.0583
MA(1)	0.44164	0.09855
MA(2)	0.68876	0.07449
MA(3)	−0.37918	0.09753
Dof	13.418	7.498
GJR GARCH(1,1), Student-t distribution		
ω_0	0.000024	0.000022
ω_1	0.85545	0.10459
ω_2	0.05522	0.05242
ω_3	0.00972	0.05459
Degree of freedom	13.418	7.498
AIC	−1983.73	
AIC (center method)	−1978.51	

source: calculation

convex combination and the center method, we find that the AIC of convex combination is lower than center method for both ARMA(3.4)-GJR-GARCH(1,1) and ARMA(3,4)-EGARCH(1,1). This result indicates the superiority of convex combination method over the center method.

3.3   In-Sample Forecast and Volatility

Then, the best fit GARCH model is used to predict the return of intervals and volatility of Comex and Nymex as shown in Figs. 3 and 4. These figures illustrate the accuracy of the predicted return against actual interval return (upper panel)

Table 4. Estimated results form EGARCH of Nymex

ARIMA(4,0,3) with Student-t distribution		
Parameter	Value	Standard error
Constant	0.0009	0.00001
AR(1)	-0.31467	0.04812
AR(2)	0.77496	0.03358
AR(3)	0.75047	0.02937
AR(4)	-0.324847	0.04747
MA(1)	0.64736	0.01685
MA(2)	-0.60512	0.01967
MA(3)	-0.95722	0.01944
Dof.	13.0926	0.00009
EGARCH(1,1), Student-t distribution		
ω_0	0.00246	0.00598
ω_1	0.99951	0.00002
ω_2	-0.03632	0.03329
ω_3	-0.11246	0.0202
Degree of freedom	13.0926	0.00009
AIC	-1567.34	
AIC (center method)	-1547.61	

source: calculation

and closing price returns (bottom panel) to see the performance of GJR-GARCH and EGARCH with convex combination. In addition, the predicted volatility σ_t^2 is also plotted in the middle panel. From the graph, it is obvious that the performance is satisfactory. Different results of predicted volatilities are shown in the middle of Figs. 3 and 4. We observed that the volatility of Comex is high during 2013–2014 corresponding to the Greek crisis. For Nymex, we observed that the volatility is high during 2015–2016. We expected that the increasing doubts about the success of the oil producers meeting and rising production as well as the record US and global crude oil inventories have put a high pressure on crude oil prices.

3.4   Results of Copulas

In this section, copula model is employed to measure the dependency of Comex and Nymex. The obtained standardized residuals from the best fit GARCH process are used to compute the dependence in the copula model. First of all, we present the scatter plot of copula between Comex and Nymex in Fig. 5. We observed an unclear relationship between Comex and Nymex, thus the various families of copulas are proposed to capture the relationship between these two variables.

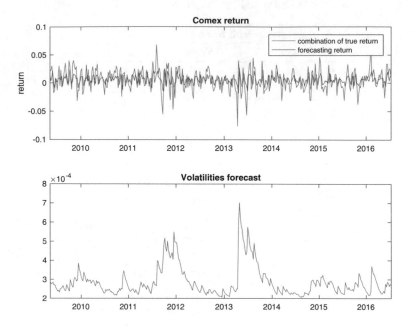

Fig. 3. Volatilities forecast and interval return of Comex

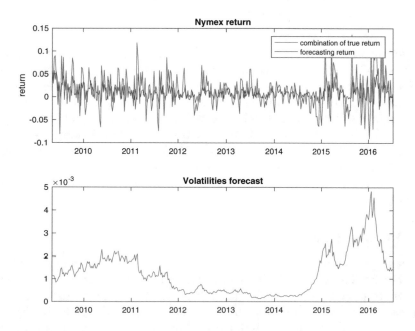

Fig. 4. Volatilities forecast and interval return of Nymex

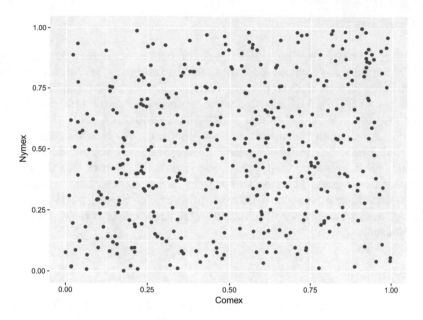

Fig. 5. Scatter plot between Comex and Nymex prices

Table 5. Estimation results of copulas

Copula	AIC	Copula	AIC
Gaussian	−17.2708	BB6	−15.0331
Student-t	−19.5115*	BB7	−16.7158
Frank	−13.4916	BB8	−15.3787
Clayton	−17.0466	Rotated Joe	−14.6511
Gumbel	−16.1062	Rotated BB1	−16.2804
Joe	−12.5512	Rotated BB7	−11.1828
BB1	−17.0757	Rotated BB8	−17.2221
ρ of student-t	0.2479	Lower-Upper tail dependence	"[0.0392 0.0392]"
Degree of freedom	8.4774	Kendall tau	0.1594

source: calculation

Finally, the results of copula model are presented in Table 5. We found that among 14 copula families, Student-t copula function shows the lowest value of AIC (−19.5115). Thus, we selected Student-t copula function to explain the dependence between Comex and Nymex. The result indicates that there exists a weak positive dependence between Comex and Nymex ($\rho = 0.2479$, degree of freedom $= 8.4774$). Moreover, we found the tail dependence between these two, where the upper and lower tail dependence was found to be 0.0392. We can conclude that there exists a dependence between Comex and Nymex not only in the normal event, but also in the extreme event.

4   Conclusion

This study investigates the performance of convex combination via various GARCH families and copula-based approach. In this study, we consider crude oil and gold with interval data as the application study. The results confirm that the EGARCH and GJR-GARCH with convex combination method improve the estimation. We also used the obtained standardized residuals from GARCH process with the copula model to measure the dependency of crude oil and gold. We found that among the various copula families, Student-t copula shows the lowest value of AIC, thus we used a Student-t copula function to join the marginal of crude oil and gold. The result of copula model showed that there exists a positive dependence between these two variables. Moreover, we also found the positive tail dependence which indicates that there exists a dependence in the extreme event.

References

1. Akaike, H.: Information theory and an extension of the maximum likelihood principle. In: Proceedings of 2nd International Symposium on Information Theory, Budapest, pp. 267–281 (1973)
2. Abbate, A., Marcellino, M.: Point, interval and density forecasts of exchange rates with time-varying parameter models. Discussion paper, Deutsche Bundesbank, No. 19/2016 (2016)
3. Billard, L., Diday, E.: Regression analysis for interval-valued data. In: Data Analysis, Classification and Related Methods, Proceedings of the 7th Conference for the IFCS, IFCS 2002, pp. 369–374. Springer, Berlin (2000)
4. Blanco Fernndez, A., Corral, N., Gonzlez Rodrguez, G.: Estimation of a flexible simple linear model for interval data based on set arithmetic. Comput. Stat. Data Anal. 55(9), 2568–2578 (2011). North-Holland
5. Bollerslev, T.: Generalized autoregressive conditional heteroskedasticity. J. Econ. 31, 307–327 (1986)
6. Hansen, B.E.: Reversion. Econometrics. www.ssc.wisc.edu/bhansen (2013)
7. Chanaim, S., Sriboonchitta, S., Rungruang, C.: A convex combination method for linear regression with interval data. In: Integrated Uncertainty in Knowledge Modelling and Decision Making, 5th International Symposium, IUKM 2016, Da Nang, Vietnam, 30 November–2 December 2016, Proceedings, pp. 469-480. Springer (2016)
8. Engle, R.F.: Dynamic conditional correlation - a simple class of multivariate GARCH models. J. Bus. Econ. Stat. 20(3), 339–350 (2002)
9. Glosten, L., Jagannathan, R., Runkle, D.: On the relation between the expected value and the volatility of the nominal excess return on stocks. J. Financ. 1993, 1779–1801 (1993)
10. Joe, H.: Multivariate Models and Multivariate Dependence Concepts. CRC Press, Boca Raton (1997)
11. Neto, L.E.A., de Carvalho, F.A.T.: Centre and range method for fitting a linear regression model to symbolic interval data. Comput. Stat. Data Anal. 52, 1500–1515 (2008)

12. Phochanachan, P., Pastpipatkul, P., Yamaka, W., Sriboonchitta, S.: Threshold regression for modeling symbolic interval data. Int. J. Appl. Bus. Econ. Res. **15**(7), 195–207 (2017)
13. Hentschel, L.: All in the family nesting symmetric and asymmetric GARCH models. J. Financ. Econ. **39**(1995), 71–104 (1994). Simon. Elsevier

Simultaneous Confidence Intervals for All Differences of Means of Normal Distributions with Unknown Coefficients of Variation

Warisa Thangjai[✉], Sa-Aat Niwitpong, and Suparat Niwitpong

Department of Applied Statistics, Faculty of Applied Science, King Mongkut's University of Technology North Bangkok, Bangkok 10800, Thailand
wthangjai@yahoo.com, {sa-aat.n,suparat.n}@sci.kmutnb.ac.th

Abstract. This paper presents a procedure for simultaneous confidence interval estimation for the differences of means of several normal populations with unknown coefficients of variation. The proposed approaches are a generalized confidence interval approach (GCI approach) and method of variance estimates recovery approach (MOVER approach). A Monte Carlo simulation was used to evaluate the performance in terms of coverage probability, average width and standard error. The simulation results indicated that the GCI and MOVER approaches are satisfactory in terms of the coverage probability, but the average widths of the MOVER approach are slightly shorter than the average widths of the GCI approach. The proposed approaches are illustrated by an example.

Keywords: Simultaneous confidence intervals · Mean · GCI approach MOVER approach · Coefficient of variation

1 Introduction

The problem of estimating the normal population mean when the coefficient of variation is known has been considered continuously in many fields such as medicine, biology, and chemistry. For example, Searls [20] provided the minimum mean squared error (MMSE) estimator to estimate mean of normal distribution when the coefficient of variation is known. Khan [4] discussed the mean estimation with known coefficient of variation in one sample case. Gleser and Healy [3] considered the normal mean estimation when the coefficient of variation is known using the minimum quadratic risk scale invariant estimator. Bhat and Rao [1] considered the problem of testing the normal mean with known coefficient of variation. Niwitpong et al. [12] and Niwitpong and Niwitpong [14] proposed confidence interval for the difference between normal means with known coefficients of variation. The confidence intervals for the normal mean with known coefficient of variation were considered by Niwitpong [10], Niwitpong and Niwitpong [13], and Niwitpong [11]. The confidence intervals for common mean of normal distributions with known coefficient of variation were proposed by Sodanin et al. [24].

© Springer International Publishing AG 2018
V. Kreinovich et al. (eds.), *Predictive Econometrics and Big Data*, Studies in Computational Intelligence 753, https://doi.org/10.1007/978-3-319-70942-0_48

In practice, the coefficient of variation is often unknown. This is because the population mean and population variance are unknown. Several researchers have considered the problem of estimating the normal population mean with unknown coefficient of variation. For example, Srivastava [22] provided uniformly minimum variance unbiased (UMVU) estimator for the normal mean estimation with unknown coefficient of variation using the minimum mean squared error estimator of Searls [20]. Srivastava and Singh [23] derived the UMVU estimator of the exact relative efficiency ratio of the estimator of Srivastava [22]. For more details about the mean of normal distribution with unknown coefficient of variation, see the research papers of Sahai [16], Sahai and Acharya [17], and Sodanin et al. [24].

The problem of multiple comparisons of parameters has received considerable attention in previous literature. Malley [9] constructed simultaneous confidence intervals (SCIs) for ratios of means of normal distributions. Kharrati-Kopaei et al. [5] proposed simultaneous fiducial generalized confidence intervals (SFG-CIs) for the successive differences of exponential location parameters under heteroscedasticity. Zhang [31] presented SFGCIs for several inverse Gaussian populations. Sadooghi-Alvandi and Malekzadeh [15] proposed a parametric bootstrap approach to construct SCIs for ratios of means of several lognormal distributions. Li et al. [8] provided a parametric bootstrap SCIs for differences of means from several two-parameter exponential distributions. Kharrati-Kopaei [6] described SCIs for the differences of location parameters of successive exponential distributions in the unbalanced case under heteroscedasticity.

To our knowledge, there is no previous work on constructing SCIs for differences of normal means with unknown coefficients of variation using generalized confidence interval approach (GCI approach) and method of variance estimates recovery approach (MOVER approach). In this paper, the GCI and MOVER approaches to construct the SCIs are presented. Confidence interval estimation using the GCI approach was introduced by Weerahandi [29] and has been considered by Krishnamoorthy and Lu [7], Tian [26], Tian and Wu [27], and Ye et al. [30]. The MOVER approach to construct confidence interval for sum of two populations was introduced by Zou and Donner [32] and Zou et al. [33]. Also, the MOVER approach for confidence interval estimation for difference of two populations was proposed by Donner and Zou [2]. Several researchers have used the concept of MOVER approach to obtain confidence interval; see e.g., Sangnawakij et al. [18] and Sangnawakij and Niwitpong [19]. In this paper, the concepts of the GCI and MOVER approaches are extended to k populations and use to construct SCIs of differences of normal means with unknown coefficient of variation.

The remainder of this paper is organized as follows. The GCI and MOVER approaches to construct SCIs for differences of means of several normal distributions with unknown coefficients of variation are presented in Sect. 2. Section 3, simulation results are presented to evaluate the coverage probabilities, average widths, and standard errors of the GCI approach in comparison to the MOVER approach. Section 4, the GCI and MOVER approaches are illustrated with an example. And finally, the conclusions are presented in Sect. 5.

2   Simultaneous Confidence Intervals

Assume that $X = (X_1, X_2, \ldots, X_n)$ are independent random samples each having the same normal distribution with mean μ and variance σ^2. Let $\tau = \sigma/\mu$ be the coefficient of variation (CV). Let $\bar{X}$ and S be sample mean and sample standard deviation for X, respectively.

Searls [20] proposed the following minimum mean squared error (MMSE) estimator for the normal population mean

$$\theta = \frac{\sum_{j=1}^{n} X_j}{n + \tau^2} = \frac{\bar{X}}{1 + (\sigma^2/n\mu^2)} = \frac{n\bar{X}}{n + (\sigma^2/\mu^2)}. \tag{1}$$

However, the CV is usually unknown in practice. Therefore, Srivastava [21] considered the following estimate for the mean with unknown CV

$$\hat{\theta} = \frac{\sum_{j=1}^{n} X_j}{n + \hat{\tau}^2} = \frac{\bar{X}}{1 + (S^2/n\bar{X}^2)} = \frac{n\bar{X}}{n + (S^2/\bar{X}^2)}. \tag{2}$$

Let $X_i = (X_{i1}, X_{i2}, \ldots, X_{in_i})$ be a random sample from the i-th normal population $N(\mu_i, \sigma_i^2)$, $i = 1, 2, \ldots, k$. For the i-th sample, let $\bar{X}_i$ and S_i be sample mean and sample standard deviation of X_i, respectively. And let $\bar{x}_i$ and s_i be the observed values of $\bar{X}_i$ and S_i, respectively. The estimator of Srivastava [21] has the following form

$$\hat{\theta}_i = \frac{n_i \bar{X}_i}{n_i + (S_i^2/\bar{X}_i^2)}. \tag{3}$$

In this paper we are interested in constructing SCIs for $\theta_{il} = \theta_i - \theta_l$, $i, l = 1, 2, \ldots, k$, $i \neq l$. The estimate for the differences of means of normal distributions with unknown coefficients of variation, θ_{il}, $i, l = 1, 2, \ldots, k$, $i \neq l$, is

$$\theta_{il} = \theta_i - \theta_l = \frac{n_i \bar{X}_i}{n_i + (\sigma_i^2/\mu_i^2)} - \frac{n_l \bar{X}_l}{n_l + (\sigma_l^2/\mu_l^2)}. \tag{4}$$

The estimator of $\theta_{il} = \theta_i - \theta_l$ is obtained by

$$\hat{\theta}_{il} = \hat{\theta}_i - \hat{\theta}_l = \frac{n_i \bar{X}_i}{n_i + (S_i^2/\bar{X}_i^2)} - \frac{n_l \bar{X}_l}{n_l + (S_l^2/\bar{X}_l^2)}. \tag{5}$$

Definition 1: *Let $X = (X_1, X_2, \ldots, X_n)$ be a random sample of size n from normal distribution with mean μ and variance σ^2. Let $\bar{X}$ and S be sample mean and sample standard deviation for X, respectively. Let $\theta = n\bar{X} \big/ \left(n + (\sigma^2/\mu^2)\right)$ be the mean with unknown coefficient of variation. Let $\hat{\theta} = n\bar{X} \big/ \left(n + (S^2/\bar{X}^2)\right)$ be the estimated mean with unknown coefficient of variation for θ. And let τ_1*

and τ_2 be the mean and variance of $\hat{\theta}$, respectively. According to Thangjai et al. [25], the distribution of

$$\sqrt{n}\left(\hat{\theta} - \tau_1\right) \xrightarrow{D} N\left(0, n\tau_2\right),\tag{6}$$

where $\tau_1 = A \cdot B$ and $\tau_2 = C \cdot D$, where

$$A = \left(\frac{\mu}{1 + \left(\frac{\sigma^2}{n\mu^2 + \sigma^2}\right)\left(1 + \frac{2\sigma^4 + 4n\mu^2\sigma^2}{(n\mu^2 + \sigma^2)^2}\right)}\right)$$

$$B = \left(1 + \frac{\left(\frac{n\sigma^2}{n\mu^2 + \sigma^2}\right)^2\left(\frac{2}{n} + \frac{2\sigma^4 + 4n\mu^2\sigma^2}{(n\mu^2 + \sigma^2)^2}\right)}{\left(n + \left(\frac{n\sigma^2}{n\mu^2 + \sigma^2}\right)\left(1 + \frac{2\sigma^4 + 4n\mu^2\sigma^2}{(n\mu^2 + \sigma^2)^2}\right)\right)^2}\right)$$

$$C = \left(\frac{\mu}{1 + \left(\frac{\sigma^2}{n\mu^2 + \sigma^2}\right)\left(1 + \frac{2\sigma^4 + 4n\mu^2\sigma^2}{(n\mu^2 + \sigma^2)^2}\right)}\right)^2$$

$$D = \left(\frac{\sigma^2}{n\mu^2} + \frac{\left(\frac{n\sigma^2}{n\mu^2 + \sigma^2}\right)^2\left(\frac{2}{n} + \frac{2\sigma^4 + 4n\mu^2\sigma^2}{(n\mu^2 + \sigma^2)^2}\right)}{\left(n + \left(\frac{n\sigma^2}{n\mu^2 + \sigma^2}\right)\left(1 + \frac{2\sigma^4 + 4n\mu^2\sigma^2}{(n\mu^2 + \sigma^2)^2}\right)\right)^2}\right).$$

2.1   Generalized Confidence Interval Approach

Definition 2: Let $X = (X_1, X_2, \ldots, X_n)$ be a random sample from a distribution $F(x|\delta)$ and let x be the observation of X. Let $\delta = (\theta, \nu)$ be a vector of unknown parameters where θ is a parameter of interest, ν is a vector of nuisance parameters. Let $R(X; x, \delta)$ be a function of X, x, and δ. The random quantity $R(X; x, \delta)$ is called be generalized pivotal quantity (GPQ) if the following two conditions are satisfied; see Weerahandi [29]:

 (i) The distribution of $R(X; x, \delta)$, $X = x$, is free of all unknown parameters.
 (ii) The observed value of $R(X; x, \delta)$, $X = x$, is the parameter of interest θ.

Let $R(\alpha/2)$ and $R(1 - \alpha/2)$ be the $100(\alpha/2)$-th and the $100(1 - \alpha/2)$-th percentiles of $R(X; x, \delta)$, respectively. Then $(R(\alpha/2), R(1 - \alpha/2))$ becomes a $100(1 - \alpha)\%$ two-sided generalized confidence interval for θ.

It is well known that $\bar{X}_i$ and S_i^2 are independently distributed with

$$\bar{X}_i \sim N\left(\mu_i, \frac{\sigma_i^2}{n_i}\right), \frac{(n_i - 1)S_i^2}{\sigma_i^2} \sim \chi_{n_i-1}^2; i = 1, 2, \ldots, k,\tag{7}$$

where $\chi_{n_i-1}^2$ denotes a chi-square distribution with $n_i - 1$ degrees of freedom.

According to Tian and Wu [27], the generalized pivotal quantities for estimating μ_i and σ_i^2 based on the i-th sample are given by

$$R_{\mu_i} = \bar{x}_i - \frac{Z_i}{\sqrt{U_i}}\sqrt{\frac{(n_i - 1)s_i^2}{n_i}}\tag{8}$$

and

$$R_{\sigma_i^2} = \frac{(n_i - 1)\, s_i^2}{V_i}, \tag{9}$$

where Z_i denotes standard normal distribution, and U_i and V_i denote chi-square distribution with $n_i - 1$ degrees of freedom.

Therefore, the generalized pivotal quantity for θ_i based on the i-th sample is obtained by

$$R_{\theta_i} = \frac{n_i R_{\mu_i}}{n_i + \left(R_{\sigma_i^2}/R_{\mu_i}^2\right)}, \tag{10}$$

where R_{μ_i} is defined in Eq. (8) and $R_{\sigma_i^2}$ is defined in Eq. (9).

Thus the generalized pivotal quantity for $\theta_{il} = \theta_i - \theta_l$, $i, l = 1, 2, \ldots, k$, $i \neq l$ is defined by

$$R_{\theta_{il}} = R_{\theta_i} - R_{\theta_l} = \frac{n_i R_{\mu_i}}{n_i + \left(R_{\sigma_i^2}/R_{\mu_i}^2\right)} - \frac{n_l R_{\mu_l}}{n_l + \left(R_{\sigma_i^2}/R_{\mu_l}^2\right)}. \tag{11}$$

Therefore, the $100\,(1 - \alpha)\,\%$ two-sided simultaneous confidence interval for all differences of normal means with unknown coefficients of variation θ_{il} based on GCI approach is defined by

$$SCI_{il(GCI)} = \left(L_{il(GCI)}, U_{il(GCI)}\right) = \left(R_{\theta_{il}}\left(\alpha/2\right), R_{\theta_{il}}\left(1 - \alpha/2\right)\right), \tag{12}$$

where $R_{\theta_{il}}\left(\alpha/2\right)$ and $R_{\theta_{il}}\left(1 - \alpha/2\right)$ denote the $100\,(\alpha/2)$-th and $100\,(1 - \alpha/2)$-th percentiles of $R_{\theta_{il}}$, respectively.

The following algorithm is used to construct the SCI based on GCI approach:

Algorithm 1.
Step 1 Calculate the values of $\bar{x}_i$ and s_i^2, the observed values of $\bar{X}_i$ and S_i^2, $i = 1, 2, \ldots, k$.
Step 2 Generate the values of Z_i from standard normal distribution, $i = 1, 2, \ldots, k$.
Step 3 Generate the values of U_i from chi-square distribution with $n_i - 1$ degrees of freedom, $i = 1, 2, \ldots, k$.
Step 4 Compute the values of R_{μ_i} from equation (8), $i = 1, 2, \ldots, k$.
Step 5 Generate the values of V_i from chi-square distribution with $n_i - 1$ degrees of freedom, $i = 1, 2, \ldots, k$.
Step 6 Compute the values of $R_{\sigma_i^2}$ from equation (9), $i = 1, 2, \ldots, k$.
Step 7 Compute the values of R_{θ_i} from equation (10), $i = 1, 2, \ldots, k$.
Step 8 Compute the values of $R_{\theta_{il}}$ from equation (11), $i, l = 1, 2, \ldots, k$, $i \neq l$.
Step 9 Repeat Step 2 - Step 8 a large number of times, $m = 2500$, and from these m values, obtained the $R_{\theta_{il}}$.
Step 10 Compute the $100\,(\alpha/2)$-th percentile of $R_{\theta_{il}}$ defined by $R_{\theta_{il}}\left(\alpha/2\right)$ and compute the $100\,(1 - \alpha/2)$-th percentile of $R_{\theta_{il}}$ defined by $R_{\theta_{il}}\left(1 - \alpha/2\right)$.

2.2 Method of Variance Estimates Recovery Approach

First, the confidence interval for mean of normal distribution with unknown coefficient of variation is considered. Recall that the standard normal distribution with the mean 0 and variance 1 is defined by

$$Z = \frac{\bar{X} - \mu}{\sqrt{Var(\hat{\theta})}} \sim N(0,1). \tag{13}$$

The student's t-distribution with $n-1$ degrees of freedom can be defined as the distribution of the random variable T with

$$T = \frac{Z}{\sqrt{\frac{\chi^2_{n-1}}{n-1}}} = \frac{\bar{X} - \mu}{\sqrt{Var(\hat{\theta})}} \Bigg/ \sqrt{\frac{(n-1)S^2}{(n-1)\sigma^2}}$$

$$= \frac{\bar{X} - \mu}{\sqrt{Var(\hat{\theta})}} \left(\frac{\sigma}{s}\right) \approx \frac{\bar{X} - \mu}{\sqrt{Var(\hat{\theta})}} \left(\frac{\hat{\sigma}}{s}\right), \tag{14}$$

where $\frac{(n-1)S^2}{\sigma^2} \sim \chi^2_{n-1}$ and $\hat{\sigma} = S$.

Therefore, the $100(1-\alpha)\%$ two-sided confidence interval for the mean of normal distribution with unknown coefficient of variation θ_i, $i = 1, 2, \ldots, k$, is obtained by

$$l_i = \bar{x}_i - t_{1-\alpha/2} \frac{s_i}{\hat{\sigma}_i} \sqrt{Var(\hat{\theta}_i)} \tag{15}$$

and

$$u_i = \bar{x}_i + t_{1-\alpha/2} \frac{s_i}{\hat{\sigma}_i} \sqrt{Var(\hat{\theta}_i)}, \tag{16}$$

where $Var(\hat{\theta}_i) = E \cdot F$, where (from Eq. (6))

$$E = \left(\frac{\mu_i}{1 + \left(\frac{\sigma_i^2}{n_i\mu_i^2 + \sigma_i^2}\right)\left(1 + \frac{2\sigma_i^4 + 4n_i\mu_i^2\sigma_i^2}{\left(n_i\mu_i^2 + \sigma_i^2\right)^2}\right)} \right)^2$$

$$F = \left(\frac{\sigma_i^2}{n_i\mu_i^2} + \frac{\left(\frac{n_i\sigma_i^2}{n_i\mu_i^2 + \sigma_i^2}\right)^2\left(\frac{2}{n_i} + \frac{2\sigma_i^4 + 4n_i\mu_i^2\sigma_i^2}{\left(n_i\mu_i^2 + \sigma_i^2\right)^2}\right)}{\left(n_i + \left(\frac{n_i\sigma_i^2}{n_i\mu_i^2 + \sigma_i^2}\right)\left(1 + \frac{2\sigma_i^4 + 4n_i\mu_i^2\sigma_i^2}{\left(n_i\mu_i^2 + \sigma_i^2\right)^2}\right)\right)^2} \right).$$

In the case of a difference of two parameters, Donner and Zou [2] proposed the method of variance estimates recovery approach (MOVER approach) to construct a $100(1-\alpha)\%$ two-sided confidence interval (L, U) for $\theta_1 - \theta_2$ where

θ_1 and θ_2 denote two parameters of interest, L and U denote the lower limit and upper limit of the confidence interval, respectively. The (l_i, u_i) contains the parameter values for θ_i where $i = 1, 2$. The lower limit L and upper limit U are defined by

$$L = \hat{\theta}_1 - \hat{\theta}_2 - \sqrt{\left(\hat{\theta}_1 - l_1\right)^2 + \left(u_2 - \hat{\theta}_2\right)^2} \tag{17}$$

and

$$U = \hat{\theta}_1 - \hat{\theta}_2 + \sqrt{\left(u_1 - \hat{\theta}_1\right)^2 + \left(\hat{\theta}_2 - l_2\right)^2}. \tag{18}$$

It is reasonable to extend the concept of Donner and Zou [2] to construct a $100\,(1 - \alpha)\,\%$ two-sided confidence interval (L, U) for $\theta_{il} = \theta_i - \theta_l$, $i, l = 1, 2, \ldots, k$, $i \neq l$. The lower limit L and upper limit U are obtained by

$$L = \hat{\theta}_i - \hat{\theta}_l - \sqrt{\left(\hat{\theta}_i - l_i\right)^2 + \left(u_l - \hat{\theta}_l\right)^2} \tag{19}$$

and

$$U = \hat{\theta}_i - \hat{\theta}_l + \sqrt{\left(u_i - \hat{\theta}_i\right)^2 + \left(\hat{\theta}_l - l_l\right)^2}, \tag{20}$$

where $\hat{\theta}_i$ and $\hat{\theta}_l$ are defined in Eq. (3), l_i and l_l are defined in Eq. (15), and u_i and u_l are defined in Eq. (16).

Therefore, the $100\,(1 - \alpha)\,\%$ two-sided simultaneous confidence interval for all differences of normal means with unknown coefficients of variation θ_{il} based on MOVER approach is defined by

$$SCI_{il(MOVER)} = \left(L_{il(MOVER)}, U_{il(MOVER)}\right), \tag{21}$$

where

$$L_{il(MOVER)} = \hat{\theta}_i - \hat{\theta}_l - \sqrt{\left(\hat{\theta}_i - l_i\right)^2 + \left(u_l - \hat{\theta}_l\right)^2}$$

$$U_{il(MOVER)} = \hat{\theta}_i - \hat{\theta}_l + \sqrt{\left(u_i - \hat{\theta}_i\right)^2 + \left(\hat{\theta}_l - l_l\right)^2}.$$

3   Simulation Studies

In this section, results of the simulation study in which the SCIs based on GCI approach (SCI_{GCI}) are compared with the SCIs based on MOVER approach (SCI_{MOVER}) are presented. The performance of these two approaches was evaluated through the coverage probabilities, average widths, and standard errors of the confidence intervals.

The following algorithm was used to obtain the coverage probabilities of two simultaneous confidence intervals:

Algorithm 2.

Step 1 Generate X_i, a random sample of sample size n_i from normal population with parameters μ_i and σ_i^2, for $i = 1, 2, \ldots, k$. Calculate $\bar{x}_i$ and s_i (the observed values of $\bar{X}_i$ and S_i).

Step 2 Construct the two-sided SCIs based on the GCI approach $(SCI_{il(GCI)})$ from Algorithm 1 and record whether or not all the values of $\theta_{il} = \theta_i - \theta_l$, $i, l = 1, 2, \ldots, k$, $i \neq l$, are in their corresponding SCI_{GCI}.

Step 3 Construct the two-sided SCIs based on the MOVER approach $(SCI_{il(MOVER)})$ from equation (21) and record whether or not all the values of $\theta_{il} = \theta_i - \theta_l$, $i, l = 1, 2, \ldots, k$, $i \neq l$, are in their corresponding SCI_{MOVER}.

Step 4 Repeat Step 1 - Step 3 a large number of times, $M = 5000$. Then, the fraction of times that all $\theta_{il} = \theta_i - \theta_l$, $i, l = 1, 2, \ldots, k$, $i \neq l$, are in their corresponding SCIs provides an estimate of the coverage probability.

Table 1. The coverage probabilities of 95% of two-sided simultaneous confidence intervals for all differences of means of normal distributions with unknown coefficients of variation: 3 sample cases.

(n_1, n_2, n_3)	$(\sigma_1, \sigma_2, \sigma_3)$	SCI_{GCI}	SCI_{MOVER}
$(10, 10, 10)$	$(0.1, 0.1, 0.1)$	0.9606	0.9625
	$(0.1, 0.2, 0.3)$	0.9594	0.9599
$(20, 20, 20)$	$(0.1, 0.1, 0.1)$	0.9563	0.9569
	$(0.1, 0.2, 0.3)$	0.9546	0.9555
$(10, 20, 30)$	$(0.1, 0.1, 0.1)$	0.9547	0.9557
	$(0.1, 0.2, 0.3)$	0.9548	0.9551
$(30, 30, 30)$	$(0.1, 0.1, 0.1)$	0.9530	0.9547
	$(0.1, 0.2, 0.3)$	0.9523	0.9522
$(50, 50, 50)$	$(0.1, 0.1, 0.1)$	0.9505	0.9521
	$(0.1, 0.2, 0.3)$	0.9488	0.9496
$(30, 50, 100)$	$(0.1, 0.1, 0.1)$	0.9486	0.9483
	$(0.1, 0.2, 0.3)$	0.9530	0.9543
$(100, 100, 100)$	$(0.1, 0.1, 0.1)$	0.9506	0.9512
	$(0.1, 0.2, 0.3)$	0.9491	0.9493
$(200, 200, 200)$	$(0.1, 0.1, 0.1)$	0.9483	0.9485
	$(0.1, 0.2, 0.3)$	0.9499	0.9508
$(500, 500, 500)$	$(0.1, 0.1, 0.1)$	0.9500	0.9503
	$(0.1, 0.2, 0.3)$	0.9496	0.9494
$(1000, 1000, 1000)$	$(0.1, 0.1, 0.1)$	0.9473	0.9490
	$(0.1, 0.2, 0.3)$	0.9475	0.9485

Table 2. The average widths (standard errors) of 95% of two-sided simultaneous confidence intervals for all differences of means of normal distributions with unknown coefficients of variation: 3 sample cases.

(n_1, n_2, n_3)	$(\sigma_1, \sigma_2, \sigma_3)$	SCI_{GCI}	SCI_{MOVER}
$(10, 10, 10)$	$(0.1, 0.1, 0.1)$	0.1996 (0.0122)	0.2002 (0.0120)
	$(0.1, 0.2, 0.3)$	0.4254 (0.0626)	0.4207 (0.0608)
$(20, 20, 20)$	$(0.1, 0.1, 0.1)$	0.1309 (0.0056)	0.1313 (0.0055)
	$(0.1, 0.2, 0.3)$	0.2780 (0.0387)	0.2769 (0.0382)
$(10, 20, 30)$	$(0.1, 0.1, 0.1)$	0.1487 (0.0161)	0.1490 (0.0161)
	$(0.1, 0.2, 0.3)$	0.2626 (0.0196)	0.2623 (0.0192)
$(30, 30, 30)$	$(0.1, 0.1, 0.1)$	0.1049 (0.0037)	0.1051 (0.0035)
	$(0.1, 0.2, 0.3)$	0.2235 (0.0310)	0.2229 (0.0307)
$(50, 50, 50)$	$(0.1, 0.1, 0.1)$	0.0799 (0.0022)	0.0801 (0.0021)
	$(0.1, 0.2, 0.3)$	0.1701 (0.0233)	0.1698 (0.0231)
$(30, 50, 100)$	$(0.1, 0.1, 0.1)$	0.0820 (0.0073)	0.0820 (0.0073)
	$(0.1, 0.2, 0.3)$	0.1468 (0.0097)	0.1467 (0.0095)
$(100, 100, 100)$	$(0.1, 0.1, 0.1)$	0.0560 (0.0011)	0.0560 (0.0010)
	$(0.1, 0.2, 0.3)$	0.1190 (0.0161)	0.1188 (0.0161)
$(200, 200, 200)$	$(0.1, 0.1, 0.1)$	0.0394 (0.0006)	0.0394 (0.0005)
	$(0.1, 0.2, 0.3)$	0.0836 (0.0113)	0.0836 (0.0112)
$(500, 500, 500)$	$(0.1, 0.1, 0.1)$	0.0249 (0.0003)	0.0248 (0.0002)
	$(0.1, 0.2, 0.3)$	0.0527 (0.0071)	0.0527 (0.0071)
$(1000, 1000, 1000)$	$(0.1, 0.1, 0.1)$	0.0176 (0.0002)	0.0175 (0.0001)
	$(0.1, 0.2, 0.3)$	0.0373 (0.0050)	0.0372 (0.0050)

In the simulation, four configuration factors are considered to evaluate the performance of the two approaches: (1) sample cases: $k = 3$ and $k = 5$; (2) population means: $\mu_1 = \mu_2 = \ldots = \mu_k = 1$; (3) population standard deviations: $\sigma_1, \sigma_2, \ldots, \sigma_k$; (4) sample sizes: $n_1, n_2, \ldots, n_k$. See the following tables for specific combinations of configurations. The nominal confidence level was chosen to be 0.95. The simulation results from $k = 3$ and $k = 5$ are presented in the following four tables.

For $k = 3$, it is seen from Tables 1 and 2 that both the SCI_{GCI} and SCI_{MOVER} perform satisfactorily in terms of the coverage probability, average width, and standard error of the confidence interval. In almost all cases, the coverage probabilities of the SCI_{GCI} are similar to those of the SCI_{MOVER}. However, the average widths of the SCI_{MOVER} are slightly shorter than those of the SCI_{GCI}. As seen from Tables 3 and 4 for $k = 5$, the performances of SCI_{GCI} and SCI_{MOVER} are similar to those of $k = 3$. Therefore, the MOVER approach performs better than the GCI approach in terms of average width. However, the GCI approach can be considered as an alternative to construct SCIs for all pairwise differences of means from several normal distributions with unknown coefficients of variation.

Table 3. The coverage probabilities of 95% of two-sided simultaneous confidence intervals for all differences of means of normal distributions with unknown coefficients of variation: 5 sample cases.

$(n_1, n_2, n_3, n_4, n_5)$	$(\sigma_1, \sigma_2, \sigma_3, \sigma_4, \sigma_5)$	SCI_{GCI}	SCI_{MOVER}
$(10, 10, 10, 10, 10)$	$(0.1, 0.1, 0.1, 0.1, 0.1)$	0.9631	0.9645
	$(0.1, 0.1, 0.2, 0.3, 0.3)$	0.9583	0.9584
$(20, 20, 20, 20, 20)$	$(0.1, 0.1, 0.1, 0.1, 0.1)$	0.9551	0.9569
	$(0.1, 0.1, 0.2, 0.3, 0.3)$	0.9551	0.9553
$(10, 10, 20, 30, 30)$	$(0.1, 0.1, 0.1, 0.1, 0.1)$	0.9569	0.9580
	$(0.1, 0.1, 0.2, 0.3, 0.3)$	0.9578	0.9587
$(30, 30, 30, 30, 30)$	$(0.1, 0.1, 0.1, 0.1, 0.1)$	0.9532	0.9543
	$(0.1, 0.1, 0.2, 0.3, 0.3)$	0.9536	0.9533
$(50, 50, 50, 50, 50)$	$(0.1, 0.1, 0.1, 0.1, 0.1)$	0.9511	0.9521
	$(0.1, 0.1, 0.2, 0.3, 0.3)$	0.9526	0.9526
$(30, 30, 50, 100, 100)$	$(0.1, 0.1, 0.1, 0.1, 0.1)$	0.9523	0.9529
	$(0.1, 0.1, 0.2, 0.3, 0.3)$	0.9526	0.9531
$(100, 100, 100, 100, 100)$	$(0.1, 0.1, 0.1, 0.1, 0.1)$	0.9539	0.9543
	$(0.1, 0.1, 0.2, 0.3, 0.3)$	0.9496	0.9501
$(200, 200, 200, 200, 200)$	$(0.1, 0.1, 0.1, 0.1, 0.1)$	0.9509	0.9514
	$(0.1, 0.1, 0.2, 0.3, 0.3)$	0.9509	0.9514
$(500, 500, 500, 500, 500)$	$(0.1, 0.1, 0.1, 0.1, 0.1)$	0.9490	0.9490
	$(0.1, 0.1, 0.2, 0.3, 0.3)$	0.9514	0.9516
$(1000, 1000, 1000, 1000, 1000)$	$(0.1, 0.1, 0.1, 0.1, 0.1)$	0.9502	0.9503
	$(0.1, 0.1, 0.2, 0.3, 0.3)$	0.9515	0.9518

Table 4. The average widths (standard errors) of 95% of two-sided simultaneous confidence intervals for all differences of means of normal distributions with unknown coefficients of variation: 5 sample cases.

$(n_1, n_2, n_3, n_4, n_5)$	$(\sigma_1, \sigma_2, \sigma_3, \sigma_4, \sigma_5)$	SCI_{GCI}	SCI_{MOVER}
$(10, 10, 10, 10, 10)$	$(0.1, 0.1, 0.1, 0.1, 0.1)$	0.1993 (0.0082)	0.1998 (0.0081)
	$(0.1, 0.1, 0.2, 0.3, 0.3)$	0.4248 (0.0422)	0.4197 (0.0409)
$(20, 20, 20, 20, 20)$	$(0.1, 0.1, 0.1, 0.1, 0.1)$	0.1310 (0.0038)	0.1313 (0.0037)
	$(0.1, 0.1, 0.2, 0.3, 0.3)$	0.2791 (0.0258)	0.2778 (0.0254)
$(10, 10, 20, 30, 30)$	$(0.1, 0.1, 0.1, 0.1, 0.1)$	0.1515 (0.0107)	0.1518 (0.0107)
	$(0.1, 0.1, 0.2, 0.3, 0.3)$	0.2624 (0.0129)	0.2621 (0.0127)
$(30, 30, 30, 30, 30)$	$(0.1, 0.1, 0.1, 0.1, 0.1)$	0.1049 (0.0025)	0.1051 (0.0024)
	$(0.1, 0.1, 0.2, 0.3, 0.3)$	0.2229 (0.0202)	0.2222 (0.0200)
$(50, 50, 50, 50, 50)$	$(0.1, 0.1, 0.1, 0.1, 0.1)$	0.0801 (0.0015)	0.0802 (0.0014)
	$(0.1, 0.1, 0.2, 0.3, 0.3)$	0.1699 (0.0151)	0.1696 (0.0150)
$(30, 30, 50, 100, 100)$	$(0.1, 0.1, 0.1, 0.1, 0.1)$	0.0822 (0.0048)	0.0823 (0.0048)
	$(0.1, 0.1, 0.2, 0.3, 0.3)$	0.1433 (0.0064)	0.1433 (0.0063)
$(100, 100, 100, 100, 100)$	$(0.1, 0.1, 0.1, 0.1, 0.1)$	0.0560 (0.0008)	0.0560 (0.0007)
	$(0.1, 0.1, 0.2, 0.3, 0.3)$	0.1189 (0.0104)	0.1188 (0.0104)
$(200, 200, 200, 200, 200)$	$(0.1, 0.1, 0.1, 0.1, 0.1)$	0.0394 (0.0004)	0.0394 (0.0003)
	$(0.1, 0.1, 0.2, 0.3, 0.3)$	0.0836 (0.0073)	0.0836 (0.0073)
$(500, 500, 500, 500, 500)$	$(0.1, 0.1, 0.1, 0.1, 0.1)$	0.0248 (0.0002)	0.0248 (0.0001)
	$(0.1, 0.1, 0.2, 0.3, 0.3)$	0.0527 (0.0046)	0.0527 (0.0046)
$(1000, 1000, 1000, 1000, 1000)$	$(0.1, 0.1, 0.1, 0.1, 0.1)$	0.0176 (0.0001)	0.0175 (0.0001)
	$(0.1, 0.1, 0.2, 0.3, 0.3)$	0.0372 (0.0032)	0.0372 (0.0032)

4   An Empirical Application

In this section, an example is given to illustrate the usage of the GCI approach and MOVER approach in practice. This data set was provided by Walpole et al. [28]. Testing patient blood samples for HIV antibodies, a spectrophotometer determines the optical density of each sample. Ten different runs at four randomly selected laboratories were measured. The summary statistics are as follows: $\bar{x}_1 = 1.0037$, $\bar{x}_2 = 1.0579$, $\bar{x}_3 = 1.0997$, $\bar{x}_4 = 1.0224$, $s_1^2 = 0.0056$, $s_2^2 = 0.0227$, $s_3^2 = 0.0301$, $s_4^2 = 0.0238$, $n_1 = 10$, $n_2 = 10$, $n_3 = 10$, and $n_4 = 10$. For illustration, the 95% simultaneous confidence intervals for the six differences of normal means with unknown coefficients of variation, $\theta_{il} = \theta_i - \theta_l$, $i, l = 1, 2, 3, 4$, $i \neq l$, obtained by using the two approaches, are given in Table 5. The results show that the MOVER approach performs much better than the GCI approach in the sense that the widths of the MOVER approach are shorter than the widths of the GCI approach in almost all cases. Therefore, this result confirms our simulation study in the previous section for the same values of $n_i = 10$ and the different values of σ_i, $i = 1, 2, \ldots, k$.

Table 5. The 95% simultaneous confidence intervals for all pairwise differences of means of normal distributions with unknown coefficients of variation.

Parameters	CI_{GCI}		CI_{MOVER}	
	Lower	Upper	Lower	Upper
$\theta_2 - \theta_1$	−0.0705	0.1740	−0.0659	0.1745
$\theta_3 - \theta_1$	−0.0378	0.2314	−0.0390	0.2312
$\theta_4 - \theta_1$	−0.1124	0.1355	−0.1038	0.1413
$\theta_3 - \theta_2$	−0.1198	0.2143	−0.1224	0.2061
$\theta_4 - \theta_2$	−0.2011	0.1171	−0.1896	0.1186
$\theta_4 - \theta_3$	−0.2488	0.0851	−0.2434	0.0886

5   Discussion and Conclusions

In this paper, GCI and MOVER approaches were proposed to construct SCIs for the differences of normal means with unknown coefficients of variation. The concepts of the GCI and MOVER approaches are extended to k populations and used to construct the SCIs. The performances of these approaches were investigated using Monte Carlo simulations. Coverage probabilities of the GCI approach are similar to those of the MOVER approach in almost all cases. However, the average widths of the MOVER approach are slightly shorter than those of the GCI approach. Moreover, the MOVER approach is recommended for use since it is conceptually simpler than the GCI approach in practice. The coverage probability of the MOVER approach is similar to the result in research papers by Donner and Zou [2].

Acknowledgements. The first author gratefully acknowledges the financial support from Science Achievement Scholarship of Thailand.

References

1. Bhat, K.K., Rao, K.A.: On tests for a normal mean with known coefficient of variation. Inter. Stat. Rev. **75**, 170–182 (2007)
2. Donner, A., Zou, G.Y.: Closed-form confidence intervals for function of the normal standard deviation. Stat. Meth. Med. Res. **21**, 347–359 (2010)
3. Gleser, L.J., Healy, J.D.: Estimating the mean of a normal distribution with known coefficient of variation. J. Am. Stat. Assoc. **71**, 977–981 (1976)
4. Khan, R.A.: A note on estimating the mean of a normal distribution with known coefficient of variation. J. Am. Stat. Assoc. **63**, 1039–1041 (1968)
5. Kharrati-Kopaei, M., Malekzadeh, A., Sadooghi-Alvandi, M.: Simultaneous fiducial generalized confidence intervals for the successive differences of exponential location parameters under heteroscedasticity. Stat. Prob. Lett. **83**, 1547–1552 (2013)
6. Kharrati-Kopaei, M.: A note on the simultaneous confidence intervals for the successive differences of exponential location parameters under heteroscedasticity. Stat. Meth. **22**, 1–7 (2015)
7. Krishnamoorthy, K., Lu, Y.: Inference on the common means of several normal populations based on the generalized variable method. Biometrics **59**, 237–247 (2003)
8. Li, J., Song, W., Shi, J.: Parametric bootstrap simultaneous confidence intervals for differences of means from several two-parameter exponential distributions. Stat. Prob. Lett. **106**, 39–45 (2015)
9. Malley, J.D.: Simultaneous confidence intervals for ratios of normal means. J. Am. Stat. Assoc. **77**, 170–176 (1982)
10. Niwitpong, S.: Confidence intervals for the normal mean with known coefficient of variation. World Acad. Sci. Eng. Technol. **6**, 1365–1368 (2012)
11. Niwitpong, S.: Confidence intervals for the normal mean with a known coefficient of variation. Far East J. Math. Sci. **97**, 711–727 (2015)
12. Niwitpong, S., Koonprasert, S., Niwitpong, S.: Confidence interval for the difference between normal population means with known coefficients of variation. Appl. Math. Sci. **6**, 47–54 (2012)
13. Niwitpong, S., Niwitpong, S.: On simple confidence intervals for the normal mean with a known coefficient of variation. World Acad. Sci. Eng. Technol. **7**, 1444–1447 (2013)
14. Niwitpong, S., Niwitpong, S.: Confidence intervals for the difference between normal means with known coefficients of variation. Ann. Oper. Res. **247**, 1–15 (2016)
15. Sadooghi-Alvandi, S.M., Malekzadeh, A.: Simultaneous confidence intervals for ratios of means of several lognormal distributions: a parametric bootstrap approach. Comput. Stat. Data Anal. **69**, 133–140 (2014)
16. Sahai, A.: On an estimator of normal population mean and UMVU estimation of its relative efficiency. Appl. Math. Comput. **152**, 701–708 (2004)
17. Sahai, A., Acharya, R.M.: Iterative estimation of normal population mean using computational-statistical intelligence. Comput. Sci. Technol. **4**, 500–508 (2016)
18. Sangnawakij, P., Niwitpong, S., Niwitpong, S.: Confidence intervals for the ratio of coefficients of variation of the gamma distributions. Lecture Notes in Computer Science, vol. 9376, pp. 193–203 (2015)

19. Sangnawakij, P., Niwitpong, S.: Confidence intervals for coefficients of variation in two-parameter exponential distribution. Commun. Stat. Simul. Comput. **46**, 6618–6630 (2017)
20. Searls, D.T.: The utilization of a known coefficient of variation in the estimation procedure. J. Am. Stat. Assoc. **59**, 1225–1226 (1964)
21. Srivastava, V.K.: On the use of coefficient of variation in estimating mean. J. Indian Soc. Agric. Stat. **26**, 33–36 (1974)
22. Srivastava, V.K.: A note on the estimation of mean in normal population. Metrika **27**, 99–102 (1980)
23. Srivastava, V.K., Singh, R.S.: Uniformly minimum variance unbiased estimator of efficiency ratio in estimation of normal population mean. Stat. Prob. Lett. **10**, 241–245 (1990)
24. Sodanin, S., Niwitpong, S., Niwitpong, S.: Confidence intervals for common mean of normal distributions with known coefficient of variation. Lecture Notes in Computer Science, vol. 9978, pp. 574–585 (2016)
25. Thangjai, W., Niwitpong, S., Niwitpong, S.: Confidence intervals for the common mean of several normal populations with unknown coefficients of variation. Commun. Stat. Simul. Comput. (2017, Submitted)
26. Tian, L.: Inferences on the common coefficient of variation. Stat. Med. **24**, 2213–2220 (2005)
27. Tian, L., Wu, J.: Inferences on the common mean of several log-normal populations: the generalized variable approach. Biometrical J. **49**, 944–951 (2007)
28. Walpole, R.E., Myers, R.H., Myers, S.L., Ye, K.: Probability and Statistics for Engineers and Scientists. Prentice Hall, Upper Saddle River (2012)
29. Weerahandi, S.: Generalized confidence intervals. J. Am. Stat. Assoc. **88**, 899–905 (1993)
30. Ye, R.D., Ma, T.F., Wang, S.G.: Inferences on the common mean of several inverse Gaussian populations. Comput. Stat. Data Anal. **54**, 906–915 (2010)
31. Zhang, G.: Simultaneous confidence intervals for several inverse Gaussian populations. Stat. Prob. Lett. **92**, 125–131 (2014)
32. Zou, G.Y., Donner, A.: Construction of confidence limits about effect measures: a general approach. Stat. Med. **27**, 1693–1702 (2008)
33. Zou, G.Y., Taleban, J., Hao, C.Y.: Confidence interval estimation for lognormal data with application to health economics. Comput. Stat. Data Anal. **53**, 3755–3764 (2009)

Estimating the Value of Cultural Heritage Creativity from the Viewpoint of Tourists

Phanee Thipwong[1], Chung-Te Ting[2], Yu-Sheng Huang[2],
Yun-Zu Chen[3], and Wan-Tran Huang[1(✉)]

[1] Department of Business Administration, Asia University, Taichung, Taiwan
Panee_tipwong@yahoo.com, wthuangwantran7589@gmail.com
[2] Department of Tourism, Food and Beverage Management, Chang Jung
Christian University, Tainan, Taiwan
{ctting,yshuang}@mail.cjcu.edu.tw
[3] Department of Business Administration, Chang Jung Christian University,
Tainan, Taiwan
winniechen0224@gmail.com

Abstract. Creativity in cultural heritage raises expectations of added value, and helps promote local economic development through new elements introduced to the original cultural industry. The purpose of this study is to discuss the willingness to pay (WTP) for commercialization and creativity of cultural heritage. However, no general market price for cultural heritage exists. Thus, the Contingent Valuation Method (CVM) in the non-market valuation method was applied to determine whether commercialization and creativity of cultural heritage through a personally inherent attitude and through different preferences could analyze the tourists' willingness to pay, and to consider the factors that influence their WTP. In this study, 410 subjects were used for the CVM construction of the WTP based on three situations when cultural heritage and creativity are combined, as well as in the application of the double-bounded dichotomous choice model of survival analysis to estimate the WTP-influencing factors. The results showed that the higher the income of the subject, the higher of WTP for value-added services for preserving cultural heritage, participating in activities, and helping local development.

Keywords: Cultural heritage · Willingness to pay
Contingent valuation method

1 Introduction

International Council on Monuments and Sites (ICOMOS) defines heritage as a broad concept that includes both natural and the cultural environments. It involves landscapes, historic sites, environment, biodiversity, collections, previous and continuing cultural practices, knowledge, and living experiences. It records and expresses the long processes of historic development, becoming the essence of diverse national, regional, indigenous, and local identities. It is an integral part of modern life. Hence, cultural

© Springer International Publishing AG 2018
V. Kreinovich et al. (eds.), *Predictive Econometrics and Big Data*, Studies in Computational Intelligence 753, https://doi.org/10.1007/978-3-319-70942-0_49

heritage of historical significance, traditions, and artistic value reflects not only our ancestors' lifestyles and attitudes toward life, but also those of the modern people. Culture and cultural heritage are crucial to a people's identity, self-respect, and dignity [1]. Culture and cultural heritage are prominent resources in any society; thus, cultural heritage serves as a tourism attraction. Tourism leads to financial and political support for the management of this heritage. Cultural heritage is an important resource. Tourist trips typically include cultural heritage elements, ranging widely from a journey to a historical town center, to a visit to a museum or a stroll around a historic garden [2]. Therefore, a unique cultural heritage is an important attraction that helps develop the tourism industry.

For this reason, the traditional historic sites are undergoing a redefinition and reinterpretation, creating and making cultural heritage competitive and attractive [3]. Landry claims that a city's cultural heritage reflects its history, industries, and art assets, including the physical buildings and landmarks, as well as the intangible local and aboriginal life traditions, festivals, customs, legends, hobbies, and passion. Amateur cultural activities, such as the ubiquitous traditional food and cooking, leisure activities, clothing, and subculture, are also included [4]. In recent years, intangible heritage in the form of indispensable performances and visual arts have become a wave of popular emerging global culture industry. The concept of cultural industries comprises all enterprises, including self-employed individuals whose economic activities focus on production, dissemination, and intermediation of artistic and cultural products or services [5]. In the other words, cultural industry covers a broad range (e.g., music industry, publishing industry, arts, film industry, and so on).

Belova et al. defines "creative industries" as entrepreneurial activities in which economic value is linked to cultural content. Creative industries bring together the traditional strengths of classical culture with the "added value" of entrepreneurial strategies and the new knowledge-based electronic and communications skills. Thus, creative industry enhances the economic impact of the familiar cultural giants, placing particular emphasis on the contemporary. Experimented and engaged with modernity, culture is able to renew and revitalize itself. The creative industries are thus the source of the innovation and competitiveness essential to urban cultures in the fast changing and globalizing world [6]. A couple of examples are the Creative Industries Development Strategy proposed by Singapore and Creative Industries Mapping Documents 1998 by UK.

Ministry of Economic Affairs of Taiwan explains that "cultural and creative industries" are simply those industries that have their origins in innovation or cultural accretion. They have the potential to create wealth or jobs through the production and utilization of intellectual property, which can help to enhance the living environment for society as a whole [7]. While cultural and creative industry nowadays is an important sign of global economic development, the industry itself and its extended products have obtained huge profits, not only for enterprises and countries, but also for natural markets and civilizations all over the world. The marginal benefit for sightseeing promotion is apparent. Thus, many local governments around the world try their best promoting positive self-images by cultivating culture and creative industries. Due to globalization and technology trends, modern economy has transformed into an innovative knowledge-based economic structure.

In this study, Tainan City's cultural heritage is the research object. Tainan City is Taiwan's most ancient city that, during the Ming and Qing dynasties, served as the location of the government and was the political, economic, and cultural center of Taiwan. The city, as a result, possesses abundant potential cultural heritage.

The contingent valuation method (CVM) technique has great flexibility, allowing valuation of a wider variety of non-market goods and services than any of the indirect techniques possible. It is, in fact, the only method currently available for estimating nonuse values. In natural resources, the contingent valuation method (CVM) studies generally derive values through the elicitation of the willingness to pay (WTP) of respondents to prevent damage to natural resources or to restore damaged natural resources [8]. Because no general market price of cultural heritage is available, this study applies the non-market valuation of contingent valuation method (CVM) to build the situation of cultural heritage creativity and to understand the willingness to pay (WTP) of tourists to Tainan for the creation of cultural heritage goods, as well as for commercialization.

Based on this description, this study uses the double-bounded dichotomous choice as one of CVMs to find out the willingness to pay (WTP) through the factors that influence the economic value of creativity in cultural heritage and to provide the reference of the relevant units for future cultural heritage planning. Hanemann et al. indicates that the major advantage of the double-bounded dichotomous choice is that one could identify the location of the maximum willingness to pay (WTP) value from the data derived from this approach and that is an incentive compatible method. The statistical efficiency of conventional dichotomous choice contingent valuation surveys can be improved by asking each respondent a second dichotomous choice question which depends on the response to the first question—if the first response is "yes," the second bid is some amount greater than the first bid; while, if the first response is "no," the second bid is some amount smaller [9].

2 Contingent Valuation Method (CVM)

CVM refers to a hypothetical survey method of valuing the benefits of an intervention in monetary terms by estimating the maximum WTP of an individual. Currently, an issue in the application of CVM is the technique used to elicit this monetary valuation. The most obvious way to measure non-market values is through directly questioning individuals on their WTP for a product or service. CVM is a survey or questionnaire-based approach to the valuation of non-market products and services [10].

This study uses the questionnaire as a tool. The method has four non-market financial evaluation: open-ended, bidding game, payment card, and closed-ended. In this study, the closed-ended of double-bounded dichotomous was used. In other words, questionnaire design in the CVM is adopted by a closed-ended CV. The unique aspect of the dichotomous-choice questions is that subjects are asked if they would pay a fixed sum of money for the item being evaluated. Moreover, the double-bounded dichotomous choice approach is statistically more efficient than the single-bounded dichotomous choice approach [9, 11]. A typical close-ended question is a yes-no question.

A respondent simply read the statement and answer yes or no. It is not necessary for the respondent to elaborate thoughts or ideas.

According to Alberini, the questionnaire optimum amount is selected in the pre-test questionnaire. The amounts are set from low to high, removing approximately 10% of the extreme number, and selecting the amounts of 20%, 40%, 60%, and 80% as the first bids. If the response of the subjects in the first bid is "no," the first bid will cut in half to determine the amount for the second bid. Conversely, if the response of the subjects in the first bid is "yes," the amount of the first bid is increased by half the amount to set the second bid. Alberini recommends that the double-bound dichotomous choice of bid be divided into 4 to 6 groups. The most appropriate bid is divided into four groups [12].

This study focuses on three aspects whether a respondent has the willingness to pay for creativity of cultural heritage. The three questions are: (1) Are you willing to pay $___ to preserve cultural heritage? (2) Are you willing to pay $___ to participate in activities if Tainan City accedes to artistic and cultural activities that can improve the quality of leisure? (3) Are you willing to pay $___ to help if Tainan City undergoes local development. If the respondent's initial answer to each question with a "bid" of money is "yes," then a lightly increased amount is provided to see whether the respondent is willing to pay more. On the contrary, if the initial answer to each question is "no," the "bid" is slightly decreased to evaluate the willing to pay.

In addition, the application of CVM to cultural heritage, due to the complexity of the issue itself, is highly difficult. Thus, it must be designed for simple and easy situations. Subjects need to understand that the context because CVM is crucial to the study of cultural heritage creativity.

3   Empirical Result

In this study, the socioeconomic background (gender), cultural heritage creativity cognitive factors (socioemotional cognition, value-added services cognition, overall-image cognition), cultural heritage creativity cognitive cluster (leisure emotional expression group, pursuit of quality in leisure group, cold and conservative group), travel experiences (participate in archaeological activities) of the subjects were included in the model. Based on the foregoing theoretical model, an empirical model of the cultural heritage creativity WTP was set using following formula:

$$WTP1 = f(\ln income, sex, fac1, fac2, fac3, d1, d2, d3) \tag{1}$$

$$WTP2 = f(\ln income, sex, fac1, fac2, fac3, d1, d2, d3) \tag{2}$$

$$WTP3 = f(\ln income, sex, fac1, fac2, fac3, d1, d2, d3) \tag{3}$$

Here, WTP1, WTP2, and WTP3, respectively, are three aspects of willing to pay: preserving cultural heritage, participating in activities, and helping local development. The socioeconomic variables of ln income is logarithmic of the income of the subject; sex is a gender variable, male is 1, female is 0; fac1 is socioemotional cognition; fac2 is value-added services cognition; fac3 is overall-image cognition; d1 and d2 are the

Table 1. Cultural heritage creativity WTP variables

Variable name		Variable descriptions
Socioeconomic variables	ln income	Logarithmic of subject's income (TWD)
	Sex	Male as 1, Female as 0
Cultural heritage creativity cognitive factors	fac1	Socioemotional cognition
	fac2	Value-added services cognition
	fac3	Overall-image cognition
Cultural heritage creativity cognitive cluster	d1	Dummy variable of leisure emotional expression group is set to 1
		Dummy variable of pursuit of quality in leisure group is set to 0
		Dummy variable of cold and conservative group is set to 0
	d2	Dummy variable of leisure emotional expression group is set to 0
		Dummy variable of multi-cognition group is set to 1
		Dummy variable of cold and conservative group is set to 0
Travel experiences	d3	Dummy variable of participate in archaeological activities, Participation = 1, non-participation = 0

dummy variables of the cluster; the leisure emotional expression group is set to 1 and 0; the pursuit of quality in leisure group is set to 0 and 1; the cold and conservative group is set to 0 and 0; and d3 is the dummy variable of participating in archaeological activities, with participation as 1, did not participate as 0, preceding variable definition, as shown in Table 1. In this study, after establishing the evaluation model, and in accordance to the aforementioned variable data, the empirical analysis of the cultural heritage creativity WTP pay for the maximum likelihood estimation (MLE) was performed.

Preserving cultural heritage is shown in Table 2. Assessment model of the log-normal was at 10% significance, whereas the overall-image cognition (fac3) was estimated to be negative and significant. At 5% significance, the leisure emotional expression group (d1) was estimated to be positive and significant. The assessment model of the Weibull was at 10% significance in the ln income was estimated to be positive and significant. Also, at 10% significance, the value-added services cognition (fac2) value of positive and significant. At 1% significance, the overall-image cognition (fac3) value was negative and significant. The assessment model of the gamma was at 5% significance; thus, leisure emotional expression group (d1) was estimated to be positive and significant.

Participating in activities is shown in Table 3. Assessment model of the log-normal distribution at 5% significance of ln income was estimated to be positive and significant. At 5% significance, the socioemotional cognition (fac1) was estimated to be

Table 2. Preserving cultural heritage results of the evaluation function

Variable name		Evaluation function of the probability of allocation patterns		
		Log-normal	Weibull	Gamma
Intercept term		6.11 (0.67)	7.27 (0.71)	5.46 (0.61)
Socioemotional variables	ln income	0.97 (0.62)	1.36 (0.71)*	0.73 (0.54)
	Sex	0.12 (0.13)	0.17 (0.13)	0.08 (0.12)
Cultural heritage creativity cognitive factors	fac1	−0.48 (0.33)	−0.64 (0.39)	−0.39 (0.29)
	fac2	0.17 (0.18)	0.31 (0.19)*	0.04 (0.17)
	fac3	−0.34 (0.17)*	−0.51 (0.17)***	−0.19 (0.16)
Cultural heritage creativity cognitive cluster	d1	0.38 (0.17)**	0.24 (0.18)	0.40 (0.16)**
	d2	−0.02 (0.24)	−0.13 (0.25)	0.11 (0.22)
Travel experiences	d3	0.20 (0.24)	0.36 (0.24)	0.12 (0.22)
Scale		1.00 (0.05)	0.90 (0.05)	0.92 (0.07)
Log likelihood		−417.74	−435.39	−412.11
Restricted Log likelihood		−427.33	−447.10	−421.30
Log likelihood ratio		19.18	23.42	18.38

*,**,*** respectively significant at 10%, 5%, 1%. Log-likelihood ratio = (−2) × (Restricted Log likelihood − Log Likelihood), $\chi 2(7,0.95) = 14.07$.

Table 3. Participating activities of the evaluation function

Variable name		Evaluation function of the probability of allocation patterns		
		Log-normal	Weibull	Gamma
Intercept term		6.87 (0.76)	7.96(0.83)	6.13 (0.67)
Socioemotional variables	ln income	1.69 (0.69)**	1.85 (0.83)**	1.46 (0.64)**
	Sex	−0.11 (0.15)	0.00 (0.15)	−0.14 (0.13)
Cultural heritage creativity cognitive factors	fac1	−0.76 (0.37)**	−0.79 (0.45)*	−0.72 (0.34)**
	fac2	0.24 (0.21)	0.24 (0.22)	0.18 (0.19)
	fac3	−0.49 (0.19)**	−0.54 (0.21)***	−0.37 (0.18)**
Cultural heritage creativity cognitive cluster	d1	0.30 (0.20)	0.16 (0.22)	0.32 (0.18)*
	d2	−0.03 (0.27)	0.04 (0.29)	0.05 (0.24)
Travel experiences	d3	0.22 (0.27)	0.37 (0.29)	0.14 (0.24)
Scale		1.12 (0.06)	1.00 (0.05)	1.04 (0.07)
Log likelihood		−425.46	−444.68	−419.51
Restricted Log likelihood		−435.10	−454.39	−428.70
Log likelihood ratio		19.28	19.42	18.38

*,**,*** respectively significant at 10%, 5%, 1%. Log-likelihood ratio = (−2) × (Restricted Log likelihood − Log Likelihood), $\chi 2(7,0.95) = 14.07$.

negative and significant. At 5% significance, the overall-image cognition (fac3) was estimated to be negative and significant. The assessment model of the Weibull distribution at 5% significance ln income was estimated to be positive and significant. At 10% significance, the socioemotional cognition (fac1) was estimated to be negative and significant. At 1% significance, the overall-image cognition (fac3) was estimated to be positive and significant. The assessment model of the gamma at 5% significance of ln income were estimated to be positive and significant. At 5% significance, the socioemotional cognition (fac1) was estimated to be negative and significant. At 5% significant, the overall-image cognition (fac3) was estimated to be positive and significant. At 10% significance, the leisure emotional expression group (d1) was estimated to be positive and significant.

Helping local development is shown in Table 4. Assessment model of the log-normal distribution at 10% significance of value-added services cognition (fac2) were estimated to is positive and significant. At 5% significant, the overall-image cognition (fac3) estimate is negative and significant. At 10% significance, the leisure emotional expression group (d1) was estimated to be positive and significant. The assessment model of the Weibull at 10% significance of ln income was estimated to be positive and significant. At 10% significance, value-added services cognition (fac2) was estimated to be positive and significant. At 5% significance, the overall-image cognition (fac3) was estimated to be positive and significant. The assessment model of the gamma at 10% significance of overall-image cognition (fac3) was estimated to be

Table 4. Local development results of the evaluation function

Variable name		Evaluation function of the probability of allocation patterns		
		Log-normal	Weibull	Gamma
Intercept term		5.36 (0.68)	6.42 (0.71)	4.83 (0.72)
Socioemotional variables	ln income	0.97 (0.63)	1.29 (0.70)*	0.80 (0.60)
	Sex	0.13 (0.14)	0.18 (0.14)	0.12 (0.14)
Cultural heritage creativity cognitive factors	fac1	−0.39 (0.33)	−0.51 (0.38)	−0.33 (0.31)
	fac2	0.36 (0.21)*	0.42 (0.22)*	0.32 (0.21)
	fac3	−0.39 (0.19)**	−0.40 (0.19)**	−0.35 (0.18)*
Cultural heritage creativity cognitive cluster	d1	0.34 (0.19)*	0.13 (0.19)	0.42 (0.19)**
	d2	−0.17 (0.27)	−0.18 (0.27)	−0.13 (0.26)
Travel experiences	d3	−0.03 (0.25)	0.06 (0.25)	−0.08 (0.24)
Scale		1.09 (0.06)	0.93 (0.05)	1.10 (0.06)
Log likelihood		−423.24	−437.59	−421.48
Restricted Log likelihood		−432.12	−445.97	−430.89
Log likelihood ratio		17.76	16.76	18.82

*,**,*** respectively significant at 10%, 5%, 1%. Log-likelihood ratio = (−2) × (Restricted Log likelihood − Log Likelihood), $\chi 2(7,0.95) = 14.07$.

Table 5. Cultural heritage creativity WTP

WTP item	WTP (TWD/year/person)	Lower (TWD/year person)	Upper (TWD/year/person)
Preserving cultural heritage	922.9	630.5	1369.7
Participating in activities	914.2	587.1	1453.0
Help local development	620.9	416.0	941.3

positive and significant. At 5% significance, leisure emotional expression group (d1) was estimated to be positive and significant.

The results of whether a consumer is willing to pay (TWD/year/person) for creativity in cultural heritage from three aspects are show in Table 5: preserving cultural heritage, 922.9 TWD; participating in activities, 914.2 TWD; helping local development, 620.9 TWD.

4 Conclusions and Discussion

The results demonstrate that the higher the income of the subjects, the higher of WTP for Weibull, value-added services cognition (fac2) factor in preserving cultural heritage and helping local development. When the subjects strongly agree, WTP is higher for Weibull in the leisure emotional expression group (d1). When subjects can more easily express their own leisure emotional needs, WTP was higher for the log-normal and gamma distributions for the cultural heritage creativity.

The highest amount of WTP in the three situations is for preserving cultural heritage, for which subjects were willing to pay 922.9 TWD. This result shows that people with cultural heritage cognition want to maintain cultural heritage to give it more added-value. The lowest amount of WTP in the three situations is helping the local development, where subjects are willing to pay 620.9 TWD. The possible reason for this result is that the subjects are not local residents and have no local identity; thus, they do not think of the local development combined with the cultural heritage can affect WTP.

In fact, participating activities has similar amount as preserving cultural heritage. Subjects were willing to pay 914.2 TWD, possibly because subjects have positive experience of participating activities. Watching ballet dance or appreciating modern art display at a historical building can provide the subject with a unique sensation and then encourage them to participate more often. At the same time, the cultural heritage can be better known by people.

As a result, in order to develop better cultural heritage creativity and tourism, cultural heritage management is crucial. While the consumers are willing to pay for preserving cultural heritage, local governments should work on the recognition of heritage in both local residents and outsiders. Take Tainan for example. Being a city full of cultural heritage in Taiwan, its government should pay attention to making Tainan and its culture recognized by potential consumers in order to help its tourism sustain. While consumers have various activities to participate, the tour becomes more interesting and attractive to greater number of potential consumers. Hence, the creativity added in cultural heritage is the key to the sustainability of tourism. In other words, cultural heritage creativity helps the local tourism, and the profit made from the tourist industry can possibly help preserving the local cultural heritage. Such dynamic economic value will make contribution to the local development, helping the local industries to thrive.

References

1. Andersen, K.: Sustainable Tourism and Cultural Heritage: A Review of Development Assistance and its Potential to Promote Sustainability. World Bank, Washington (1999)
2. De la Torre, M.: Assessing the Values of Cultural Heritage: Research Report. The Getty Conservation Institute, Los Angeles (2002)
3. Nasser, N.: Planning for urban heritage places: reconciling conservation, tourism, and sustainable development. J. Plann. Lit. **17**(4), 467–479 (2003)
4. Landry, C.: The Creative City: A Toolkit for Urban Innovators. Earthscan Publications Ltd., London (2008)
5. Fesel, B., Söndermann, M.: Culture and creative industries in Germany. German commission for UNESCO, Grafische Werkstatt Druckerei und Verlag (2007)
6. Belova, E., Cantell, T., Causey, S., Korf, E., O'Connor, J.: Creative industries in the modern city: Encouraging enterprise and creativity in St. Petersburg. Tacis Cross Border Cooperation Small Project Facility (2002)
7. Ministry of Economic Affairs: White Paper on SMEs in Taiwan, pp. 162–165 (2004)
8. Lipton, D.W., Wellman, K., Sheifer, I.C., Weiher, R.F.: Economic valuation of natural resources: a handbook for coastal policymakers. National Oceanic and Atmospheric Administration (NOAA), USA (1995)
9. Hanemann, M., Loomis, J., Kanninen, B.: Statistical efficiency of double-bounded dichotomous choice contingent valuation. Am. J. Agric. Econ. **73**, 1255–1263 (1991)
10. Smith, R.D.: Contingent valuation: indiscretion in the adoption of discrete choice question formats? Centre for Health Program Evaluation Working paper 74 (1997)
11. Boyle, G.: Dichotomous-choice, contingent-valuation questions: functional form is important. Northeast. J. Agric. Resour. Econ. **19**(2), 25–31 (1990)
12. Alberini, A.: Efficiency vs bias of willingness-to-pay estimates: bivariate and inter-data models. J. Environ. Econ. Manag. **29**, 169–180 (1995)

A Bibliometric Review of Global Econometrics Research: Characteristics and Trends

Van-Chien Pham[1] and Man-Ling Chang[2]([⊠])

[1] Department of Business Administration, Asia University,
Taichung 41354, Taiwan
[2] Department of Business Administration, National Chung Hsing University,
Taichung 402, Taiwan
manllian@dragon.nchu.edu.tw

Abstract. Using a bibliometric analysis, this research analyses on the Social Science Citation Index (SSCI) publications from the Institute for Scientific Information (ISI), Web of Science database, based on 12965 publications from 1497 journals during 1992 to 2016. The research was assessed the research's characteristics and trends of most productive countries/regions and institutions, was pointed out the sharp increasing in China on econometrics research, Applied Economics (England) published the most econometrics articles. The research was also pointed out the temporal evolution of recent hot econometrics research issues. Global trends and characteristics was found throughout this research can give a general overview for further researches on econometrics.

Keywords: Trend research · Bibliometric analysis · Econometrics
SSCI · Web of Science

1 Introduction

Bibliometric analysis is a analyzing method based on statistics, is applied widely in sciences and also in research analysis. The research characteristics and/or trends of research in economics were also conducted by bibliometricians, such as in financial [1], accounting [2], sport-economics [3], as well as management [4]. Despite importance of researching about, there is not many approaches by bibliometric method about econometrics research so far. By bibliometric method, Baltagi [5] studied the charactersistics of econometrics research by ranking the most productive authors, institutions and countries based on the data of 16 leading international journals in the area and cover the period 1989–2005.

In this research, we will use statistical analysis approach with titles, abstracts, author keywords of articles, information about countries and institutions, in order to dissect and assess global trends and characteristics of econometrics research from 1992 to 2016, so as to bring a panorama about econometrics research to researchers and to help them to identify suitable researching orientation as well.

© Springer International Publishing AG 2018
V. Kreinovich et al. (eds.), *Predictive Econometrics and Big Data*, Studies in Computational Intelligence 753, https://doi.org/10.1007/978-3-319-70942-0_50

2 Data Sources and Research Methodology

Data for this research was obtained from the online version of the SSCI, the Web of Science. The online version of the SSCI database was searched by the keyword "*Econometrics*" as a part of the title, abstract, and keywords (author keywords) to compile all manuscripts related to econometrics research. Besides, the term "*Econometric*" was also used as supporting searching-word because of differences in the terminology of authors.

The impact factor (IF), subject category and rank in category of the journals were gathered from the 2015 Journal Citation Reports (JCR). The contributions of different countries/territories and institutes were determined by the participation of at least one author of the publication, through the author addresses. The term "independent country" was understood as all authors who are from the same country, the term "international collaborative" was used when the coauthors are from more than one country.

Due to limitations of the SSCI database, before 1992 articles were incomplete abstract information, while the analysis in this study needs to use the related information in the title, abstract and keywords. Therefore, the publications before 1992 will not be considered. All of the articles in the 25-years period 1992–2016 were evaluated by the following aspects: document type and language of publication, publication characteristics by region, country, institution, author, relevant journals and hot topic analysis.

3 Result and Discussion

3.1 Document Type and Language of Publication

Seventeen document types were found in the total 12965 publications during the 25-years study period. The journal article, as the most frequency document type, comprises 92.0% (11913) of the total production, and followed distantly by proceedings papers (802; 6.2%), book reviews (416; 3.2%), editorial materials (289; 2.2%), and reviews (244; 1.9%).

One hundred ninety articles were published in 1992, and the number of articles went up slightly to 335 in 2004, then increased rapidly in the period from 2005 to 2013 with 853 articles published in 2013, before gently to reach a pick of 909 articles in 2015 and decreased in 2016 (Fig. 1). The number of proceedings papers and book reviews decreased after 2008 and maintained a low level afterwards. Journal articles represented the majority of document types, therefore only the 11913 articles were used in further analysis.

There was 16 languages in use, ninety-seven percent of total articles in this period were published in English, other languages that were generally less appeared were Spanish (0.95%), French (0.75%), Czech (0.48%), German (0.35%).

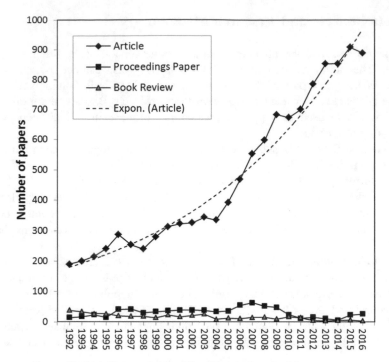

Expon. (Article): Exponential trendline of the number of articles

Fig. 1. Trends of articles, proceedings papers, and book reviews on *Econometrics* from 1992 to 2016

3.2   Countries/Continents Research Characteristics

In recent researches about the global research trends, the total of articles index (including independent and international collaborative articles) were commonly used for analyzing research trends [6, 7]. The top 20 productive countries ranked by the total number of econometrics articles with more than 150 articles published includes two North American countries (USA and Canada), thirteen Europe countries (UK, Germany, France, Italy, Spain, Netherlands, Switzerland, Greece, Belgium, Sweden, Austria, Norway, and Turkey), four Asia countries (China, Japan, South Korea, Taiwan), and Australia. USA is the most frequency country in econometrics research, with 4414 articles, accounting to 38% of total articles in the world, followed by UK (1647, 14%), Germany (848, 7.4%). Research by European scholars increased rapidly and has become the most productive continent since 2004 (see Fig. 2).

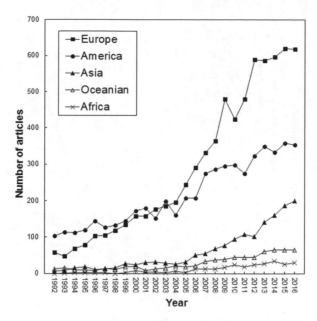

Fig. 2. Publication trends of regions

Besides, the Fig. 2 also shows that while there are the dwelling of research by Europe and American authors in recent years, research trend of econometrics in Asia continued to increase strongly. To clarify this trend line, an analysis for the Asia is conducted with five most productive countries in the continent. The result in Fig. 3 indicates the rise of Asia after 2005 until now is due to the sharply increasing of China. Research trend of econometrics in China increased sharply from 2006 onwards, with 520 articles in the priod 1992–2016 (300 articles were published in the last 4 years), China has the highest rate of publication growth in the world.

3.3 Most Relevant Journals and Subject Categories

In total, 11913 articles were published in 1445 journals. There were 10 journals with more than 100 publication articles in econometrics from 1992–2016, one-sixth of all articles was published in this ten journals (Table 1). Applied Economics (England) published the most econometrics articles (346, 2.9%), following by Journal of Econometrics (Netherlands, 331), and Econometric Theory (USA, 236).

Six of the top 10 most relevant journals were classified under *Economics* subject category by JCR in 2015, three of them were classified under the subject categories related to *Mathematics*.

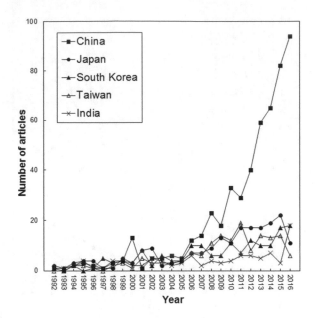

Fig. 3. Publication trends of five most active countries of Asia

3.4   Institutions' Research Characteristics

Table 2 ranks the top 11 most influential institutions by total number of articles pub-lished with more than 100 publication articles in *econometrics research* from 1992–2016, which is as many as ten institutions were in the USA, the another institution is University of Cambridge in UK. The leading institutions were University of Illinois with 131 articles (accounting for 1.14% of total publications), University of California Berkeley (128, 1.11%), and University of Wisconsin (121, 1.05%). Notably, there was the appearance of a financial institution, the World Bank, in the top 11 most influential institutions, the only institutions in the list is not a university. Most of publications of the World Bank on econometrics were published by researchers in Development Research Group.

The University of Illinois was also ranked at first place about number of first author articles, while the highest number of corresponding author articles and independent articles belong to the University of Wisconsin. In econometrics research, University of California at Berkeley had the strongest researching collaboration with outside insti-tutions, also had the highest inter-instutionally collaborative ratio among the 11 most influential institutions, along with Cornell University and Stanford University (84%). All the top 11 most influential institutions in econometrics research (actually all 39 leading institutions with the number of articles of more than 60, which the table does not shows) have the high inter-instutionally collaborative ratio of more than 60%.

Table 1. Top 10 most relevant journals ranked by total number of articles with impact factor, JCR category and rank in category.

Journal	Number of Articles (%)	Impact factor (IF)	JCR Category	Rank in Category	Country of Publisher
Applied Economics	346 (2.9)	0.586	Economics	236/345	England
Journal of Econometrics	331 (2.8)	1.611	Mathematics, Interdisciplinary Applications	30/101	Netherlands
Econometric Theory	236 (2.0)	1.162	Mathematics, Interdisciplinary Applications Statistics and Probability	48/101 50/123	USA
Economic Modelling	205 (1.7)	0.997	Economics	149/345	Netherlands
Energy Policy	163 (1.4)	3.045	Energy & Fuels Environmental Sciences	29/88 59/225	England
Energy Economics	153 (1.3)	2.862	Economics	22/345	Netherlands
American Journal of Agricultural Economics	148 (1.2)	1.436	Agricultural Economics & Policy	5/17	USA
Economics Letters	143 (1.2)	0.603	Economics	230/345	Switzerland
Applied Economics Letters	142 (1.2)	0.378	Economics	280/345	England
Journal of Applied Econometrics	108 (0.91)	1.872	Economics Social sciences, Mathematical Methods	57/345 9/49	England

3.5   Authorship Characteristics

Table 3 indicates the top 15 most productive authors with more than 19 publications in the period 1992–2016. Bhat, the most productive researcher from University of Texas at Austin (USA), published 37 articles, followed by Phillips (35), and Mcaleer (34). Bhat also published the highest number of articles as corresponding author (33).

Table 2. Top 11 most influential institutions of articles during 1992–2016

Institute	Total number of articles (%)	Rank of number of first author articles (%)	Rank of number of corresponding author articles (%)	Rank of number of independent articles (%)	Rank of number of inter-instutionally collaborative articles (%)	Inter-instutionally collaborative ratio *
University of Illinois, USA	131 (1.14)	1 (0.67)	2 (0.67)	3 (0.62)	2 (1.5)	76
University of California at Berkeley, USA	128 (1.11)	3 (0.57)	3 (0.55)	13 (0.42)	1 (1.6)	84
University of Wisconsin, USA	121 (1.05)	2 (0.66)	1 (0.73)	1 (0.96)	12 (1.1)	60
World Bank, USA	110 (0.96)	7 (0.50)	8 (0.48)	6 (0.50)	5 (1.3)	77
Cornell University, USA	110 (0.96)	5 (0.50)	4 (0.53)	22 (0.36)	3 (1.4)	84
Harvard University, USA	104 (0.90)	4 (0.51)	5 (0.51)	2 (0.70)	15 (1.1)	66
Massachusetts Institution of Technology, USA	103 (0.90)	5 (0.50)	6 (0.50)	26 (0.34)	4 (1.3)	83
University of Maryland, USA	102 (0.89)	10 (0.44)	9 (0.47)	11 (0.44)	9 (1.2)	78
University of Penn, USA	101 (0.88)	18 (0.37)	16 (0.39)	26 (0.34)	7 (1.3)	83
University of Cambridge, UK	101 (0.88)	9 (0.48)	11 (0.44)	5 (0.56)	12 (1.1)	72
Stanford University, USA	101 (0.88)	14 (0.42)	13 (0.41)	29 (0.32)	5 (1.3)	84

* the percentage of inter-instutionally collaborative articles in total institution articles

Table 3. Top 15 most productive authors on econometrics between 1992 and 2016

Author	Rank (TP)	Rank (TPF)	Rank (TPR)	Rank (TPI)
Bhat, CR	1 (37)	4 (17)	1 (33)	11 (6)
Phillips, PCB	2 (35)	1 (23)	2 (22)	1 (16)
Mcaleer, M	3 (34)	29 (8)	19 (10)	126 (2)
Franses, PH	4 (23)	22 (9)	9 (12)	126 (2)
Baltagi, BH	5 (22)	2 (22)	3 (20)	6 (7)
Hendry, DF	6 (21)	6 (15)	7 (14)	4 (9)
Fingleton, B	6 (21)	4 (17)	6 (15)	2 (12)
Anselin, L	6 (21)	3 (18)	5 (16)	6 (7)
Tsionas, EG	9 (20)	17 (10)	9 (12)	11 (6)
Plantinga, AJ	9 (20)	55 (6)	40 (7)	400 (1)
Filippini, M	9 (20)	9 (12)	157 (4)	400 (1)
Serletis, A	12 (19)	8 (14)	4 (17)	400 (1)
Egger, P	12 (19)	9 (12)	12 (11)	25 (4)
Colombo, MG	12 (19)	6 (15)	30 (8)	#N/A
Chavas, JP	12 (19)	29 (8)	30 (8)	126 (2)

TPː total publications
TPF: total publications as the first author
TPR: total publications as the corresponding author
TPI: total publications as the independent author
N/A: not available

Phillips was ranked at first on total publications as first author (23), as well as independent author (16). As noted by Chang and Ho [1], a bias can exit in authorship analysis due to the same name of two or more authors, or the different names was used by an author in his or her manuscripts.

3.6 Most Frequently Cited Articles

Table 4 lists the top 13 most frequently cited articles ranked by total citations from the publication year to 2016. Eight of these 13 articles were published in the first decade of the 21st century, 5 before 2000, and no articles published in the last 7 years (among 2010–2016). Journal of Health Economics and Journal of Econometrics are two journal in which two of their articles are listed in the top 13. The most frequently cited was "How much should we trust differences-in-differences estimates?" by Bertrand, Duflo, and Mullainathan. This articles was published in 2004 and was cited 1776 times among 2004–2016, also had the highest average time cited per year (136.6 times).

Analysing the time cited each year from the publication year until 2016 for the top 13 most frequently cited articles showns two articles with the rapidly rising in the number of time the articles was cited. That is "How much should we trust differences-in-differences estimates?" that noted above and "How to do xtabond2: An introduction to difference and system GMM in Stata" by David Roodman.

Table 4. Top 13 most frequently cited econometrics research articles ranked by total citations to 2016

Authors	Journal	PY	Title	TC	C/Y
Bertrand, M; Duflo, E; Mullainathan, S	Quarterly Journal of Economics	2004	How much should we trust differences-in-differences estimates?	1776	136.6
Brazier, J; Roberts, J; Deverill, M	Journal of Health Economics	2002	The estimation of a preference-based measure of health from the SF-36	1315	87.7
Fornell, C; Johnson, MD; Anderson, EW; Cha, JS; Bryant, BE	Journal of Marketing	1996	The American customer satisfaction index: Nature, purpose, and findings	933	44.4
Manning, WG; Mullahy, J	Journal of Health Economics	2001	Estimating log models: to transform or not to transform?	886	55.4
Goffe, WL; Ferrier, GD; Rogers, J	Journal of Econometrics	1994	Global optimization of statistical functions with simulated annealing	820	35.7
Duffie, D; Pan, J; Singleton, K	Econometrica	2000	Transform analysis and asset pricing for affine jump-diffusions	818	48.1
Papke, LE; Wooldridge, JM	Journal of Applied Econometrics	1996	Econometric methods for fractional response variables with an application to 401(k) plan participation rates	762	36.3
Heckman, JJ; Ichimura, H; Todd, P	Review of Economic Studies	1998	Matching as an econometric evaluation estimator	760	40.0
Gali, J; Gertler, M	Journal of Monetary Economics	1999	Inflation dynamics: A structural econometric analysis	744	41.3
Bai, JS; Ng, S	Econometrica	2002	Determining the number of factors in approximate factor models	728	48.5
Imbens, GW	Review of Economics and Statistics	2004	Nonparametric estimation of average treatment effects under exogeneity: A review	685	52.7
Smith, JA; Todd, PE	Journal of Econometrics	2005	Does matching overcome LaLonde's critique of nonexperimental estimators?	673	56.1
Roodman, David	Stata Journal	2009	How to do xtabond2: An introduction to difference and system GMM in Stata	632	79.0

PY: Publication year; TC: Total citations from the publication year to 2016; C/Y: Average citations per year

The manuscript was published in 2009 about application of system generalized method-of-moments estimators with xtabond2 in Stata is also the most recently articles was published in the list.

Table 5. Top 24 most frequency of author keywords used among 1992–2016 and in five 5-years periods

Author keywords	1992–2016 total number of articles (%)	1992–2016 Rank (%)	1992-1996 Rank (%)	1997-2001 Rank (%)	2002–2006 Rank (%)	2007-2011 Rank (%)	2012–2016 Rank (%)
Spatial econometrics	391	1 (4.8)	24 (0.54)	5 (2.1)	1 (4.1)	1 (5.3)	1 (5.6)
Econometrics	319	2 (3.9)	2 (7.3)	2 (4.2)	2 (3.5)	2 (3.6)	2 (3.9)
Econometric models	282	3 (3.5)	1 (7.8)	1 (5.4)	3 (3.3)	3 (3.4)	3 (2.7)
Panel data	221	4 (2.7)	24 (0.54)	4 (3.3)	4 (3.1)	4 (3.0)	5 (2.6)
Economic growth	173	5 (2.1)	43 (0.27)	12 (1.0)	6 (1.6)	5 (2.1)	4 (2.7)
Cointegration	164	6 (2.0)	4 (3.0)	3 (4.0)	5 (2.4)	6 (2.0)	8 (1.5)
China	154	7 (1.9)	24 (0.54)	30 (0.60)	7 (1.5)	7 (1.7)	6 (2.5)
Forecasting	106	8 (1.3)	3 (4.1)	6 (1.9)	9 (1.1)	8 (1.1)	11 (1.1)
Productivity	95	9 (1.2)	5 (1.9)	6 (1.9)	8 (1.3)	13 (0.97)	12 (1.0)
Econometric analysis	90	10 (1.1)	43 (0.27)	55 (0.30)	9 (1.1)	22 (0.76)	7 (1.6)
Econometric modeling	88	11 (1.1)	8 (1.6)	6 (1.9)	24 (0.61)	17 (0.84)	10 (1.2)
Foreign direct investment (FDI)	85	12 (1.0)	43 (0.27)	20 (0.75)	37 (0.52)	8 (1.1)	9 (1.3)
Innovation	79	13 (0.97)	43 (0.27)	20 (0.75)	16 (0.79)	8 (1.1)	12 (1.0)
Time series	78	14 (0.96)	5 (1.9)	20 (0.75)	13 (0.87)	11 (1.1)	16 (0.86)
Efficiency	67	15 (0.82)	16 (0.81)	9 (1.3)	37 (0.52)	12 (1.1)	21 (0.69)
Monetary policy	62	16 (0.76)	16 (0.81)	10 (1.2)	11 (1.0)	20 (0.80)	34 (0.56)
Endogeneity	61	17 (0.75)	#N/A	20 (0.75)	21 (0.70)	28 (0.67)	14 (0.89)
Instrumental variables	57	18 (0.70)	12 (1.4)	15 (0.90)	24 (0.61)	26 (0.72)	25 (0.64)
Causality	57	18 (0.70)	24 (0.54)	30 (0.60)	16 (0.79)	17 (0.84)	23 (0.67)
Unemployment	56	20 (0.69)	16 (0.81)	30 (0.60)	13 (0.87)	53 (0.46)	17 (0.81)
Convergence	53	21 (0.65)	24 (0.54)	30 (0.60)	16 (0.79)	22 (0.76)	34 (0.56)
Bayesian econometrics	53	21 (0.65)	43 (0.27)	#N/A	46 (0.44)	22 (0.76)	17 (0.81)
Inflation	52	23 (0.64)	43 (0.27)	30 (0.60)	16 (0.79)	22 (0.76)	34 (0.56)
Growth	52	23 (0.64)	16 (0.81)	30 (0.60)	24 (0.61)	17 (0.84)	40 (0.53)
Economic development	48	25 (0.59)	#N/A	55 (0.30)	24 (0.61)	31 (0.63)	23 (0.67)
Poverty	47	26 (0.58)	#N/A	69 (0.15)	37 (0.52)	15 (0.88)	40 (0.53)
Climate change	47	26 (0.58)	#N/A	69 (0.15)	73 (0.26)	53 (0.46)	14 (0.89)
International trade	46	28 (0.56)	#N/A	20 (0.75)	46 (0.44)	77 (0.34)	19 (0.78)
Bootstrap	46	28 (0.56)	43 (0.27)	12 (1.0)	24 (0.61)	38 (0.55)	46 (0.50)

* N/A: not avaiable

3.7   Hot Issues

Researching the temporal evolution of hot research issues identifies for us the most attractive matters of researchers (and also of entreprises, institutions). It is very useful for apprehending the research trends in recent years, then establishing better direction for further researches. The bibliometric approach is one of the best ways to analyze this. In analyzing hot issues, the bibliometric method uses statistical techniques to analyzing frequency of words in titles, author keywords, and occasionally in abstracts. To identify hot issues in econometrics research, author keywords, titles, and abstracts statistical analyses were used.

Analysis of single word in the titles shows some of the most used word by authors is "*analysis*" (1244 articles, 10.4% of total articles), "*econometric*" (1191, 10.0%), "*evidence*" (994, 8.3%), "*model*" (782, 6.6%), "*economic*" (647, 5.5%), "*market*" (605, 5.1%), "*spatial*" (600,5.0%), "*growth*" (558, 4.7%). "*Economic*", "*spatial*", "*impact*", "*china*", "*development*" are the words which the strong uptrends, while there is the downtrends in the use of "*estimation*", "*forecasting*", "*application*", and "*modelling*".

The most frequency of author keywords were used in the period 1992–2016 with more than 100 articles including "spatial econometrics" (391 articles), "econometrics" (319), "econometric models" (282), "panel data" (221), "economic growth" (173), "cointegration" (164), "china" (154), and "forecasting" (106) (see Table 5).

For more detailed analysis of the global trends in econometrics research, five macroeconomics and four microeconomics issues were examined. Figure 4 shows the comparison of the growth trends of five macroeconomics hot issues in econometrics research. "*Monetary policy*" (including "exchange rate") is the most interesting subject

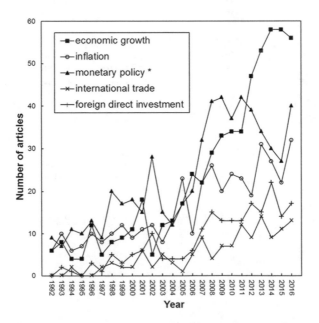

Fig. 4. Comparison the growth trends of five macroeconomic groups

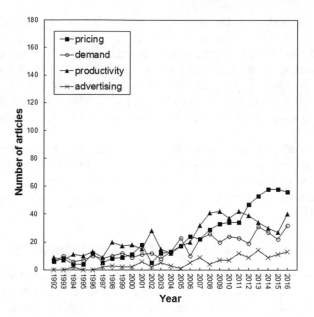

Fig. 5. Comparison the growth trends of four microeconomic groups

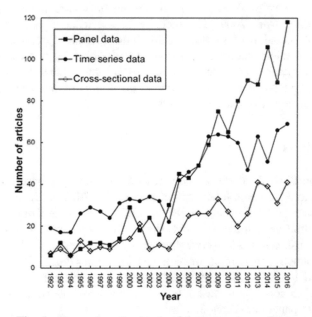

Fig. 6. Comparison the trends of three structures of data

in many years before 2012. However, from 2012, "*economic growth*" takes its place. Research on economic growth in econometrics increased rapidly from 2004 to 2014 before dwelling in the last two years. Recent researches on economic growth, through econometrics methods, provided evidences to clarify the relationship between economic growth and other social and economic issues. Detailed analysis for all 213 articles with the word "economic growth" in the title shows that energy consumption is the most interesting issue in the relationship with economic growth (with 22 in total 213 articles), which is to be found in researches for China [8], as well as Asean countries [9], Afica [10], and Central America [11]. There are also much work on *inflation, international trade, foreign direct investment* with the uptrends, and are all increased in 2016.

On the other hand, in applied econometrics in microeconomics research, pricing is the hostest issue, which is strongly cared from 2006 (Fig. 5). Involving to *"pricing"* matters includes *"asset pricing"*, *"price elasticity"*, *"hedonic pricing"*, *"option pricing"*. Research on demand and productivity although had the rising trend-lines, but decreased sharply in 2016.

3.8   Structures of Data Used in Applied Econometrics Research

Data structures of panel data (longitudinal data), time series data, and cross-sectional data are widely used in applied econometrics to describe various econometric situations [12]. Figures 6 shows the comparison of the trends of this three data structures in econometrics research. Before 2004, econometrics scholars prefer to use *time series* data best. The interest in time series econometrics continued to grow rapidly, however, showing signs of slowing down and giving up this most attractiveness place to *panel data* econometrics from 2009. The increase in publications ralated to panel data indicates the expectation of researchers for more realistic specifications and better understanding of microeconomic behaviors [13] as noted by two founder fathers of panel data econometrics in their article published in 1966: "One of the main reasons for being interested in panel data is the unique possibility of uncovering disaggregate dynamic relationships using such data sets" [14].

4   Conclusions

Researching on SSCI database of econometrics has already gained some significant results about global research trends and characteristics. Most of the articles originated from USA researchers, according to 38% of all articles, while China has the highest rate of publication growth in the world, shows the biggest attention to econometrics research. Universities in USA overwhelmingly occupy nine out of eleven places in the list of most productive institutions of econometrics articles during 1992–2016. Applied Economics and Journal of Econometrics have the most number of published articles. Recent researches concentrate on hot issues: spatial econometrics, economic growth, forecasting, and panel data econometrics.

This research indicated global trends and characteristics on econometrics research in the period 1992–2016. These trends shall orientate for further researches about econometrics.

References

1. Chang, C.-C., Ho, Y.-S.: Bibliometric analysis of financial crisis research. Afr. J. Bus. Manage. **4**(18), 3898–3910 (2010)
2. Merigó, J.M., Yang, J.-B.: Accounting research: a bibliometric analysis. Aust. Account. Rev. **27**(1), 71–100 (2017). https://doi.org/10.1111/auar.12109
3. Santos, J., Garcia, P.: A bibliometric analysis of sport economics research. Int. J. Sport Finance **6**(3), 222–244 (2011)
4. Podsakoff, P., MacKenzie, S., Podsakoff, N., Bachrach, D.: Scholarly influence in the field of management: a bibliometric analysis of the determinants of university and author impact in the management literature in the past quarter century. J. Manag. **34**(4), 641–720 (2008)
5. Baltagi, B.H.: Worldwide econometrics rankings: 1989-2005. Econometric Theory **23**, 952–1012 (2007). https://doi.org/10.1017/S026646660707051X
6. Han, J.-S., Ho, Y.-S.: Global trends and performances of acupuncture research. Neurosci. Biobehav. Rev. **35**, 680–687 (2011)
7. Li, J., Wang, M.-H., Ho, Y.-S.: Trends in research on global climate change: a Science Citation Index Expanded-based analysis. Global Planet. Change **77**, 13–20 (2011)
8. Wang, S., Zhou, C., Li, G., Feng, K.: CO2, economic growth, and energy consumption in China's provinces: investigating the spatiotemporal and econometric characteristics of China's CO2 emissions. Ecol. Ind. **69**, 164–195 (2016)
9. Heidari, H., Katircioglu, S., Saeidpour, L.: Economic growth, CO2 emissions, and energy consumption in the five Asean countries. Int. J. Electr. Power Energy Syst. **64**, 785–791 (2015)
10. Wolde-Rufael, Y.: Energy consumption and economic growth: the experience of African countries revisited. Energy Econ. **31**(2), 227–234 (2009)
11. Apergis, N., Payne, J.: Energy consumption and economic growth in central America: evidence from a panel cointegration and error correction model. Energy Econ. **31**(2), 211–216 (2009)
12. Burdisso, T., Sangiácomo, M.: Panel time series: Review of the methodological evolution. Stata J. **16**(2), 424–442 (2016)
13. Dupont-Kieffer, A., Pirotte, A.: The early years of panel data econometrics. Hist. Polit. Econ. **43**, 258–282 (2011). https://doi.org/10.1215/00182702-1158754
14. Balestra, P., Nerlove, M.: Pooling cross-section and time-series data in the estimation of a dynamic model: the demand for natural gas. Econometrica **34**, 585–612 (1966)

Macro-Econometric Forecasting for During Periods of Economic Cycle Using Bayesian Extreme Value Optimization Algorithm

Satawat Wannapan[✉], Chukiat Chaiboonsri, and Songsak Sriboonchitta

Faculty of Economics, Chiang Mai University,
Chiang Mai, Thailand
lionz1988@gmail.com, chukiat1973@gmail.com

Abstract. This paper aims to computationally analyze the extreme events which can be described as crises or unusual times-series trends among the macroeconomic variables. These data are statistically estimated by employing the optimally extreme point for supporting policy makers to specify the economic expansion target and economic warning level. The Nonstationary Extreme Value Analysis (NEVA) applying Bayesian inference and Newton-optimal method are employed to complete the researchs solutions and estimate the time-series variables such as GDP, CPI, FDI, and unemployment rate collected during 1980 to 2015. The results show there are extreme values in the trend of macroeconomic factors in Thailand economic system. This extreme estimation is presented as an interval. In addition, the empirical results from the optimization approach state that the exactly extreme points can be computationally found. Ultimately, it is clear that the computationally statistical approach, especially Bayesian statistics, is inevitably important for econometric researches in the recent era.

Keywords: Extreme event
Nonstationary Extreme Value Analysis (NEVA) · Newtons method
Differential Evolution Monte Carlo (DE-MC) · Macroeconomics

1 Introduction

Macroeconomics has been mentioned as an element that forms the part of a complex whole. In other words, the study of possible sources of economic fluctuations has been the major preoccupation in macroeconomics in recent years. As seen in Fig. 1, the examples of macroeconomic factors such as Thailand GDP, consumer price index (CPI), foreign direct investment (FDI), and rates of unemployment, evidently indicate that there are many fluctuated time-series trends found in Thai economic system during 1980 to 2015. Accordingly, this seems that to investigate these oscillated points for explaining and forecasting economic situations is extremely crucial.

Interestingly, the question is fundamental if one shall gain insight into the workings of the economy, and aid in the formulation and conduct of economic policy (Bjrnland 2000). To address this issue, the combination between a multiplex econometric method and mathematically statistical analysis, including the Bayesian extreme value model and Newton-optimal processing, was computationally employed to focus on a precise estimation, which is crucial for economic predictions. Consequently, for the former model, extreme events in the macroeconomic factors will be clarified, and these empirical results can evidence the rare situations in Thai economic system. For the latter method, the precisely extreme point will be computationally announced for providing policy recommendations.

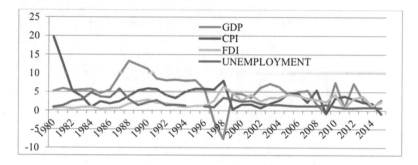

Fig. 1. Presenting the macroeconomic variables of Thailand economic system (GDP, CPI, FDI, and unemployment rate)

2   The Objective and Scope of Research

The paper aims to study the extreme events of Thailand macroeconomic variables as well as clarify the extremely optimal point to notify the alarming sign for policy makers. In addition, the scope of this research intensively focuses on mathematically computational approaches applying to the econometric analysis. The overview of the conceptual framework and Bayesian extreme value optimization algorithm under a covering of this study is presented in Fig. 2.

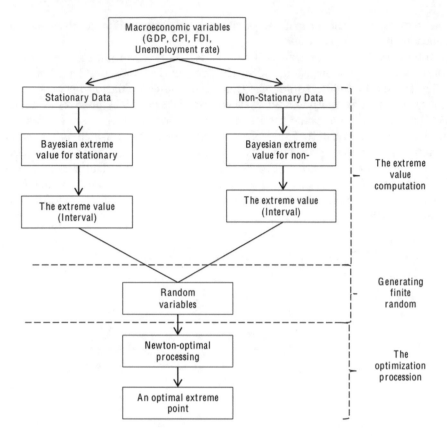

Fig. 2. The conceptual frame work and Bayesian extreme value optimization algorithm of the research

3   Literature

Undeniably, a business cycle in an economic system and macroeconomic analysis have been considerably interested and econometrically investigated by many scholars for few decades. In business cycle theory, this study has been a renewed interest since the instability of the world economy in the aftermath of the oil price shocks in the 1970's. More recently, a new branch of the classical models referred to as the Real Business Cycle (RBC) models have been developed, that emphasize real productivity shocks, as opposed to aggregated demand shocks, as a source of economic fluctuations (Kydland and Prescott 1982; Bjrnland 2000).

According to macroeconomic research papers, for instance, Chow (2001), Hrdahl et al. (2004), Visco (2014), Hall et al. (2014), and Wu and Xia (2016), this group of researchers worked on the topic that specified the way to understand and explain economic crises as well as econometrical forecasting. Also, their empirical statements are only based on normal situational data and classically statistical analyses. On the other hand, focusing on the subjective statistical analysis called Bayesian extreme value approach for macroeconomic factorial estimating, this seems that there are few researchers studying on this method. For example, Calabrese and Giudici (2015) who explored the generalized extreme value (GEV) for the basis of both macroeconomic and bank-specific microeconomic factors, and Chaiboonsri and Chaitip (2017) who investigated the extreme linkage between the financial indexes such as SGX, KLSE, SET, IDX, and PSE and the macroeconomic variable (GDP) based on Bayesian inferences. Interestingly, this paper is the extension of the Bayesian extreme value computation by employing the optimization processing called Newton Method. Moreover, this mathematically computational analysis relied on macroeconomic factors can be one of useful solutions regarding the issue of economic alarming signs, which have been difficultly defined for long time.

4   Research Methodology

4.1   The ADF Unit Root Test Based on Bayesian Inference

The ADF test analyzes the null hypothesis that a time-series index is stationary against the alternative (non-stationary data). Conditionally, dynamics in data have the autoregressive moving average model (ARMA) (Said and Dickey 1984). The ADF test is based on estimating the regression test, which is expressed by the Eq. (1),

$$\Delta y_t = c + \alpha' D_t + \varphi y_{t-1} + \sum_{j=1}^{p} \gamma_j \Delta y_{t-j} + \varepsilon_t. \tag{1}$$

The prior density of φ is formulated and expressed in the Eq. (2),

$$p(\theta) = p(\phi)p(a^*|\phi) \tag{2}$$

The marginal likelihood for ϕ is

$$l(\phi|D)\alpha \int l(\varphi|D)\phi(a^*|\phi)da^* \tag{3}$$

After setting the prior, we give the error measure and compute a single point estimate of WLS for the weight. The likelihood and the prior are specified. Thus, we calculate the *posterior distribution* over w, which explains our *belief level*, via Bayes rule (Eq. (4)),

$$p(w|t, \alpha, \sigma^2) = \frac{likelihood \times prior}{normalised\ factor} = \frac{p(t|w, \sigma^2)p(w|\alpha)}{p(t|\alpha, \sigma^2)}. \tag{4}$$

In academic researches, Bayesian statistics considers hypotheses regarding multiple parameters by adapting Bayes factor comparisons. Let M_i be the model devised in the term of the null hypothesis and M_j be the model of the alternative hypothesis. The posterior odds ratio of M_i and M_j is shown in the Eq. (5),

$$\frac{pr(M_i|y)}{pr(M_j|y)} = \frac{pr(y|M_i)}{pr(y|M_j)} \times \frac{\pi(M_i)}{M_j}. \tag{5}$$

The Bayes factor is summarized by the statistical model proposed by Jeffrey (1961). The interpretation in half-units on Jeffreys scales was simply explained as (Table 1):

Table 1. Presentation the summary of Jeffreys Guideline for model comparison

Items	The recommendations of Jeffrey's was modified by Authors
BF < 1/10	Strong evidence for M_j
1/10 < BF < 1/3	Moderate evidence for M_j
1/3 < BF < 1	Weak evidence for M_j
1 < BF < 3	Weak evidence for M_i
3 < BF < 10	Moderate evidence for M_i
10 < BF	Strong evidence for M_i

Source: modified from Jeffreys (1991) by authors

4.2   Generalized Pareto Distributions (GDP) and Non-stationary Extreme Value Theory Analysis

At the beginning, focusing on the class of issues where the behavior of the distributions over (below) a high (small) threshold is of interest, characterized as *extreme events*. A random quantity with distribution function $F(x)$ under certain conditions shown by Pickands (1975), $F(x|u) = P(X \leq u+x|X > u)$, can be approximated by a Generalized Pareto distribution (GPD), which is defined in Eq. (6).

$$G(x|\zeta, \sigma, u) = \begin{cases} 1 - (1 + \frac{\zeta(x-u)}{\sigma})^{-1/\zeta}, & if \ \zeta \neq 0 \\ 1 - exp[-\frac{(x-u)}{\sigma}], & if \ \zeta = 1 \end{cases} \qquad (6)$$

where $\sigma > 0$ and ζ are the scale and shape parameter, respectively. The data exhibit heavy tail behavior when $\zeta > 0$. To conduct the threshold come from a GPD introduced in Eq. (6), the proposed model assumes that observations under the threshold, u, is generated from a certain distribution with parameters, η. The model is here denoted as $H(.|\eta)$. Hence, the distribution function F of any observation X can be described as

$$F(x|\eta, \zeta, \sigma, u) = \begin{cases} H(x|\eta), & if \ x < u \\ H(u|\eta) + [1 - H(u|\eta)]G(x|\zeta, \sigma, u), & if \ x \geq u. \end{cases} \qquad (7)$$

For an Eq. (7) which is a sample size n, $x = (x_1, ..., x_n)$ from F, parameter vector $\theta = (\eta, \sigma, \zeta, u)$, $A = (i : x_i) < u$, and $B = (i : x_i)$, the likelihood function is defined as (Behrens et al. 2004)

$$L(\theta; x) = \prod_A h(x|\eta) \prod_B (1 - H(u|\eta))\{\frac{1}{\sigma}[1 + \frac{\zeta(x_i - u)}{\sigma}]_+^{-(1+\zeta)/\zeta}\}, \qquad (8)$$

for $\zeta \neq 0$, and $L(\theta; x) = \prod_A h(x|\eta) \prod_B (1 - H(u|\eta))[(1/\sigma)exp\{(x_i - u)/\sigma\}]$, for $\zeta = 0$.

The threshold, u, as shown in Eq. (8), is the point where the density has a discontinuity, which depends on the larger or smaller jumped density, and each case the choice of which observations will be considered as exceedances that can be more obvious or less evident. The smaller jumped density is more difficult to estimate the threshold. Thus, strong discontinuities, or large jumps, indicate separation of the data such that it is expected the parameter estimation would be easier. As shown in Fig. 3, the density model of threshold is schematically presented.

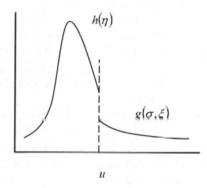

Fig. 3. Presentation the schematics of the threshold model (Modified from Behrens et al. 2004)

In terms of Nonstationary Extreme Value Analyses (NEVA) using Bayesian Inference, applying from Cheng et al. (2014), NEVA employs a Bayesian technical computation to infer the GPD distribution of parameters under stationary and non-stationary conditions. Typically, the Bayesian inference consisting of Markov chain Monte Carlo (MCMC) approach for obtaining the posterior distribution of parameters from an arbitrary distribution has been popular and used in several investigations of extremes (Coles and Powell 1996; Cheng et al. 2014). This approach integrates the knowledge brought by a prior distribution and the observational vector; $\overrightarrow{y} = (y_t)_{t=1:N_t}$; and the posterior distribution of parameters; $\theta = (\mu, \sigma, \zeta)$. In this case, N_t refers the number of observations in the observable vector $\overrightarrow{y}$. The omission prior for the location and scale parameters are non-informative normal distributions. It can be adjusted to informative priors, and other choices of distribution functions can be used in NEVA.

Assuming the linkage between observations, the Bayes theorem for estimation of GPD parameters under the non-stationary assumption can be expressed as (Renard et al. 2006; Coles 2001; Cheng et al. 2014) (see an Eq. (9))

$$p(\beta|\overrightarrow{y}, x)\alpha p(\overrightarrow{y}|\beta, x)p(\beta|x) \qquad (9)$$

$$p(\beta|\overrightarrow{y}, x) = \prod_{N_t, t=1} p(\overrightarrow{y}|\beta, x(t)) = \prod_{N_t, t=1} p(\overrightarrow{y}|\mu(t), \sigma, \zeta), \qquad (10)$$

where $\beta = (\mu_1, \mu_0, \sigma, \zeta)$ in Eq. (10) are estimated parameters. Conversely, the stationary model can be described as a special case of the above two equation without $x(t)$,

$$p(\theta|\overrightarrow{y})\alpha p(\overrightarrow{y}|\theta)p(\theta) = \prod_{N_t, t=1} p(y_t|\theta)p(\theta), \qquad (11)$$

where $x(t)$ refers to the set of all covariate values under the non-stationary assumption. The outcomes of posterior distributions $p(\theta|\overrightarrow{y})$ as well as $p(\theta|\overrightarrow{y}, x)$ contribute information about parameters under the stationary condition $\theta = (\mu, \sigma, \zeta)$ or non-stationary model $\beta = (\mu_1, \mu_0, \sigma, \zeta)$ (see an Eq. (11)). Technically, in this paper, NEVA generates a large number of realizations from the parameter joint posterior distribution. Modernly, the *Differential Evolution Markov Chain* $(DE - MC)$ (Ter Braak 2006; Cheng et al. 2014) is applied. The DE-MCMC conducts and utilizes the genetic algorithm, which is the Differential Evolution (DE) for global optimization over the parameter space with the (DE) approach.

Obviously, the DE-MC algorithm employs the fitness assignment scheme used in the corresponding Multi-objective Shuffled Complex Evolution Metropolis (MOSCEM) algorithm, which has demonstrated certain effectiveness and efficiency in the multi-objective optimization of macro-econometric models, and it is relied on the number of external non-dominated points (Zhu and Li 2010).

Inside the framework of the fitness assignment, the calculation of a population P with m individual configurations, $C1, Cm$, is based on the strength S_i of each non-dominated configuration C_i. The strength S_i is described as the proportion of configurations in P dominated by C_i. Then, the fitness of an individual C_i is represented as follows, (see an Eq. (12)),

$$fit(C_i) = \begin{cases} s_i \\ 1 + \sum_i^{i=j} s_j, \end{cases} \qquad (12)$$

where $fit(C_t) = s_i$ refers to C_t is non-dominated. Conversely, $fit(C_t) = s_i$ denotes C_t is dominated. The configurations with fitness less than 1.0 are technically non-dominated. The fitness function influences to the non-dominated solutions with less dominated configurations while those with a lot of neighbors in their niche are penalized (Vrugt 2002; Cheng et al. 2014). Modernly, the new algorithm of Monte Carlo simulated processing employs *differential Evolution* (DE) to produce new configurations based on the current population. Inside this scheme, configurations $C_i(x_1, ..., x_n)$ and mutant vectors V_i is normally set as follows, (see an Eq. (13)),

$$V_i = C_{r1} + F(C_{r2} - C_{r3}) \qquad (13)$$

where $i, rl, r2$, and $r3$ are randomly integer numbers in the interval $[1, m]$, and $F > 0$ is a tunable amplification control constant. In addition, other formations of mutant vector generating are also provided in the study explored by Storn and Price (1997). Consequently, a new confirmation $C_i'(x_1', ..., x_n')$ is simulated by the crossover operation on V_i and C_i, and these can be defined in Eq. (14).

$$x_j' = \begin{cases} v_j, & j = (k)_n, [k+1]_n, ..., [k+L+1]_n, \\ x_j, & otherwise \end{cases} \tag{14}$$

where $[.]_n$ indicates the modulo operation with modulus n, k is a randomly generated integer from the interval $[0, n-1]$, L is an integer drawn from $[0, n-1]$ with probability $\Pr(L = l) = (CR)^l$, and $CR \in [0, 1)$ is the crossover probability.

Necessarily, Monte Carlo movements apply the *Metropolis Transitions* for generating new population. Defining this transition as a probability, the metropolis transition relies on the variation of the fitness function value. The Metropolis-Hastings ratio is expressed as (see an Eq. (15)):

$$w(C_i \to C_t^i) = e^{\frac{-(fit(c_t')fit(C_t)}{T}}, \tag{15}$$

where T is the simulated macroeconomic population, which is used to control the acceptance rate of the Metropolis transitions. Also, the new form C_i' generated by DE, which is confirmed with the probability as follows (see an Eq. (16)):

$$min(1, w(c_i \to c_i')) \tag{16}$$

Ultimately, the majority of the computations of DE-MC convolutions such as the generating of new configurations, evaluation of objective functions, fitness assignments, and Metropolis transitions can be independently complemented, and this modernly simulated calculation only requires scrutinizing operations to access the shared population information, which is particularly suitable for a massively parallel computing system with shared memory (Cheng et al. 2014).

4.3   The Statistical Probability for Generating Random Variable Sets

4.3.1   Random Sets

Since the data (information) from the Bayesian extreme value estimation is resulted as an interval of the finite numerical set, random elements based on the mathematical theory of probability are used to performance experimental or observing data sets. A random element is a map $X : \Omega \to U$ (two arbitrary sets) where Ω equipped with a σ - field and U equipped with σ - field (U) such that $X^{-1}(U) \subseteq A$. When P is a probability measure on, the law of X is $P_X = PX^{-1}$ (probability measure on U). The law P_x contains all probabilistic information for studying X, and the range U of X is the space of our data where $U = \mathbb{R}^d$ is an Euclidean space and continuous function on $[0, 1]$; $U = C([0; 1])$. Hence, a random element whose range U is a space of set called a random set (roughly speaking: it is a set obtained at random) (Nguyen 2014).

Considering into random sets as sampling designs, they are the outcome of a random experiment where the results of the experiment are sets. For instance, when we wish to obtain a portion of a (finite) population, we randomly design a procedure to achieve it. Formally, as shown in Eq. (17), let U be a finite set and its power set 2^U is interested. In order to choose elements of 2^U, a probability density function $f \colon 2^U$ is $[0, 1]$, $\sum_{A \subseteq U} f(A) = 1$ is designed at random. A sampling design is a finite random set S on the population of interest, and it is a random element; $S : (\Omega, A, P) \rightarrow (2^U, 2^{2^U}, p_S)$ with probability density $f(.)(P(S = A)) = f(A)$. Its distribution F is defined as $2^u \rightarrow [0,1]$ where $P(S = A) = F(A) = \sum_{B \subseteq A} f(B)$. Thus, the degree of the separation of the design is characterized by the coverage function (Nguyen 2014). This can be described as follows

$$\pi_S : U \rightarrow [0, 1], \qquad \pi_S(u) = P(u \in S) = \sum_{u \in A} f(A) \qquad (17)$$

where the simple random sample, which is a random sample of size n, is explained as (see an Eq. (18))

$$P(A) = \begin{cases} \frac{1}{\binom{N}{n}}, & if \ |A| = n. \\ 0, & if \ |A| \neq n \end{cases} \qquad (18)$$

For the experimental data sampling, this type of data sample is created by random sampling designs. Let $p(.)$ be given and supposed that $A \subseteq V$ such that (see an Eq. (19))

$$\sum_{K \in A} X_k \rightarrow \sum_{K \in V} X_k \qquad (19)$$

4.4   Investigating the Extreme Point by Using Newton Optimization Approach

Following the process of random variable generating, the interesting point that can be noticed in the randomly extreme variable analysis is which point is the exact extreme position for macroeconomic factors using in this study. To mathematically solve the issue, a computationally operational method, namely *Newtons method* is employed to investigate the optimization for clarifying an extreme spot in Thai macroeconomic variables. Typically, the basic idea of Newtons method is used as a basis on linearization. Given $R^1 \rightarrow R^1$ is a differential function. Thus, we are trying to solve the equation as (see Eq. (20))

$$F(x) = 0 \qquad (20)$$

Technically, the initial point x_0 is made to be the starting points, and assembled the linear approximation of $F(x)$ in the neighbor of $x_0 : F(x_0+h) \approx F(x_0)h$ and resolve the consequent linear formation $F(x_0) + F'(x_0)h = 0$ As a result, the recurrent method is expressed as follows (Polyak 2007):

$$x_{k+1} = x_k - F'(x_k)^{-1} F(x_k), \qquad k = 0, 1, ... \qquad (21)$$

The above equation (Eq. (21)) is the method invented by Newton in 1669. More precisely, this arising linear equation is dealt with polynomial which is expressed of $F(x_0 + h)$. Moreover, the progress in improvement of the method is related to such famous mathematicians such as Fourier who proved the quadratic method convergences in the neighborhood of a root in an Eq. 18 as well as Cauchy provided the multidimensional extension and used the method to prove the existence of a root of an equation in 1829 (Polyak 2007). In this paper, the authors employ the values of random variables, which are the results calculating from the random sampling variables, to be the data set for lag optimizing.

Theoretically, the quadratic convergence in Newtons method is based on two fundamentals.

Theorem 1. The main convergence result is defined as continuously differentiable on the data set: $B = |x : |x - x_0| \leq r|$ which is the visible linear operation model $(F'(x_0))$, where $(x \in B)$, expressed as $|F'(x_0)^{-1}F(x_0)| \leq \eta$ and $|F'(x_0)^{-1}F'(x)| \leq K$ such that $h = K\eta < \frac{1}{2}$ and $r \geq \frac{\eta}{h2^k}(2h)^{2^k}$. Consequently, the process in an Eq. (21) is well clarified and it will converge to x^* with a quadratic rate as follows (Polyak 2007),

$$|x_k - x^*| \leq \frac{\eta}{h2^k}(2h)^{2^k}. \tag{22}$$

Theorem 2. Suppose F is defined and twice continuously differentiable on the balls $B - [x : |x - x_0| \leq r]$ Differentially, the linear function $F'(x)$ is invertible on B and $|F'(x)^{-1}| \leq \beta$, $|F'(x)| \leq K$, $|F'(x)| \leq \eta$, where $x \in B$. Now, we can achieve linear parameters as follows: $h = K\beta^2\eta < 2$ and $r \geq \beta\eta \sum_{n=0}^{\infty}(\frac{h}{2})^{2^k-1}$. Thus, the solution $x^* \in B$ and the progress converge to x^* with quadratic rate can be defined as (see an Eq. (22))

$$|x_k - x^*| \leq \frac{\beta\eta(h/2)^{2^k-1}}{1 - (h/2)^{2^k}}. \tag{23}$$

Graphically, as presented in Fig. 4, the convergence processing of Newtons method approximates $F'(x)$ by its tangent line at $x^{(0)}$ whose root, $x^{(1)}$, serves as the next approximation of the true, x^*. The next similar step yields $x^{(2)}$, which is already gone closer to the root at x^*

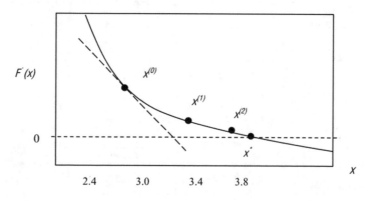

Fig. 4. Presentation the convergence processing in Newton-optimal approach

5   Empirical Results

5.1   Descriptive Information

Generally, the basic information, including average values, max-min points, and standard deviation are displayed in Table 2. Technically, collected variables such as GDP, CPI, FDI, and unemployment rate are transformed into growth rates.

Table 2. Descriptive data of Thailand macroeconomic variables

	GDP	CPI	FDI	UNEM
Mean	5.196206	3.993222	2.360140	1.948571
Median	5.533828	3.323594	2.278402	1.500000
Maximum	13.28811	19.70425	6.434801	5.800000
Minimum	−7.633734	−0.898069	0.419531	0.700000
Std.Dev	4.032781	3.755795	1.483663	1.221468
Observations	35	35	35	35

Source: From computed

5.2   The Test of Data Stationary Conditions

Obviously, macroeconomic factors especially used to do the investigation of extreme dependences are the time-series information. Before computationally statistical analyses, these variables should be checked the data stationary condition. Technically, the Bayesian inference is employed to verify the stable condition. The unit-root test is shown in Table 3, and the details have both stationary $(I(0))$ and non-stationary $(I(1))$ variables.

Table 3. The ADF unit root test based on Bayesian inference in yearly data of Thailand macroeconomic factors during 1980 to 2015

Variables	Bayesian factor model	Hypotheses	Numbers of MCMC regress iterations	Bayesian factor ratios (M1/M2)	Interpretation of the Bayesian factor	Result
Thai GDP	Model 1 Model 2	H0 (Mi): Non-stationary H1 (Mj): Stationary	101,000 74,000	0.204	Moderate evidence for Mj	I(0)
CPI	Model 1 Model 2	H0 (Mi): Non-stationary H1 (Mj): Stationary	101,000 92,000	0.000	Strong evidence for Mj	I(0)
FDI	Model 1 Model 2	H0 (Mi): Non-stationary H1 (Mj): Stationary	101,000 92,000	0.205	Strong evidence for Mj	I(0)
Unemployment rate	Model 1 Model 2	H0 (Mi): Non-stationary H1 (Mj): Stationary	101,000 101,000	68	Strong evidence for Mi	I(0)

Source : From computed

5.3   The Investigation of the Extreme Value Estimation Using Bayesian Inference

As expressed in Table 4, with the Bayesian extreme value analysis called Non-stationary Extreme Value Analysis (NEVA), its advantage is this approach can computationally explore the extreme events of the observable variables, even though the set of data contains both stationary and non-stationary conditions (Cheng et al. 2014). Empirically, the results deduced from the Bayesian approach found that the values of each factor are displayed as an interval. The details are shown in the Figs. 5, 6, 7 and 8 in the Appendix (part 1), respectively. Descriptively, the interval of the extreme value regarding Thailand GDP is between 8% and 13%. For the consumer price index (CPI), the length of the extreme interval is during 5.8% to 12.5%. Considering the extreme result of yearly foreign direct investments (FDI), the interval is the length between 2 and 4%. For the factor of Thailand unemployment rates, the non-stationary estimated outcome shows that the extreme interval is between 2.3 and 4.4%. Lastly, for the general Thailand stock value, the extremely calculated interval is between 200 and 500%, respectively.

5.4   The Newtons Method for Optimizing the Extreme Values

To extend the details of the results estimated from Bayesian extreme value processing, the Newton-optimization method can be the efficiently computational statistical tool for providing the precisely optimal position to clarify that which an extreme value should be announced to be the alarming point. By comparing with the general mean, the process of the Newton-optimal approach is more complex and reliable since the simulated iterations were employed. As seen the details in Table 3, the empirical results presented that there are the differences between the Newton-optimal value and general mean. Descriptively, the optimal point for GDP is 10.55 (10.50%), CPI is 8.96 (9.15%), FDI is 3.18 (3.00%), and unemployment rate is 3.496 (3.35%), respectively. In addition, the optimal extreme point of each macroeconomic factor is graphically displayed in the Figs. 9, 10, 11 and 12 in the Appendix (part 2).

Table 4. Presentation the optimally extreme value calculation of Thailand macroeconomic factors

Variables (Growth rate)	The extreme interval (Bayesian extreme estimation) (Percent)	General mean (Percent)	Optimal value (Newtons method) (Percent)
GDP	8 13	10.5	10.55
CPI	5.8 12.5	9.15	8.96
FDI	2–4	3	3.18
Unemployment rate	2.3–4.4	3.35	3.496

Source: From computed

6   Conclusion

This paper successfully clarifies the extreme events among five macroeconomic factors such as the expansion rates of Thailand GDP, CPI, FDI, unemployment, and stock values. The main propose of this study is to apply the computationally statistical approaches, including the nonstationary extreme value analysis (NEVA) using Bayesian inference, random variable processing, and Newton-optimal method, to solve the alarming point based on extreme events. With time-series data collected during 1980 to 2015 (35 observations), the empirically estimated results state that the NEVA method can clearly provide the interval of extreme value measurement of each variable. Moreover, the Newton optimization method can computationally investigate the exact position for the specification of economic alarming levels.

Following the previous statement, to precisely forecasting the alarming sign for the economic system, especially in the macroeconomic level, is very necessary. The empirical outcomes in this paper such as the interval of extreme value calculations and the optimal point can be usefully supported policy makers to simultaneously decide the economic expansion target and economic alarm level. Accordingly, it is obvious that mathematical theorems and computationally statistical methods are inevitably crucial for econometric researches in the recent era.

Appendix A: The Bayesian extreme value estimation (Part 1)

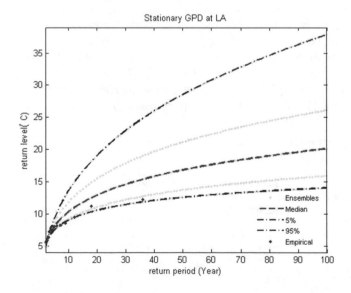

Fig. 5. Presentation the Bayesian extreme result regarding Thailand GDP

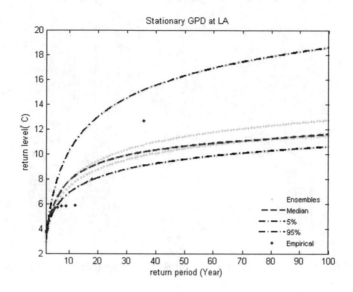

Fig. 6. Presentation the Bayesian extreme result regarding CPI

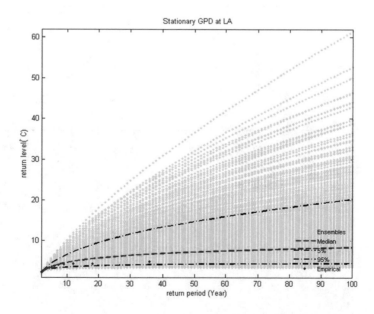

Fig. 7. Presentation the Bayesian extreme result regarding FDI

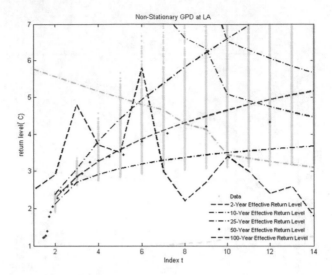

Fig. 8. Presentation the Bayesian extreme result regarding unemployment rate

Appendix B: The Newton-optimal processing (Part 2)

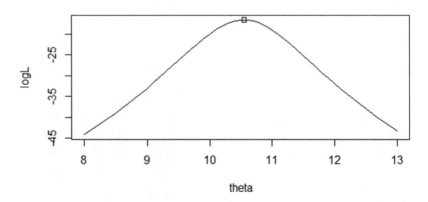

Fig. 9. Presentation the Newton-optimal point for the growth rate of GDP

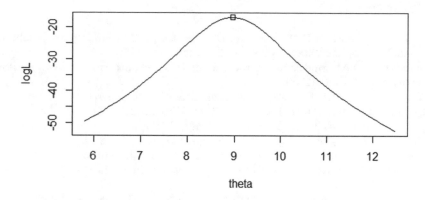

Fig. 10. Presentation the Newton-optimal point for the growth rate of CPI

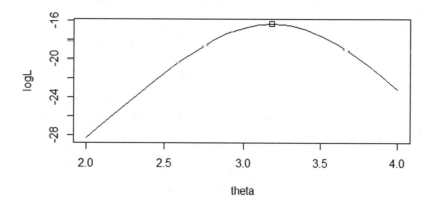

Fig. 11. Presentation the Newton-optimal point for the growth rate of FDI

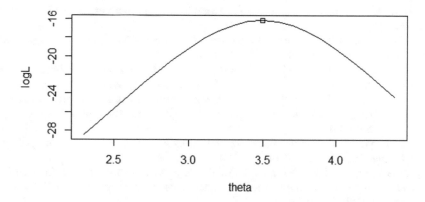

Fig. 12. Presentation the Newton-optimal point for the growth rate of unemployment

References

Behrens, C.N., Lopes, H.F., Gamerman, D.: Bayesian analysis of extreme events with threshold estimation. Stat. Modell. **4**, 227–244 (2004)

Bjrnland, H.C.: VAR Models in Macroeconomic Research. Statistics Norway Research Department, Norway (2000)

Calabrese, R., Giudici, P.: Estimating bank default with generalised extreme value regression models. J. Oper. Res. Soc. **66**(11), 1783–1792 (2015)

Cheng, L., AghaKouchak, A., Gilleland, E., Katz, R.W.: Non-stationary extreme value analysis in a changing climate. Clim. Change **127**, 353–369 (2014)

Chaiboonsri, C., Chaitip, P.: Forecasting methods for safeguarding ASEAN-5 stock exchanges during extreme volatility. Int. J. Trade Global Markets **10**(1), 123–130 (2017)

Chow, G.C.: Econometric and economic policy. Stat. Sin. **11**, 631–660 (2001)

Coles, S.: Introduction to Statistical Modeling of Extreme Values. Springer, London (2001)

Coles, S.G., Powell, E.A.: Bayesian methods in extreme value modelling. Int. Stat. **64**, 114–193 (1996)

Hall, S.G., Roudoi, A., Albu, L.L., Lupu, R., Călin, A.C.: Lawrence R. Klein and the economic forecasting a survey. Roman. J. Econ. Forecast. **17**(1), 5–14 (2014)

Hrdahl, P., Tristani, O., Vestin, D.: A joint econometric model o f macroeconomic and term structure dynamics. Working paper number 405. European Central Bank (2004). http://www.ecb.int

Jeffreys, H.: Theory of Probability, 3rd edn. Oxford University Press, New York (1961)

Kydland, F.E., Prescott, E.C.: Time to build and aggregate fluctuations. Econometrica **50**, 1345–1370 (1982)

Nguyen, H.T. Probability for statistics in econometrics. Center of Excellence in Econometrics, Faculty of Economics, Chiang Mai University, Thailand (2014). http://old.viasm.edu.vn/wp-content/uploads/2014/11/VIASMWorkshop.pdf

Pickands, J.: Statistical inference using extreme order statistics. Ann. Stat. **3**, 119–131 (1975)

Polyak, B.T.: Newtons method and its use in optimization. Eur. J. Oper. Res. **181**, 1086–1096 (2007)

Renard, B., et al.: An application of Bayesian analysis and Markov chain Monte Carlo methods to the estimation of a regional trend in annual maxima. Water Resour. Res. **42** (2006)

Said, S.E., Dickey, D.: Testing for unit roots in autoregressive moving-average models with unknown order. Biometrika **71**, 599–607 (1984)

Storn, R., Price, K.: Differential evolution – a simple and efficient heuristic for global optimization over continuous spaces. J. Glob. optim. **11**, 341–359 (1997)

Ter Braak, C.J.F.: A Markov chain Monte Carlo version of the genetic algorithm differential evolution: easy Bayesian computing for real parameter spaces. Stat. Comput. **16**, 239–249 (2006)

Visco, I.: Lawrence R. Klein: macroeconomics, econometrics and economic policy. J. Pol. Model. **36**, 605–628 (2014)

Vrugt, J.A., Gupta, H.V., Bastidas, L.A., Boutem, W., Sorooshian, S.: Effective and efficient algorithm for multiobjective optimization of hydrologic models. Water Resour. Res. **39**(8), 1214–1232 (2002)

Wu, J.C., Xia, F.D.: Measuring the macroeconomic impact of monetary policy at the zero lower bound. J. Money Credit Bank. **48**(2–3), 254–291 (2016)

Zhu, W., Li, Y.: GPU-accelerated differential evolutionary Markov chain Monte Carlo Method for multi-objective optimization over continuous space. In: Proceedings of the 2nd Workshop on Bio-Inspired Algorithms for Distributed Systems, BADS 2010, pp. 1–8 (2010)

Forecasting of VaR in Extreme Event Under Economic Cycle Phenomena for the ASEAN-4 Stock Exchange

Satawat Wannapan[✉], Pattaravadee Rakpuang, and Chukiat Chaiboonsri

Faculty of Economics, Chiang Mai University, Chiang Mai, Thailand
lionz1988@gmail.com, kaiimook1994@gmail.com, chukiat1973@gmail.com

Abstract. This paper was proposed to computationally investigate the cycling details and risk management of the ASEAN-4 financial stock indexes, including Bangkok Bank (BBL), Development Bank of Singapore Limited (DBS), Commerce International Merchant Bankers (CIMB), and Bank Mandiri (Mandiri). These daily time-series data were observed during 2012 to 2017. Technically, this paper employed the econometric tool called Markov Switching Model (MS-model), the extreme value application called Generalized Pareto Distribution (GPD-model), and the risk management method called Value at Risk (VaR) to provide the estimated solutions and recommendations for investing in these financial stocks. Empirically, the switching regime estimation resulted that these four financial indexes obviously contain real business cycling movements, which were described as bull and bear regimes. Additionally, the results estimated by the GPD model confirmed that there were extreme events inside the trends of the four stock indexes. Ultimately, the outcomes calculated by the risk measurement for extreme cases, which were economic crises, stated that there was an enormously high risk to considerably invest only in short earnings within these four financial stock indexes. Consequently, long-run investment should be mentioned.

Keywords: Financial stock index · ASEAN · MS-model
GPD estimation · Value at Risk (VaR)

1 Introduction

In the new era of modern economic developments, the financial sectors, especially in the ASEAN Economic Community (AEC), are on which one of the driving economic factors that many researchers are now focusing. Generally, most countries in ASEAN have been defined as the emerging market, which can interestingly be the target for investing. However, not every country gives priority to financial sectors intensively, even though this field is necessary. As a result, to clarify the details of permanent financial issues in ASEAN countries is inevitable. Thus, the examination on the financial sectors in ASEAN are really significant due to

© Springer International Publishing AG 2018
V. Kreinovich et al. (eds.), *Predictive Econometrics and Big Data*, Studies in Computational Intelligence 753, https://doi.org/10.1007/978-3-319-70942-0_52

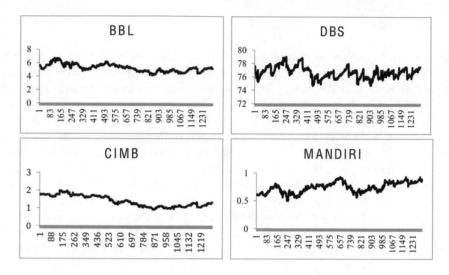

Fig. 1. Presentation the descriptive data of ASEAN financial stock indexes

speculations and long-term investments. In particular, the risk forecasting in the financial market is one part that is important to decide a suitable choice for the portfolios of investors.

Consideration into the financial index of major countries in ASEAN, the details were graphically shown in Fig. 1. BBL - Bangkok Bank is one of the biggest financial companies in Thailand. DBS - The Development Bank of Singapore Limited is the major Singaporean banking service. For Malaysia, CIMB - Commerce International Merchant Bankers Bhd is defined to be the financial leader, and Mandiri Bank Mandiri is the main banking company in Indonesia. Moreover, AEC countries have been recently improving their power on bargaining international transactions in the world financial market. Undeniably, many investors are very interested to invest in the various types of business tasks. Consequently, this paper is contributed to statistically investigate the rare situations, which can be also expressed as extreme dependences, in the financial sectors of ASEAN countries and conclude suitable recommendations for addressing the issue.

2   The Objective of Research

The main objective of this paper is to explore extreme events and provide choices for risk reduction in the financial indexes which are the banking stock indicators in four countries of AEC for recommending suitable choices of financial investments. The data was collected as daily time-series information, which was during 2012 to 2016.

3   Literature

Because the Real Business Cycle theory (RBC) based on microeconomics can be an adequate basis for understanding business cycles, the theoretical research conducted by King and Rebelo (2000) stated that the real business cycle approach as the new orthodoxy of macroeconomics (New Keynesian school) is now commonly applied, employing to the tasks in monetary economics, public finance, asset pricing and so on. The empirical outcome of this remeasurement suggested the nature of macroeconomic fluctuations originated from real causes (shocks) as regimes, which is a prematurely dismissed idea ignored by economists that needs to be considered. In addition, Ernst and Stockhammer (2003) analyzed the Austrian business cycle model to compare the Keynesian and classical economic thoughts, using the result to develop the dynamics of European economic fluctuations. Interestingly, for the new era of econometric research, real business cycle theory is becoming more and more crucial for describing economic fluctuations in real circumstances.

To extend the issue regarding economic cycle in this econometric investigation, extreme value theory has been considered to predict risks in financial factors (for example, banking and insurance) and to discover probability of risks in the future. The exploration on the testing of GEV conditions as well as the theoretical review in the statistical tools by Embrechts et al. (1999) and Neves and Alves (2008) revealed the results needed to forecast the probability on each situation in a teletraffic data set. Furthermore, the example studied by Einmahl and Magnus (2008) examined the statistics of athletic performance using the theory of estimation to specify suitable instruments for athletes and beyond. Moreover, Andulai (2011) used the theory of risk estimation in the financial institutions in Europe which relied on the underlying theory analysis to predict the risk level in European stock markets. Lastly, Garrido and Lezaud (2013) showed the overview of probability and statistical tools on the theory to estimate and provide a real parameter on the GEV index. Interestingly, this paper would apply the GPD extreme analysis to improve the restriction of GEV calculation.

In order to reduce the loss deviation, the value at risk (VaR) adapted in this paper is important to specify risk management. Manganell and Engle (2001) employed the Historical Simulation VaR method to evaluate the risk level and showed the overview. Furthermore, the exploration on the conditional Value-at-risk (VaR) explained financial risks based on the distribution of random sampling data by Rockafellar and Uryasev (2002). In addition, Sampera et al. (2013) used the GlueVaR, which is a new measurement of risk in financial instruments and insurance instruments, explaining the relationship in GlueVaR, VaR, and TVaR. The paper showed the terms of GlueVaR as an overall risk measure of non-financial problems. Lastly, Perez and Murphy (2015) used the conditional VaR to evaluate risks of financial instruments as well comparing unfiltered variables and filtered variables of the VaR that are considered alternatives in the upcoming future (Fig. 2).

4   The Framework of Research

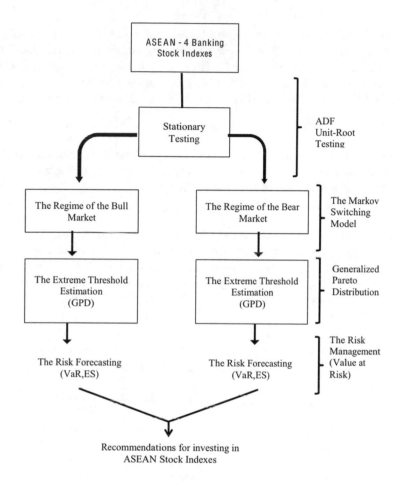

Fig. 2. The conceptual framework of the research

5   Methodology

5.1   The Markov Switching Model (MS-Model)

Theoretically, the Markov Switching Model (MS-model) explains the institutional changes on business and economic cycles (boom periods and recession periods), which can be effectively employed to investigate and clarify the movement of price indexes in stock markets. In particular, this paper focused on the yearly bank stock indicators during 2012–2017. As stated by Hamilton (2005), Markov regime switching models are a type of specification of which the selling point is the flexibility in handling processes driven by heterogeneous states of the world. Technically, this model is defined as follows,

$$y_t = c_{s_t} + \phi y_{t-1} + \varepsilon_t. \tag{1}$$

As the detail in Eq. (1), the value $C_{s_t} = 1$ for $t = 1, 2, 3, ..., n$ and $C_{s_t} = 2$ for $t = t_0 + 1, t_0 + 2, ..., t_0 + n$ follow the assumption of a normal distribution with zero mean and variance $u(0, \sigma_{s_t}^n)$ (Perlin 2010). In addition, the theory of probability is applied for MS model, which is supposed that the boom period is $C_{s_t} = 1$ and the recession period is $C_{s_t} = 2$. Thus, this can be written as

$$Pr(s_t = j | s_{t-1} = i, \ s_{t-2} = k, ..., y_{t-1}, y_{t-2}, ...) \ = \ Pr(s_t = j | s_{t-1} = i) \ = \ p_{ij}. \tag{2}$$

Assuming that the econometrician observes y_t directly but can only make an inference about the value of s_t based on what we see happening with y_t. This inference will take the form of two probabilities

$$\xi_{it} = Pr(s_t = j | \Omega_t, \theta), \tag{3}$$

for $j = 1, 2$, these two probabilities sum to 1 by construction. Here, $\Omega_t = [y_t, y_{t-1}, ..., y_1, y_0]$ denotes the set of observations obtained as of date t and θ, which are a vector of population parameters. Thus, the above example would be $\theta = (\sigma, \phi, c_1, c_2, p_{11}, p_{22})$ and it presumes to be known with certainty. The inference is performed iteratively for $t = 1, 2, ..., T$ with step t accepting as input the values in Eq. (4),

$$\xi_{it} = Pr(s_{t-1} = j | \Omega_t, \theta). \tag{4}$$

The key of magnitudes is in order to perform iterations of the densities under the two regimes. The details are expressed as Eq. (5),

$$\eta_{jt} = f(y_t | s_t = j, \Omega_{t-1}; \theta) \ = \ \frac{1}{\sqrt{2\pi}\sigma} exp[-\frac{(y_t - c_j - \phi y_{t-1})^2}{2\sigma^2}], \tag{5}$$

where $j = 1, 2$. Specifically, we can calculate the conditional density of the observation from Eq. (6),

$$f(y_t | \Omega_{t-1}; \theta) \ = \ \sum_{i=1}^{2} \sum_{i=1}^{2} p_{ij} \xi_{i,t-1} \eta_{jt}, \tag{6}$$

and the desired output is then

$$\xi_{ij} = \frac{\sum_{i=1}^{2} p_{ij} \xi_{i,t-1} \eta_j t}{f(y_t | \Omega_{t-1}; \theta)}. \tag{7}$$

As a result of executing of iterations, we will have succeeded in evaluating the sample conditional log likelihood of the observed data, which is described as follows

$$log f(y_1, y_2, ..., y_T | y_0; \theta) = \sum_{t=1}^{T} log f(y_t | \Omega_{t-1}; \theta). \tag{8}$$

For the specified value of θ, the estimation of the value θ can be then obtained by maximizing by numerical optimization.

5.2   The Generalized Pareto Distribution Model for Extreme Value Estimating

The Generalize Pareto Distribution model (GPD) includes the extreme value theory that explained about the case of extreme value information on the behavior of the distribution high threshold exceedances. Studying on extreme cases, the distribution function that a random quantity on $F(x)$ beneath condition is shown by Pickands (1975),

$$F(x|u) = P(X \leq u + X \leq u). \tag{9}$$

From Mierlus-Mazilu (2010), this paper examines the Univariate Generalized Pareto Distribution, where X is a random variable, u is a designed threshold, $y = x - u$ are the overabundances and x_F is the final process of F. Then, F_u is verified and written as,

$$F_u(Y) = \frac{F(u+y) - F(u)}{1 - F(u)} = \frac{F(X) - F(u)}{1 - F(u)}. \tag{10}$$

The Eq. (10) is can approximated by a Generalized Pareto Distribution (GPD), which is defined as

$$G(x|\zeta, \sigma, u) = \begin{cases} 1 - (1 + \frac{\zeta(x-u)}{u})^{-1/\zeta}, & if \ \zeta \neq 0 \\ 1 - exp[-\frac{(x-u)}{\sigma}], & if \ \zeta = 0. \end{cases} \tag{11}$$

For the Eq. (11), where $\sigma > 0$ and ξ is the scale and shape parameter, respectively. The information of parameter is heavy when $\xi > 0$. The Eq. (12) showed and conducted the threshold of GPD model. Assuming in the object GDP that under the threshold, u is generalized from a certain distribution with parameters, η. The model is represented as $H|(.|\eta)$. Consequently, the distribution function F of any observation X able to described as

$$F(x|\eta, \zeta, \sigma, u) = \begin{cases} H(x|n), & if \ x < u \\ H(u|\eta) + [1 - H(u|\eta)]G(x|\zeta, \sigma, u), & if \ x \geq 0. \end{cases} \tag{12}$$

Considering the Eqs. (13) and (14), there is a sample size $n, x = (x_1, ..., x_n)$ from F, parameter vector $\theta = (\eta, \sigma, \xi, u)$, $A = (i : x_i < u)$ and $B = (i : x_i \geq u)$ the likelihood function is defined as (Behrens et al. 2004),

$$L(\theta; x) = \prod_A h(x|\eta) \prod_B h(1 - H(u|\eta))[\frac{1}{\sigma}(1 + \frac{\xi(x_i - u)}{\sigma})^{-\frac{1+\xi}{\xi}}], \tag{13}$$

for $\xi \neq 0$ and

$$L(\theta; x) = \prod_A h(x|\eta) \prod_B h(1 - H(u|\eta))[(\frac{1}{\sigma})exp(\frac{x_i - u}{\sigma})], \quad \xi = 0. \tag{14}$$

Obviously, the Eq. (14) represented the threshold, where u is the matter density having discontinuity which depends on the larger or smaller jump density.

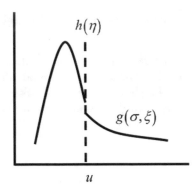

Fig. 3. Presentation the schematics of the threshold model (Applying from Behrens et al. 2004)

In each event observation, it studies the analysis that is able to be more obvious and less evident. However, it is more difficult to predict the threshold. Subsequently, strong discontinuities or large jumps show the separation of information that is easier for evaluating the parameter estimation. Hence, the density model of threshold is schematically presented in Fig. 3.

5.3   The Value at Risk Model (VaR)

This theory is generated to estimate risk reduction, especially in this paper. This risk management method is employed to investigate the risks of bank stocks indexes of top banking companies in each of the four countries. At the beginning, the reliability result of VaR has the procedure that consists of the comparison of the VaR estimation with actual realized loss in the next period and the statistical accuracy in the result of VaR. In the test, the first process is called unconditional coverage test proposed by Kupiec (1995). This conditional estimation specifies the frequency exceedances, which are in the estimated line level of VaR, and the violation of independence property is employed to check the accuracy of the VaR measurement. Conditionally, the violations are also randomly distributed, and the likelihood ratio test is statistically used to evaluate as follows,

$$LR_u = 2[log((\frac{N}{T})^N)(1 - \frac{N}{T})^{T-N} - log(p^N(1 - p)^{T-N})]. \quad (15)$$

Considering the Eq. (15), $H_0 : \frac{N}{T} = p$ and $H_1 : \frac{N}{T} \neq p$ are the correct null and alternatives, respectively. If $LR_u \to x^2$ under H_0 that implies good specification. Let p as the evaluated failure rate ($p = 1-q$ where the confidence level for VaR is q). When T is the total number of such trials, N is the number of failures that is able to be modeled with a binominal distribution with probability of occurrence equals to α.

Furthermore, the second test is called conditional coverage test proposed by Chirtofersen (1998). This conditional calculation explains the checking accuracy of the VaR results whether the conditional coverage test and independence property were used. The null hypothesis that indicated the failure process is independent and the expected proportion of violations equals p. This can be expressed as follows

$$LR_{cc} = -2log[(1-p)^{T-N}p^N] + 2log[(1-\pi_{01})^{n_{00}}]\pi_{01}^{01}(1-\pi_{11})^{n_{10}}\pi_{11}^{n_{11}}] \to x^2. \quad (16)$$

The number of observations with value i followed by j is n_{ij} for $i, j = 0, 1$ respectively, and the corresponding probabilities are $\pi_{ij} = \frac{n_{ij}}{\sum_j n_{ij}}$. The value $i, j = 1$ denote that a violation is made, while $i, j = 0$ indicates the opposite. To estimate the value at risk for the peak over threshold in GPD, it is able to be written as Generalized Pareto Distribution estimating the value at risk (Kisacik 2006), which is expressed as Eq. (17).

$$VaR_{GPD} = u + (\frac{\sigma}{\xi})[(\frac{n}{N_u}\alpha)]^{-\xi}, \ \xi \neq 0, \quad (17)$$

where u : threshold
         σ : scale parameter
         ξ : shape parameter
         α : 1-p
         N_u : number of exceedances
         n : sample size

$$ES_q = E(X|X > VaR_q) = VaR_q + E(X - VaR_q|X > VaR_q), \quad (18)$$

from the Eq. (18), this is the expected shortfall (ES), which is described as the expected loss size and it is able to be written as GPD distribution,

$$\widehat{ES_q} = \frac{\widehat{VaR}}{1-\xi} + \frac{(\hat{\tau}(h) - \xi h)}{1-\xi}. \quad (19)$$

6   Empirical Results

6.1   Descriptive Data

Generally, the descriptively statistical information is represented in Table 1. Each variable collected from ASEANs stocks such as BBL (Thailand), DBS (Singapore), CIMB (Malaysia), and Mandiri (Indonesia) has been shown as a basic average, median, and max-min calculation and checked for stationarity condition to ensure the conversion to long-run equilibriums. Consequently, the results indicate that the Singapore banking stock has the biggest volume of banking stock index.

Table 1. Descriptive data of the collected stock indexes

	BBL (Thailand)	DBS (Singapore)	CIMB (Malaysia)	Mandiri (Indonesia)
Maximum	6.72	79.02	1.99	0.93
Minimum	4.13	74.66	0.91	0.49
Average	5.30	76.67	1.44	0.72
Medien	5.28	76.64	1.46	0.73

Noted: From computation

6.2   Expected Durations of Regimes

As seen in the details of Table 2, the Markov Switching Model found the number of expected duration regimes, which are expansions (Bull markets) and depressions (Bear markets). For the BBL index, the result contains 1,022 days for the bull period and 281 times for recessions. Speaking to the DBS index, it seems the bull time equals 1,079, and 113 days for regimes of recession periods. In the case of CIMB index, it is approximately 1,011 times for bull periods and nearly 291 days for bear periods. Lastly, the boom period and recession period of Mandiri index are 762 and 533 times, respectively.

Table 2. Presentation the expected duration of regimes of financial stock indexes

	BBL (Thailand)	DBS (Singapore)	CIMB (Malaysia)	Mandiri (Indonesia)
Bull (Daily time periods)	1,022	1,079	1,011	762
Bear (Daily time periods)	281	113	291	533

Noted: From computation

6.3   The Extremely Estimated Parameters of the GPD Model

In Table 3, to demonstrate the value of estimated parameters using the value at risk of Generalized Pareto Distribution (VaR_{GPD}) on each situation in the economy (expansion and depression periods), three parameters are defined as a scale parameter (σ), shape parameter (ξ), and threshold (μ). At the beginning, BBL index includes 1.0233 for the scale estimator (σ), 0.1882 for the shape (ξ), and 0.02 (μ) is the threshold in expansion periods. On the other hand, the scale, shape, and threshold parameters are estimated and respectively expressed as 1.0767 (σ), 0.2275 (ξ), 0.03 (μ) in depression periods. Considering the DBS index, in boom periods, there are 1.0498 (σ), 0.1496 (ξ), and 0.025 (μ) for the GPD estimators, respectively. On the other side, for decreasing periods, there are 1.0332 (σ), 0.3226 (ξ), 0.01 (μ), respectively. In the matter of the CIMB index, the results of estimated parameters in bull periods are 0.9613 (σ), 0.2016

Table 3. Presentation of GPD estimators for financial stock indexes

	BBL (Thailand)		DBS (Singapore)		CIMB (Malaysia)		Mandiri (Indonesia)	
	Bull	*Bear*	Bull	*Bear*	Bull	*Bear*	Bull	*Bear*
Scale parameter	1.0233	*1.0767*	1.0498	*1.0332*	0.9613	*0.9699*	1.0131	*1.0169*
Shape parameter	0.1882	*0.2275*	0.1496	*0.3226*	0.2016	*0.1662*	0.2302	*0.2802*
Threshold	0.02	*0.03*	0.0025	*0.01*	0.01	*0.025*	0.02	*0.05*

Noted: From computation

(ξ), 0.01 (μ). Conversely, the bear periods are 0.9699 (σ), 0.1662 (ξ), 0.025 (μ), respectively. Lastly, the Mandiri index contains 1.0131 for the scale parameter (σ), 0.2302 for the shape estimator (ξ), and 0.05 for the threshold (μ). On the other hand, 1.0169 (σ), 0.2802 (ξ), and 0.05 (μ) are the GPD parameters in recession times.

6.4 Results of the Value at Risk (VaR) and Expected Shortfall (ES) Estimations

Empirically, Table 4 reveals two values: risk value of least loss and the percentage return in the short-run (Expected shortfall: ES). The data were represented as two periods of the economic system (expansion period and recession period). As estimated results of risk values, the BBL index in expansion periods contains 1.57% for risk measurement and 2.68% for the risk in the bear period. The risk values of DBS are 1.92% in the boom periods and 2.88% in recession times. In addition, the results of the CIMB index approximately provide 2.30% and 3.09% in boom periods and recession periods, respectively. Lastly, the Mandiri index contains 1.40% in boom periods and 1.61% in bear periods.

Additionally, Table 4 showed the percentage of expected shortfalls, which are estimated by using normally situational data. Firstly, the BBL index includes 2% and 1% in the bull and bear markets, respectively. For the DBS index, the result indicates 0.26% and 1%, which are in the boom and depression periods, respectively. For CIMB index, the most minimum return of these indexes is 1.00% and 2.00%, respectively. Lastly, the Mandiri index has the values of expected shortfalls, which are approximately 2.00 perncet in boom periods and 5.00% in recession periods.

The Table 4 also displayed the comparison between risk measurements (VaR) and expected shortfalls. The empirical outcomes can be implied that there are two groups of the financial stock indexes, including a positively financial yield (BBL and Mandiri) and negatively financial (DBS and IMB) group. The results show BBL and Mandiri contain the optimistic trend for investing in both bull and bear periods, which indicates the expected shortfalls are more than risk measurements. Conversely, the results of DBS and CIMB provide negative signs, which are not suitable for speculating in both bull and bear periods.

Table 4. Presentation the results of VaR and ES

	BBL (Thailand)	DBS (Singapore)	CIMB (Malaysia)	Mandiri (Indonesia)
Bull period				
VaR value	1.57%	1.92%	2.30%	1.40%
Expected shortfall	2.00%	0.26%	1.00%	2.00%
Bull period				
VaR value	2.68%	2.88%	3.09%	1.61%
Expected shortfall	3.00%	1.00%	2.00%	5.00%

Noted: From computation

7   Conclusion

The purpose of this paper was to computationally investigate the volume of risks in four major financial stock indexes of AEC countries such as BBL (Bangkok Bank, Thailand), DBS (The Development Bank of Singapore, Singapore), CIMB (Commerce International Merchant Bankers Bhd, Malaysia) and Mandiri (Bank Mandiri, Indonesia). The daily time-series data was observed during the period of 2012 to 2017. Methodologically, this paper focused on forecasting econometric methods. For example, the Markov Switching Model (MS) for economical regime specification, Generalized Pareto Distribution (GPD) for the extreme value theory, and the calculation of values at risks (VaR) were all employed.

Empirically, from the investigation of economic cycles, the results estimated from the switching model representing four financial stock indexes were computationally separated into two states, including the boom and recession periods (see details in Table 2). These separated regimes can clarify the details of economic cycles efficiently. Moreover, this application of the switching model can successfully confirm that the fluctuations of ASEAN banking stock indexes obviously occur and consist of the business cycle theory.

Based on the eventual analysis relating to the separated regimes estimated by the MS-model, the results of the GPD models displayed that there are evidently unusual events among the trends of ASEAN banking stocks (see the details in Table 3). The computational results calculated from the GPD model were then employed to clarify risk measurement by using Value at Risk forecasting (VaR). Demonstrably, the results confirm that two indexes of ASEAN banking stock indexes are not suitable for short-run investing. For example, in the case of DBS and CIMB indexes, the computational results indicated that these financial stock indexes contain the higher chance to loose rather than gaining profits. This can be implied that the indexes in Singapore and Malaysia are not suitable for speculation, which the risk measurement estimated by the Value at Risk (VaR) (Seen the details in Table 4) showed that the chance to get lost is higher than the chance to get earning profits. To address this issue, it obviously seems that a long-run strategy to achieve positive outcomes from these banking stocks should be intensively focused. On the other hand, in the case of BBL and Mandiri indexes, the results showed these indexes can be financially speculated. To deal with this issue, an appropriate time to invest and accurately computational prediction are intensively required.

Ultimately, it is undeniable that investments in banking systems are not always suitable for short-term speculations. The empirical results from this investigation also confirm that the stability of banking systems is inevitably related to the banking stock indexes. Long-run investment in banking stocks is crucial, and this should be carefully considered by authorities. As a result, this study can strongly conclude the uniqueness of the banking system in Asean countries really needs somethings to clarify a suitable prediction, for instance, econometrical and computational tools.

References

Avdulai, K.: The Extreme Value Theory as a Tool to Measure Market Risk. Working paper 26/2011.IES FSV. Charles University (2011). http://ies.fsv.cuni.cz

Behrens, C.N., Lopes, H.F., Gamerman, D.: Bayesian analysis of extreme events with threshold estimation. Stat. Modell. **4**, 227–244 (2004)

Chaithep, K., Sriboonchitta, S., Chaiboonsri, C., Pastpipatkul, P.: Value at risk analysis of gold price return using extreme value theory. EEQEL **1**(4), 151–168 (2012)

Christoffersen, P.: Evaluating interval forecasts. Int. Econ. Rev. **39**, 841–862 (1998)

Einmahl, J.H.J., Magnus, J.R.: Records in athletics through extreme-value theory. J. Am. Stat. Assoc. **103**, 1382–1391 (2008)

Embrechts, T., Resnick, S.T., Samorodnitsky, G.: Extreme value theory as a risk management tool. North Am. Actuarial J. **3**(2), 30–41 (1999)

Ernst, E., Stockhammer, E.: Macroeconomic Regimes: Business Cycle Theories Reconsidered. Working paper No. 99. Center for Empirical Macroeconomics, Department of Economics, University of Bielefeld (2003). http://www.wiwi.uni-bielefeld.de

Garrido, M.C., Lezaud, P.: Extreme value analysis: an introduction. J. de la Socit Franaise de Statistique, 66–97 (2013). https://hal-enac.archivesouvertes.fr/hal-00917995

Hamilton, J.D.: Regime Switching Models. Palgrave Dictionary of Economics (2005)

Jang, J.B.: An extreme value theory approach for analyzing the extreme risk of the gold prices. J. Financ. Rev. **6**, 97–109 (2007)

King, R.G., Rebelo, S.T.: Resuscitating Real Business Cycles. Working paper 7534, National Bureau of Economic Research (2000). http://www.nber.org/papers/w7534

Kisacik, A.: High volatility, heavy tails and extreme values in value at risk estimation. Institute of Applied Mathematics Financial Mathematics/Life Insurance Option Program Middle East Technical University, Term Project (2006)

Kupiec, P.: Techniques for verifying the accuracy of risk management models. J. Derivat. **3**, 73–84 (1995)

Manganelli, S., Engle, F.R.: Value at Risk Model in Finance. Working paper No. 75. European Central Bank (2001)

Marimoutou, V., Raggad, B., Trabelsi, A.: Extreme Value Theory and Value at Risk: Application to oil Market (2006). https://halshs.archives-ouvertes.fr/halshs-00410746

Mierlus-Mazilu, I.: On generalized Pareto distribution. Roman. J. Econ. Forecast. **1**, 107–117 (2010)

Mwamba, J.W.M., Hammoudeh, S., Gupta, R.: Financial Tail Risks and the Shapes of the Extreme Value Distribution: A Comparison between Conventional and Sharia-Compliant Stock Indexes. Working paper No. 80. Department of Economics Working Paper Series, University of Pretoria (2014)

Neves, C., Alves, M.I.F.: Testing extreme value conditions an overview and recent approaches. Stati. J., REVSTAT **6**(1), 83–100 (2008)

Perez, P.G., Murphy, D.: Filtered Historical Simulation Value-at-Risk Models and Their Competitors. Working paper No. 525. Bank of England (2015). http://www.bankofengland.co.uk/research/Pages/workingpapers/default.aspx

Perlin, M.: MS Regress - The MATLAB Package for Markov Regime Switching Models (2010). Available at SSRN: http://ssrn.com/abstract=1714016

Pickands, J.: Statistical inference using extreme order statistics. Ann. Stat. **3**, 110–131 (1975)

Rockafellar, R.T., Uryasev, S.: Conditional value-at-risk for general loss distributions. J. Bank. Financ. **26**, 1443–1471 (2002). http://www.elsevier.com/locate/econbase

Sampara, J.B., Guillen, M., Santolino, M.: Beyond value-at-risk: glue VaR distortion risk measures. Working paper No. 2. Research Institute of Applied Economics, Department of Econometrics, Riskcenter - IREA University of Barcelona (2013)

Taghipour, A.: Banks, stock market and economic growth: the case of Iran. J. Iran. Econ. Rev. **14**(23) (2009)

Interval Forecasting on Big Data Context

Berlin Wu[⊠]

Department of Mathematical Sciences, National ChengChi University,
Taipei, Taiwan
berlin@nccu.edu.tw

Abstract. Purpose: The object of this research is to construct an optimal internal forecasting method in big data context.

Design/methodology/approach: An intelligent model construction, including consumer behavior and market information, structural changes detection, nonlinear pattern recognition, spatial causality, semantic processing mode is presented.

Findings: The major drawback in forecasting field is that the statistical forecasting result is derived from historical data but it often encounters non-realistic problem when people predict future trends or market changes in real world.

Practical Applications: Construction of Big Data platform will be a new technique provides to solve the structured change and uncertain problems. According to the artificial intelligence evolution and on line improvement to the market conditions, it will do a better performance to prevailing future event.

Originality: We efficiently integrate the idea of structure change, entropy and market behavior in the forecasting process.

Conclusion: Since historical time series analysis has difficult to prove the relationship/causality with future events. Especially in the case of a structural change, the future is full of high uncertainty, ambiguity and unexpected.

Keywords: Structure change · Interval forecasting methods · Big data
Fuzzy entropy

1 Introduction

Despite current time series analysis have reached an increasingly sophisticated level, researchers on this field still feel dilemma in the forecasting performance. Economic forecasting mostly comes after economic model construction, while model construction is based on historic data without information from the future events. The major drawback in econometric forecasting is that econometric model based on the result of historical trends, it encounters difficult phenomenon in practice [13, 16]. With these well-constructed models people still cannot predict future changes or market trends. Since historical time series analysis has difficult to prove the relationship/causality with

---

It is not the strongest nor the most intelligent who will survive, but those who can best manage change.

© Springer International Publishing AG 2018
V. Kreinovich et al. (eds.), *Predictive Econometrics and Big Data*, Studies in Computational Intelligence 753, https://doi.org/10.1007/978-3-319-70942-0_53

future events. Especially with the present of a structural change, the future is full of high uncertainty and ambiguity [1–3].

In this paper we introduce an integrated process to construct and forecasting a time series model with structure changed characteristic on the big data context. The technique of changing period detection is used for an effective procedures instead of unit root test in the model construction.

According to its artificial intelligence evolution and on line improvement to the market conditions, it will do a quick response to prevailing future event. A Big Data platform construction will be a new technique provides to solve the structured change, market trend and uncertain problem.

Time series analysis and classical analysis biggest difference is that in order to detect the structural transformation of substituted stable type with a single test. In the model construction process, it is often assumed that the data type is stationary or by a single test [8, 9]. But structural changes in the market economy, steady type of hypothesis is difficult to set up. Especially in the number of multiple columns situations, consistent smooth lines and structural change is almost impossible. Therefore, we need to more carefully consider the leadership behind the relationship between them. We consider the data base in a large background of macroeconomic factors, A Forecasting model selection will be proceeded by the following analysis process:

1. System variables and related domain are examined
2. Apply the transition interval test, structural changes in the latest issue of the model section,
3. Large model within the entire kinetic energy and the kinetic energy of the external integration of consumer behavior and market trends and forecast
4. The forecasting result will take the big data context for forecasting with the new system parameters of consumer behavior and market trends

Finally, an empirical study will be demonstrated in an efficient way on the big data context with the applied concept of fuzzy entropy property, the market behavior and forecasting trend.

2 Dynamic Trend with Fuzzy Entropy

2.1 Trend with Fuzzy Entropy

The concept of structural change in a time series should be based on the dynamic trend of variable. It should be a gradually-emerging change interval and not a change point whereby the change happens abruptly at a certain point of time.

When we use set theory to examine whether there is any change point in a time series, first we cluster the time series, find out the cluster center, and then use the fuzzy membership degree, fuzzy entropy and other relevant concepts to perform classification. The definition of fuzzy membership degree is as follows:

Definition 2.1 *Fuzzy Membership Degree. Let a time series be* $\{x_t, t = 1, 2, \ldots, N\}$, *with C_1 and C_2 being two cluster centers of the time series,* $\mu_{it}, i = 1, 2.$ *to represent* the membership degree of an element x_t in the time series X_t to C_1 and C_2, the membership degree is thus defined as: $\mu_{it} = 1 - \frac{|x_t - C_i|}{\sum_{i=1}^{2}|x_t - C_i|}$.

Entropy is a concept in the Thermodynamics study [10, 11, 15], whose original meaning is the degree at which work can be transformed. The Statistical Physics provides another definition to it: the measure to describe the random motion. In addition, the Probability Theory and Information Theory give it a more common definition: measure of the unboundedness of a random variable, or the measure of the amount of missing information. So fuzzy entropy is used to measure the uncertainty of fuzzy sets, and is an important tool for the processing of fuzzy data, while the membership degree is used to characterize elements that do not clearly belong to some particular sets.

So the change interval that studies variables, compared with the classical method of investigation for time series, has better descriptive power. [12] suggested that the use of fuzzy entropy is effective in identifying whether a structural change happens in a time series. Besides, it can also be used together with the mean cumulated fuzzy entropy of t times, to observe the change in message of fuzzy entropy, based upon which a standard for the classification of change model can be established [4–7].

Definition 2.2 *Fuzzy Entropy. Let a time series be* $\{x_t, t = 1, 2, \ldots, N\}$, *with* μ_{it} *being the membership degree of* x_t *to the cluster centers Ci (i = 1,2,...,k), the fuzzy entropy is thus defined as:*

$$\delta(x_t) = -\left(\frac{1}{k}\right) \sum_{i=1}^{k} [\mu_{it} ln(\mu_{it}) + (1 - \mu_{it}) ln(1 - \mu_{it})]$$

Definition 2.1. *Mean Cumulated Fuzzy Entropy for one-variable time series.* Let $\{x_t, \ t = 1, 2, \cdots, N\}$ be a time series with fuzzy entropy $\delta(x_t)$. The mean cumulated fuzzy entropy is thus defined as:

$$MS\delta(x_t) = \frac{1}{t} \sum_{i=1}^{t} \delta(x_i) \tag{2.1}$$

There is usually a threshold level λ set up for fuzzy classification, because no matter it is for nature or humanities, the determination for classification is very subjective and often non-unanimous. Hence, an objective measure is needed. According to empirical experience, λ cannot take a value too huge or too small, otherwise the classification cannot be done or too many classes will be created. So a value for λ between 0.001 and 0.1 will be ideal.

Here we wish to find two cluster centers. This is determined based on common experience of empirical analysis and trend of time series. The procedures are as follows:

Step 1: *Use the k-means method (Sharma, 1996) to find out two cluster centers C_1 and C_2 in time series $\{x_t\}$, and determine the membership degree μ_{it}, $i = 1, 2$ of $\{x_t\}$ to the two cluster centers.*

Step 2:   *Compute the fuzzy entropy $\delta(x_t)$, mean cumulated fuzzy entropy $MS\delta(x_t) =$*

$$\frac{1}{t}\sum_{i=1}^{t}\delta(x_i)$$ *and the median of this series $Median(MS\delta(x_t))$ that correspond to x_t.*

Step 3:   *Take a suitable threshold value λ, classify the mean cumulated fuzzy entropy $MS\delta(x_t)$ series that correspond to x_t. If the mean cumulated fuzzy entropy $MS\delta(x_t)$ falls into the interval $[0, Median(MS\delta(x_t)) - \lambda)$, we will use 1 to represent Group 1; if $MS\delta(x_t)$ falls into the interval $[Median(MS\delta(x_t)) - \lambda, Median(MS\delta(x_t)) + \lambda)$, we use 2 to represent Group 2; and if $MS\delta(x_t)$ falls into the interval $[Median(MS\delta(x_t)) + \lambda, 1]$, 3 will be used to represent Group 3.*

Step 4:   *If the result of classification is inconsistent, we then make an adjustment to the result. If it is consistent, go to Step 5.*

Step 5:   *Select an appropriate determination level. If the number of consecutive samples is greater than [], then these consecutive samples belong to the same group. During classification, if more than one group is found, we know that structural change happens in this time series. Thereafter, find the change interval.*

2.2   Data Mining for Variable Interaction

Under the big data context, the impact of macroeconomic variables is very much complex. Each factor leads a pattern with a nonlinear trend and have certain causality property with other variables. Therefore, we must first clarify the main economic variables in considerations. Then by real-time intelligence, market consumption, investment, and trade data for economic growth to make better predictions. Since the of macroeconomic data is highly correlated with leading or lag causation, the activities of a single economic forecast market performance conditions is not enough and t is the inevitable choice of the statistical agency.

"Big" big data does not lie in how much amount of data, but rather a kind of "information as to the large and pluralism" methodology. The mobile Internet, the next generation Internet, cloud computing, networking, social networks, there are many large and small departments and units built up financial systems, human resources systems and client systems, their mutual combination brings new wave of expansion and development of its business intelligence will greatly enrich statistics pipeline sources, affecting the quality of statistics sources. People pick through the data for the past, reveal the law to face the future, predict trends, will greatly improve the ability of the government's macro decision-making judgments and standards.

Based on the new economic statistics of data mining and processing, we can see that the current macroeconomic variables analysis facing a major information inno-vation. In order to promote the quality of forecasting, the macroeconomic factors need to be reviewed before data processing to improve the macroeconomic accounts.

In order to reform the data process of macroeconomic accounting method, "big data" concept requires us to extract information from the mass of "gold mine" from which to create a new accounting value. Their reformation breaks the traditional thinking using new statistical techniques, and actively improve the various types of cargo bulk products, commodity production, sale, transport statistical monitoring system platform. Using various types of commodity trading platform, the Internet of Things, mobile Internet, cloud computing, enterprise financial systems and other data information. By extraction, transformation, integration of information, establishment of GDP accounting data warehouse. Variable economic census database, the basic unit of directory database, database agricultural census, census basic information database, the tertiary industry census database, industrial census database, input-output survey data and other dynamic information. Accounting information reduce uncertainty and difficult to capture judge. Table 1 demonstrates the main macroeconomic mining projects and interaction.

In GDP accounting, for example, economists do GDP accounting process often feel confused or troubled. It is because that the data items and weights are not so precise and consistent, information collected is incomplete, delayed, inadequate representation and other issues. Message bias will result in variable effectiveness analysis, significant deviations modeling and conclusions. Such as armaments security value, the value of real estate was often played down or taken into account, the value of tourism, entertainment, education and cultural, medicine and underground economic accounting, still cannot be fully calculated in GDP accounting process. Therefore, how to optimize the GDP accounting items of information sources and the weight in order to improve data quality is very important. Under the new "National Economic Industry Classification Benchmark, China, 2011 Edition" improving the GDP accounting industry classification, improve the relevance and usefulness of the industry reflects the problem; management, assessment and GDP-related performance and coordination of relevant variables. Therefore, in the background of big data, whether accounting, modeling, forecasting, and should have a new basis for supporting measures.

2.3 Residuals Evaluation with Interval Data

An interval-valued fuzzy set can be viewed as a continuous fuzzy set, which further represents uncertain matters. Take "test results" as an example. In foreign countries, A, B, C and D are used to evaluate a student's result, whereby A represents 100–80 marks, B represents 79–70 marks, C represents 69–60 marks, D represents 59–50 marks, in place of numerical scores. In the past we think that obtaining a high score means learning well. However, does a student who gets 85 marks have better learning ability than another who scores 80 marks? Not necessarily so. That's why the interval-valued fuzzy set resolves the phenomena of uncertainty.

When a sample of interval-valued data is available, we have to consider its operations However, there is still no complete definition for the measure of interval distance (see Wu and Nguyen [13]). How to define a well-defined interval distance? First we represent the interval with $(c_i; r_i)$ with c being the center, r being radius. This way, the interval distance can be considered as the difference of the center plus the difference of the radius. The difference of the center can be seen as the difference in location, and the difference of the radius can be seen as the difference in scale. However, in order to lower the impact of the scale difference on the location difference, we take the *ln* value of the scale difference, and then plus 1 to avoid the *ln* natural log value becoming negative.

Table 1. The main macroeconomic variable and its interaction

	GDP	Exchange rate	Interest	Energy	Stock market	Inflation	Unemployment rate	Real estate	Population	Tempreture
CPI	1	1	-1	-1	1	2	1	2	0	1
Raw material	1	-1	1	1	0	1	-2	0	1	0
Labor	2	-1	1	1	0	1	1	0	1	0
R&D	2	0	1	1	0	0	-1	0	0	0
Investment	2	1	0	1	1	1	1	1	1	1
Import/export	2	2	-2	-1	2	1	1	-1	0	1
Medicine/health	1	1	1	1	0	0	1	0	1	1
Tourism, entertainment	1	1	1	1	1	0	1	-1	1	1
Cultural & education	1	1	1	0	0	0	1	0	1	1
National defence	-1	-2	1	1	-1	0	1	1	1	0

Once such a transformation has been selected, instead of the original trapezoid-data, we have a new value y = f(x). In the ideal situation, this new quantity y is normally distributed. (In practice, a normal distribution for y may be a good first approximation.) When selecting the transformation, we must take into account that, due to the possibility of a rescaling, the numerical values of the quantity x is not uniquely determined.

Definition 2.3 Scaling for an interval fuzzy number on R. *Let $A = [a, b]$ be an interval fuzzy number on U with its center cA = $(b - a)/2$. Then the defuzzification number RA of A = [a, b] is defined as*

$$RA = cA + \frac{\|A\|}{2\ln(e + \|A\|)} \quad where, \; \|A\| \; is \; the \; length \; of \; the \; interval.$$

However, there are few literatures and definitions appear on the measurement system. In this section, a well-defined distance for interval data will be presented.

Example 2.1. *Let $x_1 = [2,3]$, $x_2 = [1.5,3.5]$, $x_3 = [1,4]$, $x_4 = [2,3.5]$, Then,*

$$Rx_1 = 2.5 + \frac{1}{2 \times \ln(e+1)} = 2.89; \quad Rx_2 = 2.5 + \frac{2}{2 \times \ln(e+2)} = 3.14$$

$$Rx_3 = 2.5 + \frac{3}{2 \times \ln(e+3)} = 3.36; \quad Rx_4 = 2.75 + \frac{1.5}{2 \times \ln(e+1.5)} = 3.27$$

Definition 2.4. *Let $A_i = [a_i, b_i]$ (i = 1, 2, n) is a sequence of interval fuzzy number on U with its center cA = (b-a)/2. Then the distance between A_i and A_j is defined as*

$$d(A_i, A_j) = |cA_i - cA_j| + \left| \frac{\|A_i\|}{2\ln(e + \|A_i\|)} - \frac{\|A_j\|}{2\ln(e + \|A_i\|)} \right|$$

The traditional five-point scale used in scoring is often five options. The sequence score from small to large is often 1, 2, 3, 4, 5 points then to calculate the sum. According to this principle, the scoring method can also apply in discrete type of fuzzy numbers. It can let the distance of option value become larger, and it might help achieve a significant level after testing.

Definition 2.5. *Interval means square error (IMSE) of prediction for a sample of interval-valued fuzziness. Let $\{s_i = [a_i, b_i], i = 1, \ldots, N\}$ be an interval time series, with prediction interval being $\hat{s}_i = [\hat{a}_i, \hat{b}_i]$ and $\varepsilon_i = d(s_i, \hat{s}_i)$ beging the error between the prediction interval and the actual interval, thus:*

$$IMSE = \frac{1}{l} \sum_{i=N+1}^{N+l} \varepsilon_i^2$$

where l is the forecasted expectancy value.

Example 2.2. *Let* $\hat{A} = \{[3,6],[2,6],[5,8],[3,8]\}$ be an interval time series of forecasting data with respect to the real time series. $A = \{[3,4],[3,5],[5,7],[4,5]\}$. Table 2 illustrates the distance from forecasting values and real values.

Table 2. Distance for the interval data

Sample	Forecasting value	Actual value	Distance $d(A_i, A_j) = \lvert cx_i - cx_j \rvert + \left\lvert \dfrac{\lVert A_i \rVert}{2\ln(e + \lVert A_i \rVert)} - \dfrac{\lVert A_j \rVert}{2\ln(e + \lVert A_i \rVert)} \right\rvert$
	[3,6]	[3,4]	$d(s_1, \hat{s}_1) = \lvert 3.5 - 4.5 \rvert + \lvert 0.86 - 0.38 \rvert. = 1.48$
3	[2,6]	[3,5]	$d(s_3, \hat{s}_3) = \lvert 4 - 4 \rvert + \lvert 1.05 - 0.64 \rvert = 0.41$
4	[5,8]	[5,7]	$d(s_4, \hat{s}_4) = \lvert 6 - 6.5 \rvert + \lvert 0.86 - 0.64 \rvert = 0.72$
5	[3,8]	[4,5]	$d(s_5, \hat{s}_5) = \lvert 4.5 - 5.5 \rvert + 0.86 - 0.38 = 1.48$
$IMSE = \frac{1}{4} \times (1.48^2 + 0.41^2 + 0.72^2 + 1.48^2) = 1.15$			

Example 2.3. The distribution of 27 sets of interval data of unemployment rate (unit: percent) is shown as Fig. 1, as follows:

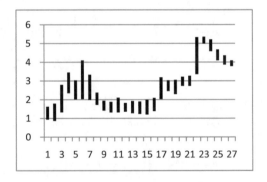

Fig. 1. Chart of interval time series

If we wish to divide the data into two groups, using Definition 2.5, we can obtain two interval clusters $I_1 = (1.83, 2.5)$ and $I_2 = (3.71, 5.2)$ The result of clustering is as follows (Fig. 2):

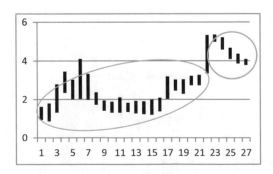

Fig. 2. Result of interval clustering

3   An Integrated Decision System with Forecasting

3.1   Model Construction and Forecasting

In the traditional research, people observe the microeconomic index from the financial point of view, try to find the internal market and marcoeconomic relevance. However, we find that it is not sufficient to make projections based on internal economic statistics analysis. The performance of individual economic statistics is in global reflecting internal (local) individual economic trends and is closely related to the individual economic and economic situation of the global environment. Therefore, we think that in the overall economic prosperity (overall economic indicators) and the internal operation of the company (stocks earnings data), in order to more accurate forecasting economic trends.

The traditional research for the econometric trend is mostly focus on the GDP investigation. Hence the association variables computation are considered highly on these financial indicators. But we think that only use these variables is not sufficient, since the demonstration on the microeconomics data is not only influenced by its local economic condition. It is influebced by the big environment, the global economic condition. Therefore we take these two function into considerations. It will make our foresting result more realistic and robust.

Let y_t be a time series in a dynamic system modeled by with

$$y_{t+1} = a_1 + \phi_1 y_t + sign(\Omega_{1t})f(\text{big data}\,\Omega_{1t}) + sign(\Omega_{2t})g(\text{big data}\,\Omega_{2t})$$
$$+ sign(\Delta y_i) \sum\nolimits_{i=0}^{t} \frac{y_i}{y_{i+1}} e^{-|\Delta y_i|} + \varepsilon_{t+1}$$

Be an integrated transferred time series, f is the function with a set of corresponding exogenous variables $\Omega_{1t} = \{o_1, o_2, \ldots, o_{50}\}$, Δy_t is the variation of entropy y_t is the national miroeconomic variables, $\Omega_{2t} = \{w_1, w_2, \ldots, w_{50}\}$ a set of corresponding indogenous variables, $\sum_{i=0}^{t} \Delta y_i e^{-|\Delta y_i|}$ is the cumulative dynamic entropy.

For the interval data type, we extend last equation into the following

$$\begin{cases} x_{t+1} = a_1 + \phi_1 x_t + sign(\Omega_t)f_1(\Omega_{1t}) + sign(\Delta x_i)g_1(\Omega_{2t}) + sign(\Delta x_i)\sum_{i=0}^{t}\frac{x_i}{x_{i+1}}e^{-|\Delta x_i|} + \varepsilon_{t+1} \\ l_{t+1} = a_2 + \phi_2 l_t + sign(\Omega_t)f_2(\Omega_t) + sign(\Delta l_i)g_2(\Omega_{2t}) + sign(\Delta l_i)\sum_{i=0}^{t}\frac{l_i}{l_{i+1}}e^{-|\Delta l_i|} + \delta_{t+1} \end{cases}$$

Where $x_t = center\ of\ y_t$, $l_t = length\ of\ y_t$, $\varepsilon_{t+1} \sim WN(0, \sigma_\varepsilon^2)$, $\delta_{t+1} \sim WN(0, \sigma_\delta^2)$.

3.2   Forecasting with Financial Entropy

The concept of entropy originates from machine engineering, where in a process involving heat it is a measure of the portion of heat becoming unavailable for doing work. In information theory, Entropy is a measure of *unpredictability* or *information content* in a random variable. Since entropy can be measured in price values or range of prices, in this context, the term will be referred to the *ratio entropy*, which ratio quantifies of the time series. The role of entropy rate for a dynamic process is one bit per value. For exmple, if the process demonstrates upward trend, and hence the entropy rate, is lower. This is because, if asked to predict the next outcome, we could choose the most frequent result and be right more often than wrong. The entropy of a message multiplied by the length of that message is a measure of how much information the message contains. If some messages come out smaller, at least one must come out larger. In practical use, this is generally not a problem, because we are usually only interested in compressing certain types of messages, for example English documents as opposed to gibberish text, or digital photographs rather than noise, and it is unimportant if our compression algorithm makes certain kinds of sequences larger. However the problem can still arise even in every day use when applying a compression algorithm to an already compressed data.

Wu [12] suggested that, the use of fuzzy entropy is effective in identifying whether a structural change happens in a time series. Besides, it can also be used together with the mean cumulated fuzzy entropy of t times, to observe the change in message of fuzzy entropy, based upon which a standard for the classification of change model can be established.

$$\begin{cases} x_{t+1} = a_1 + \phi x_t + \varepsilon_{t+1} & \varepsilon_{t+1} \sim WN\left(0, \sigma_\varepsilon^2\right) \\ l_{t+1} = a_1 + l_t + \delta_{t+1} & \delta_{t+1} \sim WN\left(0, \sigma_\delta^2\right) \end{cases} \tag{3.1}$$

Let $\hat{x}_t(i)$ and $\hat{l}_t(i)$ be the one step forecasting value of the interval time series as in Eq. (2.1), $xf(i)$ and $lf(i)$ We denote the revised forecasting value as

$$\begin{cases} xf(1) = \hat{x}_t(1) + sign(l_t - l_{t-1}) \frac{l_t}{l_{t-1}} \\ lf(1) = \hat{l}_t(1) + sign(x_t - x_{t-1}) \frac{x_t}{x_{t-1}} \end{cases} \tag{3.2}$$

$$\begin{cases} xf(i) = \hat{x}_t(i) + sign\left(\hat{l}_t(i) - \hat{l}_t(i-1)\right) \frac{\hat{l}_t(i)}{\hat{l}_t(i-1)} \\ lf(i) = \hat{l}_t(i) + sign(\hat{x}_t(i) - \hat{x}_t(i-1)) \frac{\hat{x}_t(i)}{\hat{x}_t(i-1)} \end{cases} \tag{3.3}$$

3.3   An Integrated Decision System for Fuzzy Time Series Analysis and Forecasting

Here we wish to find two cluster centers. This is determined based on common experience of empirical analysis and trend of time series. The procedures are as follows:

Step 1: *Use the k-means method to find out two cluster centers C_1 and C_2 in time series $\{x_t\}$, and determine the membership degree $\mu_{it}, i = 1, 2$ of $\{x_t\}$ to the two cluster centers.*

Step 2: *Compute the fuzzy entropy $\delta(x_t)$, mean cumulated fuzzy entropy $MS\delta(x_t) = \frac{1}{t}\sum_{i=1}^{t}\delta(x_i)$, and Median $(MS\delta(x_t))$ of this series, that correspond to x_t.*

Step 3: *Take a suitable threshold value λ, classify the mean cumulated fuzzy entropy $MS\delta(x_t)$ series that correspond with x_t.*

> *If the mean cumulated fuzzy entropy $MS\delta(x_t)$ falls into the interval $[0,$ Median $(MS\delta(x_t)) - \lambda)$, we assign 1 to represent Group 1;*
> *if $MS\delta(x_t)$ falls into the interval $[$Median $(MS\delta(x_t)) - \lambda,$ Median $(MS\delta(x_t)) + \lambda)$, we assign 2 to represent Group 2;*
> *if $MS\delta(x_t)$ falls into the interval $[$Median $(MS\delta(x_t)) + \lambda, 1]$, we assigne 3 to represent Group 3.*

Step 4: *If it is consistent, go to Step 5. Otherwise adjust the level λ and go to step 3.*

Step 5: *Step 5: Select an appropriate determination level α. If the number of consecutive samples is greater than $[\alpha N]$, then these consecutive samples belong to the same group. During classification, if more than one group is found, we know that structural change happens in this time series. Thereafter, find the change interval.*

Step 7: *Construct a system of AR(1) model for the center and length of the dynamic process.*

Step 8: *Forecasting the time series as in the Eq. (3.2).*

4   Empirical Studies

We consider the daily gold price time series during January 2006 to October 2015. Data source is from New York Mercantile Exchange. The weekly interval time series (with minimum and maximum) are derived from these 2418 daily data into 500 interval data with the form [center, radius] by the maximum price and the minimum price weekly.

The trend of time series are illustrated at Fig. 3(a). Figure 3(b) shows the trend with groups. We use the threshold fuzzy classification method (*TFC* method) to perform classification and construct a more comprehensive model. The statistical softwares, R and Minitab, are perform to do the computational work.

Firstly, we establish some models including general time series model, fuzzy time series model, and our threshold fuzzy classification model, by means of 487 weekly data from January 2006 to July 2015. We retain 13 weekly data from August 2015 to October 2015 to make forecasting performance comparison.

From the computation, classification is successful and structural change happens. So we need to consider the radius of gold prices from weeks 145, the first week of Jan. 2009. Also, we can see weeks 289 to weeks 467 (the second week of Oct. 2011 to the second week of March 2015) as well as weeks 468 to weeks 487 (the third week of March 2015 to the fourth week of July 2015) are two groups with consecutive samples greater than 19. This means the classification is successful and structural change happens. So we need to consider the center of gold prices from weeks 468 (the third week of March 2015). While for the radius of gold price is from 144 on.

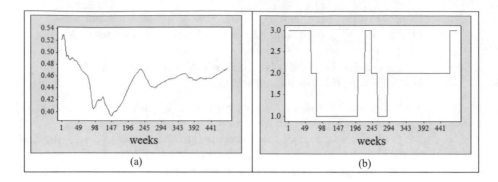

Fig. 3. (a) Chart of mean cumulated fuzzy entropy of gold price center, (b) The classification diagram for Fig. 3, obtained by setting $\lambda = 0.01$

We construct a new threshold fuzzy model only using the last 20 weeks data. The best-fitted model for the center is as follows:

$$\left(1 - 1.31B + 0.41B^2\right)c_t = 115.97 + \varepsilon_t, t \geq 468 \tag{4.1}$$

The best-fitted model for the radius is as follows:

$$(1 - 0.92B)r_t = 1.19 + (1 - 0.75B)\varepsilon_t, t \geq 145 \tag{4.2}$$

Forecasting (fittings) Comparison with other models.

After constructing the model, we look at the core interest of this study – the forecasting ability. Table 3 is the comparison results for the forecast of gold price. Compared with Moving-Average model, Exponential smoothing model, ARIMA model and threshold fuzzy classification model (TFC model), we can see TFC model is much better than other four models based on three statistical criteria.

Table 3. Forecasting comparison with other models gold price (Center)

Model	Performance indices		
	MSE	MAE	MAPE
Moving average model	2399.92	43.4	3.78%
Exponential smoothing	2120.61	40.54	3.53%
ARIMA	1506.61	33.93	2.96%
TFC model with B-data	265.04	13.56	1.19%

Average model, Exponential-Smoothing model, Exponential smoothing model, ARIMA model and TFC model, we can see TFC model is better than other four models based on the MSE (Table 4).

Table 4. Forecasting comparison with other models for Gold Price (Radius)

Model	Performance indices		
	MSE	MAE	MAPE
Moving average model	48.65	5.46	40.00%
Exponential smoothing	31.59	4.07	32.57%
ARIMA	24.95	3.83	35.01%
TFC model	23.78	4.1	40.67%

Table 5 is the comparison results for the forecast of gold price intervals. It reveals that *IMSE* and *IMAE* of TFC model are much smaller than those of the traditional models. Furthermore, if we take the Coverage Rate into consideration, the TFC model outperforms the ARIMA model. In addition, the forecasted interval value of ARIMA and TFC models for test data (from 468 to 487) is illustrated at Table 5.

Table 5. Forecasting Performance of Gold Price (Interval, from 468–487)

	Performance indices		
	IMSE	IMAE	Fitting performance
Moving average model	2469.90	44.17	64.4%
Exponential smoothing	2169.42	41.12	64.3%
ARIMA	1544.95	34.48	68.5%
TFC model	283.32	14.14	82.6%

Overall, the forecasts of gold price we made, especially for weekly center and fuzzy interval are excellent. In addition, the forecast interval of gold prices using the fuzzy classification method will provide a better forecasting ability for the uncertainty in forecasting gold prices effectively. At the end, we also find that if the change period of its structural change can be determined for a time series, better results on the model construction and forecasting ability can be produced.

Forecasting result

The result of 1. overall external economic fortunes (overall economic indicators) 2. internal company operating conditions (stocks earnings data) and 3. the entropy state are shown at table are demonstrated at Table 6.

Table 6. The result for the contribution value of center forecasting

	Step 1		Step 2		Step 3	
	Center	Radius	Center	Radius	Center	Radius
1.macroeconomics	6	1	8	−1	7	0
2.internal effect	7	−1	9	2	−2	2
3.entropy state	10	1	9	2	7	2
Total	23	1	26	3	12	4

The forecasting for out of data are showing at Table 7.

Table 7. Weakly gold price forecasting for out of data

Period	Real value			Forecasting		
	Center	Radius	Interval	Center	Radius	Interval
488 (2015/8/3-7)	1089.9	4.2	(1086,1094)	1073 + 23	5.7	(1062,1084)
489 (2015/8/10-14)	1113.7	9.5	(1104,1123)	1097 + 26	7.0	(1070,1092)
490 (2015/8/17-21)	1139.8	11.4	(1117,1163)	1122 + 12	8.2	(1062,1084)
491 (2015/8/24-28)	1137.9	15.5	(1122,1153)			

5   Conclusion

In this research we use big data context and new measurement system to do the forecasting work. Big data can be summarized into real number data and non-real data. Especially for non-real data, such as semantic data, interval data, trapezoid data, graphics, digital information encoded it, and so we will present a new statistical analysis of new technologies, new metrics and decision evaluation.

These proposed methods illustrate a innovative applications. Clarify the properties of measurement system for fuzzy data and their statistical analysis; extension of linear model building to more general random elements. Propose strategy for selecting an optimal models based on the fitted density and procedure to check the stability of the models' parameters.

Since the model specification testing and forecasting is particularly important, the experience from the inherent economic significance and statistical significance of uncertainty needs to be considerate. The main idea of Interval data operation is to transform the data into centroid and length/area reduce the vague set of supported by randomly selecting a subset of samples. In addition the soft ranking techniques are proposed to measure the order of separation margin and to be effective in the gray zone.

In the empirical study there are a number of distinct artifuture intelligent theories to explore the phenomenon in the economic model construction. Empirical studies will demonstrate that our frontier and efficient forecasting techniques will be more efficient information comparing to the point value forecasting either in academic or the realistic field.

Some topics that are worth researching for future study are:

1. Is there a better way to determine the cluster centers for intervals, so that the relationship between intervals can be described more clearly?
2. This study mainly discusses the analysis of interval data with regards to Z-type, Λ-type, pi-type and S-type fuzzy data. This seems to be worth further research.
3. How does a change interval differ when the threshold value λ and significance level α differ? Meaning that, how to define a threshold value λ and significance level α that are more effective, and find out a more meaningful structural change time.

References

1. Kim, D.-W., Lee, K.H., Lee, D.: On cluster validity index for estimation of the optimal number of fuzzy clusters. Pattern Recogn. **37**(10), 2009–2025 (2004)
2. Kumar, K., Wu, B.: Detection of change points in time series analysis with fuzzy statistics. Int. J. Syst. Sci. **32**(9), 1185–1192 (2001)
3. Lai, W., Wu, B.: Overcome the sort problem of low discrimination by interval fuzzy number. In: Innovative Management, Information and Production Nonlinear Mathematics for Uncertainty and its Applications, pp. 249–258. Springer, Berlin (2013)
4. Pakhira, M.K., Bandyopadhyay, S., Maulik, U.: A study of some fuzzy cluster validity indices, genetic clustering and application to pixel classification. Fuzzy Sets Syst. **15**(2), 191–214 (2005)
5. Windham, M.P.: Cluster validity for fuzzy clustering algorithms. Fuzzy Sets Syst. **5**(2), 177–185 (1981)
6. Namwong, J., Wattanakul, K.: Measuring of happiness of people in Tippanate's community, Chiang Mai province. Empirical Econometrics Quant. Econ. Lett. **1**(2), 59–70 (2012)
7. San, O.M., Huynh, V., Nakamori, Y.: An alternative extension of the K-means algorithm for clustering categorical data. Int. J. Appl. Math. Comput. Sci. **14**(2), 241–247 (2004)
8. Sudtasan, T.: Detection of regime switching in stock prices before "window dressing" at the yearend using genetic algorithm. Int. J. Intell. Technol. Appl. Stat. **5**(2), 143–155 (2012)
9. Sudtasan, T., Suriya, K.: Making profit in stock investment before XD dates by using genetic algorithm. In: Watada, J., et al (eds.) Innovative Management in Information and Production. Springer Science and Business Media, New York (2013)
10. Tseng, F., Tzeng, G.: A fuzzy seasonal ARIMA model for forecasting. Fuzzy Sets Syst. **126**, 367–376 (2002)
11. Tseng, F., Tzeng, G., Yu, H., Yuan, B.: Fuzzy ARIMA model for forecasting the foreign exchange market. Fuzzy Sets Syst. **158**, 609–624 (2005)
12. Wu, B., Chen, M.: Use fuzzy statistical methods in change periods detection. Appl. Math. Comput. **99**, 241–254 (1999)
13. Wu, B., Nguyen, H.: New statistical analysis on the marketing research and efficiency evaluation with fuzzy data. Manag. Decis. **52**, 1330–1342 (2014)
14. Yang, C.C., Cheng, Y.-T., Wu, B., Sriboonchitta, S.: Hold a mirror up to nature: A new approach on correlation evaluation with fuzzy data. In: Quantitative Modeling in Marketing and Management, pp. 285–299. World Scientific (2012)
15. Zeng, W., Li, H.: Relationship between similarity measure and entropy of interval valued fuzzy sets. Fuzzy Sets Syst. **157**(11), 1477–1484 (2006)
16. Zhou, H.D.: Nonlinearity or structural break? - data mining in evolving financial data sets from a Bayesian model combination perspective. In: Proceedings of the 38th Hawaii International Conference on System Sciences (2005)

Bayesian Empirical Likelihood Estimation for Kink Regression with Unknown Threshold

Woraphon Yamaka$^{(\boxtimes)}$, Pathairat Pastpipatkul, and Songsak Sriboonchitta

Centre of Excellence in Econometrics, Faculty of Economics,
Chiang Mai University, Chiang Mai, Thailand
woraphon.econ@gmail.com

Abstract. Bayesian inference provides a flexible way of combining data with prior information from our knowledge. However, Bayesian estimation is very sensitive to the likelihood. We need to evaluate the likelihood density, which is difficult to evaluate, in order to use MCMC. Thus, this study considers using the Bayesian empirical likelihood(BEL) approach to kink regression. By taking the empirical likelihood into a Bayesian framework, the simulation results show an acceptable bias and MSE values when compared with LS, MLE, and Bayesian when the errors are generated from both normal and non-normal distributions. In addition, BEL can outperform the competing methods with quite small sample sizes under various error distributions. Then, we apply our approach to address a question: Has the accumulation of foreign reserves effectively protected the Thai economy from the financial crisis? The results demonstrate that foreign reserves provide both positive and negative effects on economic growth for high and low growth regimes of foreign reserve, respectively. We also find that foreign reserves seem to have played a role in offsetting the effect of the crisis when the growth rate of foreign reserves is less than 2.48%.

Keywords: Bayesian empirical likelihood · Kink regression
GDP · Foreign reserves

1   Introduction

There is a lot of evidence that suggest that the structure of the economic data might exhibit nonlinear behavior during the last few decades, especially after the economic globalization. Thus, a number of nonlinear models have been proposed in the past 30–40 years. The purpose of this study is to propose a modern method of time series analysis in the evaluation of econometric models, in particular non-linear model. Recently, the classical methods, for example Ordinary Least Squares, Bayesian, and Maximum likelihood estimators, are almost always used and perhaps the most commonly and widely accepted estimators in parameter estimation in nonlinear model. However, these methods often face the problem of poor estimation such as bias and inconsistency of the results since the data are assumed to have normal distribution. In the estimation, the time

© Springer International Publishing AG 2018
V. Kreinovich et al. (eds.), *Predictive Econometrics and Big Data*, Studies in Computational Intelligence 753, https://doi.org/10.1007/978-3-319-70942-0_54

series approach to modeling typically involves a set of strong assumptions. It might not be appropriate to assume that error is a sequence of independent and identically normal distributed random variables [7]. Moreover, [11,15,17] also confirm that the explanatory variable is also conditionally non-normally distributed in many applications, thus the assumption of normality may then yield biased estimation for model parameters of interest. In addition, many researchers are also concerned about the limited data that bring about an underdetermined, or ill-posed problem for the observed data, in which case the traditional estimation techniques have difficulty obtaining the optimal solution. As is widely understood, the larger sample size of data can bring the higher probability of finding a statistically significant result [14]. [1] suggested that when the samples are limited, it is often hard to get meaningful results.

Thus, in this study, we propose an alternative estimation called Bayesian empirical likelihood (BEL) since it can relax a strong assumption of normality and the limited data. Typically, Bayesian estimation analyses do not assume large samples, as is the case with maximum likelihood and least squares estimation. A smaller data size can be estimated without losing any power while retaining precision [16]. [10] proposed that, Bayesian estimation requires only 1:3 ratio of parameters to observations. For least squares estimation, [18] showed that it is inadmissible when the number of coefficients in the model is larger. Thus, it is reasonable here to conduct a Bayesian estimation for constructing distributions based on limited information.

As we mentioned above, although the Bayesian estimation is common and reasonable with small data size, it still produces a biased estimation or even misleading results when the true distribution of error may not be normally distributed especially if it is heavy-tailed or skewed [21]. We know that for the Bayesian estimation, the important step is the simulation from the posterior distribution by using Markov Chain Monte Carlo (MCMC). However, Bayesian estimation is very sensitive to the likelihood; we need to evaluate the likelihood density, which is difficult to do, in order to use MCMC. Due to the complexity of the estimation and model, the likelihood is normally assumed to have normal or other parametric likelihood density. In these cases, it is required to develop a more robust likelihood to make statistical inference on Bayesian estimation. To this end, throughout this paper, we incorporate an empirical likelihood into a Bayesian framework.

The Empirical Likelihood as a robust an alternative to classical likelihood approaches was first introduced by [13]. It has many interesting properties and is efficient as parametric likelihood [12,19]. The main idea of Empirical Likelihood is to use a maximum-entropy discrete distribution supported by the observed data and constrained by nonlinear equations related with the parameters of the model. In brief, its a non-parametric likelihood, which is fundamental for the likelihood-based statistical methodology. This study uses empirical likelihood to approximate the likelihood in the Bayesian computation to be followed by Bayesian inference.

In the recent decade, the BEL method has been discussed by [2,4]. The BEL inference has been applied to various time series models. For example, [20,21] considered BEL estimation in Quantile model; [3] developed a BEL approach for linear regression model. However, to our knowledge, there is no work done on the BEL inference on kink regression model. Here we develop a BEL inference on nonlinear kink regression of [5] based on the constructed EL and the prior distribution of parameters including a kink or threshold parameter. The Markov Chain Monte Carlo (MCMC) method is presented to make Bayesian inference on parameters by using the Metropolis-Hastings algorithm.

The rest of the paper is organized as follows. In Sect. 2, we introduce the kink regression, and discuss BEL approach. The simulation study is conducted to demonstrate the finite sample performance of the BEL approach through Monte Carlo simulations in Sect. 3. In Sect. 4, we use microeconomic real data example to show that the BEL approach can be used as alternative method in nonlinear model. In this section, we try to addresses a question: Has the accumulation of foreign reserves effectively protected the Thai economy from the financial crisis? Concluding remarks are given in Sect. 5.

2   Kink Regression Model

2.1   Model Structure

In this study, the following two-regime kink regression model is considered

$$
\begin{aligned}
Y_t = \beta_1^-(x'_{1,t} \le \gamma_1)_- + \beta_1^+(x'_{1,t} > \gamma_1)_+ +, ..., +\beta_K^-(x'_{K,t} \le \gamma_K)_- \\
+ \beta_K^+(x'_{K,t} > \gamma_K)_+ + \alpha Z + \varepsilon_t
\end{aligned}
\tag{1}
$$

where Y_t is $[T \times 1]$ sequence of response variable at time t, $x'_{k,t}$ is a matrix of $(T \times K)$ predictor variables at time t. Here the relationship of $x'_{k,t}$ and Y_t changes at the unknown location or kink point γ_K, thus β is a matrix of $(T \times K \times 2)$ unknown parameters where $(\beta_1^-, ..., \beta_K^-)$ and $(\beta_1^+, ..., \beta_K^+)$ are the coefficients with respect to variable $x'_{k,t}$ for value of $x'_{k,t} \le \gamma_k$ and with respect to variable $x'_{k,t}$ for value of $x'_{k,t} > \gamma_k$, respectively. Z is a vector of covariates whose relationship with Y_t is linear. Following [5], the predictor variables are subject to regime-change at unknown kink point or threshold variable $(\gamma_1, ..., \gamma_K)$ and thereby separating this regressor into two regimes. These threshold variables are compact and strictly in the interior of the support of $(x_{1,t}, ..., x_{K,t})$. In addition, the error term of the model ε_t is a vector of error for which we do not assume any distribution, but we only assume that $E(\varepsilon_t) = 0$.

2.2   Constructing Empirical Likelihood

Since the distribution of errors ε_t in Eq. (1) is unspecified, the likelihood function is unavailable. Therefore, it is necessary to find an appropriate likelihood. In this study, we adopt empirical likelihood (EL) of [13] as an alternative parametric likelihood in our Bayesian estimation. In this section, we will briefly discuss the concept of empirical likelihood, and its relationship with estimating functions. Let $p_1, ..., p_k$ be the set of implied probability weights allocated to the data and $\theta \in \{\beta_1^-, ..., \beta_K^-, \beta_1^+, ..., \beta_K^+, \alpha, \gamma_1, ..., \gamma_K\}$. It carries a lot of information about the stochastic properties of the data. Then, let $X^- = (x'_t \leq \gamma)_-$ and $X^+ = (x'_t > \gamma)_+$, the empirical likelihood for estimated parameters in Eq. (1), in the spirit of [13], is

$$EL(\theta) = \max \prod_{t=1}^{T} p_i. \tag{2}$$

By taking logarithm Eq. (2), we have

$$EL(\beta) = \max \sum_{t=1}^{T} \log p_t, \tag{3}$$

where the maximization is subject to the constraints

$$\sum_{t=1}^{T} p_t \frac{\partial m(X_{it}^-, X_{it}^+; \theta)}{\partial \theta} (y_t - m(X_{it}^-, X_{it}^+; \theta)) = 0, \tag{4}$$

$$\sum_{t=1}^{T} p_t = 1, \tag{5}$$

where, $m(X_{it}^-, X_{it}^+; \theta) = \beta_1^-(x'_{1,t} \leq \gamma_1)_- + \beta_1^+(x'_{1,t} > \gamma_1)_+, ..., \beta_K^-(x'_{K,t} \leq \gamma_K)_- + \beta_K^+(x'_{K,t} > \gamma_K)_+ + \alpha Z$. Sometime the high dimensionality of the parameter space $(\theta, p_1, ..., p_T)$ makes the above maximization problem difficult to solve, and leads to expressions which are hard to maximize. Instead of maximizing $EL(\theta)$ with respect to the parameters $(\theta, p_1, ..., p_T)$ jointly, we use a profile likelihood.

The empirical likelihood, Eq. (3), is essentially a constrained profile likelihood, with constraints Eqs. (4) and (5). In getting the empirical likelihood at each candidate parameter value θ, this optimization problem can be solved for the optimal p_t. Suppose, we know θ, then we can write the empirical likelihood as

$$EL(\theta, p_1, ..., p_T) = EL(p_1, ..., p_T) \tag{6}$$

We maximize this profile empirical likelihood to obtain $(p_1, ..., p_T)$. By conducting the Lagrange multipliers, we can maximize the empirical likelihood in Eq. (3) subject to the constraints in Eqs. (4) and (5) as

$$L(p, \lambda_0, \lambda_1) = \sum_{t=1}^{T} \log(p_t) + \lambda_0 \left(\sum_{t=1}^{T} p_t - 1 \right)$$
$$+ \lambda'_1 \sum_{t=1}^{T} p_t \frac{\partial m(X_{it}^-, X_{it}^+; \theta)}{\partial A_k} (Y_t - m(X_{it}^-, X_{it}^+; \theta)), \tag{7}$$

where $\lambda \in \mathbb{R}$ is the Lagrange multipliers. It is a straightforward exercise to show that with the first order condition for L with respect to p_t, and setting the derivative to zero, we can find that $\lambda_0 = -T$, and by defining $\lambda = -T\lambda_1$, we obtain the optimal p_t as

$$p_t = \frac{1}{T}\left(1 + \lambda' \frac{\partial m(X_{it}^-, X_{it}^+; \theta)}{\partial \theta}(Y_t - m(X_{it}^-, X_{it}^+; \theta))\right)^{-1} \tag{8}$$

Then, substituting the optimal p_t into the empirical likelihood in Eq. (2) we obtain

$$EL(\theta) = \max \prod_{t=1}^{T} \frac{1}{T}\left(1 + \lambda' \frac{\partial m(X_{it}^-, X_{it}^+; \theta)}{\partial \theta}(Y_t - m(X_{it}^-, X_{it}^+; \theta))\right)^{-1}. \tag{9}$$

By taking logarithm, we get

$$\log EL(\theta) = \sum_{t=1}^{T} \log(1 + \lambda' \frac{\partial m(X_{it}^-, X_{it}^+; \theta)}{\partial \theta}(Y_t - m(X_{it}^-, X_{it}^+; \theta)) - T log(T). \tag{10}$$

To compute the profile empirical likelihood at a θ, it consists of two estimation steps. Firstly, it is important to solve a nonlinear optimization to obtain p_t, λ, and $EL(\theta_i)$ which depends on θ_i. Second step, the profile empirical likelihood is then maximized with respect to candidate θ_i. Then, we propose another candidate θ_i to repeat the first step again. After $EL(\theta_i)$ is computed for all candidates θ_i, the maximum value of $EL(\tilde{\theta}_i)$ is selected. The maximization problem can now be represented as the problem of minimizing $Q(\lambda)$

$$Q(\lambda) = -\sum_{t=1}^{T} \log(1 + \lambda' \frac{\partial m(X_{it}^-, X_{it}^+; \theta)}{\partial \theta}(Y_t - m(X_{it}^-, X_{it}^+; \theta)). \tag{11}$$

subject to $0 \leq p_t \leq 1$, that is

$$1 + \lambda' \frac{\partial m(X_{it}^-, X_{it}^+; \theta)}{\partial \theta}(Y_t - m(X_{it}^-, X_{it}^+; \theta)) \geq 1/T \tag{12}$$

To compute θ, one use a nested optimization algorithm where the outer maximization loop with respect to θ encloses the inner minimization loop with respect to λ. Some comments on the inner loop and the outer loop are in order. In the application study, the number of all possible θ_i can be so large that it becomes infeasible and insensible to evaluate them all. Thus, we can employ a standard least squares estimator to get the estimated θ_{LS} and specify the sensible range of candidate $\theta_i = [-2\theta_{LS}, 2\theta_{LS}]$.

2.3 Bayesian Empirical Likelihood for Kink Regression

The posterior distribution consists of the estimation of empirical likelihood function and the prior distribution. It is derived using Bayes rule. Here, we briefly remind how it is derived. Let (Ω, A, P) be a probability space. Let $A_n, n \geq 1$, be a countable, measurable partition of Ω, and $B \in A$ be an event with $P(B) > 0$. Then, for any $n \geq 1$,

$$P(A_n \,|B) = \frac{P(B\,|A_n)P(A_n)}{\sum\limits_{j=1}^{\infty} P(B\,|A_j)P(A_j)} \tag{13}$$

Indeed, we have

$$P(A_n \,|B) = \frac{P(A_n \cap B)}{P(B)} = \frac{P(B\,|A_n\,P(A_n))}{P(B)} \tag{14}$$

And writing

$$B = B \cap \Omega = B \cap (\cup_{j=1}^{\infty} A_j) = U_{j=1}^{\infty}(B \cap A_j) \tag{15}$$

We have

$$P(B) = \sum_{j=1}^{\infty} P(B \cap A_j) = \sum_{j=1}^{\infty} P(B\,|A_j)P(A_j). \tag{16}$$

Consider the discrete case, let X, Y be discrete random variables. Then

$$P(X = x\,|Y = y) = \frac{P(Y = y\,|X = x)P(X = x)}{\sum_{x'} P(Y = y\,|X = x')P(X = x')} \tag{17}$$

is the conditional density of X given Y. Here, $P(X = x)$ denotes prior density of X, while $P(Y = y)$ denotes empirical likelihood density. Thus, $P(X = x\,|Y = y)$ is the posterior density and we can rewrite Eq. 17 as

$$P(\theta\,|Y, X) \propto EL(\theta) \cdot \pi(\theta), \tag{18}$$

where $\pi(\theta)$ denotes a prior density of each estimated parameters. To estimate the posterior distribution in kink regression models, [20] suggested that the value of the empirical likelihood is relatively easy to compute given θ which makes the Metropolis-Hastings algorithm of [6] feasible for sampling from the posterior. However, it remains for this study to derive the conditional posterior distribution for unknown parameter θ. If we select a proper prior, the posterior in Eq. 18 is also proper. In this study, we choose the priors as follows.

We take $\theta = \{\alpha, \beta^-, \beta^+\}$ to be normally distributed with mean θ_0 and variance 0.01, γ_k is assumed to have uniform distribution. Hence, the conditional posteriors of θ and γ_k can be computed as in the following:

(1) The conditional posterior distribution for θ is

$$\theta^* = \left(\frac{X'X}{(\varepsilon'\varepsilon/n)^2} + 0.01\right)^{-1} \times \left(\frac{X'X}{(\varepsilon'\varepsilon/n)^2}\tilde{\theta} + 0.01(\theta_0)\right), \tag{19}$$

where $\tilde{\theta} = (X'X)^{-1}X'Y$ and $X = \{x'_{k,t} \leq \gamma_k, x'_{k,t} > \gamma_k\} = \left\{x^-_{k,t}, x^+_{k,t}\right\}$.

(2) The conditional posterior distribution for γ_k can be written as

$$P(\gamma_k \,|\theta, Y, X) = \sum_1 EL(\theta \,|Y, X) \cdot \pi(\theta), \qquad (20)$$

To sample all of these parameters based on conditional posterior distribution, we employ the Markov Chain Monte Carlo, Metropolis-Hastings algorithm, which is especially useful in extracting marginal distributions from fully conditional distributions. We run the Gibbs sampler for 20,000 iterations where the first 5,000 iterations serve as a burn-in period. For Metropolis-Hastings algorithm, we apply it to find kink value Θ, γ_k where the acceptance ratio is

$$r = \frac{EL(\theta^* \,|Y, X)\pi(\theta_{i-1} \,|\theta^*)}{EL(\theta_{i-1} \,|Y, X) \, \pi(\theta^* \,|\theta_{i-1})} \qquad (21)$$

Then, we set

$$\theta_j = \begin{cases} \theta_{j-1} & U < r \\ \theta_j^* & U > r \end{cases} \qquad (22)$$

where θ_{j-1} is the estimated vector of parameter at $(j-1)^{th}$ draw and θ_j^* is candidate vector of parameters which random from $N(\theta_{j-1}, 0.01)$. U is Uniform(0,1). This means that if the candidate θ_j^* looks good, keep it; otherwise, keep the current value θ_{j-1}. By using a MetropolisHastings algorithm, we estimate the parameters using the average of the Markov chain on θ as an estimate of θ, when the posterior density is likely to be close to normal and the trace of θ_j looks stationary.

2.4   Testing for a Kink Effect

Since the non-linear structure of the model has been considered in this study, we develop Bayes factor which is a reliable testing procedures in model comparison and kink effect test in the Bayesian approach. The purpose of this Bayes factor is to check whether or not kink regression model is significant relative to the linear regression model. It can be used to assess the models of interest namely linear and kink regression, so that the best fit model will be identified given a data set and a possible model set. In short, Bayes factor is a useful tool for selecting a possible model [8]. Bayes factor is used for the ratio of the posterior under one model to another model. In this study, we consider the linear model to be a null model denoted by M_1 and the switching model to be an alternative model denoted by M_2. More specifically, Bayes factor BF is given by

$$BF = \frac{P(Y, X \,|M_1)}{P(Y, X \,|M_2)} = \frac{\int EL(Y, X \,|\theta_1)\pi(\theta_1 \,|M_1)d\theta_1}{\int EL(Y, X \,|\theta_2)\pi(\theta_2 \,|M_2)d\theta_2}, \qquad (23)$$

where $P(Y, X \,|M_1)$ and $P(Y, X \,|M_2)$ are the posterior density of the null model and alternative model, respectively. θ_1 and θ_2 are the vector of parameters of M_1 and M_2, respectively. For choosing the appropriate model, we follow the idea of [9] which can be summarized in Table 1.

Table 1. Interpretation

$\log(BF)$	<2	$2-6$	$6-10$	>10
	Supports M_2	Positive to M_1	Strong support to M_1	Very strong support M_1

Source: Kass and Raftery [9]

3   Simulation Study

In this section, we use Monte Carlo simulations to investigate the performance of the BEL method in terms of coverage probability and estimation efficiency. We focus on the asymptotic properties of the posterior distribution in Eq. 18, to check whether the empirical likelihood is valid for posterior inference based on the criteria provided in [20]. Moreover, we also compare the performance of our BEL estimation on kink regression model with the Bayesian with normal likelihood and prior, least squares (LS) and maximum likelihood (ML) estimations. To compare these methods, the study used bias and Mean Squared Error (MSE) approaches.

In the simulation, we use the following equation to generate the dataset Y_t

$$Y_t = \alpha + \beta_1^-(x'_{1,t} \le \gamma_1)_- + \beta_1^+(x'_{1,t} > \gamma_1)_+ + \varepsilon_t \tag{24}$$

where the true value for parameters α, β_1^-, and β_1^+ are $\alpha = 1$, $\beta_1^- = -2$, and $\beta_1^+ = 0.5$, respectively. The threshold value is $\gamma_1 = 6$. The covariate $x'_{1,t}$ is independently generated from the standard normal distribution $N(\gamma_1, 1)$ to guarantee that γ_1 is located in $x'_{1,t}$. To make a fair comparison, we consider the following random errors ε_t: (1) $\chi^2(2)$, (2)$N(0,1)$, (3)$t(0,1,4)$, (4)$Unif(-2,2)$, and (5) non-standard normal scale mixture, $N(0,1)$ with respect to a uniform distribution supported on $(-1,1)$, denoted by NUnif. In this Monte Carlo simulation, we consider sample size $n = 20$ and $n = 40$. Then, we assess the performance of our proposed method through the bias and Mean Squared Error (MSE) of each parameter in which the bias and MSE of each parameter are given by

$$bias = \left| N^{-1} \sum_{r=1}^{N} (\tilde{\theta}_r - \theta_r) \right|,$$

and

$$MSE = N^{-1} \sum_{r=1}^{N} (\tilde{\theta}_r - \theta_r)^2.$$

where $N = 100$ is the number of bootstrapping; and $\tilde{\theta}_r$ and θ_r are the estimated value and true value, respectively.

Tables 2, 3, 4, 5 and 6 report the results of the sampling experiments for the chi-squared, normal, student-t, uniform and mixture normal-uniform error distributions, respectively. We found that BEL is not the best method when compared with LS, ML and Bayesian methods. Although some of the bias and MSE values of estimated parameters from BEL are lower than other methods,

Table 2. Kink regressions with $\chi^2(2)$ error

n = 20	BEL		LS		ML		Bayesian	
	bias	MSE	bias	MSE	bias	MSE	bias	MSE
α	0.7703	0.7878	0.9571	8.3447	1.8247	3.6852	1.4142	2.5118
$\beta_1{}^-$	0.7745	2.0034	0.1165	0.2790	0.0292	0.0364	−0.1464	0.1174
$\beta_1{}^+$	2.7387	8.7729	0.2414	0.6313	0.0139	0.0424	0.0012	0.0388
γ_1	0.3120	1.8453	0.3776	4.0403	0.1610	0.2899	0.1821	0.1979
n = 100	**BEL**		**LS**		**ML**		**Bayesian**	
	bias	MSE	bias	MSE	bias	MSE	bias	MSE
α	0.7255	0.9069	2.029	4.1938	2.0254	4.1782	1.9125	3.6989
$\beta_1{}^-$	0.7078	0.8438	0.0111	0.0059	0.0103	0.0057	0.0217	0.0033
$\beta_1{}^+$	1.4887	3.4708	0.0057	0.0047	0.0066	0.0047	0.0209	0.0026
γ_1	0.1023	0.8919	0.0155	0.0672	0.0103	0.0641	0.104	0.0145

Source: Calculations

Table 3. Kink regressions with $N(0,1)$ error

n = 20	BEL		LS		ML		Bayesian	
	bias	MSE	bias	MSE	bias	MSE	bias	MSE
α	0.0947	0.3337	0.1217	0.2334	0.1025	0.2392	0.0883	0.214
$\beta_1{}^-$	0.2124	0.5303	0.0088	0.0151	0.0004	0.0148	0.0021	0.0181
$\beta_1{}^+$	0.2036	0.5025	0.0374	0.0376	0.0521	0.0338	0.0423	0.0173
γ_1	0.1874	0.2389	0.0122	0.1479	0.0623	0.1092	0.141	0.2192
n = 100	**BEL**		**LS**		**ML**		**Bayesian**	
	bias	MSE	bias	MSE	bias	MSE	bias	MSE
α	0.0303	0.0217	0.0728	0.0709	0.0727	0.0709	0.0195	0.0321
$\beta_1{}^-$	0.0093	0.0099	0.0191	0.0052	0.0191	0.0051	0.006	0.0018
$\beta_1{}^+$	0.0181	0.0062	0.0050	0.0022	0.0050	0.0022	0.0225	0.0021
γ_1	0.0146	0.0693	0.0476	0.0388	0.0476	0.0388	0.0657	0.0155

Source: Calculations

the BEL does not show evidence of completely outperforming other methods. However, it still provides a good estimation and could be an alternative estimation for kink regression model, especially when the sample size is small ($n = 20$). We found that the performance of BEL estimator is different when $n = 20$ and $n = 40$ since the values of bias and MSE are lower when the sample size increases. In the case of small sample size, we compared the results of BEL estimator with other competing methods and found that the bias and MSE of its estimated parameters are mostly lower than other methods. In contrast, the larger sample size cannot bring about lower bias and MSE of the parameters when compared

Table 4. Kink regressions with $t(0, 1, 4)$ error

	BEL		LS		ML		Bayesian	
n = 20	bias	MSE	bias	MSE	bias	MSE	bias	MSE
α	0.5109	0.6225	0.3432	0.3984	0.2062	0.3262	0.2558	0.2631
$\beta_1{}^-$	0.0784	0.0258	0.1035	0.0280	0.0444	0.0143	0.066	0.0082
$\beta_1{}^+$	0.0176	0.0078	0.0347	0.0116	0.0069	0.0041	0.0327	0.0108
γ_1	0.1887	0.1661	0.2073	0.1813	0.0576	0.0966	0.1728	0.0484
	BEL		LS		ML		Bayesian	
n = 100	bias	MSE	bias	MSE	bias	MSE	bias	MSE
α	0.3661	0.39	0.1013	0.0651	0.0747	0.0425	0.0719	0.0483
$\beta_1{}^-$	0.051	0.0536	0.0004	0.0038	0.0102	0.0034	0.0101	0.0027
$\beta_1{}^+$	0.4207	0.9698	0.0093	0.0019	0.0019	0.0011	0.0289	0.0056
γ_1	0.1607	0.0723	0.0474	0.0197	0.0278	0.0116	0.0219	0.0375

Source: Calculations

Table 5. Kink regressions with $Unif(-2, 2)$ error

	BEL		LS		ML		Bayesian	
n = 20	bias	MSE	bias	MSE	bias	MSE	bias	MSE
α	0.0911	0.4862	0.0655	0.2768	0.0656	0.2769	0.1392	0.2179
$\beta_1{}^-$	0.0233	0.0236	0.0239	0.0255	0.0239	0.0255	0.0824	0.0247
$\beta_1{}^+$	0.0457	0.2012	0.0007	0.024	0.0001	0.024	0.0161	0.0295
γ_1	0.1793	0.1976	0.1352	0.1481	0.1352	0.1481	0.1328	0.4431
	BEL		LS		ML		Bayesian	
n = 100	bias	MSE	bias	MSE	bias	MSE	bias	MSE
α	0.0332	0.0524	0.0279	0.0735	0.0279	0.0735	0.0453	0.0696
$\beta_1{}^-$	0.0163	0.033	0.0131	0.003	0.0131	0.003	0.0061	0.0039
$\beta_1{}^+$	0.0228	0.0686	0.0288	0.0047	0.0288	0.0047	0.0148	0.0032
γ_1	0.0038	0.0136	0.0634	0.0331	0.0635	0.0331	0.0253	0.0218

Source: Calculations

with LS, ML, and Bayesian. This indicates that BEL is useful and more accurate when the sample size is small and even if non-normal distribution is assumed.

In summary, we can conclude that the overall performance of the BEL estimator applied to kink regression is good over a wide range of error distributions. The BEL estimators produce acceptable bias and MSE values compared with LS, ML, and Bayesian when the errors are generated from both normal and non-normal distributions. In addition, BEL can outperform the competing methods with small sample sizes under various error distributions.

Table 6. Kink regressions with $N(0,1) - Unif(-1,1)$ error

$n = 20$	BEL		LS		ML		Bayesian	
	bias	MSE	bias	MSE	bias	MSE	bias	MSE
α	0.3699	0.2559	0.4297	0.2618	0.4297	0.2618	0.4204	0.2327
β_1^-	0.069	0.7991	0.0078	0.0049	0.0078	0.0049	0.0129	0.005
β_1^+	0.0129	0.6453	0.0126	0.0016	0.0126	0.0016	0.0051	0.0034
γ_1	0.0079	0.0138	0.0551	0.0185	0.0552	0.0185	0.0525	0.0179
$n = 100$	BEL		LS		ML		Bayesian	
	bias	MSE	bias	MSE	bias	MSE	bias	MSE
α	0.361	0.3221	0.3136	0.1233	0.3136	0.1232	0.3418	0.1531
β_1^-	0.0059	0.0018	0.0003	0.0013	0.0003	0.0013	0.062	0.002
β_1^+	0.0035	0.0018	0.0056	0.0009	0.0056	0.0009	0.0023	0.0012
γ_1	0.0015	0.0099	0.0159	0.015	0.0159	0.015	0.0256	0.027

Source: Calculations

4   Empirical Analysis

This study investigates the basic question whether the accumulation of foreign reserves protected the Thai economy from the financial crisis. Motivated by this, we fit a Kink model for answering this basic question, which is given as follows:

$$GDP_t = \alpha + \beta_1^-(FR_t \leq \gamma)_- + \beta_1^+(FR_t > \gamma)_+ + \varepsilon_t,$$

where GDP_t is the growth of Gross Domestic product and FR_t is the growth of foreign exchange reserves of Thailand. The data is quarterly time series data and the sample period is Q1/1993-Q4/2016. Table 7 gives summary statistics, Jarque-Bera (JB) test, and Augmented Dickey-Fuller (ADF) test, to better understand the characteristics of the series. It can be seen that our data is stationary with non-normality. This is confirmed by the p value < 0.01 of the ADF and JB tests, respectively, in Table 7. Prior to estimating the kink regression model, we also had a concern about the nonlinear behavior in our model. In this study, therefore, we conduct a nonlinear structure test for our data. Bayes factor (BF) test is used to determine whether our model appears to have a nonlinear behavior. Using the Bayes factor formula, the results shown in Table 8 provide the values of Bayes factor of these two models in which we find that the value of $\log(BF)$ is equal to -0.03997. This result means the model M_2 is more anecdotally supported by the data under consideration than the model M_1, and hence, the data is more likely to have the nonlinear structure. This result led us to reject the null hypothesis of linear regression (M_1) for the relationship between foreign exchange reserves and Gross Domestic product for the Thai economy, which means that the kink regression model is better to describe this non-linear relationship.

Table 7. Data description

	FR growth	GDP growth
Mean	0.023169	0.016913
Median	0.023414	0.016678
Maximum	0.257449	0.107358
Minimum	−0.150070	−0.083100
Std. Dev.	0.054383	0.025234
Skewness	0.589845	−0.514270
Kurtosis	6.405468	7.101790
JB Probability	0	0
ADF-Probability	0	0

Source: Calculations

Table 8. Bayes factor of kink effect

	$P(Y, X\,|M_1)$	$P(Y, X\,|M_2)$	BF	$\log(BF)$	Interpret
Regime 1 vs. Regime 2	91.2986	−95.02217	0.9608	−0.03997	Support M2

Source: Calculations

Table 9. Coefficients (standard errors) from kink regression

Parameter	BEL
α	0.0278 (0.0003)
$\beta_1{}^-$ (regime 1)	0.3670 (0.0155)
$\beta_1{}^+$ (regime 2)	0.5222 (0.0218)
γ	0.0248 (0.0028)
Acceptance rate	0.4624

Source: Calculations

Note: () is standard deviation

4.1   Application Results

The kink regression model is then estimated by BEL estimator, and the estimated results are shown in Table 9. We can observe a different sign of effect of foreign reserves on the GDP_t in these two regimes. In this study, we interpret regime 1 and 2 as low and high foreign reserves growth regimes, respectively. The results demonstrate that the growth of foreign exchange reserves shows positive coefficient (0.3670) in regime 1 and negative coefficient (−0.5222) in regime 2. Consider the kink or threshold point (γ), we can observe that there exists a statistically significant kink effect in our model.

We now can plot a fitted kink regression line in Fig. 2. The result illustrates a steep positive slope for foreign reserves with a kink or threshold point (γ_1) around 0.0248, switching to a low negative slope above that point. For regime 1, ($FR_t \leq 0.0248$), the growth of FR by one percentage is found to increase

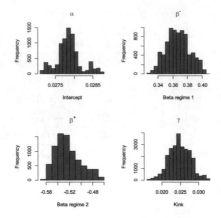

Fig. 1. Histograms based on the MCMC draws.

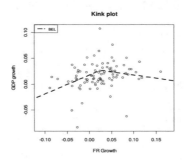

Fig. 2. Kink plot of GDP growth and Debt/GDP

the growth rate of GDP by 0.1755%. This result indicates the usefulness of foreign reserves during the crisis if the growth rate of foreign reserves is less than 2.34%. In contrast, the growth of FR by one percent is found to decrease the growth rate of GDP by 0.0037%, when $FR_t > 0.0248$. We expect that the increase in foreign reserves in this regime can reduce domestic savings and thereby decreasing domestic investment and economic growth.

Figure 2 depicts the histograms based on the MCMC draws, which gives some basic insight into the geometry of the posteriors obtained in the empirical analysis. The results demonstrate a good convergence behaviour and it seems to converge to the normal distribution; thus we can get accurate posterior inference for parameters that appear to have good mixing (Fig. 1).

5   Concluding Remarks

In this study we propose using empirical likelihood as a working likelihood for kink regression with unknown threshold in Bayesian estimation. The Empirical Likelihood can be employed as a robust alternative to classical likelihood. The

BEL approach can relax a strong assumption of normality and the limited data. Although the idea of BEL approach is not new, our work provides an important addition to the literature by employing this approach for non-linear kink regression model. Moreover, we also propose using a Bayes factor to check whether or not kink regression model is significant relative to the linear regression model. It can be used to assess the models of interest namely linear and kink regression, so that the best fit model will be identified given a data set and a possible model set. In short, Bayes factor is a useful tool for selecting a possible model.

We then conduct a simulation study to show the performance and accuracy of BEL estimation for kink regression model Simulation results validate that our BEL approach can provide accurate estimates for all unknown parameters including a kink parameter. We find that our BEL estimator applied to kink regression performs well over a wide range of error distributions. The BEL estimators produce acceptable bias and MSE values compared with LS, ML, and Bayesian when the errors are generated from both normal and non-normal distributions. In addition, BEL can outperform the competing methods with quite small sample sizes under various error distributions.

Finally, the empirical results demonstrate that foreign reserves provide both positive and negative effects on economic growth for high and low growth regimes of foreign reserves, respectively. We also find that foreign reserves seem to have played a role in offsetting the effect of the crisis when the growth rate of foreign reserves is less than 2.48%.

Acknowledgements. The authors are grateful to Puay Ungphakorn Centre of Excellence in Econometrics, Faculty of Economics, Chiang Mai University for the financial support.

References

1. Button, K.S., Ioannidis, J.P., Mokrysz, C., Nosek, B.A., Flint, J., Robinson, E.S., Munaf, M.R.: Power failure: why small sample size undermines the reliability of neuroscience. Nat. Rev. Neurosci. **14**(5), 365–376 (2013)
2. Chang, I.H., Mukerjee, R.: Bayesian and frequentist confidence intervals arising from empirical type likelihoods. Biometrika **95**, 139–147 (2008)
3. Chib, S., Shin, M., Simoni, A.: Bayesian empirical likelihood estimation and comparison of moment condition models. arXiv preprint arXiv:1606.02931 (2016)
4. Fang, K.T., Mukerjee, R.: Empirical-type likelihoods allowing posterior credible sets with frequentist validity: higher-order asymptotics. Biometrika **93**, 723–733 (2006)
5. Hansen, B.E.: Regression kink with an unknown threshold. J. Bus. Econ. Stat. **35**(2), 228–240 (2017)
6. Hastings, W.K.: Monte Carlo sampling methods using Markov chains and their applications. Biometrika **57**, 97–109 (1970)
7. Howrey, E.P.: The role of time series analysis in econometric model evaluation. In: Evaluation of Econometric Models, pp. 275–307. Academic Press (1980)
8. Kwon, Y.: Bayesian analysis of threshold autoregressive models (2003)

9. Kass, R.E., Raftery, A.E.: Bayes factors. J. Am. Stat. Assoc. **90**(430), 773–795 (1995)
10. Lee, S.Y., Song, X.Y.: Evaluation of the Bayesian and maximum likelihood approaches in analyzing structural equation models with small sample sizes. Multivar. Behav. Res. **39**(4), 653–686 (2004)
11. Lourens, S., Zhang, Y., Long, J.D., Paulsen, J.S.: Bias in estimation of a mixture of normal distributions. J. Biometrics Biostatistics **4** (2013)
12. Mykland, P.A.: Bartlett identities and large deviations in likelihood theory. Ann. Stat. 1105–1117 (1999)
13. Owen, A.B.: Empirical Likelihood. Wiley, New York (1998)
14. Peto, R., Pike, M.C., Armitage, P., Breslow, N.E., Cox, D.R., Howard, S.V., Smith, P.G.: Design and analysis of randomized clinical trials requiring prolonged observation of each patient. II. analysis and examples. Br. J. Cancer **35**(1), 1 (1977)
15. Sriboochitta, S., Yamaka, W., Maneejuk, P., Pastpipatkul, P.: A generalized information theoretical approach to non-linear time series model. In: Robustness in Econometrics, pp. 333–348. Springer International Publishing (2017)
16. Van de Schoot, R., Broere, J.J., Perryck, K.H., Zondervan-Zwijnenburg, M., Van Loey, N.E.: Analyzing small data sets using Bayesian estimation: the case of posttraumatic stress symptoms following mechanical ventilation in burn survivors. Eur. J. Psychotraumatology **6**(1), 25216 (2015)
17. Yamaka, W., Pastpipatkul, P., Sriboonchitta, S.: Has the accumulation of foreign reserves protect the thai economy from financial crisis?: An approach of empirical likelihood. Int. J. Econ. Res. **14**(6), 275–285 (2017)
18. Stein, C.: Inadmissibility of the usual estimator for the mean of multivariate normal distribution. In: Proceedings of the Third Berkeley Symposium on Mathematical Statistics and Probability, vol. 1, pp. 197–206 (1955)
19. Variyath, A.M., Chen, J., Abraham, B.: Empirical likelihood based variable selection. J. Stat. Plann. Infer. **140**(4), 971–981 (2010)
20. Yang, Y., He, X.: Bayesian empirical likelihood for quantile regression. Ann. Stat. **40**(2), 1102–1131 (2012)
21. Zhang, Y.Q., Tang, N.S.: Bayesian empirical likelihood estimation of quantile structural equation models. J. Syst. Sci. Complex. **30**(1), 122–138 (2017)

Spatial Choice Modeling Using the Support Vector Machine (SVM): Characterization and Prediction

Yong Yoon[⊠]

Faculty of Economics, Chulalongkorn University, Bangkok, Thailand
yong.y@chula.ac.th

Abstract. We take a cursory look at the support vector machine (SVM) as a useful and effective algorithm for characterizing and predicting spatial choice problems in economics. Beginning with a discussion of the SVM for the linearly separable case as well as the nonlinear non-separable case using the soft margin SVM and kernels, we then describe how the SVM can be used to characterize and predict spatial choice models, which can be seen as a special case of discrete choice models, using examples from a simple 1-D to the more complex multi-dimensional features space.

Keywords: Support vector machine · Spatial choice modeling
Characterization and prediction

1 Introduction

Big data and statistical or machine learning has arrived in economics in a big way [12,13]. Amongst various tools, the support vector machine (SVM) originally developed by Vapnik [17] and Alexey Ya. Chervonenkis in the early 1960s has become widely used not only for machine learning problems like pattern recognition, but also in other areas such as genetics, politics [11], and marketing [3]. Here, we take a look at how the SVM can be a useful and effective algorithm for characterizing and predicting spatial choice models in economics.

As a way of introducing the spatial choice problem, a good place to begin perhaps is with the "two ice cream vendors on a beach" example inspired by Hotelling [10]. The basic setting is to imagine a stretch of beach, say, a hundred meters long, on which two ice cream vendors are selling ice cream, each located at either end of the beach. Assuming that both offer the same flavors and the prices they charge are the same, it is reasonable to assume that a sunbather located somewhere in between the two vendors would choose to buy from the nearest one. From this basic setting, we abstract from the physical distance of Hotelling's paper by considering instead attributes of a good or service and individual preferences.

For example, suppose instead of ice cream vendors we have two Thai restaurants, $i \in \{1, 2\}$, selling Thai food with spiciness $x_i \in [0, 10]$, denoting the degree

© Springer International Publishing AG 2018
V. Kreinovich et al. (eds.), *Predictive Econometrics and Big Data*, Studies in Computational
Intelligence 753, https://doi.org/10.1007/978-3-319-70942-0_55

of spiciness ranging from "not at all spicy" to "very very spicy". Say we have $y \in \{1, 2, \cdots, j\}$ potential customers that have individual preferences for spiciness $S^{\{j\}} \in [0, 10]$. Then we can define the utility of the j−th consumer in choosing to visit restaurant i as:

$$u_i^{\{j\}} = C_i^{\{j\}} - |x_i - S^{\{j\}}|$$

where $C_i^{\{j\}}$ is the opportunity cost of the j−th consumer reaching restaurant i. For simplicity, we may assume that $C_i^{\{j\}}$ is constant and the same for all consumers, such that utility matters only on how "close" a restaurant's spiciness is to an individual's ideal taste.

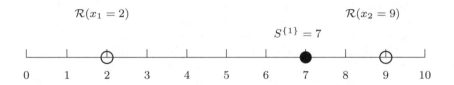

Fig. 1. Simple 1-D linear spatial choice model

Figure 1 illustrates the simple 1-D linear spatial choice model with two choices of restaurants $\mathcal{R}$ offering spiciness of $x_1 = 2$ and $x_1 = 9$. Suppose a certain individual has taste $S^{\{1\}} = 7$. Then because restaurant offering spiciness of $x_1 = 2$ is further away to the individual's preferred taste compared to restaurant offering spiciness of $x_1 = 9$, i.e., since $|x_1 - S^{\{1\}}| > |x_2 - S^{\{1\}}|$ or $|9 - 7| > |2 - 7|$, the individual will choose the restaurant offering spiciness of $x_1 = 9$ resulting in larger utility for the consumer than if restaurant offering spiciness of $x_1 = 2$ was chosen.

Rather than considering the closeness of attributes to a consumer's ideal point as the decision rule for such spatial choice problems, we propose the modeling of such a spatial choice problem by the SVM algorithm as an alternative and effective way to characterizing and predicting choice, and one that is also easily applicable to the multi-dimensional attribute space. Essentially, the SVM can easily construct a hyperplane in a high- or infinite-dimensional space, which can then be used for classification, regression, or other tasks. We exploit the idea that the SVM separates the feature space by a hyperplane that has the largest distance to the nearest training-data point(s) of any class (i.e. so-called functional margin), thus the large margin classifier effectively divides the space such that it meaningfully matches characterization and prediction of an economic agents' spatial choice.

For our simple 1-D linear spatial model, referring to Fig. 2, the SVM algorithm finds $\mathcal{H} = 5.5$, the separating hyperplane (or rather a point in this case as we lack sufficient dimensions for this simple example) such that any individual with $S^{\{i\}} < (>) 5.5$ would choose restaurant offering spiciness of $x_1 = 2$ ($x_2 = 9$). What is interesting to note is that the SVM splits the feature space

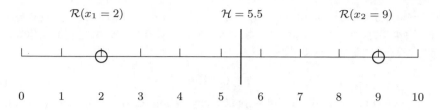

Fig. 2. Splitting the linear space with a 'hyperplane' by SVM

right in the middle of the two choices by the hyperplane, such that the choice alternatives on either side of the hyperplane will be at an equal distance from the hyperplane. A consumer positioned on the hyperplane will be indifferent to the two restaurants.

2 Support Vector Machine (SVM)

Assume a two-dimensional space with a number of points represented by $\mathbf{x}$ and labeled $y_i \in \{1, -1\}$. The SVM constructs a hyperplane $\mathcal{H}$ defined as $\mathbf{w} \cdot \mathbf{x} = 0$ (i.e. with parameters (w, b)), where the vector $\mathbf{w}$ is perpendicular to the hyperplane.

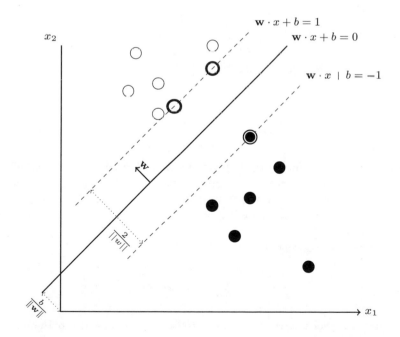

Fig. 3. Maximum-margin hyperplane and margins for a SVM

For any arbitrary point represented by a vector $\mathbf{u}$ which could be on either side of the hyperplane $\mathcal{H}$, projecting $\mathbf{u}$ onto $\mathbf{w}$ such that $\mathbf{w} \cdot \mathbf{u} \geq c$ and defining $c = -b$ gives the following decision rule:

$$\mathbf{w} \cdot \mathbf{u} + b \geq 0, \quad \text{then} \quad +ve \tag{1}$$

which defines one side of the hyperplane, say, where characterization or choices are $+ve$. We have now to define some constraints to establish b and $\mathbf{w}$ exactly. Say, given a sample point x, then

$$\mathbf{w} \cdot \mathbf{x}_+ + b \geq 1 \quad \text{and} \quad \mathbf{w} \cdot \mathbf{x}_- + b \leq -1$$

To simplify the two equations above, multiplying both equations by y_i gives $y_i(\mathbf{w} \cdot x_i + b) \geq 1$ (the same for both equations), which can be re-written as $y_i(\mathbf{w} \cdot x_i + b) - 1 \geq 0$.

As shown in Fig. 3 we could also draw two separate and parallel boundaries, which we may call margins, running parallel on either side of the hyperplane by

$$y_i(\mathbf{w} \cdot x_i + b) - 1 = 0 \tag{2}$$

Taking a point on each of the margins, say $\mathbf{x}_+$ and $\mathbf{x}_-$, say for *positive* and *negative* examples (depicted by *white* and *black* circles in Fig. 3), we can measure their distance apart as

$$(\mathbf{x}_+ - \mathbf{x}_-) \cdot \frac{\mathbf{w}}{\|w\|} = \frac{2}{\|w\|} \tag{3}$$

since $\mathbf{x}_+ = 1 - b$ and $\mathbf{x}_- = 1 + b$, by Eq. 2. And what the SVM does, because it looks for the hyperplane with the largest margin, is to maximize the margin $2/\|w\|$, or what is the same thing, $\forall i \in \{1, \ldots, n\}$, as

$$\min \frac{1}{2}\|w\|^2 \tag{4}$$

subject to

$$y_i(\mathbf{w} \cdot x_i + b) \geq 1$$

which is a rather simple quadratic programming (QP) problem for which there are readily available algorithms of complexity $O(n^3)$.[1] But, when n and d are large (tens of thousands) even the best QP methods fail. However, one very desirable characteristics of the SVM is that most of the data points end up being irrelevant.[2]

Setting up the Lagrangian,

$$\mathcal{L} = \frac{1}{2}\|w\|^2 - \sum_{i=1}^{n} \alpha_i \left[y_i(\mathbf{w} \cdot \mathbf{x}_i + b) - 1 \right] \tag{5}$$

---

[1] For example, so-called interior point algorithms that are variations of the Karmarkar algorithm for linear programming.

[2] The relevant data are only the points that end up exactly on the margin of the optimal classifier and these are often a very small fraction of n.

the F.O.C gives,

$$\nabla_{\mathbf{w}}\mathcal{L} = 0, \text{i.e., } \mathbf{w} = \sum_{i=1}^{n} \alpha_i y_i \mathbf{x}_i \tag{6}$$

$$\frac{\partial \mathcal{L}}{\partial b} = 0, \text{i.e., } \sum_{i=1}^{n} \alpha_i y_i = 0 \tag{7}$$

$$\alpha_i \left[y_i(\mathbf{w}\mathbf{x}_i + b) - 1 \right] = 0, \text{ for all } i \leq n. \tag{8}$$

The above Karush-Kuhn-Tucker (KKT) conditions for optimality provide a complete characterization of the optimal hyperplane. The normal $\mathbf{w}$ must be a linear combination of the observed vectors $\mathbf{x}$, given by Eq. 6; the coefficients of this linear combination must add up to 0, which is Eq. 7, and; finally the complementarity slackness conditions Eq. 8 tell us that the only non-zero Lagrange multipliers α_i are those associated to the vectors $\mathbf{x}$ exactly on the margin, i.e., corresponding to Eq. 2 which are the so-called *support vectors* and they are the only ones needed in Eq. 6. The support vectors are the observations $\mathbf{x}_i^{(s)}$ at the exact distance $1/||w||$ from the separating hyperplane, i.e., satisfying Eq. 2. And as mentioned above, the number of such vectors is usually much smaller than n which makes it possible to consider very large numbers of examples points and higher dimensions with $\mathbf{x}_i$ having many coordinates.

3   Duality: Soft-Margin SVM and Kernels

Of course, more realistically, the feature space may not be as easily linearly separable as assumed so far. The SVM literature proposes two further extensions to tackle not-so-easily linear separable spaces, (1) by using soft margins, and, (2) by transformation using kernels.

The soft margin SVM largely due to Cortes and Vapnik [4] introduces extra "slackening" to which instance $\xi_i \geq 0$ can be thought of as instances that are misclassified, and 0 otherwise. The optimization problem is then reformulated as

$$\min \frac{1}{2}||w||^2 + C\sum \xi_i \tag{9}$$

subject to

$$y_i(\mathbf{w} \cdot x_i + b) \geq 1 - \xi_i$$

with $\forall i \in \{1, \ldots, n\}$ and $\xi \geq 0$, and where $C\sum \xi_i$ serves as some penalty for slackness. Then given a new instance x_i, the classifier $f(x) = sign(\mathbf{w} \cdot x_i + b)$.[3] The corresponding Lagrangian is then,

$$\mathcal{L} = \frac{1}{2}||w||^2 + C\sum \xi_i - \sum_{i=1}^{n} \alpha_i \left[y_i(\mathbf{w} \cdot \mathbf{x}_i + b) - 1 + \xi_i \right] - \sum \beta_i \xi_i \tag{10}$$

---

[3] Note also that for the hard-margin form, we can set C to a very large value close to infinity resulting in a huge penalty.

where the constraint $\sum \beta_i \xi_i$ is included because of the non-negative constraint on $\xi \in [0, \infty]^n$.

The F.O.C give $\nabla_{\mathbf{w}} \mathcal{L} = 0$, i.e., $\mathbf{w} = \sum_{i=1}^{n} \alpha_i y_i \mathbf{x}_i$, (as in Eq. 6), $\partial \mathcal{L}/\partial b = 0$, i.e., $\sum_{i=1}^{n} \alpha_i y_i = 0$, (as in Eq. 7), and $\mathcal{L}/\partial \xi = 0$, i.e., $\alpha = C - \beta$ or $\beta = C - \alpha$.

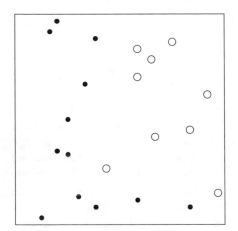 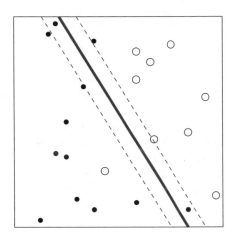

Fig. 4. Soft-margin SVM

Figure 4 shows the case where data points are not easily linearly separable (left panel). By applying the soft margin SVM (right panel), separation is found with all points in the margin and those classified incorrectly serving as support vectors.

For the primal problem (Eqs. 4 and 9), it is easy to show that its dual problem is

$$\max \sum_{i=1}^{n} \alpha_i - \frac{1}{2} \sum_{i,j=1}^{n} \alpha_i \alpha_j y_i y_j (\mathbf{x}_i \cdot \mathbf{x}_j) \tag{11}$$

subject to

$$\sum \alpha_i y_i = 0, \text{ and } \alpha_i \in [0, C]$$

and, by the strong duality property for such quadratic programming problems, the solution to this dual problem yields the same solution as the primal (see [2,5,6]).

What is important to note here, however, is that the dual tells us that the optimization problem essentially depends on the dot product (inner product of data vectors), $\mathbf{x}_i \cdot \mathbf{x}_j$. Thus, we have the decision rule for any $\mathbf{u}_j$ that

$$\sum \alpha_i y_i \mathbf{x}_i \cdot \mathbf{u}_i + b \geq 0 \text{ then } +ve$$

This leads us to the "kernel trick", which is essentially a way of transforming the original data points with a kernel function in such as a way that allows the SVM algorithm to fit the maximum-margin hyperplane by linear separation in the higher-dimensional transformed feature space. The idea is to apply a kernel or an appropriate mapping function, or more specifically a Grim function which follows the so-called Mercer's condition, on the dot product, i.e., $K(\mathbf{x}_i, \mathbf{x}_j)$, which would allow the linear SVM to find a separating hyperplane in the higher-dimension feature space which can then be translated back into decision boundaries in the original data space.[4] The objective function Eq. 11 can be re-written as

$$\max \sum_{i=1}^{n} \alpha_i - \frac{1}{2} \sum_{i,j=1}^{n} \alpha_i \alpha_j y_i y_j K(\mathbf{x}_i \cdot \mathbf{x}_j) \tag{12}$$

and with the dual representation of the optimal weight vector $\mathbf{w} = \sum_{i=1}^{n} \alpha_i y_i \phi(\mathbf{x}_i)$ for some appropriate non-linear mapping function, $\phi : I = \mathcal{R}^2 \rightarrow F = \mathcal{R}^d$ from, say, a 2-dimensional input space I into a higher d-dimensional feature space F, the equation for the optimal separating hyperplane can be written as

$$\mathbf{w} \cdot \mathbf{x} + b = \sum \alpha_i y_i K(\mathbf{x}_i \cdot \mathbf{u}_i) + b = 0$$

where α_i are the optimal Lagrange multipliers obtained by maximizing Eq. 12 and b is the optimal perpendicular distance from the origin now with $\mathbf{w}$ and support vectors $\mathbf{x}^{(s)}$ in the higher dimension feature space F.

Figure 5 shows the decision boundary for a non-linear case (left panel) using the Gaussian kernel, which is represented with the kernel transformation on a higher-dimensional feature space (right panel) that is linearly separable. Although many useful kernel functions correspond to an infinite-dimensional ϕ vector, which is equivalent to SVM with infinite number of features, yet the algorithm does not bear much additional computational cost as long as K itself is easy to compute (i.e. the Gram matrix will be in $O(m^2)$ dimension).[5]

---

[4] If $K(\mathbf{x}_i, \mathbf{x}_j)$ the satisfies the Mercer's condition, i.e. $\int_a \int_b K(a, b) g(a) g(b) da db \geq 0$, then $K(a, b) = \phi(a) \cdot \phi(b)$ for some transformation ϕ and the similarity function K is referred to as a kernel or mercer kernel. The Mercer condition simply asserting that the Gram matrix K has a positive-definite structure for any possible data set, i.e., $g' \cdot K \cdot g \geq 0$, for all x_i.

[5] Some of the most commonly used kernels are (1) the dth-degree polynomial kernel, $(\mathbf{x} \cdot x_i + \theta)^d$, (2) the Gaussian or radial basis kernel, $exp(-\gamma ||\mathbf{x} - x_i||^2)$, (3) the sigmoid kernel, $tanh(\kappa_1 (\mathbf{x} \cdot x_i) + \kappa_2)$, and (4) the inverse multi-quadratic, $(\sqrt{||\mathbf{x} - x_i||^2 2\sigma^2 + c^2})^{-1}$.

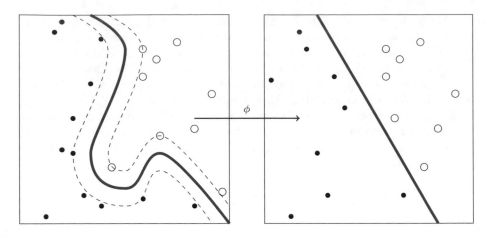

Fig. 5. Separation by the Gaussian kernel: $exp(-\gamma||\mathbf{x} - x_i||^2)$

4 Characterizing and Predicting Spatial Choice Models Using the SVM

In economics, spatial choice models can be viewed as a special case of the broader class of discrete choice models, which were popularized by the Economics Nobel Prize laureate McFadden [7–9], amongst others. More specifically, spatial choice models considered here satisfy the conditions for discrete choice, namely, that the choices or set of alternatives are collectively exhaustive, mutually exclusive, and finite.[6] As another application, let's go back to Hotelling's model mentioned in the introduction by assuming this time two attributes, x_1 and x_2, spanning a 2-dimensional feature space and having *black* and *white* labels or examples positioned as shown in Fig. 6 (top panel).

Clearly the linear SVM will struggle to fit a separating hyperplane on this nonlinear non-separable feature space. Hence, we can employ the "kernel trick" by adding another dimension z (shown in the bottom panel) that allows for the linearly separation by the SVM in this higher-dimensional feature space. This translated back into the original space results in four separate quadrants that neatly predicts *black* (or *white*) labels depending on whether the attributes x_1 and x_2 taken together lie in the 'northwest' or 'southeast' ('northeast' or 'southwest') quadrants. More formally, for a certain individual with preferences $S^{\{x_1\}}$ and $S^{\{x_2\}}$ for attributes $x^{\{1\}}$ and $x^{\{2\}}$ respectively that fully characterize two mutually independent and distinct choices, $y_i \in \{black, white\}$, then we may predict that the individual will choose either *black* or *white* depending on

---

[6] See Train (2003) [16]. Empirically discrete choice models employ many interesting forms such as the Binary/Multinomial Logit and Probit, Conditional Logit, Nested Logit, Generalized Extreme Value Models, Mixed Logit, and Exploded Logit. Discrete choice models have been used extensively in transportation, psychology, energy, housing, marketing, voting, and many more.

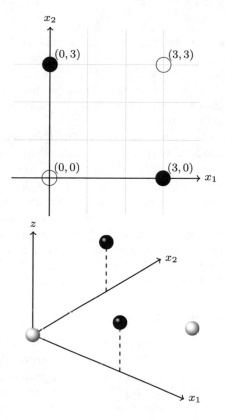

Fig. 6. The kernel trick

the 'closeness' of features describing *black* and *white* to his or her preferences. Furthermore, a useful measure of 'closeness' is the euclidean distance, D. That is, for the 2-D feature space, we have for label $l \in \{black, white\}$

$$D_l = \sqrt{(x_l^{\{1\}} - S^{\{x_1\}})^2 + (x_l^{\{2\}} - S^{\{x_2\}})^2}$$

and the individual will choose *black* which result in higher utility if $D_{l=black} < D_{l=white}$, and vice versa (or indifferent if $D_{l=black} = D_{l=white}$).

Characterization and prediction by the euclidean distance can easily be generalized for cases with multiple m features, i.e. $X_i \in \mathbb{R}^m$. The important point here is to note that characterization by the SVM and prediction (in contrast to other machine learning techniques such as decision trees, neural networks and the k-nearest neighbors algorithms, amongst others) is exactly the same as what we would have had with the euclidean distance method. This should not be surprising because the SVM as a large margin classifier divides the feature space by a linear hyperplane that maximizes the separation of the different classes or examples with respect to the support vectors and does this precisely by the euclidean distance. Moreover, increasing features does not necessarily increase the computational burden of SVM very much even for the nonlinear non-separable case, because by employing the kernel method when moving into a higher dimensional space often requires computation of only a few finite supporting vectors, i.e. most α_i in Eq. 12 would be zero and once an optimal set of *alpha* is learned, all other $\mathbf{x}_i$ where $\alpha_i = 0$ can be discarded and the resulting sparse solution is desirable as it reduces the required memory and computation requirements for classification.

Compared to the logistic model, which is used often in discrete choice modeling, a number of papers [3,15] have shown that the SVM outperforms characterization and prediction compared to the logistic regression. In general, parametric models, including discriminant analysis, which is another popular classification method, tend to face structural inadequacy especially when dealing with relational or nested boundaries between choices [18]. As a last point, over-fitting in high dimensional settings with a small number of examples may not be an issue with spatial choice modeling. On the contrary, it may be desirable that SVM allows us more complex and interesting boundaries for classification and prediction of spatial choice models.

5   Concluding Remarks

We have looked at the support vector machine (SVM) as a useful algorithm to modeling spatial choice problems. With linearly separable data, the SVM can be viewed as a standard quadratic programming (QP) problem, for which optimization using the Lagrangian and the dual QP involved only m variables or m^2 dot products. we also treated the soft margin variant for nonlinear non-separable data by including a margin or extra slackness that allowed for misclassification, which essentially is similar to using a simple surrogate loss term called a hinge loss and applying a form on L-2 regularization. The optimization was seen to have a dual form that requires only pair-wise similarity between the data. More generally, we showed how to exploit these dual forms to work with the SVM that can operate implicitly in a very high-dimensional feature space by defining non-linearity similarity kernels that measure the dot product in those spaces, making the classifier independent of the implicit feature dimension while remaining computationally in the quadratic to cubic in the data space. This makes SVM somewhat minimalist in terms of memory and computational time.

We also discussed first using a simple linear 1-D case, then a higher 2-D feature space, how the SVM can be exploited for spatial choice modeling. Generalization for m-dimension space is straightforward. What is important to note is that even for the more complex and higher dimensional feature space, the major advantage of the SVM is that it did not necessarily suffer from the curse of dimensionality because only computation involving a few support vectors mattered. Furthermore, the SVM has been found to be superior in terms of classification and prediction compared to other standard techniques like the logistic regression commonly used in discrete choice modeling. The treatment of SVM and demonstration of the easiness of translating the SVM to spatial choice problems in the discussion above is admittedly superficial, but we hope that we have demonstrated sufficiently that the easiness of the abstraction of attributes and the choices (or labels) of the SVM make it a natural choice for characterization and prediction of spatial choice models.

Admittedly, the utility function adopted here is a simple, deterministic one, and understanding how the SVM would apply in relation to more complex, stochastic utility models could be a further extension. Furthermore, there are further technical issues to consider. For example, although the SVM is essentially a two-class classifier, it would be interesting to see how solving multi-class spatial choice problems can be done with multi-class extensions of the basic SVM [14]. Also we have only looked at supervised learning, and unsupervised learning based on ideas of the SVM, so-called support vector clustering (SCV) due to Ben-Hur et al. [1] and others can also be explored further in the context of spatial choice modeling.

References

1. Ben-Hur, A., Horn, D., Siegelmann, H.Y., Vapnik, V.: Support vector clustering. J. Mach. Learn. Res. **2**, 125–137 (2001)
2. Burges, C.J.C.: A tutorial on support vector machines for pattern recognition. Data Min. Knowl. Disc. **2**(2), 121–167 (1998)
3. Cui, D., Curry, D.: Prediction in marketing using support vector machine. Mark. Sci. **24**(4), 595–615 (2005)
4. Cortes, C., Vapnik, V.: Support vector networks. Mach. Learn. **20**(3), 273–297 (1995)
5. Cristianini, N., Shawe-Taylor, J.: An Introduction to Support Vector Machines and Other Kernel-Based Learning Methods. Cambridge University Press, Cambridge (2000)
6. Hastie, T., Tibshirani, R., Jerome, F.: The Elements of Statistical Learning. Springer New York Inc., New York (2001)
7. McFadden, D.L.: Conditional logit analysis of qualitative choice behaviour. In: Zarembka, P. (ed.) Frontiers in Econometrics, pp. 105–142. Academic Press, New York (1973)
8. McFadden, D.L.: Econometric models of probabilistic choice. In: Manski, C.F., McFadden, D.L. (eds.) Structural Analysis of Discrete Data with Econometric Applications, pp. 198–272. MIT Press, Cambridge (1981)
9. McFadden, D.: Econometric analysis of qualitative response models. In: Griliches, Z., Intriligator, M.D. (eds.) Handbook of Econometrics, vol. 2, 1st edn, chap. 24, pp. 1395–1457. Elsevier (1984)
10. Hotelling, H.: Stability in competition. Econ. J. **39**(153), 41–57 (1929)
11. Jakulin, A., Buntine, W., La Pira, T.M., Brasher, H.: Analyzing the US senate in 2003: similarities, networks, clusters and blocs. Polit. Anal. **17**(3), 291–310 (2009)
12. Mullainathan, S., Spiess, J.: Machine learning: an applied econometric approach. J. Econ. Perspect. **31**(2), 87–106 (2017)
13. Varian, H.: Big data: new tricks for econometrics. J. Econ. Perspect. **28**(2), 3–28 (2014)
14. Shigeo, A. Multiclass support vector machines. In: Support Vector Machines for Pattern Classification, pp. 113–161. Springer, London (2010)

15. Salazar, D.A., Velez, J.I., Salazar, J.C.: Comparison between SVM and logistic regression: which one is better to discriminate? Rev. Colomb. Estadistica **35**(2), 223–237 (2012)
16. Train, K.: Discrete Choice Methods with Simulation. Department of Economics, SUNY-Oswego, Oswego (2003)
17. Vapnik, V.N.: The Nature of Statistical Learning Theory. Springer Science, Berlin (2000)
18. West, P.M., Brockett, P.L., Golden, L.L.: A comparative analysis of neural networks and statistical methods for predicting consumer choice. Mark. Sci. **16**(4), 370–391 (1997)

Author Index